Operator Theory
Advances and Applications
Vol. 87

Editor
I. Gohberg

Recent Developments in Operator Theory and Its Applications

International Conference in Winnipeg, October 2–6, 1994

Edited by

I. Gohberg
P. Lancaster
P.N. Shivakumar

Birkhäuser Verlag
Basel · Boston · Berlin

Editors' addresses:

I. Gohberg
School of Mathematical Sciences
Tel Aviv University
69978 Tel Aviv
Israel

P. Lancaster
Department of Mathematics and Statistics
University of Calgary
Calgary, Alberta T2N 1NA
Canada

P.N. Shivakumar
Department of Applied Mathematics and
Institute of Industrial Mathematical Sciences
University of Manitoba
Winnipeg, Manitoba
Canada R3T 2N2

1991 Mathematics Subject Classification 47-XX

A CIP catalogue record for this book is available from the Library of Congress, Washington D.C., USA

Deutsche Bibliothek Cataloging-in-Publication Data
Recent developments in operator theory and its applications:
international conference in Winnipeg, October 2-6, 1994 / ed.
by I. Gohberg ... - Basel; Boston; Berlin: Birkhäuser, 1996
(Operator theory; Vol. 87)
ISBN-13:978-3-0348-9878-2 e-ISBN-13:978-3-0348-9035-9
DOI: 10.1007/978-3-0348-9035-9

NE: Gochberg, Izrail' C. [Hrsg.]; GT

Softcover reprint of the hardcover 1st edition 1996

Printed on acid-free paper produced from chlorine-free pulp. TCF ∞
Cover design: Heinz Hiltbrunner, Basel

ISBN-13:978-3-0348-9878-2
e-ISBN-13:978-3-0348-9035-9

9 8 7 6 5 4 3 2 1

Table of Contents

EDITORIAL INTRODUCTION

The present volume contains the proceedings of the International Conference on Applications of Operator Theory held in Winnipeg, Canada (October 2nd to 6th, 1994), which was organized by the Institute of Industrial Mathematical Sciences (IIMS) of the University of Manitoba. At this conference 92 participants representing 15 countries participated, and 64 papers were presented. This meeting was the second of a linked pair. The first was a program of advanced instruction held at the Fields Institute, Ontario, followed by a research conference. The first of these events gave rise to the volume "Lectures on Operator Theory and its Applications", published by the American Mathematical Society for the Fields Institute in 1995.

These two events were the creation of the following Program Committee:

M. A. Dahleh (M.I.T.)
P. A. Fillmore (Dalhousie)
B. A. Francis (Toronto)
F. Ghahramani (Manitoba)
K. Glover (Cambridge)
I. Gohberg (Tel Aviv)
T. Kailath (Stanford)
P. Lancaster (Calgary), Chair
H. Langer (Vienna)
P. N. Shivakumar (Manitoba)
A. A. Shkalikov (Moscow)
B. Simon (Cal. Tech.)
H. Widom (Santa Cruz)

Both events focused on the following main topics: Infinite matrices and projection methods, linear operators on indefinite scalar product spaces, differential operators and mathematical systems theory and control.

This volume contains a selection of papers in modern operator theory and its applications. They are dedicated to recent achievements and many are written by leaders in the mentioned fields. This collection, together with the lecture volume, will be useful and interesting for a wide audience in Mathematical and Engineering Sciences.

The editors are very pleased to record their gratitude to Sharon Henderson of the IIMS for her efficient help.

I. Gohberg (Tel Aviv)

P. Lancaster (Calgary)

P. N. Shivakumar (Winnipeg)

Operator Theory:
Advances and Applications, Vol. 87

INVERSE SCATTERING PROBLEM FOR CONTINUOUS TRANSMISSION LINES WITH RATIONAL REFLECTION COEFFICIENT FUNCTION

D. ALPAY, I. GOHBERG and L. SAKHNOVICH

In this paper we obtain explicit formula for the reflexivity coefficient function (or potential) of an ordinary differential operator if its reflection coefficient is a rational matrix valued function. The solution is given in terms of a realization of the reflection coefficient function.

1 Introduction

Let H denote the differential operator

$$(Hf)(x) = -iJ\frac{df}{dx}(x) - \begin{pmatrix} 0 & k(x) \\ k(x)^* & 0 \end{pmatrix} f(x), \qquad x \geq 0. \tag{1.1}$$

In (1.1), $J = \begin{pmatrix} I_m & 0 \\ 0 & -I_m \end{pmatrix}$, the function k is called the potential and belongs to $L_1^{m\times m}[0,\infty)$ and the unknown f is $\mathbb{C}^{2m\times p}$–valued for some p.

Associated to (1.1) are a number of matrix valued functions which play an important role in the study of the operator H, in particular the scattering function $S(\lambda)$, the spectral function $W(\lambda)$ and the asymptotic equivalence matrix function $V(\lambda)$. These matrix functions are computed in terms of solutions of the eigenvalue problem

$$-iJ\frac{dX(x,\lambda)}{dx} - \begin{pmatrix} 0 & k(x) \\ k(x)^* & 0 \end{pmatrix} X(x,\lambda) = \lambda X(x,\lambda) \tag{1.2}$$

where $\lambda \in \mathbb{R}$ and where the unknown X is $\mathbb{C}^{2m\times m}$–valued and subject to various boundary conditions. Precise definitions and the main properties of the functions S, W and V are reviewed in the sequel.

The direct and inverse problem for the equation (1.2) interpreted as the model for a scattering medium has been analyzed in [6]; then the function $k(x)$ is called the local reflexivity coefficient function. Associated to this medium is its reflection coefficient function

$R(\lambda) = X_{21}(0,\lambda)X_{11}(0,\lambda)^{-1}$, where $X = \begin{pmatrix} X_{11} \\ X_{21} \end{pmatrix}$ is the $\mathbb{C}^{2m\times m}$–valued solution of (1.2) subject to the asymptotic

$$X(x,\lambda) = e^{-ix\lambda}\begin{pmatrix} I_m \\ 0 \end{pmatrix} + o(1). \qquad (x \to +\infty) \tag{1.3}$$

The direct scattering problem is to compute the functions S, W and R from the function k; the inverse scattering problem is the other way around, and consists in recovering k from one of these functions. When S, W or V is given, we refer to [12], [13], [15]. The case where the function R is given is considered in [6]. For the case where S or W are rational, explicit formulas for the potential $k(x)$ built in terms of realizations of S or W were obtained in [3]. The discrete analogues of these results were obtained in [2].

In this paper we obtain formulas for the function $k(x)$ also in the case when the reflection coefficient function is given and is rational. The solution is in the same framework as [3] and [2]. The paper consists of four sections; this introduction is the first. In section 2 we review various results on differential expressions of the form (1.1) and in section 3 we study the connections between the reflection coefficient function R and the functions S, W and V. In the fourth section we give explicit formulas for $k(x)$ when R is rational.

Finally some notations : $\mathbb{C}^{p\times m}$ denotes the space of p–rows q–columns matrices with complex entries and $\mathbb{C}^{p\times 1}$ is written as $\mathbb{C}^p$; for a matrix A, the symbol A^T denotes the transpose of A and A^* the adjoint of A; the matrix whose entries are the complex conjugates of the entries of A is denoted by $\overline{A}$. The operator norm of the matrix A is denoted by $||A||$. If $\lambda \to R(\lambda)$ is a $\mathbb{C}^{m\times m}$–valued rational function, R^* denotes the function $\lambda \to R(\overline{\lambda})^*$.

2 Some results on differential expressions

In this section we review some results on the solutions of the differential equation (1.2).

Theorem 2.1 *Let $k \in L_1^{m\times m}[0,+\infty)$. The differential equation* (1.2) *has a unique $\mathbb{C}^{2m\times 2m}$–valued solution $U(x,\lambda)$ defined for $x \geq 0$ and $\lambda \in \mathbb{R}$ and subject to the asymptotic*

$$U(x,\lambda) = e^{-i\lambda xJ} + o(1) \tag{2.1}$$

as $x \to +\infty$. Furthermore,

$$U(x,\lambda)^*JU(x,\lambda) = J, \qquad x \geq 0, \ \ \lambda \in \mathbb{R}. \tag{2.2}$$

Proof: For the existence of $U(x,\lambda)$ see e.g. [9, Theorem 9.2 p. 213–214]. Here is a simple argument that shows that (2.2) holds. We differentiate the function $x \to U(x,\lambda)^*JU(x,\lambda)$ with respect to x. Since U is a solution of (1.2) we have

$$\frac{d}{dx}(U(x,\lambda)^*JU(x,\lambda)) \;=\; (\frac{d}{dx}U(x,\lambda)^*JU(x,\lambda) + U(x,\lambda)^*J(\frac{d}{dx}U(x,\lambda))$$

$$\begin{aligned} &= \left(i\lambda U^*(x,\lambda)J - U^*(x,\lambda)\begin{pmatrix} 0 & k(x) \\ k^*(x) & 0 \end{pmatrix}\right) JU(x,\lambda) + \\ &\quad +U(x,\lambda)^*J(-i\lambda U(x,\lambda) - \begin{pmatrix} 0 & k(x) \\ k(x)^* & 0 \end{pmatrix} U(x,\lambda)) \\ &= -U(x,\lambda)^*\begin{pmatrix} 0 & k(x) \\ k(x)^* & 0 \end{pmatrix} JU(x,\lambda) \\ &\quad -U(x,\lambda)^*J\begin{pmatrix} 0 & k(x) \\ k(x)^* & 0 \end{pmatrix} U(x,\lambda) \end{aligned}$$

which is equal to zero, since

$$\begin{pmatrix} 0 & k(x) \\ k(x)^* & 0 \end{pmatrix} J + J\begin{pmatrix} 0 & k(x) \\ k(x)^* & 0 \end{pmatrix} = 0.$$

For a fixed λ, and preassigned ε, there is $x_\lambda \geq 0$ such that:

$$x \geq x_\lambda \Longrightarrow ||U(x,\lambda) - e^{-i\lambda xJ}|| \leq \varepsilon.$$

We set $U_0(x,\lambda) = e^{-i\lambda xJ}$. From

$$\begin{aligned} U(x,\lambda)^*JU(x,\lambda) - J &= U(x,\lambda)^*JU(x,\lambda) - U_0(x,\lambda)^*JU_0(x,\lambda) \\ &= (U(x,\lambda)^* - U_0(x,\lambda)^*)JU_0(x,\lambda) + \\ &\quad +U(x,\lambda)J(U(x,\lambda) - U_0(x,\lambda)) \end{aligned}$$

we obtain for $x \geq x_\lambda$,

$$||U(x,\lambda)^*JU(x,\lambda) - J|| \leq \varepsilon + (1+\varepsilon)\varepsilon,$$

and so $U(x,\lambda)^*JU(x,\lambda) = J$.

The uniqueness of $U(x,\lambda)$ is proved as follows. If $U_1(x,\lambda)$ and $U_2(x,\lambda)$ are two $\mathbb{C}^{2m\times 2m}$–valued solutions of (1.2) satisfying the condition (2.1), the same computations as before show that

$$\frac{d}{dx}U_1(x,\lambda)JU_2(x,\lambda)^* = 0$$

and so,

$$U_1(x,\lambda)JU_2(x,\lambda)^* = J$$

because of the estimate (2.1). From the J-unitarity of U_1 and U_2 follows that $U_1 = U_2$. □

We note that the $\mathbb{C}^{2m\times m}$–valued function X solution to (1.2) and subject to (1.3) is equal to

$$X(x,\lambda) = U(x,\lambda)\begin{pmatrix} I_m \\ 0 \end{pmatrix}.$$

The matrix function $U(0,\lambda)$ is J–unitary on $\mathbb{R}$. In fact, it has much more properties, as is explained in Theorem 2.2, which is proved in [14, Sections 2]. To state the results of

Melik–Adamyan we first need some definitions. The Wiener algebra $\mathcal{W}^{m\times m}$ consists of the $\mathbb{C}^{m\times m}$–valued functions W which can be written as

$$W(\lambda) = D + \int_{-\infty}^{+\infty} e^{i\lambda x}k(x)dx$$

where $D \in \mathbb{C}^{m\times m}$ and where $k \in L^1_{m\times m}(\mathbb{R})$ and where the variable $\lambda \in \mathbb{R}$. We note that

$$D = \lim_{\lambda\to\pm\infty} W(\lambda),$$

and will use the notation $W(\infty) = D$.

The subalgebras $\mathcal{W}_+^{m\times m}$ and $\mathcal{W}_-^{m\times m}$ consist of the elements of $\mathcal{W}^{m\times m}$ for which the support of k is in $\mathbb{R}^+$ and $\mathbb{R}^-$ respectively. Note that an element W in $\mathcal{W}_+^{m\times m}$

$$W(\lambda) = D + \int_{0}^{+\infty} e^{i\lambda x}k(x)dx$$

is analytic and bounded in the open upper half-plane: if $\lambda = \alpha + i\beta$ with $\beta \geq 0$, we have

$$\begin{aligned} \|W(\lambda)\| &\leq \|D\| + \int_0^{+\infty} e^{-\beta u}\|k(u)\|du \\ &\leq \|D\| + \int_0^{+\infty} \|k(u)\|du. \end{aligned}$$

In particular, the maximum modulus principle for a strip [8, p .231] gives

$$\sup_{\lambda\in\mathbb{C}_+\cup\mathbb{R}} \|W(\lambda)\| = \sup_{\lambda\in\mathbb{R}} \|W(\lambda)\|.$$

We also note that $D = \lim_{\mathrm{Im}\,\lambda\to+\infty} W(\lambda), \quad \lambda \in \mathbb{C}_+$. Similar remarks hold for elements of $\mathcal{W}_-^{m\times m}$.

If $W \in \mathcal{W}^{m\times m}$ can be written as $W = W_+W_-$ where W_+ and its inverse are in $\mathcal{W}_+^{m\times m}$ and W_- and its inverse are in $\mathcal{W}_-^{m\times m}$ we say that W admits a left Wiener–Hopf (or left spectral) factorization. Similarly, $W = W_-W_+$ with W_- and W_+ as above, is called a right Wiener–Hopf (or right spectral) factorization.

Theorem 2.2 *Let $k \in L_1^{m\times m}[0,+\infty)$ and let $U(x,\lambda)$ be the $\mathbb{C}^{2m\times 2m}$–valued solution of the differential expression* (1.1) *satisfying the asymptotic* (2.1). *Denote by $V(\lambda)$ the function $U(0,\lambda)$. Then V has the following properties:*
(a) *It can be represented in the form*

$$V(\lambda) = I_{2m} + \int_0^{+\infty} M(t) \begin{pmatrix} e^{-i\lambda t}I_m & 0 \\ 0 & e^{i\lambda t}I_m \end{pmatrix} dt$$

where $M \in L^1_{2m\times 2m}[0,+\infty)$. This is equivalent to the following: if $V = (V_{ij})$ is the decomposition of V into four $\mathbb{C}^{m\times m}$–valued blocks, V_{11} and V_{21} belong to $\mathcal{W}_-^{m\times m}$ and V_{22} and V_{12} belong to $\mathcal{W}_+^{m\times m}$.
(b) *The function V_{11} is invertible in $\mathcal{W}_-^{m\times m}$ and the function V_{22} is invertible in $\mathcal{W}_+^{m\times m}$.*

For more on the class of matrix valued functions which satisfy conditions (a) and (b) in the above theorem we refer to [14, Definition 1.2 p. 50]; we denote this class by $\mathcal{K}$. The matrix valued function $V(\lambda)$ is called the asymptotic equivalence matrix, or the A–matrix (see [7], [9, p. 156], [14, p. 65]).

The next lemma concerns with the case $k(x)^T = k(x)$. It includes in particular the case where $k(x)$ is scalar.

Lemma 2.3 *Let us suppose that the local reflexivity function $k(x)$ satisfies $k(x)^T = k(x)$. Then the $\mathbb{C}^{2m\times 2m}$–valued solution of* (1.2) *which satisfies the asymptotic* (2.1) *is such that*

$$U(x,\lambda) = j\overline{U(x,\lambda)}j \tag{2.3}$$

where

$$j = \begin{pmatrix} o & I_m \\ I_m & 0 \end{pmatrix}.$$

Proof: Since $k(x) = k(x)^T$ we have

$$iJ\frac{d}{dx}\overline{U(x,\lambda)} = \lambda\overline{U(x,\lambda)} + \begin{pmatrix} o & \overline{k(x)} \\ k(x) & 0 \end{pmatrix}\overline{U(x,\lambda)}. \tag{2.4}$$

Furthermore, we obtain from (2.1)

$$\overline{U(x,\lambda)} = e^{i\lambda xJ} + o(1). \tag{2.5}$$

Since

$$j\begin{pmatrix} 0 & \overline{k(x)} \\ k(x) & 0 \end{pmatrix}j = \begin{pmatrix} 0 & k(x) \\ \overline{k(x)} & 0 \end{pmatrix}, \quad jJj = -J$$

and $je^{i\lambda xJ}j = e^{-i\lambda xJ}$, multiplying both sides of (2.4) and (2.5) on the left and on the right by j leads to

$$\begin{aligned} ijJj\frac{d}{dx}j\overline{U(x,\lambda)}j &= \lambda j\overline{U(x,\lambda)}j + j\begin{pmatrix} 0 & \overline{k(x)} \\ k(x) & 0 \end{pmatrix}jj\overline{U(x,\lambda)}j \\ &= \lambda(j\overline{U(x,\lambda)}j) + \begin{pmatrix} o & k(x) \\ \overline{k(x)} & 0 \end{pmatrix}j\overline{U(x,\lambda)}j \\ &= \lambda(j\overline{U(x,\lambda)}j) + \begin{pmatrix} 0 & k(x) \\ k(x)^* & 0 \end{pmatrix}j\overline{U(x,\lambda)}j, \end{aligned}$$

and

$$j\overline{U(x,\lambda)}j = e^{-i\lambda xJ} + o(1).$$

Thus, $j\overline{U(x,\lambda)}j$ is a solution of (1.2) which satisfies the asymptotic (2.1); by uniqueness of such a solution, we obtain (2.3). □

We note that equation (2.3) can be rewritten as

$$\overline{U_{11}(x,\lambda)} = U_{22}(x,\lambda) \tag{2.6}$$

$$\overline{U_{21}(x,\lambda)} = U_{12}(x,\lambda). \tag{2.7}$$

3 The reflection coefficient function

In this section we study the reflection coefficient function R and relate it to the scattering function S and to the spectral function W. These relationships will be used later in solving the inverse spectral problem associated to R. Let us first recall the definition of S (see [14, p. 64-65]): the differential equation (1.2) has a unique $\mathbb{C}^{2m\times m}$–valued solution $Z(x,\lambda)$ which satisfies the conditions

$$(I_m \ -I_m)Z(0,\lambda)=0 \tag{3.1}$$

and

$$(I_m \ \ 0)Z(x,\lambda)=e^{-i\lambda x}I_m+o(1) \ \ (x\rightarrow+\infty). \tag{3.2}$$

Then there exists a $\mathbb{C}^{m\times m}$–valued function S, called the scattering function, such that

$$(0 \ \ I_m)Z(x,\lambda)=S(\lambda)e^{i\lambda x}+o(1) \ \ (x\rightarrow+\infty). \tag{3.3}$$

The scattering function S is in $\mathcal{W}^{m\times m}$; it takes unitary values and admits a Wiener–Hopf factorization $S(\lambda) = S_-(\lambda)S_+(\lambda)$, where S_- and its inverse are in $\mathcal{W}_-^{m\times m}$ and S_+ and its inverse are in $\mathcal{W}_+^{m\times m}$.

Before relating S and R we give a number of properties of R.

Theorem 3.1 *Let us consider a differential expression* (1.1) *and let $U(x,\lambda)$ be the $\mathbb{C}^{2m\times 2m}$–valued solution of* (1.2) *with asymptotic* (2.1) *and let $V(\lambda) = U(0,\lambda)$ be the asymptotic equivalence matrix. Assume also that $R(\lambda)$ is the reflection coefficient function (defined by* (1.3)*). Then*
(1) *The function $R(\lambda)$ is in $\mathcal{W}_-^{m\times m}$ and can be rewritten as*

$$R(\lambda)=(V_{12}(\lambda)V_{22}(\lambda)^{-1})^*, \quad \lambda\in\mathbb{R} \tag{3.4}$$

and

$$R(\infty)=I_m. \tag{3.5}$$

It holds that

$$I_m-R(\lambda)^*R(\lambda)=V_{11}(\lambda)^{-*}V_{11}(\lambda)^{-1}, \quad \lambda\in\mathbb{R} \tag{3.6}$$

$$I_m-R(\lambda)R(\lambda)^*=V_{22}(\lambda)^{-*}V_{22}(\lambda)^{-1}, \quad \lambda\in\mathbb{R}. \tag{3.7}$$

Equality (3.6) *is a left Wiener–Hopf factorization and* (3.7) *is a right Wiener–Hopf factorization.*
(2) *We have $\sup_{\lambda\in\mathbb{C}_-\cup\mathbb{R}}||R(\lambda)||<1$. The function $V_{11}-V_{21}$ is invertible in $\mathcal{W}_-^{m\times m}$ and the function $V_{12}-V_{22}$ is invertible in $\mathcal{W}_+^{m\times m}$.*

Proof: First note that $V_{11}(\lambda)$ and $V_{22}(\lambda)$ are invertible for real λ since V_{11} is invertible in $\mathcal{W}_-^{m\times m}$ and V_{22} is invertible fo in $\mathcal{W}_+^{m\times m}$; hence $R(\lambda)=V_{21}(\lambda)V_{11}(\lambda)^{-1}$ is well defined; since V belongs to the class $\mathcal{K}$, both V_{21} and V_{11}^{-1} are in $\mathcal{W}_-^{m\times m}$ and so $R\in\mathcal{W}_-^{m\times m}$. From the J–unitarity of V on the real line we have

$$V_{11}(\lambda)^*V_{12}(\lambda)=V_{21}(\lambda)^*V_{22}(\lambda) \tag{3.8}$$

and so

$$V_{11}(\lambda)^{-*}V_{21}(\lambda)^*=V_{12}(\lambda)V_{22}(\lambda)^{-1}$$

from which we obtain (3.4). To prove (3.5) it suffices to remark that $V_{21}(\infty) = 0$ and $V_{11}(\infty) = I_m$. The J–unitarity of V also gives the equalities

$$\begin{aligned} V_{22}(\lambda)^* V_{22}(\lambda) - V_{12}(\lambda)^* V_{12}(\lambda) &= I_m \\ V_{11}(\lambda)^* V_{11}(\lambda) - V_{21}(\lambda)^* V_{21}(\lambda) &= I_m. \end{aligned}$$

Dividing both sides of the first equality by $V_{22}(\lambda)^*$ on the left and by $V_{22}(\lambda)$ on the right, we obtain (3.7); equality (3.6) is obtained by dividing the second equality by $V_{11}(\lambda)^*$ on the right and by $V_{11}(\lambda)$ on the left. Since $V_{11}^{\pm} \in \mathcal{W}_-^{m\times m}$, (3.6) is a left Wiener–Hopf factorization and, similarly, (3.7) is a right Wiener–Hopf factorization since $V_{22}^{\pm 1} \in \mathcal{W}_+^{m\times m}$.

We now turn the proof of (2). Since $R \in \mathcal{W}_-^{m\times m}$, the maximum modulus gives that

$$\sup_{\lambda\in\mathbb{C}_-\cup\mathbb{R}} ||R(\lambda)|| = \sup_{\lambda\in\mathbb{R}} ||R(\lambda)||. \tag{3.9}$$

From (3.6) we have $||R(\lambda)|| < 1$ for every $\lambda \in \mathbb{R}$; in view of (3.5), it follows that

$$\sup_{\lambda\in\mathbb{R}} ||R(\lambda)|| < 1$$

and so $\sup_{\lambda\in\mathbb{C}_-\cup\mathbb{R}} ||R(\lambda)|| < 1$.

Since V is in the class $\mathcal{K}$, the function $V_{11} - V_{21}$ belongs to $\mathcal{W}_-^{m\times m}$; to prove that its inverse is also in $\mathcal{W}_-^{m\times m}$, it suffices to check that $\det(V_{11}(\lambda) - V_{21}(\lambda)) \neq 0$ for $\lambda \in \mathbb{C}_- \cup \mathbb{R}$; this in turn follows from the equality

$$V_{11}(\lambda) - V_{21}(\lambda) = (I_m - R(\lambda))V_{11}(\lambda) \tag{3.10}$$

and the fact that $||R(\lambda)|| < 1$ for $\lambda \in \mathbb{C}_- \cup \mathbb{R}$.

The claim on $V_{22} - V_{12}$ is proved in a similar manner. □

Theorem 3.2 *Let us consider a differential expression of the form* (1.1), *let* $V(\lambda)$ *be its asymptotic equivalence matrix function and let* S *be its scattering matrix function. Then* S *is given by the formulas*

$$S(\lambda) = (V_{22}(\lambda) - V_{12}(\lambda))^{-1}(V_{11}(\lambda) - V_{21}(\lambda)) \tag{3.11}$$

and

$$S(\lambda) = (V_{22}(\lambda) - V_{12}(\lambda))^*(V_{11}(\lambda) - V_{21}(\lambda))^{-*}. \tag{3.12}$$

Furthermore, $S(\infty) = I_m$ *and* (3.11) (*resp.* (3.12)) *is a right spectral factorization* (*resp is a right spectral factorization*).

Proof: Since the columns of the function $U(x,\lambda)$ span all the solutions of (1.2), there are two matrix valued functions $C(\lambda)$ and $D(\lambda)$ such that

$$Z(x,\lambda) = U(x,\lambda)\begin{pmatrix} C(\lambda) \\ D(\lambda) \end{pmatrix}$$

(where Z has been defined as the $\mathbb{C}^{2m\times m}$–solution of (1.2) subject to (3.1) and (3.2)).

From (3.2) and (3.3) we obtain $C(\lambda) = I_m$ and $D(\lambda) = S(\lambda)$. Since (3.1) holds, we have

$$(V_{11}(\lambda) - V_{21}(\lambda)) + (V_{12}(\lambda) - V_{22}(\lambda))S(\lambda) = 0$$

and so we obtain (3.11) and in particular $S(\infty) = I_m$. That (3.1) is a left Wiener–Hopf factorization follows from (2) of Theorem 3.1. To obtain (3.12), we note that the J–unitarity of the asymptotic equivalence matrix on $\mathbb{R}$ leads to

$$(I_m \; -I_m)V(\lambda)JV(\lambda)^* \begin{pmatrix} I_m \\ -I_m \end{pmatrix} = 0$$

and thus,

$$(V_{11}(\lambda) - V_{21}(\lambda))(V_{11}(\lambda) - V_{21}(\lambda))^* = (V_{12}(\lambda) - V_{22}(\lambda))(V_{12}(\lambda) - V_{22}(\lambda))^*,$$

which leads to (3.12). □

For theorems of this type and more discussion on the scattering matrix S we refer to [14, Section 3].

We note that (3.16) and (3.12) can be rewritten as

$$S(\lambda) = V_{22}(\lambda)^{-1}(I_m - R(\lambda)^*)^{-1}(I_m - R(\lambda))V_{11}(\lambda) \tag{3.13}$$

and

$$S(\lambda) = V_{22}(\lambda)^*(I_m - R(\lambda))(I_m - R(\lambda)^*)^{-1}V_{11}(\lambda)^{-*} \tag{3.14}$$

This last equation allows us to compute S when R is given and to get the Wiener–Hopf factorization $S = S_-S_+$ of the scattering matrix, with

$$S_-(\lambda) = V_{22}(\lambda)^*(I_m - R(\lambda)) \tag{3.15}$$

and

$$S_+(\lambda) = (I_m - R(\lambda)^*)^{-1}V_{11}(\lambda)^{-*} \tag{3.16}$$

Note that $S_-(\infty) = S_+(\infty) = I_m$.

We now turn to the spectral function W and first recall its definition. Let us consider the set D_H of functions $f \in L_2^{2m}[0, +\infty)$ which are absolutely continuous and for which $(I_m \; -I_m)f(0) = 0$. The operator H (defined in (1.1)) restricted to D_H is selfadjoint; a $\mathbb{C}^{m\times m}$–valued function W defined for $\lambda \in \mathbb{R}$ and such that $W(\lambda) > 0$, for all $\lambda \in \mathbb{R}$ is called a spectral function of H if there is a unitary map $U : L_2^{2m}[0, +\infty) \to L_2^m(W)$ such that

$$(UHf)(\lambda) = \lambda(Uh)(\lambda)$$

for $f \in D_H$. The function $W(\lambda) = S_-(\lambda)^{-1}S_-(\lambda)^{-*}$ is a spectral function of H and the map U is computed in terms of continuous orthogonal polynomials (see [14], [9]).

Proposition 3.3 *Consider a differential expression of the form* (1.1) *with reflection coefficient function* $R(\lambda)$ *and spectral matrix function* $W(\lambda)$. *Then* W *and* R *are related by*

$$W(\lambda) = (I_m - R(\lambda))^{-1}(I_m - R(\lambda)R(\lambda)^*))(I_m - R(\lambda)^*)^{-1}. \tag{3.17}$$

Proof: Using the formula (3.15) for the factor S_- we have

$$\begin{aligned} W(\lambda) &= S_-(\lambda)^{-1}S_-(\lambda)^{-*} \\ &= (I_m - R(\lambda))^{-1}V_{22}(\lambda)^{-*}V_{22}(\lambda)^{-1}(I_m - R(\lambda)^*)^{-1} \end{aligned}$$

which leads to (3.17) since as proved in Theorem 3.1, $I_m - R(\lambda)R(\lambda)^* = V_{22}(\lambda)^{-*}V_{22}(\lambda)^{-1}$. □

We now consider the case where $k(x) = k(x)^T$; then (2.6) and (2.7) hold and thus:

Proposition 3.4 *Let us consider a differential expression of the form* (1.2) *and suppose that* $k(x) = k(x)^T$. *Then the scattering function* S, *the reflection coefficient function* R, *and the spectral function* W *are symmetric, i.e.*

$$R(\lambda) = R(\lambda)^T, \quad S(\lambda) = S(\lambda)^T, \quad W(\lambda) = W(\lambda)^T. \tag{3.18}$$

and the asymptotic equivalence matrix is j*-symmetric, i.e.*

$$V(\lambda) = j\overline{V(\lambda)}j \tag{3.18}$$

for $\lambda \in \mathbb{R}$, (*with* $j = \begin{pmatrix} 0 & I_m \\ I_m & 0 \end{pmatrix}$).

Proof: To obtain (3.18) we set $x = 0$ in (2.3). Then,

$$\overline{V_{11}(\lambda)} = V_{22}(\lambda) \tag{3.19}$$

and

$$\overline{V_{21}(\lambda)} = V_{12}(\lambda). \tag{3.20}$$

Then, (3.4) leads to

$$\begin{aligned} R(\lambda)^T &= \overline{V_{12}(\lambda)V_{22}(\lambda)}^{-1} \\ &= V_{21}(\lambda)V_{11}(\lambda)^{-1} \\ &= R(\lambda) \end{aligned}$$

and the claim on W follows from (3.17). Similarly, from

$$V_{11}(\lambda) - V_{12}(\lambda) = (V_{22}(\lambda) - V_{21}(\lambda))S(\lambda)$$

we obtain, using (3.19) and (3.20)

$$(V_{22}(\lambda) - V_{21}(\lambda)) = (V_{11}(\lambda) - V_{12}(\lambda))\overline{S(\lambda)}$$

and so

$$S(\lambda) = \overline{S(\lambda)}^{-1}$$

Since $S(\lambda)S(\lambda)^* = I_m$, it follows that $S(\lambda) = S(\lambda)^T$. □

4 The rational case

In this section we assume that the reflection coefficient R is rational; then, since the limit $\lim_{|\mathrm{Im}\ \lambda|\to\infty} R(\lambda)$ exists and is equal to zero, R is analytic at infinity and $R(\infty) = 0$. Hence, it follows that $\sup_{\mathbb{R}} \|R(\lambda)\| < 1$. As we will show in the sequel, the scattering function S, the spectral function W and the asymptotic equivalence matrix function V are also rational and analytic at infinity. Now, any $\mathbb{C}^{p\times p}$-valued rational function analytic at infinity admits a representation of the form

$$D + C(\lambda I_n - A)^{-1}B, \tag{4.1}$$

called its realization. In (4.1), the matrices A, B, C and D are in $\mathbb{C}^{n\times n}$, $\mathbb{C}^{n\times m}$, $\mathbb{C}^{m\times n}$, $\mathbb{C}^{m\times m}$ respectively. The realization is called minimal if n in (4.1) is minimal. For general information on these realization we refer to [4], [10, chapter 7] and [16].

We take $R(\lambda) = -C(\lambda I_n - A)^{-1}B$ a minimal realization of the reflection coefficient function R and compute (in general nonminimal) realizations of the spectral matrix W and of the factor S_- in the Wiener–Hopf factorization $S = S_-S_+$ of the scattering function. We then use the results of [3] to obtain expressions of the local reflexivity coefficient function in terms of these realizations. Finally, we compute R in terms of a minimal realization of the asymptotic equivalence matrix, using results of [1], and express $k(x)$ in terms of this realization.

We first quote a result on Wiener–Hopf factorizations; for minimal realizations this theorem appears in [4]; the general case of nonminimal realizations was considered in [5].

Theorem 4.1 *Let $W(\lambda) = I_m + C(\lambda I_n - A)^{-1}B$ be a realization of the $\mathbb{C}^{m\times m}$-valued rational function W and assume that A has no real spectrum. Then W admits a Wiener–Hopf factorization relative to the real line if and only if the following two conditions are fulfilled:*

(i) $A^\times = A - BC$ has no real eigenvalues

(ii) $\mathbb{C}^n = M \oplus M^\times$

where M (resp. $M^\times$) is the space spanned by the eigenvectors and generalized eigenvectors corresponding to the eigenvalues of A (resp. $A^\times$) in the upper (resp. lower) half plane. Furthermore, in that case, W admits a canonical factorization $W = W_-W_+$ with

$$W_-(\lambda) = I_m + C(\lambda I_n - A)^{-1}(I - \pi)B \tag{4.2}$$

$$W_+(\lambda) = I_m + C\pi(\lambda I_n - A)^{-1}B \tag{4.3}$$

$$W_-(\lambda)^{-1} = I_m - C(I - \pi)(\lambda I_n - A^\times)^{-1}B \tag{4.4}$$

$$W_+(\lambda)^{-1} = I_m - C(\lambda I_n - A^\times)^{-1}\pi B \tag{4.5}$$

where π is the projection of $\mathbb{C}^n$ along M onto $M^\times$.

Theorem 4.2 *Let R be the reflection coefficient function of a differential expression of the form* (1.2). *Assume that R is rational and let $R(\lambda) = -C(\lambda I_n - A)^{-1}B$ be a minimal realization of R. Then the local reflexivity coefficient function is given by the formula*

$$k(x) = -2ce^{2ixa}(I_m + \Omega(Y - e^{-2ixa^*}Ye^{2ixa}))^{-1}(b + i\Omega c^*). \tag{4.6}$$

The rest of the theorem explains the entries in this formula: define

$$\mathcal{A} = \begin{pmatrix} A & BB^* \\ 0 & A^* \end{pmatrix} \quad \text{and} \quad \mathcal{A}^\times = \begin{pmatrix} A & BB^* \\ C^*C & A^* \end{pmatrix} \tag{4.7}$$

and

$$a = \begin{pmatrix} T & 0 \\ 0 & I \end{pmatrix} \begin{pmatrix} (I-\pi)\mathcal{A}^\times(I-\pi) & (I-\pi)\begin{pmatrix} 0 \\ C^*C \end{pmatrix} \\ 0 & A \end{pmatrix} \begin{pmatrix} T^{-1} & 0 \\ 0 & I \end{pmatrix}, \tag{4.8}$$

$$b = \begin{pmatrix} T & 0 \\ 0 & I \end{pmatrix} \begin{pmatrix} (I-\pi)\begin{pmatrix} 0 \\ C^* \end{pmatrix} \\ B \end{pmatrix}, \tag{4.9}$$

$$c = ((C\ \ 0)(I-\pi)\ \ C)\begin{pmatrix} T^{-1} & 0 \\ 0 & I \end{pmatrix}. \tag{4.10}$$

The matrices $\mathcal{A}$ and $\mathcal{A}^\times$ have no real spectrum. Denote by Q (resp. $Q^\times$) the Riesz projection corresponding to the eigenvalues of $\mathcal{A}$ (resp. $\mathcal{A}^\times$) in $\mathbb{C}_+$; in (4.8), (4.9), (4.10), *π is the projection defined by* $\operatorname{Ker} \pi = \operatorname{Im} Q$ *and* $\operatorname{Im} \pi = \operatorname{Ker} Q^\times$, *and T is a linear bijection from* $\operatorname{Im}(I-\pi)$ *onto $\mathbb{C}^q$, with $q = \dim \operatorname{Im}(I-\pi)$.*

The spectrum of the matrices a and $a^\times = a - bc$ is in $\mathbb{C}_+$ and Ω and Y are the unique solutions of the Lyapunov equations

$$i(\Omega a^{\times *} - a^\times \Omega) = bb^* \tag{4.11}$$

and

$$i(Ya - a^*Y) = -c^*c \tag{4.12}$$

Proof: Let $\mathcal{A}$ and $\mathcal{A}^\times$ be defined by (4.7). The equality

$$I_m - R(\lambda)R(\overline{\lambda})^* = I_m - (C\ \ 0)(\lambda I_{2n} - \mathcal{A})^{-1}\begin{pmatrix} 0 \\ C^* \end{pmatrix}$$

is a (in general non minimal) realization of $I_m - RR^*$. The operator $\mathcal{A}$ has clearly no real spectrum. Furthermore,

$$(I - R(\lambda)R(\overline{\lambda})^*)^{-1} = I_m + (C\ \ 0)(\lambda I_{2n} - \mathcal{A}^\times)^{-1}\begin{pmatrix} 0 \\ C^* \end{pmatrix}$$

is a (in general nonminimal) realization of $(I_m - RR^*)^{-1}$. The proof that $\mathcal{A}^\times$ has no real spectrum follows from [11], where it is proved (Proof of Step 1 of Theorem 3.2 p.232) that the real poles of $(I_m - RR^*)^{-1}$ are the real eigenvalues of $\mathcal{A}^\times$ with same multiplicities. In the present case, the function R is analytic and strictly contractive on the real line and so $(I_m - RR^*)^{-1}$ has no real poles. Thus, $\mathcal{A}^\times$ has no real eigenvalues. It follows that Q and $Q^\times$ are well defined; the projection π is well defined since $I_m - R(\lambda)R(\overline{\lambda})^* > 0$ for $\lambda \in \mathbb{R}$. Since $(I_m - R(\lambda)R(\overline{\lambda})^*) = V_{22}(\overline{\lambda})^{-*}V_{22}(\lambda)^{-1}$ is a right Wiener–Hopf factorization (as was shown in Theorem 3.1), Theorem 4.1 gives

$$V_{22}(\overline{\lambda})^* = I_m + (C\ \ 0)(I-\pi)(\lambda I - \mathcal{A}^\times)^{-1}\begin{pmatrix} 0 \\ C^* \end{pmatrix}.$$

Since $(I-\pi)\mathcal{A}^\times = (I-\pi)\mathcal{A}^\times(I-\pi)$, we can rewrite (4.13) as

$$V_{22}(\overline{\lambda})^* = I_m + (C\ \ 0)T^{-1}(\lambda I_q - T(I-\pi)\mathcal{A}^\times(I-\pi)T^{-1})^{-1}T(I-\pi)\begin{pmatrix} 0 \\ C^* \end{pmatrix}.$$

Writing

$$\begin{aligned} V_{22}(\overline{\lambda})^* - V_{12}(\overline{\lambda})^* &= V_{22}(\overline{\lambda})^*(I_m - V_{22}(\overline{\lambda})^{-*}V_{12}(\overline{\lambda})^*) \\ &= V_{22}(\overline{\lambda})^*(I_m - R(\lambda)) \end{aligned}$$

we obtain the realization

$$V_{22}(\overline{\lambda})^* - V_{12}(\overline{\lambda})^* = I_m + c(\lambda I - a)^{-1}b$$

where a, b, c are defined by (4.8)–(4.10). The spectrum of the matrix a is clearly in $\mathbb{C}_+$. To check that the spectrum of $a^\times$ is inside $\mathbb{C}_+$, it suffices to note that

$$a^\times = \begin{pmatrix} T & 0 \\ 0 & I \end{pmatrix}\begin{pmatrix} (I-\pi)\mathcal{A}(I-\pi) & 0 \\ -B(C\ 0)(I-\pi) & A-BC \end{pmatrix}\begin{pmatrix} T^{-1} & 0 \\ 0 & I \end{pmatrix}.$$

The operator $A - BC$ has its spectrum inside $\mathbb{C}_+$ since $I_m - R$ is invertible in $\mathbb{C}_- \cup \mathbb{R}$ and since

$$(I - R(\lambda))^{-1} = I + C(\lambda I_n - (A - BC))^{-1}B$$

is a minimal realization. Similarly, from Theorem 4.1,

$$\begin{aligned} V_{22}(\overline{\lambda})^{-*} &= I_m - (C\ \ 0)(\lambda I - \mathcal{A})^{-1}(I-\pi)\begin{pmatrix} 0 \\ C^* \end{pmatrix} \\ &= I_m - (C\ \ 0)(I-\pi)(\lambda - (I-\pi)\mathcal{A}(I-\pi))^{-1}(I-\pi)\begin{pmatrix} 0 \\ C^* \end{pmatrix} \end{aligned}$$

and so,

$$V_{22}(\overline{\lambda})^{-*} = I_m - (C\ \ 0)(I-\pi)T^{-1}(\lambda I_q - T(I-\pi)\mathcal{A}(I-\pi)T^{-1})^{-1}T(I-\pi)\begin{pmatrix} 0 \\ C^* \end{pmatrix}, \tag{4.14}$$

since $(I-\pi)\mathcal{A}(I-\pi) = \mathcal{A}(I-\pi)$. The realization (4.14) is minimal and so $T(I-\pi)\mathcal{A}(I-\pi)T^{-1}$ has no spectrum in $\mathbb{C}_- \cup \mathbb{R}$ since $V_2(\overline{\lambda})^{-*}$ is analytic there. □

The matrix $\mathcal{A}^\times$ defined in (4.7) was introduced in [11, p. 228], and is called the state characteristic matrix of R associated to the given minimal realization. The fact that $R(\lambda) = -C(\lambda I_n - A)^{-1}B$ is a minimal realization is not needed; it is sufficient to assume that both A and $A - BC$ have no spectrum in $\mathbb{C}_- \cup \mathbb{R}$. We now compute the function $k(x)$ using formula (3.17) for the spectral function.

Theorem 4.3 *Let R be the reflection coefficient function of the differential expression* (1.1) *and assume that R is rational. Let $R(\lambda) = -C(\lambda I_n - A)^{-1}B$ be a minimal realization of R. Then*

$$W(\lambda) = I_m + \mathbf{C}(\lambda I - \mathbf{A})^{-1}\mathbf{B} \tag{4.15}$$

is a realization of the spectral function W, and the local reflexivity coefficient function is given by

$$k(x) = 2\mathbf{C}(\mathbf{P}e^{-2ix\mathbf{A}^\times}|\mathrm{Im}\,\mathbf{P})^{-1}\mathbf{P}\mathbf{B}. \tag{4.16}$$

In these expressions

$$\mathbf{A} = \begin{pmatrix} A - BC & -BC & 0 & -BB^* \\ 0 & A & BB^* & 0 \\ 0 & 0 & A^* & -C^*B^* \\ 0 & 0 & 0 & (A-BC)^* \end{pmatrix} \tag{4.17}$$

$$\mathbf{B} = \begin{pmatrix} B \\ \begin{pmatrix} 0 \\ C^* \end{pmatrix} \\ C^* \end{pmatrix} \tag{4.18}$$

and

$$\mathbf{C} = -(C \;\; (C \;\; 0) \;\; B^*) \tag{4.19}$$

The operators $\mathbf{A}$ and $\mathbf{A}^\times$ have no real spectrum and $\mathbf{P}$ denotes the Riesz projection corresponding to the eigenvalues of $\mathbf{A}$ in $\mathbb{C}_+$.

Proof: If $W_i(\lambda) = D_i + C_i(\lambda I_{n_i} - A_i)B_i$, $i = 1, 2, 3$ are given realizations of the matrix valued functions W_i, $i = 1, 2, 3$, then

$$W_1(\lambda)W_2(\lambda)W_3(\lambda) = D + C(\lambda I_n - A)^{-1}B$$

is a realization of the product $W_1W_2W_3$, where

$$A = \begin{pmatrix} \begin{pmatrix} A_1 & B_1C_2 \\ 0 & A_2 \end{pmatrix} & \begin{pmatrix} B_1D_2C_3 \\ B_2C_3 \end{pmatrix} \\ 0 & A_3 \end{pmatrix}$$

$$B = \begin{pmatrix} B_1D_2D_3 \\ B_2D_3 \\ B_3 \end{pmatrix}, \quad C = (C_1 \;\; D_1C_2 \;\; D_1C_3)$$

and $D = D_1D_2D_3$.

We apply this formula with

$$W_1(\lambda) = (I - R(\lambda))^{-1} = I - C(\lambda I_n - (A - BC))^{-1}B$$

$$W_3(\lambda) = (I - R(\overline{\lambda})^*)^{-1} = I - B^*(\lambda I_n - (A - BC)^*)^{-1}C^*$$

and

$$\begin{aligned} W_2(\lambda) &= I - R(\lambda)R(\overline{\lambda})^* \\ &= I_m - (C \; 0)(\lambda I_{2n} - A)^{-1} \begin{pmatrix} 0 \\ C^* \end{pmatrix} \end{aligned}$$

and obtain the realization (4.15) of $W(\lambda)$.
We have

$$\mathbf{BC} = \begin{pmatrix} -BC & -BC & 0 & -BB^* \\ 0 & 0 & 0 & 0 \\ -C^*C & -C^*C & 0 & -C^*B^* \\ -C^*C & -C^*C & 0 & -C^*B^* \end{pmatrix}$$

and thus

$$\mathbf{A}^\times = \mathbf{A} - \mathbf{BC} = \begin{pmatrix} A & 0 & 0 & 0 \\ 0 & A & BB^* & 0 \\ C^*C & C^*C & A^* & 0 \\ C^*C & C^*C & 0 & A^* \end{pmatrix}.$$

Since as already noted $\mathcal{A}^\times = \begin{pmatrix} A & BB^* \\ C^*C & A^* \end{pmatrix}$ has no real spectrum, the matrix $\mathbf{A}^\times$ has no real spectrum. The formula for $k(x)$ follows then from [3]. □

We conclude with a result which gives a realization of R when the asymptotic equivalence matrix is given.

Proposition 4.4 *Let R be the reflection coefficient function of the differential expression (1.1) and assume that R is rational. Then the asymptotic equivalence matrix function $V(\lambda)$ is rational, it is analytic at infinity and $V(\infty) = I_{2m}$. A realization of R is of the form*

$$R(\lambda) = (C_2 - C_1)\left(\lambda I_{2n} - \begin{pmatrix} A & -B_1C_1 \\ 0 & A - B_1C_1 \end{pmatrix}\right)^{-1} \begin{pmatrix} B_1 \\ B_1 \end{pmatrix} \tag{4.20}$$

where

$$V(\lambda) = I_{2m} + \begin{pmatrix} C_1 \\ C_2 \end{pmatrix} (\lambda I_n - A)^{-1}(B_1 \ \ B_2) \tag{4.21}$$

is a minimal realization of V *(with* C_1 *and* $C_2 \in \mathbb{C}^{m\times n}$ *and* $B_1, B_2 \in \mathbb{C}^{n\times m}$*).*

Proof: From (3.6) we obtain that V_{11} is rational; it follows then from the fact that $V_{11} \in \mathcal{K}$ that V_{11} is analytic at infinity and $V_{11}(\infty) = I_m$; $V_{21}(\lambda) = R(\lambda)V_{11}(\lambda)$ is then also rational and analytic at infinity, with $V_{21}(\infty) = 0$. The case of V_{22} and V_{12} is treated similarly using (3.7) and (3.4). We have from (4.21)

$$V_{11}^{-1}(\lambda) = I_m - C_1(\lambda I_n - (A - B_1C_1))^{-1}B_1$$

and

$$V_{21}(\lambda) = C_2(\lambda I_n - A)^{-1}B_1$$

and hence, the realization (4.20). □

Minimal realization of rational matrix valued functions analytic at infinity and J–unitary on the real line were studied in [1], from which we obtain the following: the realization (4.20) is the realization of a function J-unitary on the real line if and only if there is an invertible hermitian matrix H (uniquely defined from the realization) and such that

$$A^*H - HA = i(C_1^*C_1 - C_2^*C_2) \tag{4.22}$$

and

$$(B_1 \ \ B_2) = -iH^{-1}(C_1^* - C_2^*). \tag{4.23}$$

Thus, (4.20) can be rewritten as

$$R(\lambda) = (C_2 - C_1)\left(\lambda I_{2n} - \begin{pmatrix} A & -iH^{-1}C_1^*C_1 \\ 0 & A + iH^{-1}C_1^*C_1 \end{pmatrix}\right)\begin{pmatrix} -iH^{-1}C_1^* \\ -iH^{-1}C_1^* \end{pmatrix} \tag{4.24}$$

where H, A, C_1, C_2 are related by (4.22).

References

[1] D. Alpay and I. Gohberg. *Unitary rational matrix functions*, volume 33 of *Operator Theory: Advances and Applications*, pages 175–222. Birkhäuser Verlag, Basel, 1988.

[2] D. Alpay and I. Gohberg. Inverse spectral problems for difference operators with rational scattering matrix function. *Integral Equations and Operator Theory*, 20:125–170, 1994.

[3] D. Alpay and I. Gohberg. Inverse spectral problem for differential operators with rational scattering matrix functions. *Journal of Differential Equations*, 118:1–19, 1995.

[4] H. Bart, I. Gohberg, and M. Kaashoek. *Minimal factorization of matrix and operator functions*, volume 1 of *Operator Theory: Advances and Applications*. Birkhäuser Verlag, Basel, 1979.

[5] H. Bart, I. Gohberg, and M. Kaashoek. *Invariants for Wiener–Hopf equivalence of analytic operator functions*, volume 21 of *Operator Theory: Advances and Applications*, pages 317–356. Birkhäuser Verlag, Basel, 1986.

[6] A. Bruckstein, B. Levy, and T. Kailath. Differential methods in inverse scattering. *SIAM journal of applied mathematics*, 45:312–335, 1985.

[7] Yu. L. Daleckii and M.G. Kreĭn. *Stability solutions of differential equations in Banach spaces*, volume 43 of *Translations of mathematical monographs*. American mathematical society, Providence, Rhode Island, 1974.

[8] N. Dunford and J. Schwartz. *Linear operators*, volume 1. Interscience, 1957.

[9] H. Dym and A. Iacob. *Positive definite extensions, canonical equations and inverse problems*, volume 12 of *Operator Theory: Advances and Applications*, pages 141–240. Birkhäuser Verlag, Basel, 1984.

[10] I. Gohberg, P. Lancaster, and L. Rodman. *Invariant subspaces of matrices and applications.* Wiley, New–York, 1986.

[11] I. Gohberg and S. Rubinstein. *Proper contractions and their unitary minimal completions*, volume 34 of *Operator Theory: Advances and Applications*, pages 223–247. Birkhäuser Verlag, Basel, 1988.

[12] M.G. Kreĭn. On the determination of a potential of a particle from its s–function. *Dokl. Akad. Nauk. SSSR*, 105:637–640, 1955.

[13] M.G. Kreĭn. *Topics in differential and integral equations and operator theory*, volume 7 of *Operator theory: Advances and Applications.* Birkhäuser Verlag, 1983.

[14] F.E. Melik-Adamyan. Canonical differential operators in Hilbert space. *Izvestya Akademii Nauk. Armyanskoi SSR Matematica*, 12:10–31, 1977.

[15] F.E. Melik-Adamyan. On a class of canonical differential operators. *Izvestya Akademii Nauk. Armyanskoi SSR Matematica*, 24:570–592, 1989. English translation in: Soviet Journal of Contemporary Mathematics, vol. 24, pages 48–69 (1989).

[16] M.W. Wonham. *Linear Multivariable Control: Geometric Approach.* Springer–Verlag, New–York, 1979.

Daniel Alpay
Department of Mathematics
Ben–Gurion University of the Negev
POB 653. 84105 Beer-Sheva
Israel

Israel Gohberg
School of Mathematical Sciences
The Raymond and Beverly Sackler Faculty
of Exact Sciences
Tel–Aviv University
Tel–Aviv, Ramat–Aviv 69989, Israel

Lev Sakhnovich
Pr. Dobrovolskogo 154 app. 199
Odessa 270111
Ukraine

MSC: 34L25, 81U40, 47A56

Operator Theory:
Advances and Applications, Vol. 87

THE BAND METHOD AND GRASSMANNIAN APPROACH FOR COMPLETION AND EXTENSION PROBLEMS

J.A. Ball*, I. Gohberg and M.A. Kaashoek

The Grassmannian approach is used to develop a new addition of the band method. This addition allows one to obtain a linear fractional representation of all solutions of a completion problem from special extensions that are not necessarily band extensions (for the positive case) or triangular extensions (for the contractive case). Also linear fractional representations are obtained for all solutions of a completion problem of non-band type.

1. INTRODUCTION

The band method originates in work of Dym and Gohberg (see [4], [5]) and consequently has been developed and refined, primarily by Gohberg, Kaashoek and Woerdeman (see [8], [9], [10], [11]). For the following discussion we follow the treatment in Chapter XXXIV of the recent monograph [7] which is based on a presentation of the band method by Ellis, Gohberg and Lay [6].

We consider an algebra $\mathcal{M}$ with a unit e and an involution $*$ which has a band structure. The existence of the involution means that, for each $x \in \mathcal{M}$, there corresponds an element $x^* \in \mathcal{M}$ such that

(a) $(x+y)^* = x^* + y^*$,

(b) $(\alpha x)^* = \bar{\alpha} x^*$,

(c) $(xy)^* = y^* x^*$

(d) $x^{**} = x$.

The band structure means that $\mathcal{M}$ admits a direct sum decomposition

$$\mathcal{M} = \mathcal{M}_1 \oplus \mathcal{M}_2^0 \oplus \mathcal{M}_d \oplus \mathcal{M}_3^0 \oplus \mathcal{M}_4, \tag{1.1}$$

* The first author was partially supported by National Science Foundation grant DMS-9500912.

where $\mathcal{M}_1$, $\mathcal{M}_2^0$, $\mathcal{M}_d$, $\mathcal{M}_3^0$ and $\mathcal{M}_4$ are linear manifolds of $\mathcal{M}$ such that

(E1) the unit element e is in $\mathcal{M}_d$,

(E2) $\mathcal{M}_1^* = \mathcal{M}_4$, $(\mathcal{M}_2^0)^* = \mathcal{M}_3^0$, $\mathcal{M}_d^* = \mathcal{M}_d$,

(E3) the following multiplication table describes some additional restrictions on the multiplication in $\mathcal{M}$:

	$\mathcal{M}_1$	$\mathcal{M}_2^0$	$\mathcal{M}_d$	$\mathcal{M}_3^0$	$\mathcal{M}_4$
$\mathcal{M}_1$	$\mathcal{M}_1$	$\mathcal{M}_1$	$\mathcal{M}_1$	$\mathcal{M}_+^0$	$\mathcal{M}$
$\mathcal{M}_2^0$	$\mathcal{M}_1$	$\mathcal{M}_+^0$	$\mathcal{M}_2^0$	$\mathcal{M}_c$	$\mathcal{M}_-^0$
$\mathcal{M}_d$	$\mathcal{M}_1$	$\mathcal{M}_2^0$	$\mathcal{M}_d$	$\mathcal{M}_3^0$	$\mathcal{M}_4$
$\mathcal{M}_3^0$	$\mathcal{M}_+^0$	$\mathcal{M}_c$	$\mathcal{M}_3^0$	$\mathcal{M}_-^0$	$\mathcal{M}_4$
$\mathcal{M}_4$	$\mathcal{M}$	$\mathcal{M}_-^0$	$\mathcal{M}_4$	$\mathcal{M}_4$	$\mathcal{M}_4$

(1.2)

where

$$\mathcal{M}_+^0 := \mathcal{M}_1 \oplus \mathcal{M}_2^0, \quad \mathcal{M}_-^0 := \mathcal{M}_3^0 \oplus \mathcal{M}_4, \quad \mathcal{M}_c := \mathcal{M}_2^0 \oplus \mathcal{M}_d \oplus \mathcal{M}_3^0.$$

We are now ready to state the extension problem studied in this paper. Let $\mathcal{M}$ be an *algebra with band structure* (1.1) *in a unital C^*-algebra $\mathcal{R}$*. The latter means that $\mathcal{M}$ is a $*$-subalgebra of a unital C^*-algebra $\mathcal{R}$ and the unit e of $\mathcal{M}$ is also the unit of $\mathcal{R}$. Let k be an element in $\mathcal{M}_{\ell+} := \mathcal{M}_d \oplus \mathcal{M}_2^0$. An element f of $\mathcal{M}_+ := \mathcal{M}_d \oplus \mathcal{M}_2^0 \oplus \mathcal{M}_1$ is said to be an *$\mathcal{R}$-positive real part completion* of k if f has the form

$$f = k + m_1 \tag{1.3}$$

for an $m_1 \in \mathcal{M}_1$ and $b = f + f^*$ is positive definite in $\mathcal{R}$ (i.e., $b = a^*a$ for some invertible a in $\mathcal{R}$). Sometimes it is convenient to speak of positive extensions b of $k + k^*$ rather than positive real part extensions of k; by an *$\mathcal{R}$-positive extension* b of $k + k^*$ we simply mean an element $b \in \mathcal{M}$ of the form $m_1 + k + k^* + m_4$ with $m_1 \in \mathcal{M}_1$ and $m_4 \in \mathcal{M}_4$ such that b is positive definite in $\mathcal{R}$. Note that in this situation necessarily $m_4 = m_1^*$ since both b and $k + k^*$ are self-adjoint.

In this paper we are interested in describing all $\mathcal{R}$-positive real part completions of the given element $k \in \mathcal{M}_{\ell+}$, or equivalently, all $\mathcal{R}$-positive extensions of the element

$k + k^* \in \mathcal{M}_c = \mathcal{M}_2^0 \oplus \mathcal{M}_d \oplus \mathcal{M}_3^0$. This problem and variations on it have been discussed at length in a number of publications (see [4], [5], [6], [8], [9], [10], [11]). Here we follow the version which is presented in Chapter XXXIV of the book [7]. There it is shown how to get a linear fractional parametrization for the set of all solutions once one has a band extension b which admits both a right and a left spectral factorization relative to (1.1). Before stating this result we need to introduce some more definitions.

Let $\mathcal{M}$ be an algebra with band structure (1.1), and let $b \in \mathcal{M}$. We say that b admits a *right spectral factorization* (relative to the decomposition (1.1)) if b factors as $b = b_+^* b_+$, where b_+ is an invertible element of $\mathcal{M}$ such that b_+ and its inverse b_+^{-1} are both in $\mathcal{M}_+$. Analogously, b is said to have a *left spectral factorization* (relative to (1.1)) if $b = b_-^* b_-$ with b_- an invertible element of $\mathcal{M}$ and $b_-^{\pm 1}$ in $\mathcal{M}_-$. From the symmetry of the multiplication table one can see that b admits a right spectral factorization if b^{-1} admits a left factorization, and conversely. In the sequel we write a^{-*} for $(a^{-1})^*$ or $(a^*)^{-1}$.

Suppose now that the $b \in \mathcal{M}$ is a $\mathcal{R}$-positive completion of $k + k^*$. We say that b is a *band extension* of $k + k^*$ if b has the additional property that b^{-1} is in $\mathcal{M}_c$. Band extensions which also have right and left spectral factorizations are of particular interest. Such special extensions can be found by solving linear equations (cf., Theorems XXXIV.1.1 and 1.2 in [7]). The next result (see Theorem XXXIV.2.1 in [7]) indicates how the set of all solutions can be described once one has found a band extension with right and left spectral factorization.

The statement of this result requires some extra structure. In the sequel $\|\cdot\|_{\mathcal{R}}$ denotes the norm of the C^*-algebra $\mathcal{R}$. We shall assume that the following axiom holds.

AXIOM (A). *If $g \in \mathcal{M}_+$ and $\|g\|_{\mathcal{R}} < 1$, then $(e-g)^{-1} \in \mathcal{M}_+$.*

THEOREM 1.1. *Let $\mathcal{M}$ be an algebra with band structure* (1.1) *in a unital C^*-algebra $\mathcal{R}$, and assume that axiom* (A) *holds. Let $k \in \mathcal{M}_{\ell+}$, and suppose that $k + k^*$ has a band extension b which admits both a right and a left spectral factorization relative to* (1.1):

$$b = u^{-*}u^{-1} = v^{-*}v^{-1}, \quad u^{\pm 1} \in \mathcal{M}_+, v^{\pm 1} \in \mathcal{M}_-. \tag{1.4}$$

(a) *Then each $\mathcal{R}$-positive extension of $k + k^*$ is of the form*

$$\tilde{\mathcal{F}}(g) = (vg + u)^{-*}(e - g^*g)(vg + u)^{-1}, \tag{1.5}$$

where the free parameter g is an arbitrary element in $\mathcal{M}_1$ such that $\|g\|_{\mathcal{R}} < 1$. Moreover, *the map $\tilde{\mathcal{F}}$ in (1.5) provides a one-to-one correspondence between all such g and all $\mathcal{R}$-positive extensions of $k + k^*$.*

(b) *Write b in the form $b = c + c^*$ where $c \in \mathcal{M}_+$. Then each $\mathcal{R}$-positive real part extension of k is of the form*

$$\mathcal{F}(g) = (-c^*vg + cu)(vg + u)^{-1}, \tag{1.6}$$

where the free parameter g is as in part (a). Moreover the map $\mathcal{F}$ in (1.6) provides a one-to-one correspondence between all such g and all $\mathcal{R}$-positive real part extensions of k.

A parallel result holds for strictly contractive extensions of a given element k of $\mathcal{M}_\ell := \mathcal{M}_4 \oplus \mathcal{M}_3^0 \oplus \mathcal{M}_d \oplus \mathcal{M}_2^0$. Here we say that $f \in \mathcal{M}$ is a *strictly contractive extension* (or *completion*) of the given element $k \in \mathcal{M}_\ell$ if $k - f \in \mathcal{M}_1$ and $\|f\|_{\mathcal{R}} < 1$. A problem of interest is to describe all strictly contractive extensions.

In a way which is completely analogous to the case for positive real part completions, it is shown in [7] how one can obtain a parametrization of the set of all completely contractive extensions of k once one has a particular completely contractive extension which has special properties. One such special property is that the completely contractive extension c be a *triangular extension* of k, i.e., that $c(e - c^*c)^{-1}$ belongs to $\mathcal{M}_\ell$. In addition it is required that $e - c^*c$ admit a right spectral factorization relative to (1.1) and that $e - cc^*$ admit a left spectral factorization relative to (1.1). Then the following result appears in [7], Section XXXIV.3.

THEOREM 1.2. *Let $\mathcal{M}$ be an algebra with band structure (1.1) in a unital C^*-algebra $\mathcal{R}$, and assume that axiom (A) holds. Let $k \in \mathcal{M}_\ell$, and suppose that k has a triangular extension c such that $e - c^*c$ admits a right and $e - cc^*$ admits a left spectral factorization relative to (1.1), i.e.*

$$e - c^*c = u^{-*}u^{-1} \ (u^{\pm 1} \in \mathcal{M}_+), \quad e - cc^* = v^{-*}v^{-1} \ (v^{\pm 1} \in \mathcal{M}_-). \tag{1.7}$$

Then each strictly contractive extension of k in $\mathcal{M}$ is given by

$$\mathcal{F}(g) = (vg + cu)(c^*vg + u)^{-1}, \tag{1.8}$$

where the free parameter g is an arbitrary element in $\mathcal{M}_1$ such that $\|g\|_{\mathcal{R}} < 1$. Moreover, the map $\mathcal{F}$ in (1.8) provides a one-to-one correspondence between all such g and all strictly contractive extensions of k.

In this paper we show that the linear fractional representations in Theorems 1.1 and 1.2 also can be obtained from extensions that are not band extensions (for the positive case) or triangular extensions (for the contractive case) and have less restrictive conditions on the factors u and v appearing in (1.4) and (1.7). We shall prove the following theorems.

THEOREM 1.3. *Let $\mathcal{M}$ be an algebra with band structure (1.1) in a unital C^*-algebra $\mathcal{R}$, and assume that axiom (A) holds. Let $k \in \mathcal{M}_{\ell+}$, and suppose that $k + k^*$ has an $\mathcal{R}$-positive extension f such that*

(i) $f = u^{-*}u^{-1}$, *where* $u^{\pm 1} \in \mathcal{M}_+$;

(ii) $f = v^{-*}v^{-1}$, *where* $v \in \mathcal{R}$ *and* $v^*\mathcal{M}_1 = \mathcal{M}_1$;

(iii) $f^{-1}\mathcal{M}_1 \subset \mathcal{M}_+$.

(a) *Then each $\mathcal{R}$-positive extension of $k + k^*$ is of the form*

$$\tilde{\mathcal{F}}(g) = (vg + u)^{-*}(e - g^*g)(vg + u)^{-1}, \tag{1.9}$$

where the free parameter g is an arbitrary element of $\mathcal{M}_1$ such that $\|g\|_{\mathcal{R}} < 1$. Moreover, the map $\tilde{\mathcal{F}}$ in (1.9) provides a one-to-one correspondence between all such g and all $\mathcal{R}$-positive extensions of $k + k^{\times}$.

(b) *Write f in the form $f = c + c^*$ where $c \in \mathcal{M}_+$. Then each $\mathcal{R}$–positive real part extension of k is of the form*

$$\mathcal{F}(g) = (-c^*vg + cu)(vg + u)^{-1}, \tag{1.10}$$

where the free parameter g is an arbitrary element in $\mathcal{M}_1$ such that $\|g\|_{\mathcal{R}} < 1$. Moreover the map $\mathcal{F}$ in (1.10) provides a one-to-one correspondence between all such g and all $\mathcal{R}$-positive real part extensions of k.

THEOREM 1.4. *Let $\mathcal{M}$ be an algebra with band structure (1.1) in a unital C^*-algebra $\mathcal{R}$, and assume that axiom (A) holds. Let $k \in \mathcal{M}_\ell$, and suppose that k has a strictly contractive extension c such that*

(i) $e - c^*c = u^{-*}u^{-1}$, *where* $u^{\pm 1} \in \mathcal{M}_+$;

(ii) $e - cc^* = v^{-*}v^{-1}$, *where* $v^{\pm 1} \in \mathcal{R}$ *and* $v^*\mathcal{M}_1 = \mathcal{M}_1$;

(iii) $(e - c^*c)^{-1}c^*\mathcal{M}_1 \subset \mathcal{M}_+$.

Then each strictly contractive extension of k in $\mathcal{M}$ is given by

$$\mathcal{F}(g) = (vg + cu)(c^*vg + u)^{-1}, \tag{1.11}$$

where the free parameter g is an arbitrary element in $\mathcal{M}_1$ such that $\|g\|_{\mathcal{R}} < 1$. Moreover the map $\mathcal{F}$ in (1.11) provides a one-to-one correspondence between all such g and all strictly contractive extensions of k.

Note that the construction of the coefficient matrix Θ for the linear fractional map parametrizing all solutions of a positive real part or strictly contractive completion problem does not require the full strength of a band extension or triangular extension with left and right spectral factorizations. All that is required is a special positive or strictly contractive completion satisfying the collection of properties (i), (ii) and (iii) listed in Theorem 1.3 or Theorem 1.4, respectively. This suggests we introduce some definitions.

DEFINITION 1.5 (a) *Let* k *be a given element of* $\mathcal{M}_{\ell+}$, *and let* $c \in \mathcal{M}_+$ *be an* $\mathcal{R}$*-positive real part extension of* k. *We say that* c *is a canonical* $\mathcal{R}$*-positive real part extension of* k *if* $f = c + c^*$ *satisfies conditions* (i), (ii) *and* (iii) *of Theorem* (1.3). *In this case we also refer to* f *as a canonical* $\mathcal{R}$*-positive extension of* $k + k^*$.

(b) *Let* k *be an element of* $\mathcal{M}_\ell$, *and let* $c \in \mathcal{M}$ *be a strictly contractive extension of* k. *If* c *satisfies conditions* (i), (ii) *and* (iii) *of Theorem* (1.4), *then* c *is said to be a canonical strictly contractive extension of* k.

The assertion of Theorem 1.3(b) then is: If c is a canonical $\mathcal{R}$-positive real part extension of $k \in \mathcal{M}_{\ell+}$, then formula (1.10), with the free parameter g in $\mathcal{M}_1$ with $\|g\|_{\mathcal{R}} < 1$, gives a parametrization of the set of all solutions of the positive real part completion problem with data k. In a similarly way Theorem 1.4 asserts that all strictly contractive extensions of k in $\mathcal{M}_\ell$ may be obtained from a canonical one.

Theorems 1.3 and 1.4 contain Theorems 1.1 and 1.2, respectively, as special cases. Indeed, let b be a band extension of $k + k^*$ such that (1.4) holds. Then conditions (i), (ii) and (iii) in Theorem 1.3 are fulfilled with $f = b$. For statement (i) this holds true trivially. Since $(v^*)^{\pm 1} \in \mathcal{M}_+$, the multiplication table (1.2) yields $v^*\mathcal{M}_1 = \mathcal{M}_1$. The fact that b is a band extension means that $b^{-1} \in \mathcal{M}_c$, and thus, using the multiplication table again, $b^{-1}\mathcal{M}_1 \subset \mathcal{M}_+^0 \subset \mathcal{M}_+$. Similarly, if c is a triangular extension such that (1.7) holds, then (i), (ii) and (iii) in Theorem 1.4 are fulfilled. For (i) this is obviously true. Statement (ii) follows from the multiplication table and the fact that $(v^*)^{\pm 1} \in \mathcal{M}_+$. Finally, since for a triangular extension c we have

$$(e - c^*c)^{-1}c^* \in \mathcal{M}_3^0 \oplus \mathcal{M}_d \oplus \mathcal{M}_2^0 \oplus \mathcal{M}_1,$$

the multiplication table (1.2) yields $(e - c^*c)^{-1}c^*\mathcal{M}_1 \subset \mathcal{M}_+^0$, and hence (iii) is satisfied.

As a first illustration of Theorems 1.3 and 1.4 let us consider the positive extension problem for the partially given $(n+1) \times (n+1)$ matrix

$$A = \begin{pmatrix} a_{0,0} & a_{0,1} & \cdots & a_{0,n-1} & ? \\ a_{1,0} & a_{1,1} & \cdots & a_{1,n-1} & a_{1,n} \\ \vdots & \vdots & & \vdots & \vdots \\ a_{n-1,0} & a_{n-1,1} & \cdots & a_{n-1,n-1} & a_{n-1,n} \\ ? & a_{n,1} & \cdots & a_{n,n-1} & a_{n,n} \end{pmatrix}. \tag{1.12}$$

Here the entries a_{ij} are given complex numbers such that $a_{ij} = \bar{a}_{ji}$ for $|i-j| \leq n-1$. We want to determine all positive extensions of A assuming one exists, i.e., we seek all positive definite $(n+1) \times (n+1)$ matrices $F = (f_{ij})_{i,j=0}^{n}$ such that $f_{ij} = a_{ij}$ for $|i-j| \leq n-1$. To put this problem in the context of the band method, let $\mathcal{M}$ be the set of all $(n+1) \times (n+1)$ matrices M with complex entries and consider the following subsets

$$\mathcal{M}_1 = \{M = (m_{ij})_{i,j=0}^{n} \mid m_{ij} = 0 \text{ for } j-i \leq n-1\},$$
$$\mathcal{M}_2^0 = \{M = (m_{ij})_{i,j=0}^{n} \mid m_{ij} = 0 \text{ for } j-i > n-1 \text{ or } j-i \leq\},$$
$$\mathcal{M}_d = \{M = (m_{ij})_{i,j=0}^{n} \mid m_{ij} = 0 \text{ for } i \neq j\},$$
$$\mathcal{M}_3^0 = \{M = (m_{ij})_{i,j=0}^{n} \mid m_{ij} = 0 \text{ for } j-i \geq 0 \text{ or } j-i < -(n-1)\},$$
$$\mathcal{M}_4 = \{M = (m_{ij})_{i,j=0}^{n} \mid m_{ij} = 0 \text{ for } j-i \geq -(n-1)\}.$$

Then $\mathcal{M}$ is a unital C^*-algebra in its own right,

$$\mathcal{M} = \mathcal{M}_1 \oplus \mathcal{M}_2^0 \oplus \mathcal{M}_d \oplus \mathcal{M}_3^0 \oplus \mathcal{M}_4, \tag{1.13}$$

and with the decomposition (1.13) $\mathcal{M}$ is an algebra with band structure. Notice that in this case $\mathcal{M}_+$ is the subalgebra of all upper triangular $(n+1) \times (n+1)$ matrices and $\mathcal{M}_1$ is the subalgebra of all lower triangular $(n+1) \times (n+1)$ matrices. Replacing the question marks in (1.12) by the number zero, we may view the partially matrix A as an element in $\mathcal{M}_c$ which in this case coincides with the set of all $(n-1)$-band matrices. According to Theorem 1.1, to find all positive extensions of A we have first to determine a band extension B of A (i.e., a positive extension B with the additional property that B^{-1} is an $(n-1)$-band matrix), and next one has to compute a left and a right spectral factorization of B. According to Theorem 1.3 one may start with any positive extension F of A and use spectral factorizations of F. In fact, as we shall see, in this example each positive real part extension is a canonical one and hence can be used to obtain all positive extensions. Indeed, assume F is a positive extension of A, and consider factorizations

$$F = U^{-*}U^{-1}, \qquad F = V^{-*}V^{-1}, \tag{1.14}$$

where $U^{\pm 1}$ are upper triangular and $V^{\pm 1}$ are lower triangular. Then condition (i) in Theorem 1.3 is fulfilled. Since V^* and $(V^*)^{-1}$ are both upper triangular, we have $V^*\mathcal{M}_1 = \mathcal{M}_1$, and hence condition (ii) is fulfilled. Finally, notice that $\mathcal{M}\mathcal{M}_1$ consists of all $(n+1)\times(n+1)$ matrices of which all entries are zero except those in the last column. In particular, $F^{-1}\mathcal{M}_1 \subset \mathcal{M}_+$, and condition (iii) in Theorem 1.3 holds. It follows that all positive extensions of A are given by

$$(VG+U)^{-*}(I-G^*G)(VG+U)^{-1},$$

where U and V are as in (1.14) and the free parameter G is an $(n+1)\times(n+1)$ matrix of which all entries are zero except the element in the right upper corner which is an arbitrary complex number in the open unit disc.

The proofs of Theorems 1.3 and 1.4 are based on the Grassmannian approach to interpolation and extension problems originating in [3] as adapted to a more abstract setting in [1]. These techniques are presented and developed further in the second and third section of the paper. In fact, as a result of finding relations between the band method and the Grassmannian approach, we derive two general theorems: (1) about linear fractional representations of positive and contractive extensions and (2) about the construction of the corresponding coefficient matrices. These results are stated in Section 2; their proofs appear in Section 3. When specified for the band structure the results of Section 2 yield Theorems 1.3 and 1.4 above as immediate corollaries. In Section 4 we consider the Carathéodory and Nehari extension problems, and we illustrate for these classical problems the difference between band extensions and triangular extensions on the one hand and canonical positive real part extensions and canonical strictly contractive extensions on the other hand.

In the last section we show that the Grassmannian approach of Section 2 also can be used to obtain linear fractional representations of all extensions for a positive extension problem of non-band type, where the multiplication table differs from the one appearing in (1.2). The latter theme will be pursued further for interpolation problems of Sarason and Nevanlinna-Pick type in a future publication.

2. THE GRASSMANNIAN APPROACH

Throughout this section we will use the following set up. We have given a unital C^*-algebra $\mathcal{R}$ with unit e, a $*$-subalgebra $\mathcal{N}$ of $\mathcal{R}$, a subalgebra $\mathcal{N}_+$ of $\mathcal{N}$, and a linear submanifold $\mathcal{N}_1$ of $\mathcal{N}$. We assume that the unit e of $\mathcal{R}$ is in $\mathcal{N}_+$ and that $\mathcal{N}_1$ is a right module over $\mathcal{N}_+$. In other words,

$$e \in \mathcal{N}_+, \qquad \mathcal{N}_1\mathcal{N}_+ \subset \mathcal{N}_1. \tag{2.1}$$

Furthermore, we fix an element $k \in \mathcal{N}$. The above notation will be kept fixed throughout this section. Note that we do not require $\mathcal{N}_1$ to be a subset of $\mathcal{N}_+$.

An element $g \in \mathcal{R}$ is said to be *positive definite* in $\mathcal{R}$ (notation: $g >_{\mathcal{R}} 0$) if $g = a^*a$ for some invertible element $a \in \mathcal{R}$. We call $g \in \mathcal{R}$ *strictly contractive* if $e - g^*g$ is positive definite or, equivalently, if $\|g\|_{\mathcal{R}} < 1$, where $\|\cdot\|_{\mathcal{R}}$ denotes the norm on $\mathcal{R}$.

We say that $f \in \mathcal{R}$ is an *$\mathcal{R}$-positive real part extension* of the given element $k \in \mathcal{N}$ if $f - k \in \mathcal{N}_1$ and $f + f^*$ is positive definite in $\mathcal{R}$. We call $f \in \mathcal{R}$ a *strictly contractive extension* of k if $k - f \in \mathcal{N}_1$ and $\|f\|_{\mathcal{R}} < 1$. Notice that the condition $k \in \mathcal{N}$ and $f - k \in \mathcal{N}_1$ automatically implies that $f \in \mathcal{N}$, because $\mathcal{N}_1$ is contained in $\mathcal{N}$. Thus $\mathcal{R}$-positive real part extensions and strictly contractive extensions of k are elements of $\mathcal{N}$. Therefore, the role of the algebra $\mathcal{N}$ is minor, and in what follows we may as well take $\mathcal{N} = \mathcal{R}$.

Our aim is to derive linear fractional representations of all $\mathcal{R}$-positive real part extensions of k and of all strictly contractive extensions of k. For this purpose we need an additional connection between $\mathcal{N}_+$ and $\mathcal{R}$, which is expressed by the following axiom.

AXIOM (A). *If $g \in \mathcal{N}_+$ and $\|g\|_{\mathcal{R}} < 1$, then $(e - g)^{-1} \in \mathcal{N}_+$.*

Assume axiom (A) holds, and let g be an element of $\mathcal{N}_+$ with spectral radius $r_{\text{spec}}(g)$ strictly less than one. Then $(e - g)^{-1} \in \mathcal{N}_+$. To see this, notice that $r_{\text{spec}}(g) < 1$ implies that $\|g^n\|_{\mathcal{R}} < 1$ for some n. Hence, by axiom (A), we have $(e - g^n)^{-1} \in \mathcal{N}_+$. Also $e + g + \cdots + g^{n-1} \in \mathcal{N}_+$, and therefore

$$(e - g)^{-1} = (e - g^n)(e + g + \cdots + g^{n-1}) \in \mathcal{N}_+.$$

Let x be an invertible element in $\mathcal{R}$, and consider the set $\mathcal{N}'_+ = x\mathcal{N}_+x^{-1}$. Notice that $\mathcal{N}'_+$ is a subalgebra of $\mathcal{R}$. From the remark made in the previous paragraph it follows that axiom (A) holds for $\mathcal{N}'_+$ if and only if this axiom holds for $\mathcal{N}_+$. Indeed, assume axiom (A) holds for $\mathcal{N}_+$, and let $g' \in \mathcal{N}'_+$ be strictly contractive. Then $g = x^{-1}g'x \in \mathcal{N}_+$, and $r_{\text{spec}}(g) < 1$. Thus, by the remark made in the previous paragraph, $(e - x^{-1}g'x)^{-1} \in \mathcal{N}_+$, and hence

$$(e - g')^{-1} = x(e - x^{-1}g'x)^{-1}x^{-1} \in \mathcal{N}'_+.$$

Another consequence of axiom (A) is the following lemma.

LEMMA 2.1. *Assume that axiom* (A) *holds. Let* $a \in \mathcal{N}_+$ *be such that* $a + a^*$ *is positive definite in* $\mathcal{R}$. *Then* a *is invertible and* $a^{-1} \in \mathcal{M}_+$.

For the proof of Lemma 2.1 we refer to the proof of Lemma XXXIV.2.2 in [7].

To describe all $\mathcal{R}$-positive real part extensions and all strictly contractive extensions of our given k, consider a 2×2 block matrix Θ with entries in $\mathcal{R}$,

$$\Theta = \begin{pmatrix} \Theta_{11} & \Theta_{12} \\ \Theta_{21} & \Theta_{22} \end{pmatrix} \in \mathcal{R}^{2\times 2}. \tag{2.2}$$

We view Θ as the coefficient matrix of a linear fractional map, i.e., with Θ we associate the map

$$\mathcal{F}_\Theta(h) = (\Theta_{11}h + \Theta_{12})(\Theta_{21}h + \Theta_{22})^{-1}. \tag{2.3}$$

We shall consider coefficient matrices Θ satisfying the following additional conditions:

(CM1) $\Theta\begin{pmatrix} \mathcal{N}_1 \\ \mathcal{N}_+ \end{pmatrix} = \begin{pmatrix} e & k \\ 0 & e \end{pmatrix}\begin{pmatrix} \mathcal{N}_1 \\ \mathcal{N}_+ \end{pmatrix}$,

(CM2) Θ_{22} is invertible and $\Theta_{22}^{-1} \in \mathcal{N}_+$,

(CM3a) $\Theta^*\begin{pmatrix} 0 & e \\ e & 0 \end{pmatrix}\Theta = \begin{pmatrix} -p & 0 \\ 0 & q \end{pmatrix}$,

or

(CM3b) $\Theta^*\begin{pmatrix} -e & 0 \\ 0 & e \end{pmatrix}\Theta = \begin{pmatrix} -p & 0 \\ 0 & q \end{pmatrix}$.

Here p and q are positive definite elements in $\mathcal{R}$, and k, $\mathcal{N}_+$ and $\mathcal{N}_1$ are as in the beginning of this section. We shall prove the following theorems.

THEOREM 2.2. *Assume that axiom* (A) *holds.*

(a) *If* Θ *in* (2.2) *satisfies conditions* (CM1), (CM2) *and* (CM3a), *then each* $\mathcal{R}$-*positive real part extension* f *of* k *is of the form*

$$f = \mathcal{F}_\Theta(h), \tag{2.4}$$

where the free parameter h *is an arbitrary element in* $\mathcal{N}_1$ *such that* $q - h^*ph$ *is positive definite in* $\mathcal{R}$. *Moreover, the map* $\mathcal{F}_\Theta$ *provides a one-to-one correspondence between all such* h *and all* $\mathcal{R}$-*positive real part extensions* f *of* k.

(b) *If* Θ *in* (2.2) *satisfies conditions* (CM1), (CM2) *and* (CM3b), *then each strictly contractive extension* f *of* k *is of the form* (2.4), *where the free parameter* h *is as in part* (a), *and in this case the map* $\mathcal{F}_\Theta$ *provides a one-to-one correspondence between all such* h *and all strictly contractive extensions* f *of* k.

THEOREM 2.3. *Assume that axiom* (A) *holds.*

(a) *There exists a* 2×2 *block matrix* Θ *as in* (2.2) *satisfying conditions* (CM1), (CM2) *and* (CM3a) *if and only if* k *has a* $\mathcal{R}$*-positive real part extension* c *with the following properties:*

(i) $c + c^* = u^{-*}qu^{-1}$ *with* $u^{\pm 1} \in \mathcal{N}_+$,

(ii) *there exists* $v \in \mathcal{R}$ *such that* $v^*(c + c^*)v = p$, *and*

$$(c + c^*)^{-1}\mathcal{N}_1 = v\mathcal{N}_1 \subset \mathcal{N}_+. \tag{2.5}$$

In fact, if Θ *has the properties* (CM1), (CM2) *and* (CM3a), *then* $c = \Theta_{12}\Theta_{22}^{-1}$ *is an* $\mathcal{R}$*-positive real part extension of* k *and statements* (i) *and* (ii) *hold with* $u = \Theta_{22}$ *and* $v = \Theta_{21}$. *Conversely, if* c *is an* $\mathcal{R}$*-positive real part extension of* k *satisfying* (i) *and* (ii), *then*

$$\Theta = \begin{pmatrix} -c^*v & cu \\ v & u \end{pmatrix} \tag{2.6}$$

has the properties (CM1), (CM2) *and* (CM3a).

(b) *There exists a* 2×2 *block matrix* Θ *as in* (2.2) *satisfying conditions* (CM1), (CM2) *and* (CM3b) *if and only if* k *has a strictly contractive extension* c *with the following properties*

(j) $e - c^*c = u^{-*}qu^{-1}$ *with* $u^{\pm 1} \in \mathcal{N}_+$,

(jj) *there exists* $v \in \mathcal{R}$ *such that* $v^*(e - cc^*)v = p$ *and*

$$(e - cc^*)^{-1}\mathcal{N}_1 = v\mathcal{N}_1, \quad c^*v\mathcal{N}_1 \subset \mathcal{N}_+. \tag{2.7}$$

In fact, if Θ *has the properties* (CM1), (CM2) *and* (CM3b), *then* $c = \Theta_{12}\Theta_{22}^{-1}$ *is a strictly contractive extension of* k *and statements* (j) *and* (jj) *hold with* $u = \Theta_{22}$ *and* $v = \Theta_{11}$. *Conversely, if* c *is a strictly contractive extension of* k *satisfying* (j) *and* (jj), *then*

$$\Theta = \begin{pmatrix} v & cu \\ c^*v & u \end{pmatrix} \tag{2.8}$$

has the properties (CM1), (CM2) *and* (CM3b).

Theorem 2.3 becomes more transparant if one requires additionally that the coefficient matrix Θ in (2.2) is invertible in $\mathcal{R}^{2\times 2}$.

THEOREM 2.4. *Assume that axiom* (A) *holds.*

(a) *There exists a* 2×2 *block matrix* Θ *as in* (2.2) *such that* Θ *is invertible in* $\mathcal{R}^{2\times 2}$ *and* Θ *satisfying conditions* (CM1), (CM2) *and* (CM3a) *if and only if* k *has a* $\mathcal{R}$*-positive real part extension* c *with the following properties:*

(1a) $c+c^* = u^{-*}qu^{-1}$ *with* $u^{\pm 1} \in \mathcal{N}_+$,

(2a) $c+c^* = v^{-*}pv^{-1}$ *with* $v^{\pm 1} \in \mathcal{R}$ *and* $p^{-1}v^*\mathcal{N}_1 = \mathcal{N}_1$,

(3a) $(c+c^*)^{-1}\mathcal{N}_1 \subset \mathcal{N}_+$.

In fact, if Θ *is invertible and satisfies* (CM1), (CM2) *and* (CM3a), *then* $c = \Theta_{12}\Theta_{22}^{-1}$ *is an* $\mathcal{R}$*-positive real part extension of* k *and statements* (1a), (2a) *and* (3a) *hold with* $u = \Theta_{22}$ *and* $v = \Theta_{21}$. *Conversely, if* c *is an* $\mathcal{R}$*-positive real part extension of* k *satisfying* (1a) *and* (2a) *and* (3a), *then*

$$\Theta = \begin{pmatrix} -c^*v & cu \\ v & u \end{pmatrix} \tag{2.9}$$

is invertible in $\mathcal{R}^{2\times 2}$ *and has the properties* (CM1), (CM2) *and* (CM3a).

(b) *There exists a* 2×2 *block matrix* Θ *as in* (2.2) *such that* Θ *is invertible in* $\mathcal{R}^{2\times 2}$ *and* Θ *satisfies conditions* (CM1), (CM2) *and* (CM3b) *if and only if* k *has a strictly contractive extension* c *with the following properties*

(1b) $e - c^*c = u^{-*}qu^{-1}$ *with* $u^{\pm 1} \in \mathcal{N}_+$,

(2b) $e - cc^* = v^{-*}pv^{-1}$ *with* $v^{\pm 1} \in \mathcal{R}$ *and* $p^{-1}v^*\mathcal{N}_1 = \mathcal{N}_1$,

(3b) $(e-c^*c)^{-1}c^*\mathcal{N}_1 \subset \mathcal{N}_+$.

In fact, if Θ *is invertible and satisfies the properties* (CM1), (CM2) *and* (CM3b), *then* $c = \Theta_{12}\Theta_{22}^{-1}$ *is a strictly contractive extension of* k *and statements* (1b), (2b) *and* (3b) *hold with* $u = \Theta_{22}$ *and* $v = \Theta_{11}$. *Conversely, if* c *is a strictly contractive extension of* k *satisfying* (1b), (2b) *and* (3b), *then*

$$\Theta = \begin{pmatrix} v & cu \\ c^*v & u \end{pmatrix} \tag{2.10}$$

is invertible in $\mathcal{R}^{2\times 2}$ *and has the properties* (CM1), (CM2) *and* (CM3b).

Theorems 1.3 and 1.4 in Section 1 are immediate corollaries of Theorems 2.2 and 2.4. For example, to prove Theorem 1.3 we apply first Theorem 2.4(a) with $\mathcal{N}_1 = \mathcal{M}_1$, $\mathcal{N}_+ = \mathcal{M}_+$, $\mathcal{N} = \mathcal{M}$, $p = q = e$, and with $k \in \mathcal{M}_{\ell+}$. We may write f in Theorem 1.3 as $f = c + c^*$ with $c \in \mathcal{M}_+$. Thus c is an $\mathcal{R}$-positive real part extension of k. According to Theorem 2.4(a) conditions (i), (ii) and (iii) in Theorem 1.3 imply that

$$\Theta = \begin{pmatrix} -c^*v & cu \\ v & u \end{pmatrix}$$

has the properties (CM1), (CM2) and (CM3a). But we can apply Theorem 2.2(a) to prove part (b) of Theorem 1.3. Since $\tilde{\mathcal{F}}(g) = \mathcal{F}(g) + \mathcal{F}(g)^*$, part (a) also follows. Theorem 1.4 is proved in a similar way using Theorems 2.4(b) and 2.2(b).

In the remaining part of this section we prove Theorem 2.4, assuming Theorems 2.2 and 2.3 hold. The latter two theorems are proved in Section 3. We begin with two auxiliarly results.

LEMMA 2.5. *Let Θ be as in (2.2), and assume that Θ_{22} is invertible in $\mathcal{R}$. Put $c = \Theta_{12}\Theta_{22}^{-1}$.*

(a) *If, in addition, condition* (CM3a) *holds, then $\Theta_{11} = -c^*\Theta_{21}$ and*

$$c + c^* = \Theta_{22}^{-*}q\Theta_{22}^{-1}, \qquad \Theta_{21}^*(c + c^*)\Theta_{21} = p. \tag{2.11a}$$

(b) *If, in addition, condition* (CM3b) *holds, then $\Theta_{21} = c^*\Theta_{11}$ and*

$$e - c^*c = \Theta_{22}^{-*}q\Theta_{22}^{-1}, \qquad \Theta_{11}^*(e - cc^*)\Theta_{11} = p. \tag{2.11b}$$

PROOF. (a) Condition (CM3a) may be restated as

$$\begin{pmatrix} \Theta_{21}^* & \Theta_{11}^* \\ \Theta_{22}^* & \Theta_{12}^* \end{pmatrix} \begin{pmatrix} \Theta_{11} & \Theta_{12} \\ \Theta_{21} & \Theta_{22} \end{pmatrix} = \begin{pmatrix} -p & 0 \\ 0 & q \end{pmatrix}. \tag{2.12a}$$

It follows that

$$c + c^* = \Theta_{12}\Theta_{22}^{-1} + \Theta_{22}^{-*}\Theta_{12}^* = \Theta_{22}^{-*}(\Theta_{22}^*\Theta_{12} + \Theta_{12}^*\Theta_{22})\Theta_{22}^{-1} = \Theta_{22}^{-*}q\Theta_{22}^{-1},$$

which proves the first identity in (2.11a). Next,

$$-c^*\Theta_{21} = -\Theta_{22}^{-*}\Theta_{12}^*\Theta_{21} = \Theta_{22}^{-*}\Theta_{22}^*\Theta_{11} = \Theta_{11}.$$

Here we used that $\Theta_{22}^*\Theta_{11} + \Theta_{12}^*\Theta_{21} = 0$, because of (2.12a). We proceed with

$$\begin{aligned} \Theta_{21}^*(c + c^*)\Theta_{21} &= \Theta_{21}^*\Theta_{22}^{-*}q\Theta_{22}^{-1}\Theta_{21} \\ &= \Theta_{21}^*\Theta_{22}^{-*}(\Theta_{22}^*\Theta_{12} + \Theta_{12}^*\Theta_{22})\Theta_{22}^{-1}\Theta_{21} \\ &= \Theta_{21}^*\Theta_{12}\Theta_{22}^{-1}\Theta_{21} + \Theta_{21}^*\Theta_{22}^{-*}\Theta_{12}^*\Theta_{21}. \end{aligned}$$

From (2.12a) we see that $\Theta_{21}^*\Theta_{12} = -\Theta_{11}^*\Theta_{22}$. Thus

$$\begin{aligned} \Theta_{21}^*(c + c^*)\Theta_{21} &= -\Theta_{11}^*\Theta_{22}\Theta_{22}^{-1}\Theta_{21} - \Theta_{21}^*\Theta_{22}^{-*}\Theta_{22}^*\Theta_{11} \\ &= -\Theta_{11}^*\Theta_{21} - \Theta_{21}^*\Theta_{11} = p, \end{aligned}$$

which proves the second identity in (2.11a).

(b) Condition (CM3b) may be restated as

$$\begin{pmatrix} -\Theta_{11}^* & \Theta_{21}^* \\ -\Theta_{12}^* & \Theta_{22}^* \end{pmatrix} \begin{pmatrix} \Theta_{11} & \Theta_{12} \\ \Theta_{21} & \Theta_{22} \end{pmatrix} = \begin{pmatrix} -p & 0 \\ 0 & q \end{pmatrix}. \tag{2.12b}$$

It follows that $-\Theta_{12}^*\Theta_{12} + \Theta_{22}^*\Theta_{22} = q$, and hence

$$e - c^*c = e - \Theta_{22}^{-*}\Theta_{12}^*\Theta_{12}\Theta_{22}^{-1} = \Theta_{22}^{-*}q\Theta_{22}^{-1},$$

and the first identity in (2.11b) is proved. Next,

$$c^*\Theta_{11} = \Theta_{22}^{-*}\Theta_{12}^*\Theta_{11} = \Theta_{22}^{-*}\Theta_{22}^*\Theta_{21} = \Theta_{21},$$

and finally $\Theta_{11}^*(e - cc^*)\Theta_{11} = \Theta_{11}^*\Theta_{11} - \Theta_{21}^*\Theta_{21} = p$. □

LEMMA 2.6. *Let Θ be as in* (2.2), *and assume that Θ_{22} is invertible.*

(a) *Let condition* (CM3a) *be fulfilled. Then Θ is invertible in $\mathcal{R}^{2\times 2}$ if and only if Θ_{21} is invertible in $\mathcal{R}$.*

(b) *Let condition* (CM3b) *be fulfilled. Then Θ is invertible in $\mathcal{R}^{2\times 2}$ if and only if Θ_{11} is invertible in $\mathcal{R}$.*

PROOF. (a) Recall that condition (CM3a) is equivalent to (2.12a). Assume that Θ is invertible. Then we see from (2.12a) that

$$\begin{pmatrix} \Theta_{11} & \Theta_{12} \\ \Theta_{21} & \Theta_{22} \end{pmatrix} \begin{pmatrix} -p^{-1} & 0 \\ 0 & q^{-1} \end{pmatrix} \begin{pmatrix} \Theta_{21}^* & \Theta_{11}^* \\ \Theta_{22}^* & \Theta_{12}^* \end{pmatrix} = \begin{pmatrix} e & 0 \\ 0 & e \end{pmatrix}.$$

In particular, $\Theta_{21}p^{-1}\Theta_{21}^* = \Theta_{22}q^{-1}\Theta_{22}^*$. Since Θ_{22} is invertible, this implies that Θ_{21}^* has a left inverse. On the other hand, from (2.11a) we see that $\Theta_{21}^*\Theta_{22}^{-*}q\Theta_{22}^{-1}\Theta_{21}p^{-1} = e$. Hence Θ_{21}^* has also a right inverse. Thus Θ_{21} is invertible.

Next, assume Θ_{21} is invertible. Since $\Theta_{11} = -c^*\Theta_{21}$ and $\Theta_{12} = c\Theta_{22}$, we have

$$\begin{pmatrix} \Theta_{11} & \Theta_{12} \\ \Theta_{21} & \Theta_{22} \end{pmatrix} = \begin{pmatrix} e & c \\ 0 & e \end{pmatrix} \begin{pmatrix} -(c+c^*)\Theta_{21} & 0 \\ \Theta_{21} & \Theta_{22} \end{pmatrix}. \tag{2.13}$$

Notice that the first equality in (2.11a) implies that $c + c^*$ is invertible. Also, Θ_{21} and Θ_{22} are invertible. Thus the second factor in the right hand side of (2.13) is invertible in $\mathcal{R}^{2\times 2}$. The same is true for the first factor. Thus Θ is invertible in $\mathcal{R}^{2\times 2}$.

(b) We are assuming that Θ_{22} is invertible. Thus Θ is invertible if and only if $\Theta_{11} - \Theta_{12}\Theta_{22}^{-1}\Theta_{21}$ is invertible.

Suppose now that Θ_{11} is invertible. Then

$$\Theta_{11}^*(\Theta_{11} - \Theta_{12}\Theta_{22}^{-1}\Theta_{21}) = \Theta_{11}^*\Theta_{11} + \Theta_{21}^*\Theta_{21} \geq \Theta_{11}^*\Theta_{11}$$

(where we used $\Theta_{11}^*\Theta_{12} = \Theta_{21}^*\Theta_{22}$), and hence $\Theta_{11} - \Theta_{12}\Theta_{22}^{-1}\Theta_{21}$ is invertible as wanted.

Conversely, suppose that $\Theta_{11} - \Theta_{12}\Theta_{22}^{-1}\Theta_{21}$ is invertible. Then, from the identity derived in the preceding paragraph we have

$$\Theta_{11}^* = (\Theta_{11}^*\Theta_{11} + \Theta_{21}^*\Theta_{21})(\Theta_{11} - \Theta_{12}\Theta_{22}^{-1}\Theta_{21})^{-1}.$$

Note that the left factor on the right hand side is

$$\begin{aligned}\Theta_{11}^*\Theta_{11} + \Theta_{21}^*\Theta_{21} &= \Theta_{11}^*\Theta_{11} - \Theta_{21}^*\Theta_{21} + 2\Theta_{21}^*\Theta_{21}\\ &= e + 2\Theta_{21}^*\Theta_{21}\\ &\geq e,\end{aligned}$$

and hence $\Theta_{11}^*\Theta_{11} + \Theta_{21}^*\Theta_{21}$ is invertible. We conclude that Θ_{11}^* and hence also Θ_{11} is invertible. □

PROOF OF THEOREM 2.4. Assume Θ is invertible and satisfies (CM1), (CM2) and (CM3a). Put $c = \Theta_{12}\Theta_{22}^{-1}$, and set $u = \Theta_{22}$ and $v = \Theta_{21}$. By Theorem 2.3(a), the element c is an $\mathcal{R}$-positive real part extension of k and conditions (i) and (ii) in Theorem 2.3(a) are fulfilled. In particular, with this choice of c, u and v, conditions (1a) and (3a) are satisfied. Notice that $v = \Theta_{21}$ is invertible, by Lemma 2.6(a). So, we see from (ii) in Theorem 2.3(a) that $c + c^* = v^{-*}pv^{-1}$. By using the latter identity in (2.5), we obtain $vp^{-1}v^*\mathcal{N}_1 = v\mathcal{N}_1$, and hence $p^{-1}v^*\mathcal{N}_1 = \mathcal{N}_1$. Thus (2a) also holds.

Conversely, let c be an $\mathcal{R}$-positive real part extension of k satisfying (1a), (2a) and (3a). Then conditions (i) and (ii) in Theorem 2.3(a) are fulfilled. It follows that Θ in (2.9) has the properties (CM1), (CM2) and (CM3a). Since v is invertible, Lemma 2.6(a) implies that Θ is invertible in $\mathcal{R}^{2\times 2}$.

(b) Assume Θ is invertible and satisfies (CM1), (CM2) and (CM3b). Put $c = \Theta_{12}\Theta_{22}^{-1}$, and set $u = \Theta_{22}$ and $v = \Theta_{11}$. From Theorem 2.3(b) we know that c is a strictly contractive extension of k and conditions (j) and (jj) in Theorem 2.3(b) are fulfilled. In particular, with this choice of c, u and v, condition (1b) holds. Notice that

$$(e - c^*c)^{-1}c^*\mathcal{N}_1 = c^*(e - c^*c)^{-1}\mathcal{N}_1 = c^*v\mathcal{N}_1 \subset \mathcal{N}_+,$$

by (2.7). Hence (3b) is fulfilled. Since Θ is invertible, Lemma 2.6(b) gives that $v = \Theta_{11}$ is invertible. So we see from (jj) in Theorem 2.3(b) that $e - cc^* = v^{-*}pv^{-1}$. By using the latter identity in the first part of (2.7), we obtain $vp^{-1}v^*\mathcal{N}_1 = v\mathcal{N}_1$, and hence $p^{-1}v^*\mathcal{N}_1 = \mathcal{N}_1$. Thus (2b) also holds.

Conversely, let c be a strictly contractive extension of k satisfying (1b), (2b) and (3b). Then $(e - cc^*)^{-1}\mathcal{N}_1 = vp^{-1}v^*\mathcal{N}_1 = v\mathcal{N}_1$, and thus

$$c^*v\mathcal{N}_1 = c^*(e - cc^*)^{-1}\mathcal{N}_1 = (e - c^*c)^{-1}c^*\mathcal{N}_1 \subset \mathcal{N}_+.$$

We conclude that conditions (j) and (jj) in Theorem 2.3(b) are fulfilled. It follows that Θ in (2.10) has the properties (CM1), (CM2) and (CM3b). Since $\Theta_{11} = v$ is invertible, Lemma 2.6(b) implies that Θ is invertible in $\mathcal{R}^{2\times 2}$. □

3. PROOFS OF THEOREMS 2.2 AND 2.3

Throughout this section we use the notations and terminology introduced in Section 2. In particular, $\mathcal{R}$ is a unital C^*-algebra, $\mathcal{N}_+$ is a subalgebra of $\mathcal{R}$ which contains the unit e, $\mathcal{N}_1$ is a linear submanifold of $\mathcal{R}$ which is a right module over $\mathcal{N}_+$, and k is an element of $\mathcal{R}$. (Actually, $k \in \mathcal{N}$, where $\mathcal{N}$ is a $*$-subalgebra of $\mathcal{R}$ containing $\mathcal{N}_+$ and $\mathcal{N}_1$.) Our aim is to prove Theorems 2.2 and 2.3. We begin with the following proposition, which is typical for the Grassmannian approach.

PROPOSITION 3.1. *Let f be an element of $\mathcal{R}$, and consider the linear manifold*

$$G = \begin{pmatrix} f \\ e \end{pmatrix} \mathcal{N}_+. \tag{3.1}$$

Then $f - k \in \mathcal{N}_1$ if and only if

$$G \subset \begin{pmatrix} e & k \\ 0 & e \end{pmatrix} \begin{pmatrix} \mathcal{N}_1 \\ \mathcal{N}_+ \end{pmatrix}. \tag{3.2}$$

PROOF. Assume (3.2). Since $e \in \mathcal{N}_+$, we have $\begin{pmatrix} f \\ e \end{pmatrix}$ is in G. Thus there exist $n_1 \in \mathcal{N}_1$ and $x \in \mathcal{N}_+$ such that

$$\begin{pmatrix} f \\ e \end{pmatrix} = \begin{pmatrix} e & k \\ 0 & e \end{pmatrix} \begin{pmatrix} n_1 \\ x \end{pmatrix} = \begin{pmatrix} n_1 + kx \\ x \end{pmatrix}.$$

Obviously, $x = e$, and thus $f = n_1 + k$. So $f - k$ is in $\mathcal{N}_1$.

Conversely, assume $f = k + n_1$ for some $n_1 \in \mathcal{N}_1$. Since $\mathcal{N}_1\mathcal{N}_+ \subset \mathcal{N}_1$, we have $n_1\mathcal{N}_+ \subset \mathcal{N}_1$, and thus

$$G = \begin{pmatrix} f \\ e \end{pmatrix} \mathcal{N}_+ = \begin{pmatrix} e & k \\ 0 & e \end{pmatrix} \begin{pmatrix} n_1 \\ e \end{pmatrix} \mathcal{N}_+ \subset \begin{pmatrix} e & k \\ 0 & e \end{pmatrix} \begin{pmatrix} \mathcal{N}_1 \\ \mathcal{N}_+ \end{pmatrix},$$

which proves (3.2). □

PROOF OF THEOREM 2.2(a). We split the proof into five parts.

Part (α). First we show that condition (CM1) implies that

$$\Theta_{21}\mathcal{N}_1 \subset \mathcal{N}_+, \qquad \Theta_{22} \in \mathcal{N}_+. \tag{3.3}$$

Indeed, from (CM1) it follows that

$$(\,0 \quad e\,) \begin{pmatrix} \Theta_{11} & \Theta_{12} \\ \Theta_{21} & \Theta_{22} \end{pmatrix} \begin{pmatrix} \mathcal{N}_1 \\ \mathcal{N}_+ \end{pmatrix} = (\,0 \quad e\,) \begin{pmatrix} e & k \\ 0 & e \end{pmatrix} \begin{pmatrix} \mathcal{N}_1 \\ \mathcal{N}_+ \end{pmatrix} = \mathcal{N}_+.$$

In particular,

$$\Theta_{21}\mathcal{N}_1 = (\,\Theta_{21} \quad \Theta_{22}\,) \begin{pmatrix} \mathcal{N}_1 \\ 0 \end{pmatrix} \subset (\,0 \quad e\,) \begin{pmatrix} \Theta_{11} & \Theta_{12} \\ \Theta_{21} & \Theta_{22} \end{pmatrix} \begin{pmatrix} \mathcal{N}_1 \\ \mathcal{N}_+ \end{pmatrix} = \mathcal{N}_+,$$

which proves the first inclusion in (3.3). In a similar way one shows that $\Theta_{22}\mathcal{N}_+ \subset \mathcal{N}_+$. Since $e \in \mathcal{N}_+$, we conclude that $\Theta_{22} \in \mathcal{N}_+$.

Part (β). We prove that

$$\Theta_{21}^*\mathcal{N}_1 \subset p\mathcal{N}_1 \qquad \Theta_{22}^*\mathcal{N}_1 \subset q\mathcal{N}_+. \tag{3.4}$$

From (2.12a) and condition (CM1) we conclude that

$$\begin{pmatrix} -p & 0 \\ 0 & q \end{pmatrix} \begin{pmatrix} \mathcal{N}_1 \\ \mathcal{N}_+ \end{pmatrix} = \begin{pmatrix} \Theta_{21}^* & \Theta_{11}^* \\ \Theta_{22}^* & \Theta_{12}^* \end{pmatrix} \begin{pmatrix} e & k \\ 0 & e \end{pmatrix} \begin{pmatrix} \mathcal{N}_1 \\ \mathcal{N}_+ \end{pmatrix}.$$

In particular,

$$\Theta_{21}^*\mathcal{N}_1 = (\,\Theta_{21}^* \quad \Theta_{11}^*\,) \begin{pmatrix} e & k \\ 0 & e \end{pmatrix} \begin{pmatrix} \mathcal{N}_1 \\ 0 \end{pmatrix} \subset p\mathcal{N}_1,$$

$$\Theta_{22}^*\mathcal{N}_1 = (\,\Theta_{22}^* \quad \Theta_{12}^*\,) \begin{pmatrix} e & k \\ 0 & e \end{pmatrix} \begin{pmatrix} \mathcal{N}_1 \\ 0 \end{pmatrix} \subset q\mathcal{N}_+.$$

Part (γ) Let h be an element of $\mathcal{N}_1$ such that $q - h^*ph$ is positive definite in $\mathcal{R}$. In this part we show that

$$f := \mathcal{F}_\Theta(h) = (\Theta_{11}h + \Theta_{12})(\Theta_{21}h + \Theta_{22})^{-1} \tag{3.5}$$

is well-defined and f is a $\mathcal{R}$-positive real part extension of k. From Lemma 2.5(a) we know that $\Theta_{21}^*\Theta_{22}^{-*}q\Theta_{22}^{-1}\Theta_{21} = p$. Since $q >_{\mathcal{R}} 0$, we may consider its square root $q^{\frac{1}{2}} \in \mathcal{R}$. Hence

$$\begin{aligned} e - (q^{\frac{1}{2}}\Theta_{22}^{-1}\Theta_{21}hq^{-\frac{1}{2}})^*(q^{\frac{1}{2}}\Theta_{22}^{-1}\Theta_{21}hq^{-\frac{1}{2}}) &= e - q^{-\frac{1}{2}}h^*\Theta_{21}^*\Theta_{22}^{-*}q\Theta_{22}^{-1}\Theta_{21}hq^{-\frac{1}{2}} \\ &= e - q^{-\frac{1}{2}}h^*phq^{-\frac{1}{2}} \\ &= q^{-\frac{1}{2}}(q - h^*ph)q^{-\frac{1}{2}} >_{\mathcal{R}} 0. \end{aligned}$$

Therefore $\|q^{\frac{1}{2}}\Theta_{22}^{-1}\Theta_{21}hq^{-\frac{1}{2}}\|_{\mathcal{R}} < 1$. It follows that $q^{\frac{1}{2}}\Theta_{22}^{-1}\Theta_{21}hq^{-\frac{1}{2}} + e$ is invertible. But then

$$\Theta_{21}h + \Theta_{22} = \Theta_{22}q^{-\frac{1}{2}}\{q^{\frac{1}{2}}\Theta_{22}^{-1}\Theta_{21}hq^{-\frac{1}{2}} + e\}q^{\frac{1}{2}}$$

is invertible. Thus f is well-defined by (3.5).

Put $x = \Theta_{21}h + \Theta_{22}$. We shall show that x and x^{-1} are in $\mathcal{N}_+$. Since $h \in \mathcal{N}_1$, the first inclusion in (3.3) yields $\Theta_{21}h \in \mathcal{N}_+$. From (CM2) we know that $\Theta_{22}^{-1}\Theta_{21}h \in \mathcal{N}_+$. Also $e \in \mathcal{N}_+$, and, by the second part of (3.3), we have $\Theta_{22} \in \mathcal{N}_+$. Thus

$$x = \Theta_{21}h + \Theta_{22} = \Theta_{22}[\Theta_{22}^{-1}\Theta_{21}h + e] \in \mathcal{N}_+\mathcal{N}_+ \subset \mathcal{N}_+.$$

Recall that $\|q^{\frac{1}{2}}\Theta_{22}^{-1}\Theta_{21}hq^{-\frac{1}{2}}\|_{\mathcal{R}} < 1$ This implies that $r_{\text{spec}}(\Theta_{22}^{-1}\Theta_{21}h) < 1$. It follows (use the remark made in the paragraph after Axiom (A) in Section 2) that $(\Theta_{22}^{-1}\Theta_{21}h + e)^{-1} \in \mathcal{N}_+$. Thus

$$x^{-1} = (\Theta_{22}^{-1}\Theta_{21}h + e)^{-1}\Theta_{22}^{-1} \in \mathcal{N}_+\mathcal{N}_+ \subset \mathcal{N}_+.$$

From (3.5) we see that

$$\begin{pmatrix} f \\ e \end{pmatrix} = \Theta \begin{pmatrix} h \\ e \end{pmatrix} x^{-1}. \tag{3.6}$$

Since $x^{-1}\mathcal{N}_+ \subset \mathcal{N}_+\mathcal{N}_+ \subset \mathcal{N}_+$ and $h\mathcal{N}_+ \subset \mathcal{N}_1\mathcal{N}_+ \subset \mathcal{N}_1$, formula (3.5) yields

$$\begin{pmatrix} f \\ e \end{pmatrix} \mathcal{N}_+ \subset \Theta \begin{pmatrix} \mathcal{N}_1 \\ \mathcal{N}_+ \end{pmatrix} = \begin{pmatrix} e & k \\ 0 & e \end{pmatrix} \begin{pmatrix} \mathcal{N}_1 \\ \mathcal{N}_+ \end{pmatrix},$$

where the second equality comes from (CM1). But then we can apply Proposition 3.1 to show that $f - k \in \mathcal{N}_1$.

Finally, using (3.5) again, we have

$$\begin{aligned} f + f^* = (f^* \quad e) \begin{pmatrix} 0 & e \\ e & 0 \end{pmatrix} \begin{pmatrix} f \\ e \end{pmatrix} &= x^{-*}(h^* \quad e)\Theta^* \begin{pmatrix} 0 & e \\ e & 0 \end{pmatrix} \Theta \begin{pmatrix} h \\ e \end{pmatrix} x^{-1} \\ = x^{-*}(h^* \quad e) \begin{pmatrix} -p & 0 \\ 0 & q \end{pmatrix} \begin{pmatrix} h \\ e \end{pmatrix} x^{-1} &= x^{-*}(q - h^*ph)x^{-1} >_{\mathcal{R}} 0. \end{aligned}$$

Thus f is a $\mathcal{R}$-positive real part extension of k.

Part (δ). This part concerns the uniqueness of the representation (3.5). Take $h \in \mathcal{N}_1$ with $q - h^*ph >_{\mathcal{R}} 0$. Define f by (3.5). We want to show that h is uniquely determined by f. As before, put $x = \Theta_{21}h + \Theta_{22}$. Since

$$\begin{pmatrix} f \\ e \end{pmatrix} = \Theta \begin{pmatrix} h \\ e \end{pmatrix} (\Theta_{21}h + \Theta_{22})^{-1},$$

we can use (2.12a) to show that

$$\begin{pmatrix} \Theta_{21}^* f + \Theta_{11}^* \\ \Theta_{22}^* f + \Theta_{12}^* \end{pmatrix} = \begin{pmatrix} -ph \\ q \end{pmatrix} x^{-1}.$$

Now, put $w = f - \Theta_{12}\Theta_{22}^{-1}$. Then

$$\begin{aligned} \Theta_{21}^* f + \Theta_{11}^* &= \Theta_{21}^* w + \Theta_{21}^*\Theta_{12}\Theta_{22}^{-1} + \Theta_{11}^* \\ &= \Theta_{21}^* w + \{\Theta_{21}^*\Theta_{12} + \Theta_{11}^*\Theta_{22}\}\Theta_{22}^{-1} = \Theta_{21}^* w. \end{aligned}$$

Here we used that $\Theta_{21}^*\Theta_{12} + \Theta_{11}^*\Theta_{22} = 0$, which follows from (2.12a). Similarly,

$$\begin{aligned} \Theta_{22}^* f + \Theta_{12}^* &= \Theta_{22}^* w + \Theta_{22}^*\Theta_{12}\Theta_{22}^{-1} + \Theta_{12}^* \\ &= \Theta_{22}^* w + \{\Theta_{22}^*\Theta_{12} + \Theta_{12}^*\Theta_{22}\}\Theta_{22}^{-1} \\ &= \Theta_{22}^* w + q\Theta_{22}^{-1}. \end{aligned}$$

We conclude that

$$\Theta_{21}^* w = -phx^{-1}, \qquad \Theta_{22}^* w + q\Theta_{22}^{-1} = qx^{-1}. \tag{3.7}$$

Thus

$$h = -p^{-1}\Theta_{21}^* w(\Theta_{22}^* w + q\Theta_{22}^{-1})^{-1} q. \tag{3.8}$$

Since w is uniquely determined by f, we conclude that the same holds true for h.

Part (ε). Suppose f is a $\mathcal{R}$-positive real part extension of k. In this part we show that $f = \mathcal{F}_\Theta(h)$ for some $h \in \mathcal{N}_1$ with $q - h^*ph$ positive definite in $\mathcal{R}$. Put $w = f - \Theta_{12}\Theta_{22}^{-1}$. From Part ($\gamma$) we know that $c := \Theta_{12}\Theta_{22}^{-1}$ is also a $\mathcal{R}$-positive real part extension of k. Thus $w = f - c \in \mathcal{N}_1$. From the second inclusion in (3.4) we see that $\Theta_{22}^* w \in \mathcal{N}_+$. Thus $\Theta_{22}^* w\Theta_{22} + q$ belongs to $q\mathcal{N}_+$.

We first show that $\Theta_{22}^* w\Theta_{22} + q$ is invertible. Since c and f are both $\mathcal{R}$-positive real part extensions, we have

$$2\Theta_{22}^*(c + c^*)\Theta_{22} + \Theta_{22}^*(w + w^*)\Theta_{22} = \Theta_{22}^*(c + c^* + f + f^*)\Theta_{22} >_{\mathcal{R}} 0.$$

From Lemma 2.5(a) we know that $c + c^* = \Theta_{22}^{-*} q \Theta_{22}^{-1}$. Thus

$$(q + \Theta_{22}^* w \Theta_{22}) + (q + \Theta_{22}^* w \Theta_{22})^* >_{\mathcal{R}} 0.$$

Put $a := q^{-\frac{1}{2}}(q + \Theta_{22}^* w \Theta_{22}) q^{-\frac{1}{2}}$. Then $a + a^* >_{\mathcal{R}} 0$ and $a \in \mathcal{M}_+ := q^{\frac{1}{2}} \mathcal{N}_+ q^{-\frac{1}{2}}$. Notice that $\mathcal{M}_+$ is a subalgebra of $\mathcal{R}$. From the remark made in the second paragraph after Axiom (A) in Section 2, we know that this axiom is also fulfilled for $\mathcal{M}_+$. But then we can use Lemma 2.1 (with $\mathcal{M}_+$ in place of $\mathcal{N}_+$). Therefore $q + \Theta_{22}^* w \Theta_{22} = q^{\frac{1}{2}} a q^{\frac{1}{2}}$ is invertible and

$$(\Theta_{22}^* w \Theta_{22} + q)^{-1} = q^{-\frac{1}{2}} a^{-1} q^{-\frac{1}{2}} \in \mathcal{N}_+ q^{-1}. \tag{3.9}$$

Inspired by (3.8) we take

$$h := -p^{-1} \Theta_{21}^* w \Theta_{22} (\Theta_{22}^* w \Theta_{22} + q)^{-1} q. \tag{3.10}$$

Since $\Theta_{22} \in \mathcal{N}_+$ and $(\Theta_{22}^* w \Theta_{22} + q)^{-1} q \in \mathcal{N}_+$, the rule $\mathcal{N}_+ \mathcal{N}_+ \subset \mathcal{N}_+$ yields

$$\Theta_{22}(\Theta_{22}^* w \Theta_{22} + q)^{-1} q \in \mathcal{N}_+.$$

Recall that $w \in \mathcal{N}_1$ and $\mathcal{N}_1 \mathcal{N}_+ \subset \mathcal{N}_1$. Thus $w\Theta_{22}(\Theta_{22}^* w \Theta_{22} + q)^{-1} q \in \mathcal{N}_1$. But then we can use the first inclusion in (3.4) to show that $h \in \mathcal{N}_1$.

From (2.12a) and (CM1) it follows that

$$\begin{pmatrix} \Theta_{11} & \Theta_{12} \\ \Theta_{21} & \Theta_{22} \end{pmatrix} \begin{pmatrix} -p^{-1} & 0 \\ 0 & q^{-1} \end{pmatrix} \begin{pmatrix} \Theta_{21}^* & \Theta_{11}^* \\ \Theta_{22}^* & \Theta_{12}^* \end{pmatrix} \begin{pmatrix} e & k \\ 0 & e \end{pmatrix} \begin{pmatrix} n_1 \\ n_+ \end{pmatrix} = \begin{pmatrix} e & k \\ 0 & e \end{pmatrix} \begin{pmatrix} n_1 \\ n_+ \end{pmatrix},$$

for $n_1 \in \mathcal{N}_1$, $n_+ \in \mathcal{N}_+$. This implies among other things that

$$(-\Theta_{21} p^{-1} \Theta_{21}^* + \Theta_{22} q^{-1} \Theta_{22}^*) n_1 = 0, \quad n_1 \in \mathcal{N}_1; \tag{3.11}$$

$$(-\Theta_{11} p^{-1} \Theta_{21}^* + \Theta_{12} q^{-1} \Theta_{22}^*) n_1 = n_1, \quad n_1 \in \mathcal{N}_1. \tag{3.12}$$

Recall that $w\Theta_{22}(\Theta_{22}^* w \Theta_{22} + q)^{-1} q \in \mathcal{N}_1$. Thus (3.10) and (3.11) yield

$$\begin{aligned} \Theta_{21} h + \Theta_{22} &= -\Theta_{21} p^{-1} \Theta_{21}^* w (\Theta_{22}^* w + q \Theta_{22}^{-1})^{-1} q + \Theta_{22} \\ &= -\Theta_{22} q^{-1} \Theta_{22}^* w (\Theta_{22} q^{-1} \Theta_{22}^* w + e)^{-1} \Theta_{22} + \Theta_{22} \\ &= (\Theta_{22} q^{-1} \Theta_{22}^* w + e)^{-1} \Theta_{22}. \end{aligned}$$

Thus $\Theta_{21} h + \Theta_{22}$ is invertible and

$$(\Theta_{21} h + \Theta_{22})^{-1} = q^{-1} (\Theta_{22}^* w + q \Theta_{22}^{-1}). \tag{3.13}$$

Similarly, using (3.10), (3.12) and (3.13) we have

$$\begin{aligned}(\Theta_{11}h+\Theta_{12})(\Theta_{21}h+\Theta_{22})^{-1} &= \{-\Theta_{11}p^{-1}\Theta_{21}^{*}w\Theta_{22}(\Theta_{22}^{*}w\Theta_{22}+q)^{-1}q+\Theta_{12}\}\cdot \\ &\qquad \cdot q^{-1}(\Theta_{22}^{*}w+q\Theta_{22}^{-1}) \\ &= -\Theta_{11}p^{-1}\Theta_{21}^{*}w+\Theta_{12}q^{-1}(\Theta_{22}^{*}w+q\Theta_{22}^{-1}) \\ &= -\Theta_{12}q^{-1}\Theta_{22}^{*}w+w+\Theta_{12}q^{-1}\Theta_{22}^{*}w+\Theta_{12}\Theta_{22}^{-1} \\ &= w+c=f.\end{aligned}$$

Hence $f=\mathcal{F}_{\Theta}(h)$.

It remains to prove that $q-h^{*}ph$ is positive definite in $\mathcal{R}$. To do this notice that

$$\begin{aligned}0<_{\mathcal{R}} f+f^{*} &= (\,f^{*} \quad e\,)\begin{pmatrix}0 & e\\ e & 0\end{pmatrix}(\,f \quad e\,) \\ &= (\Theta_{21}h+\Theta_{22})^{-*}(\,h^{*} \quad e\,)\Theta^{*}\begin{pmatrix}0 & e\\ e & 0\end{pmatrix}\begin{pmatrix}h\\ e\end{pmatrix}(\Theta_{21}h+\Theta_{22})^{-1} \\ &= (\Theta_{21}h+\Theta_{22})^{-*}(\,h^{*} \quad e\,)\begin{pmatrix}-p & 0\\ 0 & q\end{pmatrix}\begin{pmatrix}h\\ e\end{pmatrix}(\Theta_{21}h+\Theta_{22})^{-1} \\ &= (\Theta_{21}h+\Theta_{22})^{-*}(q-h^{*}ph)(\Theta_{21}h+\Theta_{22})^{-1},\end{aligned}$$

which yields the desired result. □

PROOF OF THEOREM 2.3(a). We split the proof in two parts.

Part (α). In this part we assume that Θ in (2.2) satisfies conditions (CM1), (CM2) and (CM3a). Put

$$c=\Theta_{12}\Theta_{22}^{-1}, \quad u=\Theta_{22}, \quad v=\Theta_{21}. \tag{3.14}$$

We shall show that with this choice of c, u and v conditions (i) and (ii) in Theorem 2.3(a) are fulfilled.

Notice that $c=\mathcal{F}_{\Theta}(0)$, and hence we know from Theorem 2.2(a) that c is an $\mathcal{R}$-positive real part extension of k. From Lemma 2.5(a) we know that $c+c^{*}=u^{-*}qu^{-1}$ and $v^{*}(c+c^{*})v=p$. By condition (CM2) the element $u^{-1}\in\mathcal{N}_{+}$. From Part (α) of the proof of Theorem 2.2(a) we know that $u\in\mathcal{N}_{+}$ (see (3.3)). Thus (i) is fulfilled. The first part of formula (3.3) gives $v\mathcal{N}_{1}\subset\mathcal{N}_{+}$. It remains to prove the equality in (2.5).

According to Lemma 2.5(a) we have $\Theta_{11}=-c^{*}v$. Clearly, $cu=\Theta_{12}$. Thus our 2×2 block matrix Θ is given by (2.6), where c, u and v are defined by (3.14). Recall that

$c = k + n_1$ for some $n_1 \in \mathcal{N}_1$. It follows that

$$\begin{aligned}\begin{pmatrix} e & -k \\ 0 & e \end{pmatrix} \Theta &= \begin{pmatrix} e & -k \\ 0 & e \end{pmatrix} \begin{pmatrix} -c^*v & cu \\ v & u \end{pmatrix} \\ &= \begin{pmatrix} e & -k \\ 0 & e \end{pmatrix} \begin{pmatrix} e & c \\ 0 & e \end{pmatrix} \begin{pmatrix} -(c+c^*)v & 0 \\ v & u \end{pmatrix} \\ &= \begin{pmatrix} e & n_1 \\ 0 & e \end{pmatrix} \begin{pmatrix} -(c+c^*)v & 0 \\ v & u \end{pmatrix}. \end{aligned} \tag{3.15}$$

By combining this with (CM1) we obtain

$$\begin{pmatrix} e & n_1 \\ 0 & e \end{pmatrix} \begin{pmatrix} -(c+c^*)v & 0 \\ v & u \end{pmatrix} \begin{pmatrix} \mathcal{N}_1 \\ \mathcal{N}_+ \end{pmatrix} = \begin{pmatrix} \mathcal{N}_1 \\ \mathcal{N}_+ \end{pmatrix}.$$

Since $n_1\mathcal{N}_+ \subset \mathcal{N}_1$, by (2.1), we have

$$\begin{pmatrix} e & -n_1 \\ 0 & e \end{pmatrix} \begin{pmatrix} \mathcal{N}_1 \\ \mathcal{N}_+ \end{pmatrix} = \begin{pmatrix} \mathcal{N}_1 \\ \mathcal{N}_+ \end{pmatrix}, \tag{3.16}$$

and therefore

$$\begin{pmatrix} -(c+c^*)v & 0 \\ v & u \end{pmatrix} \begin{pmatrix} \mathcal{N}_1 \\ \mathcal{N}_+ \end{pmatrix} = \begin{pmatrix} \mathcal{N}_1 \\ \mathcal{N}_+ \end{pmatrix}. \tag{3.17}$$

Now, let y_1 be an arbitrary element of $\mathcal{N}_1$. Then according to (3.17),

$$(c+c^*)vy_1 = \begin{pmatrix} -(c+c^*)v & 0 \end{pmatrix} \begin{pmatrix} -y_1 \\ 0 \end{pmatrix} \in \mathcal{N}_1.$$

Thus $(c+c^*)v\mathcal{N}_1 \subset \mathcal{N}_1$. On the other hand, using (3.17) again, there exist $x_1 \in \mathcal{N}_1$ and $x_+ \in \mathcal{N}_+$ such that

$$\begin{pmatrix} -(c+c^*)v & 0 \\ v & u \end{pmatrix} \begin{pmatrix} x_1 \\ x_+ \end{pmatrix} = \begin{pmatrix} -y_1 \\ 0 \end{pmatrix}.$$

This yields $(c+c^*)vx_1 = y_1$, and thus $(c+c^*)v\mathcal{N}_1 = \mathcal{N}_1$, which proves the equality in (2.5).

Part (β). Let c be an $\mathcal{R}$-positive real part extension of k satisfying (i) and (ii) in Theorem 2.3(a). Define Θ by (2.6). In this part we show that Θ has the properties (CM1), (CM2) and (CM3a).

As $\Theta_{22} = u$, condition (CM2) is fulfilled trivially, because of (i). Property (CM3a) follows from the following calculation:

$$\begin{aligned}\Theta^* \begin{pmatrix} 0 & e \\ e & 0 \end{pmatrix} \Theta &= \begin{pmatrix} -v^*c & v^* \\ u^*c^* & u^* \end{pmatrix} \begin{pmatrix} 0 & e \\ e & 0 \end{pmatrix} \begin{pmatrix} -c^*v & cu \\ v & u \end{pmatrix} \\ &= \begin{pmatrix} v^* & -v^*c \\ u^* & u^*c^* \end{pmatrix} \begin{pmatrix} -c^*v & cu \\ v & u \end{pmatrix} \\ &= \begin{pmatrix} -v^*(c+c^*)v & 0 \\ 0 & u^*(c+c^*)u \end{pmatrix} = \begin{pmatrix} -p & 0 \\ 0 & q \end{pmatrix}. \end{aligned}$$

To prove (CM1), write $c = k + n_1$ with $n_1 \in \mathcal{N}_1$. By repeating the earlier calculation in (3.15), we obtain

$$\begin{pmatrix} e & -k \\ 0 & e \end{pmatrix} \Theta = \begin{pmatrix} e & n_1 \\ 0 & e \end{pmatrix} \begin{pmatrix} -(c+c^*)v & 0 \\ v & u \end{pmatrix}.$$

Since $n_1 \in \mathcal{N}_1$, we have $n_1\mathcal{N}_+ \subset \mathcal{N}_1$, by (2.1), and hence (3.16) holds true. Thus in order to establish (CM1) it suffices to show that

$$\begin{pmatrix} -(c+c^*)v & 0 \\ v & u \end{pmatrix} \begin{pmatrix} \mathcal{N}_1 \\ \mathcal{N}_+ \end{pmatrix} = \begin{pmatrix} \mathcal{N}_1 \\ \mathcal{N}_+ \end{pmatrix}. \tag{3.18}$$

To prove (3.18) we first show that

$$x_1 = (c + c^*)vp^{-1}v^*x_1, \qquad x_1 \in \mathcal{N}_1. \tag{3.19}$$

Consider the linear transformations

$$\sigma : \mathcal{N}_1 \to \mathcal{N}_1, \quad \sigma x_1 = (c + c^*)vx_1,$$
$$\tau : \mathcal{N}_1 \to \mathcal{N}_1, \quad \tau x_1 = p^{-1}v^*x_1.$$

From (2.5) we know that $(c + c^*)v\mathcal{N}_1 = \mathcal{N}_1$. The equality $v^*(c + c^*)v = p$ implies that for $x_1 \in \mathcal{N}_1$ we have

$$\tau x_1 = p^{-1}v^*x_1 \in p^{-1}v^*\mathcal{N}_1 = p^{-1}v^*(c + c^*)v\mathcal{N}_1 = \mathcal{N}_1.$$

Thus τ is well-defined. Furthermore, $\tau\sigma x_1 = p^{-1}v^*(c + c^*)vx_1 = x_1$, and thus $\tau\sigma = I_{\mathcal{N}_1}$ (the identity operator on $\mathcal{N}_1$). Since σ is surjective, the latter happens only when σ is the inverse of τ. Thus $\sigma\tau = I_{\mathcal{N}_1}$, and (3.19) is proved.

From our hypotheses we know that

$$v\mathcal{N}_1 \subset \mathcal{N}_+, \quad u \in \mathcal{N}_+, \quad (c + c^*)v\mathcal{N}_1 = \mathcal{N}_1.$$

Thus the left hand side of (3.18) is a subset of the right hand side of (3.18). To prove the equality, take $y_1 \in \mathcal{N}_1$ and $y_+ \in \mathcal{N}_+$ arbitrary, and set

$$x_1 = -p^{-1}v^*y_1, \qquad x_+ = u^{-1}[y_+ + vp^{-1}v^*y_1].$$

Then $x_1 = -\tau y_1$ (where τ is the map defined in the previous paragraph). Thus $x_1 \in \mathcal{N}_1$. Furthermore, $vp^{-1}v^*y_1 = -vx_1 \in v\mathcal{N}_1 \subset \mathcal{N}_+$, and thus $x_+ \in \mathcal{N}_+$. Finally, we have

$$\begin{pmatrix} -(c+c^*)v & 0 \\ v & u \end{pmatrix} \begin{pmatrix} x_1 \\ x_+ \end{pmatrix} = \begin{pmatrix} (c+c^*)vp^{-1}v^*y_1 \\ -vp^{-1}v^*y_1 + y_+ + vp^{-1}v^*y_1 \end{pmatrix} = \begin{pmatrix} y_1 \\ y_+ \end{pmatrix},$$

because of (3.19). Thus (3.18) holds, and (CM1) is proved. □

We proceed with the proofs of Theorems 2.2(b) and 2.3(b). The general structure of the proofs is the same as that of the proofs of Theorems 2.2(a) and 2.3(a); the differences are only in the details.

PROOF OF THEOREM 2.2(b). Let Θ be as in (2.2), and assume that conditions (CM1), (CM2) and (CM3b) are satisfied. We have to prove that all strictly contractive extensions are given by $\mathcal{F}_\Theta(h)$, where h is an arbitrary element of $\mathcal{N}_1$ such that $q-h^*ph >_{\mathcal{R}} 0$. We split the proof into four parts.

Part (α). In this part we show that

$$\Theta_{21}\mathcal{N}_1 \subset \mathcal{N}_+, \qquad \Theta_{22} \in \mathcal{N}_+, \tag{3.20}$$

$$\Theta_{11}^*\mathcal{N}_1 \subset p\mathcal{N}_1, \qquad \Theta_{12}^*\mathcal{N}_1 \subset q\mathcal{N}_+. \tag{3.21}$$

Since (CM1) holds, (3.20) is proved in the same way as (3.3). To prove (3.21) notice that (CM1) and (CM3b) imply that

$$\begin{pmatrix} -p & 0 \\ 0 & q \end{pmatrix} \begin{pmatrix} \mathcal{N}_1 \\ \mathcal{N}_+ \end{pmatrix} = \begin{pmatrix} -\Theta_{11}^* & \Theta_{21}^* \\ -\Theta_{12}^* & \Theta_{22}^* \end{pmatrix} \begin{pmatrix} e & k \\ 0 & e \end{pmatrix} \begin{pmatrix} \mathcal{N}_1 \\ \mathcal{N}_+ \end{pmatrix}. \tag{3.22}$$

So we can use the same arguments as in Part (β) of the proof of Theorem 2.2(a) to show that the inclusions in (3.21) hold.

Part (β). Let h be an element of $\mathcal{N}_1$ such that $q - h^*ph$ is positive definite in $\mathcal{R}$. In this part we show that $f := \mathcal{F}_\Theta(h)$ is well-defined and f is a strictly contractive extension of k.

Put

$$\hat{\Theta} = \begin{pmatrix} \hat{\Theta}_{11} & \hat{\Theta}_{12} \\ \hat{\Theta}_{21} & \hat{\Theta}_{22} \end{pmatrix} = \Theta \begin{pmatrix} p^{-\frac{1}{2}} & 0 \\ 0 & q^{-\frac{1}{2}} \end{pmatrix}, \quad J = \begin{pmatrix} e & 0 \\ 0 & -e \end{pmatrix}.$$

Then (CM3b) can be rewritten as $\hat{\Theta}^* J \hat{\Theta} = J$, and, by (CM2), the element $\hat{\Theta}_{22}$ is invertible. But then we can apply Theorem 2.1 in [2] to show that $\hat{\Theta}_{22}^{-1}\hat{\Theta}_{21}$ is strictly contractive. In other words, $\|q^{\frac{1}{2}}\Theta_{22}^{-1}\Theta_{21}p^{-\frac{1}{2}}\|_{\mathcal{R}} < 1$. Since $q - h^*ph$ is positive definite in $\mathcal{R}$, we also have $\|p^{\frac{1}{2}}hq^{-\frac{1}{2}}\|_{\mathcal{R}} < 1$. It follows that $q^{\frac{1}{2}}\Theta_{22}^{-1}\Theta_{21}hq^{-\frac{1}{2}}$ is strictly contractive. Hence

$$\Theta_{21}h + \Theta_{22} = \Theta_{22}q^{-\frac{1}{2}}\{q^{\frac{1}{2}}\Theta_{22}^{-1}\Theta_{21}hq^{-\frac{1}{2}} + e\}q^{\frac{1}{2}}$$

is invertible, and $f = \mathcal{F}_\Theta(h)$ is well-defined.

Put $x = \Theta_{21}h + \Theta_{22}$. We claim that x and x^{-1} are in $\mathcal{N}_+$. To prove this claim one can use the same arguments as in the second paragraph of Part (γ) of the proof of Theorem 2.2(a). In fact, one only has to replace the reference to (3.3) by a reference to (3.20).

Since $f = \mathcal{F}_\Theta(h)$, we have

$$\begin{pmatrix} f \\ e \end{pmatrix} = \Theta \begin{pmatrix} h \\ e \end{pmatrix} x^{-1}, \quad x = \Theta_{21}h + \Theta_{22}. \tag{3.23}$$

Now we proceed as in the third paragraph of Part (γ) of the proof of Theorem 2.2(a) to show that $f - k \in \mathcal{N}_1$. Thus f is an extension of k. To see that f is strictly contractive we use (3.22) and (CM3b) in the following computation:

$$\begin{aligned} e - f^*f &= (\, f^* \quad e\,) \begin{pmatrix} -e & 0 \\ 0 & e \end{pmatrix} \begin{pmatrix} f \\ e \end{pmatrix} \\ &= x^{-*}\,(\, h^* \quad e\,)\,\Theta^* \begin{pmatrix} -e & 0 \\ 0 & e \end{pmatrix} \Theta \begin{pmatrix} h \\ e \end{pmatrix} x^{-1} \\ &= x^{-*}\,(\, h^* \quad e\,) \begin{pmatrix} -p & 0 \\ 0 & q \end{pmatrix} \begin{pmatrix} h \\ e \end{pmatrix} x^{-1} \\ &= x^{-*}(q - h^*ph)x^{-1} >_{\mathcal{R}} 0. \end{aligned}$$

Thus $\|f\|_{\mathcal{R}} < 1$.

Part (γ). This part concerns the uniqueness of the representation. Assume $f = \mathcal{F}_\Theta(h)$ with $h \in \mathcal{N}_1$ and such that $q - h^*ph >_{\mathcal{R}} 0$. We want to show that h is uniquely determined by f.

As before, put $x = \Theta_{21}h + \Theta_{22}$. Using (3.23) and (CM3b), it follows that

$$\begin{pmatrix} -\Theta_{11}^* f + \Theta_{21}^* \\ -\Theta_{12}^* f + \Theta_{22}^* \end{pmatrix} = \begin{pmatrix} -ph \\ q \end{pmatrix} x^{-1}. \tag{3.24}$$

Put $w = f - \Theta_{12}\Theta_{22}^{-1}$. Using (2.12b), which is a reformulation of (CM3b), we see that

$$\begin{aligned} -\Theta_{11}^* f + \Theta_{21}^* &= -\Theta_{11}^* w - \Theta_{11}^*\Theta_{12}\Theta_{22}^{-1} + \Theta_{21}^* \\ &= -\Theta_{11}^* w + (-\Theta_{11}^*\Theta_{12} + \Theta_{21}^*\Theta_{22})\Theta_{22}^{-1} = -\Theta_{11}^* w, \end{aligned}$$

and

$$\begin{aligned} -\Theta_{12}^* f + \Theta_{22}^* &= -\Theta_{12}^* w - \Theta_{12}^*\Theta_{12}\Theta_{22}^{-1} + \Theta_{22}^* \\ &= -\Theta_{12}^* w + (-\Theta_{12}^*\Theta_{12} + \Theta_{22}^*\Theta_{22})\Theta_{22}^{-1} = -\Theta_{12}^* w + q\Theta_{22}^{-1}. \end{aligned}$$

By inserting these identities in (3.24) we obtain

$$-\Theta_{11}^* w = phx^{-1}, \qquad -\Theta_{12}^* w + q\Theta_{22}^{-1} = qx^{-1},$$

and thus

$$h = p^{-1}\Theta_{11}^{*}w(q\Theta_{22}^{-1} - \Theta_{12}^{*}w)^{-1}q. \tag{3.25}$$

Since w is uniquely determined by f, we conclude that the same holds for h.

Part (δ). Suppose that f is a strictly contractive extension of k. In this part we show that $f = \mathcal{F}_\Theta(h)$ for some $h \in \mathcal{N}_1$ with $q - h^*ph$ positive definite in$\mathcal{R}$. Put $c = \Theta_{12}\Theta_{22}^{-1}$, and set $w = f - c$. From Part (β) we know that c is a strictly contractive extension of k. In fact, $c = \mathcal{F}_\Theta(0)$. Thus $f - c \in \mathcal{N}_1$. From the second inclusion in (3.21) we see that $\Theta_{12}^{*}w \in q\mathcal{N}_+$. It follows that $q\Theta_{22}^{-1} - \Theta_{12}^{*}w \in q\mathcal{N}_+$.

We first show that $q - \Theta_{12}^{*}w\Theta_{22}$ is invertible. Since $f = w + c$, we have

$$e - f^*f = e - c^*c - w^*w - c^*w - w^*c.$$

Recall that $c = \Theta_{12}\Theta_{22}^{-1}$, by definition, and $e - c^*c = \Theta_{22}^{-*}q\Theta_{22}^{-1}$, by Lemma 2.5(b). Therefore,

$$\begin{aligned}\Theta_{22}^{*}(e - f^*f)\Theta_{22} + \Theta_{22}^{*}w^*w\Theta_{22} + q &= 2q - \Theta_{12}^{*}w\Theta_{22} - \Theta_{22}^{*}w^*\Theta_{12}\\ &= (q - \Theta_{12}^{*}w\Theta_{22}) + (q - \Theta_{12}^{*}w\Theta_{22})^*.\end{aligned}$$

We know that $e - f^*f$ and q are positive definite in $\mathcal{R}$. It follows that

$$(q - \Theta_{12}^{*}w\Theta_{22}) + (q - \Theta_{12}^{*}w\Theta_{22})^* >_{\mathcal{R}} 0. \tag{3.26}$$

Thus $q - \Theta_{12}^{*}w\Theta_{22}$ is invertible.

Put $a := q^{-\frac{1}{2}}(q - \Theta_{12}^{*}w\Theta_{22})q^{-\frac{1}{2}}$. Then $a + a^* >_{\mathcal{R}} 0$, and $a \in \mathcal{M}_+ := q^{\frac{1}{2}}\mathcal{N}_+q^{-\frac{1}{2}}$. By repeating arguments in the second paragraph of Part (ε) of the proof of Theorem 2.2(a), we may conclude that $a^{-1} \in \mathcal{M}_+$ and therefore,

$$(q - \Theta_{12}^{*}w\Theta_{22})^{-1} = q^{-\frac{1}{2}}a^{-1}q^{-\frac{1}{2}} \in \mathcal{N}_+q^{-1}. \tag{3.27}$$

Inspired by (3.25) we take

$$h := p^{-1}\Theta_{11}^{*}w\Theta_{22}(q - \Theta_{12}^{*}w\Theta_{22})^{-1}q. \tag{3.28}$$

Since $\Theta_{22} \in \mathcal{N}_+$, by (3.20), and $(q - \Theta_{12}^{*}w\Theta_{22})^{-1}q$ belongs to $\mathcal{N}_+$, by (3.28), the rule $\mathcal{N}_+\mathcal{N}_+ \subset \mathcal{N}_+$ yields $\Theta_{22}(q - \Theta_{12}^{*}w\Theta_{22})^{-1}q \in \mathcal{N}_+$. Recall that $w \in \mathcal{N}_1$ and $\mathcal{N}_1\mathcal{N}_+ \subset \mathcal{N}_1$. Thus $w\Theta_{22}(q - \Theta_{12}^{*}w\Theta_{22})^{-1}q \in \mathcal{N}_1$. But then we can use the first inclusion of (3.21) to show that $h \in \mathcal{N}_1$.

From (2.12b) and (CM1) it follows that

$$\begin{pmatrix} \Theta_{11} & \Theta_{12} \\ \Theta_{21} & \Theta_{22} \end{pmatrix} \begin{pmatrix} -p^{-1} & 0 \\ 0 & q^{-1} \end{pmatrix} \begin{pmatrix} -\Theta_{11}^* & \Theta_{21}^* \\ -\Theta_{12}^* & \Theta_{22}^* \end{pmatrix} \begin{pmatrix} e & k \\ 0 & e \end{pmatrix} \begin{pmatrix} n_1 \\ n_+ \end{pmatrix}$$
$$= \begin{pmatrix} e & k \\ 0 & e \end{pmatrix} \begin{pmatrix} n_1 \\ n_+ \end{pmatrix}, \quad n_1 \in \mathcal{N}_1, \quad n_+ \in \mathcal{N}_+.$$

In particular, we have

$$(\Theta_{21}p^{-1}\Theta_{11}^* - \Theta_{22}q^{-1}\Theta_{12}^*)n_1 = 0, \quad n_1 \in \mathcal{N}_1, \tag{3.29}$$
$$(\Theta_{11}p^{-1}\Theta_{11}^* - \Theta_{12}q^{-1}\Theta_{12}^*)n_1 = n_1, \quad n_1 \in \mathcal{N}_1. \tag{3.30}$$

Recall that $w\Theta_{22}(q - \Theta_{12}^* w\Theta_{22})^{-1}q \in \mathcal{N}_+$. Thus (3.28) and (3.29) yield

$$\begin{aligned} \Theta_{21}h + \Theta_{22} &= \Theta_{21}p^{-1}\Theta_{11}^* w\Theta_{22}(q - \Theta_{12}^* w\Theta_{22})^{-1}q + \Theta_{22} \\ &= \Theta_{22}q^{-1}\Theta_{12}^* w\Theta_{22}(q - \Theta_{12}^* w\Theta_{22})^{-1}q + \Theta_{22} \\ &= \Theta_{22}(q - \Theta_{12}^* w\Theta_{22})^{-1}q. \end{aligned}$$

It follows that $\Theta_{21}h + \Theta_{22}$ is invertible, and thus (using (3.28) and 3.30)) we obtain

$$\begin{aligned} (\Theta_{11}h + \Theta_{12})(\Theta_{21}h + \Theta_{22})^{-1} &= \{\Theta_{11}p^{-1}\Theta_{11}^* w\Theta_{22}(q - \Theta_{12}^* w\Theta_{22})^{-1}q + \Theta_{12}\}\cdot \\ &\qquad \cdot q^{-1}(q - \Theta_{12}^* w\Theta_{22})\Theta_{22}^{-1} \\ &= \Theta_{12}q^{-1}\Theta_{12}^* w + w + \Theta_{12}q^{-1}(q - \Theta_{12}^* w\Theta_{22})\Theta_{22}^{-1} \\ &= w + \Theta_{12}\Theta_{22}^{-1} = f. \end{aligned}$$

Hence $f \in \mathcal{F}_\Theta(h)$. Finally,

$$\begin{aligned} 0 <_{\mathcal{R}} e - f^* f &= (f^* \quad e) \begin{pmatrix} -e & 0 \\ 0 & e \end{pmatrix} \begin{pmatrix} f \\ e \end{pmatrix} \\ &= (\Theta_{21}h + \Theta_{22})^{-*} (h^* \quad e) \Theta^* \begin{pmatrix} -e & 0 \\ 0 & e \end{pmatrix} \Theta \begin{pmatrix} h \\ e \end{pmatrix} (\Theta_{21}h + \Theta_{22})^{-1} \\ &= (\Theta_{21}h + \Theta_{22})^{-*} (h^* \quad e) \begin{pmatrix} -p & 0 \\ 0 & q \end{pmatrix} \begin{pmatrix} h \\ e \end{pmatrix} (\Theta_{21}h + \Theta_{22})^{-1} \\ &= (\Theta_{21}h + \Theta_{22})^{-*}(q - h^* ph)(\Theta_{21}h + \Theta_{22})^{-1}. \end{aligned}$$

Thus $q - h^* ph$ is positive definite in $\mathcal{R}$. □

PROOF OF THEOREM 2.3(b). We split the proof in two parts.

Part (α). In this part we assume that Θ in (2.2) satisfies conditions (CM1), (CM2) and (CM3b). Put

$$c = \Theta_{12}\Theta_{22}^{-1}, \qquad u = \Theta_{22}, \qquad v = \Theta_{11}. \tag{3.31}$$

We shall show that with this choice of c, u and v conditions (j) and (jj) in Theorem 2.3(b) are fulfilled.

Notice that $c = \mathcal{F}_\Theta(0)$, and hence Theorem 2.2(b) yields that c is a strictly contractive extension of k. From Lemma 2.5(b) we know that $e - c^*c = u^{-*}qu^{-1}$ and $v^*(e - cc^*)v = p$. By condition (CM2) we have $u^{-1} \in \mathcal{N}_+$. From (3.20), which holds if (CM1) is fulfilled, we know that $u^{-1} \in \mathcal{N}_+$. Thus (j) is fulfilled. According to Lemma 2.5(b), we have $c^*v = \Theta_{21}$. Thus the first part of (3.20) implies that $c^*v\mathcal{N}_1 \subset \mathcal{N}_+$. It remains to prove the equality in (2.7).

Since $\Theta_{21} = c^*v$ and $\Theta_{12} = cu$ our 2×2 block matrix Θ is given by (2.8), where c, u and v are defined by (3.31). Recall that $c = k + n_1$, for some $n_1 \in \mathcal{N}_1$. It follows that

$$\begin{pmatrix} e & -k \\ 0 & e \end{pmatrix} \Theta = \begin{pmatrix} e & n_1 \\ 0 & e \end{pmatrix} \begin{pmatrix} (e - cc^*)v & 0 \\ c^*v & u \end{pmatrix}. \tag{3.32}$$

By combining this with (CM1) and using (3.16) we obtain

$$\begin{pmatrix} (e - cc^*)v & 0 \\ c^*v & u \end{pmatrix} \begin{pmatrix} \mathcal{N}_1 \\ \mathcal{N}_+ \end{pmatrix} = \begin{pmatrix} \mathcal{N}_1 \\ \mathcal{N}_+ \end{pmatrix}. \tag{3.33}$$

Argueing as in the last paragraph of Part (α) of the proof of Theorem 2.3(a) we see that $(e - cc^*)v\mathcal{N}_1 = \mathcal{N}_1$, which yields the equality in (2.7).

Part (β). Let c be a strictly contractive extension of k satisfying (j) and (jj) in Theorem 2.3(b). Define Θ by (2.8). In this part we show that Θ has the properties (CM1), (CM2) and (CM3b).

As $\Theta_{22} = u$, condition (CM2) is fulfilled trivially, because of (j). Property (CM3b) follows from

$$\begin{aligned} \Theta^* \begin{pmatrix} -e & 0 \\ 0 & e \end{pmatrix} \Theta &= \begin{pmatrix} -v^* & v^*c \\ -u^*c^* & u^* \end{pmatrix} \begin{pmatrix} v & cu \\ c^*v & u \end{pmatrix} = \begin{pmatrix} -v^*(e - cc^*)v & 0 \\ 0 & u^*(e - c^*c)u \end{pmatrix} \\ &= \begin{pmatrix} -p & 0 \\ 0 & q \end{pmatrix}. \end{aligned}$$

To prove (CM1), write $c = k + n_1$ with $n_1 \in \mathcal{N}_1$. Then

$$\begin{pmatrix} e & -k \\ 0 & e \end{pmatrix} \Theta = \begin{pmatrix} e & n_1 \\ 0 & e \end{pmatrix} \begin{pmatrix} (e - cc^*)v & 0 \\ c^*v & u \end{pmatrix}.$$

Since $n_1\mathcal{N}_+ \subset \mathcal{N}_1$, the identity (3.16) holds true. Thus in order to establish (CM1) it satisfies to show that

$$\begin{pmatrix} (e - cc^*)v & 0 \\ c^*v & u \end{pmatrix} \begin{pmatrix} \mathcal{N}_1 \\ \mathcal{N}_+ \end{pmatrix} = \begin{pmatrix} \mathcal{N}_1 \\ \mathcal{N}_+ \end{pmatrix}. \tag{3.34}$$

To prove (3.34) we first show that

$$x_1 = (e - cc^*)vp^{-1}v^*x_1, \quad x_1 \in \mathcal{N}_1. \tag{3.35}$$

Consider the linear transformations

$$\tilde{\sigma} : \mathcal{N}_1 \to \mathcal{N}_1, \quad \tilde{\sigma}x_1 = (e - cc^*)vx_1; \quad \tilde{\tau} : \mathcal{N}_1 \to \mathcal{N}_1, \quad \tilde{\tau}x_1 = p^{-1}v^*x_1.$$

From the first part of (2.7) we know that $(e - cc^*)v\mathcal{N}_1 = \mathcal{N}_1$. Thus $\tilde{\sigma}$ is well-defined and $\operatorname{Im}\tilde{\sigma} = \mathcal{N}_1$. The equality $v^*(e - cc^*)v = p$ implies that for each $x_1 \in \mathcal{N}_1$ we have

$$\tilde{\tau} = p^{-1}v^*x_1 \in p^{-1}v^*\mathcal{N}_1 = p^{-1}v^*(e - cc^*)v\mathcal{N}_1 = \mathcal{N}_1.$$

Thus $\tilde{\tau}$ is well-defined. Furthermore, $\tilde{\tau}\tilde{\sigma}x_1 = p^{-1}v^*(e - cc^*)vx_1 = x_1$. Thus $\tilde{\sigma}$ is a right inverse of $\tilde{\tau}$. Since $\tilde{\sigma}$ is surjective, we conclude that $\tilde{\sigma}$ is invertible and $\tilde{\sigma}^{-1} = \tilde{\tau}$. Therefore, $\tilde{\sigma}\tilde{\tau}$ is the identity operator on $\mathcal{N}_1$, and (3.35) is proved.

From our hypotheses we know that

$$c^*v\mathcal{N}_1 \subset \mathcal{N}_+, \quad u \in \mathcal{N}_+, \qquad (e - c^*c)v\mathcal{N}_1 = \mathcal{N}_1.$$

Thus the left hand side of (3.34) is a subset of the right hand side of (3.34). To prove the equality, take $y_1 \in \mathcal{N}_1$, and $y_+ \in \mathcal{N}_+$ arbitrary, and set

$$x_1 = p^{-1}v^*y_1, \qquad x_+ = u^{-1}(y_+ - c^*vp^{-1}v^*y_1).$$

Thus $x_1 = \tilde{\tau}y_1$, and hence $x_1 \in \mathcal{N}_1$. Furthermore, $c^*vp^{-1}v^*y_1 = c^*vx_1 \in c^*v\mathcal{N}_1 \subset \mathcal{N}_+$, and so $x_+ \in \mathcal{N}_+$. Finally, we have

$$\begin{pmatrix} (e - cc^*)v & 0 \\ c^*v & u \end{pmatrix} \begin{pmatrix} x_1 \\ x_+ \end{pmatrix} = \begin{pmatrix} (e - cc^*)vp^{-1}v^*y_1 \\ c^*vp^{-1}v^*y_1 + y_+ - c^*vp^{-1}v^*y_1 \end{pmatrix} = \begin{pmatrix} y_1 \\ y_+ \end{pmatrix},$$

because of (3.35). Thus (3.34) holds, and (CM1) is proved. □

4. THE CARATHEODORY AND NEHARI EXTENSION PROBLEMS

In this section we describe the canonical positive real part extensions and canonical strictly contractive extensions for the classical extension problems of Carathéodory and Nehari, respectively. We also point out the difference with the usual band or triangular extension.

4.1. **The Carathéodory problem.** Let $\mathbb{D}$ be the unit disk with boundary $\partial\mathbb{D} = \mathbb{T}$. As usual, $L^\infty(\mathbb{T})$ denotes the space of essentially bounded Lebesgue measurable functions on $\mathbb{T}$, and $H^\infty(\mathbb{T})$ denotes the standard Hardy space consisting of those functions $f \in L^\infty(\mathbb{T})$ with Fourier series of the form $f \sim \sum_{j=0}^\infty f_j\zeta^j$. In the sequel, we omit the symbol $\mathbb{T}$, and simply write L^∞ and H^∞. We also let $\mathcal{W}$ be the Wiener algebra of all functions in L^∞ with absolutely convergent Fourier series, $\mathcal{W}_+$ the subalgebra $\mathcal{W} \cap H^\infty$, and $\mathcal{P}_n$ the subspace of polynomials $p(\zeta) = \sum_{j=0}^n p_j\zeta^j$ of polynomials in ζ of degree at most n.

Given $n+1$ complex numbers $c_0, \dots, c_n$, the *Carathéodory problem* is to find a function φ analytic on $\mathbb{D}$, with positive real part on the unit disk $\mathbb{D}$, such that

$$\frac{1}{j!}\varphi^{(j)}(0) = c_j, \qquad j = 0, \dots, n. \tag{4.1}$$

If for convenience we insist also that φ be in the Wiener algebra $\mathcal{W}_+$, we can see the problem as a special case of the positive real part completion problem for an abstract band. Indeed, to put the Carathéodory problem in the framework of the band method we let $\mathcal{C}$ be the C^*-algebra of all complex-valued continuous functions on $\mathbb{T}$, which has $1_\mathbb{T}$ as its unit, we put $\mathcal{M} = \mathcal{W}$, and we set

$$\mathcal{M}_1 = \zeta^{n+1}\mathcal{W}_+, \quad \mathcal{M}_2^0 = \zeta\mathcal{P}_{n-1}, \quad \mathcal{M}_d = \mathbb{C}1_\mathbb{T}, \quad \mathcal{M}_3^0 = (M_2^0)^-, \quad \mathcal{M}_4 = M_1^-,$$

where the bar denotes complex conjugate. We have

$$\mathcal{M} = \mathcal{M}_1 \oplus \mathcal{M}_1^0 \oplus \mathcal{M}_d \oplus \mathcal{M}_3^0 \oplus \mathcal{M}_4, \tag{4.2}$$

and the multiplication table (1.2) is fulfilled. Put

$$k(\zeta) = c_0 + c_1\zeta + \dots + c_n\zeta^n \in \mathcal{M}_{\ell_+} = \mathcal{M}_d \oplus \mathcal{M}_2^0. \tag{4.3}$$

Then $\varphi \in \mathcal{W}_+$ is a solution of the Carathéodory problem for $c_0, \dots, c_n$ if and only if φ is a $\mathcal{C}$-positive real part extension of k in (4.3). Thus, according to Theorem 1.3, to describe all solutions of the Carathéodory problem we have first to find a solution which is a canonical $\mathcal{C}$-positive real part extension of k.

PROPOSITION 4.1. *Let $\varphi \in \mathcal{W}_+$ be a solution of the Carathéodory problem for $c_0, \dots, c_n$. Then φ is a canonical $\mathcal{C}$-positive real part extension of k if and only if $(\varphi + \overline{\varphi})^{-1}$ is a trigonometric polynomial of degree $n+1$, i.e.,*

$$\left(\varphi(\zeta) + \overline{\varphi(\zeta)}\right)^{-1} = \sum_{j=n+1}^{n+1} \alpha_j\zeta_j, \quad \zeta \in \mathbb{T}. \tag{4.4}$$

Furthermore, $f = \varphi + \overline{\varphi}$ is a band extension relative to (4.2) *if and only if $(\varphi + \overline{\varphi})^{-1}$ is a trigonometric polynomial of degree n, i.e.,*

$$\left(\varphi(\zeta) + \overline{\varphi(\zeta)}\right)^{-1} = \sum_{j=-n}^{n} \alpha_j \zeta_j, \quad \zeta \in \mathbb{T}. \tag{4.5}$$

PROOF. Since $\varphi \in \mathcal{W}_+$ and $\varphi(\zeta) + \overline{\varphi(\zeta)}$ is uniformly positive definite on $\mathbb{T}$, it is well-known that for $f = \varphi + \overline{\varphi}$ a factorization as in condition (i) of Theorem 1.3 automatically exists. Also, $f = \varphi + \overline{\varphi}$ factorizes as $f = v^{-*}v^{-1}$ with $(v^*)^{\pm 1} \in \mathcal{W}_+$. Since the multiplication table (1.2) holds, $(v^*)^{\pm 1} \in \mathcal{W}_+$ implies $v^*\mathcal{M}_1 = \mathcal{M}_1$, and hence condition (ii) in Theorem 1.3 is also fulfilled. Finally, for the case considered here condition (iii) in Theorem 1.3 may be restated as $f^{-1}\mathcal{W}_+ \subset \zeta^{-n-1}\mathcal{W}_+$. As f^{-1} is selfadjoint, this shows that condition (iii) in Theorem 1.3 is equivalent to (4.4). The statement about the band extension follows directly from the definition of a band extension. □

4.2. **The Nehari problem.** In the Nehari problem one has given a sequence $a_0, a_{-1}, a_{-2}, \dots$ of complex numbers, and one seeks $f \in L^\infty$ with $\|f\|_\infty < 1$ such that

$$\frac{1}{2\pi}\int_{-\pi}^{\pi} f(e^{it})e^{-int}\,dt = a_n, \quad n = 0, -1, -2, \dots.$$

We are interested in solutions that belong to the Wiener algebra $\mathcal{W}$, and therefore we assume additionally that $\sum_{j=0}^{\infty} |a_{-j}| < \infty$.

To put the Nehari problem into the framework of the band method we let $\mathcal{R}$ be the C^*-algebra L^∞, which has $1_\mathbb{T}$ as its unit, we take $\mathcal{M} = \mathcal{W}$, and we set

$$\mathcal{M}_1 = \zeta\mathcal{W}_+, \quad \mathcal{M}_2^0 = \mathcal{M}_3^0 = \{0\}, \quad \mathcal{M}_d = \mathbb{C}1_\mathbb{T}, \quad \mathcal{M}_4 = \mathcal{M}_1^-,$$

where the bar denotes complex conjugate. Then

$$\mathcal{M} = \mathcal{M}_1 \oplus \mathcal{M}_2^0 \oplus \mathcal{M}_d \oplus \mathcal{M}_3^0 \oplus \mathcal{M}_4, \tag{4.6}$$

and again the multiplication table (1.2) is fulfilled. Furthermore, $f \in \mathcal{W}$ is a solution of the Nehari problem for $a_0, a_{-1}, a_{-2}, \dots$ if and only if relative to the band structure (4.6) the element f is a strictly contractive extension of k, where

$$k(\zeta) = \sum_{j=-\infty}^{0} \zeta^j a_j \in \mathcal{M}_\ell = \mathcal{W}_+^-. \tag{4.7}$$

Thus, according to Theorem 1.4, to describe all solutions of the Nehari problem we have first to find a solution which is a canonical strictly contractive extension of k in (4.7). In what follows $\mathcal{W}_- = \mathcal{W}_+^-$, the set of co-analytic functions in $\mathcal{W}$. Notice that $\mathcal{M}_\ell = \mathcal{W}_+$.

PROPOSITION 4.2. *Let $f \in \mathcal{W}$ be a solution of the Nehari problem for the sequence $a_0, a_{-1}, a_{-2}, \ldots$. Then f is a canonical strictly contractive extension of k in* (4.7) *if and only if*

$$(1_{\mathbb{T}} - |f|^2)^{-1}\overline{f} \in \zeta\mathcal{W}_-, \tag{4.8}$$

and f is a triangular extension of k if and only if

$$(1_{\mathbb{T}} - |f|^2)^{-1}\overline{f} \in \mathcal{W}_-. \tag{4.9}$$

PROOF. Since $f \in \mathcal{W}$ and $1 - \overline{f(\zeta)}f(\zeta)$ is uniformly positive definite on $\mathbb{T}$, it is well-known that for $1_{\mathbb{T}} - \overline{f}f$ a factorization as in condition (i) of Theorem 1.4 automatically exists. Also, $1 - f\overline{f} = v^{-*}v^{-1}$ with $v^{\pm 1} \in \mathcal{W}_-$. Then $(v^*)^{\pm 1} \in \mathcal{W}_+$, and hence $v^*\mathcal{M}_1 = \mathcal{M}_1$. Thus condition (ii) in Theorem 1.4 is also fulfilled. Next, notice that for the case considered here condition (iii) in Theorem 1.4 may be restated as $(1_{\mathbb{T}} - \overline{f}f)^{-1}\overline{f}\mathcal{W}_+ \subset \zeta^{-1}\mathcal{W}_+$, and hence condition (iii) in Theorem 1.4 is fulfilled if and only if (4.8) holds. Finally, the statement about the triangular extension follows directly from the definition of a triangular extension. □

5. OPERATOR MATRIX EXTENSION PROBLEMS

In this section we consider two positive extension problems for partially given operator matrices. The first problem is a classical one, and concerns the case when the given entries form a band. For this band problem we identify the canonical positive extensions.

In the second problem the pattern of the given entries is a bordered band (cf., [11]). This second problem is not of band type and the corresponding multiplication table does not satisfy (1.2). We show that for this non-band problem the result of Section 3 can be used to derive a linear fractional representation of all positive extensions.

5.1. **Band operator matrices.** Let $\mathcal{X} = H_1 \oplus \cdots \oplus H_n$ be a Hilbert space direct sum. Each element $A \in \mathcal{L}(X)$, the algebra of bounded linear operators on $\mathcal{X}$, has a canonical representation as an $n \times n$ operator matrix $A = (A_{ij})_{i,j=1}^n$. Fix $0 \leq m < n$. We say that A is an *m-band operator matrix* if $A_{ij} = 0$ for $|i - j| > m$. Let A be such an operator. An operator $F = (F_{ij})_{i,j=1}^n$ in $\mathcal{L}(\mathcal{X})$ is called an *extension* (or *completion*) of A

if

$$F_{ij} = A_{ij}, \qquad |i-j| \leq m. \tag{5.1}$$

The positive extension problem in this setting is the problem of constructing such an F which is also strictly positive as an operator on $\mathcal{L}(\mathcal{X})$.

This problem can be put in the band method setting as follows. We let $\mathcal{M}$ be the unital C^*-algebra $\mathcal{M}(\mathcal{X})$, and we set

$$\mathcal{M}_1 = \{A = (A_{ij})_{i,j=1}^n \mid A_{ij} = 0 \text{ for } j-i \leq m\}, \tag{5.2a}$$

$$\mathcal{M}_2^0 = \{A \in (A_{ij})_{i,j=1}^n \mid A_{ij} = 0 \text{ for } j-i > m \text{ or } j-i \leq 0\}, \tag{5.2b}$$

$$\mathcal{M}_d = \{A = (A_{ij})_{i,j=1}^n \mid A_{ij} = 0 \text{ for } j \neq i\}, \tag{5.2c}$$

$$\mathcal{M}_3^0 = \{A = (A_{ij})_{i,j=1}^n \mid A_{ij} = 0 \text{ for } j-i \geq 0 \text{ or } j-i < -m\}, \tag{5.2d}$$

$$\mathcal{M}_4 = \{A = (A_{ij})_{i,j=1}^n \mid A_{ij} = 0 \text{ for } j-i \geq -m\}. \tag{5.2e}$$

Then

$$\mathcal{M} = \mathcal{M}_1 \oplus \mathcal{M}_2^0 \oplus \mathcal{M}_d \oplus \mathcal{M}_3^0 \oplus \mathcal{M}_4, \tag{5.3}$$

and (5.3) defines a band structure in $\mathcal{M}$. In particular, the multiplication table (1.2) is satisfied.

Thus, by Theorem 1.3, the problem of constructing all (strictly) positive extensions of the m-band operator matrix A reduces to the problem of finding a canonical positive extension. The next proposition identifies such extensions.

PROPOSITION 5.1. *Let F be a positive extension of the m-band operator matrix A. Then F is a canonical positive extension if and only if F^{-1} is an $(m+1)$-band operator matrix.*

PROOF. Since F is a positive extension, F is strictly positive as an operator on $\mathcal{L}(\mathcal{X})$. It follows (cf., Section XXII.1 in [7]) that F factorizes as $F = U^{-*}U^{-1}$ and $F = V^{-*}V^{-1}$, where $U^{\pm 1}$ is block upper triangular and $V^{\pm 1}$ block lower triangular, both relative to the decomposition $\mathcal{X} = H_1 \oplus \cdots \oplus H_n$. Notice that in this case $\mathcal{M}_+$ coincides with the space of all block upper triangular operator matrices in $\mathcal{L}(\mathcal{X})$. Thus condition (i) in Theorem 1.3 is fulfilled. Since V^* is block upper triangular and invertible, the multiplication table (1.2), which is valid in this case, shows that $V^*\mathcal{M}_1 = \mathcal{M}_1$. Hence condition (ii) in Theorem 1.3 is also fulfilled. It is straightforward to check that condition (iii) in Theorem 1.3 is equivalent to the requirement that F^{-1} is an $(m+1)$-band operator matrix. □

Recall that a positive extension F of the m-band operator matrix A is a band extension if F^{-1} is an m-band operator matrix. Thus a band extension is a canonical positive extension but the set of all canonical positive extensions is strictly larger. For example, if $m = n-1$, then each positive extension is a canonical one.

5.2. **A non-band extension problem.** Consider the Hilbert space direct sum $\mathcal{Y} = H_0 \oplus H_1 \oplus \cdots \oplus H_n$, and let $\mathcal{N} = \mathcal{L}(\mathcal{Y})$ be the C^*-algebra of all bounded linear operators on $\mathcal{Y}$. The unit of $\mathcal{N}$ will be denoted by E. We write the elements of $\mathcal{N}$ as 2×2 operator matrices relative to the partioning $\mathcal{Y} = H_0 \oplus \mathcal{X}$, where $\mathcal{X}$ is the Hilbert space direct sum $\mathcal{X} = H_1 \oplus \cdots \oplus H_n$. In what follows we assume that $\mathcal{M} = \mathcal{L}(\mathcal{X})$ is endowed with the band structure considered in the previous section. Thus

$$\mathcal{M} = \mathcal{M}_1 \oplus \mathcal{M}_2^0 \oplus \mathcal{M}_d \oplus \mathcal{M}_3^0 \oplus \mathcal{M}_4, \tag{5.4}$$

with the subspaces being given by (5.2a)–(5.2e). Consider the spaces

$$\mathcal{N}_+ = \left\{ \begin{pmatrix} \alpha & \beta \\ 0 & \delta \end{pmatrix} \in \mathcal{L}(\mathcal{Y}) \mid \alpha \in \mathcal{L}(H_0), \quad \beta \in \mathcal{L}(\mathcal{X}, H_0), \quad \delta \in \mathcal{M}_+ \right\},$$
$$\mathcal{N}_1 = \left\{ \begin{pmatrix} 0 & 0 \\ 0 & \delta \end{pmatrix} \in \mathcal{L}(\mathcal{Y}) \mid \delta \in \mathcal{M}_1 \right\}.$$

Here $\mathcal{M}_+ = \mathcal{M}_1 \oplus \mathcal{M}_2^0 \oplus \mathcal{M}_d$, and thus $\mathcal{N}_+$ is just the algebra of all upper triangular $(n+1)\times(n+1)$ operator matrices relative to the decomposition $\mathcal{Y} = H_0 \oplus H_1 \oplus \cdots \oplus H_n$. Notice that $E \in \mathcal{N}_+$. Since the multiplication table (1.2) holds for the band structure (5.4), we have $\mathcal{N}_1\mathcal{N}_+ \subset \mathcal{N}_1$. It follows that $\mathcal{N}_+$ and $\mathcal{N}_1$ satisfy the conditions laid down in the first paragraph of Section 2. Fix

$$K = \begin{pmatrix} \alpha_K & \beta_K \\ 0 & \delta_K \end{pmatrix} \in \mathcal{N}_+. \tag{5.5}$$

We want to describe all positive real part extensions of K. Notice that in this case $F \in \mathcal{L}(\mathcal{Y})$ is a positive real part extension of K if and only if the following three conditions are fulfilled:

(a) $F = \begin{pmatrix} \alpha_K & \beta_K \\ 0 & \delta \end{pmatrix} \in \mathcal{N}_+$,

(b) $\delta - \delta_K \in \mathcal{M}_1$,

(c) $F + F^*$ is strictly positive as an operator on $\mathcal{Y}$.

From (b) we see that without loss of generality we may assume that

$$\delta_K \in \mathcal{M}_2 = \mathcal{M}_2^0 \oplus \mathcal{M}_d. \tag{5.6}$$

Thus $K + K^*$ has the form of an m-band operator matrix bordered by a full operator column on the left and a full operator row at the top. From (c) it follows that for the above extension problem to be solvable it is necessary that $\alpha_K + \alpha_K^*$ is a strictly positive operator on H_0. Therefore in the sequel we assume that

$$\alpha_K + \alpha_K^* = \alpha^*\alpha, \qquad \alpha^{\pm 1} \in \mathcal{L}(H_0). \tag{5.7}$$

We shall prove the following theorem.

THEOREM 5.2. *Let K be as in (5.5), and assume that (5.7) is satisfied. Put*

$$\Delta_K = \delta_K + \delta_K^* - \beta_K^*(\alpha_K + \alpha_K^*)^{-1}\beta_K, \tag{5.8}$$

and let $\Delta_{K,c}$ be the m-band operator matrix which one obtains if the entries in Δ_K outside the m-band are set to zero. Then K has a positive real part extension if and only if the m-band operator matrix $\Delta_{K,c}$ has a strictly positive completion. Furthermore, in this case all positive real part completions of K are obtained in the following way. Let b be the unique band extension of $\Delta_{K,c}$, factorize b as

$$b = u^{-*}u^{-1}, \qquad b = v^{-*}v^{-1}, \tag{5.9}$$

where $u^{\pm 1}$ are upper triangular $m \times m$ operator matrices and $v^{\pm 1}$ are lower triangular $m \times m$ operator matrices, both relative to the decomposition $\mathcal{X} = H_1 \oplus \cdots \oplus H_n$, and set

$$U = \begin{pmatrix} \alpha^{-1} & -\alpha^{-1}\alpha^{-*}\beta_K u \\ 0 & u \end{pmatrix}, \qquad V = \begin{pmatrix} \alpha^{-1} & -\alpha^{-1}\alpha^{-*}\beta_K v \\ 0 & v \end{pmatrix}, \tag{5.10}$$

$$C = \begin{pmatrix} \alpha_K & \beta_K \\ 0 & \delta_K + \rho \end{pmatrix}, \tag{5.11}$$

where ρ is the $m \times m$ operator matrix which one obtains by setting all (i,j)-th operator entries of $b - \Delta_K$ to zero except those with indices $j - i > m$, which remain unchanged. Then all positive real part extensions F of K are given by

$$F = (-C^*VG + CU)(VG + U)^{-1}. \tag{5.12}$$

where the free parameter G is an arbitrary element of $\mathcal{N}_1$ satisfying $\|G\| < 1$. Moreover, the map (5.12) provides a one-to-one correspondence between all such G and all positive real part extensions F of K.

PROOF. We split the proof into four parts.

Part (a). Assume K has a positive real part extension F. In this part we show that $\Delta_{K,c}$ has a strictly positive completion. We know that

$$F + F^* = \begin{pmatrix} \alpha_K + \alpha_K^* & \beta_K \\ \beta_K^* & \delta + \delta^* \end{pmatrix},$$

where $\delta - \delta_K \in \mathcal{M}_1$, with $\mathcal{M}_1$ being given by (5.2a). Using (5.7) we have the following factorization:

$$F + F^* = \begin{pmatrix} \alpha^* & 0 \\ \beta_K^* \alpha^{-1} & e \end{pmatrix} \begin{pmatrix} e & 0 \\ 0 & \tau \end{pmatrix} \begin{pmatrix} \alpha & \alpha^{-*}\beta_K \\ 0 & e \end{pmatrix}, \tag{5.13}$$

where

$$\tau = \delta + \delta^* - \beta_K^* (\alpha_K + \alpha_K^*)^{-1} \beta_K. \tag{5.14}$$

Since $F + F^*$ is strictly positive on $\mathcal{Y}$, the operator τ is strictly positive on $\mathcal{X}$. The fact that $\delta - \delta_K \in \mathcal{M}_1$ implies that $\tau - \Delta_K \in \mathcal{M}_1 + \mathcal{M}_4$. Also, $\Delta_K - \Delta_{K,c} \in \mathcal{M}_1 + \mathcal{M}_4$. Thus $\tau - \Delta_{K,c} \in \mathcal{M}_1 + \mathcal{M}_4$, and therefore τ is a strictly positive completion of $\Delta_{K,c}$.

Part (b). Assume that $\Delta_{K,c}$ has a strictly positive completion. Let C be as in (5.11). In this part we show that C is a positive real part completion of K. By definition,

$$C - K = \begin{pmatrix} 0 & 0 \\ 0 & \rho \end{pmatrix},$$

where

$$\rho = P_1 (b + \beta_K^* (\alpha_K + \alpha_K^*)^{-1} \beta_K). \tag{5.15}$$

Here P_1 is the projection of $\mathcal{M} = \mathcal{L}(\mathcal{X})$ onto $\mathcal{M}_1$ along the other spaces in the decomposition (5.4). It follows that $C - K \in \mathcal{N}_1$. Notice that

$$C + C^* = \begin{pmatrix} \alpha^* & 0 \\ \beta_K^* \alpha^{-1} & e \end{pmatrix} \begin{pmatrix} e & 0 \\ 0 & \Delta_K + \rho + \rho^* \end{pmatrix} \begin{pmatrix} \alpha & \alpha^{-*}\beta_K \\ 0 & e \end{pmatrix}. \tag{5.16}$$

We claim that $\Delta_K + \rho + \rho^* = b$. To see this, recall that b is a completion of $\Delta_{K,c}$. Thus $b - \Delta_{K,c}$ is in $\mathcal{M}_1 + \mathcal{M}_4$. Also, $\Delta_K - \Delta_{K,c} \in \mathcal{M}_1 + \mathcal{M}_4$. Therefore, using the selfadjointness of b and Δ_K, we may write $b - \Delta_K = m_1 + m_1^*$, where $m_1 = P_1(b - \Delta_K) = \rho$. Thus $\Delta_K + \rho + \rho^* = b$. The operator b is strictly positive on $\mathcal{X}$. Thus (5.16) shows that $C + C^*$ is strictly positive on $\mathcal{Y}$. In fact, using the identity (5.16), the equality $b = \Delta_K + \rho + \rho^*$, and the factorizations of b in (5.10), we see that

$$C + C^* = U^{-*} U^{-1}, \qquad C + C^* = V^{-*} V^{-1}, \tag{5.17}$$

where U and V are as in (5.10).

Part (c). Again we assume that $\Delta_{K,c}$ has a strictly positive extension. In this part we show that

$$U^{\pm 1} \in \mathcal{N}_+, \quad V^*\mathcal{N}_1 = \mathcal{N}_1, \quad U^*\mathcal{N}_1 \subset \mathcal{N}_+. \tag{5.18}$$

Since $u^{\pm 1} \in \mathcal{M}_+$, the operators $U^{\pm 1} \in \mathcal{N}_+$ by definition.

Notice that

$$V^{-*} = \begin{pmatrix} \alpha & \alpha^{-*}\beta_K \\ 0 & v^{-1} \end{pmatrix}^* = \begin{pmatrix} \alpha^* & 0 \\ \beta_K^*\alpha^{-1} & v^{-*} \end{pmatrix}$$

Let Z be an arbitrary element of $\mathcal{N}_1$. Thus

$$Z = \begin{pmatrix} 0 & 0 \\ 0 & z \end{pmatrix}, \tag{5.19}$$

for some $z \in \mathcal{M}_1$. It follows that

$$V^{-*}Z = \begin{pmatrix} 0 & 0 \\ 0 & v^{-*}z \end{pmatrix} \in \mathcal{N}_1.$$

Here we use that $v^{-*} \in \mathcal{M}_+$ and $\mathcal{M}_+\mathcal{M}_1 \subset \mathcal{M}_1$, because of the multiplication table (1.2), which holds for the band structure defined by (5.4). Next, put $\delta = v^*z$. Since $v^* \in \mathcal{M}_+$, we may use the multiplication table (2.1) again, to show that $\delta \in \mathcal{M}_+\mathcal{M}_1 \subset \mathcal{M}_1$, and thus

$$Y := \begin{pmatrix} 0 & 0 \\ 0 & \delta \end{pmatrix} \in \mathcal{N}_1, \qquad V^{-*}Y = Z$$

This proves the second part of (5.18).

Finally, notice that

$$U^* = \begin{pmatrix} \alpha^{-*} & 0 \\ u^*\beta_K^*\alpha^{-1}\alpha^{-*} & u^* \end{pmatrix}.$$

Since b is a band extension, we know (see, e.g., Lemma XXXIV.1.4 in [7]) that $u \in \mathcal{M}_2 = \mathcal{M}_2^0 \oplus \mathcal{M}_d$. Thus $u^* \in \mathcal{M}_3 = \mathcal{M}_3^0 \oplus \mathcal{M}_d$. Again, let Z in (5.19) be an arbitrary element of $\mathcal{N}_1$. We have

$$U^*Z = \begin{pmatrix} 0 & 0 \\ 0 & u^*z \end{pmatrix} \in \mathcal{N}_+.$$

Indeed, $u^*z \in \mathcal{M}_3\mathcal{M}_1 \subset \mathcal{M}_+$ according to the multiplication table (1.2). This proves the third part of (5.18).

Part (d). We complete the proof of Theorem 5.2. Assume that $\Delta_{K,c}$ has a strictly positive completion. From (5.17) and (5.18) it follows that conditions (1a), (2a) and (3a) in Theorem 3.4 are fulfilled. Since Axiom (A) holds for $\mathcal{N}_+$ with respect to $\mathcal{R} = \mathcal{L}(\mathcal{Y})$, we may apply Theorem 2.4(a) to show that the 2×2 operator matrix

$$\Theta = \begin{pmatrix} -C^*V & CU \\ V & U \end{pmatrix}$$

has the properties (CM1), (CM2) and (CM3a). But then we can use Theorem 2.2(a) to finish the proof. □

In addition to $\mathcal{N}_1$ let us consider the following subspaces of $\mathcal{N} = \mathcal{L}(\mathcal{Y})$:

$$\mathcal{N}_2^0 = \left\{ \begin{pmatrix} 0 & \beta \\ 0 & \delta \end{pmatrix} \in \mathcal{L}(\mathcal{Y}) \mid \beta \in \mathcal{L}(\mathcal{X}, H_0), \quad \delta \in \mathcal{M}_2^0 \right\},$$
$$\mathcal{N}_d = \left\{ \begin{pmatrix} \alpha & 0 \\ 0 & \delta \end{pmatrix} \in \mathcal{L}(\mathcal{Y}) \mid \alpha \in \mathcal{L}(H_0), \quad \delta \in \mathcal{M}_d \right\},$$
$$\mathcal{N}_3^0 = (\mathcal{N}_2^0)^*, \qquad \mathcal{N}_4 = \mathcal{N}_1^*.$$

Then

$$\mathcal{N} = \mathcal{N}_1 \oplus \mathcal{N}_2^0 \oplus \mathcal{N}_d \oplus \mathcal{N}_3^0 \oplus \mathcal{N}_4. \tag{5.20}$$

The decomposition in (5.20) does not define a band structure on $\mathcal{N}$, because in this case the multiplication table (1.2) does not hold. For example, $\mathcal{N}_2^0\mathcal{N}_1$ is not contained in $\mathcal{N}_1$ (as would be required by (1.2)). Nevertheless we can define the notions of band extension and canonical positive extension as in Section 1. Indeed, let K be as in (5.5), and assume that (5.6) holds. Then $K + K^* \in \mathcal{N}_c := \mathcal{N}_2^0 \oplus \mathcal{N}_d \oplus \mathcal{N}_3^0$. We say that $G \in \mathcal{L}(\mathcal{Y})$ is a *band extension* of $K + K^*$ if G is strictly positive on $\mathcal{Y}$,

$$G - (K + K^*) \in \mathcal{N}_1 \oplus \mathcal{N}_4, \tag{5.21}$$

and $G^{-1} \in \mathcal{N}_c$. Similarly, G is said to be a *canonical positive extension* of $K + K^*$ if (5.21) holds and G factorizes as $G = U^{-*}U^{-1}$ and $G = V^{-*}V^{-1}$, where U and V are invertible and satisfy the conditions in (5.18). For a band structure with the properties described in Section 1 a band extension is always a canonical positive extension. This statement also holds in the present context.

PROPOSITION 5.3. *Let K be as in* (5.5), *and assume* (5.6) *holds. Let B be a band extension of $K + K^*$ relative to* (5.20). *Then B is a canonical positive extension of $K + K^*$.*

PROOF. Write $B = F + F^*$, where F is a positive real part extension of K. Then $B = F + F^*$ admits a factorization as in (5.13), where τ is a strictly positive completion of $\Delta_{K,c}$. From (5.13) it follows that B^{-1} is of the form

$$B^{-1} = \begin{pmatrix} * & * \\ * & \tau^{-1} \end{pmatrix}.$$

Since $B^{-1} \in \mathcal{N}_c$, this implies that $\tau^{-1} \in \mathcal{M}_c$, and hence τ is a band extension of $\Delta_{k,c}$ relative to the band structure (5.4). But then we know from the proof of Theorem 5.2 that $B = F + F^*$ factorizes as $B = U^{-*}U^{-1}$ and $B = V^{-*}V^{-1}$, where U and V are invertible and satisfy the conditions in (5.18). Thus B is a canonical positive extension of $K + K^*$. □

We conclude with two examples. The first shows that in the present setting we cannot use the recipe of Theorem 1.1 to obtain a linear fractional representation of all positive extensions. The second shows that a canonical positive extension does not have to be a band extension.

Example 5.4. In this example $H_0 = H_1 = H_2 = H_3 = \mathbb{C}$, and thus $\mathcal{N}$ is the algebra of 4×4 matrices over $\mathbb{C}$ with

$$\mathcal{N}_1 = \left\{ \begin{pmatrix} 0 & 0 & 0 & 0 \\ 0 & 0 & a & b \\ 0 & 0 & 0 & c \\ 0 & 0 & 0 & 0 \end{pmatrix} : a, b, c \in \mathbb{C} \right\},$$

$$\mathcal{N}_2^0 = \left\{ \begin{pmatrix} 0 & a & b & c \\ 0 & 0 & 0 & 0 \\ 0 & 0 & 0 & 0 \\ 0 & 0 & 0 & 0 \end{pmatrix} : a, b, c \in \mathbb{C} \right\},$$

$$\mathcal{N}_d = \left\{ \begin{pmatrix} a & 0 & 0 & 0 \\ 0 & b & 0 & 0 \\ 0 & 0 & c & 0 \\ 0 & 0 & 0 & d \end{pmatrix} : a, b, c, d \in \mathbb{C} \right\},$$

$$\mathcal{N}_3^0 = (\mathcal{N}_2^0)^*, \qquad \mathcal{N}_4 = (\mathcal{N}_1)^*.$$

Take

$$K = \begin{pmatrix} \frac{1}{2} & \frac{1}{2} & 0 & 0 \\ 0 & \frac{1}{2} & 0 & 0 \\ 0 & 0 & \frac{1}{2} & 0 \\ 0 & 0 & 0 & \frac{1}{2} \end{pmatrix} \in \mathcal{N}_2 = \mathcal{N}_2^0 \oplus \mathcal{N}_d.$$

We want to find all positive extensions of $K + K^*$, i.e., all positive definite $G \in \mathcal{M}$ such that

$$G - (K + K^*) \in \mathcal{N}_1 + \mathcal{N}_4.$$

In this case $B = K + K^*$ is such an extension. This extension in fact is a band extension. Indeed,

$$B^{-1} = \begin{pmatrix} \frac{3}{4} & -\frac{2}{3} & 0 & 0 \\ -\frac{2}{3} & \frac{4}{3} & 0 & 0 \\ 0 & 0 & 1 & 0 \\ 0 & 0 & 0 & 1 \end{pmatrix} \in \mathcal{N}_c = \mathcal{N}_2^0 \oplus \mathcal{N}_d \oplus \mathcal{N}_3^0.$$

Next, we check that B is a canonical positive extension. By Gaussian elimination, we know that B has a factorization $B = U^{-*}U^{-1}$ where $U^{\pm 1} \in \mathcal{N}_+$ since in this case $\mathcal{N}_+$ consists of the usual upper triangular matrices. We need in addition that $U^*\mathcal{N}_1 \subset \mathcal{N}_+$; for an upper triangular

$$U = \begin{pmatrix} u_{11} & u_{12} & u_{13} & u_{14} \\ 0 & u_{22} & u_{23} & u_{24} \\ 0 & 0 & u_{33} & u_{34} \\ 0 & 0 & 0 & u_{44} \end{pmatrix}.$$

A computation shows that this is equivalent to $u_{24} = 0$. From the form of B and B^{-1}, it is clear that U has the form

$$U = \begin{pmatrix} u_{11} & u_{12} & 0 & 0 \\ 0 & u_{22} & 0 & 0 \\ 0 & 0 & 1 & 0 \\ 0 & 0 & 0 & 1 \end{pmatrix},$$

so the condition $U^*\mathcal{N}_1 \subset \mathcal{N}_+$ also is automatically satisfied.

Next we require a factorization $B = V^{-*}V^{-1}$ with $(V^*)^{\pm 1}\mathcal{N}_1 = \mathcal{N}_1$. A computation shows that a 4×4 matrix

$$Y = \begin{pmatrix} y_{11} & y_{12} & y_{13} & y_{14} \\ y_{21} & y_{22} & y_{23} & y_{24} \\ y_{31} & y_{32} & y_{33} & y_{34} \\ y_{41} & y_{42} & y_{43} & y_{44} \end{pmatrix}$$

is invertible and has $Y^{\pm 1}\mathcal{N}_1 = \mathcal{N}_1$ if and only if the set of conditions

$$\begin{aligned} &y_{12} = 0, \quad y_{13} = 0, \quad y_{32} = 0, \quad y_{42} = 0, \quad y_{43} = 0, \\ &y_{22} \neq 0, \quad y_{33} \neq 0, \quad y_{11}y_{44} - y_{14}y_{41} \neq 0 \end{aligned} \tag{5.21}$$

is satisfied. Thus

$$Y = \begin{pmatrix} 0 & 0 & 0 & 1 \\ 0 & \frac{\sqrt{3}}{2} & 0 & \frac{1}{2} \\ 0 & 0 & 1 & 0 \\ 1 & 0 & 0 & 0 \end{pmatrix}$$

meets all these requirements. Moreover a computation shows that $Y^*Y = B$. Thus, if we set $V = Y^*$, we have a pair of operators U, V meeting all the conditions for a canonical positive extension of $K + K^*$, and we can apply Theorems 2.2(a) and 2.4(a) to show that all positive real part extensions F of K are given by

$$F = (-K^*VG + KU)(VG + U)^{-1},$$

where the free parameter G is an arbitrary element of $\mathcal{N}_1$ satisfying $\|G\| < 1$.

The point that we want to make here is that such a linear fractional parametrization cannot be obtained by applying the recipe from Theorem 1.1 based on a band extension using spectral factorizations of B. Indeed, if we choose Y to be upper triangular with $Y^{\pm 1}\mathcal{N}_1 = \mathcal{N}_1$, so that we have

$$y_{21} = 0, \quad y_{31} = 0, \quad y_{41} = 0 \tag{5.22}$$

in addition to (5.21), then YY^* has the form

$$YY^* = \begin{pmatrix} y_{11} & 0 & 0 & y_{14} \\ 0 & y_{22} & y_{23} & y_{24} \\ 0 & 0 & y_{33} & y_{34} \\ 0 & 0 & 0 & y_{44} \end{pmatrix} \begin{pmatrix} \overline{y}_{11} & 0 & 0 & 0 \\ 0 & \overline{y}_{22} & 0 & 0 \\ 0 & \overline{y}_{23} & \overline{y}_{33} & 0 \\ \overline{y}_{14} & \overline{y}_{24} & \overline{y}_{34} & \overline{y}_{44} \end{pmatrix}.$$

Then the $(4,1)$ entry of YY^* is $(YY^*)_{41} = y_{44}\overline{y}_{41} = b_{41} = 0$ (where B is *any* positive definite extension of $K + K^*$ for which $B = YY^*$). As necessarily $y_{44} \neq 0$ since Y is invertible and upper triangular, necessarily $y_{14} = 0$. But then

$$(YY^*)_{12} = y_{14}\overline{y}_{24} = 0$$

while $b_{12} = k_{12} = 1/2$ and we have a contradiction.

Example 4.2. We take $\mathcal{N} = \mathcal{N}_1 \oplus \mathcal{N}_2^0 \oplus \mathcal{N}_d \oplus \mathcal{N}_3^0 \oplus \mathcal{N}_4$ as in the previous example, and we set

$$K = \begin{pmatrix} \frac{1}{2} & 0 & y & z \\ 0 & \frac{1}{2} & 0 & 0 \\ 0 & 0 & \frac{1}{2} & 0 \\ 0 & 0 & 0 & \frac{1}{2} \end{pmatrix} \in \mathcal{N}_2 = \mathcal{N}_2^0 \oplus \mathcal{N}_d.$$

Here y and z are real parameters satisfying $0 < y < z < 1$. Consider

$$B = \begin{pmatrix} 1 & 0 & y & z \\ 0 & 1 & 0 & 0 \\ y & 0 & 1 & \frac{y}{z} \\ z & 0 & \frac{y}{z} & 1 \end{pmatrix}.$$

We claim that B is a canonical positive extension of $K+K^*$ but not a band extension. To prove this we need a factorization $B = V^{-*}V^{-1}$ with $V^*\mathcal{M}_1 = \mathcal{M}_1$. As in the previous example, V^{-*} must have the form

$$V^{-*} = \begin{pmatrix} y_{11} & 0 & 0 & y_{14} \\ y_{21} & y_{22} & y_{23} & y_{24} \\ y_{31} & 0 & y_{33} & y_{34} \\ y_{41} & 0 & 0 & y_{44} \end{pmatrix}$$

with $y_{22} \neq 0, \quad y_{33} \neq 0, \quad y_{11}y_{44} - y_{14}y_{41} \neq 0$. Note that

$$V^{-*} = \begin{pmatrix} \sqrt{1-z^2} & 0 & 0 & z \\ 0 & 1 & 0 & 0 \\ 0 & 0 & \sqrt{z^2-y^2}/z & y/z \\ 0 & 0 & 0 & 1 \end{pmatrix}$$

meets these specifications, and in addition

$$V^{-*}V^{-1} = \begin{pmatrix} \sqrt{1-z^2} & 0 & 0 & z \\ 0 & 1 & 0 & 0 \\ 0 & 0 & \sqrt{z^2-y^2}/z & y/z \\ 0 & 0 & 0 & 1 \end{pmatrix} \begin{pmatrix} \sqrt{1-z^2} & 0 & 0 & 0 \\ 0 & 1 & 0 & 0 \\ 0 & 0 & \sqrt{z^2-y^2}/z & 0 \\ z & 0 & y/z & 1 \end{pmatrix} = B.$$

Next set,

$$\tilde{U} = \begin{pmatrix} 1 & 0 & -y & (y^2-z^2)/(1-y^2)z \\ 0 & 1 & 0 & 0 \\ 0 & 0 & 1 & y(z^2-1)/(1-y^2)z \\ 0 & 0 & 0 & 1 \end{pmatrix}$$

Then $\tilde{U}$ is upper triangular, and a direct computation shows that $B\tilde{U}$ is lower triangular; indeed

$$B\tilde{U} = \begin{pmatrix} 1 & 0 & y & z \\ 0 & 1 & 0 & 0 \\ y & 0 & 1 & y/z \\ z & 0 & y/z & 1 \end{pmatrix} \begin{pmatrix} 1 & 0 & -y & (y^2-z^2)/(1-y^2)z \\ 0 & 1 & 0 & 0 \\ 0 & 0 & 1 & y(z^2-1)/(1-y^2)z \\ 0 & 0 & 0 & 1 \end{pmatrix}$$

has

$$(B\tilde{U})_{12} = 0, \quad (B\tilde{U})_{13} = -y + y = 0,$$
$$(B\tilde{U})_{14} = (1/(1-y^2)z)[y^2 - z^2 + y^2z^2 - y^2 + z^2 - y^2z^2] = 0,$$
$$(B\tilde{U})_{23} = 0, \quad (B\tilde{U})_{24} = 0,$$
$$(B\tilde{U})_{34} = (1/(1-y^2)z)[y^3 - yz^2 + yz^2 - y + y - y^3] = 0.$$

Hence if $U = \tilde{U}D$ for an appropriate diagonal matrix D, then we have the factorization $B = U^{-*}U^{-1}$ where $U^{\pm 1}\mathcal{M}_+ = \mathcal{N}_+$. As in Example 4.1, the condition $U^*\mathcal{N}_1 \subset \mathcal{N}_+$

is equivalent to $u_{24} = 0$; as $\tilde{u}_{24} = 0$, this holds as well. Thus the pair of operators U and V meets all the requirements of (5.17) and (5.18) for $B = C + C^*$. Thus B is a canonical positive extension of $K + K^*$. However, this extension is not a band extension. A computation shows that $B^{-1} = UU^* = \tilde{U}DD^*\tilde{U}^*$ is equal to

$$B^{-1} = \begin{pmatrix} d_{11} & 0 & -yd_{33} & \big((y^2 - z^2)/(1 - y^2)z\big)d_{44} \\ 0 & d_{22} & 0 & 0 \\ 0 & 0 & d_{33} & \big(y(z^2 - 1)\big)/\big((1 - y^2)z\big)d_{44} \\ 0 & 0 & 0 & d_{44} \end{pmatrix}$$

$$\times \begin{pmatrix} 1 & 0 & 0 & 0 \\ 0 & 1 & 0 & 0 \\ -y & 0 & 1 & 0 \\ (y^2 - z^2)/(1 - y^2)z & 0 & y(z^2 - 1)/(1 - y^2)z & 1 \end{pmatrix}$$

where d_{11}, d_{22}, d_{33} and d_{44} are the (positive) diagonal entries of DD^*. Thus

$$(B^{-1})_{34} = \big(y(z^2 - 1)/(1 - y^2)z\big)d_{44} \neq 0$$

and B is not a band extension.

REFERENCES

[1] J.A. Ball, Nevanlinna-Pick interpolation: generalizations and applications, in *Recent Results in Operator Theory Vol. I* (Ed. J.B. Conway and B.B. Morrel), Longman Scientific and Tech., Essex, 1988, pp. 51–94.

[2] J.A. Ball, I. Gohberg and M.A. Kaashoek, Nevanlinna-Pick interpolation for time-varying input-output maps: The discrete case, in: *Time-variant systems and interpolation* (Ed. I. Gohberg), OT 56, Birkhäuser Verlag, Basel, 1992, pp. 1–51

[3] J.A. Ball and J.W. Helton, A Beurling-Lax theorem for the Lie group $U(m, n)$ which contains most classical interpolation, *J. Operator Theory* 9 (1983), 107–142.

[4] H. Dym and I. Gohberg, Extensions of band matrices with band inverses, *Linear Alg. Appl.* 36 (1981), 1–24.

[5] H. Dym and I. Gohberg, Extensions of matrix valued functions and block matrices, *Indiana Univ. Math. J.* 31 (1982), 733–765.

[6] R.L. Ellis, I. Gohberg and D.C. Lay, Extensions with positive ral part, a new version of the abstract band method with applications, *Integral Equations and Operator Theory* 16 (1993), 360–384.

[7] I. Gohberg, S. Goldberg and M.A. Kaashoek, *Classes of Linear Operators Vol. II*, OT 63, Birkhauser, Basel, 1993.

[8] I. Gohberg, M.A. Kaashoek and H.J. Woerdeman, The band method for positive and contractive extension problems, *J. Operator Theory* 22 (1989), 109–155.

[9] I. Gohberg, M.A. Kaashoek and H.J. Woerdeman, The band method for positive and contractive extension problems: An alternative version and new applications, *Integral Equations and Operator Theory* 12 (1989), 343–382.

[10] I. Gohberg, M.A. Kaashoek and H.J. Woerdeman, The band method for several positive extension problems of non-band type, *J. Operator Theory* 16 (1991), 191–218.

[11] I. Gohberg, M.A. Kaashoek and H.J. Woerdeman, The band method for bordered algebras, in *Contributions to Operator Theory and its Applications: Tsuyoshi Ando Anniversary Volume* (Ed. T. Furuta, I. Gohberg and T. Nakazi), OT 62, Birkhauser, Basel, 1993, pp. 85–98.

MSC: primary 47A57, secondary 30E05

J.A. Ball
Department of Mathematics
Virginia Tech
Blacksburg, Virginia 24061-0123, U.S.A.

I. Gohberg
Raymond and Beverly Sackler Faculty of Exact Sciences
School of Mathematical Sciences
Tel-Aviv University
Ramat-Aviv, 69978 Tel-Aviv, Israel

M.A. Kaashoek
Faculteit Wiskunde en Informatica
Vrije Universiteit
De Boelelaan 1081a
1081 HV Amsterdam, The Netherlands

Operator Theory:
Advances and Applications, Vol. 87

POLAR DECOMPOSITIONS IN FINITE DIMENSIONAL INDEFINITE SCALAR PRODUCT SPACES: SPECIAL CASES AND APPLICATIONS

Y. Bolshakov, C. V. M. van der Mee,[1] A. C. M. Ran,
B. Reichstein, L. Rodman[2]

Polar decompositions $X = UA$ of real and complex matrices X with respect to the scalar product generated by a given indefinite nonsingular matrix H are studied in the following special cases: (1) X is an H-contraction, (2) X is an H-plus matrix, (3) H has only one positive eigenvalue, and (4) U belongs to the connected component of the identity in the group of H-unitary matrices. Applications to linear optics are presented.

1 Introduction

Let F be either the field of real numbers $\mathbf{R}$ or the field of complex numbers $\mathbf{C}$. Fix a real symmetric (if $F = \mathbf{R}$) or complex hermitian (if $F = \mathbf{C}$) invertible $n \times n$ matrix H. Consider the scalar product induced by H by the formula $[x, y] = \langle Hx, y\rangle$, $x, y \in F^n$. Here $\langle \cdot , \cdot \rangle$ stands for the standard scalar product in F^n defined by $\langle x, y\rangle = \sum_{j=1}^{n} x_j \bar{y}_j$, where $(x_1, \cdots, x_n)^T$ and $(y_1, \cdots, y_n)^T$ are column vectors in F^n. (Of course, $\bar{y}_j = y_j$ if $F = \mathbf{R}$.) The scalar product $[\cdot , \cdot]$ is nondegenerate ($[x, y] = 0$ for all $y \in F^n$ implies $x = 0$), but is indefinite in general. In other words, the real number $[x, x]$ can be positive, negative, or zero for various $x \in F^n$ (unless H is definite). The vector $x \in F^n$ is called *positive* if $[x, x] > 0$, *neutral* if $[x, x] = 0$, and *negative* if $[x, x] < 0$.

---

[1] The work of this author was performed under the auspices of C.N.R.–G.N.F.M. and partially supported by the research project, "Nonlinear problems in analysis and its physical, chemical, and biological applications: Analytical, modelling and computational aspects," of the Italian Ministry of Higher Education and Research (M.U.R.S.T.)

[2] The work of this author partially supported by an NSF grant DMS 9123841 and by an NSF International Cooperation Grant.

Similarly, a subspace $\mathcal{M} \subset F^n$ is called *positive* (resp. *negative*) if all non-zero vectors $x \in \mathcal{M}$ are positive (resp. negative). We write H-positive or H-negative if we wish to emphasize the dependence of these definitions on H. If all vectors in a subspace are neutral, we say that the subspace is *isotropic* (or *H-isotropic*).

Well-known concepts related to the scalar product $[\,\cdot\,,\,\cdot\,]$ are defined in obvious ways. Thus, given an $n \times n$ matrix A over F, the adjoint $A^{[*]}$ is defined by $[Ax, y] = [x, A^{[*]}y]$ for all $x, y \in F^n$. The formula $A^{[*]} = H^{-1}A^*H$ is verified immediately (here and elsewhere we denote by A^* the conjugate transpose of A, so that $A^* = A^T$ if $F = \mathbf{R}$). A matrix A is called *H-selfadjoint* if $A^{[*]} = A$, or equivalently, if HA is hermitian. In particular, if HA is positive semidefinite hermitian, we say that the matrix A is *H-positive*. An $n \times n$ matrix U is called *H-unitary* if $[Ux, Uy] = [x, y]$ for all $x, y \in F^n$, or, equivalently, $U^*HU = H$. Observe that for every H-unitary matrix U we have $|\det U| = 1$; in particular, $\det U = \pm 1$ if $F = \mathbf{R}$.

In this article we continue to study decompositions of an $n \times n$ matrix X over F of the form

$$X = UA, \tag{1.1}$$

where U is H-unitary and A is H-selfadjoint (with or without additional restrictions). We call the decomposition (1.1) without additional restrictions on U and A an *H-polar decomposition* of X. Given non-negative integers p, q, (1.1) is called an *(H, p, q)-polar decomposition* if the number of positive (resp., negative) eigenvalues, when counted with multiplicities, of HA does not exceed p (resp., q). A general theory of H-polar decompositions has been developed in a preceding article [BMRRR1]; it is devoted to the problems of existence, uniqueness (up to equivalence) and basic properties of H-polar and (H, p, q)-polar decompositions, and to the existence of H-polar decompositions of H-normal matrices. Most of the concepts and notations used here are introduced in [BMRRR1]. In the present article we study H-polar decompositions of the type (1.1), where various constraints are imposed on the matrices X, U, A and H, and discuss its applications in linear optics.

We shall now briefly discuss these various subjects, some of their history, and the contents of the sections.

Motivated by the theory of the H-modulus for H-nonexpansive operators, i.e., operators X for which $H - X^*HX$ is positive semidefinite (see [P1,P2,AI]), and the theory of the H-modulus for H-plus operators, i.e., operators mapping positive vectors into positive or neutral vectors (see [AI,KS1,KS2]), we reprove and refine well-known results on the existence of an H-polar decomposition of H-contractions in Section 2 and of H-plus matrices in Section 3. In the case when H has precisely one positive eigenvalue, more specific H-polar decomposition results are obtained for H-plus matrices. Necessary and sufficient conditions are given for a matrix to be an H-plus matrix. In Section 4 we give a full description of all matrices X that allow an H-polar decomposition in the case when H has only one positive eigenvalue. The constraints that the structure of H imposes on the Jordan structure of A, make it possible to give a much more complete description than is given in [BMRRR1]. In Section 5 we seek H-polar decompositions where the H-unitary factor is required to belong to a prescribed connected component of the group of H-unitary matrices. For $F = \mathbf{R}$ and $n = 2, 3$, we give examples of matrices X having an H-polar decomposition, but where U cannot be chosen in just any prescribed connected component. For $F = \mathbf{R}$ and $n \geq 4$, such selections turn out to be always possible if H has only one positive eigenvalue and an H-polar decomposition exists. In Section 6 we apply our results on H-polar decomposition to linear optics, where we must study the case $F = \mathbf{R}$, $n = 4$, $H = \operatorname{diag}(1, -1, -1, -1)$, and U in the connected component of the identity. The polarization matrices involved are real H-plus matrices with respect to $H = \text{diag}(1,-1,-1,-1)$. Well-known results on two classes of polarization matrices, namely those satisfying the so-called Stokes criterion (see [K,MH,N,M]) and the so-called weighted sums of pure Mueller matrices (see [C,M]), are generalized. We indicate when matrices belonging to the larger one of these two classes (i.e., the class of matrices satisfying the Stokes criterion) have an H-polar decomposition. Necessary and sufficient conditions are given for a real 4×4 matrix to belong to either of these two classes, thus improving upon results given in [M].

The following notations will be used. The number of positive (negative, zero) eigenvalues of a hermitian matrix A is denoted by $\pi(A)$ $(\nu(A), \delta(A))$. F^n (where $F = \mathbf{R}$ or $F = \mathbf{C}$) stands for the vector space of n-dimensional columns over F. We denote by $F^{m\times n}$ the vector space of $m \times n$ matrices over F. The standard matrices are $J_k(\lambda)$ (the $k \times k$ upper triangular Jordan block with $\lambda \in \mathbf{C}$ on the main diagonal), I_m the $m \times m$ identity matrix, O_m the $m \times m$ zero matrix, and $Q_m = [\delta_{i+j,m+1}]_{i,j=1}^m$ the $m \times m$ matrix with 1's on the southwest–northeast diagonal and zeros elsewhere. The block diagonal matrix with matrices $Z_1, \dots, Z_k$ on the main diagonal is denoted by $Z_1 \oplus \cdots \oplus Z_k$ or $\text{diag}(Z_1, \dots, Z_k)$. The set of eigenvalues (including nonreal eigenvalues for real matrices) of a matrix X is denoted by $\sigma(X)$. $\text{Ker}\, A$ and $\text{Im}\, A$ stand for the null space and range of a matrix A. The symbol $\mathcal{M} \oplus \mathcal{N}$ denotes the direct sum of the subspaces $\mathcal{M}$ and $\mathcal{N}$.

Although we have sought to write the present paper in a self-contained way, occasionally we will draw on concepts and results from the previous paper [BMRRR1] and from the paper [BMRRR2] on H-unitary extensions and H-polar decompositions with HA positive semidefinite hermitian. The canonical form of an ordered pair $\{A, H\}$ where A is H-selfadjoint, is described in Section 2 of [BMRRR1] (as well as in Section I.3.2 of [GLR] and many other sources) and will not be redefined here. We will use freely the canonical form, in particular the sign characteristic, of the pair $\{A, H\}$.

For the reader's convenience, we quote here one result from [BMRRR1] (Theorem 4.4):

THEOREM 1.1. ($F = \mathbf{C}$ *or* $F = \mathbf{R}$) *An* $n \times n$ *matrix* X *admits* H*-polar decomposition if and only if all the conditions (i), (ii), and (iii) below are satisfied.*

(i) For each negative eigenvalue λ *of* $X^{[*]}X$ *the part of the canonical form of* $\{X^{[*]}X, H\}$ *corresponding to* λ *can be presented in the form*

$$\{\text{diag}\,(A_i)_{i=1}^m, \quad \text{diag}\,(H_i)_{i=1}^m\},$$

where, for $i = 1, \dots, m$,

$$A_i = J_{k_i}(\lambda) \oplus J_{k_i}(\lambda), \qquad H_i = Q_{k_i} \oplus -Q_{k_i}.$$

(ii) The part of the canonical form of $\{X^{[*]}X, H\}$ *corresponding to the zero eigenvalue can be presented in the form*

$$\{\text{diag}\,(B_i)_{i=0}^m, \quad \text{diag}\,(H_i)_{i=0}^m\},$$

where $B_0 = O_{k_0}$, $H_0 = I_{p_0} \oplus -I_{n_0}$, $p_0 + n_0 = k_0$ *and, for each* $i = 1, \dots, m$, *the pair* $\{B_i, H_i\}$ *is of one of the following two forms:*

$$B_i = J_{k_i}(0) \oplus J_{k_i}(0), \qquad H_i = Q_{k_i} \oplus -Q_{k_i}, \quad k_i > 1,$$

or

$$B_i = J_{k_i}(0) \oplus J_{k_i-1}, \qquad H_i = \varepsilon_i(Q_{k_i} \oplus Q_{k_i-1}),$$

with $\varepsilon = \pm 1$, *and* $k_i > 1$.

Assume that (ii) holds and denote the corresponding basis in $Ker(X^{[*]}X)^n$ *in which this is achieved by*

$$\{e_{i,j}\}_{i=0}^{m}{}_{j=1}^{l_i}$$

where $l_0 = k_0$ *and* $l_i = 2k_i$ *in case* B_i *is an even size matrix, and* $l_i = 2k_i - 1$ *in case* B_i *is an odd size matrix.*

(iii) There is a choice of basis $\{e_{i,j}\}_{i=0}^{m}{}_{j=1}^{l_i}$ *such that (ii) holds and*

$$\begin{aligned} KerX &= \operatorname{span}\{e_{i,1} + e_{i,\ k_i+1} \mid l_i = 2k_i,\ i = 1, \dots, m\} \oplus \\ &\oplus \operatorname{span}\{e_{i,1} \mid l_i = 2k_i - 1,\ i = 1, \dots, m\} \oplus \operatorname{span}\ \{e_{0,j}\}_{j=1}^{k_0}. \end{aligned}$$

2 H-Contractive Matrices

Let $F = \mathbf{R}$ or $F = \mathbf{C}$. We consider F^n together with the indefinite scalar product defined by the invertible hermitian matrix H over F.

An $n \times n$ matrix X (over F) is called *H-nonexpansive* if it does not increase the indefinite scalar product of two vectors, i.e., $[Xv, Xv] \le [v, v]$ for all $v \in F^n$, or, equivalently, the matrix $H - X^*HX = H(I - X^{[*]}X)$ is positive semidefinite hermitian. It was proved by Potapov [P2] (for the case $F = \mathbf{C}$) that every H-nonexpansive matrix X admits an H-polar decomposition. As a matter of fact, he showed that X can be factored as $X = UA$, where U is H-unitary and A is H-selfadjoint, with the additional conditions $\operatorname{Ker} A = \operatorname{Ker} A^2$ and $\sigma(A) \subset [0, \infty)$; such a matrix A was called an *H-modulus* of X. This result was later extended to the infinite dimensional case by Ju. P. Ginzburg ([Gi1,Gi2]), and by M. G. Krein and Ju. L. Shmul'jan ([KS1,KS2]). We will adopt the term *H-contraction* instead of H-nonexpansive matrix.

Given an arbitrary $n \times n$ matrix X, the question naturally arises when it is possible to find an H-polar decomposition $X = UA$ of X with the additional property that $\sigma(A) \subset \{\lambda \mid \lambda \ge 0\}$. The following theorem provides a complete answer to this question.

THEOREM 2.1. *($F = \mathbf{C}$ or $F = \mathbf{R}$) An $n \times n$ matrix X admits an H-polar decomposition $X = UA$ with $\sigma(A) \subset \{\lambda \mid \lambda \ge 0\}$ if and only if the condition* (i) *below and the conditions* (ii), *and* (iii) *of Theorem 1.1 hold.*

(i) $\sigma(X^{[*]}X) \subset \{\lambda \mid \lambda \ge 0\}$.

Proof. First let $F = \mathbf{C}$. If X has an H-polar decomposition $X = UA$ with $\sigma(A) \subset \{\lambda | \lambda \ge 0\}$ then $X^{[*]}X = A^2$ has all its eigenvalues in $\{\lambda | \lambda \ge 0\}$. Thus (i) holds. Now apply Theorem 1.1. The converse follows as in the proof of Theorem 1.1 (see the proof of Theorem 4.4 of [BMRRR1]).

The proof for $F = \mathbf{R}$ is essentially the same, especially since Theorem 1.1 applies to both $F = \mathbf{C}$ and $F = \mathbf{R}$. □

Let us see how the theory of H-polar decompositions applies to the particular class of matrices studied by Potapov, i.e., the class of H-contractions. To do this we need some preliminary material on H-contractions.

THEOREM 2.2. *Let X be an $n \times n$ matrix which is an H-contraction. Then the following hold:*

(a) $\sigma(X^{[*]}X) \subset \{\lambda | \lambda \geq 0\}$;

(b) *let $\mathcal{M}_+$ be the spectral invariant subspace of $X^{[*]}X$ corresponding to eigenvalues in $[0,1)$, then $\mathcal{M}_+$ is H-positive; let $\mathcal{M}_-$ denote the spectral invariant subspace of $X^{[*]}X$ corresponding to eigenvalues in $(1,\infty)$ then $\mathcal{M}_-$ is H-negative; in other words, every Jordan block of $X^{[*]}X$ with eigenvalue $\lambda > 1$ (resp. $\lambda < 1$) is of order 1 and the corresponding sign in the sign characteristic of $\{X^{[*]}X, H\}$ is -1 (resp. $+1$).*

(c) *on $\operatorname{Ker}(X^{[*]}X - I)^n$ the matrix $X^{[*]}X$ has Jordan blocks of order at most two. The signs in the sign characteristic of $\{X^{[*]}X, H\}$ corresponding to blocks with eigenvalue one of order one may be both $+1$ and -1, the signs corresponding to blocks with eigenvalue one of order two are all -1.*

Conversely, if a matrix X is such that the canonical form of $\{X^{[*]}X, H\}$ satisfies (a)–(c), then X is an H-contraction.

Proof. Assume that X is an H-contraction. Part (a) was proved in [P2], where also the first part of (c) was observed.

We provide a full independent proof of the conditions (a), (b) and (c). If B is an H-selfadjoint matrix, the canonical form of the pair $\{B, H\}$ shows that, after reduction to the canonical form, for every non-real eigenvalue λ of B the matrix $H - HB$ has a 2×2 principal submatrix of the form $\begin{bmatrix} 0 & 1-\overline{\lambda} \\ 1-\lambda & * \end{bmatrix}$; this 2×2 matrix is never positive semidefinite. Applying this observation to $B = X^{[*]}X$, we obtain that all eigenvalues of $X^{[*]}X$ are real. Furthermore, let $X^{[*]}X = S^{-1}JS$ and $H = S^*H_0S$, where the pair $\{J, H_0\}$ is the canonical form of the pair $\{X^{[*]}X, H\}$. Then J is an H_0-contraction, and

$$H - X^*HX = H - HX^{[*]}X = S^*(H_0 - H_0J)S \geq 0;$$

so $H_0 - H_0J \geq 0$. Now $H_0 - H_0J$ is block diagonal. Suppose λ is an eigenvalue of $X^{[*]}X$ (hence also an eigenvalue of J), and let k be the order of one of the Jordan blocks in J with eigenvalue λ. Then $H_0 - H_0J$ contains a block of the form $\varepsilon(Q_k - Q_kJ_k(\lambda))$, where ε is the sign in the sign characteristic of $\{J, H_0\}$ corresponding to this block. Clearly, this block can only be positive semidefinite if (c) holds and every Jordan block of $X^{[*]}X$ with eigenvalue $\lambda > 1$ (resp. $\lambda < 1$) is of order 1 with the sign -1 (resp. $+1$).

It remains to prove that $X^{[*]}X$ has no negative eigenvalues. Let $\mathcal{M}_+$ be the spectral invariant subspace of $X^{[*]}X$ corresponding to the eigenvalues which are less than 1. By the already proved parts of (a), (b) and (c), $\mathcal{M}_+$ is H-positive. In other words, the scalar product induced by H on $\mathcal{M}_+$ is positive definite. On such a subspace $X^{[*]}X$ cannot have negative eigenvalues. This completes the proof of the properties (a), (b) and (c).

The converse statement follows easily from the canonical form of $\{X^{[*]}X, H\}$. □

The opposite concept is the concept of an H-expansive matrix. An $n \times n$ matrix X (over F) is called an *H-expansion* if $[Xv, Xv] \geq [v, v]$ for all $v \in F^n$. Using an obvious observation that a matrix is an H-expansion if and only if it is a $(-H)$-contraction, the result analogous to Theorem 2.2 holds for H-expansions. To obtain the statement of this result, replace in Theorem 2.2 "H-contraction" by "H-expansion," replace the signs in (b) and (c) by their opposites, and interchange "H-positive" and "H-negative" in (b).

We say that a matrix X is *H-monotone* if it is an H-expansion or an H-contraction. Another piece of information we need for H-monotone matrices, is the following.

LEMMA 2.3. *If X is H-monotone, then* $\operatorname{Ker} X^{[*]}X = \operatorname{Ker} X$.

Proof. Assume first that X is an H-contraction. It is proved in [BR] (Lemma 4.4) that

$$\operatorname{rank}(X^{[*]}X) \leq \operatorname{rank}(X) \leq \operatorname{rank}(X^{[*]}X) + d, \tag{2.1}$$

where

$$d = \min\{\pi(H) - \pi(HX^{[*]}X), \quad \nu(H) - \nu(HX^{[*]}X)\}.$$

$$\pi(H) - \pi(HX^{[*]}X) = p_1(0), \quad \nu(H) - \nu(HX^{[*]}X) = 0. \tag{2.2}$$

Here $p_1(0)$ is the number of 1×1 nilpotent blocks in the canonical form of $X^{[*]}X$ with the sign $+1$ in the sign characteristic of $\{X^{[*]}X, H\}$. Clearly, (2.1) and (2.2) yield $\operatorname{Ker} X^{[*]}X = \operatorname{Ker} X$. The case of X an H-expansion is considered analogously. □

Combining the results above easily yields the following theorem of Potapov [P2].

THEOREM 2.4. ($F = \mathbf{C}$ *or* $F = \mathbf{R}$) *Let X be H-monotone. Then X admits a unique H-polar decomposition $X = UA$, with the additional property that $\sigma(A) \subset \{\lambda | \lambda \geq 0\}$. For this H-polar decomposition we also have* $\operatorname{Ker} A = \operatorname{Ker} A^2$.

Proof. Let X be an H-contraction. Then $\sigma(X^{[*]}X) \subset [0, \infty)$ and there are no Jordan blocks of order ≥ 2 corresponding to any zero eigenvalue. Thus Theorem 2.1 implies that X has an H-polar decomposition with $\sigma(A) \subset [0, \infty)$. More precisely, if we write

$$X^{[*]}X = O_{k_0} \oplus \bigoplus_{i=1}^{m} Y_i$$

with $0 < \lambda_1 < \cdots < \lambda_m$ and $\sigma(Y_i) = \{\lambda_i\}$, then

$$A = O_{k_0} \oplus \bigoplus_{i=1}^{m} Z_i,$$

where $Z_i^2 = Y_i$ and $\sigma(Z_i) = \{\sqrt{\lambda_i}\}$, is an H-selfadjoint matrix such that $X^{[*]}X = A^2$ and $\operatorname{Ker} A = \operatorname{Ker} A^2 = \operatorname{Ker} X^{[*]}X$. Further, A is a real matrix if X is a real matrix.

Since any matrix with only positive eigenvalues has a unique square root that is a matrix with only positive eigenvalues, there exists a unique H-modulus A such that $X^{[*]}X = A^2$, where X is a given H-contraction. □

3 H-plus Matrices

Let $F = \mathbf{R}$ or $F = \mathbf{C}$. We consider F^n together with the indefinite scalar product $[\cdot,\cdot]$ defined by the invertible hermitian matrix H over F.

Krein and Shmul'jan [KS1,KS2] have developed a theory of plus operators, which are operators on an indefinite scalar product space that transform nonnegative vectors into nonnegative vectors. The results they obtained were formulated in an infinite dimensional setting and hence their term "plus *operator*" is appropriate. However, since we are working exclusively in a finite dimensional context, we will adopt the term "H-plus matrix" instead, where the matrix H generating the scalar product has been attached to our terminology.

An $n\times n$ matrix X (over F) will be called an *H-plus matrix* if $[X^{[*]}Xu,u] = [Xu,Xu] \geq 0$ whenever $[u,u] \geq 0$. Clearly, X is an H-plus matrix if $[X^{[*]}Xu,u] \geq 0$ whenever $[u,u] > 0$. Thus defining

$$\mu(X) = \inf_{[u,u]=1} [X^{[*]}Xu,u], \tag{3.1}$$

we see that X is an H-plus matrix if and only if $\mu(X) \geq 0$. Then

$$[X^{[*]}Xz,z] \geq \mu(X)[z,z], \qquad z \in F^n. \tag{3.2}$$

In the complex case the formula (3.2) is well-known (see [Bo], Theorem II.8.1); one can prove (3.2) in the complex case also using convexity of numerical ranges (see Theorem 1.6 in [A]). We relegate the proof of (3.2) for the case $F = \mathbf{R}$ to the appendix of this section. We call X a *strict H-plus matrix* if $\mu(X) > 0$. Finally, we call X a *doubly H-plus matrix* if both X and $X^{[*]}$ are H-plus matrices. As we will indicate below, every strict H-plus matrix is a doubly H-plus matrix. However, there exist H-plus matrices which are not doubly H-plus matrices.

Example 3.1. Let $H = \mathrm{diag}\,(1,-1)$ and $X = \begin{bmatrix} 0 & 1 \\ 0 & 0 \end{bmatrix}$. Then $X^{[*]} = \begin{bmatrix} 0 & 0 \\ -1 & 0 \end{bmatrix}$. We now easily check that $[X^{[*]}Xu,u] = |u_2|^2$ and $[XX^{[*]}u,u] = -|u_1|^2$, where $u = (u_1,u_2)$. Thus X is an H-plus matrix, but $X^{[*]}$ is not. □

If H is positive definite, every $n\times n$ matrix is an H-plus matrix and $\mu(X)$ is the smallest (automatically nonnegative) eigenvalue of $X^{[*]}X$. Then X is a strict H-plus matrix if and only if it is invertible. On the other hand, if H is negative definite, every $n\times n$ matrix X is an H-plus matrix. Since the definition of $\mu(X)$ does not make sense in this case, we cannot even define strict H-plus matrices in the above way. In the rest of this section we will therefore assume that H is indefinite.

The next result allows us to derive the spectral properties of strict H-plus matrices from those of H-contractions.

LEMMA 3.1. *An $n\times n$ matrix X is a strict H-plus matrix if and only if, for some $c > 0$, cX is a $(-H)$-contraction. One may choose $c = \mu(X)^{-1/2}$. Moreover, every strict H-plus matrix is a doubly H-plus matrix.*

Proof. One easily checks the following string of implications: X is a strict H-plus matrix, if and only if there exists $\mu > 0$ such that $X^{[*]}X - \mu I$ is H-positive, if and only if there exists $c > 0$ such that $(cX)^{[*]}(cX) - I$ is H-positive, if and only if there exists $c > 0$

such that $I-(cX)^{[*]}(cX)$ is $(-H)$-positive, if and only if cX is a $(-H)$-contraction for some $c>0$. Obviously, $c=\mu^{-1/2}$.

The second part of the proposition follows from the fact that $Y^{[*]}$ is an H-contraction whenever Y is an H-contraction [P2]. □

We now discuss the spectral properties of $X^{[*]}X$ valid for H-plus matrices X. Although their infinite-dimensional versions (for $F=\mathbf{C}$) can be found in [KS1,KS2,Bo], for convenience we give full and concise proofs which apply to both the real and the complex cases.

PROPOSITION 3.2. *Let X be an H-plus matrix. Then the following hold:*

(a) $\sigma(X^{[*]}X)\subset\mathbf{R}$;

(b) *$X^{[*]}X$ does not have Jordan blocks of order ≥ 2 corresponding to eigenvalues different from $\mu(X)$, while there are no Jordan blocks of order ≥ 3 corresponding to the eigenvalue $\mu(X)$;*

(c) *The eigenvectors corresponding to the eigenvalues larger (resp. smaller) than $\mu(X)$ are positive (resp. negative);*

(d) *The Jordan blocks of size 2 corresponding to the eigenvalue $\mu(X)$ have the positive sign in the sign characteristic of $\{X^{[*]}X,H\}$;*

(e) *If X is a doubly H-plus matrix, then $\sigma(X^{[*]}X)\subset[0,\infty)$.*

Proof. If X be a strict H-plus matrix, parts (a), (b), (c) and (d) are immediate from Lemma 3.1 and the corresponding parts of Theorem 2.2. If X is an H-plus matrix, then $X^{[*]}X-\mu(X)I$ being H-positive implies parts (a), (b), (c) and (d). It remains to prove part (e).

Let X be an H-plus matrix such that $\mu(X)=0$. If $\lambda\in\sigma(X^{[*]}X)\cap(-\infty,0)$, then there exists u such that $X^{[*]}Xu=\lambda u$ and $[u,u]<0$. Using $X[\mathrm{Ker}\,(X^{[*]}X-\lambda I)]=\mathrm{Ker}\,(XX^{[*]}-\lambda I)$ and writing $v=Xu$, we find $[v,v]>0$ and $XX^{[*]}v=\lambda v$, which contradicts part (c) if $X^{[*]}$ is an H-plus matrix. Hence, if X is a doubly H-plus matrix, then $\sigma(X^{[*]}X)\subset[0,\infty)$. □

The following result has been proved in [M] in the case $F=\mathbf{R}$ and $n=4$. The main result in [M] can be simplified, since its author failed to observe that $\sigma(X^{[*]}X)\subset[0,\infty)$.

PROPOSITION 3.3. *Let X be an $n\times n$ matrix. Then X is a strict H-plus matrix if and only if $X^{[*]}X$ has the following properties:*

(a) $\sigma(X^{[*]}X)\subset[0,\infty)$;

(b) *There exists $\mu>0$ such that there are no Jordan blocks of order exceeding 1 corresponding to the eigenvalues of $X^{[*]}X$ different from μ. The eigenvectors corresponding to the eigenvalues smaller than μ are negative; those corresponding to the eigenvalues larger than μ are positive;*

(c) *There do not exist Jordan blocks of order exceeding 2 corresponding to the eigenvalue μ of $X^{[*]}X$; the blocks of size 2 have the positive sign in the sign characteristic of $\{X^{[*]}X,H\}$.*

X is a non-strict H-plus matrix if and only if $X^{[*]}X$ fails to satisfy at least one of (a), (b) *and* (c), *and has, in addition, the following properties:*

(d) $\sigma(X^{[*]}X) \subset \mathbf{R}$ *and* $0 \in \sigma(X^{[*]}X)$;

(e) *There are no Jordan blocks of order exceeding* 1 *corresponding to the eigenvalues of $X^{[*]}X$ different from* 0. *The eigenvectors corresponding to the negative eigenvalues are negative; those corresponding to the positive eigenvalues are positive;*

(f) *There do not exist Jordan blocks of order exceeding* 2 *corresponding to the zero eigenvalue of $X^{[*]}X$; the blocks of size* 2 *have the positive sign in the sign characteristic of* $\{X^{[*]}X, H\}$.

Proof. A strict H-plus matrix has the above properties (a)-(c) and a non-strict plus matrix has the above properties (d)-(f), as a consequence of Proposition 3.2. Conversely, suppose X is an $n \times n$ matrix having the above properties (a)-(c). Then with no loss of generality, we may assume that

$$H = \operatorname{diag}(\underbrace{+1,\cdots,+1,}_{p-\nu \text{ entries}}\ \underbrace{+1,-1,\cdots,+1,-1,}_{\nu \text{ pairs}}\ \underbrace{-1,\cdots,-1}_{k-\nu \text{ entries}}) \tag{3.3}$$

and

$$X^{[*]}X = \operatorname{diag}(\lambda_1,\cdots,\lambda_{p-\nu}, D(\varepsilon_1,\mu),\cdots,D(\varepsilon_\nu,\mu),\lambda_{p+\nu+1},\cdots,\lambda_n),$$

where $\lambda_j \geq \mu > 0$ if $j = 1,\cdots,p-\nu$; $0 \leq \lambda_j \leq \mu$ if $j = p+\nu+1,\cdots,n$; $\varepsilon_1,\cdots,\varepsilon_\nu > 0$; and

$$D(\varepsilon,\mu) = \begin{bmatrix} \mu+\varepsilon & \varepsilon \\ -\varepsilon & \mu-\varepsilon \end{bmatrix}.$$

Then one easily verifies that for any vector $z = (z_1,\cdots,z_n)$

$$\begin{aligned}
[X^{[*]}Xz,z] &= \sum_{j=1}^{p-\nu} \lambda_j|z_j|^2 - \sum_{j=p+\nu+1}^{n} \lambda_j|z_j|^2 \\
&+ \mu\sum_{j=1}^{\nu}\left(|z_{p-\nu+2j-1}|^2 - |z_{p-\nu+2j}|^2\right) \\
&+ \sum_{j=1}^{\nu} \varepsilon_j|z_{p-\nu+2j-1} + z_{p-\nu+2j}|^2.
\end{aligned}$$

As a result,

$$\begin{aligned}
[X^{[*]}Xz,z] - \mu[z,z] &= \sum_{j=1}^{p-\nu}(\lambda_j-\mu)|z_j|^2 + \sum_{j=p+\nu+1}^{n}(\mu-\lambda_j)|z_j|^2 \\
&+ \sum_{j=1}^{\nu} \varepsilon_j|z_{p-\nu+2j-1} + z_{p-\nu+2j}|^2 \geq 0,
\end{aligned}$$

where $\mu > 0$, which implies that X is a strict H-plus matrix.

Now let X be an $n \times n$ matrix having the properties (d)-(f). Then with no loss of generality, we may assume that H has the form (3.3) and

$$X^{[*]}X = \operatorname{diag}(\lambda_1, \cdots, \lambda_{p-\nu}, D(\varepsilon_1, 0), \cdots, D(\varepsilon_\nu, 0), \lambda_{p+\nu+1}, \cdots, \lambda_n),$$

where $\lambda_j \geq 0$ if $j = 1, \cdots, p-\nu$; $\lambda_j \leq 0$ if $j = p+\nu+1, \cdots, n$; and $\varepsilon_1, \cdots, \varepsilon_\nu > 0$. Then one easily verifies that for any vector $z = (z_1, \cdots, z_n)$

$$[X^{[*]}Xz, z] = \sum_{j=1}^{p-\nu} \lambda_j |z_j|^2 + \sum_{j=p+\nu+1}^{n} (-\lambda_j)|z_j|^2 + \sum_{j=1}^{\nu} \varepsilon_j |z_{p-\nu+2j-1} + z_{p-\nu+2j}|^2 \geq 0,$$

which implies that X is an H-plus matrix with $\mu(X) = 0$. □

Let us now characterize the H-plus matrices allowing an H-polar decomposition. The part pertaining to strict H-plus matrices has been proved before by Krein and Shmul'jan [KS2] in an infinite-dimensional setting under conditions that are satisfied in the finite-dimensional case. In fact, in [KS2] the existence of a unique H-modulus for strict H-plus matrices (and for a certain class of strict H-plus operators) is proved. Indeed, let X be a strict H-plus matrix. Then for $c = \mu(X)^{-1/2}$, the matrix cX is a $(-H)$-contraction and hence has a unique $(-H)$-modulus A_0. But then $A = \mu(X)^{1/2}A_0$ is an H-modulus for X. The uniqueness of A_0 is equivalent to the uniqueness of A. Of course, the proof using Proposition 3.1 breaks down in the infinite dimensional case, and therefore [KS2] needed a different proof for the existence of a unique H-modulus. Note that Condition (c) of Theorem 3.4 is redundant if X is a doubly H-plus matrix.

THEOREM 3.4. *Let X be an H-plus matrix. Then X has an H-polar decomposition if and only if the following conditions are satisfied:*

(a) *$X^{[*]}X$ is invertible, or $0 \in \sigma(X^{[*]}X)$ and there are at least as many linearly independent positive eigenvectors corresponding to the zero eigenvalue as there are Jordan blocks of order 2; in other words, the part of the canonical form of $\{X^{[*]}X, H\}$ corresponding to the zero eigenvalue of $X^{[*]}X$ can be presented in the form*

$$\{O_k \oplus B \oplus \cdots \oplus B,\ G \oplus K \oplus \cdots \oplus K\}, \tag{3.4}$$

where $G = I_p \oplus -I_q$ $(p+q=k)$, $B = \begin{bmatrix} 0 & 1 \\ 0 & 0 \end{bmatrix} \oplus (0)$, $K = \begin{bmatrix} 0 & 1 \\ 1 & 0 \end{bmatrix} \oplus (1)$, and the summands B and K are repeated m times each in (3.4).

(b) *In case $0 \in \sigma(X^{[*]}X)$, there is a basis $\{e_{ij}\}_{i=0}^{m}{}_{j=1}^{l_i}$ $(l_0 = k, l_i = 3$ for $i = 1, \ldots, m)$ in $\operatorname{Ker}(X^{[*]}X)^n$ with respect to which the canonical form (3.4) is achieved and which has the additional property that*

$$\operatorname{Ker} X = \operatorname{span}\{e_{01}, \ldots, e_{0k}\} \oplus \operatorname{span}\{e_{11}, \ldots, e_{m1}\}.$$

(c) *$X^{[*]}X$ does not have negative eigenvalues.*

In particular, a strict H-plus matrix has an H-polar decomposition.

Proof. Lemma 3.1 and Theorem 2.4 imply that any strict H-plus matrix X has an H-polar decomposition $X = UA$ where A has only nonnegative eigenvalues and $\operatorname{Ker} A = \operatorname{Ker} A^2$.

To determine if a given H-plus matrix X has an H-polar decomposition, it suffices to examine the nilpotent part of $X^{[*]}X$ and the part corresponding to the negative eigenvalues. The theorem then is a straightforward application of Theorem 1.1, taking into account that (a) all Jordan blocks of order 2 corresponding to the zero eigenvalue have the positive sign in the sign characteristic of $\{X^{[*]}X, H\}$ and that there are no Jordan blocks of order exceeding 2, and (b) for each negative eigenvalue, all Jordan blocks are of order 1 and the corresponding eigenvectors are negative. □

If H has exactly one positive eigenvalue, then $X^{[*]}X$ and $XX^{[*]}$ are similar if X is an H-plus matrix. Indeed, if $X^{[*]}X$ and $XX^{[*]}$ were to have a different Jordan structure, then their nilpotent parts would not be similar (see [F] for a complete description of the relationships between the Jordan form of AB and that of BA). In view of Proposition 3.3 the only way in which this is possible is when one of the two matrices $X^{[*]}X$ and $XX^{[*]}$ has exactly one Jordan block of order 2 corresponding to the zero eigenvalue while the other matrix has all Jordan blocks of order 1 corresponding to the zero eigenvalue. But then in view of Theorem 3.4 one of X and $X^{[*]}$ would have an H-polar decomposition, whereas the other does not, which is impossible. Indeed, if $X = UA$ is an H-polar decomposition of X, then $X^{[*]} = U^{-1} \cdot UAU^{-1}$ is an H-polar decomposition of $X^{[*]}$.

The similarity of $X^{[*]}X$ and $XX^{[*]}$ for X an H-plus matrix can be used to refine Proposition 3.3. Namely, if H has exactly one positive eigenvalue and X is an $n \times n$ matrix, then X is a doubly H-plus matrix if and only if X has the following properties:

(g) $\sigma(X^{[*]}X) \subset [0, \infty)$;

(h) There exists $\mu \geq 0$ such that there are no Jordan blocks of order exceeding 1 corresponding to the eigenvalues of $X^{[*]}X$ different from μ. The eigenvectors corresponding to the eigenvalues smaller than μ are negative; those corresponding to the eigenvalues larger than μ are positive;

(i) There do not exist Jordan blocks of order exceeding 2 corresponding to the eigenvalue μ of $X^{[*]}X$; the blocks of size 2 have the positive sign in the sign characteristic of $\{X^{[*]}X, H\}$.

Indeed, suppose (g)-(i) hold with $\mu = 0$. (The case $\mu > 0$ implies that X is a strict H-plus matrix and hence a doubly H-plus matrix). Since (g)-(i) imply that X is an H-plus matrix (see (d)-(f) in Proposition 3.3), we have $X^{[*]}X$ and $XX^{[*]}$ similar. But then $XX^{[*]}$ satisfies (g)-(i), and therefore $X^{[*]}$ is an H-plus matrix.

In connection with the remark made two paragraphs ago, observe that in general the matrices $XX^{[*]}$ and $X^{[*]}X$ need not be similar:

Example 3.2 (based on the formula (7.1) in [BR].) Let

$$H = \begin{bmatrix} 0 & 1 \\ 1 & 0 \end{bmatrix} \oplus \begin{bmatrix} 0 & -1 \\ -1 & 0 \end{bmatrix} \oplus (1) \oplus (-1),$$

and let

$$X = \begin{bmatrix} 1 & 0 & 0 & 0 & 0 & 0 \\ 0 & 0 & 0 & 0 & 0 & 0 \\ 0 & 0 & 0 & 0 & 0 & 1 \\ 0 & 0 & 0 & 0 & 0 & 0 \\ 0 & 1 & 0 & 0 & 0 & 0 \\ 0 & 0 & 0 & 1 & 0 & 0 \end{bmatrix}.$$

A calculation shows that

$$X^{[*]}X = \begin{bmatrix} 0 & 1 \\ 0 & 0 \end{bmatrix} \oplus \begin{bmatrix} 0 & 1 \\ 0 & 0 \end{bmatrix} \oplus (0) \oplus (0),$$

whereas the matrix $XX^{[*]}$ has 1 only in the positions (1,5), (3,4) and (5,2), all other positions in $XX^{[*]}$ being zero. Clearly, $X^{[*]}X$ and $XX^{[*]}$ are not similar (they have different ranks). □

Let $F = \mathbf{R}$ and $n \geq 2$, and let H have exactly one positive eigenvalue. With no loss of generality, we take $H = \text{diag}\,(1, -1, \cdots, -1)$. Then the set

$$\mathcal{C} = \{x = (x_1, \cdots, x_n) \in \mathbf{R}^n : [x, x] \geq 0,\ x_1 \geq 0\}$$

is a positive cone in $\mathbf{R}^n$, i.e., $u + v \in \mathcal{C}$ and $\lambda u \in \mathcal{C}$ whenever $u, v \in \mathcal{C}$ and $\lambda \in [0, \infty)$. In fact, if $u = (u_1, \cdots, u_n)$ and $v = (v_1, \cdots, v_n)$ belong to $\mathcal{C}$, then Schwarz' inequality implies that

$$\begin{aligned}[u+v, u+v] &= \{u_1^2 - (u_2^2 + \cdots + u_n^2)\} + \{v_1^2 - (v_2^2 + \cdots + v_n^2)\} \\ &+ 2\{u_1 v_1 - (u_2 v_2 + \cdots + u_n v_n)\} \geq 0.\end{aligned}$$

The cone $\mathcal{C}$ has the following properties [BP,Kr]:

1. The interior of $\mathcal{C}$, relative to the usual topology of $\mathbf{R}^n$, coincides with the set of positive vectors in $\mathbf{R}^n$ with positive first component, and it is obviously nonempty.

2. $e = (1, 0, \cdots, 0)$ is an order unit relative to the partial order of $\mathbf{R}^n$ generated by the cone $\mathcal{C}$ (i.e., $x \geq y$ if and only if $x - y \in \mathcal{C}$). Indeed, if $x = (x_1, \cdots, x_n) \in \mathcal{C}$, then $\lambda_+ e - x \in \mathcal{C}$ and $x - \lambda_- e \in \mathcal{C}$ where $\lambda_\pm = x_1 \pm \sqrt{x_2^2 + \cdots + x_n^2}$. As a result, a real $n \times n$ matrix X maps the interior of $\mathcal{C}$ into itself if and only if $(Xe)_1 > 0$ and Xe is a positive vector.

3. The dual cone, i.e., the set of all vectors $x \in \mathbf{R}^n$ satisfying $[x, y] \geq 0$ for every $y \in \mathcal{C}$ (where we note that we have defined duality with respect to the indefinite scalar product rather than with respect to the usual scalar product of $\mathbf{R}^n$), coincides with $\mathcal{C}$. Indeed, if $[x, y] \geq 0$ for every $y \in \mathcal{C}$, then (using $y = (\sqrt{x_2^2 + \cdots + x_n^2}, x_2, \cdots, x_n)$ if $(x_2, \cdots, x_n)$ is nontrivial, and using $y = e$ if $x_2 = \cdots = x_n = 0$) we have $x_1 \geq \sqrt{x_2^2 + \cdots + x_n^2}$. Thus $x \in \mathcal{C}$. Conversely, if $x, y \in \mathcal{C}$, then $[x, y] = x_1 y_1 - (x_2 + \cdots + x_n y_n) = x_1 y_1 - [x_2^2 + \cdots + x_n^2]^{1/2}[y_2^2 + \cdots + y_n^2]^{1/2} \geq 0$. As a result, if a real $n \times n$ matrix X satisfies $X[\mathcal{C}] \subset \mathcal{C}$, this is also true for $X^{[*]}$.

PROPOSITION 3.5. *Let $F = \mathbf{R}$ and $H = \operatorname{diag}(1, -1, \cdots, -1)$. Then a real $n \times n$ matrix $X = [X_{ij}]_{i,j=1}^{n}$ satisfies $X[\mathcal{C}] \subset \mathcal{C}$ if and only if it is a doubly H-plus matrix with $X_{11} \geq 0$.*

Proof. Let X be a real $n \times n$ matrices leaving invariant the cone $\mathcal{C}$. Suppose $u \in \mathbf{R}^n$ and $[u, u] \geq 0$. If $u_1 \geq 0$, then $u \in \mathcal{C}$ and hence $Xu \in \mathcal{C}$, so that $[Xu, Xu] \geq 0$. On the other hand, if $u_1 \leq 0$, then $(-u) \in \mathcal{C}$ and hence $(-Xu) \in \mathcal{C}$, so that $[Xu, Xu] = [-Xu, -Xu] \geq 0$. Thus X is an H-plus matrix. Further, $X^{[*]}[\mathcal{C}] \subset \mathcal{C}$ (see item 3 above), and therefore X is a doubly H-plus matrix. Finally, since X_{11} is the first component of Xe and $Xe \in \mathcal{C}$, we get $X_{11} \geq 0$.

Conversely, let X be a doubly H-plus matrix with $X_{11} \geq 0$. First note that every $x = (x_1, \cdots, x_n) \in \mathcal{C}$ can be written as the sum of three vectors from the boundary of $\mathcal{C}$:

$$x = \frac{x_1 - \sigma}{2} \begin{pmatrix} 1 \\ q_2 \\ \vdots \\ q_n \end{pmatrix} + \frac{x_1 - \sigma}{2} \begin{pmatrix} 1 \\ -q_2 \\ \vdots \\ -q_n \end{pmatrix} + \begin{pmatrix} \sigma \\ x_2 \\ \vdots \\ x_n \end{pmatrix},$$

where $\sigma = \sqrt{x_2^2 + \cdots + x_n^2}$ and $(q_2, \cdots, q_n) \in \mathbf{R}^{n-1}$ has unit length. Thus in order to prove $X[\mathcal{C}] \subset \mathcal{C}$ it suffices to prove that $Xu \in \mathcal{C}$ for every vector u of the form $u = \operatorname{col}(1, q)$ where $q \in \mathbf{R}^{n-1}$ has unit length. First of all, $X^{[*]}$ being an H-plus matrix implies $[X^{[*]}e, X^{[*]}e] \geq 0$. With $X_{11} \geq 0$ this yields $X_{11} \geq [X_{12}^2 + \cdots + X_{1n}^2]^{1/2}$. Now applying X to $u = \operatorname{col}(1, q)$ with $q = (q_2, \cdots, q_n) \in \mathbf{R}^{n-1}$ having unit length, we obtain $[Xu, Xu] \geq 0$ because X is an H-plus matrix, as well as

$$[Xu]_1 = X_{11} + X_{12}q_2 + \cdots + X_{1n}q_n \geq X_{11} - \sqrt{X_{12}^2 + \cdots + X_{1n}^2} \geq 0,$$

which implies $Xu \in \mathcal{C}$. Hence $X[\mathcal{C}] \subset \mathcal{C}$. □

The next result is immediate from the Perron-Fröbenius theory (e.g., [BP]).

PROPOSITION 3.6. *Suppose $H = \operatorname{diag}(1, -1, \cdots, -1)$, X is a nontrivial doubly H-plus matrix, and σ_0 is the sign of X_{11}. Then the following statements hold:*

(a) *The spectral radius ρ of X is an eigenvalue of $\sigma_0 X$ and there exists a corresponding eigenvector in $\mathcal{C}$.*

(b) *Let m be the order of the largest Jordan block corresponding to the eigenvalue $\sigma_0\rho$. Then the Jordan blocks corresponding to any eigenvalue λ of X of absolute value ρ have orders not exceeding m.*

(c) *Let either $X_{11}^2 > X_{21}^2 + \cdots + X_{n1}^2$ or $X_{11}^2 > X_{12}^2 + \cdots + X_{1n}^2$. Then*

(c_1) *$\rho > 0$;*

(c_2) *$\sigma_0\rho$ is an algebraically simple eigenvalue of both X and $X^{[*]}$ to which correspond positive eigenvectors;*

(c_3) *X and $X^{[*]}$ have no other eigenvalues on the spectral circle $|z| = \rho$.*

Proof. Parts (a) and (b) follow from Theorem 1.3.2 of [BP]. Part (c) follows from Theorem 1.3.26 of [BP], because under the additional assumption either Xe or $X^{[*]}e$ belongs to the interior of $\mathcal{C}$. □

If $F = \mathbf{R}$ and $H = \mathrm{diag}\,(1, -1, \cdots, -1)$, then any real $n \times n$ matrix X leaving invariant the cone $\mathcal{C}$ has an H-polar decomposition $X = UA$ where U is H-unitary, A is H-selfadjoint, and both U and A leave invariant $\mathcal{C}$, unless $X^{[*]}X$ is nilpotent and different from the zero matrix. This follows almost immediately from Theorem 3.4 and Proposition 3.5, provided we can prove, in the cases where X allows an H-polar decomposition, that one may choose U such that U (and hence also $U^{-1} = U^{[*]}$) leaves invariant $\mathcal{C}$. However, any H-unitary matrix U has the property that either $U_{11} \geq 1$ or $U_{11} \leq -1$; thus an H-unitary matrix U leaves invariant $\mathcal{C}$ if and only if $U_{11} \geq 1$. If $U_{11} \leq 1$ for the H-unitary matrix U in the H-polar decomposition $X = UA$, we can simply replace $X = UA$ by $X = (-U)(-A)$. Since in that case both X and $(-U)^{-1}(= (-U)^{[*]})$ map $\mathcal{C}$ into itself, this is also the case for $(-A)$.

Appendix: Proof of (3.2) in the real case.

Throughout the appendix (with the exception of Corollary 3.7) we assume that $F = \mathbf{R}$, and that X is a real H-plus matrix of order n. Without loss of generality we can (and do) assume that the pair $\{X^{[*]}X, H\}$ is in the canonical form (see, e.g., Theorem I.5.3 in [GLR], or Section 2 of [BMRRR1]).

We denote by $J_k(\lambda \pm i\mu)$ the $k \times k$ real Jordan block with complex conjugate eigenvalues $\lambda \pm i\mu$ (the integer k is necessarily even). More explicitly, $J_k(\lambda \pm i\mu)$ is a block $k/2 \times k/2$ matrix with 2×2 blocks, where the block diagonal consists of the blocks $\begin{bmatrix} \lambda & \mu \\ -\mu & \lambda \end{bmatrix}$, the first block superdiagonal consists of the blocks I_2, and all other blocks are zeros.

Observe that all eigenvalues of $X^{[*]}X$ are real. Indeed, if $\lambda \pm i\mu$ is a pair of non-real complex conjugate eigenvalues of $X^{[*]}X$, then the pair $\{X^{[*]}X, H\}$ contains a pair of blocks $\{J_k(\lambda \pm i\mu), Q_k\}$. Denote by e_p the standard unit vector having 1 in the p-th position and zeros elsewhere. If $m = k/2$ is odd, then, denoting $Z = J_k(\lambda \pm i\mu)^{[*]}J_k(\lambda \pm i\mu)$, we have

$$[Ze_m, e_m] = -\mu, \quad [Ze_{m+1}, e_{m+1}] = \mu, \quad [e_m, e_m] = [e_{m+1}, e_{m+1}] = 0;$$

if m is even, then

$$[Z(e_{m-1} + e_{m+1}), e_{m-1} + e_{m+1}] = -2\mu, \quad [Z(e_m + e_{m+2}), e_m + e_{m+2}] = 2\mu,$$

$$[e_{m-1} + e_{m+1}, e_{m-1} + e_{m+1}] = [e_m + e_{m+2}, e_m + e_{m+2}] = 0;$$

and in both cases a contradiction to X being an H-plus matrix is obtained.

Assume that $\{X^{[*]}X, H\}$ contains a pair of blocks $\{J_{2m}(\lambda), \varepsilon Q_{2m}\}$. The set of vectors $x = \sum_{j=1}^{m} x_j e_j \in \mathbf{R}^{2m}$ satisfying the equation $\langle \varepsilon Q_{2m} x, x\rangle = 1$ is described by the formula

$$2\varepsilon(x_1 x_{2m} + x_2 x_{2m-1} + \cdots + x_m x_{m+1}) = 1. \tag{3.5}$$

A calculation shows that if (3.5) holds, then

$$\langle \varepsilon Q_m J_{2m}(\lambda)^{[*]} J_{2m}(\lambda)x, x\rangle = \lambda + 2\varepsilon(x_2 x_{2m} + \cdots + x_m x_{m+2}) + \varepsilon x_{m+1}^2.$$

Since X is H-plus, the equality

$$\lambda + 2\varepsilon(x_2x_{2m} + \cdots + x_mx_{m+2}) + \varepsilon x_{m+1}^2 \geq 0 \tag{3.6}$$

holds for every x satisfying (3.5). If $m > 1$, then letting $x_3 = \cdots = x_{2m-1} = 0$, $x_{2m} = 1$, x_2 arbitrary (and x_1 determined from (3.5)), it follows that $\lambda + 2\varepsilon x_2 \geq 0$ for all $x_2 \in \mathbf{R}$, which is impossible. Thus, $m = 1$. Now (3.6) takes the form $\lambda + \varepsilon x_2^2 \geq 0$, and since this inequality must be satisfied for all non-zero x_2 (because for any such x_2 the value of x_1 can be determined from (3.5)), we obtain $\varepsilon = +1$ and $\lambda \geq 0$. Conclusion: the even size Jordan blocks of $X^{[*]}X$ must have size 2, their signs in the sign characteristic are all $+1$, and their eigenvalues are all nonnegative.

Analogously we verify that $X^{[*]}X$ cannot have odd size Jordan blocks of size larger than 1. Thus:

$$X^{[*]}X = J_2(\lambda_1) \oplus \cdots \oplus J_2(\lambda_k) \oplus (\mu_1) \oplus \cdots \oplus (\mu_s),$$
$$H = Q_2 \oplus \cdots \oplus Q_2 \oplus (\varepsilon_1) \oplus \cdots \oplus (\varepsilon_s),$$

where $\lambda_j \geq 0$; $\mu_1, \cdots, \mu_s$ are real, and $\varepsilon = \pm 1$. Assume first $k \neq 0$ (i.e., $X^{[*]}X$ is not diagonalizable). Let $u = (x_1, y_1, \cdots, x_k, y_k, z_1, \cdots, z_s)^T$, where x_j, y_j, z_j are real numbers. We have

$$[u, u] = \sum_{i=1}^{k} 2x_jy_j + \sum_{j=1}^{s} \varepsilon_j z_j^2, \tag{3.7}$$

$$[X^{[*]}Xu, u] = \sum_{i=1}^{k} (2\lambda_i x_i y_i + y_i^2) + \sum_{j=1}^{s} \varepsilon_j \mu_j z_j^2. \tag{3.8}$$

Therefore, if $[u, u] = 1$, then

$$[X^{[*]}Xu, u] = \lambda_k + \sum_{i=1}^{k-1} 2(\lambda_i - \lambda_k)x_iy_i + \sum_{i=1}^{k} y_i^2 + \sum_{j=1}^{s} (\mu_j - \lambda_k)\varepsilon_j z_j^2 \geq 0 \tag{3.9}$$

by the H-plus property of X. Since the parameters $x_1, \cdots, x_{k-1}, y_1 \cdots, y_k, z_1, \cdots, z_s$ can be chosen arbitrarily in (3.7) (as long as $y_k \neq 0$, to ensure existence of $x_k \in \mathbf{R}$ such that $[u, u] = 1$), we conclude that $\lambda_i - \lambda_k = 0$ for $i = 1, \cdots, k-1$ and $(\mu_j - \lambda_k)\varepsilon_j \geq 0$ for $j = 1, \cdots, s$. In other words, all λ_i's are equal to the same number, call it λ, and $\varepsilon_j = +1$ (resp. $\varepsilon_j = -1$) for every eigenvalue $\mu_j > \lambda$ (resp. $\mu_j < \lambda$). The formula (3.9) shows also that $\mu(X) = \lambda$, where $\mu(X)$ is defined by (3.1); indeed, it suffices to take $z_j = 0$ and y_i as close to zero as we wish in (3.9). Now

$$[X^{[*]}Xu, u] - \mu(X)[u, u] = \sum_{i=1}^{k} y_i^2 + \sum_{j=1}^{s} (\mu_j - \lambda)\varepsilon_j z_j^2 \geq 0$$

for all u, which proves (3.2) in the case when $k \neq 0$.

Finally, assume $k = 0$, i.e., $X^{[*]}X$ is diagonalizable. Write

$$X^{[*]}X = (\mu_1) \oplus \cdots \oplus (\mu_s), \quad H = (\varepsilon_1) \oplus \cdots \oplus (\varepsilon_s),$$

where $\varepsilon_j = 1$ for $j = 1, \cdots, p$; $\varepsilon_j = -1$ for $j = p+1, \cdots, s$ $(1 \leq p < s)$. For $u = (x_1, \cdots, x_s)^T$ we have

$$[u, u] = x_1^2 + \cdots + x_p^2 - x_{p+1}^2 - \cdots - x_s^2, \tag{3.10}$$

$$[X^{[*]}Xu,u] = \mu_1 x_1^2 + \cdots + \mu_p x_p^2 - \mu_{p+1}x_{p+1}^2 - \cdots - \mu_s x_s^2. \tag{3.11}$$

If $[u,u]=1$, then for a fixed index j, $(1 \le j \le p)$, we have

$$[X^{[*]}Xu,u] = \mu_j + (\mu_k - \mu_j)x_2^2 + \cdots + (\mu_p - \mu_j)x_p^2 + (\mu_j - \mu_{p+1})x_{p+1}^2 + \cdots + (\mu_j - \mu_s)x_s^2, \tag{3.12}$$

which must be nonnegative. Letting here $x_2 = \cdots = x_p = 0$, and observing that $x_{p+1}, \cdots, x_s$ can attain arbitrary real values independently, it follows that $\mu_j \ge 0$ and $\mu_j \ge \mu_k$ for $j = 1, \cdots, p$ and $k = p+1, \cdots, s$. Applying (3.12) with $\mu_j = \min(\mu_1, \cdots, \mu_p)$ yields the value of $\mu(X)$: $\mu(X) = \min\{\mu_j \in \sigma(X^{[*]}X) |$ there is a nonnegative eigenvector of $X^{[*]}X$ corresponding to $\mu_j\}$. Using this formula for $\mu(X)$, the equalities (3.10) and (3.11) easily yield the inequality (3.2) for every $u \in \mathbf{R}^n$. This concludes the proof of formula (3.2) in the real case.

We remark that the above proof can be adapted to the complex case as well.

As a byproduct of the above proof, a characterization of $\mu(X)$ is obtained:

COROLLARY 3.7. ($F = \mathbf{R}$ or $F = \mathbf{C}$.) *Let X be an H-plus matrix. Then $\mu(X)$ coincides with the minimal eigenvalue of $X^{[*]}X$ for which there exists an eigenvector v satisfying* $[v,v] \ge 0$.

4 Indefinite Scalar Products with Only One Positive Square

In this section $H = H^*$ is an invertible $n \times n$ matrix with only one positive eigenvalue and $n-1$ negative eigenvalues. We are interested in H-polar decompositions.

As H has only one positive eigenvalue the possibilities for H-selfadjoint matrices A are rather restricted. We shall list them below in terms of the canonical forms (as in Theorem 2.1 of [BMRRR1], or Section I.3.2 of [GLR], for example) of $\{A, H\}$. Let $F = \mathbf{C}$. Given an H-selfadjoint matrix A, there exists an invertible matrix S such that either one of the following six alternatives occurs:

(a)

$$S^{-1}AS = \operatorname{diag}(\lambda, \bar{\lambda}) \oplus \operatorname{diag}(\lambda_i^2)_{i=1}^{n-2},$$

where $\lambda \notin \mathbf{R}$, $\lambda_i \in \mathbf{R}$ (not necessarily distinct), and

$$S^*HS = \begin{bmatrix} 0 & 1 \\ 1 & 0 \end{bmatrix} \oplus -I_{n-2}.$$

In this case $(S^{-1}AS)^2$ is given by

$$S^{-1}A^2S = \operatorname{diag}(\lambda^2, \bar{\lambda}^2) \oplus \operatorname{diag}(\lambda_i^2)_{i=1}^{n-2}.$$

It is important to distinguish between $\lambda \in i\mathbf{R}$ and $\lambda \notin i\mathbf{R}$. In the first case $\lambda^2 = \bar{\lambda}^2 < 0$.

(b)

$$S^{-1}AS = (\lambda) \oplus \operatorname{diag}(\lambda_i)_{i=1}^{n-1},$$

with $\lambda \in \mathbf{R}$, $\lambda_i \in \mathbf{R}$ not necessarily distinct, and

$$S^*HS = (1) \oplus -I_{n-1}.$$

In this case

$$S^{-1}A^2S = (\lambda^2) \oplus \operatorname{diag}(\lambda_i^2)_{i=1}^{n-1}.$$

(c)

$$S^{-1}AS = \begin{bmatrix} \lambda & 1 \\ 0 & \lambda \end{bmatrix} \oplus \operatorname{diag}(\lambda_i)_{i=1}^{n-2},$$

where $\lambda \in \mathbf{R} \setminus \{0\}$, $\lambda_i \in \mathbf{R}$, not necessarily distinct, and

$$S^*HS = \pm \begin{bmatrix} 0 & 1 \\ 1 & 0 \end{bmatrix} \oplus -I_{n-2}.$$

In this case,

$$S^{-1}A^2S = \begin{bmatrix} \lambda^2 & 2\lambda \\ 0 & \lambda^2 \end{bmatrix} \oplus \operatorname{diag}(\lambda_i^2)_{i=1}^{n-2}.$$

(d)

$$S^{-1}AS = \begin{bmatrix} \lambda & 1 & 0 \\ 0 & \lambda & 1 \\ 0 & 0 & \lambda \end{bmatrix} \oplus \operatorname{diag}(\lambda_i)_{i=1}^{n-3},$$

where $\lambda \in \mathbf{R} \setminus \{0\}$, $\lambda_i \in \mathbf{R}$, not necessarily distinct, and

$$S^*HS = \begin{bmatrix} 0 & 0 & -1 \\ 0 & -1 & 0 \\ -1 & 0 & 0 \end{bmatrix} \oplus -I_{n-3}.$$

In this case,

$$S^{-1}A^2S = \begin{bmatrix} \lambda^2 & 2\lambda & 1 \\ 0 & \lambda^2 & 2\lambda \\ 0 & 0 & \lambda^2 \end{bmatrix} \oplus \operatorname{diag}(\lambda_i^2)_{i=1}^{n-3}.$$

(e)

$$S^{-1}AS = \begin{bmatrix} 0 & 1 \\ 0 & 0 \end{bmatrix} \oplus \operatorname{diag}(\lambda_i)_{i=1}^{n-2},$$

where $\lambda_i \in \mathbf{R}$, not necessarily distinct, and

$$S^*HS = \pm \begin{bmatrix} 0 & 1 \\ 1 & 0 \end{bmatrix} \oplus -I_{n-2}.$$

In this case,

$$S^{-1}A^2S = O_2 \oplus \operatorname{diag}(\lambda_i^2)_{i=1}^{n-2}.$$

(f)

$$S^{-1}AS = \begin{bmatrix} 0 & 1 & 0 \\ 0 & 0 & 1 \\ 0 & 0 & 0 \end{bmatrix} \oplus \operatorname{diag}(\lambda_i)_{i=1}^{n-3},$$

where $\lambda \in \mathbf{R} \setminus \{0\}$, $\lambda_i \in \mathbf{R}$, not necessarily distinct, and

$$S^*HS = \begin{bmatrix} 0 & 0 & -1 \\ 0 & -1 & 0 \\ -1 & 0 & 0 \end{bmatrix} \oplus -I_{n-3}.$$

In this case,

$$S^{-1}A^2S = \begin{bmatrix} 0 & 0 & 1 \\ 0 & 0 & 0 \\ 0 & 0 & 0 \end{bmatrix} \oplus \operatorname{diag}(\lambda_i^2)_{i=1}^{n-3}.$$

In the case $F = \mathbf{R}$, the same classification (a)-(f) is valid, with the only exception that in (a), $\operatorname{diag}(\lambda, \bar{\lambda})$ is replaced by $\begin{bmatrix} \lambda & \mu \\ -\mu & \lambda \end{bmatrix}$, $\lambda, \mu \in \mathbf{R}$, $\mu \neq 0$.

We now state the main result of this section.

THEOREM 4.1. *($F = \mathbf{C}$ or $F = \mathbf{R}$) Let $H = H^*$ be an invertible $n \times n$ matrix with one positive eigenvalue. An $n \times n$ matrix X allows H-polar decomposition if and only if X has precisely one of the following mutually exclusive properties:*

(i) *$X^{[*]}X$ has a non-real eigenvalue,*

(ii) *$X^{[*]}X$ has a negative eigenvalue λ of algebraic and geometric multiplicity two, and H is indefinite on* $\operatorname{Ker}(X^{[*]}X - \lambda)$,

(iii) *$X^{[*]}X$ has all its eigenvalues in $\{\lambda \in \mathbf{R} |\ \lambda \geq 0\}$, and there is a positive λ such that* $\operatorname{Ker}(X^{[*]}X - \lambda)^n$ *is H-indefinite,*

(iv) *$X^{[*]}X$ has all its eigenvalues in $\{\lambda \in \mathbf{R} |\ \lambda \geq 0\}$, is diagonalizable and* $\operatorname{Ker} X$ *contains a k-dimensional H-nonpositive subspace and a p-dimensional H-nonnegative subspace, where k (respectively, p) is the number of negative (respectively, positive) signs in the sign characteristic of $\{X^{[*]}X, H\}$ corresponding to the zero eigenvalue of $X^{[*]}X$ (observe that $p \leq 1$),*

(v) *$X^{[*]}X$ has all its eigenvalues in $\{\lambda \in \mathbf{R} |\ \lambda \geq 0\}$,*

$$\operatorname{rank}(X^{[*]}X)|_{\operatorname{Ker}(X^{[*]}X)^n} = 1, \quad \dim \operatorname{Ker} X^{[*]}X \geq 2,$$

in the canonical form of $\{X^{[*]}X, H\}$ there is a block of the form

$$\left(\begin{bmatrix} 0 & 1 \\ 0 & 0 \end{bmatrix}, \begin{bmatrix} 0 & -1 \\ -1 & 0 \end{bmatrix} \right),$$

and $\operatorname{Ker} X$ *is the direct sum of an $(r-3)$-dimensional strictly H-negative subspace and the subspace $[H(\operatorname{Ker}(X^{[*]}X))]^{\perp}$, where $r = \dim \operatorname{Ker}[(X^{[*]}X)^n]$.*

Proof. We give the proof here only in the complex case. Suppose $X = UA$ where U is H-unitary and A is H-selfadjoint. Then A is as in one of the cases (a)-(f) described above. In case (a) precisely one of (i) and (ii) in the statement of the theorem holds. In case (b), (iv) holds, in cases (c) and (d), (iii) holds. In case (e), (iv) holds, and finally, in case (f), (v) holds.

Conversely, suppose precisely one of (i)-(v) holds for X. Because of Lemma 4.3 of [BM-RRR1], we may assume either $\sigma(X^{[*]}X) = \{\lambda\}$ with $\lambda \in \mathbf{R}$ or $\sigma(X^{[*]}X) = \{\lambda, \bar{\lambda}\}$ with $\lambda \notin \mathbf{R}$, and H has at most one positive eigenvalue.

Assume $\sigma(X^{[*]}X) = \{\lambda, \bar{\lambda}\}$ with $\lambda \notin \mathbf{R}$. As H has only one positive eigenvalue, in this case there is an S such that

$$S^{-1}X^{[*]}XS = \begin{bmatrix} \lambda & 0 \\ 0 & \bar{\lambda} \end{bmatrix}, \quad S^*HS = \begin{bmatrix} 0 & 1 \\ 1 & 0 \end{bmatrix}.$$

Let μ be such that $\mu^2 = \lambda$, and put $A = S \text{ diag } (\mu, \bar{\mu})\, S^{-1}$. Then A is H-selfadjoint and $A^2 = X^{[*]}X$. So X admits H-polar decomposition by Theorem 4.1(e) of [BMRRR1].

Assume $\sigma(X^{[*]}X) = \{\lambda\}$, $\lambda < 0$, then (ii) holds. So there is an S such that

$$S^{-1}X^{[*]}XS = \text{diag}\,(\lambda, \lambda), \quad S^*HS = \text{diag}\,(1, -1).$$

Then there is also a V such that

$$V^{-1}X^{[*]}XV = X^{[*]}X = \begin{bmatrix} \lambda & 0 \\ 0 & \lambda \end{bmatrix}, \quad V^*HV = \begin{bmatrix} 0 & 1 \\ 1 & 0 \end{bmatrix}.$$

Take A as defined by

$$A = V \text{ diag } \left(i\sqrt{-\lambda}, -i\sqrt{-\lambda}\right) V^{-1}.$$

Then A is H-selfadjoint and $A^2 = \lambda I = X^{[*]}X$.

Now assume $\sigma(X^{[*]}X) = \{\lambda\}$, $\lambda > 0$, and $\text{Ker}\,(X^{[*]}X - \lambda)^n$ is H-indefinite, i.e., (iii) holds. Then for some invertible S there are three possibilities:

$$S^{-1}X^{[*]}XS = \lambda I_n, \quad S^*HS = (1) \oplus -I_{n-1},$$

or

$$S^{-1}X^{[*]}XS = \begin{bmatrix} \lambda & 1 \\ 0 & \lambda \end{bmatrix} \oplus \lambda I_{n-2}, \quad S^*HS = \pm \begin{bmatrix} 0 & 1 \\ 1 & 0 \end{bmatrix} \oplus -I_{n-2},$$

or

$$S^{-1}X^{[*]}XS = \begin{bmatrix} \lambda & 1 & 0 \\ 0 & \lambda & 1 \\ 0 & 0 & \lambda \end{bmatrix} \oplus \lambda I_{n-3}, \quad S^*HS = \begin{bmatrix} 0 & 0 & -1 \\ 0 & -1 & 0 \\ -1 & 0 & 0 \end{bmatrix} \oplus -I_{n-3}.$$

In the first case, put $A = \sqrt{\lambda}\, I_n$. Then $A^2 = X^{[*]}X$ and A is H-selfadjoint. In the second case, put

$$A = S \left(\begin{bmatrix} \sqrt{\lambda} & \frac{1}{2\sqrt{\lambda}} \\ 0 & \sqrt{\lambda} \end{bmatrix} \oplus \sqrt{\lambda}\, I_{n-2} \right) S^{-1}.$$

Then A is H-selfadjoint and $A^2 = X^{[*]}X$. In the third case, put

$$A = S\left(\begin{bmatrix} \sqrt{\lambda} & \frac{1}{2\sqrt{\lambda}} & \frac{-1}{8\sqrt{\lambda^3}} \\ 0 & \sqrt{\lambda} & \frac{1}{2\sqrt{\lambda}} \\ 0 & 0 & \sqrt{\lambda} \end{bmatrix} \oplus \sqrt{\lambda}\, I_{n-3}\right) S^{-1}.$$

Then A is H-selfadjoint and $A^2 = X^{[*]}X$.

Next, assume $\sigma(X^{[*]}X) = \{\lambda\}$, $\lambda > 0$, and $\operatorname{Ker}(X^{[*]}X - \lambda)^n$ is H-definite. Then

$$S^{-1}X^{[*]}XS = \lambda I_n, \quad S^*HS = -I_n$$

for some S (recall that in this part of the proof we assume that H has at most one positive eigenvalue; in particular, the case of a negative definite H is not excluded). Taking $A = \sqrt{\lambda}\, I_n$ we find that A is H-selfadjoint and $A^2 = X^{[*]}X$.

Finally, assume $\sigma(X^{[*]}X) = \{0\}$. Then either (iv) or (v) holds. In case (iv) holds we can apply Theorem 5.3 of [BMRRR2] to show that X admits an $(H, 0, n)$-polar decomposition. So it remains to consider case (v).

In case (v) holds and $\sigma(X^{[*]}X) = \{0\}$, observe that $\operatorname{rank} X^{[*]}X = 1$, $\dim \operatorname{Ker} X^{[*]}X \geq 2$ and $X^{[*]}X$ has one Jordan block of order 2, and we have $r = n \geq 3$. Put $M = \operatorname{Ker} X$ and denote $\operatorname{Ker} X^{[*]}X$ by N. Then $N \cap (HN)^\perp = (HN)^\perp$, as both are one dimensional, and $(HN)^\perp \subset N$ (these facts can be easily verified using the canonical form of $\{X^{[*]}X, H\}$). Because of the hypothesis on $\operatorname{Ker} X$, we also have $(HN)^\perp \subset M$. Let e_0 be a vector such that $\operatorname{span}\{e_0\} = (HN)^\perp$ and choose a basis $e_0, e_1, \ldots, e_{n-3}$ for M such that

$$\begin{aligned} \langle He_i, e_j \rangle &= 0 \quad \text{for } i \neq j \quad (\text{and for } i = j = 0) \\ \langle He_i, e_i \rangle &= -1 \text{ for } i = 1, \ldots, n-3. \end{aligned}$$

As one sees from the canonical form of $\{X^{[*]}X, H\}$ (or proves quite easily directly), $\operatorname{Im} X^{[*]}X = (HN)^\perp$. Choose any f_0 such that $X^{[*]}Xf_0 = e_0$, $\langle Hf_0, e_0 \rangle = -1$, and $\langle Hf_0, f_0 \rangle = 0$. (Note that this choice is possible by the hypothesis (v) and by the canonical form of $\{X^{[*]}X, H\}$.)

Next, we choose a vector $g \in (HM)^\perp \cap N$ such that $g \notin M$. To see that such a choice is possible, argue as follows. Since

$$\operatorname{Ker} X \subset \operatorname{Ker} X^{[*]}X,$$

and $(\dim \operatorname{Ker} X^{[*]}X) - (\dim \operatorname{Ker} X) = 1$, there is $g_0 \in (\operatorname{Ker} X^{[*]}X) \setminus (\operatorname{Ker} X)$. Put

$$g = g_0 + \sum_{j=1}^{n-3} \langle Hg_0, e_j \rangle e_j.$$

Clearly, $g \in \operatorname{Ker} X^{[*]}X$ and $g \notin \operatorname{Ker} X$. Also, for $i = 1, \ldots, n-3$,

$$\langle Hg, e_i \rangle = \langle Hg_0, e_i \rangle - \langle Hg_0, e_i \rangle = 0,$$

and

$$\langle Hg, e_0 \rangle = 0,$$

because $e_0 \in (H \operatorname{Ker} X^* X)^{\perp}$. It follows that $g \in (HM)^{\perp}$.

We note that $\langle Hg, g\rangle < 0$, otherwise span $\{e_0, g\}$ would be a two-dimensional H-nonnegative subspace, and an H-nonnegative subspace can have dimension at most one as H has only one positive eigenvalue. Scaling g we may assume that $\langle Hg, g\rangle = -1$.

Consider

$$f = f_0 + \sum_{j=1}^{n-3} \langle Hf_0, e_j\rangle e_j + \langle Hf_0, g\rangle g.$$

Then $X^{[*]}Xf = X^{[*]}Xf_0 = e_0$, and for $i \geq 1$,

$$\begin{aligned} \langle Hf, e_i\rangle &= \langle Hf_0, e_i\rangle + \sum_{j=1}^{n-3} \langle Hf_0, e_j\rangle \langle He_j, e_i\rangle \\ &+ \langle Hf_0, g\rangle \langle Hg, e_i\rangle = 0 \end{aligned}$$

as $\langle He_j, e_j\rangle = -\delta_{ij}$ and $\langle Hg, e_i\rangle = 0$, since $g \in (HM)^{\perp}$. Likewise $\langle Hf, g\rangle = 0$. Furthermore, we have $\langle Hf, e_0\rangle \neq 0$. Indeed, suppose $\langle Hf, e_0\rangle = 0$. Then (since $N = M + \operatorname{span}\{g\}$) $\langle Hf, x\rangle = 0$ for all $x \in \operatorname{Ker} X^{[*]}X$, so $f \in (H \operatorname{Ker} X^{[*]}X)^{\perp} = \operatorname{span}\{e_0\}$. But then, $X^{[*]}Xf = e_0 = 0$ and $N = \operatorname{Ker} X^{[*]}X$. Contradiction. As $\langle Hf, e_0\rangle = \langle Hf, X^{[*]}Xf\rangle = \langle HX^{[*]}Xf, f\rangle$ and $HX^{[*]}X \leq 0$ (considering the canonical form of $\{X^{[*]}X, H\}$ the latter fact is easily seen), we have $\langle Hf, e_0\rangle < 0$. Observe also $\langle Hf, e_0\rangle = \langle Hf_0, e_0\rangle = -1$. Next,

$$\langle Hf, f\rangle = \langle Hf, f_0\rangle = \langle Hf_0, f_0\rangle = 0.$$

Take as a basis in F^n the vectors $e_0, g, f, e_1, e_1, \dots, e_{n-3}$, and let S be the $n \times n$ matrix with these vectors as its columns in the order in which they appear here. Then

$$S^{-1}X^{[*]}XS = \begin{bmatrix} 0 & 0 & 1 \\ 0 & 0 & 0 \\ 0 & 0 & 0 \end{bmatrix} \oplus 0,$$

and

$$S^*HS = \begin{bmatrix} 0 & 0 & -1 \\ 0 & -1 & 0 \\ -1 & 0 & 0 \end{bmatrix} \oplus -I_{n-3}.$$

Take A as follows:

$$A = S\left(\begin{bmatrix} 0 & 1 & 0 \\ 0 & 0 & 1 \\ 0 & 0 & 0 \end{bmatrix} \oplus 0 \right) S^{-1}.$$

Then $A^2 = X^{[*]}X$, and

$$\operatorname{Ker} A = \operatorname{span}\{e_0, e_1, \dots, e_{n-3}\} = \operatorname{Ker} X.$$

Clearly, also A is H-selfadjoint. By Theorem 4.1(e) of [BMRRR1], X allows an H-polar decomposition. □

5 Polar Decompositions with Special Unitary Factors

The H-polar decomposition (1.1) will be called *special* if $\det U = 1$ and *connected* if U belongs to the connected component of I in the group of H-unitary $n \times n$ matrices (over F). If $F = \mathbf{C}$, then every H-polar decomposition is connected; if H is definite and $F = \mathbf{R}$, then the classes of connected and of special H-polar decompositions coincide. In this section we study special and connected H-polar decompositions.

First, we find the possible values of $\det U$ in H-polar decompositions $X = UA$ of a given $n \times n$ matrix X. Two H-polar decompositions $X = UA$ and $X = \tilde{U}\tilde{A}$ of X are called *equivalent* if the matrices A and $\tilde{A}$ are H-unitarily similar, i. e., $\tilde{A} = W^{-1}AW$ for some H-unitary matrix W. A complete description of the equivalence classes is given in [BMRRR1]. Clearly, if two H-polar decompositions are equivalent, then they are (H,p,q)-polar decompositions for the same (p,q), but the converse is false in general: Two (H,p,q)-polar decompositions with the same (p,q) need not be equivalent. Fix an H-polar decomposition $X = \tilde{U}\tilde{A}$. By Proposition 7.1 of [BMRRR1], all values of $\det U$ in equivalent H-polar decompositions $X = UA$ are given by the formula

$$\{\det(VW) \cdot \det \tilde{U} \mid V \text{ and } W \text{ are } H\text{-unitary such that } X = VXW\}. \tag{5.1}$$

It is convenient to study the formula (5.1) in two steps, by considering separately premultiplication and postmultiplication by H-unitary matrices.

We fix an invertible hermitian $n \times n$ matrix H (over F). For any $X \in F^{n \times n}$, denote by $\mathcal{U}_\ell(X)$ (resp., $\mathcal{U}_r(X)$) the group of H-unitary matrices U such that $UX = X$ (resp., $XU = X$). Denote also by $\mathcal{DU}_\ell(X)$ (resp., $\mathcal{DU}_r(X)$) the set $\{\det U \mid U \in \mathcal{U}_\ell(X)\}$ (resp., $\{\det U \mid U \in \mathcal{U}_r(X)\}$) of values of the determinant function on $\mathcal{U}_\ell(X)$ (resp., $\mathcal{U}_r(X)$). As usual, we distinguish the two cases $F = \mathbf{R}$ and $F = \mathbf{C}$. We denote by $d(V)$ the *defect* of a subspace V with respect to $[\,\cdot\,,\,\cdot\,]$ (the indefinite scalar product induced by H), i.e., the number of zero eigenvalues of the Gram matrix (relative to $[\,\cdot\,,\,\cdot\,]$) of any basis in V. The defect is zero precisely when the subspace is H-*nondegenerate*. It is well known that a subspace $\mathcal{M}$ is H-nondegenerate if and only if its orthogonal companion

$$\mathcal{M}^{[\perp]} = \{x \in F^n \mid [x,y] = 0 \text{ for all } y \in \mathcal{M}\}$$

is actually a direct complement of $\mathcal{M}$ in F^n.

The invertibility of H easily implies

$$\dim V + d(V) \le n$$

for every subspace $V \subset F^n$.

THEOREM 5.1.

(i) $(F = \mathbf{C})$. *$\mathcal{DU}_\ell(X)$ coincides with the unit circle if and only if*

$$\dim(\operatorname{Im} X) + d(\operatorname{Im} X) < n; \tag{5.2}$$

otherwise, $\mathcal{DU}_\ell(X) = \{1\}$.

(ii) $(F = \mathbf{R})$. $\mathcal{DU}_\ell(X) = \{1, -1\}$ *if and only if* (5.2) *holds; otherwise,* $\mathcal{DU}_\ell(X) = \{1\}$.

Proof. $U \in \mathcal{U}_\ell(X)$ if and only if U is H-unitary and $Ux = x$ for every $x \in \operatorname{Im} X$. In other words, U is a Witt extension (in the terminology of [BMRRR2]) of the identity linear transformation on $\operatorname{Im} X$. The formula for the Witt extensions (given in Theorem 2.3 of [BMRRR2]) shows that all such Witt extensions either have a constant determinant (if $\dim(\operatorname{Im} X) + d(\operatorname{Im} X) = n$) or can have an arbitrary value of the determinant on the unit circle (if $F = \mathbf{C}$) or in the set $\{1, -1\}$. □

THEOREM 5.2. *$\mathcal{DU}_r(X)$ coincides with the unit circle (if $F = \mathbf{C}$) or with $\{1,-1\}$ (if $F = \mathbf{R}$) if and only if* $\operatorname{Ker} X$ *is not H-isotropic. Otherwise, i.e., if* $\operatorname{Ker} X$ *is H-isotropic, then $\mathcal{DU}_r(X) = \{1\}$ in both of the cases $F = \mathbf{C}$ and $F = \mathbf{R}$.*

Proof. Using the obvious equality

$$\mathcal{DU}_r(X) = \mathcal{DU}_\ell(X^{[*]}),$$

and applying Theorem 5.1 (with X replaced by $X^{[*]}$) we see that $\mathcal{DU}_r(X) \neq \{1\}$ if and only if

$$\dim(\operatorname{Im} X^{[*]}) + d(\operatorname{Im} X^{[*]}) < n. \tag{5.3}$$

It is well known (and easy to verify) that

$$\operatorname{Im} X^{[*]} = (\operatorname{Ker} X)^{[\perp]},$$

and that $\dim(\operatorname{Ker} X)^{[\perp]} = n - \dim \operatorname{Ker} X$. Also,

$$d(\mathcal{M}) = d(\mathcal{M}^{[\perp]}) \tag{5.4}$$

for every subspace $\mathcal{M} \subset F^n$ (to verify (5.4), simply observe that $d(\mathcal{M}) = \dim(\mathcal{M} \cap \mathcal{M}^{[\perp]})$). Using all these observations, we see that (5.3) is equivalent to $d(\operatorname{Ker} X) < \dim(\operatorname{Ker} X)$, i.e., $\operatorname{Ker} X$ is not H-isotropic. □

Combining Theorems 5.1 and 5.2 with formula (5.1), the following result is obtained.

THEOREM 5.3. *Let $X = \tilde{U}\tilde{A}$ be an H-polar decomposition. If*

$$d(\operatorname{Ker} X) = d(\operatorname{Im} X) = \dim(\operatorname{Ker} X), \tag{5.5}$$

then $\det U = \det \tilde{U}$ for every H-polar decomposition of X which is equivalent to $X = \tilde{U}\tilde{A}$. If at least one of the equalities (5.5) fails, then for every $\alpha \in F$, $|\alpha| = 1$, there exists an H-polar decomposition $X = UA$ with $\det U = \alpha$ and which is equivalent to $X = \tilde{U}\tilde{A}$.

Observe that (5.5) holds for every invertible X.

We emphasize that Theorem 5.3 holds only for equivalent H-polar decompositions (in particular, the inertia of HA and that of $H\tilde{A}$ must be the same). If one considers H-polar decompositions $X = UA$ irrespective of the inertia of HA, then there is considerably more freedom in the values of $\det U$. For example, in the case when $F = \mathbf{R}$ and n is odd, the trivial equality $X = UA = (-U)(-A)$ shows that both 1 and -1 appear as values of $\det U$.

In the remainder of this section we assume $F = \mathbf{R}$ and focus on the more subtle problem of having U in a prescribed connected component of the group $\mathcal{U}(H;\mathbf{R})$ of H-unitary matrices (since the group $\mathcal{U}(H;\mathbf{C})$ is connected, this question is trivial for $F = \mathbf{C}$). We start with two examples.

Example 5.1. Let

$$X = \begin{bmatrix} 0 & 0 \\ 0 & 1 \end{bmatrix}, \quad H = \begin{bmatrix} 0 & 1 \\ 1 & 0 \end{bmatrix}.$$

We have an H-polar decomposition

$$X = H \begin{bmatrix} 0 & 1 \\ 0 & 0 \end{bmatrix} = U \cdot A$$

with $U = H$. Then if $X = U_1 A_1$ is another H-polar decomposition there is a real H-unitary V such that $A_1 = VA$ and $U_1 = UV^{-1}$. Write $V = (v_{ij})_{i=1}^{2}\,_{j=1}^{2}$. Then $VA = \begin{bmatrix} 0 & v_{11} \\ 0 & v_{21} \end{bmatrix}$ is H-selfadjoint if and only if $v_{21} = 0$. Hence $V = \begin{bmatrix} v_{11} & v_{12} \\ 0 & v_{22} \end{bmatrix}$. From the H-unitarity of V (and V being real), it follows that also $v_{12} = 0$ and $v_{11}^{-1} = v_{22}$. So $V = \operatorname{diag}(v_{11}, v_{11}^{-1})$, for some nonzero real number v_{11}. We see that for any H-polar decomposition of X the H-unitary factor is of the form

$$\begin{bmatrix} 0 & 1 \\ 1 & 0 \end{bmatrix} \begin{bmatrix} v_{11}^{-1} & 0 \\ 0 & v_{11} \end{bmatrix} = \begin{bmatrix} 0 & v_{11} \\ v_{11}^{-1} & 0 \end{bmatrix};$$

in particular, any H-unitary factor has determinant -1, and hence cannot be in the connected component of I in $\mathcal{U}(H;\mathbf{R})$. □

Example 5.2. As a second example, let

$$X = \begin{bmatrix} 0 & 0 & 1 \\ 0 & 1 & 0 \\ 1 & 0 & 0 \end{bmatrix} \begin{bmatrix} 0 & 1 & 0 \\ 0 & 0 & 1 \\ 0 & 0 & 0 \end{bmatrix} = UA$$

with $H = U$. Let V be a real H-unitary matrix such that VA is H-selfadjoint. Put

$$V = \begin{bmatrix} v_{11} & v_{12} & v_{13} \\ v_{21} & v_{22} & v_{23} \\ v_{31} & v_{32} & v_{33} \end{bmatrix}.$$

Then the H-selfadjointness of VA implies $v_{31} = v_{32} = 0$ and $v_{11} = v_{22}$, as one checks easily, while the H-unitarity of V implies that V must be of the form

$$V = \begin{bmatrix} v_{11} & v_{12} & -\frac{1}{2} v_{11} v_{12}^2 \\ 0 & v_{11} & -v_{12} \\ 0 & 0 & v_{11} \end{bmatrix} \quad \text{with} \quad v_{11}^2 = 1.$$

Any H-unitary matrix U_1 that is the H-unitary component in an H-polar decomposition of X is then necessarily of the form HV^{-1}:

$$U_1 = \begin{bmatrix} 0 & 0 & v_{11} \\ 0 & v_{11} & v_{12} \\ v_{11} & -v_{12} & -\frac{1}{2}v_{11}v_{12}^2 \end{bmatrix}.$$

Put

$$S = \begin{bmatrix} \frac{1}{\sqrt{2}} & 0 & \frac{1}{\sqrt{2}} \\ 0 & 1 & 0 \\ \frac{1}{\sqrt{2}} & 0 & -\frac{1}{\sqrt{2}} \end{bmatrix}.$$

Then $S^*HS = \mathrm{diag}\,(1,1,-1)$, and S^*U_1S is of the form

$$\begin{bmatrix} \frac{1}{4}v_{11}(4-v_{12}^2) & -\frac{v_{12}}{\sqrt{2}} & \frac{1}{4}v_{11}v_{12}^2 \\ \frac{v_{12}}{\sqrt{2}} & v_{11} & -\frac{v_{12}}{\sqrt{2}} \\ \frac{1}{4}v_{11}v_{12}^2 & \frac{v_{12}}{\sqrt{2}} & -\frac{1}{4}v_{11}(4+v_{12}^2) \end{bmatrix}.$$

As $\det \begin{bmatrix} \frac{1}{4}v_{11}(4-v_{12}^2) & -\frac{1}{2}v_{12}\sqrt{2} \\ \frac{1}{2}v_{12}\sqrt{2} & v_{11} \end{bmatrix} = 1 + \frac{1}{4}v_{12}^2$ and the sign of $-\frac{1}{4}v_{11}(4+v_{12}^2)$ depends on v_{11}, it is clear that U_1 can only be in two of the four connected components of $\mathcal{U}(H;\mathbf{R})$; see Theorem 5.2. In particular, U_1 can be in the connected component of the identity for this choice of X. However, for $X = S\,\mathrm{diag}\,(1,-1,1)\,S^* \cdot A$, the same argument shows that for this choice of X, there is no H-polar decomposition with the unitary factor in the connected component of I. □

The following results give sufficient conditions for existence of H-polar decompositions with the H-unitary factor in an arbitrary connected component of $\mathcal{U}(H;\mathbf{R})$.

THEOREM 5.4. *Suppose $n \geq 4$, and let H be a real invertible selfadjoint $n \times n$ matrix with exactly one positive eigenvalue. Assume X admits an H-polar decomposition. Then X has an H-polar decomposition with the H-unitary factor in any preselected connected component of the group of real H-unitary matrices.*

Observe that Theorem 5.4 is sharp in the sense that it does not hold for $n = 2$ or $n = 3$, as shown by Examples 5.1 and 5.2.

THEOREM 5.5. *Let H be a real symmetric indefinite invertible $n \times n$ matrix, and suppose X is an $n \times n$ matrix which allows an $(H, n, 0)$-polar decomposition. Suppose either*

(a) *n is odd, or*

(b) *$X^{[*]}X \neq 0$, or*

(c) *n is even and* $\dim[\operatorname{Ker} X \cap (H \operatorname{Ker} X)^{\perp}] \neq \frac{n}{2}$.

Then, for any preselected connected component of the group of real H-unitary matrices X admits an $(H, n, 0)$-polar decomposition with the unitary factor in this connected component.

In a way Theorem 5.5 is sharp, as testified by Example 5.2.

Both Theorems 5.4 and 5.5 will be deduced from Theorem 5.6 below, which is of independent interest.

THEOREM 5.6. *Let H be a real symmetric invertible $n \times n$ matrix, and suppose X is a real $n \times n$ matrix which allows an H-polar decomposition $X = UA$. Suppose the canonical form of $\{A, H\}$ is of the form*

$$H = \varepsilon_1 P_1 \oplus (\varepsilon_2) \oplus H_3, \quad A = J_1 \oplus (\lambda_2) \oplus A_3, \tag{5.6}$$

where J_1 is a real Jordan block of order n_1, $\varepsilon_1 = \pm 1$ if $\sigma(J_1)$ is real, and $\varepsilon_1 = 1$ if $\sigma(J_1)$ is nonreal (in which case n_1 is necessarily even), $\varepsilon_2 = \pm 1$, $\lambda_2 \in \mathbf{R}$, and A_3 is some real $n_3 \times n_3$ matrix. Furthermore, assume that one of the following three properties are satisfied:

(i) $\sigma(J_1) = \{\lambda_1\} \subset \mathbf{R}$, *$n_1$ is even but not divisible by* 4;

(ii) $\sigma(J_1) = \{\lambda_1\} \subset \mathbf{R}$, *$n_1$ is odd, and either $\varepsilon_2 = 1$ and $\dfrac{n_1 + \varepsilon_1}{2}$ is even, or $\varepsilon_2 = -1$ and $\dfrac{n_1 + \varepsilon_1}{2}$ is odd;*

(iii) $\sigma(J_1) = \{\alpha \pm i\beta\}$, $\beta \neq 0$, *in which case $\varepsilon_1 = 1$ and n_1 is necessarily even, but we assume that $\dfrac{n_1}{2}$ is odd.*

Then, for any preselected connected component of the group of real H-unitary matrices, there is an H-polar decomposition of X such that its unitary factor is in this preselected connected component.

Proof. Put
$$V = aI_{n_1} \oplus (b) \oplus I_{n_3},$$
where $a^2 = b^2 = 1$. Then V is H-unitary and VA is H-selfadjoint. Hence $X = (UV^{-1})(VA)$ is an alternative H-polar decomposition. We may assume without loss of generality that $H_3 = I_{p_3} \oplus -I_{m_3}$, where $p_3 + m_3 = n_3$. Let S_1 be an arbitrary invertible matrix such that $S_1^*(\varepsilon_1 P_1)S_1 = I_{p_1} \oplus -I_{m_1}$, where $p_1 + m_1 = n_1$. Then

$$\begin{bmatrix} S_1^* & 0 \\ 0 & I_{n_3+1} \end{bmatrix} H \begin{bmatrix} S_1 & 0 \\ 0 & I_{n_3+1} \end{bmatrix} = I_{p_1} \oplus -I_{m_1} \oplus (\varepsilon_2) \oplus I_{p_3} \oplus -I_{m_3}.$$

Observe: $p_1 = m_1 = n_1/2$ if n_1 is even; $p_1 = (n_1 + \varepsilon_1)/2$, $m_1 = (n_1 - \varepsilon_1)/2$ if n_1 is odd.

Consider $S^{-1}UV^{-1}S$ and $S^{-1}US$, where $S = S_1 \oplus I_{n_3+1}$, and partition

$$S^{-1}US = (U_{ij})_{i,j=1}^5, \quad S^{-1}UV^{-1}S = (W_{ij})_{i,j=1}^5,$$

where U_{11}, W_{11} are $p_1 \times p_1$; U_{22}, W_{22} are $m_1 \times m_1$; U_{33}, W_{33} are 1×1; U_{44}, W_{44} are $p_3 \times p_3$; and U_{55}, W_{55} are $m_3 \times m_3$.

First consider the case n_1 is even, $n_1/2$ odd, and $\varepsilon_2 = +1$. Then to see in which connected component UV^{-1} is we have to look at

$$\det \begin{bmatrix} W_{11} & W_{13} & W_{14} \\ W_{31} & W_{33} & W_{34} \\ W_{41} & W_{43} & W_{44} \end{bmatrix} = \det \begin{bmatrix} aU_{11} & bU_{13} & U_{14} \\ aU_{31} & bU_{33} & U_{34} \\ aU_{41} & bU_{43} & U_{44} \end{bmatrix} = a^{n_1/2} b \det \begin{bmatrix} U_{11} & U_{13} & U_{14} \\ U_{31} & U_{33} & U_{34} \\ U_{41} & U_{43} & U_{44} \end{bmatrix},$$

and at

$$\det \begin{bmatrix} W_{22} & W_{25} \\ W_{52} & W_{55} \end{bmatrix} = \det \begin{bmatrix} aU_{22} & U_{25} \\ aU_{52} & U_{55} \end{bmatrix} = a^{n_1/2} \det \begin{bmatrix} U_{22} & U_{25} \\ U_{52} & U_{55} \end{bmatrix}.$$

Taking $a, b = \pm 1$ we can choose the signs of these determinants arbitrarily, independently of each other. This proves the theorem in this case. The case when n_1 even, $n_1/2$ odd, and $\varepsilon_2 = -1$ is handled the same way.

Next, consider case (ii), n_1 is odd. Assume first $\varepsilon_2 = 1$, $(n_1 + \varepsilon_1)/2$ is even. Then to see in which connected component UV^{-1} is we have to consider again

$$\det \begin{bmatrix} W_{11} & W_{13} & W_{14} \\ W_{31} & W_{33} & W_{34} \\ W_{41} & W_{43} & W_{44} \end{bmatrix} = a^{(n_1+\varepsilon_1)/2} b \det \begin{bmatrix} U_{11} & U_{13} & U_{14} \\ U_{31} & U_{33} & U_{34} \\ U_{41} & U_{43} & U_{44} \end{bmatrix}, \tag{5.7}$$

and

$$\det \begin{bmatrix} W_{22} & W_{25} \\ W_{52} & W_{55} \end{bmatrix} = a^{(n_1-\varepsilon_1)/2} \det \begin{bmatrix} U_{22} & U_{25} \\ U_{52} & U_{55} \end{bmatrix}. \tag{5.8}$$

Again, choosing $a, b = \pm 1$ we can obtain any signs for these determinants, independently of each other, because $a^{(n_1+\varepsilon_1)/2} b = b$ and $a^{(n_1-\varepsilon_1)/2} = a$. This proves (ii) in case $\varepsilon_2 = 1$. The case when $\varepsilon_2 = -1$ and $(n_1 + \varepsilon_1)/2$ is odd is done likewise.

Finally, consider case (iii), and again assume first $\varepsilon_2 = 1$. The argument now is the same as the one for case (i). The case $\varepsilon_2 = -1$ is treated in the same manner. □

Proof of Theorem 5.4. In the case when $n \geq 4$ and H has only one positive eigenvalue there is a limited number of possibilities for H-selfadjoint matrices A. Each of them is listed below, and it will turn out that all possibilities fall under one of the three cases (i), (ii), (iii) of Theorem 5.6.

Case 1. A is diagonalizable and $\sigma(A) \subset \mathbf{R}$. In this case we are in case (ii) of Theorem 5.6 for $n_1 = 1$, $\varepsilon_1 = 1$, $\varepsilon_2 = -1$, $H_3 = -I_{n-2}$.

Case 2. A has precisely one Jordan block of order 2 with a real eigenvalue, all eigenvalues are real. We are in case (i) of Theorem 5.6 with $n_1 = 2$, $\varepsilon_2 = -1$, $H_3 = -I_{n-3}$.

Case 3. A has precisely one Jordan block of order 3 with a real eigenvalue, the corresponding sign in the sign characteristic is -1, and $\sigma(A) \subset \mathbf{R}$. We are in case (ii) of Theorem 5.6 with $n_1 = 3$, $\varepsilon_1 = -1$, $\varepsilon_2 = -1$. Note that $(n_1 + \varepsilon_1)/2 = 1$ is odd indeed.

Case 4. A has precisely one pair of complex conjugate eigenvalues $\alpha \pm i\beta$ with multiplicity one, all other eigenvalues are real. We are in case (iii) of Theorem 5.6 with $n_1 = 2$.

As in all cases Theorem 5.6 is applicable, Theorem 5.4 is proved. □

Proof of Theorem 5.5. Let $X = UA$ be an $(H, n, 0)$-polar decomposition. Then A is H-nonnegative, and by assumptions (a)-(c) the canonical form of $\{A, H\}$ is as in (5.6), and by the indefiniteness of H either case (i) of Theorem 5.6 holds (with $\lambda_1 = 0$ and $n_1 = 2$) or case (ii) holds with $n_1 = 1$ (here the indefiniteness of H plays a crucial role). The result now follows from Theorem 5.6. □

6 Applications: Linear Optics

In this section $F = \mathbf{R}$ and all vectors and matrices are real.

In linear optics, a beam of light may be described by a vector $I = (i, q, u, v)^T$ where i ($i > 0$) denotes intensity, q/i, u/i, and v/i describe the state of polarization, and the degree of polarization $p = (q^2 + u^2 + v^2)^{1/2}/i$ belongs to $[0, 1]$. In linear optics, transformations of one beam of light into another are described by 4×4 matrices that transform vectors $I_0 = (i_0, q_0, u_0, v_0)^T$ satisfying

$$i_0 \geq \sqrt{q_0^2 + u_0^2 + v_0^2} \tag{6.1}$$

into vectors $I = (i, q, u, v)^T$ satisfying the same inequality, a property of 4×4 matrices which we call the *Stokes criterion*. It is therefore of interest to give necessary and sufficient conditions on a 4×4 matrix to satisfy the Stokes criterion. This problem has been studied extensively, see, e.g., [K,MH,N,M]. In [K,MH,N] the Stokes criterion is studied by minimizing a quadratic form under a quadratic constraint. In [M,N] the eigenvalue structure of the matrix is exploited.

Clearly, the matrices satisfying the Stokes criterion are precisely the real 4×4 matrices that leave invariant the positive cone of vectors $I_0 = (i_0, q_0, u_0, v_0)^T$ satisfying (6.1). Using this point of view, in [K] necessary and sufficient conditions were obtained for a (special) direct sum of two 2×2 matrices to satisfy the Stokes criterion; a different proof of this result was given in [MH]. To generalize this result to general real 4×4 matrices, the indefinite scalar product generated by $H = \operatorname{diag}(1, -1, -1, -1)$ had to be employed. This has led to necessary and sufficient conditions for general real 4×4 matrices to satisfy the Stokes criterion (see [M]). These results have been sharpened and generalized in Section 3 of the present paper.

The Stokes criterion is obviously satisfied for the matrices U belonging to the orthochronous Lorentz group, i.e., those matrices U orthogonal with respect to $H = \operatorname{diag}(1, -1, -1, -1)$ such that $U_{11} > 0$. Given a 4×4 matrix M and two elements U_1 and U_2 of the orthochronous Lorentz group, M satisfies the Stokes criterion if and only if $U_1 M U_2$ does. It is therefore useful to know which 4×4 matrices allow an H-polar decomposition where the H-unitary factor belongs to the orthochronous Lorentz group. The H-selfadjoint factor can then easily be analyzed through its eigenvalue structure [M]. This problem may also be solved using the eigenvalue structure of $M^{[*]}M$ (cf., [M]). The idea to diagonalize either M or $M^{[*]}M$ appears also in [N] and [X]. In [N], the Minkowski space of special relativity (in mathematical terms, $\mathbf{R}^4$ equipped with the indefinite scalar product generated by $H = \operatorname{diag}(1, -1, -1, -1)$) is employed.

In the most important problems of linear optics, polarization matrices are obtained as weighted sums of so-called pure Mueller matrices. Pure Mueller matrices are derived from the complex 2×2 transformation matrix for the associated electric field vectors and hence transform fully polarized beams represented by real vectors $I_0 = (i_0, q_0, u_0, v_0)^T$ satisfying $i_0 = \sqrt{(q_0^2 + u_0^2 + v_0^2)}$ into fully polarized beams (represented by real vectors $I = (i, q, u, v)^T$

satisfying $i = \sqrt{(q^2+u^2+v^2)}$. In other words, pure Mueller matrices are exactly the matrices of the form cU where $c \geq 0$ and U belongs to the proper Lorentz group of all real H-unitary matrices U for which $U_{11} > 0$ and $\det U = +1$ (Note that the proper Lorentz group coincides with the connected component of I in the group of H-unitary matrices). Thus every pure Mueller matrix satisfies the Stokes criterion. The weighted sums of pure Mueller matrices are exactly the matrices belonging to the set

$$\mathcal{W} = \left\{ \sum_{j=1}^{n} c_j U_j : n \in \mathbf{N},\ c_1, \cdots, c_n \geq 0,\ U_1, \cdots, U_n \in \mathcal{G} \right\},$$

where $\mathcal{G}$ is the proper Lorentz group. Thus $\mathcal{W}$ is a subset of the set of matrices satisfying the Stokes criterion. In particular, the elements of the Lorentz group that belong to $\mathcal{W}$ are precisely the elements of the proper Lorentz group. Moreover, given a 4×4 matrix M and two elements U_1 and U_2 of the proper Lorentz group, $M \in \mathcal{W}$ if and only if $U_1 M U_2 \in \mathcal{W}$.

Necessary and sufficient conditions for a 4×4 matrix to belong to $\mathcal{W}$ have been given in [C], where a bijective linear transformation $T : \mathbf{C}^{4\times4} \to \mathbf{C}^{4\times4}$ was constructed mapping the real matrices bijectively onto the complex hermitian matrices. Then $M \in \mathcal{W}$ if and only if the so-called coherency matrix $T[M]$ is positive semidefinite (see [C,M] for two different proofs). On the other hand, there exists a criterion in terms of the eigenvalue structure of the given matrix M if it is H-selfadjoint (cf. [M]). In order to generalize this criterion to arbitrary 4×4 matrices M, it is therefore useful to have an H-polar decomposition of M where the H-unitary factor belongs to the connected component of I. Contrary to what is sometimes suggested in the literature [X], there exist matrices in $\mathcal{W}$ that do not allow an H-polar decomposition (see [M] for an example); this circumstance will slightly complicate the proof of Theorem 6.4 below.

In [M] criteria have been given for a 4×4 matrix to satisfy the Stokes criterion and, for an H-selfadjoint matrix, to belong to $\mathcal{W}$.

THEOREM 6.1. *Let M be an H-selfadjoint matrix. Then M satisfies the Stokes criterion (resp. $M \in \mathcal{W}$) if and only if one of the following two situations occurs:*

(1) *M has the one nonnegative eigenvalue λ_0 corresponding to a positive eigenvector and three real eigenvalues λ_1, λ_2 and λ_3 corresponding to negative eigenvectors, and $\lambda_0 \geq \max(|\lambda_1|, |\lambda_2|, |\lambda_3|)$ (resp. both of $\lambda_0 \pm \lambda_1 \geq |\lambda_2 \pm \lambda_3|$);*

(2) *M has the real eigenvalues λ, μ and ν but is not diagonalizable. The eigenvectors corresponding to μ and ν are negative, whereas to the double eigenvalue λ there corresponds one Jordan block of order 2 with the positive sign in the sign characteristic of $\{M^{[*]}M, H\}$. Moreover, $\lambda \geq \max(|\mu|, |\nu|)$ (resp. both $\mu = \nu$ and $\lambda \geq |\mu|$).*

For general 4×4 matrices the situation is more complicated. Since a real 4×4 matrix M satisfies the Stokes criterion if and only if M is a doubly H-plus matrix with $M_{11} \geq 0$, necessary and sufficient conditions for M to satisfy the Stokes criterion follow from Proposition 3.3 (with $F = \mathbf{R}$ and $H = \mathrm{diag}(1, -1, -1, -1)$), the third paragraph following the proof of Theorem 3.4, and Proposition 3.5 (with $n = 4$). Necessary and sufficient conditions for such a matrix to have an H-polar decomposition follow from Theorem 3.4 (with $F = \mathbf{R}$

and $H = \operatorname{diag}(1, -1, -1, -1)$). We will formulate these results below in the context of this section and then go on to characterize the matrices belonging to $\mathcal{W}$.

THEOREM 6.2. *Let M be a 4×4-matrix satisfying $M_{11} \geq 0$. Then M satisfies the Stokes criterion if and only if one of the following two situations occurs:*

(1) *$M^{[*]}M$ has the one nonnegative eigenvalue λ_0 corresponding to a positive eigenvector and three nonnegative eigenvalues λ_1, λ_2 and λ_3 corresponding to negative eigenvectors, and $\lambda_0 \geq \max(\lambda_1, \lambda_2, \lambda_3)$;*

(2) *$M^{[*]}M$ has the nonnegative eigenvalues λ, μ and ν but is not diagonalizable. The eigenvectors corresponding to μ and ν are negative, whereas to the double eigenvalue λ there corresponds one Jordan block of order 2 with the positive sign in the sign characteristic of $\{M^{[*]}M, H\}$. Moreover, $\lambda \geq \max(\mu, \nu)$.*

To derive conditions for a 4×4 matrix to belong to $\mathcal{W}$, we need a result on H-polar decomposition of matrices satisfying the Stokes criterion. The result is immediate from Theorem 3.4.

PROPOSITION 6.3. *A matrix M satisfying the Stokes criterion allows an H-polar decomposition, unless all of the eigenvalues of $M^{[*]}M$ vanish and $M^{[*]}M$ has one Jordan block of order 2 with the positive sign in the sign characteristic of $\{M^{[*]}M, H\}$.*

Suppose M satisfies the Stokes criterion and is invertible, and let S be an H-unitary matrix S such that either

$$S^{-1}M^{[*]}MS = \operatorname{diag}(\lambda_0, \lambda_1, \lambda_2, \lambda_3), \tag{6.2}$$

or

$$S^{-1}M^{[*]}MS = \begin{bmatrix} \lambda + \varepsilon & -\varepsilon \\ \varepsilon & \lambda - \varepsilon \end{bmatrix} \oplus (\mu) \oplus (\nu), \tag{6.3}$$

where $\varepsilon > 0$. Then one may choose S such that $S_{11} > 0$ and $\det S = +1$. Further, there exists an H-selfadjoint matrix A such that $M^{[*]}M = A^2$ and $M = UA$ for some H-unitary matrix U. Writing $A = S^{-1}DS$ and $N = MS^{-1}$, we obtain

$$US^{-1} = ND^{-1},$$

where N satisfies the Stokes criterion. In the case of (6.2), D^{-1} is a diagonal matrix with nonnegative entries and hence $U_{11} > 0$. In the case of (6.3), we find for the $(1,1)$-component of US^{-1}

$$[US^{-1}]_{11} = \frac{N_{11}}{\sqrt{\lambda}} + \frac{\varepsilon(N_{11} - N_{12})}{2\lambda\sqrt{\lambda}} > 0,$$

because $\varepsilon > 0$ and $N_{11} \geq |N_{12}|$. Hence in either case the H-unitary factor U satisfies $U_{11} > 0$, because $U_{11} \geq [US^{-1}]_{11}S_{11} - \sqrt{\sum_{j=2}^4 [US^{-1}]_{jj}^2}\sqrt{\sum_{j=2}^4 S_{jj}^2} > 0$. Moreover, the identities $M^{[*]}M = A^2$, $M = UA$ and $\det A > 0$ imply that $\det U = +1$ if $\det M > 0$ and $\det U = -1$ if $\det M < 0$.

We have the following result.

THEOREM 6.4. *Let M be a 4×4-matrix satisfying $M_{11} \geq 0$. Let σ be the sign ± 1 of the product of the nonzero eigenvalues of M. Then $M \in \mathcal{W}$ if and only if one of the following four situations occurs:*

(1) *$M^{[*]}M$ has the positive eigenvalue λ_0 corresponding to a positive eigenvector and a positive and two nonnegative eigenvalues λ_1, λ_2 and λ_3 (with $\lambda_1 \geq \lambda_2 \geq \lambda_3$) corresponding to negative eigenvectors, and $\sqrt{\lambda_0} \pm \sqrt{\lambda_1} \geq |\sqrt{\lambda_2} \pm \sigma\sqrt{\lambda_3}|$;*

(2) *$M^{[*]}M$ is diagonalizable with one positive and three zero eigenvalues, and $\sigma = +1$;*

(3) *$M^{[*]}M$ has the positive eigenvalue λ and the nonnegative eigenvalues μ and ν but is not diagonalizable. The eigenvectors corresponding to μ and ν are negative, whereas to the double eigenvalue λ there corresponds one Jordan block of order 2 with the positive sign in the sign characteristic of $\{M^{[*]}M, H\}$. Moreover, $\sigma = +1$, $\mu = \nu$* and *$\lambda \geq \mu$;*

(4) *$M^{[*]}M$ has only zero eigenvalues. When $M^{[*]}M$ is not diagonalizable, it has one Jordan block of order 2 with the positive sign in the sign characteristic of $\{M^{[*]}M, H\}$.*

Proof. Suppose M is an invertible matrix satisfying the Stokes criterion. Then either of the two cases of Theorem 6.2 applies. Further, according to Proposition 6.3 and Theorem 5.4, M has an H-polar decomposition of the form $M = UA$ where U belongs to the connected component of I in the group of H-unitary matrices and

$$A = S^{-1}\text{diag}\left(\sqrt{\lambda_0}, \sqrt{\lambda_1}, \sqrt{\lambda_2}, \sigma\sqrt{\lambda_3}\right) S$$

in the first case and

$$A = S^{-1} \begin{bmatrix} \sqrt{\lambda} + \frac{1}{2}\varepsilon\lambda^{-1/2} & \frac{1}{2}\varepsilon\lambda^{-1/2} & 0 & 0 \\ -\frac{1}{2}\varepsilon\lambda^{-1/2} & \sqrt{\lambda} - \frac{1}{2}\varepsilon\lambda^{-1/2} & 0 & 0 \\ 0 & 0 & \sqrt{\mu} & 0 \\ 0 & 0 & 0 & \sigma\sqrt{\nu} \end{bmatrix} S \tag{6.4}$$

in the second case, where S is an H-unitary matrix in the connected component of I. The theorem follows by applying Theorem 6.1 to the matrix SAS^{-1}.

Now let M satisfy the Stokes criterion, be singular, and have the H-polar decomposition $H = UA$ constructed in the paragraph following the proof of Proposition 6.3. Then $\operatorname{Ker} A = \operatorname{Ker} M^{[*]}M = \operatorname{Ker} M$. Further, there exists a nondegenerate A-invariant subspace L complementary to $\operatorname{Ker} A$ on which A is invertible. Then U is a Witt extension of the restriction U_0 of U to L acting as a matrix from L onto $U[L]$. One easily sees from Theorem 2.6 of [BMRRR2] (in the case $n = 4$, $m = 4 - d$, $m_+ = 1$, $m_0 = 0$, $m_- = 3 - d$, $p = 0$ and $q = d$, where $d = \dim \operatorname{Ker} A$) that there are two connected components of Witt extensions of U_0 and that there exist in fact U with $\det U = +1$ and U with $\det U = -1$. So let us choose U with $\det U = +1$, so that U belongs to the connected component of the identity in the group of H-unitary matrices. Now let σ be the sign of the nonzero eigenvalues of M. If $M^{[*]}M$ is diagonal, we find SAS^{-1} equal to $\text{diag}(\sqrt{\lambda_0}, \sqrt{\lambda_1}, \sigma\sqrt{\lambda_2}, 0)$ if $\lambda_0 \geq \max(\lambda_1, \lambda_2) > \lambda_3 = 0$, to $\text{diag}(\sqrt{\lambda_0}, \sigma\sqrt{\lambda_1}, 0, 0)$ if $\lambda_0 \geq \lambda_1 > \lambda_2 = \lambda_3 = 0$, and to diag $(\sigma\sqrt{\lambda_0}, 0, 0, 0)$ if $\lambda_0 > \lambda_1 = \lambda_2 = \lambda_3 = 0$, where S is an H-unitary matrix in the connected component of I. Thus, applying Theorem 6.1 to these A, we conclude that $M \in \mathcal{W}$ if and only if $\sqrt{\lambda_0} \pm \sqrt{\lambda_1} \geq \sqrt{\lambda_2}$, $\sqrt{\lambda_0} \pm \sigma\sqrt{\lambda_1} \geq 0$, and $\sigma\sqrt{\lambda_0} \geq 0$, respectively. On the other hand, if $M^{[*]}M$ has a Jordan block of order 2 corresponding to $\lambda > 0$ with the positive sign in the sign characteristic of $\{M^{[*]}M, H\}$, then, writing the 2×2 block in the left upper corner of the matrix in the right-hand side of (6.4) as B, SAS^{-1} is equal to diag $(B, \sigma\sqrt{\mu}, 0)$ if $\lambda \geq \mu > \nu = 0$, and to diag $(\sigma B, 0, 0)$ if $\lambda > \mu = \nu = 0$, where S is an H-unitary matrix in the connected component of I. Thus, applying Theorem 6.1 to these A, we conclude that $M \in \mathcal{W}$ if and only if $\lambda > \mu = \nu = 0$ and $\sigma = 1$.

When $M^{[*]}M$ has only zero eigenvalues and one Jordan block of order 2 with the positive sign in the sign characteristic of $\{M^{[*]}M, H\}$, there exist an H-unitary S in the connected component of I and $\varepsilon > 0$ such that

$$M^{[*]}M = S\left(\begin{bmatrix} \varepsilon & -\varepsilon \\ \varepsilon & -\varepsilon \end{bmatrix} \oplus \begin{bmatrix} 0 & 0 \\ 0 & 0 \end{bmatrix}\right) S^{-1}.$$

Further, $MM^{[*]}$ has the same Jordan structure as $M^{[*]}M$ (see the paragraph following the proof of Theorem 3.4). Now let D be an H-unitary matrix in the connected component of I mapping Ker M onto Ker $M^{[*]}$; such a matrix exists. Then $D^{[*]}$ maps Im M onto Im $M^{[*]}$. Put $M_\delta = M + \delta D$ for $\delta > 0$. Then M_δ satisfies the Stokes criterion whenever M does. Further,

$$M_\delta^{[*]}M_\delta = M^{[*]}M + \delta^2 I + \delta\left(M^{[*]}D + D^{[*]}M\right),$$

so that $M_\delta^{[*]}M_\delta x = \delta^2 x$ for every $x \in \operatorname{Ker} M$. If dim Ker $M = 3$ and hence Ker M = Ker $M^{[*]}M$, then $M_\delta^{[*]}M_\delta$ must have the same Jordan structure as $M^{[*]}M$, but then at the eigenvalue δ^2. Further, if ξ and η are vectors such that $M^{[*]}M\eta = \xi$ and $M^{[*]}M\xi = 0$, we have $(M^{[*]}D + D^{[*]}M)\eta$ proportional to ξ but nonzero. Thus the vectors ξ, η form a Jordan chain for $M_\delta^{[*]}M_\delta$ at the eigenvalue δ^2. Hence the Jordan block of $M_\delta^{[*]}M_\delta$ of order 2 has the positive sign in the sign characteristic of $\{M_\delta^{[*]}M_\delta, H\}$. Thus $M_\delta \in \mathcal{W}$. On the other hand, if dim Ker $M = 2$ and hence Ker $M = S[(0) \oplus \mathbf{R}^2]$, then besides δ^2 the eigenvalues of $M_\delta^{[*]}M_\delta$ consist of two numbers $\geq \delta^2$, because M_δ satisfies the Stokes criterion whenever M does. Then $M_\delta^{[*]}M_\delta$ is either diagonalizable or has a Jordan block of order 2 with the positive sign in the sign characteristic of $\{M_\delta^{[*]}M_\delta, H\}$. As a result, $M_\delta \in \mathcal{W}$.

Note that $\mathcal{W}$ is closed, since there exists a linear invertible transformation $T: \mathbf{C}^{4\times4} \to \mathbf{C}^{4\times4}$ mapping $\mathcal{W}$ onto the set of hermitian matrices with nonnegative eigenvalues, which is closed in $\mathbf{C}^{4\times4}$ (actually, T maps polarization matrices onto their corresponding coherency matrices; cf. [C,M]). Since M can be perturbed by an arbitrarily close element of $\mathcal{W}$ and $\mathcal{W}$ is a closed set, $M \in \mathcal{W}$. □

References

[A] T. Ando. *Linear Operators on Krein Spaces.* Lecture Notes, Sapporo, Japan, 1979.

[AI] T. Ja. Azizov and E. I. Iohvidov. The development of some of V. P. Potapov's ideas in the geometric theory of operators in spaces with indefinite metric.

in: *Matrix and Operator Valued Functions* OT 72 , eds. I. Gohberg and L.A. Sakhnovich, Birkhäuser, Basel, 1994, 17-27.

[BP] A. Berman, and R. J. Plemmons, *Nonnegative Matrices in the Mathematical Sciences*, Academic Press, New York, 1979; also: SIAM, Philadelphia, 1994.

[Bo] J. Bognár. *Indefinite Inner Product Spaces,* Springer, Berlin, 1974.

[BMRRR1] Yu. Bolshakov, C. V. M. van der Mee, A. C. M. Ran, B. Reichstein, L. Rodman. Polar decompositions in finite dimensional indefinite scalar product spaces: General theory, *Linear Algebra and its Applications*, submitted for publication.

[BMRRR2] Yu. Bolshakov, C. V. M. van der Mee, A. C. M. Ran, B. Reichstein, L. Rodman. Extension of isometries in finite dimensional indefinite scalar product spaces and polar decompositions, paper in preparation.

[BR] Yu. Bolshakov, B. Reichstein. Unitary equivalence in an indefinite scalar product: An analogue of singular value decomposition, *Linear Algebra and its Applications,* to appear.

[C] S. R. Cloude, Group theory and polarisation algebra, *Optik* **75**, 26–36 (1986).

[F] H. Flanders, Elementary divisors of AB and BA, *Proceedings AMS* **2** (1951), 871–874.

[Gi1] Ju. P. Ginzburg, On J-nonexpansive operator functions, *Dokl. Akad. Nauk SSSR* **117, No 2** (1957), 171–173 [Russian].

[Gi2] Ju. P. Ginzburg, On J-nonexpansive operators in a Hilbert space, *Nauchnye Zap. Fak. Fiziki i Matematiki*, Odesskogo Gosud. Pedagog. Instituta **22, No 1** (1958), 13–20 [Russian].

[GLR] I. Gohberg, P. Lancaster, L. Rodman. *Matrices and Indefinite Scalar Products*, OT8, Birkhäuser, Basel, 1983.

[K] N. V. Konovalov, Polarization matrices corresponding to transformations in the Stokes cone, Preprint **171**, *Keldysh Inst. Appl. Math., Acad. Sci. USSR*, Moscow, 1985 [Russian].

[Kr] M. A. Krasnoselskii, *Positive Solutions of Operator Equations.* Noordhoff, Groningen, 1964. [Translation from Russian].

[KS1] M. G. Krein and Ju. L. Shmul'jan, On plus operators in a space with an indefinite metric, *Mat. Issled.* **1**, No. **2** (1966), 131–161 [Russian]; English translation: *AMS Translations*, Series **2, 85** (1969), 93–113.

[KS2] M. G. Krein and Ju. L. Shmul'jan, J-polar representation of plus operators, *Mat. Issled.* **1**, No. **2** (1966), 172–210 [Russian]; English translation: *AMS Translations*, Series **2, 85** (1969), 115-143.

[M] C. V. M. van der Mee, An eigenvalue criterion for matrices transforming Stokes parameters, *J. Math. Phys.* 34 (1993), 5072–5088.

[MH] C. V. M. van der Mee and J. W. Hovenier, Structure of matrices transforming Stokes parameters, *J. Math. Phys.* **33** (1992), 3574–3584.

[N] D. I. Nagirner, Constraints on matrices transforming Stokes vectors, *Astron. Astrophys.*, **275** (1993), 318–324.

[P1] V. P. Potapov. Multiplicative structure of J-nonexpansive matrix functions. *Trudy Mosk. Math. Ob.* **4** (1955), 125–236 [Russian].

[P2] V. P. Potapov. A theorem on the modulus, I. Main concepts. The modulus. *Theory of Functions, Functional Analysis and its Applications* **38** (1982), 91–101, 129, Kharkov [Russian]. Translated in English: AMS Translations, Series **2**, Vol. **138**, 55–65.

[X] Zhang-Fan Xing, On the deterministic and non-deterministic Mueller matrix, *J. Mod. Opt.* **39** (1992), 461–484.

Yuri Bolshakov
Department of Mathematics
Yaroslavl State University
Yaroslavl, Russia

Cornelis V. M. van der Mee
Dipartimento di Matematica
Università di Cagliari
Via Ospedale 72, 09124 Cagliari, Italy

André C. M. Ran
Faculteit Wiskunde en Informatica
Vrije Universiteit Amsterdam
De Boelelaan 1081
1081 HV Amsterdam, The Netherlands

Boris Reichstein
Department of Mathematics
The Catholic University of America
Washington, DC 20064 USA

Leiba Rodman
Department of Mathematics
The College of William and Mary
Williamsburg, VA 23187-8795 USA

AMS Classification numbers: 15A23, 15A63, 47B50

Operator Theory:
Advances and Applications, Vol. 87

POSITIVE DIFFERENTIAL OPERATORS IN KREIN SPACE $L^2(\mathbb{R})$

Branko Ćurgus and Branko Najman

Consider the weighted eigenvalue problem

$$Lu = \lambda\,(\operatorname{sgn} x)u, \tag{1}$$

on the whole real line $\mathbb{R}$ where $L = p(D)$ is a positive symmetric differential operator with constant coefficients. This problem is a model problem for a more general problem $Lu = \lambda\, wu$ with L a differential operator and w a function taking both positive and negative values.

Our starting point is the observation that the operator $A = (\operatorname{sgn} x)L$ is symmetric and positive with respect to the indefinite inner product $[u, v] = \int u(x)\overline{v(x)}\operatorname{sgn} x\,dx$. The space $L^2(\mathbb{R})$ with this inner product is a Krein space. Once we prove that the resolvent set $\rho(A)$ is nonempty, H. Langer's spectral theory can be applied. This spectral theory shows that the spectrum of A is real and its properties on bounded open intervals not containing 0 are the same as the corresponding properties of a selfadjoint operator in a Hilbert space. In particular, A has a spectral function defined on open intervals in $\mathbb{R}$ with the endpoints different from 0 and ∞. The positive (negative, respectively) spectral points are of positive (negative, resp.) type. Therefore 0 and ∞ are the only possible critical points. A critical point λ is *regular* if the spectral function is bounded near λ. In that case the spectral function can be extended to intervals with an endpoint λ. A critical point is *singular* if it is not regular. If neither 0 nor ∞ is a singular critical point, then A is similar to a selfadjoint operator in $L^2(\mathbb{R})$. We used this fact in [5] to prove that A is similar to a selfadjoint operator in the case $p(t) = t^2$.

In this paper we generalize this result to more general polynomials p. The results of this paper are used in the forthcoming paper [6] to extend the results of [5] to a class of partial differential operators. For example, in [6] for $n > 1$ we prove the following.
The operator $(\operatorname{sgn} x_n)\Delta$ *defined on* $H^2(\mathbb{R}^n)$ *is similar to a selfadjoint operator in* $L^2(\mathbb{R}^n)$.

The question of nonsingularity of the critical point ∞ has been considered in [4]. This question leads to the investigation of the domain of A. In the present case the operator

A is positive (not uniformly positive as in [4]) and this is why the critical point at 0 may appear as a critical point of infinite type. If the spectrum of A accumulates at 0 from both sides, then 0 is a critical point of A. To determine whether it is singular or regular we are led to investigate the range of A. This question is harder than the investigation of the domain. In Section 1 we give a necessary and sufficient condition for $\mathcal{R}(B) = \mathcal{R}(C)$ for multiplication operators B, C in $L^2(\mathbb{R})$. We also prove several stability theorems for the regularity of the critical points 0 and ∞ of positive definitizable operators in a Krein space. As a consequence we get a stability theorem for the similarity to a selfadjoint operator in a Hilbert space. For related results in this direction see [7]. In Section 2 we consider the differential operators with constant coefficients in $L^2(\mathbb{R})$. We give a precise description of the spectrum of the operator A. Under some additional restrictions on p, we prove that A is similar to a selfadjoint operator in $L^2(\mathbb{R})$. It follows from the general operator theory in Krein spaces that an operator which is positive in the Krein space $(L^2(\mathbb{R}), [\cdot|\cdot])$ and similar to a selfadjoint operator in the Hilbert space $L^2(\mathbb{R})$ has the half-range completeness property. We use this fact in Section 3 to show that our results in Section 2 give sufficient conditions for the half-range completeness property for the problem (1).

The Sturm-Liouville problem with indefinite weight has attracted considerable attention; we mention the references quoted in [3, 4] for a partial list. The problem of nonsingularity of the critical points of definitizable operators in Krein spaces has been investigated in [2, 7, 8, 10]. For differential operators with indefinite weights the study of this problem has been motivated by the investigation of the half-range completeness property, cf. [1, 3]. The regularity of the critical point 0 has been considered in [5].

For definitions and basic results of the theory of definitizable operators see [9].

1 Abstract Results

In this section we use the method of [2, Lemma 1.8, Corollary 3.3 and Theorem 3.9] to investigate the regularity of the critical points 0 and ∞ of a positive definitizable operator A in the Krein space $(\mathcal{K}, [\cdot|\cdot])$.

The following two lemmas are restatements of [2, Theorem 3.9 and Corollary 3.3] in terms of the critical point 0. We prove the first. The proof of the second one is analogous.

LEMMA 1.1 *Let $A = JP$ be a positive definitizable operator in the Krein space $(\mathcal{K}, [\cdot|\cdot])$ such that 0 is not an eigenvalue of P. Assume that $\nu > 0$ and the operator JP^ν is definitizable. Then the following statement are equivalent:*

(a) *The point 0 is not a singular critical point of the operator JP.*

(b) *The point 0 is not a singular critical point of the operator JP^ν.*

PROOF The point 0 is not a singular critical point of JP if and only if it is not a singular critical point of the operator PJ which is similar to JP. Further, 0 is not a singular critical point of PJ if and only if ∞ is not a singular critical point of the operator JP^{-1}. It follows from [2, Theorem 3.9] that ∞ is not a singular critical point of JP^{-1} if and only if ∞ is not a singular critical point of $JP^{-\nu}$. Clearly, ∞ is not a singular critical point of $JP^{-\nu}$ if and only if 0 is not a singular critical point of $P^{\nu}J$. Because of the similarity of the operators, 0 is not a singular critical point of $P^{\nu}J$ if and only if 0 is not a singular critical point of JP^{ν}. This sequence of equivalent statements proves the lemma. □

It follows from [2, Lemma 1.8] that the operator $JP^{-\nu}$ is definitizable for $\nu = 2^m$ with m being a positive integer.

LEMMA 1.2 *Let A and B be definitizable operators in the Krein space $\mathcal{K}$ such that 0 is neither an eigenvalue of A nor of B. Assume that $\mathcal{R}(A) = \mathcal{R}(B)$. Then the following statements are equivalent.*

(a) *The point 0 is not a singular critical point of A.*

(b) *The point 0 is not a singular critical point of B.*

LEMMA 1.3 *Let g and h be nonnegative measurable functions on $\mathbb{R}$.*

(a) *The following statements are equivalent:*

(i) $\mathcal{D}(M_g) = \mathcal{D}(M_h)$

(ii) *The functions $\frac{h}{1+g}$ and $\frac{g}{1+h}$ are essentially bounded.*

(b) *The following statements are equivalent:*

(i) $\mathcal{R}(M_g) = \mathcal{R}(M_h)$.

(ii) *There exists a constant $C \geq 0$ such that*

$$g \leq Ch(1+g)\ \mu\text{-a.e.} \quad \textit{and} \quad h \leq Cg(1+h)\ \ \mu\text{-a.e.}\ . \tag{2}$$

PROOF The statement (a) is evident.
(b) For a μ-measurable function f denote the set $\{x \in \mathbb{R} | f(x) = 0\}$ by N_f. Note that each of the conditions (a) and (b) implies that $N_g = N_h = N$. Therefore $\mathcal{N}(M_g) = \mathcal{N}(M_h)$ consists of functions $f \in L^2(\mathbb{R}, \mu)$ with the support contained in N. Let

$$G(x) = H(x) = 0\ (x \in N)\ , G(x) = \frac{1}{g(x)},\ H(x) = \frac{1}{h(x)}\ (x \in \mathbb{R} \setminus N)\ .$$

It follows from (a) that the condition (ii) is equivalent to $\mathcal{D}(M_G) = \mathcal{D}(M_H)$. Since $\mathcal{D}(M_G) = \mathcal{R}(M_g) \oplus \mathcal{N}(M_g)$, we conclude that (i) and (ii) are equivalent. □

A polynomial p is *nonnegative* if $p(x) \geq 0$ for all $x \in \mathbb{R}$.

EXAMPLE 1 Let h be a nonnegative polynomial of degree $2k$ in one variable. If $g(t) = t^{2k}$, then h and g satisfy the conditions of Lemma 1.3 (a).

EXAMPLE 2 Let h be a nonnegative polynomial. Then $h(t) = ag(t)\tilde{h}(t)$, where $a > 0$, $\tilde{h}$ is a positive polynomial without real roots and $g(t) = (t - r_1)^{2k_1} \cdots (t - r_m)^{2k_m}$. Then h and g satisfy the condition (ii) of Lemma 1.3 (b).

THEOREM 1.4 *Let S be a selfadjoint operator in the Hilbert space $(\mathcal{K}, (\cdot|\cdot))$ such that JS^2 is a definitizable operator in the Krein space $(\mathcal{K}, [\cdot|\cdot])$. Let $\nu > 0$ and let h be a nonnegative continuous function. Assume that the operators $J|S|^\nu$ and $Jh(S)$ are definitizable.*

(a) *Assume that the functions $g(t) = |t|^\nu$ and h satisfy the conditions of Lemma 1.3 (a). Then the following statements are equivalent.*

   (i) *The point ∞ is not a singular critical point of JS^2.*

   (ii) *The point ∞ is not a singular critical point of $Jh(S)$.*

(b) *Assume that 0 is not an eigenvalue of S and that the functions $g(t) = |t|^\nu$ and h satisfy the condition (2). Then the following statements are equivalent.*

   (i) *The point 0 is not a singular critical point of JS^2.*

   (ii) *The point 0 is not a singular critical point of $Jh(S)$.*

PROOF We prove (b). The proof of (a) is similar. Lemma 1.1 implies that 0 is not a singular critical point of JS^2 if and only if it is not a singular critical point of $J|S|^\nu$.

It follows from Lemma 1.3 (b) that for any Borel measure μ the multiplication operators M_g and M_h in $L^2(\mathbf{R}, \mu)$ have the same range. The Spectral Theorem, see [11, Theorem 7.18], implies $\mathcal{R}(|S|^\nu) = \mathcal{R}(h(S))$. Therefore, $\mathcal{R}(J|S|^\eta) = \mathcal{R}(Jh(S))$. The conclusion follows from Lemma 1.2. □

COROLLARY 1.5 *Let S be a selfadjoint operator in the Hilbert space $(\mathcal{K}, (\cdot|\cdot))$ such that 0 is not an eigenvalue of S and such that JS^2 is a definitizable operator in the Krein space $(\mathcal{K}, [\cdot|\cdot])$. Let η and ν be positive numbers and let h be a nonnegative continuous function. Let $g_1(t) = |t|^\eta$ and $g_2(t) = |t|^\nu$. Assume that the functions g_1 and h satisfy the conditions of Lemma 1.3 (a) and that the functions g_2 and h satisfy the condition (2). Assume that the operators $J|S|^\eta, J|S|^\nu$ and $Jh(S)$ are definitizable. Then the following statements are equivalent.*

(i) *The operator JS^2 is similar to a selfadjoint operator in $(\mathcal{K}, (\cdot|\cdot))$.*

(ii) *The operator $Jh(S)$ is similar to a selfadjoint operator in $(\mathcal{K}, (\cdot|\cdot))$.*

2 Differential Operators with Constant Coefficients

In this section we apply the results from Section 1 to a class of positive ordinary differential operators with constant coefficients.

In the following, a root of multiplicity m of a polynomial is counted as m roots. Denote by $\mathbb{C}_+$ (respectively $\mathbb{C}_-$) the set of all complex numbers z such that $\operatorname{Im} z > 0$ (respectively $\operatorname{Im} z < 0$).

We consider an even order polynomial

$$p(z) = a_0 z^{2n} + a_1 z^{2n-1} + \cdots + a_{2n-1} z + a_{2n} \,. \tag{3}$$

with real coefficients a_j.

For the reader's convenience we give a proof of the following lemma.

LEMMA 2.1 *Let p be a polynomial of degree $2n$ with real coefficients. Let α be a complex number.*

(a) *If α is nonreal, then the polynomial equation*

$$p(z) - \alpha = 0 \tag{4}$$

has exactly n solutions in $\mathbb{C}_+$ and exactly n solutions in $\mathbb{C}_-$.

(b) *If α is real, then the equation (4) has at most n solutions in $\mathbb{C}_+$.*

PROOF (a) Let $n_+(\alpha)$ be the number of solutions of (4) in $\mathbb{C}_+$. Since (4) has no real solutions, it follows that $n_+(\alpha)$ is constant for $\alpha \in \mathbb{C}_+$. Note that the equation $a_0 z^{2n} = \alpha$ has exactly n solutions with positive imaginary parts, an application of Rouche's theorem shows that $n_+(\alpha) = n$ for $|\alpha|$ sufficiently large.

The claim (b) is evident. □

Denote $D = -i\frac{d}{dx}$. We consider the spectral problem

$$p(D)f(x) = \lambda(\operatorname{sgn} x)f(x), \quad x \in \mathbb{R}, \tag{5}$$

For a polynomial q of degree k, $q(D)$ denotes the constant coefficient differential operator in the Hilbert space $L^2(\mathbb{R})$ defined on the Sobolev space $H^k(\mathbb{R})$.

Let J be the multiplication operator defined by

$$(Jf)(x) = (\operatorname{sgn} x)f(x), \; x \in \mathbb{R} \,.$$

Then the problem (5) can be written in terms of operators as

$$p(D)f = \lambda J f, \quad f \in H^{2n}(\mathbb{R}) \,, \tag{6}$$

or, equivalently,

$$Jp(D)f = \lambda f, \quad f \in H^{2n}(\mathbb{R})\,. \tag{7}$$

It is natural to study the problem (7) in the Krein space $\mathcal{K} = L^2(\mathbb{R})$ with the scalar product $[f,g] = \int_{\mathbb{R}} f(x)\overline{g(x)}\operatorname{sgn} x\, dx$. The multiplication operator J is a fundamental symmetry on $\mathcal{K}$ and the corresponding positive definite scalar product is the standard scalar product in $L^2(\mathbb{R})$.

Since p has real coefficients the operator $p(D)$ is selfadjoint in the Hilbert space $L^2(\mathbb{R})$. Therefore, the operator $Jp(D)$ is selfadjoint in the Krein space $\mathcal{K}$. A selfadjoint operator in a Krein space may have empty resolvent set. In the next theorem we show that this is not the case for the operator $Jp(D)$.

THEOREM 2.2 *Let p be an even order polynomial with real coefficients. Let $A = Jp(D)$.*

(a) *The spectrum of the operator A is real.*

(b) *The operator A has no eigenvalues. Its residual spectrum is empty.*

(c) *The continuous spectrum of A is given by*

$$\sigma_c(A) = (-\infty, -m_p] \cup [m_p, +\infty), \quad \textit{where } m_p = \min\{p(x) : x \in \mathbb{R}\}\,. \tag{8}$$

PROOF (a) Let ζ be an arbitrary nonreal complex number. We have to prove that the operator $A - \zeta I$ has a bounded inverse. Since the operators J and $p(D)$ are closed, it is sufficient to prove that $p(D) - \zeta J$ is a bijection of $H^{2n}(\mathbb{R})$ onto $L^2(\mathbb{R})$. Let $g \in L^2(\mathbb{R})$. The special restriction of $p(D)$ defined in $L^2(\mathbb{R}_\mp)$ with the domain consisting of all functions f in $H^{2n}(\mathbb{R}_\mp)$ such that $f^{(j)}(0) = 0$, $j = 0, \ldots, n-1$, is selfadjoint in the Hilbert space $L^2(\mathbb{R}_\mp)$. Therefore, the boundary value problems

$$\begin{aligned}(p(D)y)(x) \pm \zeta\, y(x) = g(x), \quad x \in \mathbb{R}_\mp, \quad & y \in H^{2n}(\mathbb{R}_\mp)\\ & y^{(j)}(0) = 0, \quad j = 0, \ldots, n-1\end{aligned}$$

have unique solutions $y_\mp$ in $H^{2n}(\mathbb{R}_\mp)$.

Now consider the homogeneous equation

$$p(D)y - \zeta\, y = 0, \quad y \in H^{2n}(\mathbb{R}_+). \tag{9}$$

In order to find the fundamental set of solutions of (9) we have to solve the polynomial equation $p(-iz) - \zeta = 0$. Since ζ is nonreal, we can apply Lemma 2.1 (a) and conclude that this equation has n roots $z_j^+, j = 1, \ldots, n$, with negative real parts. These roots in the standard way lead to n linearly independent solutions ψ_j^+, $j = 1, \ldots, n$ of (9) which are in $H^{2n}(\mathbb{R}_+)$.

To find the fundamental set of solutions of the homogeneous equation

$$p(D)y + \zeta\, y = 0, \; y \in H^{2n}(\mathbb{R}_-). \tag{10}$$

we have to find the roots of $p(-iz)+\zeta = 0$ with positive real parts. By Lemma 2.1 (a) there are n such roots; denote them by $z_j^-, j = 1, \dots, n$. These roots in the standard way lead to n linearly independent solutions ψ_j^-, $j = 1, \dots, n$ of (10) which are in $H^{2n}(\mathbb{R}_-)$. Since the set $\{z_j^+, j = 1, \dots, n\}$ is disjoint from the set $\{z_j^-, j = 1, \dots, n\}$, the set $\{\psi_j^+, \psi_j^-, j = 1, \dots, n\}$ is linearly independent and moreover it is a basis of solutions of the homogeneous equation $q(D)y = 0$, where $q(t) = \prod_{j=1}^{n}(t+iz_j^+)(t+iz_j^-)$. Therefore the Wronskian of $\{\psi_j^+, \psi_j^-, j = 1, \dots, n\}$ does not have zeros.

Every solution $f \in H^{2n}(\mathbb{R})$ of the equation

$$p(D)f - \zeta J f = g \tag{11}$$

must satisfy

$$f(x) = \begin{cases} y_-(x) + \sum_{j=1}^{n} c_j^- \psi_j^-(x), & x \in \mathbb{R}_- \\ y_+(x) + \sum_{j=1}^{n} c_j^+ \psi_j^+(x), & x \in \mathbb{R}_+ \end{cases}$$

for some complex numbers c_j^-, c_j^+, $j = 1, \dots, n$. The continuity of $f^{(j)}$, $j = 0, 1, \dots, 2n-1$ at 0 leads to a system of $2n$ linear equations in c_j^-, c_j^+, $j = 1, \dots, n$. The determinant of this system is the Wronskian of the functions ψ_j^+, ψ_j^-, $j = 1, \dots, n$ evaluated at 0. Since this determinant is not 0, the system has unique solution. Therefore, the equation (11) has a unique solution, i.e., $p(D) - \zeta J$ is bijection of $H^{2n}(\mathbb{R})$ onto $L^2(\mathbb{R})$. Consequently, ζ is in the resolvent set of A.

(b) Let $\zeta \in \mathbb{R}$ and let $y \in H^{2n}(\mathbb{R})$ be a solution of the equation

$$p(D)y - \zeta J y = 0 .$$

The restriction y_+ (y_-, resp.) of y to $\mathbb{R}_+$($\mathbb{R}_-$, resp.) satisfies the equation (9) ((10), respectively). Applying Lemma 2.1 (b) and arguing as in the proof of (a), we conclude that the equation (9)((10), respectively), has $k_+ \le n$ ($k_- \le n$, resp.) linearly independent solutions ψ_j^+, $j = 1, \dots, k_+$ (ψ_j^-, $j = 1, \dots, k_-$, resp.). Moreover, the Wronskian of

$$\{\psi_1^+, \dots \psi_{k_+}^+, \psi_1^-, \dots, \psi_{k_-}^-\}$$

is nowhere 0. Since y_+ (y_-, resp.) is a linear combination of ψ_j^+, $j = 1, \dots, k_+$ (ψ_j^-, $j = 1, \dots, k_-$, respectively) the continuity of $y^{(m)}$ for $m = 0, 1, \dots, k_+ + k_- - 1$ at 0 implies $y_+ = 0$ and $y_- = 0$. Hence $y = 0$. Since A is selfadjoint in $\mathcal{K}$ it cannot have real numbers in residual spectrum.

(c) We use I. M. Glazman's decomposition method. Define $A_\pm$ in $L^2(\mathbb{R}_\pm)$ by $\mathcal{D}(A_\pm) = H^{2n}(\mathbb{R}_\pm) \cap H_0^n(\mathbb{R}_\pm)$ and $A_\pm y = \pm p(D)y$, $y \in \mathcal{D}(A_\pm)$. The operator A_- (A_+, respectively) is a selfadjoint operator in $L^2(\mathbb{R}_-)$ ($L^2(\mathbb{R}_+)$, resp.). The continuous spectrum of A_- (A_+, respectively) is $(-\infty, -m_p]$ ($[m_p, +\infty)$, resp.). The operator $A_- \oplus A_+$ is selfadjoint in $L^2(\mathbb{R})$ and its continuous spectrum is the union of the continuous spectra of A_- and A_+. The operators A and $A_- \oplus A_+$ have the same continuous spectrum. Therefore, by (b), $\sigma(A) = \sigma_c(A) = \sigma_c(A_-) \cup \sigma_c(A_+)$. □

THEOREM 2.3 *Let p be a nonnegative polynomial. Let $A = Jp(D)$.*

(a) *The operator A is a positive definitizable operator.*

(b) *The point ∞ is a regular critical point of A.*

(c) *The point 0 is a critical point of A if and only if $0 \in \sigma(A)$, or equivalently, if and only if $m_p = \min\{p(x) | x \in \mathbb{R}\} = 0$.*

PROOF (a) The definitizability of the positive operator A follows from Theorem 2.2.

The positivity of A and the equality (8) imply the statement (c) and the fact that ∞ is a critical point of $Jp(D)$.

Since the operators $A = Jp(D)$ and JD^{2n} are definitizable the operator D satisfies all the assumptions for S in Theorem 1.4 (a). By [5], ∞ is not a singular critical point of JD^2. By Example 1 the functions $h = p$ and $g(t) = t^{2n}$ satisfy the conditions of Lemma 1.3 (a). Therefore we can apply Theorem 1.4 (a) to conclude that ∞ is not a singular critical point of A. □

It follows from Theorem 2.3 that A is similar to a selfadjoint operator in $L^2(\mathbb{R})$ if $m_p > 0$. The same is true if $m_p = 0$ and 0 is a regular critical point of A. In the next theorem we give a sufficient condition for p under which 0 is a regular critical point of A.

Let a be an arbitrary real number. Denote by $V(a)$ the multiplication operator on $L^2(\mathbb{R})$ defined by $(V(a)f)(x) = e^{iax}f(x)$, $x \in \mathbb{R}$. Simple calculations show that the following proposition holds.

PROPOSITION 2.4 *The operators JD^{2n} and $J(D + aI)^{2n}$ are similar:*

$$V(a)^{-1}JD^{2n}V(a) = J(D + aI)^{2n}.$$

THEOREM 2.5 *Let p be a nonnegative polynomial with exactly one real root. Then 0 is a regular critical point of $A = Jp(D)$. The operator A is similar to a selfadjoint operator in $L^2(\mathbb{R})$.*

PROOF Let a be the single real root of p. By Proposition 2.4 the operators JD^2 and $J(D - aI)^2$ are similar. Therefore the operator $J(D - aI)^2$ is similar to a selfadjoint operator in $L^2(\mathbb{R})$. Put $S = D - aI$ and $q(x) = p(x + a)$.

Then q satisfies all the assumptions for h in Corollary 1.5 and $Jq(S) = Jp(D)$. Since JS^2 is similar to a selfadjoint operator in $L^2(\mathbb{R})$, Corollary 1.5 implies that $Jq(S) = Jp(D)$ is similar to a selfadjoint operator in $L^2(\mathbb{R})$. □

3 Half-range Completeness

Let A be a positive operator in the Krein space $\mathcal{K} = (L^2(\mathbb{R}), [\cdot|\cdot])$. Assume that A has a nonempty resolvent set. Let $\mathcal{K}_\pm$ be the set of all functions f in $L^2(\mathbb{R})$ which vanish on the set $\mathbb{R}_\mp$. Then $\mathcal{K} = \mathcal{K}_+ \oplus \mathcal{K}_-$ is a fundamental decomposition of $\mathcal{K}$.

Assume that neither 0 nor ∞ are singular critical points of A. Let E be the spectral function of A. Then the operator A is a selfadjoint operator in the Hilbert space $(\mathcal{K}, [(E(\mathbb{R}_+)-E(\mathbb{R}_-))\cdot, \cdot])$; see [9, Theorem 5.7]. The corresponding fundamental decomposition is $\mathcal{K} = \mathcal{L}_+ \oplus \mathcal{L}_-$, where $\mathcal{L}_\pm = E(\mathbb{R}_\pm)\mathcal{K}$. This fundamental decomposition reduces A. Let $P_\pm$ be the orthogonal projection in $\mathcal{K}$ to $\mathcal{K}_\pm$. Then the restriction

$$T_\pm := P_\pm|_{\mathcal{L}_\pm} : \mathcal{L}_\pm \longrightarrow \mathcal{K}_\pm$$

is a bounded and boundedly invertible bijection of $\mathcal{L}_\pm$ onto $\mathcal{K}_\pm$. Let $f_\pm \in \mathcal{K}_\pm$. Then $T_\pm^{-1} f_\pm \in \mathcal{L}_\pm$. Therefore

$$T_\pm^{-1} f_\pm = \int_{\mathbb{R}_\pm} dE(t) T_\pm^{-1} f_\pm \ .$$

Since $P_\pm$ is continuous we get

$$f_\pm = \int_{\mathbb{R}_\pm} dP_\pm E(t) T_\pm^{-1} f_\pm = \int_{\mathbb{R}_\pm} dF_\pm(t) f_\pm \ ,$$

where $F_\pm(\imath) = P_\pm E(\imath) T_\pm^{-1}$, for $\imath$ an open interval in $\mathbb{R}_\pm$. Then $F_\pm$ is a projection valued measure on $\mathbb{R}_\pm$.

We have proved that the elements $f_\pm$ from $\mathcal{K}_\pm$ can be represented as integrals over $\mathbb{R}_\pm$ with respect to the measure $F_\pm(\cdot) f_\pm$ which is obtained by orthogonally projecting the spectral measure $E(\cdot) T_\pm^{-1} f_\pm$ onto $\mathcal{K}_\pm$. This is exactly the continuous analogue of the familiar concept of half-range completeness property in the discrete spectrum case; see [1].

This property holds in particular for the operators from Theorem 2.5.

References

[1] Beals, R.: Indefinite Sturm-Liouville problems and half-range completeness. J. Differential Equations 56 (1985), 391-407.

[2] Ćurgus, B.: On the regularity of the critical point infinity of definitizable operators. Integral Equations Operator Theory 8 (1985), 462-488.

[3] Ćurgus, B., Langer, H.: A Krein space approach to symmetric ordinary differential operators with an indefinite weight function. J. Differential Equations 79 (1989), 31-61.

[4] Ćurgus, B., Najman, B.: A Krein space approach to elliptic eigenvalue problems with indefinite weights. Differential and Integral Equations 7 (1994), 1241-1252.

[5] Ćurgus, B., Najman, B.: The operator $(\operatorname{sgn} x)\frac{d^2}{dx^2}$ is similar to a selfadjoint operator in $L^2(\mathbb{R})$. Proc. Amer. Math. Soc. 123 (1995), 1125-1128.

[6] Ćurgus, B., Najman, B.: Differential operators in Krein space $L^2(\mathbb{R}^n)$. Preprint.

[7] Jonas, P.: Compact perturbations of definitizable operators. II. J. Operator Theory 8 (1982), 3-18.

[8] Jonas, P.: On a problem of the perturbation theory of selfadjoint operators in Krein spaces. J. Operator Theory 25(1991),183-211.

[9] Langer, H.: Spectral function of definitizable operators in Krein spaces. Functional Analysis, Proceedings, Dubrovnik 1981. Lecture Notes in Mathematics 948, Springer-Verlag, Berlin, 1982, 1-46.

[10] Veselić, K.: On spectral properties of a class of J-selfadjoint operators. I. Glasnik Mat. Ser. III 7(27) (1972), 229-248.

[11] Weidmann, J.: Linear Operators in Hilbert Spaces. Springer-Verlag, Berlin, 1980.

B. Ćurgus
Department of Mathematics,
Western Washington University,
Bellingham, WA 98225, USA
curgus@cc.wwu.edu

B. Najman
Department of Mathematics,
University of Zagreb,
Bijenička 30, 41000 Zagreb, Croatia
najman@cromath.math.hr

Mathematical Reviews 1991 Mathematics Subject Classification 47B50 47E05

Operator Theory:
Advances and Applications, Vol. 87

ELLIPTIC PROBLEMS INVOLVING AN INDEFINITE WEIGHT FUNCTION

M. Faierman and H. Langer

We consider an elliptic boundary value problem defined on a region $\Omega \subset \mathbb{R}^n$ and involving an indefinite weight function ω. We also suppose that the problem under consideration admits a variational formulation. Then by appealing to the theory of selfadjoint operators acting in a Krein space, we derive various spectral properties for the problem. In particular, when Ω is bounded we show that the principal vectors of our problem form a Riesz basis in $L^2(\Omega^\dagger;\ |\omega(x)|dx)$, where $\Omega^\dagger = \{x \in \Omega | \omega(x) \neq 0\}$, and also establish some results concerning their half-range completeness.

1. INTRODUCTION

The object of this paper is to generalize some recent results given in [8], [18] concerning the spectral theory of an elliptic boundary value problem involving an indefinite weight function for the case where the problem admits a variational formulation, and also to extend some of the results given in [9]. Accordingly, we will be concerned here with a selfadjoint elliptic eigenvalue problem, which is derived formally from a variational problem, of the form

$$Lu = \lambda\omega(x)u \quad \text{in} \quad \Omega, \tag{1.1}$$

$$B_j u = 0 \quad \text{on } \Gamma \text{ for } j = 1, \ldots, m \text{ if } \Omega \neq \mathbb{R}^n, \tag{1.2}$$

where L is a linear elliptic operator of order $2m$ defined in a non-empty region $\Omega \subset \mathbb{R}^n$, $n \geq 2$, having boundary Γ if $\Omega \neq \mathbb{R}^n$, the B_j are linear differential operators defined on Γ, and ω is a real-valued measurable function defined in Ω which assumes both positive and negative values. The formulation of the problem will be made precise in §2.

When Ω is bounded, it was shown in [9] that under certain assumptions concerning Ω, ω, and the operators involved, the principal vectors of the problem (1.1–2) are complete in a certain function space determined by Ω and the weight function $\omega(x)$; we will now show that under some further restrictions, they actually form a Riesz basis in that space. We will also derive some results concerning the half-range completeness of certain systems of principal vectors of the problem (1.1–2) and derive as well analogous results for the case where Ω is not bounded.

Elliptic eigenvalue problems of the form (1.1–2), but not necessarily admitting a variational formulation, were also considered in [8] and [18]. Under the supposition that ω and ω^{-1} are essentially bounded in Ω (as well as under other assumptions), these authors were able to establish important information concerning the spectral theory for the problem (1.1–2) by appealing to results from interpolation theory. We will show that when the problem under consideration admits a variational formulation, analogous results can be established without the restriction that $\omega^{-1} \in L^\infty(\Omega)$ (however, we do require certain assumptions concerning the behaviour of ω and these will be presented below). In fact our theory includes the case where ω vanishes in a set of positive measure (as in [9]). As examples we might mention that when $\Omega = \mathbb{R}^n$, our theory covers those cases where: (i) ω is positive on one side of a hyperplane in $\mathbb{R}^n$ and negative on the other side, (ii) Ω is positive in the interior of a hypersphere in $\mathbb{R}^n$ and negative in the exterior (or vice-versa), and (iii) ω is positive in one of the three regions in $\mathbb{R}^n$ determined by two parallel, but distinct hyperplanes, negative in one of the two remaining regions, and identically zero in the remaining region, provided that for case (i), (ii), or (iii) our assumptions concerning the behaviour of ω in a neighbourhood of the hyperplane, of the hypersphere, or of each of the hyperplanes, respectively, as well as our remaining assumptions, are satisfied.

Our method for dealing with the above problem will be via Krein space theory, that is, we will reduce the eigenvalue problem for the system (1.1–2) to one for a selfadjoint operator acting in a Krein space, and then we will employ the results of [7] and [14] to arrive at our final results. However, in order to arrive at a position where we can use the results of [7] and [14], some intermediate steps are required. To explain the first of these steps, let us mention that in the sequel we shall show that there is associated with the operator L and, if $\Omega \neq \mathbb{R}^n$, the boundary conditions (1.2) a symmetric sesquilinear form $B(\,.\,,\,.\,)$ defined on a certain Sobolev space V. If A denotes the selfadjoint operator in $L^2(\Omega)$ associated with the form B, then our method requires that when $0 \in \sigma(A)$ we decompose V into a direct sum of a finite dimensional space and a space in which we can treat our problem by means of Krein space theory. Thus our first step is involved with showing that such a decomposition is possible, and this of course requires that we impose certain restrictions upon the operator L and the boundary conditions (1.2) (see Assumptions 2.1–2). The second intermediate step occurs in the proof of Theorem 3.1 and involves showing that two positive definite inner products defined in a vector space determined by V and the weight function ω induce equivalent norms on this space, and here we make use of some ideas of [5]. Of course, in order to use the ideas of [5] we have had to introduce certain assumptions concerning the behaviour of $\omega(x)$ (see conditions (1) – (7) concerning $\omega(x)$ given in §3).

Finally, in §2 we make precise the meaning of the eigenvalue problem (1.1–2), introduce our basic definitions as well as those assumptions which we require in order to successfully deal with those problems arising in the first intermediate step cited above, and then introduce the selfadjoint operator S, whose spectral properties are connected with those of the problem (1.1–2), acting in a certain Krein space (see Theorems 2.3–4). In §3 we introduce the aforementioned restrictions on ω which we require for a successful resolution of the problems arising in the second intermediate step cited above, and then show that under these restrictions the operator S is actually fundamentally reducible (see Theorem 3.1). This last fact, together with the known results, enable us immediately to arrive at our main results (see Corollaries 3.1–3). Lastly, in §4, we compare our results with those of [8] and [18].

2. PRELIMINARIES

We let $x = (x_1, \ldots, x_n)$ denote a generic point in $\mathbb{R}^n$, $n \geq 2$, and use the notation $D_j = \partial/\partial x_j$, $D = (D_1, \ldots, D_n)$, $D^\alpha = D_1^{\alpha_1} \cdots D_n^{\alpha_n}$, where α stands for a multi-index whose length $\sum_{j=1}^n \alpha_j$ is denoted by $|\alpha|$. For G an open set in $\mathbb{R}^n$, we let $H^k(G)$, $k \in \mathbb{N} \cup \{0\}$, denote the usual Sobolev space of order k related to $L^2(G)$ and denote by $(\,.\,,\,.\,)_{k,G}$ and $\|\;\|_{k,G}$ the inner product and norm, respectively, in this space. Now let Ω denote a fixed, non-empty region in $\mathbb{R}^n$ and let us firstly focus our attention upon the case where $\Omega \neq \mathbb{R}^n$. Then for this case we shall henceforth suppose that Ω is of class C^m, $m \in \mathbb{N}$, let Γ denote the boundary of Ω, and let V denote a closed subspace of $H^m(\Omega)$ containing $C_0^\infty(\Omega)$ determined by a system of boundary operators (possibly void) of the form $B_j(x, D) = \sum_{|\alpha| \leq m_j} b_\alpha^j(x) D^\alpha$ for $j = 1, \ldots, p \leq m$, where $0 \leq m_j < m$ and the coefficients $b_\alpha^j(x)$ have continuous bounded derivatives on Γ up to order $m - m_j$. We will denote this system by $\{B_j\}_1^p$, $0 \leq p \leq m$, where $p = 0$ means that the system is void, and suppose from now on that if $p > 0$, then the system is normal, i.e., (i) the orders m_j of the operators are all distinct, and (ii) Γ is non-characteristic to B_j, $j = 1, \ldots, p$, at each point (we will also see in a moment precisely how this system is connected with the boundary conditions (1.2)). If $\Omega = \mathbb{R}^n$, then we let $V = H^m(\mathbb{R}^n)$. In $\mathcal{H} = L^2(\Omega)$ we now introduce the symmetric sesquilinear form

$$a(u, v) = \int_\Omega \sum_{|\alpha|, |\beta| \leq m} a_{\alpha\beta}(x) D^\beta u \overline{D^\alpha v} dx$$

with domain V, where the $a_{\alpha\beta}$ are complex-valued functions in $L^\infty(\Omega)$ satisfying $\overline{a_{\alpha\beta}} = a_{\beta\alpha}$, $a_{\alpha\beta}$ is uniformly continuous in Ω if $|\alpha| = |\beta| = m$, and where it is also supposed that $a(u, v)$ is coercive over V, i.e., there exist constants $c_0 > 0$ and $c_1 \geq 0$ such that $a(u, u) \geq c_0 \|u\|_{m,\Omega}^2 - c_1 \|u\|_{0,\Omega}^2$ for every $u \in V$ (we refer to [1], [2] and [9] for sufficient conditions for this to be the case when Ω is bounded). Note also that $|a(u, v)| \leq c_2 \|u\|_{m,\Omega} \|v\|_{m,\Omega}$, where c_2 denotes a positive constant. If $\Omega \neq \mathbb{R}^n$, and $p < m$, then we shall also consider a second symmetric sesquilinear form $a^\#(u, v)$ with domain V which we define as follows. Let $F_j(x, D) = \sum_{|\alpha| \leq \mu_j} f_\alpha^j(x) D^\alpha$, $0 \leq \mu_j < m$, $j = (p+1), \ldots, m$, be a system of boundary operators which together with the $B_j(x, D)$ form a Dirichlet system of order m on Γ, where the f_α^j are complex-valued functions of class $L^\infty(\Gamma)$ and where by a Dirichlet system of order m on Γ we mean a normal system of m boundary operators on Γ each of order less than m. Then we put

$$a^\#(u, v) = \int_\Gamma \sum_{j=p+1}^m \phi_j F_j u \overline{F_j v} d\sigma \quad \text{for } u, v \in V,$$

where the ϕ_j denote real-valued functions in $L^\infty(\Gamma)$, σ denotes surface measure on Γ, and, referring to (1.43) of [13, p.319] for terminology, we suppose henceforth that the form $a^\#$ is relatively bounded with respect to the form a, with a-bound less than c_0/c_2 (it is clear that this condition is always satisfied if Ω is bounded).

NOTATION. In the sequel $B(u, v)$ will denote either the sesquilinear form $a(u, v)$ or the sesquilinear form $a(u, v) + a^\#(u, v)$.

It follows immediately that B is a densely defined symmetric form in $\mathcal{H} = L^2(\Omega)$ which is coercive and continuous on V and it is easy to show that B is closed. We henceforth let γ denote the lower bound of B and A the selfadjoint operator in $\mathcal{H}$ that is associated with B.

ASSUMPTION 2.1. If $\gamma \leq 0$, then we suppose from now on that there exists a $\delta > 0$ such that $\sigma(A) \cap [\gamma, \delta)$ consists of only a finite number of points and that these points are eigenvalues of A of finite multiplicity (note that this assumption is always satisfied if Ω is bounded).

Next we observe that if we identify $\mathcal{H}$ with its antidual, then the inclusion $V \subset \mathcal{H}$ leads naturally to the inclusion $V \subset \mathcal{H} \subset V'$ and the inner product $(.,.)_{0,\Omega}$ extends to a pairing of V and its antidual V': $(u,v)_{0,\Omega}$, $u \in V'$, $v \in V$, and where V' is the completion of $\mathcal{H}$ with respect to the norm $\|u\|_{-1} = \sup|(u,v)_{0,\Omega}|$ and the supremum is taken over the set $v \in V$ for which $\|v\|_{m,\Omega} \leq 1$. Thus we have $B(u,v) = (Lu,v)_{0,\Omega}$ for $u, v \in V$, where $L \in \mathcal{L}(V,V')$, and observing that $V' \subset \mathcal{D}'(\Omega)$ (the space of distributions on Ω), we see that Lu is the distribution $\sum_{|\alpha|,|\beta| \leq m} (-1)^{|\alpha|} D^\alpha a_{\alpha\beta}(x) D^\beta u$. Finally, if we let $D(A)$ denote the domain of A, then it is clear that $D(A) = \{u \in V \big| Lu \in \mathcal{H}\}$ and $Au = Lu$ for $u \in D(A)$.

Let $\omega(x)$ denote a real-valued measurable function defined on Ω which assumes both positive and negative values and let

$$\Omega^+ = \{x \in \Omega | \omega(x) > 0\},\ \Omega^- = \{x \in \Omega | \omega(x) < 0\},\ \text{ and }\ \Omega^0 = \{x \in \Omega | \omega(x) = 0\}.$$

Then we suppose from now on that $|\Omega^\pm| > 0$ and $|\overline{\Omega^\pm} \setminus \operatorname{int} \Omega^\pm| = 0$, where $|\ |$ denotes n-dimensional Lebesgue measure, $^-$ denotes closure, and int = interior. If $\Omega_1 = \operatorname{int} \Omega^+ \cup \operatorname{int} \Omega^-$, then we also assume that $\omega \in L^1_{\mathrm{loc}}(\Omega_1)$, from which it follows that $C_0^\infty(\Omega_1)$ is dense in $\mathcal{H}_{|\omega|} = L^2(\Omega_1; |\omega(x)|dx)$. If T denotes the operator of multiplication in $\mathcal{H}$ induced by $\omega(x)$ (observe that T is selfadjoint), then we suppose that $V \hookrightarrow D(|T|^{1/2})$, where this last space is equipped with its graph norm and $\hookrightarrow$ denotes continuous imbedding. We observe from [18] that the mapping T: $D(T) \to V'$ extends by continuity to a mapping of $D(|T|^{1/2})$ into V', and hence as a mapping from V into V'. We shall again denote these extensions by T.

We are now in a position to make precise the definition of the eigenvalue problem under consideration here. Accordingly, if we firstly suppose that Ω is bounded, then we define our problem as follows: determine pairs $\{\lambda, u\}$, where $\lambda \in \mathbb{C}$ and $0 \neq u \in V$, for which

$$B(u,v) = \lambda(\omega u, v)_{0,\Omega} \text{ for every } v \in V. \tag{2.1}$$

However we know from [9] (see also §4 below) that when Ω, the $a_{\alpha\beta}$, and b^j_α, the f^j_α, and the ϕ_j are sufficiently smooth, and under suitable conditions on ω, the variational problem (2.1) corresponds to the selfadjoint regular elliptic boundary value problem (1.1–2), where for $p < m$, the B_j, $j = (p+1), \ldots, m$, are a complementary system of boundary operators of $\{B_j\}_1^p$ relative to B (see [1] for terminology and note also that if $B(u,v) = a(u,v) + a^\#(u,v)$, then we are to take $\{B_j\}_1^p$, $\{F_j\}_{p+1}^m$ as the Dirichlet system of order m used in determining the B_j, $j = (p+1), \ldots, m$). Thus in spite of the fact that we shall not assume at this stage the regularity required to arrive at (1.1–2), we shall still say that (1.1–2) is the elliptic eigenvalue problem that is associated with the (coercive) variational problem (2.1) (see [15, p.203]). On the other hand, in view of our above discussion and the fact that V' is isometrically isomorphic to V, it is clear that the spectral problem for (2.1), and hence for (1.1–2), is equivalent to the spectral problem for the pencil $\mathcal{L}(\lambda) = L - \lambda T$ acting from V to V' (in dealing with this pencil we shall use the terminology usually employed when the pencil acts from V to V — see [16]). Similarly, when Ω is unbounded, then the spectral problem for (1.1–2) is to be interpreted as that for the pencil $\mathcal{L}(\lambda)$. Thus in the sequel, when we refer to various spectral properties of the problem (1.1–2), then this is to be considered only in a formal sense (unless otherwise stated), and is to be interpreted as that for the pencil $\mathcal{L}(\lambda)$. Consequently, for the remainder of this section we will fix our attention upon the spectral theory for the pencil $\mathcal{L}(\lambda)$ and establish some results which will be used in §3 to arrive at our main results for the problem (1.1–2). Finally, let us note that $\rho(\mathcal{L})$ is an open subset of $\mathbb{C}$, and hence $\sigma(\mathcal{L})$ is closed.

If $0 \in \rho(A)$, then it is not difficult to verify that $0 \in \rho(\mathcal{L})$, and hence let us now fix our attention upon the case $0 \in \sigma(A)$ and let $N_0 = \ker A$. Then it is clear that $\ker L = N_0$, and hence $\mu = 0$ is an eigenvalue of $\mathcal{L}(\lambda)$. In order to proceed further with this case we require some further terminology.

NOTATION. For X a non-empty subset of V (resp. V') we let X^0 (resp. ${}^0X$) denote the set of all $u \in V'$ (resp. the set of all $u \in V$) such that $(u,v)_{0,\Omega} = 0$ (resp. $(v,u)_{0,\Omega} = 0$) for every $v \in X$. Also if μ is an eigenvalue of the pencil $\mathcal{L}(\lambda)$, then we henceforth denote by M_μ the span of the eigenvectors and associated vectors of $\mathcal{L}(\lambda)$ for the eigenvalue μ.

Continuing with our discussion concerning the case $0 \in \sigma(A)$, it is of importance for the success of our method to ensure that M_0 and TM_0 are of finite and equal dimension and that $M_0 \cap {}^0(TM_0) = 0$. In order to ensure that this will be the case, we require

ASSUMPTION 2.2. We suppose from now on that if $|\Omega^0| > 0$, $0 \in \sigma(A)$, and $N_0 \cap {}^0(TN_0) \neq 0$, then L has the unique continuation property, i.e., if $u \in V$, if $Lu = 0$ in the sense of distributions on Ω, and if u vanishes almost everywhere in a non-empty open subset of Ω, then $u = 0$ (we refer to [9], [11, Chapter 8], [12], and [19, Chapter 3] for sufficient conditions for this property to hold).

THEOREM 2.1. *$0 \in \sigma(\mathcal{L})$ if and only if $0 \in \sigma(A)$. Furthermore, if $0 \in \sigma(\mathcal{L})$, then* $\dim M_0 = \dim TM_0 < \infty$ *and* $M_0 \cap {}^0(TM_0) = 0$.

PROOF. The first assertion of the theorem follows from what has already been said. Let us now fix our attention upon the case $0 \in \sigma(\mathcal{L})$. Then before passing to the proof of the remaining assertions, let us make the following observations. Firstly, by appealing to the fact that $R(A)$ is closed in $\mathcal{H}$, where R denotes range, it is not difficult to verify that $R(L)$ is closed in V'.

Secondly, if Φ denotes the isometric isomorphism cited above between V' and V, then we may appeal to the fact that ΦL is a bounded selfadjoint operator in V with kernel N_0 to show that $(N_0)^0 = R(L)$. Thirdly, by appealing to the facts that $(Tu,v)_{0,\Omega} = \int_\Omega \omega u\bar{v}dx$ for $u, v \in V$ and that $C_0^\infty(\Omega_1)$ is dense in $\mathcal{H}_{|\omega|}$, we can readily verify that if $u \in V$ and $(Tu,\phi)_{0,\Omega} = 0$ for every $\phi \in C_0^\infty(\Omega_1)$, then $u = 0$ almost everywhere in Ω_1.

In light of these observations we can now argue in a manner somewhat similar to that in [9, §2] or [17, §4] to prove the remaining assertions of the theorem. ∎

Assuming still that $0 \in \sigma(A)$, it follows from Theorem 2.1 that

$$V = M_0 \dotplus {}^0(TM_0), \qquad V' = TM_0 \dotplus (M_0)^0, \tag{2.2}$$

where $\dotplus$ denotes direct sum. Moreover, if we let $V_0 = {}^0(TM_0)$, $L_0 = L|V_0$, and $T_0 = T|V_0$, then it is easy to show that L_0 maps V_0 isomorphically onto $(M_0)^0$ and that $R(T_0) \subset (M_0)^0$. Hence in V_0 we can now introduce the bounded operator $K_0 = L_0^{-1}T_0$.

NOTATION. If $0 \in \rho(A)$, then let us henceforth write V_0 for V, $(M_0)^0$ for V', L_0 for L, T_0 for T, and let $K_0 = L^{-1}T$.

Recalling that $\Omega_1 = \text{int } \Omega^+ \cup \text{ int } \Omega^-$, we now let $\mathcal{R}$ denote the restriction operator mapping $\mathcal{H}$ onto $L^2(\Omega_1)$ defined by $\mathcal{R}u = u|\Omega_1$, let $V^\dagger = \mathcal{R}V_0$, and equip $V^\dagger$ with the norm $\|u\|_{V^\dagger} = \inf \|v\|_{m,\Omega}$, where the infimum is taken over all $v \in V_0$ for which $\mathcal{R}v = u$ (it is not difficult to verify that $V^\dagger$, equipped with this norm, is a Banach space — see the proof of Theorem 2.2 below). It is clear that $V^\dagger \hookrightarrow \mathcal{H}_{|\omega|}$, while if $0 \in \rho(A)$, then the imbedding is dense. On the other hand, if $0 \in \sigma(A)$, then it is easy to show that the mapping $\mathcal{R}\colon M_0 \to \mathcal{R}M_0$ is a bijection, and if we equip $\mathcal{R}M_0$ with the norm $\|u\|_{\mathcal{R}M_0} = \|u^\wedge\|_{m,\Omega}$, where $\mathcal{R}u^\wedge = u$, then $\mathcal{R}M_0 \hookrightarrow \mathcal{H}_{|\omega|}$. Furthermore, $\mathcal{R}M_0$ and $V^\dagger$ are linearly independent subspaces of $\mathcal{H}_{|\omega|}$ whose direct sum is continuously and densely imbedded in $\mathcal{H}_{|\omega|}$ (the norm on the direct sum is defined in an obvious way). We henceforth let V_1 denote the closure of $V^\dagger$ in $\mathcal{H}_{|\omega|}$, so that $\mathcal{H}_{|\omega|} = \mathcal{M}_0 \dotplus V_1$, where $\mathcal{M}_0 = \mathcal{R}M_0$ if $0 \in \sigma(A)$ and is zero otherwise.

Let $(.,.)$ denote the inner product in $\mathcal{H}_{|\omega|}$. Then when $\mathcal{H}_{|\omega|}$ is considered only as a vector space, we may also introduce into this space a second inner product $[u,v] = \int_{\Omega_1} \omega u\bar{v}dx$, and we shall denote $\mathcal{H}_{|\omega|}$, equipped with this inner product, by $\mathcal{H}_\omega$. It is clear that $\mathcal{H}_\omega$ is a Krein space whose rank of indefiniteness is infinite. It is also clear that when V_1, considered only as a vector space, is equipped with the inner product $[.,.]$, then it becomes a Krein space whose rank of indefiniteness is infinite, and we have

$$\mathcal{H}_\omega = \mathcal{M}_0[\dotplus]V_1, \tag{2.3}$$

where $\mathcal{M}_0$ is non-degenerate and $[\dotplus]$ denotes orthogonal direct sum in $\mathcal{H}_\omega$. In the sequel we shall often write $(V_1,(.,.))$ (resp. $(V_1,[.,.])$) when we wish to consider V_1 as a Hilbert space (resp. Krein space).

We now introduce in $V^\dagger$ the bounded invertible operator $K^\dagger$ defined by $K^\dagger u = \mathcal{R}K_0 v$ for $v \in V_0$ with $\mathcal{R}v = u$, and it is clear that $[K^\dagger u, v] = [u, K^\dagger v]$ for u, $v \in V^\dagger$. Then guided by future requirements we are now going to show the connection between the operator $K^\dagger$ and the problem (1.1–2), and to this end we require

NOTATION. If S is a linear operator in the Banach space X and μ is an eigenvalue of S, then we let $\mathcal{G}_\mu(S, X)$ denote the principal subspace for the eigenvalue μ of S.

At times in the sequel we shall speak of a point $\mu \in \mathbb{C}$ as being in the point, continuous, or residual spectrum of $\mathcal{L}(\lambda)$ and by this we mean that $\mathcal{L}(\mu)$ is not invertible, $\mathcal{L}(\mu)$ is invertible and its range is dense, or $\mathcal{L}(\mu)$ is invertible and its range is not dense, respectively.

THEOREM 2.2. *$0 \neq \mu \in \sigma(\mathcal{L})$ if and only if $1/\mu \in \sigma(K^\dagger)$. Furthermore, $\mu \neq 0$ is in the point, continuous, or residual spectrum of $\mathcal{L}(\lambda)$ if and only if $1/\mu$ is in the point, continuous, or residual spectrum of $K^\dagger$, respectively, and if $\mu \in \sigma_p(\mathcal{L}) \cup \sigma_r(\mathcal{L})$, then $R(\mathcal{L}(\mu))$ is closed in V' if and only if $R(K^\dagger - \mu^{-1} I)$ is closed in $V^\dagger$, where R denotes range. Finally, if $\mu \neq 0$ is an eigenvalue of $\mathcal{L}(\lambda)$, then $\mathcal{R}$ maps M_μ injectively onto $\mathcal{G}_{1/\mu}(K^\dagger, V^\dagger)$.*

PROOF. If $0 \in \sigma(A)$, then it is easy to show that $(M_0)^0$ is isomorphic to the antidual of V_0, and hence it makes sense to consider the spectral problem for the pencil $\mathcal{L}_0(\lambda) = L_0 - \lambda T_0$ acting from V_0 to $(M_0)^0$. Moreover, since it is not difficult to verify that for $\mu \neq 0$, $\mathcal{L}(\mu)$ maps M_0 injectively onto TM_0, we see that we need only prove the theorem under the modification that $\mathcal{L}(\lambda)$ has been replaced throughout by $\mathcal{L}_0(\lambda)$ and V' by $(M_0)^0$.

Fixing our attention firstly upon the pencil $\mathcal{L}_0(\lambda)$, direct calculations show that all of the assertions of the theorem (in the modified form just cited), except the last one, are certainly true when $K^\dagger$ and $V^\dagger$ there are replaced by K_0 and V_0, respectively, and that if $\mu \neq 0$ is an eigenvalue of $\mathcal{L}_0(\lambda)$, then $M_\mu = \mathcal{G}_{1/\mu}(K_0, V_0)$ (see the proof of Theorem 3.4 of [10]).

Suppose next that $|\Omega^0| > 0$ and let us fix our attention upon the operator K_0. Recalling from above that the mapping $T\colon V \to V'$ is continuous, let $Z = \ker(T|V)$. Then it follows from (2.2) that $Z \subset V_0$, and hence $Z = \ker T_0$. Let π denote the quotient map of V_0 onto V_0/Z (we equip this latter space with the quotient norm), and in V_0/Z introduce the bounded operator $K_0^\sim$ by putting $K_0^\sim \pi(u) = \pi(K_0 u)$ for $u \in V_0$. Then it follows from direct calculations that $0 \neq \mu \in \sigma(K_0)$ if and only if $\mu \in \sigma(K_0^\sim)$, that $\mu \neq 0$ is in the point, continuous, or residual spectrum of K_0 if and only if μ is in the point, continuous, or residual spectrum of $K_0^\sim$, respectively, that for $0 \neq \mu \in \sigma_p(K_0) \cup \sigma_r(K_0)$, $R(K_0 - \mu I)$ is closed in V_0 if and only if $R(K_0^\sim - \mu I)$ is closed in V_0/Z, and lastly that for $\mu \neq 0$ an eigenvalue of K_0, $\mathcal{G}_\mu(K_0^\sim, V_0/Z) = \pi(\mathcal{G}_\mu(K_0, V_0))$. On the other hand, since $\mathcal{R}$ induces an isometric isomorphism, say $\mathcal{R}_\pi$, between V_0/Z and $V^\dagger$ and since $K^\dagger = \mathcal{R}_\pi K_0^\sim \mathcal{R}_\pi^{-1}$, it follows immediately that $0 \neq \mu \in \sigma(K_0)$ if and only if $\mu \in \sigma(K^\dagger)$, that $\mu \neq 0$ is in the point, continuous, or residual spectrum of K_0 if and only if μ is in the point, continuous, or residual spectrum of $K^\dagger$, respectively, that for $0 \neq \mu \in \sigma_p(K_0) \cup \sigma_r(K_0)$, $R(K_0 - \mu I)$ is closed in V_0 if and only if $R(K^\dagger - \mu I)$ is closed in $V^\dagger$, and lastly that for $\mu \neq 0$ an eigenvalue of K_0, $\mathcal{R}$

maps $\mathcal{G}_\mu(K_0, V_0)$ injectively onto $\mathcal{G}_\mu(K^\dagger, V^\dagger)$.

In light of these results, all the assertions of the theorem now follow. ∎

Let $u \in V^\dagger$, let $v \in V_0$ with $\mathcal{R}v = u$, and let $\| \ \|$ denote the norm in $\mathcal{H}_{|\omega|}$. Then it follows from the foregoing results and [18] that $\|K^\dagger u\|^2 = \| |T|^{1/2} K_0 v\|^2_{0,\Omega} \le c_0 \|K_0 v\|^2_{m,\Omega} \le c_1 \|T_0 v\|^2_{-1} \le c_2 \| |T|^{1/2} v\|^2_{0,\Omega} = c_2 \|u\|^2$, where the c_j denote the positive constants and we refer to the paragraph following Assumption 2.1 for terminology, and hence we conclude that $K^\dagger$ is a bounded operator in $V^\dagger$ when $V^\dagger$ is considered as a subspace of $\mathcal{H}_{|\omega|}$ equipped with the relative topology. It follows immediately that $K^\dagger$ extends by continuity to a bounded selfadjoint operator in $(V_1, [\cdot, \cdot])$ which will be denoted by K, and it is not difficult to verify that K is invertible.

Let $S = K^{-1}$. Then

THEOREM 2.3. *S is a boundedly invertible selfadjoint operator in $(V_1, [\cdot, \cdot])$ which is positive when $\gamma \ge 0$, but is not positive when $\gamma < 0$, $|\Omega^0| = 0$, and either $0 \in \rho(A)$ or $0 \in \sigma(A)$ and $N_0 \cap {}^0(TN_0) = 0$. Furthermore, even if S is not positive it is still definitizable.*

REMARK 2.1. In the course of proving the theorem we shall give some indication of why we are unable to establish the positivity or non-positivity of S for the cases: (i) $\gamma < 0$ and $|\Omega^0| > 0$ and (ii) $\gamma < 0$, $|\Omega^0| = 0$, $0 \in \sigma(A)$, and $N_0 \cap {}^0(TN_0) \ne 0$.

PROOF OF THEOREM 2.3. That S is a boundedly invertible selfadjoint operator in $(V_1, [\cdot, \cdot])$ follows immediately from the above results. Turning to the assertions concerning the positivity or non-positivity of S made in the first sentence of the theorem, let us firstly fix our attention upon the case $\gamma < 0$, $0 \in \rho(A)$, and suppose that S is positive. Thus if we let $u_0 \in D(S)$ such that $u_0 = Ku_1$ for $u_1 \in V^\dagger$ and choose $v_1 \in V_0$ such that $\mathcal{R}v_1 = u_1$, then it follows that $B(z_1, z_1) \ge 0$, where $z_1 = L_0^{-1} T_0 v_1$. However if $|\Omega^0| = 0$, then it is easy to show that $L_0^{-1} T_0 V_0$ is dense in V_0, and hence the assumption that S is positive leads to the contradiction that $B(u, u) \ge 0$ for every $u \in V_0$. On the other hand (and bearing in mind Remark 2.1), if $|\Omega^0| > 0$, then it is easy to show that $L_0^{-1} T_0 V_0$ is not dense in V_0, and hence the assumption that S is positive leads to the conclusion that $B(u, u) \ge 0$ for every u belonging to a closed proper subspace of V_0. Consequently, if $|\Omega^0| > 0$, then it is possible that $[Su, u] \ge 0$ for every $u \in D(S)$. We have thus shown that when $\gamma < 0$, $0 \in \rho(A)$, and $|\Omega^0| = 0$, then S can not be positive, and the remaining assertions concerning the positivity or non-positivity of S can be established using similar arguments.

Turning to the final assertion of the theorem, we need only fix our attention upon the case where S is not a positive operator and show that the form $[S\cdot, \cdot]$, with domain $D(S)$, has a finite number of negative squares (see [14]). Accordingly, let $\mathcal{M}$ be a negative definite subpsace of $D(S)$ with respect to the inner product $[S\cdot, \cdot]$ and put $\mathcal{N} = S\mathcal{M}$. Then $\mathcal{N}$ is a subspace of V_1 which is negative definite with respect to the inner product $[K\cdot, \cdot]$. Let $\{w_j\}_1^p$, $p \in \mathbb{N}$, denote a system of vectors in $\mathcal{N}$ satisfying $[Kw_j, w_k] = -\delta_{jk} (\delta_{jk} =$ Kronecker delta) and let $\| \ \|$ denote the norm in $\mathcal{H}_{|\omega|}$. Then for $\epsilon > 0$ and sufficiently small, it is not difficult to verify that there exists the linearly independent set of vectors in $V^\dagger$, $\{u_j\}_1^p$, such that $\|w_j - u_j\| < \epsilon$ for $j = 1, \ldots, p$, and $\mathcal{N}_1 =$ span $\{u_j\}_1^p$ is negative definite with respect to the inner product $[K\,.\,,\,.]$. Hence if $z_j = L_0^{-1}T_0 v_j$ for $j = 1, \ldots, p$, where $v_j \in V_0$ with $\mathcal{R}v_j = u_j$, then it follows that $\{z_j\}_1^p$ is a linearly independent set of vectors in V_0 whose span $\mathcal{N}_2$ is negative definite with respect to the inner product $B(\,.\,,\,.\,)$. On the other hand, if ℓ^- denotes the number of negative eigenvalues of A counted according to multiplicity, then we know from [9] that $p \leq \ell^-$, and thus it follows that $\mathcal{N}$, and hence $\mathcal{M}$, is finite dimensional with dimension not exceeding ℓ^-. ∎

Guided by future requirements we are now going to show the connection between the operator S and the problem (1.1–2) at least under certain restrictions.

THEOREM 2.4. *Suppose that $V^\dagger$, when considered only as a vector space, is also endowed with a positive definite inner product $\langle\,.\,,\,.\,\rangle$ having the following properties: (1) the norm induced in $V^\dagger$ by $\langle\,.\,,\,.\,\rangle$ is equivalent to the norm $\|\ \|_{V^\dagger}$, (2) $\langle K^\dagger u, v\rangle = \langle u, K^\dagger v\rangle$ for u, $v \in V^\dagger$, (3) $\langle |K^\dagger|\,.\,,\,.\,\rangle$ and $(\,.\,,\,.\,)$ induce equivalent norms on $V^\dagger$, where $|K^\dagger|$ denotes the modulus of $K^\dagger$ with respect to the inner product $\langle\,.\,,\,.\,\rangle$, and (4) $\langle K^\dagger u, v\rangle = [u, v]$ for u, $v \in V^\dagger$. Suppose in addition that S is fundamentally reducible in $(V_1, [\,.\,,\,.\,])$. Then the non-zero spectrum of $\mathcal{L}(\lambda)$ is situated on the real axis and is precisely the spectrum of S. Furthermore, $\sigma_r(\mathcal{L}) = \sigma_r(S) = \emptyset$. Also, $\mu \neq 0$ is in the point or continuous spectrum of $\mathcal{L}(\lambda)$ if and only if μ is in the point or continuous spectrum of S, respectively, and if $\mu \in \sigma_p(\mathcal{L})$, then $R(\mathcal{L}(\mu))$ is closed in V' if and only if $R(S)$ is closed in V_1, where R denotes range. Finally, if $\mu \neq 0$ is an eigenvalue of $\mathcal{L}(\lambda)$, then every chain in M_μ has length 1, μ is a semi-simple eigenvalue of S, and $\mathcal{R}$ maps M_μ injectively onto a subspace of $\mathcal{G}_\mu(S, V_1)$ whose closure in V_1 is precisely $\mathcal{G}_\mu(S, V_1)$.*

PROOF. Bearing in mind Theorem 2.2, let us firstly show that $0 \neq \mu \in \rho(K)$ if and only if $\mu \in \rho(K^\dagger)$. Accordingly, suppose that $0 \neq \mu \in \rho(K^\dagger)$ and let $v \in V_1$. Then there exists the sequence $\{u_j\}_1^\infty$ in $V^\dagger$ such that $(K^\dagger - \mu I)u_j \to v$ in V_1. Thus $\{(K^\dagger - \mu I)u_j\}_1^\infty$ is a Cauchy sequence in V_1 and so $\{|K^\dagger|^{1/2}(K^\dagger - \mu I)u_j\}_1^\infty$ is a Cauchy sequence in $V^\dagger$. It follows immediately that $\{|K^\dagger|^{1/2}u_j\}_1^\infty$ is a Cauchy sequence in $V^\dagger$, and so $\{u_j\}_1^\infty$ is a Cauchy sequence in V_1. Thus there is a $u \in V_1$ such that $(K - \mu I)u = v$. Let us suppose next that $0 \neq u \in V_1$ and $(K - \mu I)u = 0$. Then there exists the sequence $\{u_j\}_1^\infty$ in $V^\dagger$ such that $\{(K - \mu I)u_j\}_1^\infty$ is a null sequence in V_1. Thus $\{|K^\dagger|^{1/2}(K^\dagger - \mu I)u_j\}_1^\infty$, and hence $\{|K^\dagger|^{1/2}u_j\}_1^\infty$, is a null sequence in $V^\dagger$. Consequently we arrive at the contradiction that $\{u_j\}$ is a null sequence in V_1.

Conversely, suppose that $0 \neq \mu \in \rho(K)$ and let $v \in V^\dagger$. Then since $\ker(K^\dagger - \mu I) = 0$, $R(K^\dagger - \mu I)$ is dense in $V^\dagger$, and so there is a sequence $\{u_j\}_1^\infty$ in $V^\dagger$ such that $(K^\dagger - \mu I)u_j \to v$ in $V^\dagger$, and hence in V_1. Thus $\{u_j\}_1^\infty$ is a Cauchy sequence in V_1 and so $\{|K^\dagger|^{1/2}u_j\}_1^\infty$, and hence $\{K^\dagger u_j\}_1^\infty$, is a Cauchy sequence in $V^\dagger$. If z denotes the limit of this latter sequence, then it follows that $(K^\dagger - \mu I)z = K^\dagger v$. Hence $z \in R(K^\dagger)$, and so if $z = K^\dagger u$, then $(K^\dagger - \mu I)u = v$.

We conclude from the foregoing results that $\mu \in \sigma(S)$ if and only if $1/\mu \in \sigma(K^\dagger)$, and hence in light of Theorem 2.2 and [6, Corollary 6.2, p.133] and bearing in mind that $\sigma(K^\dagger) \subset \mathbb{R}^1$

and $\sigma_r(K^\dagger) = \emptyset$, the first three assertions of the theorem follow immediately.

Similar arguments show that $\mu \neq 0$ is in the point or continuous spectrum of $K^\dagger$ if and only if μ is in the point or continuous spectrum of K, respectively, and if $\mu \in \sigma_p(K^\dagger)$, then $R(K^\dagger - \mu I)$ is closed in $V^\dagger$ if and only if $R(K - \mu I)$ is closed in V_1. In light of Theorem 2.2, all but the final assertions of the theorem now follow.

Turning to the final assertions of the theorem, it is not difficult to verify that the operator $|K^\dagger|^{1/2}$ extends by continuity to a bounded operator in V_1, and if we denote this extension by $\mathcal{K}$, then $\mathcal{K}$ is an invertible selfadjoint operator in $(V_1, [\,.\,,\,.\,])$ which commutes with K and such that $R(\mathcal{K}) \subset V^\dagger$. Furthermore, if $\mu \neq 0$ is an eigenvalue of K, then clearly $\ker(K^\dagger - \mu I) \subset \ker(K - \mu I)$, while direct calculations show that $\mathcal{K}(\ker(K - \mu I)) \subset \ker(K^\dagger - \mu I)$. Still fixing our attention upon this eigenvalue μ, it is obvious that $\mathcal{G}_\mu(K^\dagger, V^\dagger) = \ker(K^\dagger - \mu I)$, and we now assert that $\mathcal{G}_\mu(K, V_1) = \ker(K - \mu I)$. To see this suppose that $0 \neq u_0$ and u_1 are vectors in V_1 such that $(K - \mu I)u_0 = 0$ and $(K - \mu I)u_1 = u_0$. Then $(K - \mu I)^2 \mathcal{K} u_1 = 0$, and hence

$$0 = \left[(K - \mu I)\mathcal{K}u_1, (K - \mu I)\mathcal{K}u_1\right] = \left\langle K^\dagger(K^\dagger - \mu I)\mathcal{K}u_1,\, K^\dagger(K^\dagger - \mu I)\mathcal{K}u_1\right\rangle = \mu\langle \mathcal{K}u_0, \mathcal{K}u_0\rangle.$$

Since this implies that $u_0 = 0$, we arrive at a contradiction, and hence the assertion is proved. Moreover, we also assert that $\mathcal{G}_\mu(K, V_1)$ is precisely the closure in V_1 of $\mathcal{G}_\mu(K^\dagger, V^\dagger)$. Indeed, if this is not the case, then there is a $v \neq 0$ in $\mathcal{G}_\mu(K, V_1)$ such that $(u, v) = 0$ for every $u \subset \mathcal{G}_\mu(K^\dagger, V^\dagger)$, and hence if J denotes the fundamental symmetry in $(V_1, [\,.\,,\,.\,])$ which commutes with S and if we observe that $Jv \in \mathcal{G}_\mu(K, V_1)$, then it follows that $0 = \mu^{-1}[K^\dagger u, Jv] = |\mu|^{-1}[u, \mathcal{K}^2 Jv] = |\mu|^{-1}\langle K^\dagger u, \mathcal{K}^2 Jv\rangle = \operatorname{sgn}\mu\langle u, \mathcal{K}^2 Jv\rangle$. Since $\mathcal{K}^2 Jv \in \mathcal{G}_\mu(K^\dagger, v^\dagger)$, we arrive at the contradiction that $v = 0$, and hence the assertion is proved. In light of these results and those of Theorem 2.2, the proof of the theorem is complete. ∎

REMARK 2.2. Under the hypotheses of Theorem 2.4, we see that the spectral problem for (1.1-2) has been reduced to that for S.

3. MAIN RESULTS

In this section we shall introduce some further assumptions concerning the form B and the weight function ω and under these assumptions we shall, with the aid of the results of §2, establish the main results of this paper. Accordingly, let us again return to the space V_0 introduced above and let $B_0 = B|V_0$. Then we know from [9] that if $\gamma \geq 0$, then B_0 is positive definite on V_0, that if $\gamma < 0$ and $0 \in \rho(A)$ or if $0 \in \sigma(A)$ and $N_0 \cap {}^0(TN_0) = 0$, then B_0 is indefinite on V_0, and that if $\gamma < 0$, $0 \in \sigma(A)$, and $N_0 \cap {}^0(TN_0) \neq 0$, then B_0 is either positive definite or indefinite on V_0. Furthermore, we know from [9] that if B_0 is positive definite on V_0, then the inner products $B_0(\,.\,,\,.\,)$ and $(\,.\,,\,.\,)_{m,\Omega}$ induce equivalent norms on V_0.

We are now going to impose some additional assumptions on B and ω. To begin with we suppose henceforth that B_0 is positive definite on V_0, that $\omega \in L^\infty(\Omega)$, and if $|\Omega^0| > 0$, then we let $\Omega_0 = \Omega \backslash \overline{\Omega_1}$ and suppose that ω has been modified on a set of measure zero, if necessary, so that $\omega(x) = 0$ for $x \in \Omega_0$. Next we require some further assumptions concerning ω which pertain to the geometric nature of the sets $\Omega^{\pm}$, Ω_0, and hence we suppose from now on that:

(1) $\operatorname{int}\Omega^+$, $\operatorname{int}\Omega^-$, and Ω_0, if $|\Omega^0| > 0$, are the unions of the non-empty disjoint regions $\{\Omega_r^+\}_1^{n^+}$, $\{\Omega_r^-\}_1^{n^-}$, and $\{\Omega_r^0\}_1^{n^0}$, respectively, where the $n^{\pm}$ or n^0 are either natural numbers or infinity, but we only permit n^+, n^-, or n^0 to be infinity if Ω^+, Ω^-, or Ω^0, respectively, is unbounded;

(2) each component Γ_{rj}^+ (resp. Γ_{rj}^-) of $\partial\Omega_r^+$ (resp. $\partial\Omega_r^-$), where $\partial =$ boundary, is either a component of Γ or is contained in Ω and is either a component of $\partial\Omega_s^+$ (resp. $\partial\Omega_s^-$) for some $s \neq r$, or a component of $\partial\Omega_s^-$ (resp. $\partial\Omega_s^+$) for some s, or a component of $\partial\Omega_s^0$ for some s if $|\Omega^0| > 0$ (note from the construction that if $|\Omega^0| > 0$, then any component of $\partial\Omega_r^0$ is either a component of Γ or lies in Ω and is a component of $\partial\Omega_s^{\pm}$ for some s);

(3) if a Γ_{rj}^+ is also a component of $\partial\Omega_s^-$ for some s, then Γ_{rj}^+ is of class $C^{m,1}$ and there is a $d_0 > 0$, not depending upon r, j, nor s, such that for any $x^0 \in \Gamma_{rj}^+$, and with $\nu(x^0)$ denoting the interior normal to Γ_{rj}^+ at x^0, we have

$$\left\{x \in \mathbb{R}^n \middle| x = x^0 + t\nu(x^0),\ 0 < t < 2d_0\right\} \subset \Omega_r^+, \tag{3.1}$$

$$\left\{x \in \mathbb{R}^n \middle| x = x^0 + t\nu(x^0),\ -2d_0 < t < 0\right\} \subset \Omega_s^-; \tag{3.2}$$

(4) if $|\Omega^0| > 0$ and a Γ_{rj}^+ is also a component of $\partial\Omega_s^0$ for some s, then Γ_{rj}^+ is of class $C^{m,1}$ and for $x^0 \in \Gamma_{rj}^+$ we have (3.1) and

$$\left\{x \in \mathbb{R}^n \middle| x = x^0 + t\nu(x^0),\ -2d_0 < t < 0\right\} \subset \Omega_s^0, \tag{3.3}$$

where $\nu(x^0)$ and d_0 are defined above;

(5) if $|\Omega^0| > 0$ and a Γ_{rj}^- is also a component of $\partial\Omega_s^0$ for some s, then Γ_{rj}^- is of class $C^{m,1}$ and for $x^0 \in \Gamma_{rj}^-$ we have (3.1), with Ω_r^+ replaced by Ω_r^-, and (3.3), where now $\nu(x^0)$ denotes the interior normal to Γ_{rj}^- at x^0 and d_0 is defined above.

We also require an additional assumption of an analytic nature concerning ω, and accordingly we suppose from now on that:

(6) if a Γ_{rj}^+ is also a component of $\partial\Omega_s^-$ for some s, and, bearing in mind (3.1–2), if we let $U_{rj} = \left\{x \in \mathbb{R}^n \middle| d_{rj}(x) = \operatorname{dist}\{x, \Gamma_{rj}^+\} < d_0\right\}$, then we also have

$$\omega(x) = \big(\operatorname{sgn}\,\omega(x)\big) d_{rj}(x)^{\sigma_{rj}} g_{rj}(x) \quad \text{for} \quad x \in U_{rj},$$

where g_{rj} is of class $C^{m-1,1}$ in U_{rj} and in this set $g_{rj}(x) \geq M^{-1}$, $|D^\alpha g_{rj}(x)| \leq M$ for $|\alpha| \leq m-1$, $|D^\alpha g_{rj}(x^\dagger) - D^\alpha g_{rj}(x^\#)|/|x^\dagger - x^\#| \leq M$ for $|\alpha| = m-1$ and $x^\dagger \neq x^\#$, while $0 \leq \sigma_{rj} \leq M$, and where M denotes a positive constant not depending upon r, j, nor s.

Finally, we remark at this point that the last of our assumptions concerning ω will be given in the following paragraph.

Let $\Gamma_{rj}^{\pm}$ be any one of the manifolds considered in conditions (3), (4), or (5) above and let $x^0 \in \Gamma_{rj}^{\pm}$. Then by hypothesis there is an open set $U \subset \mathbb{R}^n$ and a real-valued function ϕ of

$n-1$ variables such that the following conditions hold: **(i)** There is a Cartesian coordinate system $(y', y_n) = (y_1, \ldots, y_n)$ in $\mathbb{R}^n$ about x^0, where the y_n-axis is directed along the inward normal to $\Gamma_{rj}^{\pm}$ at x^0 (i.e., pointing into $\Omega_r^{\pm}$) and the $y_1-, \ldots, y_{n-1}$-axes lie in the tanget plane to $\Gamma_{rj}^{\pm}$ at x^0 such that $U = \left\{(y', y_n) | y' \in U', \left|y_n - \phi(y')\right| < \rho_1\right\}$, where U' is the open ball $|y'| < \rho_0$ and ρ_0, ρ_1 are positive constants not exceeding d_0; **(ii)** $\phi \in C^{m,1}(U')$, and **(iii)** $U \cap \Omega_r^{\pm} = \left\{(y', y_n) \in U | y_n > \phi(y')\right\}$, $U \cap \Gamma_{rj}^{\pm} = \left\{(y', y_n) \in U | y_n = \phi(y')\right\}$, and $U \cap \Omega_s^-$ or $U \cap \Omega_s^0$ is the set $\{(y', y_n) \in U | y_n < \phi(y')\}$. We call U a neighbourhood and (y', y_n) a system of coordinates connected with the point x^0 and we suppose henceforth (this is the last of our assumptions concerning ω) that:

(7) there are positive constants d_1 and M_1, not depending upon x^0 nor $\Gamma_{rj}^{\pm}$, such that ρ_0 and ρ_1 are not less than d_1, $|D^\alpha \phi| \le M_1$ in U' for $|\alpha| \le m$, and $\left|D^\alpha \phi(y_1') - D^\alpha \phi(y_2')\right| / |y_1' - y_2'| \le M_1$ for $|\alpha| = m$ and $y_1' \ne y_1'$.

Now let us relable the $y_1, \ldots, y_{n-1}$ coordinates by $\eta_1, \ldots, \eta_{n-1}$, respectively, let $\eta' = (\eta_1, \ldots, \eta_{n-1})$, denote that portion of $\Gamma_{rj}^{\pm}$ described by $(y', \phi(y'))$, $|y'| < \rho_0$, by $(\eta', \theta(\eta'))$, $|\eta'| < \rho_0$, and let $\nu(\eta')$ denote the interior normal to $\Gamma_{rj}^{\pm}$ at $(\eta', \theta(\eta'))$. If we now let $\mathcal{U} = \left\{\eta = (\eta', \eta_n) \in \mathbb{R}^n \middle| |\eta'| < \rho_0/2, |\eta_n| < \rho_2\right\}$, where $\rho_2 < d_1$ is a certain constant depending only upon M_1, n, d_0, and d_1, then $\mathcal{U}$ is diffeomorphic to a subset $U^{\#}$ of U under the mapping $y = (y_1, \ldots, y_n) = (\eta', \theta(\eta')) + \eta_n \nu(\eta')$. From now on we will refer to (η', η_n), as just defined, as local coordinates of $\Gamma_{rj}^{\pm}$ at the point x^0. Furthermore, observing that in $U^{\#}$ the x_j and y_j coordinates are connected by an affine transformation, we henceforth let Φ denote the diffeomorphism: $\mathcal{U} \to U^{\#} \subset \mathbb{R}_x^n$ (this means $\mathbb{R}^n$ in the x_j coordinates) induced by the transformations just cited and let $J(\eta)$ denote the Jacobian determinant of the transformation Φ. Clearly there is no loss of generality in assuming henceforth that $J(\eta) > 0$ for $\eta \in \mathcal{U}$.

Finally, for $0 < d < (1/2n^{1/2}) \min\{d_1/2, \rho_2\}$ and $\alpha \in \mathbb{Z}^n$ let $Q_\alpha = \left\{x \in \mathbb{R}^n \middle| d(\alpha_j - 1) < x_j < d(\alpha_j + 1) \text{ for } j = 1, \ldots, n\right\}$ and put $Q = \left\{x \in \mathbb{R}^n \middle| |x_j| < 1 \text{ for } j = 1, \ldots, n\right\}$. Let $\psi \in C_0^\infty(Q)$ such that $0 \le \psi \le 1$, $\psi(x) = 1$ for $x \in \frac{1}{2}Q$, supp $\psi \subset \frac{3}{4}Q$, where supp = support, and for $\alpha \in \mathbb{Z}^n$ let $\psi_\alpha(x) = \psi((x - \alpha d)/d) / \sum_{\beta \in \mathbb{Z}^n} \psi((x - \beta d)/d)$. Then $\{Q_\alpha\}_{\alpha \in \mathbb{Z}^n}$ is a locally finite covering of $\mathbb{R}^n$ (note that each $x \in \mathbb{R}^n$ lies in at most 2^n of the Q_α) and $\{\psi_\alpha\}_{\alpha \in \mathbb{Z}^n}$ is a partition of unity subordinate to this covering.

In light of these further assumptions concerning B and ω, we can now show that

THEOREM 3.1. *S is fundamentally reducible in $(V_1, [\cdot, \cdot])$.*

PROOF. It is clear that V_0 is a Hilbert space with respect to the inner product $B_0(\cdot, \cdot)$ and for the rest of the proof we suppose that V_0 is equipped with this inner product. Now let us fix our attention upon the case $|\Omega^0| > 0$ and let $Z = \ker T \cap V$. Then it follows from (2.2) that $Z \subset V_0$ and we let $Z^\perp$ denote the orthogonal complement of Z in V_0. It is not difficult to verify that $Z^\perp$ is isomorphic to $V^\dagger$ and is invariant under K_0, while $K_0 | Z^\perp$ is a bounded, invertible selfadjoint operator whose numerical range contains both positive and negative values and which is compact if Ω is bounded. Hence if $u \in V^\dagger$ and $u^\wedge$ denotes the vector in $Z^\perp$ for which $\mathcal{R}u^\wedge = u$, then we see that the topology of $V^\dagger$ can be defined by the Hilbert topology induced by the inner product $\langle u,v\rangle = B_0(u^\wedge,v^\wedge)$, and if we denote $V^\dagger$, equipped with this topology, by $\big(V^\dagger,\langle\,.\,,\,.\,\rangle\big)$ and bear in mind that $(K^\dagger u)^\wedge = K_0 u^\wedge$, then it follows from what we have said above that $K^\dagger$ is a bounded, invertible selfadjoint operator in $\big(V^\dagger,\langle\,.\,,\,.\,\rangle\big)$ whose numerical range contains both positive and negative values and which is compact if Ω is bounded. Let us also observe for later use that if T_1 denotes the operator of multiplication in $L^2(\Omega_1)$ induced by $\omega|\Omega_1$, then

$$
\begin{aligned}
&D\big((K^\dagger)^{-1}\big) = \Big\{u \in V^\dagger \big| L_0 u^\wedge \in \mathcal{H},\, L_0 u^\wedge = 0 \ \text{ in } \ L^2(\Omega\backslash\Omega_1),\, \mathcal{R}L_0 u^\wedge \in T_1(V^\dagger)\Big\}\\
&\text{and}\\
&(K^\dagger)^{-1}u = T_1^{-1}\mathcal{R}L_0 u^\wedge \ \text{ for } \ u \in D\big((K^\dagger)^{-1}\big).
\end{aligned}
\tag{3.4}
$$

Finally, analogous results hold if $|\Omega^0| = 0$; in this case we are to take $Z^\perp = V_0$.

If $|K^\dagger|$ denotes the modulus of $K^\dagger$ (with respect to $(V^\dagger,\langle\,.\,,\,.\,\rangle)$), then let us show that $(\,.\,,\,.\,)$ and $\langle|K^\dagger|\,.\,,\,.\,\rangle$ induce equivalent norms on $V^\dagger$ when this latter space is considered only as a vector space. To this end we need only prove, in light of the proof of Proposition 1 of [5] and the fact that the assumptions hitherto made concerning ω are also assumptions concerning $-\omega$, that there exist the mappings X_1, Y_1 in $\mathcal{L}(\mathcal{H})$ which are continuous from V to V and satisfy $X_1 u = u$ in $L^2(\operatorname{int}\Omega^+)$, $Y_1^\star u = u$ in $L^2(\Omega\backslash\operatorname{int}\Omega^-)$, and $|T|X_1 u = Y_1^\star T u$ in $\mathcal{H}$, where $Y_1^\star$ denotes the adjoint of Y_1 in $\mathcal{H}$ and $|T|$ denotes the absolute value of T (note that in using the arguments given in the proof just cited, we are to take H, H_A, and T there to be $L^2(\Omega_1)$, $(V^\dagger,\langle\,.\,,\,.\,\rangle)$, and T_1, respectively, let $Xu = \mathcal{R}X_1 u^\wedge$ and $Yu = \mathcal{R}Y_1 u^\wedge$ for $u \in V^\dagger$, let $(u_1+u_2)^\wedge = u_1^\wedge + u_2^\wedge$ for $u_1 \in \mathcal{M}_0$, $u_2 \in V^\dagger$ (see (2.2–3)), and make use of (3.4)). Accordingly, letting $X_1 u = u$ in $\operatorname{int}\Omega^+$ and $Y_1^\star u = u$ in $\operatorname{int}\Omega^+$ if $|\Omega^0| = 0$, $Y_1^\star u = u$ in $\operatorname{int}\Omega^+ \cup \Omega_0$ if $|\Omega^0| > 0$ for $u \in \mathcal{H}$, we are now going to show how to extend $X_1 u$ and $Y_1^\star u$ to all of Ω so that the above requirements are fulfilled; and with $Y_1^\star$ thus defined, we can then obtain Y_1 by duality. Hence to begin with, let us firstly suppose that $|\Omega^0| = 0$ and fix our attention upon an Ω_s^-. If no component of $\partial\Omega_s^-$ coincides with a Γ_{rj}^+ for any r, j, then let us put $X_1 u = Y_1^\star u = 0$ in Ω_s^-. On the other hand, suppose that there is a component of $\partial\Omega_s^-$, say Γ_{sk}^- which coincides with a Γ_{rj}^+. Let $\{Q_\alpha\}_{\alpha\in I_k}$ denote those members of the family $\{Q_\alpha\}_{\alpha\in\mathbb{Z}^n}$ having non-empty intersection with Γ_{sk}^-, where I_k denotes an index set, and let $u_\alpha(x) = \psi_\alpha(x)u(x)$ for $\alpha \in I_k$. Fixing our attention upon a particular $\alpha \in I_k$, let $x_\alpha^0 \in \Gamma_{sk}^- \cap Q_\alpha$. Then passing to local coordinates at x_α^0, letting $\eta^\alpha = (\eta^{\alpha\prime},\eta_n^\alpha)$, $\mathcal{U}_\alpha$, $U_\alpha^\#$, Φ_α, and $J_\alpha(\eta^\alpha)$ denote the analogues of $\eta = (\eta',\eta_n)$, $\mathcal{U}$, $U^\#$, Φ, and $J(\eta)$, respectively, when x^0 is replaced by x_α^0 (see above), and bearing in mind assumptions (3) and (6) above concerning ω, let $v_\alpha(\eta^\alpha) = u_\alpha\big(x(\eta^\alpha)\big)$ and $g_\alpha(\eta^\alpha) = g_{rj}\big(x(\eta^\alpha)\big)$ for $\eta \in \mathcal{U}_\alpha$, $\sigma = \sigma_{rj}$, and let us suppose that $X_1 u_\alpha$ and $Y_1^\star u_\alpha$ go over to $X_\alpha v_\alpha$ and $Y_\alpha^\star v_\alpha$, respectively, in this transformation. Guided by the above cited requirements for X_1 and $Y_1^\star$ and putting $v_\alpha(\eta^\alpha) = 0$, $J_\alpha(\eta^{\alpha\prime},-\eta_n^\alpha/p)J_\alpha(\eta^\alpha)^{-1} = g_\alpha(\eta^\alpha)g_\alpha(\eta^{\alpha\prime},-\eta_n^\alpha/p)^{-1} = 0$

in $\mathbb{R}^n_{\eta^\alpha} \setminus \mathcal{U}_\alpha$ for $p = 1, \ldots, 2m$, we set

$$(X_\alpha v_\alpha)(\eta^\alpha) = (Y_\alpha^\star v_\alpha)(\eta^\alpha) = v_\alpha(\eta^\alpha) \quad \text{for} \quad \eta_n^\alpha > 0,$$

$$(X_\alpha v_\alpha)(\eta^\alpha) = \chi(\eta_n^\alpha) J_\alpha(\eta^\alpha)^{-1} \sum_{p=1}^{2m} c_p (v_\alpha J_\alpha)(\eta^{\alpha\prime}, -\eta_n^\alpha/p) \quad \text{for} \quad \eta_n^\alpha < 0,$$

$$(Y_\alpha^\star v_\alpha)(\eta^\alpha) = -\chi(\eta_n^\alpha) g_\alpha(\eta^\alpha) J_\alpha(\eta^\alpha)^{-1} \sum_{p=1}^{2m} p^\sigma c_p (v_\alpha J_\alpha / g_\alpha)(\eta^{\alpha\prime}, -\eta_n^\alpha/p) \quad \text{for} \quad \eta_n^\alpha < 0,$$

where $\chi(t) \in C_0^\infty(\mathbb{R}^1)$, $0 \leq \chi \leq 1$, $\chi(t) = 1$ for $|t| < d/16$, $\chi(t) = 0$ for $|t| > d/8$, and the c_p are constants which we now determine. Indeed, the requirements that $Y_1^\star$ be the adjoint of Y_1 in $\mathcal{H}$ and that X_1 and Y_1 map V continuously into V lead to the equations (see below as well as the proofs of Lemma 1 of [5] and Lemma 3.1 of [18]):

$$\sum_{p=1}^{2m} p^j c_p = (-1)^j \quad \text{for} \quad j = (-m+1), \ldots, 0,$$

$$\sum_{p=1}^{2m} p^{\sigma + j} c_p = (-1)^{j+1} \quad \text{for} \quad j = 1, \ldots, m,$$

which determine the c_p uniquely, and hence we have now defined $X_\alpha v_\alpha$ and $Y_\alpha^\star v_\alpha$ in all of $\mathbb{R}^n_{\eta^\alpha}$. If $(X_\alpha v_\alpha)^\#$ and $(Y_\alpha^\star v_\alpha)^\#$ denote the pullbacks of $X_\alpha v_\alpha$ and $Y_\alpha^\star v_\alpha$, respectively, to $\mathbb{R}^n_x$ (by this we mean that $(X_\alpha v_\alpha)^\# | U_\alpha^\#$ and $(Y_\alpha^\star v_\alpha)^\# | U_\alpha^\#$ are the pullbacks of $X_\alpha v_\alpha | \mathcal{U}_\alpha$ and $Y_\alpha^\star v_\alpha | \mathcal{U}_\alpha$, respectively, to $U_\alpha^\#$ and $(X_\alpha v_\alpha)^\#$, $(Y_\alpha^\star v_\alpha)^\#$ are defined to be zero in $\mathbb{R}^n_x \setminus U_\alpha^\#$) and if we define $X_1 u$ and $Y_1^\star u$ in Ω_s^- by putting here $X_1 u = \sum_k \sum_{\alpha \in I_k} (X_\alpha v_\alpha)^\# | \Omega_s^-$, $Y_1^\star u = \sum_k \sum_{\alpha \in I_k} (Y_\alpha^\star v_\alpha)^\# | \Omega_s^-$, where k runs over that set of integers for which the component Γ_{sk}^- of $\partial\Omega_s^-$ coincides with a component of $\partial\Omega_p^+$ for some p, then we claim that this gives the required extensions of $X_1 u$ and $Y_1^\star u$ to Ω_s^-. A repetition of the above arguments to all of the components of $\operatorname{int}\Omega^-$ we claim gives the required result. Indeed, to justify these claims, it is clear from the construction that the only assertion that remains to be proved is that Y_1 maps V continuously into itself. With this in mind, let us observe from the construction that $Y_1 u = u$ in $L^2(\Omega^+(d))$ and that $Y_1 u = 0$ in $L^2(\Omega^-(d))$, where $\Omega^+(d) = \{x \in \operatorname{int}\Omega^+ | \operatorname{dist}\{x, \operatorname{int}\Omega^-\} \geq d/8\}$, $\Omega^-(d) = \{x \in \operatorname{int}\Omega^- | \operatorname{dist}\{x, \operatorname{int}\Omega^+\} \geq d/8\}$. Now let us again fix our attention upon the case where the component Γ_{sk}^- of $\partial\Omega_s^-$ coincides with a Γ_{rj}^+, and bearing in mind assumption (6) above concerning ω, let $U_{rj}^d = \{x \in U_{rj} | d_{rj} < d/4\}$. Then with x_α^0 as above, let $U_{\alpha,d}^\# = U_\alpha^\# \cap U_{rj}^d$ and let $f \in C_0^0(U_{\alpha,d}^\#)$. Then for $u \in \mathcal{H}$ we have

$$(3.5) \qquad (u, Y_1^\star f)_{0,\Omega} = \sum_{\beta \in I_k} \left[(u, \psi_\beta f)_{0, \Omega_r^+ \cap U_{\alpha,\beta}^\#} + (u, (Y_\beta^\star h_\beta)^\#)_{0, \Omega_s^- \cap U_{\alpha,\beta}^\#} \right],$$

where I_k is defined above, $h_\beta(\eta^\beta) = (\psi_\beta f)(x(\eta^\beta))$, $U_{\alpha,\beta}^\# = U_{\alpha,d}^\# \cap U_\beta^\#$, and the summation is over those $\beta \in I_k$ for which supp $\psi_\beta \cap$ supp $f \neq \emptyset$. Hence the number of non-zero terms appearing in the summation does not exceed a fixed positive number not depending upon α, r, j, s, nor k.

Furthermore, if we pass to local coordinates and appeal to the definition of $Y_\beta^\star h_\beta$ given above, then it follows from (3.5) that

$$(u, Y_1^\star f)_{0,\Omega} = \sum_{\beta \in I_k} (Z_\beta, h_\beta J_\beta)_{0,\left[\Phi_\beta^{-1}(U^\#_{\alpha,\beta})\right]_+} \tag{3.6}$$

where

$$Z_\beta(\eta^\beta) = g_\beta(\eta^\beta)^{-1}\Big[(g_\beta u_\beta)(\eta^\beta) - \sum_{p=1}^{2m} p^{\sigma+1} c_p \chi(-p\eta_n^\beta)(g_\beta u_\beta)(\eta^{\beta\prime}, -p\eta_n^\beta)\Big],$$

$u_\beta(\eta^\beta) = u\big(x(\eta^\beta)\big)$, and $\left[\Phi_\beta^{-1}(U^\#_{\alpha,\beta})\right]_+ = \left\{\eta^\beta \in \Phi_\beta^{-1}(U^\#_{\alpha,\beta}) | 0 < \eta_n^\beta < d/4\right\}$. On the other hand, since

$$\begin{aligned}(Z_\beta, h_\beta J_\beta)_{0,\left[\Phi_\beta^{-1}(U^\#_{\alpha,\beta})\right]_+} &= \int\limits_{\left[\Phi_\alpha^{-1}(U^\#_{\alpha,\beta}\right]_+} Z_\alpha(\eta^\alpha)(\psi_\beta \bar f)\big(x(\eta^\alpha)\big) J_\alpha(\eta^\alpha) d\eta^\alpha \\ &= \int\limits_{\mathcal{U}^d_{\alpha,+}} Z_\alpha(\eta^\alpha)(\psi_\beta \bar f)\big(x(\eta^\alpha)\big) J_\alpha(\eta^\alpha) d\eta^\alpha,\end{aligned}$$

where, with $\mathcal{U}_\alpha^d = \{\eta^\alpha \in \mathcal{U}_\alpha | |\eta_n^\alpha| < d/4\}$, $\mathcal{U}^d_{\alpha,+} = \{\eta^\alpha \in \mathcal{U}_\alpha^d | \eta_n^\alpha > 0\}$ and $\bar f$ denotes the complex conjugate of f, it follows from (3.6) that

$$(u, Y_1^\star f)_{0,\Omega} = \int\limits_{\mathcal{U}^d_{\alpha,+}} Z_\alpha(\eta^\alpha) \bar f\big(x(\eta^\alpha)\big) J_\alpha(\eta^\alpha) d\eta^\alpha. \tag{3.7}$$

Similarly we can show that

$$(Y_1 u, f)_{0,\Omega} = \int\limits_{\mathcal{U}^d_\alpha} (Y_1 u)\big(x(\eta^\alpha)\big) \bar f\big(x(\eta^\alpha)\big) J_\alpha(\eta^\alpha) d\eta^\alpha. \tag{3.8}$$

It follows from (3.7–8) that almost everywhere in $\mathcal{U}_\alpha^d$, $(Y_1 u)\big(x(\eta^\alpha)\big) = Z_\alpha(\eta^\alpha)$ for $\eta_n^\alpha > 0$, $(Y_1 u)\big(x(\eta^\alpha)\big) = 0$ for $\eta_n^\alpha < 0$. Hence, in view of the definitions of the c_p above, we conclude that if $u \in V$, then $Y_1 u \in H^m(U^\#_{\alpha,d})$ and $\|Y_1 u\|_{m,U^\#_{\alpha,d}} \le c\|u\|_{m,U^\#_{\alpha,d}}$, where the constant c does not depend upon α, r, j, s, nor k. Moreover, since $Y_1 u | U_{rj}^d = \sum_{\beta \in I_k} \psi_\beta (Y_1 u | U^\#_{\beta,d})$, it follows from what we have just shown and the properties of the ψ_β cited above that $Y_1 u \in H^m(U_{rj}^d)$ and $\|Y_1 u\|_{m,U^d_{r,j}} \le c\|u\|_{m,U^d_{rj}}$, where the constant c does not depend upon r, j, s, nor k. Our assertion concerning the mapping $Y_1 : V \to V$ now follows.

Suppose next that $|\Omega^0| > 0$ and let us fix our attention upon an Ω_s^-. If no component of $\partial\Omega_s^-$ coincides with a component of $\partial\Omega_r^+$ or a component of $\partial\Omega_k^0$ for any r and k, then we put $X_1 u = Y_1^\star u = 0$ in Ω_s^-. If no component of $\partial\Omega_s^-$ coincides with a component of $\partial\Omega_r^+$ for any r, but there is a component Γ_{sk}^- which coincides with a component of $\partial\Omega_j^0$ for some j, then we put $X_1 u = 0$ in Ω_s^-. To define $Y_1^\star u$ in Ω_s^- we proceed as follows. Let $\{Q_\alpha\}_{\alpha \in I_k}$ denote those members of $\{Q_\alpha\}_{\alpha\in\mathbb{Z}^n}$ having non-empty intersection with Γ_{sk}^-, where I_k denotes an index set, and let $u_\alpha(x) = \psi_\alpha(x)u(x)$ for $\alpha \in I_k$. Fixing our attention upon a particular $\alpha \in I_k$, let $x_\alpha^0 \in \Gamma_{sk}^- \cap Q_\alpha$. Then passing to local coordinates at x_α^0 and using the same notation as in the previous paragraph, let $v_\alpha(\eta^\alpha) = u_\alpha(x(\eta^\alpha))$ for $\alpha \in I_k$ and let us suppose that $Y_1^\star u_\alpha$ goes over to $Y_\alpha^\star v_\alpha$ in this transformation. Guided by the requirements that $|T|X_1 u = Y_1^\star T u$ in $\mathcal{H}$ and that Y_1 maps V continuously into V, letting $v_\alpha(\eta^\alpha) = 0$, $J_\alpha(\eta^{\alpha\prime}, -\eta_n^\alpha/p)J_\alpha(\eta^\alpha)^{-1} = 0$ in $\mathbb{R}_{\eta^\alpha}^n \backslash \mathcal{U}_\alpha$ for $p = 1,\ldots,m$, and bearing in mind assumption (5) above concerning ω, we are led to take $(Y_\alpha^\star v_\alpha)(\eta^\alpha) = v_\alpha(\eta^\alpha)$ for $\eta_n^\alpha < 0$, $(Y_\alpha^\star v_\alpha)(\eta^\alpha) = \chi(\eta_n^\alpha)J_\alpha(\eta^\alpha)^{-1}\sum_{p=1}^m c_p(v_\alpha J_\alpha)(\eta^{\alpha\prime}, -\eta_n^\alpha/p)$ for $\eta_n^\alpha > 0$, where χ is defined above and the constants c_p are determined by the equations:

$$\sum_{p=1}^{m} p^j c_p = (-1)^j \quad \text{for } j = 1,\ldots,m.$$

If $(Y_\alpha^\star v_x)^\#$ denotes the pullback of $Y_\alpha^\star v_x$ to $\mathbb{R}_x^n$, and if we define $Y_1^\star u$ in Ω_s^- by putting here $Y_1^\star u = \sum_k \sum_{\alpha\in I_k}(Y_\alpha^\star v_\alpha)^\# |\Omega_s^-$, where k runs over the set of integers for which the component Γ_{sk}^- of $\partial\Omega_s^-$ coincides with a component of $\partial\Omega_j^0$ for some j, then we obtain the required extension of $Y_1^\star u$ to Ω_s^-. If no component of $\partial\Omega_s^-$ coincides with a component of $\partial\Omega_p^0$ for any p, but there is a component Γ_{sk}^- which coincides with a component of $\partial\Omega_r^+$ for some r, then we define $X_1 u$ and $Y_1^\star u$ in Ω_s^- by putting here $X_1 u = \sum_k \sum_{\alpha\in I_k}(X_\alpha v_\alpha)^\# |\Omega_s^-$, $Y_1^\star u = \sum_k \sum_{\alpha\in I_k}(Y_\alpha^\star v_\alpha)^\# |\Omega_s^-$ where k runs through that set of integers for which the component Γ_{sk}^- of $\partial\Omega_s^-$ coincides with a component of $\partial\Omega_r^+$ for some r, and the $(X_\alpha v_\alpha)^\#$, $(Y_\alpha^\star v_\alpha)^\#$, and I_k are defined precisely as in the previous paragraph. On the other hand, if there is a component of $\partial\Omega_s^-$, say Γ_{sk}^-, which coincides with a component of $\partial\Omega_r^+$ for some r and there is also a component, say Γ_{sj}^-, which coincides with a component of $\partial\Omega_p^0$ for some p, then we define $X_1 u$ and $Y_1^\star u$ in Ω_s^- by putting here $X_1 u = \sum_k \sum_{\alpha\in I_k}(X_\alpha v_\alpha)^\# |\Omega_s^-$,

$$Y_1^\star u = \sum_k \sum_{\alpha\in I_k}(Y_\alpha^\star v_\alpha)^\# |\Omega_s^- + \sum_j \sum_{\alpha\in I_j}(Y_\alpha^\star v_\alpha)^\# |\Omega_s^-, \tag{3.9}$$

where k (resp. j) runs over that set of integers for which the component Γ_{sk}^- (resp. Γ_{sj}^-) of $\partial\Omega_s^-$ coincides with a component of $\partial\Omega_r^+$ (resp. $\partial\Omega_p^0$) for some r (resp. p), and I_k, the $(X_\alpha v_\alpha)^\#$, and the $(Y_\alpha^\star v_\alpha)^\#$ in the first double summation on the right side of (3.9) are defined precisely as in the previous paragraph, while I_j and the $(Y_\alpha^\star v_\alpha)^\#$ in the second double summation on the right side of (3.9) are defined precisely as in the second case treated in this paragraph. Lastly, a repetition of the above arguments to all the components of $\operatorname{int}\Omega^-$ gives the required extensions of $X_1 u$ and $Y_1^\star u$ to $\operatorname{int}\Omega^-$.

It remains only to define the extension of $X_1 u$ to Ω_0, and accordingly let us fix our attention upon an Ω_s^0. If no component of $\partial\Omega_s^0$ coincides with a component of $\partial\Omega_r^+$ for any r, then we let $X_1 u = 0$ in Ω_s^0. On the other hand, if there is a component of $\partial\Omega_s^0$ which coincides with a component of $\partial\Omega_r^+$ for some r, but no component of $\partial\Omega_s^0$ coincides with a component of $\partial\Omega_k^-$ for any k, then we let $X_1 u = u$ in Ω_s^0. Finally, if there is a component of $\partial\Omega_s^0$ which coincides with a component of $\partial\Omega_r^+$, say Γ_{rj}^+, for some r and there is a component of $\partial\Omega_s^0$ which coincides with a component of $\partial\Omega_k^-$ for some k, then we define the extension of $X_1 u$ to Ω_s^0 as follows. Let $\{Q_\alpha\}_{\alpha\in I_{rj}}$ denote the members of $\{Q_\alpha\}_{\alpha\in\mathbb{Z}^n}$ having non-empty intersection with Γ_{rj}^+, where I_{rj} denotes an index set, and let $u_\alpha(x) = \psi_\alpha(x)u(x)$ for $\alpha \in I_{rj}$. Fixing our attention upon a particular $\alpha \in I_{rj}$, let $x_\alpha^0 \in \Gamma_{rj}^+ \cap Q_\alpha$. Then passing to local coordinates at x_α^0 and employing the same notation as before, let $v_\alpha(\eta^\alpha) = u_\alpha(x(\eta^\alpha))$ and let us suppose that $X_1 u_\alpha$ goes over to $X_\alpha v_\alpha$ in this transformation. Guided by the requirements that $|T|X_1 u = Y_1^* T u$ in $\mathcal{H}$ and that X_1 maps V continuously into V, letting $v_\alpha(\eta^\alpha) = 0$ in $\mathbb{R}^n_{\eta^\alpha}\backslash\mathcal{U}_\alpha$, and bearing in mind assumption (4) above concerning ω, we are led to take $(X_\alpha v_\alpha)(\eta^\alpha) = v_\alpha(\eta^\alpha)$ for $\eta_n^\alpha > 0$, $(X_\alpha v_\alpha)(\eta^\alpha) = \chi(\eta_n^\alpha)\sum_{p=1}^m c_p v_\alpha(\eta^{\alpha\prime}, -\eta_n^\alpha/p)$ for $\eta_n^\alpha < 0$, where χ is defined above and the c_p are determined by the equations:

$$\sum_{p=1}^{m} p^{-j} c_p = (-1)^j \quad \text{for } j = 0, \ldots, (m-1).$$

If $(X_\alpha v_\alpha)^\#$ denotes the pullback of $X_\alpha v_\alpha$ to $\mathbb{R}^n_x$, and if we define $X_1 u$ in Ω_s^0 by putting here $X_1 u = \sum_{r,j}\sum_{\alpha\in I_{rj}} (X_\alpha v_\alpha)^\# |\Omega_s^0$, where the first summation is over those pairs (r,j) for which a component of $\partial\Omega_s^0$ coincides with a Γ_{rj}^+, then we obtain the required extension of $X_1 u$ to Ω_s^0. A repetition of this argument to all components of Ω_0 gives the required extension of $X_1 u$ to Ω_0.

Thus we have shown that $(\,.\,,\,.\,)$ and $\langle |K^\dagger|\,.\,,\,.\,\rangle$ induce equivalent norms on $V^\dagger$. Let P_+ and P_- denote the positive and negative spectral projections, respectively, of $K^\dagger$ (acting in $(V^\dagger, \langle\,.\,,\,.\,\rangle)$) and let $V_\pm^\dagger = P_\pm V^\dagger$. Then it follows that $V_+^\dagger$ and $V_-^\dagger$ are positive definite and negative definite subspaces, respectively, of $\mathcal{H}_\omega$ and $V^\dagger = V_+^\dagger [\dot{+}] V_-^\dagger$. If we now take the completion of $V^\dagger$ with respect to the norm $\langle |K^\dagger| u, u\rangle^{1/2}$, then it follows from the definitions that this completion can be identified with $(V_1, (\,.\,,\,.\,))$ and $P_\pm$ extend by continuity to a pair of fundamental projectors on $(V_1, [\,.\,,\,.\,])$, which we will denote by $P_\pm^\#$ (see [6, Theorem 2.1, p.102]). Since $V_\pm^\dagger$ are invariant under $K^\dagger$, $P_\pm^\# V_1$ are invariant under K, and hence under S, which completes the proof of the theorem. ■

Recall that we have assumed that B_0 is positive definite in V_0. Hence it follows from arguments similar to those used in the proof of Theorem 2.3 that S is a positive operator in $(V_1, [\,.\,,\,.\,])$. Thus K is also positive, and so we conclude from [6, Theorem 1.2, p.63] that the spectrum of S is real. Furthermore, as a consequence of our above results (see in particular Theorem 2.4) and those of [7] and [14], we also have

COROLLARY 3.1. *It is the case that infinity is not a singular critical point of S. Moreover, S is similar to a selfadjoint operator in $(V_1, (\,.\,,\,.\,))$, and if Ω is bounded, then the eigenvalues of the problem (1.1-2) form a discrete subset of $\mathbb{R}^1$ and the restrictions of the corresponding principal vectors to the set $\Omega^+ \cup \Omega^-$ form a Riesz basis of $L^2(\Omega^+ \cup \Omega^-; |\omega(x)|dx)$.*

Suppose next that Ω is bounded, let $\{1/\mu_j^+\}_{j\geq 1}$ and $\{u_j^+\}_{j\geq 1}$ denote the positive eigenvalues and corresponding orthonormalized eigenvectors, respectively, of $K^\dagger$, and let $\{1/\mu_j^-\}_{j\geq 1}$ and $\{u_j^-\}_{j\geq 1}$ denote the negative eigenvalues and corresponding orthonormalized eigenvectors, respectively, of $K^\dagger$. Let $v_j^\pm = |\mu_j^\pm|^{1/2} u_j^\pm$ for $j \geq 1$. Then referring to the proof of Theorem 3.1 for details and letting $\mathcal{M}^{\pm} = P_{\pm}^{\#}V_1$, it follows from what was said in that proof that V_1 admits the fundamental decomposition

(3.10) $$V_1 = \mathcal{M}^{+}[\dot{+}]\mathcal{M}^{-}$$

and it is clear that the v_j^+ (resp. the v_j^-), $j \geq 1$, form an orthonormal basis of $\mathcal{M}^+$ (resp. $\mathcal{M}^-$) with respect to its intrinsic topology. Hence bearing in mind Theorems 2.2 and 2.4, we have

COROLLARY 3.2. *Suppose that Ω is bounded and that $0 \in \rho(A)$. Then the restrictions to Ω^+ of the eigenvectors $\{v_j^+\}_{j\geq 1}$ of the problem (1.1–2) form a Riesz basis of $L^2(\Omega^+;\ \omega(x)dx)$, while the restrictions to Ω^- of the eigenvectors $\{v_j^-\}_{j\geq 1}$ of the problem (1.1–2) form a Riesz basis of $L^2(\Omega^-;\ -\omega(x)dx)$.*

PROOF. Let $Q_{\pm}$ denote the operators in $\mathcal{H}_\omega$ defined by $(Q_{\pm}f)(x) = \chi_{\text{int}\,\Omega^{\pm}}(x)f(x)$ for $f \in \mathcal{H}_\omega$, where χ_E denotes the characteristic function of the set E, and let $\mathcal{N}^{\pm} = Q_{\pm}\mathcal{H}_\omega$. Then $\mathcal{H}_\omega$ admits the fundamental decomposition

(3.11) $$\mathcal{H}_\omega = \mathcal{N}^{+}\,[\dot{+}]\,\mathcal{N}^{-}$$

as well as the fundamental decomposition given by the right side of (3.10). Hence if we equip the $\mathcal{M}^{\pm}$ and $\mathcal{N}^{\pm}$ with their intrinsic topologies, observe from [6, Theorem 6.4, p.92, Lemma 7.1, p.93, and Lemma 4.6, p.107] that $\mathcal{M}^+$ (resp. $\mathcal{M}^-$) is isomorphic to $\mathcal{N}^+$ (resp. $\mathcal{N}^-$) under the mapping Q_+ (resp. Q_-), and bear in mind that $\mathcal{N}^+$ (resp. $\mathcal{N}^-$) is isomorphically isometric to $L^2(\text{int}\ \Omega^+;\ \omega(x)dx)$ (resp. $L^2(\text{int}\ \Omega^-;\ -\omega(x)dx)$, then all the assertions of the corollary are an immediate consequence of the foregoing results. ■

Suppose next that $0 \in \sigma(A)$ and let us fix our attention upon (2.3). Since $\mathcal{M}_0$ is not degenerate with respect to the inner product $[.\,,.]$, it follows from [6, Corollary 9.5, p.19 and Corollary 11.8, p.26] that $\mathcal{M}_0$ admits the fundamental decomposition

$$\mathcal{M}_0 = \mathcal{M}_0^+\,[\dot{+}]\,\mathcal{M}_0^-,$$

where $\mathcal{M}_0^+$ is positive definite and $\mathcal{M}_0^-$ is negative definite, and we henceforth let $\{z_j^+\}_1^{n^+}$ and $\{z_j^-\}_1^{n^-}$ denote orthonormal bases for $\mathcal{M}_0^+$ and $\mathcal{M}_0^-$, respectively, with respect to their intrinsic topologies, where $n^{\pm}$ are non-negative integers satisfying $n^+ + n^- = \dim\ \mathcal{M}_0$. It now follows from (2.3) and (3.10) that $\mathcal{H}_\omega$ admits the fundamental decomposition

$$\mathcal{H}_\omega = (\mathcal{M}_0^+\,[\dot{+}]\,\mathcal{M}^+)\,[\dot{+}]\,(\mathcal{M}_0^-\,[\dot{+}]\,\mathcal{M}^-),$$

and hence if we bear in mind (3.11) and argue in a manner similar to that in the proof of Corollary 3.2, then it is not difficult to verify that

COROLLARY 3.3. *Suppose that Ω is bounded and that $0 \in \sigma(A)$. Then the restrictions to Ω^+ of the principal vectors $\{z_j^+\}_1^{n^+}$, $\{v_j^+\}_{j\geq 1}$ of the problem (1.1–2) form a Riesz basis*

for $L^2(\Omega^+;\ \omega(x)dx)$, while the restrictions to Ω^- of the principal vectors $\{z_j^-\}_1^{n^-}$, $\{v_j^-\}_{j\geq 1}$ of the problem (1.1–2) form a Riesz basis of $L^2(\Omega^-;\ -\omega(x)dx)$.

4. EXAMPLES

Referring again to §2, let us suppose for the moment that Ω is a bounded region of class C^{2m}, the $a_{\alpha\beta}$ are uniformly continuous and bounded in Ω, and for $|\alpha| > 0$, $a_{\alpha\beta}$ has uniformly continuous, bounded derivatives in Ω of order $\leq |\alpha|$, the b_α^j, $|\alpha| \leq m_j$, $j = 1, \ldots, p$, have continuous bounded derivatives on Γ or order $\leq 2m - m_j$ if $p > 0$, and the f_α^j and ϕ_j, $|\alpha| \leq \mu_j$, $j = (p+1), \ldots, m$, have continuous bounded derivatives on Γ of order $\leq m$ if $p < m$ and $B(u,v) = a(u,v) + a^{\#}(u,v)$. Observing from [4, Theorem 7.12, p.86] that the form $B(u,v)$ of §2 satisfies Garding's inequality, it follows from regularity arguments somewhat similar to those used in [4, §9] as well as from [1, §8] and [2, Theorem 5.2] that the boundary value problem

$$Lu = \sum_{|\alpha|\leq 2m} a_\alpha^\dagger(x) D^\alpha u = f \text{ in } \Omega, \tag{4.1}$$
$$B_j u = 0 \text{ on } \Gamma \text{ for } j = 1, \ldots, m$$

is regular in the sense of [3] (here $f \in \mathcal{H}$ and L and the B_j are the same as in (1.1–2), except now we have written L in non–divergence form), and moreover, $D(A) = H^{2m}(\Omega) \cap V = \Big\{u \in H^{2m}(\Omega) \big| B_j u = 0 \text{ on } \Gamma \text{ for } j = 1, \ldots, m\Big\}$. Selfadjoint, regular elliptic boundary value problems of the form (4.1), with f replaced by $\lambda\omega(x)u$ and with $A \geq \gamma > 0$, but not necessarily admitting a variational formulation, have been considered in [18], and under the assumption that ω and ω^{-1} are in $L^\infty(\Omega)$ (as well as under other assumptions), it was shown that the eigenvectors form a Riesz basis in $L^2(\Omega^+ \cup \Omega^-;\ |\omega(x)|dx)$. What is important here is that if the problem (4.1) admits a variational formulation, that is, if $(Au,v)_{0,\Omega} = B(u,v)$ for $u,\ v \in D(A)$, where B is a symmetric sesquilinear from of the kind considered in §2, but not supposed coercive, then we know from [2, Theorem 5.1] that B must be coercive, and hence under our assumptions concerning ω our theory applies in full force and we can establish the same result without the requirement that $\omega^{-1} \in L^\infty(\Omega)$ (and also under weaker assumptions on Ω and on the coefficients involved).

Finally, the problem

$$-(\Delta + 1)u = \lambda\omega(x)u \text{ in } \mathbb{R}^n \tag{4.2}$$

was considered in [8], where Δ denotes the Laplacian in $\mathbb{R}^n$, and under the supposition that ω and ω^{-1} are in $L^\infty(\mathbb{R}^n)$ (as well as under other assumptions) it was shown that the operator S realized by the problem (4.2) is similar to a selfadjoint operator in $L^2(\Omega^+ \cup \Omega^-;\ |\omega(x)|dx)$. It is clear from [4, Remark, p.129] that the problem (4.2) admits a variational formulation, with the associated sesquilinear form $B(u,v)$ being $(u,v)_{1,\mathbb{R}^n}$, and hence under our assumptions concerning ω our theory applies in full force and we can establish the same result without the requirement that $\omega^{-1} \in L^\infty(\mathbb{R}^n)$.

REFERENCES

1. S. Agmon, *The coercive problem for integro-differential forms*, J. Analyse Math. **6** (1958), 183–223.
2. S. Agmon, *Remarks on self-adjoint and semi-bounded elliptic boundary value problems*, Proc. Internat. Sympos. on Linear Spaces (Jerusalem, 1960), Pergamon, Oxford, 1961, pp.1–13.
3. S. Agmon, *On the eigenfunctions and on the eigenvalues of general elliptic boundary value problems*, Comm. Pure Appl. Math. **15** (1962), 119–147.
4. S. Agmon, *Lectures on elliptic boundary value problems*, Van Nostrand, Princeton, N.J., 1965.
5. R. Beals, *Indefinite Sturm–Liouville problems and half-range completeness*, J. Differential Equations **56** (1985), 391–407.
6. J. Bognár, *Indefinite inner product spaces*, Springer, Berlin, 1974.
7. B. Curgus, *On the regularity of the critical point infinity of definitizable operators*, Integral Equations Operator Theory **8** (1985), 462–488.
8. B. Curgus and B. Najman, *A Krein space approach to elliptic eigenvalue problems with indefinite weights*, Differential Integral Equations **7** (1994), 1241–1252.
9. M. Faierman, *Elliptic problems involving an indefinite weight*, Trans. Amer. Math. Soc. **320** (1990), 253–279.
10. M. Faierman, *Non-selfadjoint elliptic problems involving an indefinite weight*, Comm. Partial Differential Equations **15** (1990), 939–982.
11. L. Hörmander, *Linear partial differential operators*, Springer, Berlin, 1976.
12. L. Hörmander, *Uniqueness theorems for second order elliptic equations*, Comm. Partial Differential Equations **8** (1983), 21–64.
13. T. Kato, *Perturbation theory for linear operators*, 2nd edn., Springer, Berlin, 1976.
14. H. Langer, *Spectral function of definitizable operators in Krein spaces*, Proc. Funct. Anal., Dubrovnik 1981, Lecture Notes in Mathematics **948**, Springer, Berlin, 1–46.
15. J.L. Lions and E. Magenes, *Non-homogeneous boundary value problems and applications*, Vol. I, Springer, Berlin, 1972.
16. A.S. Markus, *Introduction to the spectral theory of polynomial operator pencils*, Amer. Math. Soc., Providence, R.I., 1988.
17. M. Möller, *Orthogonal systems of eigenvectors and associated vectors for symmetric holomorphic operator functions*, Math. Nachr. **163** (1993), 45–64.
18. S.G. Pyatkov, *Some properties of eigenfunctions of linear pencils*, Siberian Math. J. **30** (1989), 587–597.
19. C. Zuily, *Uniqueness and non-uniqueness in the Cauchy problem*, Birkhäuser, Basel, 1983.

Department of Mathematics
University of the Witwatersrand
Johannesburg, WITS 2050
South Africa

Institut für Analysis
Technische Mathematik und
Versicherungsmathematik
Technische Universität Wien
Wiedner Hauptstr. 8–10
A–1040 Vienna
Austria

MSc: Primary 35P10, 47B50; Secondary 47F05

Operator Theory:
Advances and Applications, Vol. 87

THE KDV HIERARCHY AND ASSOCIATED TRACE FORMULAS

F. GESZTESY, R. RATNASEELAN, AND G. TESCHL

A natural algebraic approach to the KdV hierarchy and its algebro-geometric finite-gap solutions is developed. In addition, a new derivation of associated higher-order trace formulas in connection with one-dimensional Schrödinger operators is presented.

1. Introduction

The purpose of this paper is to advocate a most natural algebraic approach to hierarchies of completely integrable evolution equations such as the AKNS and Toda hierarchies and a systematic treatment of associated trace formulas. Specifically, we shall treat in great detail the simplest example of these completely integrable systems, the Korteweg-de Vries (KdV) hierarchy, and derive the corresponding higher-order trace formulas for one-dimensional Schrödinger operators. Even though the main ingredients of our approach to the KdV hierarchy (to be outlined below) appear to be well-known, it seems to us that no systematic attempt to combine them all into a complete description of the KdV hierarchy and its algebro-geometric solutions has been undertaken in the literature thus far. The principal aim of this paper is to fill this gap and at the same time provide the intimate connection with general higher-order trace formulas for the associated Lax operator.

The key ingredients just mentioned are a recursive approach to Lax pairs following Al'ber [1], [2] (see also [9], Ch. 12, [15]), naturally leading to the celebrated Burchnall-Chaundy polynomial [6], [7] and hence to hyperelliptic curves K_g of genus $g \in \mathbb{N}_0 (= \mathbb{N} \cup \{0\})$ and a classic representation of positive divisors of K_g of degree g due to Jacobi [27] and first applied to the KdV case by Mumford [36], Section III a).1, with subsequent extensions due to McKean [33]. Finally, following a recent series of papers on trace formulas for Schrödinger operators [16]–[19], [22]–[24] we present a new algorithm for deriving higher-order trace formulas associated with the KdV hierarchy.

In Section 2 we briefly review Al'ber's recursive approach to the KdV hierarchy. In particular, we illustrate the role of commuting differential expressions of order $2g+1$, $g \in \mathbb{N}_0$ and 2, respectively, in connection with the Burchnall-Chaundy polynomial, hyperelliptic curves K_g of genus g branched at infinity, and the equations of the stationary (i.e., time-independent) KdV hierarchy. Section 3 combines Al'ber's recursion formalism with Jacobi's representation of positive divisors of degree g of K_g as applied to the KdV case by Mumford and McKean and provides a detailed construction of the stationary KdV hierarchy and its algebro-geometric solutions. The principal new result of Section 3, summarized in (3.50)–(3.64), concern divisors of degree $g+1$ of K_g associated with Schrödinger-type operators with general boundary conditions of the type defined in (3.45). In Section 4 we present a systematic extension of this body of ideas to the time-dependent KdV hierarchy going beyond the standard treatment in the literature. Especially, our t-dependent discussion in connection with divisors of degree $g+1$ of K_g associated with the general eigenvalue problem (3.45) as presented in (4.36)–(4.50) is without precedent. Moreover, our proof of the theta function representation (4.51) of the Baker-Akhiezer function $\psi(P, x, x_0, t, t_0)$ in Theorem 4.6, based on the fundamental meromorphic function $\phi(P, x, t)$ defined in (4.15), is new. In Section 5 we turn to (higher-order) trace formulas for Schrödinger operators associated with general boundary conditions (cf. (5.3)), a key ingredient in the solution of inverse spectral problems. Unlike Sections 3 and 4, the approach in Section 5 applies to general (not necessarily algebro-geometric finite-gap) solutions of the KdV hierarchy. The principal new results of Section 5 are the (universally valid) nonlinear differential equation (5.18) for $\Gamma^\beta(z, x)$, $\beta \in \mathbb{R}$ (defined in (5.4)), the resulting recursion relation (5.21), and, in particular, our method of proof of Theorem 5.3 (i). In Appendix A we provide a brief summary on hyperelliptic curves of the KdV-type and their theta functions and establish our basic notation used in Sections 3 and 4. Finally, Appendix B provides an explicit illustration of the Riemann-Roch theorem in connection with hyperelliptic curves branched at infinity which appears to be of independent interest.

We emphasize that the methods of this paper are widely applicable to $1+1$-dimensional completely integrable systems. The corresponding account for the Toda and Kac-van Moerbeke hierarchy can be found in [5].

2. The KdV Hierarchy, Recursion Relations, and Hyperelliptic Curves

In this section we briefly review the construction of the KdV hierarchy using a recursive approach advocated by Al'ber [1], [2] (see also [9], Ch. 12, [15], [20]) and outline its connection with the Burchnall-Chaundy polynomial [6], [7] and associated hyperelliptic curves branched at infinity.

Suppose

$$V(.,t) \in C^{\infty}(\mathbb{R}),\ t \in \mathbb{R}, V(x,.) \in C^{1}(\mathbb{R}),\ x \in \mathbb{R} \tag{2.1}$$

and consider the differential expressions (Lax pair)

$$L(t) = -\frac{d^2}{dx^2} + V(x,t), \tag{2.2}$$

$$P_{2g+1}(t) = \sum_{j=0}^{g} \Big[f_j(x,t)\frac{d}{dx} - \frac{1}{2} f_{j,x}(x,t) \Big] L(t)^{g-j}, \quad g \in \mathbb{N}_0,\ (x,t) \in \mathbb{R}^2, \tag{2.3}$$

where the $\{f_j\}_{0 \le j \le g}$ satisfy the recursion relation

$$f_0 = 1,\ f_{j,x} = -\frac{1}{4} f_{j-1,xxx} + V f_{j-1,x} + \frac{1}{2} V_x f_{j-1}, \quad 1 \le j \le g. \tag{2.4}$$

Define in addition f_{g+1} by

$$f_{g+1,x} = -\frac{1}{4} f_{g,xxx} + V f_{g,x} + \frac{1}{2} V_x f_g. \tag{2.5}$$

Then one computes

$$[P_{2g+1}, L] = 2 f_{g+1,x}, \tag{2.6}$$

where $[.,.]$ denotes the commutator. The Lax equation

$$\frac{d}{dt} L(t) - [P_{2g+1}(t), L(t)] = 0, \quad t \in \mathbb{R} \tag{2.7}$$

is then equivalent to

$$\mathrm{KdV}_g(V) = V_t - 2 f_{g+1,x}, \quad t \in \mathbb{R}. \tag{2.8}$$

Varying $g \in \mathbb{N}_0$ yields the KdV hierarchy

$$\mathrm{KdV}_g(V) = 0, \quad g \in \mathbb{N}_0. \tag{2.9}$$

Explicitly, one obtains from (2.4),

$$\begin{aligned}
&f_0 = 1 = \tilde{f}_0,\\
&f_1 = \tfrac{1}{2}V + c_1 = c_1\tilde{f}_0 + \tilde{f}_1,\\
&f_2 = -\tfrac{1}{8}V_{xx} + \tfrac{3}{8}V^2 + c_1\tfrac{1}{2}V + c_2 = c_2\tilde{f}_0 + c_1\tilde{f}_1 + \tilde{f}_2, \qquad (2.10)\\
&f_3 = \tfrac{1}{32}V_{xxxx} - \tfrac{5}{16}VV_{xx} - \tfrac{5}{32}V_x^2 + \tfrac{5}{16}V^3 + c_2\tfrac{1}{2}V + c_1[-\tfrac{1}{8}V_{xx} + \tfrac{3}{8}V^2] + c_3\\
&\quad = c_3\tilde{f}_0 + c_2\tilde{f}_1 + c_1\tilde{f}_2 + \tilde{f}_3,\\
&\quad \text{etc.}
\end{aligned}$$

Hence by (2.8),

$$\begin{aligned}
&\mathrm{KdV}_0(V) = V_t - V_x = 0,\\
&\mathrm{KdV}_1(V) = V_t + \tfrac{1}{4}V_{xxx} - \tfrac{3}{2}VV_x - c_1V_x, \qquad (2.11)\\
&\mathrm{KdV}_2(V) = V_t - \tfrac{1}{16}V_{xxxxx} + \tfrac{5}{8}VV_{xxx} + \tfrac{5}{4}V_xV_{xx} - \tfrac{15}{8}V^2V_x - c_2V_x + c_1[\tfrac{1}{4}V_{xxx} - \tfrac{3}{2}VV_x],\\
&\quad \text{etc.}
\end{aligned}$$

represent the first few equations of the KdV hierarchy. Here c_ℓ denote integration constants which naturally arise when solving (2.4). Moreover, the corresponding homogeneous KdV equations, obtained by taking all integration constants equal to zero, $c_\ell \equiv 0$, $\ell \geq 1$ are then denoted by

$$\widetilde{\mathrm{KdV}}_g(V) := \mathrm{KdV}_g(V)\big|_{c_\ell\equiv 0,\, 1\leq \ell\leq g} \qquad (2.12)$$

and similarly we denote by $\tilde{P}_{2g+1} := P_{2g+1}(c_\ell \equiv 0)$, $\tilde{f}_j := f_j(c_\ell \equiv 0)$, etc. the corresponding homogeneous quantities.

Before we turn to a discussion of the stationary KdV hierarchy we briefly sketch the main steps leading to (2.3)–(2.8). Let $\mathrm{Ker}(L(t) - z)$, $z \in \mathbb{C}$ denote the two-dimensional nullspace of $L(t) - z$ (in the algebraic sense as opposed to the functional analytic one). We seek a representation of $P_{2g+1}(t)$ on $\mathrm{Ker}(L(t) - z)$ of the form

$$P_{2g+1}(t)\big|_{\mathrm{Ker}(L(t)-z)} = F_g(z,x,t)\frac{d}{dx} + G_{g-1}(z,x,t), \qquad (2.13)$$

where F_g are polynomials in z of the type

$$F_g(z,x,t) = \sum_{j=0}^{g} f_{g-j}(x,t)z^j, \tag{2.14}$$

$$G_{g-1}(z,x,t) = \sum_{j=0}^{g-1} g_{g-j}(x,t)z^j. \tag{2.15}$$

The Lax equation (2.7) restricted to $\mathrm{Ker}(L(t)-z)$ then yields

$$\begin{aligned} 0 &= \{\dot{L} - [P_{2g+1}, L]\}\Big|_{\mathrm{Ker}(L-z)} = \{\dot{L} + (L-z)P_{2g+1}\}\Big|_{\mathrm{Ker}(L-z)} \\ &= \Big\{ - [F_{g,xx} + 2G_{g-1,x}]\frac{d}{dx} + [V_t - F_g V_x - 2(V-z)F_{g,x} - G_{g-1,xx}]\Big\}\Big|_{\mathrm{Ker}(L-z)} \end{aligned} \tag{2.16}$$

implying

$$G_{g-1} = -F_{g,x}/2 \tag{2.17}$$

(neglecting a trivial integration constant) and

$$V_t = -\frac{1}{2}F_{g,xxx} + 2(V-z)F_{g,x} + V_x F_g. \tag{2.18}$$

Insertion of (2.14) into (2.18) then yields (2.8). We omit further details and just record a few of the polynomials F_g,

$$\begin{aligned} &F_0 = 1 = \tilde{F}_0, \\ &F_1 = c_1 + \tfrac{1}{2}V + z = c_1\tilde{F}_0 + \tilde{F}_1, \\ &F_2 = c_2 + c_1\tfrac{1}{2}V - \tfrac{1}{8}V_{xx} + \tfrac{3}{8}V^2 + (c_1 + \tfrac{1}{2}V)z + z^2 = c_2\tilde{F}_0 + c_1\tilde{F}_1 + \tilde{F}_2, \\ &\quad \text{etc.} \end{aligned} \tag{2.19}$$

One verifies

$$P_{2g+1} = \sum_{m=0}^{g} c_{g-m}\tilde{P}_{2m+1}, \quad c_0 = 1. \tag{2.20}$$

Finally, we specialize to the stationary KdV hierarchy characterized by $V_t = 0$ in (2.9) (respectively (2.8)), or more precisely, by commuting differential expressions

$$[P_{2g+1}, L] = 0 \tag{2.21}$$

of order $2g+1$ and 2, respectively. Eq. (2.18) then becomes

$$F_{g,xxx} - 4(V-z)F_{g,x} - 2V_x F_g = 0 \tag{2.22}$$

and upon multiplying by F_g and integrating one infers

$$\frac{1}{2}F_{g,xx}F_g - \frac{1}{4}F_{g,x}^2 - (V-z)F_g^2 = R_{2g+1}(z), \tag{2.23}$$

where $R_{2g+1}(z)$ is of the form

$$R_{2g+1}(z) = \prod_{n=0}^{2g}(z-E_n), \quad \{E_n\}_{0\le n\le 2g} \subset \mathbb{C}. \tag{2.24}$$

Because of (2.21) one computes

$$\left[P_{2g+1}\big|_{\mathrm{Ker}(L-z)}\right]^2 = -\left[\frac{1}{2}F_{g,xx}F_g - \frac{1}{4}F_{g,x}^2 - (V-z)F_g^2\right]\Big|_{\mathrm{Ker}(L-z)} = -R_{2g+1}(z). \tag{2.25}$$

Since $z \in \mathbb{C}$ is arbitrary, one obtains the Burchnall-Chaundy polynomial [6], [7] relating P_{2g+1} and L,

$$-P_{2g+1}^2 = R_{2g+1}(L) = \prod_{n=0}^{2g}(L-E_n). \tag{2.26}$$

The resulting hyperelliptic curve K_g of (arithmetic) genus g, obtained upon one-point compactification of the curve

$$y^2 = R_{2g+1}(z) = \prod_{n=0}^{2g}(z-E_n) \tag{2.27}$$

(cf. Appendix A), will be the basic ingredient in our algebro-geometric treatment of the KdV hierarchy in Sections 3 and 4.

The spectral theoretic content of the polynomials F_g, G_{g-1} is clearly displayed in (3.35), (3.37), (3.40)–(3.44).

3. The Stationary Formalism

Combining the recursion formalism of Section 2 with a polynomial approach to represent positive divisors of degree g of a hyperelliptic curve K_g of genus g originally developed by Jacobi [27] and applied to the KdV case by Mumford [36], Section III a).1 and McKean [33], we provide a detailed construction of the stationary KdV hierarchy and its algebro-geometric solutions. Our considerations (3.50)–(3.64) in connection with the general β-boundary conditions for Schrödinger-type operators in (3.45) are new.

As indicated at the end of Section 2, the stationary KdV hierarchy is intimately connected with pairs of commuting differential expressions P_{2g+1} and L of orders $2g+1$ and

2, respectively and hyperelliptic curves K_g obtained upon one-point compactification of the curve

$$y^2 = R_{2g+1}(z) = \prod_{n=0}^{2g}(z - E_n) \tag{3.1}$$

described in detail in Appendix A (whose results and notations we shall freely use in the remainder of this paper). Since we are interested in real-valued KdV solutions we now make the additional assumption

$$\{E_n\}_{0\leq n\leq 2g} \subset \mathbb{R},\ E_0 < E_1 < \cdots < E_{2g}, \quad g \in \mathbb{N}_0. \tag{3.2}$$

Writing

$$F_g(z,x) = \sum_{j=0}^{g} f_{g-j}(x)z^j = \prod_{j=1}^{g}[z - \mu_j(x)] \tag{3.3}$$

and combining (2.23) and (3.3) yields

$$\mu_j'(x)^2 = -4R_{2g+1}(\mu_j(x)) \prod_{\substack{k=1\\k\neq j}}^{g} [\mu_j(x) - \mu_k(x)]^{-2}, \quad 1 \leq j \leq g,\ x \in \mathbb{R}. \tag{3.4}$$

Integrating the nonlinear first-order system (3.4) as a vector field on the (complex) manifold $K_g \times \cdots \times K_g = K_g^g$, its solution is well-defined as long as the μ's do not collide. Since we focus on real-valued solutions V of the KdV hierarchy, we may restrict the vector field to the submanifold $\mathop{\times}_{j=1}^{g} \tilde{\pi}^{-1}([E_{2j-1}, E_{2j}])$ which is isomorphic to the torus $S^1 \times \cdots \times S^1 = T^g$. Thus

$$\mu_j'(x) = -2iR_{2g+1}^{1/2}(\hat{\mu}_j(x)) \prod_{\substack{k=1\\k\neq j}}^{g} [\mu_j(x) - \mu_k(x)]^{-1}, \quad 1 \leq j \leq g,\ x \in \mathbb{R}, \tag{3.5}$$

with the initial conditions

$$\{\hat{\mu}_j(x_0)\}_{1\leq j\leq g} \subset K_g,\ \tilde{\pi}(\hat{\mu}_j(x_0)) = \mu_j(x_0) \in [E_{2j-1}, E_{2j}], \quad 1 \leq j \leq g \tag{3.6}$$

for some fixed $x_0 \in \mathbb{R}$, has the unique solution $\{\hat{\mu}_j(x)\}_{1\leq j\leq g} \subset K_g$ satisfying

$$\hat{\mu}_j(.) \in C^\infty(\mathbb{R}, K_g),\ \tilde{\pi}(\hat{\mu}_j(x)) \in [E_{2j-1}, E_{2j}], \quad 1 \leq j \leq g,\ x \in \mathbb{R}. \tag{3.7}$$

These facts are verified using the charts (A.7), (A.8) which also shows that the solution $\hat{\mu}_j(x)$ changes sheets whenever it hits E_{2j-1} or E_{2j} and its projection $\mu_j(x) = \tilde{\pi}(\hat{\mu}_j(x))$ remains trapped in $[E_{2j-1}, E_{2j}]$ for all $x \in \mathbb{R}$.

Given (3.3), (3.5), and (2.17) one obtains

$$\begin{aligned} G_{g-1}(z,x) &= -\frac{1}{2}F_{g,x}(z,x) = \frac{1}{2}\sum_{j=1}^{g}\mu_j'(x)\prod_{\substack{k=1\\k\neq j}}^{g}[z-\mu_k(x)] \\ &= -i\sum_{j=1}^{g}R_{2g+1}^{1/2}(\hat{\mu}_j(x))\prod_{\substack{k=1\\k\neq j}}^{g}\Big(\frac{z-\mu_k(x)}{\mu_j(x)-\mu_k(x)}\Big) \end{aligned} \tag{3.8}$$

and hence

$$\begin{aligned} &R_{2g+1}^{1/2}(\hat{\mu}_j(x)) = \sigma_j(x)R_{2g+1}(\mu_j(x))^{1/2} = iG_{g-1}(\mu_j(x),x), \\ &\hat{\mu}_j(x) = (\mu_j(x), iG_{g-1}(\mu_j(x),x)), \quad 1\leq j\leq g. \end{aligned} \tag{3.9}$$

Moreover, since

$$[R_{2g+1}(z)+G_{g-1}(z,x)^2]\Big|_{z=\mu_j(x)} = 0, \quad 1\leq j\leq g, \tag{3.10}$$

one infers

$$R_{2g+1}(z)+G_{g-1}(z,x)^2 = F_g(z,x)H_{g+1}(z,x) \tag{3.11}$$

for some polynomial H_{g+1} in z of degree $g+1$,

$$H_{g+1}(z,x) = \prod_{\ell=0}^{g}[z-\nu_\ell(x)]. \tag{3.12}$$

Eqs. (3.9), (3.11), and (3.12) suggest defining $\{\hat{\nu}_\ell(x)\}_{0\leq\ell\leq g}\subset K_g$ by

$$R_{2g+1}^{1/2}(\hat{\nu}_\ell(x)) = -iG_{g-1}(\nu_\ell(x),x),\ \hat{\nu}_\ell(x) = (\nu_\ell(x), -iG_{g-1}(\nu_\ell(x),x)), \quad 0\leq\ell\leq g. \tag{3.13}$$

One verifies

$$\nu_0(x)\leq E_0,\ \nu_\ell(x)\in[E_{2\ell-1},E_{2\ell}], \quad 1\leq\ell\leq g,\ x\in\mathbb{R}. \tag{3.14}$$

Next, we define the fundamental meromorphic function $\phi(P,x)$ on K_g,

$$\begin{aligned} \phi(P,x) &= \frac{iR_{2g+1}^{1/2}(P)-G_{g-1}(\tilde{\pi}(P),x)}{F_g(\tilde{\pi}(P),x)} = \frac{iR_{2g+1}^{1/2}(P)+\frac{1}{2}F_{g,x}(\tilde{\pi}(P),x)}{F_g(\tilde{\pi}(P),x)} \\ &= \frac{-H_{g+1}(\tilde{\pi}(P),x)}{iR_{2g+1}^{1/2}(P)+G_{g-1}(\tilde{\pi}(P),x)},\ P=(\tilde{\pi}(P),R_{2g+1}^{1/2}(P)),\ x\in\mathbb{R}, \end{aligned} \tag{3.15}$$

with divisor $(\phi(.,x))$ given by

$$(\phi(.,x)) = \mathcal{D}_{\hat{\nu}_0(x)\underline{\hat{\nu}}(x)} - \mathcal{D}_{P_\infty\underline{\hat{\mu}}(x)}. \tag{3.16}$$

Here we abbreviated

$$\underline{\hat{\nu}}(x) = (\hat{\nu}_1(x),\dots,\hat{\nu}_g(x)),\ \underline{\hat{\mu}}(x) = (\hat{\mu}_1(x),\dots,\hat{\mu}_g(x)). \tag{3.17}$$

Given $\phi(P,x)$ we define the stationary Baker-Akhiezer (BA) function $\psi(P,x,x_0)$, meromorphic on $K_g\backslash\{P_\infty\}$, by

$$\psi(P,x,x_0) = \exp\Big[\int_{x_0}^{x} dy\phi(P,y)\Big], \quad (x,x_0) \in \mathbb{R}^2. \tag{3.18}$$

Properties of $V(x)$, $\phi(P,x)$, and $\psi(P,x,x_0)$ are summarized in the following

Lemma 3.1. *Let* $P = (z, \sigma R_{2g+1}(z)^{1/2}) = (\tilde{\pi}(P), R_{2g+1}^{1/2}(P)) \in K_g\backslash\{P_\infty\}$, $(z,x,x_0) \in \mathbb{C} \times \mathbb{R}^2$. *Then*

(i). $$V(x) = E_0 + \sum_{j=1}^{g}[E_{2j-1} + E_{2j} - 2\mu_j(x)]. \tag{3.19}$$

(ii). $\phi(P,x)$ *satisfies the Riccati-type equation*

$$\phi_x(P,x) + \phi(P,x)^2 = V(x) - z. \tag{3.20}$$

(iii). $\psi(P,x,x_0)$ *satisfies the Schrödinger equation*

$$-\psi_{xx}(P,x,x_0) + [V(x) - z]\psi(P,x,x_0) = 0. \tag{3.21}$$

(iv). $$\phi(P,x)\phi(P^*,x) = H_{g+1}(z,x)/F_g(z,x). \tag{3.22}$$

(v). $$\phi(P,x) + \phi(P^*,x) = -2G_{g-1}(z,x)/F_g(z,x) = F_{g,x}(z,x)/F_g(z,x). \tag{3.23}$$

(vi). $$\phi(P,x) - \phi(P^*,x) = 2iR_{2g+1}^{1/2}(P)/F_g(z,x). \tag{3.24}$$

(vii). $$\psi(P,x,x_0)\psi(P^*,x,x_0) = F_g(z,x)/F_g(z,x_0). \tag{3.25}$$

(viii). $$\psi_x(P,x,x_0)\psi_x(P^*,x,x_0) = H_{g+1}(z,x)/F_g(z,x_0). \tag{3.26}$$

(ix). $$\psi(P,x,x_0) = [F_g(z,x)/F_g(z,x_0)]^{1/2} \exp\Big[iR_{2g+1}^{1/2}(P)\int_{x_0}^{x} dyF_g(z,y)^{-1}\Big]. \tag{3.27}$$

Proof. (i). Insert (3.3) into (2.23) and compare the coefficient of z^{2g}. (ii). Combine (2.17), (2.23), and (3.15). (iii). Follows from $\psi_{xx}/\psi = \phi_x + \phi^2 = V - z$. (iv). Multiply the first and third expression in (3.15) replacing P by P^* in one of the two factors. (v), (vi) are clear from (3.15). (vii). Combine (3.18) and (3.23). (viii). Use (3.22), (3.25), and $\psi_x = \phi\psi$. (ix). Invoke (2.17), (3.15), and (3.18). □

Eq. (3.19) represents a trace formula for the finite-gap potential $V(x)$. The method of proof of Lemma 3.1 (i) indicates that higher-order trace formulas associated with the KdV hierarchy can be obtained from (3.3) and (2.23) comparing powers of z. Since we shall derive trace formulas for general potentials in Section 5, we postpone the special case of finite-gap potentials at this point and refer to Example 5.5.

We also record

Lemma 3.2. *Let* $(z,x) \in \mathbb{C} \times \mathbb{R}$. *Then*

$$\text{(i).} \quad H_{g+1}(z,x) = \frac{1}{2}F_{g,xx}(z,x) - [V(x)-z]F_g(z,x). \tag{3.28}$$

$$\text{(ii).} \quad H_{g+1,x}(z,x) = -2[V(x)-z]G_{g-1}(z,x). \tag{3.29}$$

Proof. (i). By (2.17), (2.23), and (3.11),

$$\begin{aligned} -\tfrac{1}{2}F_{g,xx} &= -\tfrac{1}{4}F_g^{-1}F_{g,x}^2 - (V-z)F_g - F_g^{-1}R_{2g+1} \\ &= -(V-z)F_g - F_g^{-1}(R_{2g+1}+G_{g-1}^2) = -(V-z)F_g - H_{g+1}. \end{aligned}$$

(ii). By (2.17), (2.22), and (3.28),

$$\begin{aligned} H_{g+1,x} &= -G_{g-1,xx} - (V-z)F_{g,x} - V_xF_g \\ &= \tfrac{1}{2}F_{g,xxx} - (V-z)F_{g,x} - V_xF_g = (V-z)F_{g,x} = -2(V-z)G_{g-1}. \quad \square \end{aligned}$$

Explicitly, one computes from (2.4), (2.14), and (3.28),

$$\begin{aligned} H_1 &= \tilde{H}_1 = -V + z, \\ H_2 &= -c_1V + \tfrac{1}{4}V_{xx} - \tfrac{1}{2}V^2 + \big(c_1 - \tfrac{1}{2}V\big)z + z^2 = c_1\tilde{H}_1 + \tilde{H}_2, \\ H_3 &= -c_2V + c_1\big(\tfrac{1}{4}V_{xx} - \tfrac{1}{2}V^2\big) - \tfrac{1}{16}V_{xxxx} + \tfrac{3}{8}V_x^2 + \tfrac{1}{2}VV_{xx} - \tfrac{3}{8}V^3 \\ &\quad + \big[c_2 - c_1\tfrac{1}{2}V + \tfrac{1}{8}V_{xx} - \tfrac{3}{8}V^2\big]z + \big[c_1 - \tfrac{1}{2}V\big]z^2 + z^3 = c_2\tilde{H}_1 + c_1\tilde{H}_2 + \tilde{H}_3, \end{aligned} \tag{3.30}$$

etc.

We also mention the following well-known result connecting Dirichlet and Neumann eigenvalues.

Lemma 3.3. [33] *Suppose* $\mu_j(x_0) \in \{E_{2j-1}, E_{2j}\}$, $1 \le j \le g$. *Then* $\nu_0(x_0) = E_0$, $\nu_j(x_0) \in \{E_{2j-1}, E_{2j}\}\backslash\{\mu_j(x_0)\}$, $1 \le j \le g$. *Conversely, suppose* $\nu_j(x_0) \in \{E_{2j-1}, E_{2j}\}$, $1 \le j \le g$. *Then* $\nu_0(x_0) = E_0$, $\mu_j(x_0) \in \{E_{2j-1}, E_{2j}\}\backslash\{\nu_j(x_0)\}$, $1 \le j \le g$.

Proof. If $\mu_j(x_0) \in \{E_{2j-1}, E_{2j}\}$, $1 \le j \le g$ then $G_{g-1}(z,x_0) = 0$ in (3.11) yields $R_{2g+1}(z) = F_g(z,x_0)H_{g+1}(z,x_0)$ and hence proves the first claim. Conversely, assuming $\nu_j(x_0) \in \{E_{2j-1}, E_{2j}\}$, $1 \le j \le g$ one infers from (3.13) that $G_{g-1}(\nu_j(x_0),x_0) = iR_{2g+1}^{1/2}(\hat{\nu}_j(x_0)) = 0$, $1 \le j \le g$, i.e., again $G_{g-1}(z,x_0) = 0$. Hence $R_{2g+1}(z) = F_g(z,x_0)H_{g+1}(z,x_0)$ also proves the second claim. $\square$

Given the bounded potential $V(x)$ in (3.19), consider the differential expression $\tau = -\frac{d^2}{dx^2} + V(x)$ and define the corresponding self-adjoint Schrödinger operator H in $L^2(\mathbb{R})$ by

$$Hf = \tau f,\ \tau = -\frac{d^2}{dx^2} + V(x), \quad x \in \mathbb{R},\ f \in \mathcal{D}(H) = H^{2,2}(\mathbb{R}), \tag{3.31}$$

with $H^{m,n}(.)$ the usual Sobolev spaces. The resolvent of H reads

$$((H-z)^{-1}f)(x) = \int_{\mathbb{R}} dx' G(z,x,x')f(x'),\ z \in \mathbb{C}\backslash\sigma(H),\ f \in L^2(\mathbb{R}), \tag{3.32}$$

where the Green's function $G(z,x,x')$ is explicitly given by

$$G(z,x,x') = W(\psi_+(z,.,x_0),\psi_-(z,.,x_0))^{-1}\begin{cases}\psi_+(z,x,x_0)\psi_-(z,x',x_0), & x \geq x'\\ \psi_+(z,x',x_0)\psi_-(z,x,x_0), & x \leq x'\end{cases}, \tag{3.33}$$

with $W(f,g) = fg' - f'g$ the Wronskian of f and g and $\psi_\pm(z,x,x_0)$ the branches of $\psi(P,x,x_0)$ in the charts $(\Pi_\pm,\tilde{\pi})$. One computes

$$W(\psi_+(z,.,x_0),\psi_-(z,.,x_0)) = (2/i)R_{2g+1}(z)^{1/2}F_g(z,x_0)^{-1} \tag{3.34}$$

and

$$G(z,x,x) = \frac{i\prod_{j=1}^{g}[z-\mu_j(x)]}{2R_{2g+1}(z)^{1/2}} = \frac{iF_g(z,x)}{2R_{2g+1}(z)^{1/2}}, \tag{3.35}$$

taking into account our convention (A.3) for $R_{2g+1}(z)^{1/2}$. In particular, the spectrum $\sigma(H)$ of H is given by

$$\sigma(H) = \bigcup_{j=0}^{g-1}[E_{2j},E_{2j+1}] \cup [E_{2g},\infty). \tag{3.36}$$

Eq. (3.35) illustrates the spectral theoretic content of the polynomial $F_g(z,x)$. Moreover, the Weyl m-functions $m_\pm(z,x_0)$, associated with the restriction of τ to $(x_0,\pm\infty)$ with a Dirichlet boundary condition at x_0, read

$$m_\pm(z,x_0) = \phi_\pm(z,x_0) = [\pm iR_{2g+1}(z)^{1/2} - G_{g-1}(z,x_0)]F_g(z,x_0)^{-1}, \tag{3.37}$$

where $\phi_\pm(z,x)$ denote the branches of $\phi(P,x)$ in the charts $(\Pi_\pm,\tilde{\pi})$. As a consequence, the

Weyl m-matrix $M(z,x_0)$ associated with H is given by (see, e.g., [32], Ch. 8)

$$
\begin{aligned}
M(z,x_0) &= [m_-(z,x_0)-m_+(z,x_0)]^{-1}\begin{pmatrix} m_-(z,x_0)m_+(z,x_0) & [m_-(z,x_0)+m_+(z,x_0)]/2 \\ [m_-(z,x_0)+m_+(z,x_0)]/2 & 1 \end{pmatrix} \\
&= \begin{pmatrix} \partial_1\partial_2 G(z,x_0,x_0) & \frac{1}{2}(\partial_1+\partial_2)G(z,x_0,x_0) \\ \frac{1}{2}(\partial_1+\partial_2)G(z,x_0,x_0) & G(z,x_0,x_0) \end{pmatrix} \\
&= \frac{i}{2R_{2g+1}(z)^{1/2}}\begin{pmatrix} H_{g+1}(z,x_0) & -G_{g-1}(z,x_0) \\ -G_{g-1}(z,x_0) & F_g(z,x_0) \end{pmatrix},
\end{aligned} \tag{3.38}
$$

where

$$
\begin{aligned}
&\partial_1 G(z,x_0,x') = \partial_x G(z,x,x')\big|_{x=x_0}, \quad \partial_2 G(z,x,x_0) = \partial_{x'} G(z,x,x')\big|_{x'=x_0}, \\
&\partial_1\partial_2 G(z,x_0,x_0) = \partial_x\partial_{x'} G(z,x,x')\big|_{x=x_0=x'}, \quad \text{etc.}
\end{aligned} \tag{3.39}
$$

The corresponding self-adjoint spectral matrix $\rho(\lambda,x_0)$, defined by

$$
M_{p,q}(z,x_0) = \int_{\mathbb{R}} (z-\lambda)^{-1} d\rho_{p,q}(\lambda,x_0), \tag{3.40}
$$

$$
\begin{aligned}
\rho_{p,q}(\lambda,x_0) - \rho_{p,q}(\mu,x_0) &= \lim_{\delta\downarrow 0}\lim_{\epsilon\downarrow 0}\pi^{-1}\int_{\mu+\delta}^{\lambda+\delta} d\nu\, \operatorname{Im}[M_{p,q}(\nu+i\epsilon,x_0)], \\
&\qquad \lambda,\mu\in\mathbb{R},\ 1\le p,q\le 2,
\end{aligned} \tag{3.41}
$$

explicitly reads (cf., e.g., [32], Ch. 8)

$$
\frac{d\rho_{1,1}(\lambda,x_0)}{d\lambda} = \begin{cases} \dfrac{H_{g+1}(\lambda,x_0)}{2\pi R_{2g+1}(\lambda)^{1/2}}, & \lambda\in\sigma(H)^o \\ 0, & \lambda\in\mathbb{R}\backslash\sigma(H) \end{cases}, \tag{3.42}
$$

$$
\frac{d\rho_{1,2}(\lambda,x_0)}{d\lambda} = \frac{d\rho_{2,1}(\lambda,x_0)}{d\lambda} = \begin{cases} \dfrac{-G_{g-1}(\lambda,x_0)}{2\pi R_{2g+1}(\lambda)^{1/2}}, & \lambda\in\sigma(H)^o \\ 0, & \lambda\in\mathbb{R}\backslash\sigma(H) \end{cases}, \tag{3.43}
$$

$$
\frac{d\rho_{2,2}(\lambda,x_0)}{d\lambda} = \begin{cases} \dfrac{F_g(\lambda,x_0)}{2\pi R_{2g+1}(\lambda)^{1/2}}, & \lambda\in\sigma(H)^o \\ 0, & \lambda\in\mathbb{R}\backslash\sigma(H) \end{cases}. \tag{3.44}
$$

(Here A^o denotes the interior of $A\subset\mathbb{R}$.)

Closely associated with H is $H^{\beta}_{x_0}$ in $L^2(\mathbb{R})$ defined by

$$\begin{aligned} &H^{\beta}_{x_0} f = \tau f, \beta \in \mathbb{R} \cup \{\infty\}, \quad x_0 \in \mathbb{R}, \\ &f \in \mathcal{D}(H^{\beta}_{x_0}) = \{g \in L^2(\mathbb{R}) | g, g' \in AC([x_0, \pm R]) \text{ for all } R > 0, \\ &\qquad \lim_{\epsilon \downarrow 0} [g'(x_0 \pm \epsilon) + \beta g(x_0 \pm \epsilon)] = 0, \ \tau g \in L^2(\mathbb{R})\}, \end{aligned} \tag{3.45}$$

with $AC_{(\text{loc})}(I)$ the set of (locally) absolutely continuous functions on I. Here, in obvious notation, $\beta = \infty$ denotes the Dirichlet Schrödinger operator $H^D_{x_0} = H^{\infty}_{x_0}$ and $\beta = 0$ the corresponding Neumann Schrödinger operator $H^N_{x_0} = H^0_{x_0}$. Moreover, $H^{\beta}_{x_0}$ decomposes into the direct sum of half-line operators

$$H^{\beta}_{x_0} = H^{\beta}_{-,x_0} \oplus H^{\beta}_{+,x_0}, \ L^2(\mathbb{R}) = L^2((-\infty, x_0]) \oplus L^2([x_0, \infty)). \tag{3.46}$$

The resolvent of $H^{\beta}_{x_0}$ reads

$$((H^{\beta}_{x_0} - z)^{-1} f)(x) = \int_{\mathbb{R}} dx' G^{\beta}_{x_0}(z, x, x') f(x'), \quad z \in \mathbb{C} \backslash \sigma(H^{\beta}_{x_0}), \ f \in L^2(\mathbb{R}), \tag{3.47}$$

$$\begin{aligned} G^{\beta}_{x_0}(z, x, x') &= G(z, x, x') - \frac{(\beta + \partial_2) G(z, x, x_0)(\beta + \partial_1) G(z, x_0, x')}{(\beta + \partial_1)(\beta + \partial_2) G(z, x_0, x_0)}, \\ &\qquad \beta \in \mathbb{R}, \ z \in \mathbb{C} \backslash \{\sigma(H^{\beta}_{x_0}) \cup \sigma(H)\}, \end{aligned} \tag{3.48}$$

$$\begin{aligned} G^{\infty}_{x_0}(z, x, x') &= G(z, x, x') - G(z, x, x_0) G(z, x_0, x') G(z, x_0, x_0)^{-1}, \\ &\qquad z \in \mathbb{C} \backslash \{\sigma(H^{\infty}_{x_0}) \cup \sigma(H)\}. \end{aligned} \tag{3.49}$$

Next we define the polynomial $K^{\beta}_{g+1}(z, x)$, $\beta \in \mathbb{R}$ of degree $g + 1$ in z,

$$K^{\beta}_{g+1}(z, x) = H_{g+1}(z, x) - 2\beta G_{g-1}(z, x) + \beta^2 F_g(z, x) = \prod_{\ell=0}^{g} [z - \lambda^{\beta}_{\ell}(x)], \quad \beta \in \mathbb{R}. \tag{3.50}$$

In particular,

$$H_{g+1}(z, x) = K^0_{g+1}(z, x), \ \nu_{\ell}(x) = \lambda^0_{\ell}(x), \quad 0 \le \ell \le g. \tag{3.51}$$

Explicitly, one computes

$$\begin{aligned}
K_1^\beta &= \beta^2 - V + z = \tilde{K}_1^\beta,\\
K_2^\beta &= c_1(\beta^2 - V) + \tfrac{1}{4}V_{xx} - \tfrac{1}{2}V^2 + \tfrac{1}{2}\beta V_x + \tfrac{1}{2}\beta^2 V\\
&\quad + \left(c_1 + \beta^2 - \tfrac{1}{2}V\right)z + z^2 = c_1\tilde{K}_1^\beta + \tilde{K}_2^\beta,\\
K_3^\beta &= c_2(\beta^2 - V) + c_1\left(\tfrac{1}{2}\beta^2 V + \tfrac{1}{2}\beta V_x + \tfrac{1}{4}V_{xx} - \tfrac{1}{2}V^2\right) - \tfrac{1}{8}\beta V_{xxx}\\
&\quad + \tfrac{3}{4}\beta V V_x - \tfrac{1}{8}\beta^2 V_{xx} + \tfrac{3}{8}\beta^2 V^2 - \tfrac{1}{16}V_{xxxx} + \tfrac{3}{8}V_x^2 + \tfrac{1}{2}VV_{xx} - \tfrac{3}{8}V^3\\
&\quad + [\tfrac{1}{2}\beta V_x + \beta^2 c_1 + \tfrac{1}{2}\beta^2 V - \tfrac{1}{2}c_1 V + c_2 + \tfrac{1}{8}V_{xx} - \tfrac{3}{8}V^2]z + [c_1 + \beta^2 - \tfrac{1}{2}V]z^2 + z^3\\
&= c_2\tilde{K}_1^\beta + c_1\tilde{K}_2^\beta + \tilde{K}_3^\beta,
\end{aligned} \tag{3.52}$$

etc.

Then (3.35) and

$$(\beta + \partial_1)(\beta + \partial_2)G(z,x,x) = \frac{iK_{g+1}^\beta(z,x)}{2R_{2g+1}(z)^{1/2}}, \quad \beta \in \mathbb{R} \tag{3.53}$$

together with (3.47) and (3.48) yield

$$\sigma(H_{x_0}^\beta) = \sigma(H) \cup \{\lambda_\ell^\beta(x_0)\}_{0\le\ell\le g}, \quad \beta \in \mathbb{R}, \tag{3.54}$$

$$\sigma(H_{x_0}^\infty) = \sigma(H) \cup \{\mu_j(x_0)\}_{1\le j\le g}, \quad \mu_j(x_0) = \lambda_j^\infty(x_0),\ 1 \le j \le g, \tag{3.55}$$

with

$$\lambda_0^\beta(x_0) \le E_0,\ \beta \in \mathbb{R},\ \lambda_\ell^\beta(x_0) \in [E_{2\ell-1}, E_{2\ell}],\ 1 \le \ell \le g, \quad \beta \in \mathbb{R} \cup \{\infty\}. \tag{3.56}$$

Next, one verifies

$$\begin{aligned}
\phi(P,x) + \beta &= \frac{iR_{2g+1}^{1/2}(P) - G_{g-1}(\tilde{\pi}(P),x) + \beta F_g(\tilde{\pi}(P),x)}{F_g(\tilde{\pi}(P),x)}\\
&= \frac{-K_{g+1}^\beta(\tilde{\pi}(P),x)}{iR_{2g+1}^{1/2}(P) + G_{g-1}(\tilde{\pi}(P),x) - \beta F_g(\tilde{\pi}(P),x)},
\end{aligned} \tag{3.57}$$

$$R_{2g+1}(z) + [G_{g-1}(z,x) - \beta F_g(z,x)]^2 = F_g(z,x)K_{g+1}^\beta(z,x), \tag{3.58}$$

$$[\phi(P,x) + \beta][\phi(P^*,x) + \beta] = K_{g+1}^\beta(z,x)/F_g(z,x), \tag{3.59}$$

$$[\psi_x(P,x,x_0) + \beta\psi(P,x,x_0)][\psi_x(P^*,x,x_0) + \beta\psi(P^*,x,x_0)] = K_{g+1}^\beta(z,x)/F_g(z,x_0). \tag{3.60}$$

The divisor of $\phi(.,x) + \beta$, $\beta \in \mathbb{R}$ is given by

$$(\phi(.,x) + \beta) = \mathcal{D}_{\hat{\lambda}_0^\beta(x)\hat{\underline{\lambda}}^\beta(x)} - \mathcal{D}_{P_\infty\hat{\underline{\mu}}(x)}, \tag{3.61}$$

with

$$R_{2g+1}^{1/2}(\hat{\lambda}_\ell^\beta(x)) = -iG_{g-1}(\lambda_\ell^\beta(x),x) + i\beta F_g(\lambda_\ell^\beta(x),x),$$
$$\hat{\lambda}_\ell^\beta(x) = (\lambda_\ell^\beta(x), -iG_{g-1}(\lambda_\ell^\beta(x),x) + i\beta F_g(\lambda_\ell^\beta(x),x)),\ 0 \le \ell \le g,\ \beta \in \mathbb{R}. \tag{3.62}$$

The first-order system of differential equations for $\lambda_\ell^\beta(x)$, $\beta \in \mathbb{R}$, i.e., the analog of (3.5) in the case $\beta = \infty$, will be derived in the next section (see (4.45) for $r = 0$). Here we only record the final result for completeness,

$$\lambda_\ell^{\beta\prime}(x) = -2i[\beta^2 - V(x) + \lambda_\ell^\beta(x)] R_{2g+1}^{1/2}(\hat{\lambda}_\ell^\beta(x)) \prod_{\substack{m=0\\ m\neq\ell}}^{g} [\lambda_\ell^\beta(x) - \lambda_m^\beta(x)]^{-1},$$
$$\tilde{\pi}(\hat{\lambda}_0^\beta(x)) = \lambda_0^\beta(x) \le E_0,\ \tilde{\pi}(\hat{\lambda}_\ell^\beta(x)) = \lambda_\ell^\beta(x) \in [E_{2\ell-1}, E_{2\ell}],\ 1 \le \ell \le g,\ (\beta, x) \in \mathbb{R}^2. \tag{3.63}$$

In particular, taking $\beta = 0$ in (3.63) then yields the first-order system of differential equations for $\nu_\ell(x)$, $0 \le \ell \le g$. (We remark that $V(x)$ in (3.63) has been used for reasons of brevity only. In order to obtain a system of differential equations for $\lambda_\ell^\beta(x)$ one needs to replace $V(x)$ by the corresponding trace formula (see, e.g., (5.23), (5.30), and (5.34)).)

We emphasize that due to our convention (A.3) for $R_{2g+1}^{1/2}(P)$, the differential equations (3.5) and (3.63) exhibit the well-known monotonicity properties of $\mu_j(x)$ and $\lambda_j^\beta(x)$, $\beta \in \mathbb{R}$, $j \ge 1$ with respect to $x \in \mathbb{R}$. For instance, Dirichlet eigenvalues corresponding to the right (left) half axis (x, ∞) ($(-\infty, x)$) associated with the decomposition (3.46) are always increasing (decreasing) with respect to $x \in \mathbb{R}$, etc.

We conclude with the θ-function representation for $\phi(P,x)$, $\psi(P,x,x_0)$, $V(x)$ to be derived in Section 4 (cf. Theorem 4.6) in the general t-dependent case.

$$\begin{aligned}\phi(P,x) = &-\beta + \frac{\theta(\Xi_{P_0} - \underline{A}_{P_0}(P_\infty) + \underline{\alpha}_{P_0}(\hat{\underline{\mu}}(x)))}{\theta(\Xi_{P_0} - \underline{A}_{P_0}(P_\infty) + \underline{\alpha}_{P_0}(\hat{\underline{\lambda}}^\beta(x)))} \bullet \\ &\bullet \frac{\theta(\Xi_{P_0} - \underline{A}_{P_0}(P) + \underline{\alpha}_{P_0}(\hat{\underline{\lambda}}^\beta(x)))}{\theta(\Xi_{P_0} - \underline{A}_{P_0}(P) + \underline{\alpha}_{P_0}(\hat{\underline{\mu}}(x)))} \exp\Big[-\int_{P_0}^P \omega^{(3)}_{P_\infty, \hat{\lambda}_0^\beta(x)}\Big],\end{aligned} \tag{3.64}$$

$$\begin{aligned}\psi(P,x,x_0) = &\frac{\theta(\Xi_{P_0} - \underline{A}_{P_0}(P) + \underline{\alpha}_{P_0}(\hat{\underline{\mu}}(x)))}{\theta(\Xi_{P_0} - \underline{A}_{P_0}(P_\infty) + \underline{\alpha}_{P_0}(\hat{\underline{\mu}}(x)))} \frac{\theta(\Xi_{P_0} - \underline{A}_{P_0}(P_\infty) + \underline{\alpha}_{P_0}(\hat{\underline{\mu}}(x_0)))}{\theta(\Xi_{P_0} - \underline{A}_{P_0}(P) + \underline{\alpha}_{P_0}(\hat{\underline{\mu}}(x_0)))} \bullet \\ &\bullet \exp\Big[-i(x-x_0)\int_{P_0}^P \omega^{(2)}_{P_\infty,0}\Big], \quad P_0 = (E_0, 0),\ \beta = \mathbb{R},\end{aligned} \tag{3.65}$$

with the linearizing property of the Abel map,

$$\underline{\alpha}_{P_0}(\underline{\hat{\mu}}(x)) = \underline{\alpha}_{P_0}(\underline{\hat{\mu}}(x_0)) + \frac{\underline{U}_0^{(2)}}{2\pi}(x - x_0), \quad (x, x_0) \in \mathbb{R}^2, \tag{3.66}$$

$$\underline{U}_0^{(2)} = (U_{0,1}^{(2)}, \ldots, U_{0,g}^{(2)}),\ U_{0,j}^{(2)} = \int_{b_j} \omega_{P_\infty,0}^{(2)},\ 1 \leq j \leq g. \tag{3.67}$$

The Its-Matveev formula [26], [4], Ch. 3, [38], Ch. II for $V(x)$ then reads

$$\begin{aligned} V(x) &= E_0 + \sum_{j=1}^{g}(E_{2j-1} + E_{2j} - 2\lambda_j) - 2\partial_x^2 \ln[\theta(\underline{\Xi}_{P_0} - \underline{A}_{P_0}(P_\infty) + \underline{\alpha}_{P_0}(\underline{\hat{\mu}}(x)))] \\ &= E_0 + \sum_{j=1}^{g}(E_{2j-1} + E_{2j} - 2\lambda_j) - 2\partial_x^2 \ln[\theta(\underline{\Xi}_{P_0} + \underline{A}_{P_0}(\hat{\lambda}_0^\beta(x)) + \underline{\alpha}_{P_0}(\underline{\hat{\lambda}}^\beta(x)))], \end{aligned} \tag{3.68}$$

where $\lambda_j \in [E_{2j-1}, E_{2j}]$, $1 \leq j \leq g$ are determined from

$$\omega_{P_\infty,0}^{(2)} = -[2R_{2g+1}^{1/2}(.)]^{-1} \prod_{j=1}^{g}(\tilde{\pi} - \lambda_j) d\tilde{\pi} \underset{\zeta \to 0}{=} [\zeta^{-2} + 0(1)]\, d\zeta \text{ near } P_\infty \tag{3.69}$$

and the second equality in (3.68) is a consequence of the equivalence $\mathcal{D}_{P_\infty \underline{\hat{\mu}}(x)} \sim \mathcal{D}_{\hat{\lambda}_0^\beta(x)\underline{\hat{\lambda}}^\beta(x)}$, i.e.,

$$\underline{A}_{P_0}(P_\infty) + \underline{\alpha}_{P_0}(\underline{\hat{\mu}}(x)) = \underline{A}_{P_0}(\hat{\lambda}_0^\beta(x)) + \underline{\alpha}_{P_0}(\underline{\hat{\lambda}}^\beta(x)),\ x \in \mathbb{R}. \tag{3.70}$$

4. The Time-Dependent Formalism

In this section we construct algebro-geometric solutions of the KdV hierarchy corresponding to g-gap initial values on the basis of a suitable time-dependent generalization of the polynomial approach developed in Chapters 2 and 3. Even though the final results (4.51)–(4.55) are well-known, in fact, classical by now, the approach presented in this section, based on the fundamental meromorphic function $\phi(P, x, t)$ in (4.15), merits attention since it easily extends to general 1 + 1-dimensional completely integrable systems such as the AKNS and Toda hierarchies. (The corresponding approach to the Toda and Kac-van Moerbeke lattices is presented in detail in [5].) The results (4.36)–(4.50) in connection with the general β-boundary condition in (3.45) and our strategy of proof of the theta function representation of the BA function $\psi(P, x, x_0, t, t_0)$ in (4.51), based on (4.15) and (4.16), are new.

Our starting point will be a g-gap solution $V^{(0)}$ of the stationary KdV$_g$ equation,

$$V^{(0)}(x) = E_0 + \sum_{j=1}^{g}[E_{2j-1} + E_{2j} - 2\mu_j^{(0)}(x)],\ \mu_j^{(0)}(x) \in [E_{2j-1}, E_{2j}], \quad 1 \leq j \leq g \tag{4.1}$$

satisfying

$$\sum_{\ell=0}^{g} c_{g-\ell} f_{\ell+1,x} = 0, \quad c_0 = 1, \tag{4.2}$$

where $f_{\ell+1}$ are given by (2.4) with $V = V^{(0)}$. Our principal aim then is to construct the KdV flow

$$\mathrm{KdV}_r(V) = 0, \; V(x,t_0) = V^{(0)}(x), \quad x \in \mathbb{R} \tag{4.3}$$

for some $r \in \mathbb{N}_0$. In terms of Lax operators this amounts to solving

$$\frac{d}{dt}L(t) - [P_{2r+1}(t), L(t)] = 0, \; t \in \mathbb{R}, \; [P_{2g+1}(t_0), L(t_0)] = 0. \tag{4.4}$$

As a consequence one then obtains that

$$[P_{2g+1}(t), L(t)] = 0, \quad t \in \mathbb{R}, \tag{4.5}$$

$$-P_{2g+1}(t)^2 = R_{2g+1}(L(t)) = \prod_{n=0}^{2g} (L(t) - E_n), \quad t \in \mathbb{R} \tag{4.6}$$

since the KdV_r flows are isospectral deformations of $L(t_0)$. In this paper we shall base the explicit solution of (4.3) not directly on (4.4) and (4.5) but instead take the following equations as our point of departure,

$$V_t = -\frac{1}{2}\hat{F}_{r,xxx} + 2(V-z)\hat{F}_{r,x} + V_x \hat{F}_r, \quad (x,t) \in \mathbb{R}^2, \tag{4.7}$$

$$F_{g,xx}F_g - \frac{1}{2}F_{g,x}^2 - 2(V-z)F_g^2 = 2R_{2g+1}(z), \quad (x,t) \in \mathbb{R}^2, \tag{4.8}$$

where

$$F_g(z,x,t) = \prod_{j=1}^{g} [z - \mu_j(x,t)] \tag{4.9}$$

(cf. (2.14), (2.18), and (2.23)). In order to stress the fact that the integration constants c_ℓ used in F_g and F_r (cf. (2.10), (2.14)) in general can differ from each other, we explicitly employ the notation F_g, G_{g-1}, H_{g+1}, K_{g+1}^{β}, etc. and $\hat{F}_r$, $\hat{G}_{r-1}$, $\hat{H}_{r+1}$, $\hat{K}_{r+1}^{\beta}$, etc. throughout this section. Similarly to (3.5)–(3.8), (3.11), and (3.12) we have

$$\mu_{j,x}(x,t) = -2iR_{2g+1}^{1/2}(\hat{\mu}_j(x,t)) \prod_{\substack{k=1 \\ k \neq j}}^{g} [\mu_j(x,t) - \mu_k(x,t)]^{-1}, \; 1 \le j \le g, \; (x,t) \in \mathbb{R}^2, \tag{4.10}$$

$$\{\hat{\mu}_j(x_0,t)\}_{1 \le j \le g} \subset K_g, \; \tilde{\pi}(\hat{\mu}_j(x_0,t)) = \mu_j(x_0,t) \in [E_{2j-1}, E_{2j}], \; 1 \le j \le g, \; t \in \mathbb{R}, \tag{4.11}$$

$$G_{g-1}(z,x,t) = -\frac{1}{2}F_{g,x}(z,x,t) = -i\sum_{j=1}^{g} R_{2g+1}^{1/2}(\hat{\mu}_j(x,t)) \prod_{\substack{k=1\\k\neq j}}^{g} \left(\frac{z-\mu_k(x,t)}{\mu_j(x,t)-\mu_k(x,t)}\right), \quad (4.12)$$

$$R_{2g+1}(z) + G_{g-1}(z,x,t)^2 = F_g(z,x,t)H_{g+1}(z,x,t), \quad (4.13)$$

$$H_{g+1}(z,x,t) = \prod_{\ell=0}^{g}[z-\nu_\ell(x,t)]. \quad (4.14)$$

In analogy to (3.15) and (3.18) one then considers the meromorphic function $\phi(P,x,t)$ on K_g,

$$\phi(P,x,t) = \frac{iR_{2g+1}^{1/2}(P) - G_{g-1}(\tilde{\pi}(P),x,t)}{F_g(\tilde{\pi}(P),x,t)} = \frac{-H_{g+1}}{iR_{2g+1}^{1/2}(P) + G_{g-1}(\tilde{\pi}(P),x,t)}, \quad (x,t)\in\mathbb{R}^2 \quad (4.15)$$

and the t-dependent BA function $\psi(P,x,x_0,t,t_0)$, meromorphic on $K_g\backslash\{P_\infty\}$,

$$\psi(P,x,x_0,t,t_0) = \exp\Big\{\int_{t_0}^{t} ds[\hat{F}_r(z,x_0,s)\phi(P,x_0,s) + \hat{G}_{r-1}(z,x_0,s)] + \int_{x_0}^{x} dy\phi(P,y,t)\Big\},$$
$$(x,x_0,t,t_0)\in\mathbb{R}^4. \quad (4.16)$$

Lemma 4.1. *Let* $P = (z,\sigma R_{2g+1}(z)^{1/2}) = (\tilde{\pi}(P), R_{2g+1}^{1/2}(P)) \in K_g\backslash\{P_\infty\}$, $(z,x,x_0,t,t_0)\in \mathbb{C}\times\mathbb{R}^4$, $r\in\mathbb{N}_0$. *Then*

(i). $$V(x,t) = E_0 + \sum_{j=1}^{g}[E_{2j-1}+E_{2j}-2\mu_j(x,t)]. \quad (4.17)$$

(ii). $\phi(P,x,t)$ *satisfies*

$$\phi_x(P,x,t) + \phi(P,x,t)^2 = V(x,t) - z, \quad (4.18)$$

$$\phi_t(P,x,t) = \partial_x[\hat{F}_r(z,x,t)\phi(P,x,t) + \hat{G}_{r-1}(z,x,t)]. \quad (4.19)$$

(iii). $\psi(P,x,x_0,t,t_0)$ *satisfies*

$$-\psi_{xx}(P,x,x_0,t,t_0) + [V(x,t)-z]\psi(P,x,x_0,t,t_0) = 0, \quad (4.20)$$

$$\psi_t(P,x,x_0,t,t_0) = \hat{F}_r(z,x,t)\psi_x(P,x,x_0,t,t_0) + \hat{G}_{r-1}(z,x,t)\psi(P,x,x_0,t,t_0). \quad (4.21)$$

(iv). $$\phi(P,x,t)\phi(P^*,x,t) = H_{g+1}(z,x,t)/F_g(z,x,t). \quad (4.22)$$

(v). $$\phi(P,x,t)+\phi(P^*,x,t) = -2G_{g-1}(z,x,t)/F_g(z,x,t) = F_{g,x}(z,x,t)/F_g(z,x,t). \quad (4.23)$$

(vi). $$\phi(P,x,t)-\phi(P^*,x,t) = 2iR_{2g+1}^{1/2}(P)/F_g(z,x,t). \quad (4.24)$$

Proof. The proof of (i), (4.18), (4.20), and (iv)–(vi) is analogous to that in Lemma 3.1. In order to prove (4.19) one can argue as follows. By (4.7) and (4.18),

$$\partial_t(\phi_x + \phi^2) = \phi_{tx} + 2\phi\phi_t = V_t = (\hat{F}_r\phi + \hat{G}_{r-1})_{xx} + 2\phi(\hat{F}_r\phi + \hat{G}_{r-1})_x,$$

which implies

$$(\partial_x + 2\phi)[\phi_t - (\hat{F}_r\phi + \hat{G}_{r-1})_x] = 0 \text{ and hence } \phi_t = (\hat{F}_r\phi + \hat{G}_{r-1})_x + Ce^{2\int^x dy\phi},$$

where C is independent of x (but may depend on P and t). The high-energy behavior of $\phi(P,x,t)$ derived from (4.15) yields $\phi(P,x,t) \underset{z\to\infty}{=} \pm i(z)^{1/2} + 0(1)$, $P \in \Pi_\pm$ uniformly in $(x,t) \in \mathbb{R}^2$ and hence $C = 0$ proving (4.19). (4.21) then immediately follows from (4.16) and (4.19). □

In analogy to (3.28) we now introduce

$$\hat{H}_{r+1}(z,x,t) = \frac{1}{2}\hat{F}_{r,xx}(z,x,t) - [V(x,t) - z]\hat{F}_r(z,x,t). \tag{4.25}$$

From (4.7) and (4.8) one then computes

$$\hat{H}_{r+1,x} = \frac{1}{2}\hat{F}_{r,xxx} - (V-z)\hat{F}_{r,x} - V_x\hat{F}_r = -V_t - 2(V-z)\hat{G}_{r-1}. \tag{4.26}$$

The t-dependence of F_g, G_{g-1}, and H_{g+1} is governed by

Lemma 4.2. *Let* $(z,x,t) \in \mathbb{C} \times \mathbb{R}^2$, $r \in \mathbb{N}_0$. *Then*

(i). $$F_{g,t}(z,x,t) = 2[F_g(z,x,t)\hat{G}_{r-1}(z,x,t) - \hat{F}_r(z,x,t)G_{g-1}(z,x,t)]. \tag{4.27}$$

(ii). $$G_{g-1,t}(z,x,t) = F_g(z,x,t)\hat{H}_{r+1}(z,x,t) - \hat{F}_r(z,x,t)H_{g+1}(z,x,t). \tag{4.28}$$

(iii). $$H_{g+1,t}(z,x,t) = 2[\hat{H}_{r+1}(z,x,t)G_{g-1}(z,x,t) - H_{g+1}(z,x,t)\hat{G}_{r-1}(z,x,t)]. \tag{4.29}$$

Proof. By (2.17), (4.19), and (4.24),

$$\begin{aligned}\phi_t(P) - \phi_t(P^*) &= -2iR_{2g+1}^{1/2}(P)F_g^{-2}F_{g,t}\\ &= \partial_x[\hat{F}_r(\phi(P) - \phi(P^*))] = 2iR_{2g+1}^{1/2}(P)(F_g\hat{F}_{r,x} - F_{g,x}\hat{F}_r)F_g^{-2},\end{aligned}$$

implying (4.27). Similarly, by (2.17) and (4.25),

$$G_{g-1,t} = -\tfrac{1}{2}F_{g,tx} = \hat{F}_rG_{g-1,x} - F_g\hat{G}_{r-1,x} = F_g\hat{H}_{r+1} - \hat{F}_rH_{g+1}.$$

Finally, (2.17), (4.25)–(4.28) yield

$$H_{g+1,t} = -G_{g-1,tx} - (V-z)F_{g,t} - V_tF_g = -F_{g,x}\hat{H}_{r+1} - F_g\hat{H}_{r+1,x} + \hat{F}_{r,x}H_{g+1} + \hat{F}_rH_{g+1,x}$$
$$-2(V-z)(F_g\hat{G}_{r-1} - \hat{F}_rG_{g-1}) - V_tF_g = 2(G_{g-1}\hat{H}_{r+1} - 2\hat{G}_{r-1}H_{g+1}). \quad \square$$

As a consequence of (4.27) one obtains the following time-dependence of $\mu_j(x,t)$.

Corollary 4.3. *Let* $(x,t) \in \mathbb{R}^2$, $r \in \mathbb{N}_0$. *Then*

$$\mu_{j,t}(x,t) = -2i\hat{F}_r(\mu_j(x,t),x)R_{2g+1}^{1/2}(\hat{\mu}_j(x,t))\prod_{\substack{k=1\\k\neq j}}^{g}[\mu_j(x,t)-\mu_k(x,t)]^{-1},$$
$$\hat{\mu}_j(x,t_0) = \hat{\mu}_j^{(0)}(x), \quad 1 \le j \le g, \tag{4.30}$$
$$\tilde{\pi}(\hat{\mu}_j(x,t)) = \mu_j(x,t) \in [E_{2j-1}, E_{2j}], \quad 1 \le j \le g. \tag{4.31}$$

Proof. Take $z = \mu_j(x,t)$ in (4.27) and observe (4.1), (4.3), (4.9), and

$$R_{2g+1}^{1/2}(\hat{\mu}_j(x,t)) = iG_{g-1}(\mu_j(x,t),x,t),\ \hat{\mu}_j(x,t) = (\mu_j(x,t), iG_{g-1}(\mu_j(x,t),x,t)), \tag{4.32}$$

the latter fact following from (4.12) (as in (3.9)). $\square$

One observes that the special case $r = 0$ (i.e., $\hat{F}_0 = 1$) in (4.30) is equivalent to (4.10), (4.11).

Next we record the remaining t-dependent analogs of Lemma 3.1 (vii)–(ix).

Lemma 4.4. *Let* $P = (z, \sigma R_{2g+1}(z)^{1/2}) = (\tilde{\pi}(P), R_{2g+1}^{1/2}(P)) \in K_g \backslash \{P_\infty\}$, $(z,x,x_0,t,t_0) \in \mathbb{C} \times \mathbb{R}^4$, $r \in \mathbb{N}_0$. *Then*

(i). $$\psi(P,x,x_0,t,t_0)\psi(P^*,x,x_0,t,t_0) = F_g(z,x,t)/F_g(z,x_0,t_0). \tag{4.33}$$

(ii). $$\psi_x(P,x,x_0,t,t_0)\psi_x(P^*,x,x_0,t,t_0) = H_{g+1}(z,x,t)/F_g(z,x_0,t_0). \tag{4.34}$$

(iii). $$\psi(P,x,x_0,t,t_0) = [F_g(z,x,t)/F_g(z,x_0,t_0)]^{1/2} \bullet$$
$$\bullet \exp\Big\{iR_{2g+1}^{1/2}(P)\Big[\int_{t_0}^{t} ds\hat{F}_r(z,x_0,s)F_g(z,x_0,s)^{-1} + \int_{x_0}^{x} dyF_g(z,y,t)^{-1}\Big]\Big\}. \tag{4.35}$$

Proof. (i). Combining (4.16), (4.23), and (4.27) yields

$$\psi(P,x,x_0,t,t_0)\psi(P^*,x,x_0,t,t_0)$$
$$= \exp\Big[\int_{t_0}^{t} dsF_{g,s}(z,x_0,s)F_g(z,x_0,s)^{-1} + \int_{x_0}^{x} dyF_{g,y}(z,y,t)F_g(z,y,t)^{-1}\Big]$$
$$= [F_g(z,x_0,t)/F_g(z,x_0,t_0)][F_g(z,x,t)/F_g(z,x_0,t)] = F_g(z,x,t)/F_g(z,x_0,t_0).$$

(ii). (4.22), (4.33) and $\psi_x = \phi\psi$ imply

$$\begin{aligned}&\psi_x(P,x,x_0,t,t_0)\psi_x(P^*,x,x_0,t,t_0)\\&=[H_{g+1}(z,x,t)/F_g(z,x,t)][F_g(z,x,t)/F_g(z,x_0,t_0)]=H_{g+1}(z,x,t)/F_g(z,x_0,t_0).\end{aligned}$$

(iii). Follows from (4.15), (4.16), and (4.27). □

Remark 4.5. We emphasize that instead of taking (4.7) and (4.8) as our starting point for solving (4.3), and subsequently deriving the first-order differential system (4.10), (4.30), one could have started directly with the system (4.10), (4.30) and derived (4.7), (4.8) and the remaining facts of this section (cf. [5]).

Next, we turn to the t-dependent analog of (3.50)–(3.63) and start by introducing

$$K_{g+1}^{\beta}(z,x,t)=H_{g+1}(z,x,t)-2\beta G_{g-1}(z,x,t)+\beta^2F_g(z,x,t)=\prod_{\ell=0}^{g}[z-\lambda_\ell^\beta(x,t)],\quad \beta\in\mathbb{R}, \tag{4.36}$$

with

$$H_{g+1}(z,x,t)=K_{g+1}^0(z,x,t),\ \nu_\ell(x,t)=\lambda_\ell^0(x,t),\quad 0\le\ell\le g. \tag{4.37}$$

One then verifies in analogy to (3.57)–(3.62) that

$$\begin{aligned}\phi(P,x,t)+\beta&=\frac{iR_{2g+1}^{1/2}(P)-G_{g-1}(\tilde\pi(P),x,t)+\beta F_g(\tilde\pi(P),x,t)}{F_g(\tilde\pi(P),x,t)}\\&=\frac{-K_{g+1}^{\beta}(\tilde\pi(P),x,t)}{iR_{2g+1}^{1/2}(P)+G_{g-1}(\tilde\pi(P),x,t)-\beta F_g(\tilde\pi(P),x,t)},\end{aligned} \tag{4.38}$$

$$R_{2g+1}(z)+[G_{g-1}(z,x,t)-\beta F_g(z,x,t)]^2=F_g(z,x,t)K_{g+1}^{\beta}(z,x,t), \tag{4.39}$$

$$[\phi(P,x,t)+\beta][\phi(P^*,x,t)+\beta]=K_{g+1}^{\beta}(z,x,t)/F_g(z,x,t), \tag{4.40}$$

$$\begin{aligned}&[\psi_x(P,x,x_0,t,t_0)+\beta\psi(P,x,x_0,t,t_0)][\psi_x(P^*,x,x_0,t,t_0)+\beta\psi(P^*,x,x_0,t,t_0)]\\&=K_{g+1}^{\beta}(z,x,t)/F_g(z,x_0,t_0),\end{aligned} \tag{4.41}$$

$$(\phi(.,x,t)+\beta)=\mathcal{D}_{\hat\lambda_0^\beta(x,t)\underline{\hat\lambda}^\beta(x,t)}-\mathcal{D}_{P_\infty\underline{\hat\mu}(x,t)}, \tag{4.42}$$

with

$$\begin{aligned}&R_{2g+1}^{1/2}(\hat\lambda_\ell^\beta(x,t))=-iG_{g-1}(\lambda_\ell^\beta(x,t),x,t)+i\beta F_g(\lambda_\ell^\beta(x,t),x,t),\\&\hat\lambda_\ell^\beta(x,t)=(\lambda_\ell^\beta(x,t),-iG_{g-1}(\lambda_\ell^\beta(x,t),x,t)+i\beta F_g(\lambda_\ell^\beta(x,t),x,t)),\quad 0\le\ell\le g.\end{aligned} \tag{4.43}$$

Eq. (4.36) and Lemma 4.2 then yield

$$\begin{aligned} K^{\beta}_{g+1,t}(z,x,t) = 2\{\hat{K}^{\beta}_{r+1}(z,x,t)[G_{g-1}(z,x,t) - \beta F_g(z,x,t)] \\ - K^{\beta}_{g+1}(z,x,t)[\hat{G}_{r-1}(z,x,t) - \beta \hat{F}_r(z,x,t)]\} \end{aligned} \tag{4.44}$$

and in analogy to Corollary 4.3 one obtains from (4.44),

$$\begin{aligned} &\lambda^{\beta}_{\ell,t}(x,t) = -2i\hat{K}^{\beta}_{r+1}(\lambda^{\beta}_{\ell}(x,t),x,t)R^{1/2}_{2g+1}(\hat{\lambda}^{\beta}_{\ell}(x,t)) \prod_{\substack{m=0\\ m\neq \ell}}^{g} [\lambda^{\beta}_{\ell}(x,t) - \lambda^{\beta}_{m}(x,t)]^{-1}, \\ &\hat{\lambda}^{\beta}_{\ell}(x,t_0) = \hat{\lambda}^{\beta,(0)}_{\ell}(x), \quad 0 \le \ell \le g, \ (x,t) \in \mathbb{R}^2, \end{aligned} \tag{4.45}$$

$$\tilde{\pi}(\hat{\lambda}^{\beta}_{0}(x,t)) = \lambda^{\beta}_{0}(x,t) \le E_0, \ \tilde{\pi}(\hat{\lambda}^{\beta}_{\ell}(x,t)) = \lambda^{\beta}_{\ell}(x,t) \in [E_{2\ell-1}, E_{2\ell}], \quad (x,t) \in \mathbb{R}^2, \tag{4.46}$$

where $\{\lambda^{\beta,(0)}_{\ell}(y)\}_{0\le\ell\le g}$ are the corresponding eigenvalues of $H^{\beta,(0)}_{y}$ (cf. (3.45), (3.54), and (3.56)) associated with the initial value $V^{(0)}(x)$ in (4.1).

In an analogous fashion one can analyze the behavior of $\lambda^{\beta}_{\ell}(x,t)$ as a function of $\beta \in \mathbb{R}$. In fact, (4.36) yields

$$\frac{\partial}{\partial\beta} K^{\beta}_{g+1}(z,x,t) = -2[G_{g-1}(z,x,t) - \beta F_g(z,x,t)] \tag{4.47}$$

and hence

$$\begin{aligned} \frac{\partial}{\partial\beta} K^{\beta}_{g+1}(z,x,t)\Big|_{z=\lambda^{\beta}_{\ell}(x,t)} &= -\Big[\frac{\partial}{\partial\beta}\lambda^{\beta}_{\ell}(x,t)\Big] \prod_{\substack{m=0\\ m\neq \ell}}^{g} [\lambda^{\beta}_{\ell}(x,t) - \lambda^{\beta}_{m}(x,t)] \\ &= -2[G_{g-1}(\lambda^{\beta}_{\ell}(x,t),x,t) - \beta F_g(\lambda^{\beta}_{\ell}(x,t),x,t)] = -2iR^{1/2}_{2g+1}(\hat{\lambda}^{\beta}_{\ell}(x,t)) \end{aligned} \tag{4.48}$$

by (4.43). This implies for $(\beta,x,t) \in \mathbb{R}^3$,

$$\frac{\partial}{\partial\beta}\lambda^{\beta}_{\ell}(x,t) = 2iR^{1/2}_{2g+1}(\hat{\lambda}^{\beta}_{\ell}(x,t)) \prod_{\substack{m=0\\ m\neq \ell}}^{g} [\lambda^{\beta}_{\ell}(x,t) - \lambda^{\beta}_{m}(x,t)]^{-1}, \quad 0 \le \ell \le g. \tag{4.49}$$

As in Section 3 we conclude with the θ-function representation of $\phi(P,x,t)$, $\psi(P,x,x_0,t,t_0)$, and $V(x,t)$.

Theorem 4.6. *Let* $P = (z, \sigma R_{2g+1}(z)^{1/2}) \in K_g \backslash \{P_\infty\}$, $(z,x,x_0,t,t_0) \in \mathbb{C}\times\mathbb{R}^4$, $P_0 = (E_0, 0)$. *Then*

$$\begin{aligned} \phi(P,x,t) = -\beta + \frac{\theta(\underline{\Xi}_{P_0} - \underline{A}_{P_0}(P_\infty) + \underline{\alpha}_{P_0}(\underline{\hat{\mu}}(x,t)))}{\theta(\underline{\Xi}_{P_0} - \underline{A}_{P_0}(P_\infty) + \underline{\alpha}_{P_0}(\underline{\hat{\lambda}}^{\beta}(x,t)))} \,\bullet \\ \bullet\, \frac{\theta(\underline{\Xi}_{P_0} - \underline{A}_{P_0}(P) + \underline{\alpha}_{P_0}(\underline{\hat{\lambda}}^{\beta}(x,t)))}{\theta(\underline{\Xi}_{P_0} - \underline{A}_{P_0}(P) + \underline{\alpha}_{P_0}(\underline{\hat{\mu}}(x,t)))} \exp\Big[-\int_{P_0}^{P} \omega^{(3)}_{P_\infty,\hat{\lambda}^{\beta}_{0}(x,t)}\Big] \end{aligned} \tag{4.50}$$

and

$$\psi(P,x,x_0,t,t_0)=\frac{\theta(\underline{\Xi}_{P_0}-\underline{A}_{P_0}(P)+\underline{\alpha}_{P_0}(\hat{\underline{\mu}}(x,t)))}{\theta(\underline{\Xi}_{P_0}-\underline{A}_{P_0}(P_\infty)+\underline{\alpha}_{P_0}(\hat{\underline{\mu}}(x,t)))}\bullet$$
$$\bullet\frac{\theta(\underline{\Xi}_{P_0}-\underline{A}_{P_0}(P_\infty)+\underline{\alpha}_{P_0}(\hat{\underline{\mu}}(x_0,t_0)))}{\theta(\underline{\Xi}_{P_0}-\underline{A}_{P_0}(P)+\underline{\alpha}_{P_0}(\hat{\underline{\mu}}(x_0,t_0)))}\exp\Big[-i(x-x_0)\int_{P_0}^{P}\omega^{(2)}_{P_\infty,0}-i(t-t_0)\int_{P_0}^{P}\Omega^{(2)}_{P_\infty,2r}\Big], \tag{4.51}$$

where (cf. (A.26)*)*

$$\Omega^{(2)}_{P_\infty,2r}=\sum_{s=0}^{r}c_{r-s}(2s+1)\omega^{(2)}_{P_\infty,2s}, \tag{4.52}$$

$$\underline{\alpha}_{P_0}(\hat{\underline{\mu}}(x,t))=\underline{\alpha}_{P_0}(\hat{\underline{\mu}}(x_0,t_0))+\frac{\underline{U}^{(2)}_0}{2\pi}(x-x_0)+\frac{\underline{U}^{(2)}_{2r}}{2\pi}(t-t_0), \tag{4.53}$$

$$\underline{U}^{(2)}_{2r}=(U^{(2)}_{2r,1},\dots,U^{(2)}_{2r,g}),\ U^{(2)}_{2r,j}=\int_{b_j}\Omega^{(2)}_{P_\infty,2r},\quad 1\le j\le g. \tag{4.54}$$

The Its-Matveev formula ([26], [4], *Ch. 3,* [38], *Ch. II) for $V(x,t)$ reads (cf.* (3.68)*)*

$$V(x,t)=E_0+\sum_{j=1}^{g}(E_{2j-1}+E_{2j}-2\lambda_j)-2\partial_x^2\ln[\theta(\underline{\Xi}_{P_0}-\underline{A}_{P_0}(P_\infty)+\underline{\alpha}_{P_0}(\hat{\underline{\mu}}(x,t)))]. \tag{4.55}$$

Sketch of Proof. Since (4.50) follows directly from (4.42) and (A.29), and (4.55) can be inferred from (4.51) and (4.20) upon expanding all quantities in (4.20) near P_∞ in a well-known manner, we first concentrate on the proof of (4.51). Let $\psi(P,x,x_0,t,t_0)$ be defined as in (4.16) and denote the right-hand-side of (4.51) by $\Psi(P,x,x_0,t,t_0)$. In order to prove that $\psi=\Psi$, one first observes from (4.10) and (4.30) that

$$\hat{F}_r(\tilde{\pi}(P),x_0,s)\phi(P,x_0,s)=\tfrac{\partial}{\partial s}\ln[\mu_j(x_0,s)-\tilde{\pi}(P)]+0(1)\text{ for }P\text{ near }\hat{\mu}_j(x_0,s) \tag{4.56}$$

and

$$\phi(P,y,t)=\tfrac{\partial}{\partial y}\ln[\mu_j(y,t)-\tilde{\pi}(P)]+0(1)\text{ for }P\text{ near }\hat{\mu}_j(y,t). \tag{4.57}$$

Hence

$$\exp\Big\{\int_{t_0}^{t}ds[\tfrac{\partial}{\partial s}\ln(\mu_j(x_0,s)-\tilde{\pi}(P))+0(1)]\Big\}$$
$$=\begin{cases}[\mu_j(x_0,t)-\tilde{\pi}(P)]0(1) & \text{for }P\text{ near }\hat{\mu}_j(x_0,t)\neq\hat{\mu}_j(x_0,t_0)\\ 0(1) & \text{for }P\text{ near }\hat{\mu}_j(x_0,t)=\hat{\mu}_j(x_0,t_0)\\ [\mu_j(x_0,t_0)-\tilde{\pi}(P)]^{-1}0(1) & \text{for }P\text{ near }\hat{\mu}_j(x_0,t_0)\neq\hat{\mu}_j(x_0,t)\end{cases} \tag{4.58}$$

and

$$\exp\Big\{\int_{x_0}^{x} dy[\tfrac{\partial}{\partial y}\ln(\mu_j(y,t)-\tilde{\pi}(P))+0(1)]\Big\}$$
$$=\begin{cases}[\mu_j(x,t)-\tilde{\pi}(P)]0(1) & \text{for } P \text{ near } \hat{\mu}_j(x,t)\neq\hat{\mu}_j(x_0,t)\\ 0(1) & \text{for } P \text{ near } \hat{\mu}_j(x,t)=\hat{\mu}_j(x_0,t)\,,\\ [\mu_j(x_0,t)-\tilde{\pi}(P)]^{-1}0(1) & \text{for } P \text{ near } \hat{\mu}_j(x_0,t)\neq\hat{\mu}_j(x,t)\end{cases} \tag{4.59}$$

where $0(1)\neq 0$ in (4.58) and (4.59). Consequently, all zeros and poles of ψ and Ψ on $K_g\backslash\{P_\infty\}$ are simple and coincide. By an application of the Riemann-Roch theorem it remains to identify the essential singularity of ψ and Ψ at P_∞. For that purpose we first recall the known fact that the diagonal Green's function $G(z,x,x,t)$ of $H(t)$ satisfies

$$G(z,x,x,t) \underset{\zeta\to 0}{=} (i/2)\zeta\sum_{j=0}^{\infty}\tilde{f}_j(x,t)\zeta^{2j}, \quad \zeta=1/\sqrt{z}, \tag{4.60}$$

with $\tilde{f}_j(x,t)$ the homogeneous coefficients as introduced in the context of (2.12) satisfying the recursion (2.4) for all $j\in\mathbb{N}$. Combining

$$G(z,x,x,t)=\frac{iF_g(z,x,t)}{2R_{2g+1}(z)^{1/2}} \tag{4.61}$$

(cf. (3.35)), (4.15), (4.16), and (4.60) then yields

$$\int_{x_0}^{x} dy\phi(P,y,t) \underset{\zeta\to 0}{=} \int_{x_0}^{x} dy\frac{iR_{2g+1}^{1/2}(P)}{F_g(\tilde{\pi}(P),y,t)}+0(\zeta^2) \underset{\zeta\to 0}{=} i(x-x_0)[\zeta^{-1}+0(1)], \tag{4.62}$$

which coincides with the singularity at P_∞ of the x-dependent term in the exponent of (4.51) taking into account (3.69). Finally, in order to identify the t-dependent essential singularity of ψ and Ψ, we may allude to (2.20) and, without loss of generality, consider the homogeneous case where $c_0=1$, $c_q=0$, $1\leq q\leq r$. Invoking (4.27) then yields from (4.15) and (4.61)

$$\begin{aligned}&\int_{t_0}^{t} ds[\tilde{F}_r(z,x_0,s)\phi(P,x_0,s)+\tilde{G}_{r-1}(z,x_0,s)]\\ &=\int_{t_0}^{t} ds\Big\{\tilde{F}_r(z,x_0,s)iR_{2g+1}^{1/2}(P)F_g(z,x_0,s)^{-1}+\tfrac{1}{2}\tfrac{\partial}{\partial s}\ln[F_g(z,x_0,s)]\Big\}\\ &\underset{\zeta\to 0}{=} -\tfrac{1}{2}\textstyle\int_{t_0}^{t} ds\tilde{F}_r(z,x_0,s)G(z,x_0,x_0,s)^{-1}+0(1), \quad \zeta=1/\sqrt{z}.\end{aligned} \tag{4.63}$$

Comparing (2.14) (in the homogeneous case) and (4.60) implies

$$-\tfrac{1}{2}\tilde{F}_r(z,x_0,s)G(z,x_0,x_0,s)^{-1} \underset{\zeta\to 0}{=} i\zeta^{-2r-1}+0(1) \tag{4.64}$$

and hence

$$\int_{t_0}^{t} ds[\tilde{F}_r(z,x_0,s)\phi(P,x_0,s)+\tilde{G}_{r-1}(z,x_0,s)] \underset{\zeta\to 0}{=} i(t-t_0)[\zeta^{-2r-1}+0(1)], \tag{4.65}$$

completing the proof of (4.51). The linearity of the Abel map with respect to x and t in (4.53) then follows by a standard argument considering the differential $\Omega(x,x_0,t,t_0) = d\ln\psi(.,x,x_0,t,t_0)$. □

5. General Trace Formulas

Following a recent series of papers on new trace formulas for Schrödinger operators [16]–[19], [22]–[24], [39], we first discuss appropriate Krein spectral shift functions, the key tool for general higher-order trace formulas. Subsequently, we develop a new method for deriving small-time heat kernel (respectively high-energy resolvent) expansion coefficients associated with the general β-boundary conditions in (5.3). Interest in these types of trace formulas stems from their crucial role in the solution of inverse spectral problems.

Unlike Sections 3 and 4, where we focused on the special case of stationary finite-gap solutions of the KdV hierarchy (the natural extension of solitons as reflectionless potentials), we now turn to the general situation and consider potentials of the type

$$V\in C^\infty(\mathbb{R}),\ V(x)\geq c,\ x\in\mathbb{R},\ V \text{ real-valued.} \tag{5.1}$$

As in Section 3 we introduce the differential expression $\tau = -\frac{d^2}{dx^2}+V(x)$, $x\in\mathbb{R}$ and define the self-adjoint operators H and $H_{x_0}^{\beta}$ in $L^2(\mathbb{R})$,

$$Hf=\tau f,\ f\in\mathcal{D}(H)=\{g\in L^2(\mathbb{R})|g,g'\in AC_{\text{loc}}(\mathbb{R});\ \tau g\in L^2(\mathbb{R})\} \tag{5.2}$$

and for $\beta\in\mathbb{R}\cup\{\infty\}$, $x_0\in\mathbb{R}$,

$$\begin{aligned} H_{x_0}^{\beta}f=\tau f,\ f\in\mathcal{D}(H_{x_0}^{\beta}) =&\{f\in L^2(\mathbb{R})|g,g'\in AC([x_0,\pm R]) \text{ for all } R>0,\\ &\lim_{\epsilon\downarrow 0}[g'(x_0\pm\epsilon)+\beta g(x_0\pm\epsilon)]=0,\quad \tau g\in L^2(\mathbb{R})\}, \end{aligned} \tag{5.3}$$

with $H_{x_0}^{\infty}=H_{x_0}^{D}$ ($H_{x_0}^{0}=H_{x_0}^{N}$) the corresponding Dirichlet (Neumann) Schrödinger operator. If $G(z,x,x')$ denotes the Green's function of H (as in (3.32), (3.33)), formulas (3.47)–(3.49) for the resolvent of $H_{x_0}^{\beta}$ apply without change in the present general situation. In particular, defining

$$\Gamma^{\beta}(z,x)=\begin{cases}(\beta+\partial_1)(\beta+\partial_2)G(z,x,x), & \beta\in\mathbb{R}\\ G(z,x,x), & \beta=\infty\end{cases} \tag{5.4}$$

(cf. the notation introduction in (3.39)) one computes for $\beta\in\mathbb{R}\cup\{\infty\}$,

$$\operatorname{Tr}[(H_x^{\beta}-z)^{-1}-(H-z)^{-1}]=-\frac{d}{dz}\ln[\Gamma^{\beta}(z,x)],\ z\in\mathbb{C}\backslash\{\sigma(H_x^{\beta})\cup\sigma(H)\}. \tag{5.5}$$

Given hypothesis (5.1), one can prove the existence of asymptotic expansions of the type

$$\operatorname{Tr}[(H_x^{\beta}-z)^{-1}-(H-z)^{-1}]\underset{z\to i\infty}{=}\sum_{j=0}^{\infty}r_j^{\beta}(x)z^{-j},\quad \beta\in\mathbb{R}\cup\{\infty\} \tag{5.6}$$

uniformly with respect to $x\in\mathbb{R}$ (cf. [24]). In particular, one can derive the heat kernel expansion

$$\operatorname{Tr}[e^{-\tau H_x^{\infty}}-e^{-\tau H}]\underset{\tau\downarrow 0}{\sim}\sum_{j=0}^{\infty}s_j^{\infty}(x)\tau^j,\quad x\in\mathbb{R}, \tag{5.7}$$

where

$$s_j^{\infty}(x)=(-1)^{j+1}(j!)^{-1}r_j^{\infty}(x),\quad j\in\mathbb{N}_0 \tag{5.8}$$

and s_j^{∞} (r_j^{∞}) are the well-known invariants of the KdV hierarchy.

In the special case of finite-gap potentials the connection of $\Gamma^{\beta}(z,s)$ in (5.4) with our polynomial approach in Section 3 is clearly demonstrated by (3.35) for $\beta=\infty$ and (3.53) for $\beta=\mathbb{R}$.

Before describing a new constructive (i.e., recursive) approach to the coefficients $r_j^{\beta}(x)$, $\beta\in\mathbb{R}$, we recall the definition of Krein's spectral shift function [30] associated with the pair (H_x^{β},H) (cf. [19], [23], [24]). The rank-one resolvent difference of H_x^{β} and H (cf. (3.47), (3.48)) is intimately connected with the fact that for each $x\in\mathbb{R}$, $\beta\in\mathbb{R}\cup\{\infty\}$,

$$\Gamma^{\beta}(z,x)\text{ is Herglotz with respect to }z \tag{5.9}$$

(i.e., a holomorphic map $\mathbb{C}_+\to\mathbb{C}_+$, where $\mathbb{C}_+=\{z\in\mathbb{C}\,|\,\mathrm{Im}(z)>0\}$). The exponential Herglotz representation for $\Gamma^{\beta}(z,x)$ (cf. [3]) then reads for each $x\in\mathbb{R}$,

$$\Gamma^{\beta}(z,x)=\exp\Big\{c^{\beta}+\int_{\mathbb{R}}[(\lambda-z)^{-1}-\lambda(1+\lambda^2)^{-1}][\xi^{\beta}(\lambda,x)+\delta^{\beta}]\,d\lambda\Big\},$$

$$c^{\beta}\in\mathbb{R},\ \beta\in\mathbb{R}\cup\{\infty\},\ \delta^{\beta}=\begin{cases}1, & \beta\in\mathbb{R}\\ 0, & \beta=\infty\end{cases}, \tag{5.10}$$

where, by Fatou's lemma,

$$\xi^\beta(\lambda,x) = \pi^{-1}\lim_{\epsilon\downarrow 0}\operatorname{Im}\{\ln[(\beta+\partial_1)(\beta+\partial_2)G(\lambda+i\epsilon,x,x)]\} - \delta^\beta, \quad \beta\in\mathbb{R}\cup\{\infty\} \tag{5.11}$$

for a.e. $\lambda\in\mathbb{R}$. Moreover,

$$\begin{aligned} &-1\le\xi^\beta(\lambda,x)\le 0,\ \xi^\beta(\lambda,x)=0,\ \lambda<\inf\sigma(H_x^\beta),\ \beta\in\mathbb{R},\\ &0\le\xi^\infty(\lambda,x)\le 1,\ \xi^\infty(\lambda,x)=0,\ \lambda<\inf\sigma(H) \end{aligned} \tag{5.12}$$

for a.e. $\lambda\in\mathbb{R}$. As a consequence, one obtains (cf. [39])

$$\operatorname{Tr}[f(H_x^\beta)-f(H)] = \int_{\mathbb{R}} d\lambda f'(\lambda)\xi^\beta(\lambda,x), \quad \beta\in\mathbb{R}\cup\{\infty\},\ x\in\mathbb{R} \tag{5.13}$$

for any $f\in C^2(\mathbb{R})$ with $(1+\lambda^2)f^{(j)}\in L^2((0,\infty))$, $j=1,2$ and for $f(\lambda)=(\lambda-z)^{-1}$, $z\in\mathbb{C}\backslash[\inf\sigma(H_x^\beta),\infty)$. In particular, (5.13) holds for traces of heat kernel and resolvent differences, i.e., for any $\beta\in\mathbb{R}\cup\{\infty\}$, $x\in\mathbb{R}$,

$$\operatorname{Tr}[e^{-\tau H_x^\beta}-e^{-\tau H}] = -\tau\int_{e_{x,0}^\beta}^\infty d\lambda e^{-\tau\lambda}\xi^\beta(\lambda,x), \quad \tau>0, \tag{5.14}$$

$$\operatorname{Tr}[(H_x^\beta-z)^{-1}-(H-z)^{-1}] = -\int_{e_{x,0}^\beta}^\infty d\lambda(\lambda-z)^{-2}\xi^\beta(\lambda,x), \quad z\in\mathbb{C}\backslash\{\sigma(H_x^\beta)\cup\sigma(H)\}, \tag{5.15}$$

where

$$e_{x,0}^\beta = \begin{cases} \inf\sigma(H_x^\beta), & \beta\in\mathbb{R}\\ \inf\sigma(H), & \beta=\infty \end{cases}. \tag{5.16}$$

Returning to a recursive approach for the expansion coefficients $r_j^\beta(x)$ in (5.6) we first consider the expansion

$$\Gamma^\beta(z,x) \underset{z\to i\infty}{=} (i/2)\sum_{j=-\delta^\beta}^\infty \gamma_j^\beta(x)z^{-j-1/2}, \quad \beta\in\mathbb{R}\cup\{\infty\}. \tag{5.17}$$

(A comparison of (5.17) and (4.60) reveals that $\gamma_j^\infty(x)=\tilde f_j(x)$, $j\in\mathbb{N}_0$ in the case $\beta=\infty$.) In order to obtain a recursion relation for $\gamma_j^\beta(x)$ one can use the following result.

Lemma 5.1. *Let* $z\in\mathbb{C}\backslash\sigma(H)$, $x\in\mathbb{R}$.
(i). Assume $\beta\in\mathbb{R}$. *Then* $\Gamma^\beta(z,x)=(\beta+\partial_1)(\beta+\partial_2)G(z,x,x)$ *satisfies*

$$\begin{aligned} &2[V(x)-\beta^2-z]\Gamma_{xx}^\beta(z,x)\Gamma^\beta(z,x)-[V(x)-\beta^2-z]\Gamma_x^\beta(z,x)^2-2V_x(x)\Gamma_x^\beta(z,x)\Gamma^\beta(z,x)\\ &-4\{[V(x)-z][V(x)-\beta^2-z]-\beta V_x(x)\}\Gamma^\beta(z,x)^2 = -[V(x)-z-\beta^2]^3. \end{aligned} \tag{5.18}$$

(ii). Assume $\beta = \infty$. Then $\Gamma^\infty(z,x) = G(z,x,x)$ satisfies

$$\Gamma^\infty_{xxx}(z,x) - 4[V(x) - z]\Gamma^\infty_x(z,x) - 2V_x(x)\Gamma^\infty(z,x) = 0 \tag{5.19}$$

and

$$-2\Gamma^\infty_{xx}(z,x)\Gamma^\infty(z,x) + \Gamma^\infty_x(z,x)^2 + 4[V(x) - z]\Gamma^\infty(z,x)^2 = 1. \tag{5.20}$$

While the results (5.19) and (5.20) in the Dirichlet case $\beta = \infty$ are well-known, see, e.g., [14], the result (5.18) (with the exception of the Neumann case $\beta = 0$ which was first presented in [21]) is new. Unfortunately, we have no reasonably short derivation of the differential equation (5.18). It can be verified (not without tears) after quite tedious though straightforward calculations (we recommend additional help in the form of symbolic computations).

Insertion of the expansion (5.17) into (5.18) and (5.20) in Lemma 5.1 yields

Lemma 5.2. *The coefficients $\gamma^\beta_j(x)$ in (5.17) satisfy the following recursion relation.*
(i). Assume $\beta \in \mathbb{R}$. Then

$$\begin{aligned}
\gamma^\beta_{-1} &= 1,\ \gamma^\beta_0 = \beta^2 - \tfrac{1}{2}V,\ \gamma^\beta_1 = \tfrac{1}{2}\beta^2 V + \tfrac{1}{2}\beta V_x - \tfrac{1}{8}V^2 + \tfrac{1}{8}V_{xx},\\
\gamma^\beta_2 &= -\tfrac{1}{16}V^3 + \tfrac{3}{8}\beta^2V^2 + \tfrac{3}{16}V_x(4\beta V + V_x) + \tfrac{1}{8}V_{xx}(V - \beta^2) - \tfrac{1}{8}\beta V_{xxx} - \tfrac{1}{64}V_{xxxx},\\
\gamma^\beta_{j+1} &= \tfrac{1}{8}\textstyle\sum_{\ell=1}^j [2(V - \beta^2)\gamma^\beta_{\ell-1}\gamma^\beta_{j-\ell,xx} - (V - \beta^2)\gamma^\beta_{\ell-1,x}\gamma^\beta_{j-\ell,x}\\
&\quad - 4\gamma^\beta_\ell\gamma^\beta_{j-\ell+1} - 4V(V - \beta^2)\gamma^\beta_{\ell-1}\gamma^\beta_{j-\ell} - 2V_x\gamma^\beta_{\ell-1}\gamma^\beta_{j-\ell,x} + \gamma^\beta_{\ell-1}\gamma^\beta_{j-\ell}]\\
&\quad + \tfrac{1}{8}\textstyle\sum_{\ell=0}^j [\gamma^\beta_{\ell,x}\gamma^\beta_{j-\ell,x} - 2\gamma^\beta_\ell\gamma^\beta_{j-\ell,xx} - 4(\beta^2 - 2V)\gamma^\beta_\ell\gamma^\beta_{j-\ell}], \quad j \geq 2.
\end{aligned} \tag{5.21}$$

(ii). Assume $\beta = \infty$. Then

$$\begin{aligned}
&\gamma^\infty_0 = 1,\ \gamma^\infty_1 = \tfrac{1}{2}V,\\
&\gamma^\infty_{j+1} = -\tfrac{1}{2}\textstyle\sum_{\ell=1}^j \gamma^\infty_\ell\gamma^\infty_{j+1-\ell} + \tfrac{1}{2}\sum_{\ell=0}^j [V\gamma^\infty_\ell\gamma^\infty_{j-\ell} + \tfrac{1}{4}\gamma^\infty_{\ell,x}\gamma^\infty_{j-\ell,x} - \tfrac{1}{2}\gamma^\infty_{\ell,xx}\gamma^\infty_{j-\ell}], \quad j \geq 1.
\end{aligned} \tag{5.22}$$

The final result for $r^\beta_j(x)$ then reads

Theorem 5.3. *The coefficients $r^\beta_j(x)$ in (5.6) satisfy the following recursion relations.*
(i). Assume $\beta \in \mathbb{R}$. Then

$$r^\beta_0(x) = -\tfrac{1}{2},\ r^\beta_1(x) = \beta^2 - \tfrac{1}{2}V(x),\ r^\beta_j(x) = j\gamma^\beta_{j-1}(x) - \textstyle\sum_{\ell=1}^{j-1}\gamma^\beta_{j-\ell-1}(x)r^\beta_\ell(x), \quad j \geq 2. \tag{5.23}$$

(ii). Assume $\beta = \infty$. Then

$$r_0^\infty = \tfrac{1}{2},\ r_1^\infty(x) = \tfrac{1}{2}V(x),\ r_j^\infty(x) = j\gamma_j^\infty(x) - \textstyle\sum_{\ell=1}^{j-1}\gamma_{j-\ell}^\infty(x)r_\ell^\infty(x), \quad j \geq 2. \tag{5.24}$$

Proof. It suffices to combine (5.5), (5.6), (5.17), and the following well-known fact on asymptotic expansions: $F(z) \underset{|z|\to\infty}{=} \sum_{j=1}^\infty c_j z^{-j}$ implies $\ln[1+F(z)] \underset{|z|\to\infty}{=} \sum_{j=1}^\infty d_j z^{-j}$, where $d_1 = c_1$, $d_j = c_j - \sum_{\ell=1}^{j-1}(\ell/j)c_{j-\ell}d_\ell$, $j \geq 2$. □

Theorem 5.3 (i) has first been derived (by using a different strategy) in [24]. The current derivation, based on the universal differential equation (5.18), is new. Combined with (5.21), Theorem 5.3 (i) yields the most efficient algorithm to date for computing $r_j^\beta(x)$, $\beta \in \mathbb{R}$.

The connection between $r_j^\beta(x)$ and $\xi^\beta(\lambda, x)$ is illustrated in the following result.

Theorem 5.4. [24] *Let $e_{x,0}^\beta = \inf\sigma(H_x^\beta)$, $\beta \in \mathbb{R}$, $e_0^\infty = \inf\sigma(H)$.*

(i). Assume $\beta \in \mathbb{R}$. Then

$$r_j^\beta(x) = -\tfrac{1}{2}(e_{x,0}^\beta)^j - \lim_{z\to i\infty}\textstyle\int_{e_{x,0}^\beta}^\infty d\lambda z^{j+1}(\lambda - z)^{-j-1}j(-\lambda)^{j-1}[\tfrac{1}{2} + \xi^\beta(\lambda, x)],\ j \in \mathbb{N}. \tag{5.25}$$

(ii). Assume $\beta = \infty$. Then

$$r_j^\infty(x) = \tfrac{1}{2}(e_0^\infty)^j + \lim_{z\to i\infty}\textstyle\int_{e_0^\infty}^\infty d\lambda z^{j+1}(\lambda - z)^{-j-1}j(-\lambda)^{j-1}[\tfrac{1}{2} - \xi^\infty(\lambda, x)],\ j \in \mathbb{N}. \tag{5.26}$$

We conclude with an example that yields the higher-order trace formulas for periodic potentials which also applies to the (quasi-periodic) finite-gap potentials of Section 3.

Example 5.5. Assume V is periodic, i.e., for some $\Omega > 0$, $V(x+\Omega) = V(x)$ for all $x \in \mathbb{R}$ in addition to (5.1). Then Floquet theory implies

$$\sigma(H) = \bigcup_{n=1}^\infty [E_{2(n-1)}, E_{2n-1}],\ E_0 < E_1 \leq E_2 < E_3 \leq \cdots \tag{5.27}$$

(i). Assume $\beta \in \mathbb{R}$. Then

$$\sigma(H_x^\beta) = \{\lambda_n^\beta(x)\}_{n\in\mathbb{N}_0} \cup \sigma(H),\ \lambda_0^\beta(x) \leq E_0,\ \lambda_n^\beta(x) \in [E_{2n-1}, E_{2n}], \quad n \in \mathbb{N}, \tag{5.28}$$

$$\xi^\beta(\lambda, x) = \begin{cases} 0, & \lambda < \lambda_0^\beta(x),\ E_{2n-1} < \lambda < \lambda_n^\beta(x), \quad n \in \mathbb{N} \\ -1, & \lambda_0^\beta(x) < \lambda < E_0,\ \lambda_n^\beta(x) < \lambda < E_{2n}, \quad n \in \mathbb{N} \\ -\dfrac{1}{2}, & E_{2(n-1)} < \lambda < E_{2n-1}, \quad n \in \mathbb{N} \end{cases}. \tag{5.29}$$

Inserting (5.29) into (5.25) then yields the higher-order periodic trace formulas

$$r_j^\beta(x) = \tfrac{1}{2}E_0^j - \lambda_0^\beta(x)^j + \tfrac{1}{2}\textstyle\sum_{n=1}^\infty[E_{2n-1}^j + E_{2n}^j - 2\lambda_n^\beta(x)^j], \quad j \in \mathbb{N}. \tag{5.30}$$

(ii). Assume $\beta = \infty$. Then

$$\sigma(H_x^\infty) = \{\mu_n(x)\}_{n\in\mathbb{N}} \cup \sigma(H),\ \mu_n(x) \in [E_{2n-1}, E_{2n}], \quad n \in \mathbb{N}, \tag{5.31}$$

$$\xi^\infty(\lambda, x) = \begin{cases} 0, & \lambda < E_0,\ \mu_n(x) < \lambda < E_{2n}, \quad n \in \mathbb{N} \\ 1, & E_{2n-1} < \lambda < \mu_n(x), \quad n \in \mathbb{N} \\ \frac{1}{2}, & E_{2(n-1)} < \lambda < E_{2n-1}, \quad n \in \mathbb{N}. \end{cases} \tag{5.32}$$

Insertion of (5.32) into (5.26) then yields

$$r_j^\infty(x) = \tfrac{1}{2}E_0^j + \tfrac{1}{2}\textstyle\sum_{n=1}^\infty [E_{2n-1}^j + E_{2n}^j - 2\mu_n(x)^j], \quad j \in \mathbb{N}. \tag{5.33}$$

The results (5.29) and (5.32) remain valid in the algebro-geometric finite-gap situation discussed in Section 3 where

$$E_{2n+1} = \lambda_n^\beta(x) = E_{2n+2}, \quad n \geq g+1,\ \beta \in \mathbb{R} \cup \{\infty\}. \tag{5.34}$$

Hence (5.30) and (5.33) apply to the stationary KdV solutions of Section 3 (e.g., (5.33) for $j = 1$ coincides with (3.19)) which, in general, are quasi-periodic with respect to x. Moreover, (5.30) and (5.33) also extend to certain classes of almost-periodic $V(x)$, see, e.g., [8], [28], [31], [32], Chs. 9, 11.

The periodic Dirichlet trace formula (5.33) for $j = 1$ has been noticed by Hochstadt [25] and later on by Dubrovin [10]. The general case $j = \mathbb{N}$ appeared in McKean and van Moerbeke [34] and Flaschka [13]. More recent accounts of (5.33) can be found in [8], [28], [31], [32], Chs. 9, 11, [40]. The Neumann case $\beta = 0$ in (5.30) is due to McKean and Trubowitz [35]. The general case $\beta \in \mathbb{R}$ was first studied in [24]. Additional references on the subject of trace formulas and their use in connection with the inverse spectral problem can be found in the papers cited in this paragraph and in the ones listed at the beginning of this section.

Appendix A. Hyperelliptic Curves of the KdV-type and Theta Functions

We briefly summarize our basic notation for hyperelliptic KdV-type curves (i.e., those branched at infinity) and their theta functions as employed in Sections 3 and 4. For details on this standard material we refer, e.g., to [11], [12], [29], [37].

Consider the points $\{E_n\}_{0\leq n\leq g} \subset \mathbb{R}$, $E_0 < E_1 < \cdots < E_{2g}$, $g \in \mathbb{N}_0$ and define the cut plane $\Pi = \mathbb{C}\backslash \bigcup_{j=0}^{g-1}[E_{2j}, E_{2j+1}] \cup [E_{2g}, \infty)$ with the holomorphic function

$$R_{2g+1}(.)^{1/2} : \begin{cases} \Pi \to \mathbb{C} \\ z \to \left[\prod_{n=0}^{2g}(z - E_n)\right]^{1/2} \end{cases} \tag{A.1}$$

on it. $R_{2g+1}(.)^{1/2}$ is extended to all of $\mathbb{C}$ by

$$R_{2g+1}(\lambda)^{1/2} = \lim_{\epsilon\downarrow 0} R_{2g+1}(\lambda + i\epsilon)^{1/2}, \quad \lambda \in \mathbb{C}\backslash\Pi, \tag{A.2}$$

with the sign of the square root chosen according to

$$R_{2g+1}(\lambda)^{1/2} = \begin{cases} (-1)^g i|R_{2g+1}(\lambda)^{1/2}|, & \lambda \in (-\infty, E_0) \\ (-1)^{g+j} i|R_{2g+1}(\lambda)^{1/2}|, & \lambda \in (E_{2j-1}, E_{2j}), \quad 1 \leq j \leq g \\ (-1)^{g+j}|R_{2g+1}(\lambda)^{1/2}|, & \lambda \in (E_{2j}, E_{2j+1}), \quad 0 \leq j \leq g-1 \\ |R_{2g+1}(\lambda)^{1/2}|, & \lambda \in (E_{2g}, \infty) \end{cases}. \tag{A.3}$$

Next we define the set

$$M = \{(z, \sigma R_{2g+1}(z)^{1/2}) | z \in \mathbb{C},\ \sigma \in \{-, +\}\} \cup \{P_\infty = (\infty, \infty)\} \tag{A.4}$$

and

$$B = \{(E_n, 0)\}_{0\leq n\leq 2g} \cup \{P_\infty = (\infty, \infty)\}, \tag{A.5}$$

the set of branch points. M becomes a compact Riemann surface upon introducing the charts (U_{P_0}, ζ_{P_0}) defined as follows

$$\begin{aligned} &P_0 = (z_0, \sigma_0 R_{2g+1}(z)^{1/2}) \text{ or } P_0 = P_\infty, \\ &P = (z, \sigma R_{2g+1}(z)^{1/2}) \in U_{P_0} \subset M,\ V_{P_0} = \zeta(U_{P_0}) \subset \mathbb{C}. \end{aligned} \tag{A.6}$$

$P_0 \in M\backslash B$:

$U_{P_0} = \{P \in M| \, |z - z_0| < C,\ \sigma R_{2g+1}(z)^{1/2}$ the branch obtained by straight line analytic continuation starting from $z_0\}$, $C = \min_n |z_0 - E_n|$,

$$\zeta_{P_0} : \begin{cases} U_{P_0} \to V_{P_0} \\ P \to (z - z_0) \end{cases}, \quad \zeta_{P_0}^{-1} : \begin{cases} V_{P_0} \to U_{P_0} \\ \zeta \to (z_0 + \zeta, \sigma R_{2g+1}(z_0 + \zeta)^{1/2}) \end{cases}. \tag{A.7}$$

$P_0 = (E_{n_0}, 0)$:

$$U_{P_0} = \{P \in M \,|\, |z - E_{n_0}| < C_{n_0}\}, \quad C_{n_0} = \begin{cases} \min\limits_{n \neq n_0} |E_n - E_{n_0}|, & g \in \mathbb{N} \\ \infty, & g = 0 \end{cases},$$

$$V_{P_0} = \{\zeta \in \mathbb{C} \,|\, |\zeta| < C_{n_0}^{1/2}\}, \quad \zeta_{P_0} : \begin{cases} U_{P_0} \to V_{P_0} \\ P \to \sigma(z - E_{n_0})^{1/2} \end{cases},$$

$$(z - E_{n_0})^{1/2} = |(z - E_{n_0})^{1/2}| e^{(i/2)\arg(z - E_{n_0})}, \ \arg(z - E_{n_0}) \in \begin{cases} [0, 2\pi), & n_0 \text{ even} \\ (-\pi, \pi], & n_0 \text{ odd} \end{cases},$$

$$\zeta_{P_0}^{-1} : \begin{cases} V_{P_0} \to U_{P_0} \\ \zeta \to (E_{n_0} + \zeta^2, \ \zeta[\Pi_{n \neq n_0}(E_{n_0} - E_n + \zeta^2)]^{1/2} \end{cases}, \tag{A.8}$$

$$\Big[\prod_{n \neq n_0} (E_{n_0} - E_n + \zeta^2)\Big]^{1/2} = (-1)^g i^{-n_0} \Big|\Big[\prod_{n \neq n_0} (E_{n_0} - E_n)\Big]^{1/2}\Big| \times$$
$$\Big[1 + 2^{-1}\zeta^2 \sum_{n \neq n_0} (E_{n_0} - E_n)^{-1} + 0(\zeta^4)\Big].$$

$P_0 = P_\infty$:

$$U_{P_0} = \{P \in M \,|\, |z| > C_\infty\}, \ C_\infty = \max_n |E_n|, \ V_{P_0} = \{\zeta \in \mathbb{C} \,|\, |\zeta| < C_\infty^{-1/2}\},$$

$$\zeta_{P_0} : \begin{cases} U_{P_0} \to V_{P_0} \\ P \to \sigma(1/z^{1/2}) \\ P_\infty \to 0 \end{cases}, \quad \begin{aligned} & z^{1/2} = |z^{1/2}| e^{(i/2)\arg(z)}, \\ & 0 \leq \arg(z) < 2\pi, \end{aligned}$$

$$\zeta_{P_0}^{-1} : \begin{cases} V_{P_0} \to U_{P_0} \\ \zeta \to \Big(\zeta^{-2}, \zeta^{-2g-1}\big[\Pi_n(1 - \zeta^2 E_n)\big]^{1/2}\Big) \\ 0 \to P_\infty \end{cases}, \tag{A.9}$$

$$\Big[\prod_n (1 - \zeta^2 E_n)\Big]^{1/2} = 1 - 2^{-1}\zeta^2 \sum_n E_n + 0(\zeta^4).$$

Upper and lower sheets $\Pi_\pm \subset M$ with associated charts $\zeta_\pm$ are defined by

$$\Pi_\pm = \{(z, \pm R_{2g+1}(z)^{1/2}) \in M | z \in \Pi\}, \ \zeta_\pm : \begin{cases} \Pi_\pm \to \Pi \\ P \to z \end{cases}. \tag{A.10}$$

The compact Riemann surface (curve) resulting from (A.4)–(A.9) is denoted by K_g. Topologically, K_g is a sphere with g handles and hence has genus g.

Next, define the holomorphic sheet exchange map (involution)

$$* : \begin{cases} K_g \to K_g \\ (z, \sigma R_{2g+1}(z)^{1/2}) \to (z, \sigma R_{2g+1}(z)^{1/2})^* = (z, -\sigma R_{2g+1}(z)^{1/2}) \end{cases} \tag{A.11}$$

and the two meromorphic projection maps

$$\tilde{\pi} : \begin{cases} K_g \to \mathbb{C} \cup \{\infty\} \\ (z, \sigma R_{2g+1}(z)^{1/2}) \to z \\ P_\infty \to \infty \end{cases}, \quad R_{2g+1}^{1/2} : \begin{cases} K_g \to \mathbb{C} \cup \{\infty\} \\ (z, \sigma R_{2g+1}(z)^{1/2}) \to \sigma R_{2g+1}(z)^{1/2} \\ P_\infty \to \infty \end{cases}. \tag{A.12}$$

$\tilde{\pi}$ has a pole of order 2 at P_∞ and two simple zeros at $(0, \pm R_{2g+1}(z)^{1/2})$ if $R_{2g+1}(0) \neq 0$ or a double zero at $(0,0)$ if $R_{2g+1}(0) = 0$ (i.e., if $0 \in \{E_n\}_{0 \le n \le 2g}$) and $R_{2g+1}^{1/2}$ as a pole of order $2g+1$ at P_∞ and $2g+1$ simple zeros at $(E_n, 0)$, $0 \le n \le 2g$. Moreover,

$$\tilde{\pi}(P^*) = \tilde{\pi}(P), \; R_{2g+1}^{1/2}(P^*) = -R_{2g+1}^{1/2}(P), \quad P \in K_g. \tag{A.13}$$

Thus K_g is a two-sheeted ramified covering of the Riemann sphere $\mathbb{C}_\infty (\cong \mathbb{C} \cup \{\infty\})$, in particular, K_g is compact and hyperelliptic.

Using our local charts one infers that for $g \in \mathbb{N}$, $d\tilde{\pi}/R_{2g+1}^{1/2}$ is a holomorphic differential on K_g with a zero of order $2(g-1)$ at P_∞ and hence

$$\eta_j = \tilde{\pi}^{j-1} d\tilde{\pi}/R_{2g+1}^{1/2}(.), \quad 1 \le j \le g \tag{A.14}$$

form a basis for the space of holomorphic differentials on K_g.

Next we introduce a canonical homology basis $\{a_j, b_j\}_{1 \le j \le g}$ for K_g as follows. The cycle a_ℓ starts near $E_{2\ell-1}$ on Π_+, surrounds $E_{2\ell}$ counterclockwise thereby changing to Π_-, and returns to the starting point encircling $E_{2\ell-1}$, changing sheets again. The cycle b_ℓ surrounds E_0, $E_{2\ell-1}$ counterclockwise (once) on Π_+. The cycles are chosen such that their intersection matrix reads $a_j \circ b_k = \delta_{j,k}$, $1 \le j,k \le g$. Introducing the invertible matrix C in $\mathbb{C}^g$,

$$\begin{aligned} c &= (c_{j,k})_{1 \le j,k \le g}, \; C_{j,k} = \int_{a_k} \eta_j = 2 \int_{E_{2k-1}}^{E_{2k}} z^{j-1} \, dz / R_{2g+1}(z)^{1/2} \in i\mathbb{R}, \\ \underline{c}(k) &= (c_1(k), \dots, c_g(k)), \quad c_j(k) = C_{j,k}^{-1}, \end{aligned} \tag{A.15}$$

the normalized differentials ω_j, $1 \le j \le g$,

$$\omega_j = \sum_{\ell=1}^{g} c_j(\ell) \eta_\ell, \; \int_{a_k} \omega_j = \delta_{j,k}, \quad 1 \le j,k \le g \tag{A.16}$$

form a canonical basis for the space of holomorphic differentials on K_g. The matrix τ in $\mathbb{C}^g$ of b-periods,

$$\tau = (\tau_{j,k})_{1\leq j,k\leq g}, \ \tau_{j,k} = \int_{b_k} \omega_j, \quad 1 \leq j,k \leq g \tag{A.17}$$

then satisfies

$$\tau_{j,k} = \tau_{k,j}, \quad 1 \leq j,k \leq g, \ \tau = iT, \quad T > 0. \tag{A.18}$$

In the chart $(U_{P_\infty}, \zeta_{P_\infty} = \zeta)$ induced by $1/\tilde{\pi}^{1/2}$ near P_∞ one infers

$$\begin{aligned} \underline{\omega} &= -2\Big\{ \sum_{j=1}^{g} \underline{c}(j)\zeta^{2(g-j)} / \Big[\prod_n (1-\zeta^2 E_n)\Big]^{1/2} \Big\} d\zeta \\ &= -2\Big\{ \underline{c}(g) + \Big[\tfrac{1}{2}\underline{c}(g) \textstyle\sum_{n=0}^{2g} E_n + \underline{c}(g-1)\Big]\zeta^2 + 0(\zeta^4) \Big\} d\zeta. \end{aligned} \tag{A.19}$$

Associated with the homology basis $\{a_j, b_j\}_{1\leq j\leq g}$ we also recall the canonical dissection of K_g along its cycles yielding the simply connected interior $\hat{K}_g$ of the fundamental polygon $\partial\hat{K}_g$ given by $\partial\hat{K}_g = a_1 b_1 a_1^{-1} b_1^{-1} a_2 b_2 a_2^{-1} b_2^{-1} \cdots a_g^{-1} b_g^{-1}$. The Riemann theta function associated with K_g is defined by

$$\theta(\underline{z}) = \sum_{\underline{n}\in\mathbb{Z}^g} \exp[2\pi i(\underline{n},\underline{z}) + \pi i(\underline{n}, \tau\underline{n})], \quad \underline{z} = (z_1, \dots, z_g) \in \mathbb{C}^g, \tag{A.20}$$

where $(\underline{u}, \underline{v}) = \sum_{j=1}^{g} u_j v_j$ denotes the scalar product in $\mathbb{C}^g$. It has the fundamental properties

$$\begin{aligned} &\theta(z_1, \dots, z_{j-1}, -z_j, z_{j+1}, \dots, z_g) = \theta(\underline{z}), \\ &\theta(\underline{z} + \underline{m} + \tau\underline{n}) = \theta(\underline{z}) \exp[-2\pi i(\underline{n},\underline{z}) - \pi i(\underline{n},\tau\underline{n})], \quad \underline{m}, \underline{n} \in \mathbb{Z}^g. \end{aligned} \tag{A.21}$$

A divisor $\mathcal{D}$ on K_g is a map $\mathcal{D}: K_g \to \mathbb{Z}$, where $\mathcal{D}(P) \neq 0$ for only finitely-many $P \in K_g$. The set of all divisors will be denoted by $\mathrm{Div}(K_g)$. With L_g we denote the period lattice

$$L_g = \{\underline{z} \in \mathbb{C}^g \,|\, \underline{z} = \underline{m} + \tau\underline{n}, \ \underline{m}, \underline{n} \in \mathbb{Z}^g\} \tag{A.22}$$

and the Jacobi variety $J(K_g)$ is defined by

$$J(K_g) = \mathbb{C}^g / L_g. \tag{A.23}$$

The Abel maps $\underline{A}_{P_0}(.)$, respectively $\underline{\alpha}_{P_0}(.)$ are defined by

$$\underline{A}_{P_0} : \begin{cases} K_g \to J(K_g) \\ P \to \underline{A}_{P_0}(P) = \int_{P_0}^{P} \underline{\omega} \mod (L_g) \end{cases}, \quad \underline{\alpha}_{P_0} : \begin{cases} \mathrm{Div}(K_g) \to J(K_g) \\ \mathcal{D} \to \underline{\alpha}_{P_0}(\mathcal{D}) = \sum_{P\in K_g} \mathcal{D}(P) \underline{A}_{P_0}(P) \end{cases}, \tag{A.24}$$

with $P_0 \in K_g$ a fixed base point. (In the main text we agree to fix $P_0 = (E_0, 0)$ for convenience.)

Next, let $\mathcal{M}(K_g)$ and $\mathcal{M}^1(K_g)$ denote the set of meromorphic functions (0-forms) and meromorphic differentials (1-forms) on K_g. The residue of a meromorphic differential $\nu \in \mathcal{M}^1(K_g)$ at a point $Q_0 \in K_g$ is defined by $\operatorname{res}_{Q_0}(\nu) = (2\pi i)^{-1} \int_{\gamma_{Q_0}} \nu$, where γ_{Q_0} is a counter-clockwise oriented smooth simple closed contour encircling Q_0 but no other pole of ν. Holomorphic differentials are also called (Abelian) differentials of the first kind (dfk); (Abelian) differentials of the second kind (dsk) $\omega^{(2)} \in \mathcal{M}^1(K_g)$ are characterized by the property that all their residues vanish. They are normalized, e.g., by demanding that all their a-periods vanish, i.e.,

$$\int_{a_j} \omega^{(2)} = 0, \quad 1 \le j \le g. \tag{A.25}$$

If $\omega^{(2)}_{P_1,n}$ is a dsk on K_g whose only pole is $P_1 \in \hat{K}_g$ with principal part $\zeta^{-n-2}\, d\zeta$, $n \in \mathbb{N}_0$ near P_1 and $\omega_j = (\sum_{m=0}^{\infty} d_{j,m}(P_1)\zeta^m)\, d\zeta$ near P_1, then

$$\int_{b_j} \omega^{(2)}_{P_1,n} = [2\pi i/(n+1)] d_{j,n}(P_1). \tag{A.26}$$

A basis for dsk's on K_g, holomorphic on $K_g \backslash \{P_\infty\}$, is provided by

$$\omega^{(2)}_n = \tilde{\pi}^{g+1+n} d\tilde{\pi} / R^{1/2}_{2g+1}(.), \quad n \in \mathbb{N}_0. \tag{A.27}$$

Any meromorphic differential $\omega^{(3)}$ on K_g not of the first or second kind is defined to be of the third kind (dtk). A dtk $\omega^{(3)} \in \mathcal{M}^1(K_g)$ is usually normalized by the vanishing of its a-periods, i.e.,

$$\int_{a_j} \omega^{(3)} = 0, \quad 1 \le j \le g. \tag{A.28}$$

A normal dtk $\omega^{(3)}_{P_1,P_2}$ associated with two points $P_1, P_2 \in \hat{K}_g$, $P_1 \neq P_2$ by definition has simple poles at P_1 and P_2 with residues $+1$ at P_1 and -1 at P_2 and vanishing a-periods. If $\omega^{(3)}_{P,Q}$ is a norma l dtk associated with P, $Q \in \hat{K}_g$, holomorphic on $K_g \backslash \{P, Q\}$, then

$$\int_{b_j} \omega^{(3)}_{P,Q} = 2\pi i \int_Q^P \omega_j, \quad 1 \le j \le g, \tag{A.29}$$

where the path from Q to P lies in $\hat{K}_g$ (i.e., does not touch any of the cycles a_j, b_j).

We shall always assume (without loss of generality) that all poles of dsk's and dtk's on K_g lie on $\hat{K}_g$ (i.e., not on $\partial \hat{K}_g$).

For $f \in \mathcal{M}(K_g)\backslash\{0\}$, $\omega \in \mathcal{M}^1(K_g)\backslash\{0\}$ the divisors of f and ω are denoted by (f) and (ω), respectively. Two divisors $\mathcal{D}$, $\mathcal{E} \in \mathrm{Div}(K_g)$ are called equivalent, denoted by $\mathcal{D} \sim \mathcal{E}$, if and only if $\mathcal{D} - \mathcal{E} = (f)$ for some $f \in \mathcal{M}(K_g)\backslash\{0\}$. The divisor class $[\mathcal{D}]$ of $\mathcal{D}$ is then given by $[\mathcal{D}] = \{\mathcal{E} \in \mathrm{Div}(K_g) | \mathcal{E} \sim \mathcal{D}\}$. We recall that

$$\deg((f)) = 0,\ \deg((\omega)) = 2(g-1), \quad f \in \mathcal{M}(K_g)\backslash\{0\},\ \omega \in \mathcal{M}^1(K_g)\backslash\{0\}, \tag{A.30}$$

where the degree $\deg(\mathcal{D})$ of $\mathcal{D}$ is given by $\deg(\mathcal{D}) = \sum_{P \in K_g} \mathcal{D}(P)$. One calls (f) (respectively (ω)) a principal (respectively canonical) divisor.

Introducing the complex linear spaces

$$\mathcal{L}(\mathcal{D}) = \{f \in \mathcal{M}(K_g) | f = 0 \text{ or } (f) \geq \mathcal{D}\},\ r(\mathcal{D}) = \dim_{\mathbb{C}} \mathcal{L}(\mathcal{D}), \tag{A.31}$$

$$\mathcal{L}^1(\mathcal{D}) = \{\omega \in \mathcal{M}^1(K_g) | \omega = 0 \text{ or } (\omega) \geq \mathcal{D}\},\ i(\mathcal{D}) = \dim_{\mathbb{C}} \mathcal{L}^1(\mathcal{D}) \tag{A.32}$$

($i(\mathcal{D})$ the index of speciality of $\mathcal{D}$), one infers that $\deg(\mathcal{D})$, $r(\mathcal{D})$, and $i(\mathcal{D})$ only depend on the divisor class $[\mathcal{D}]$ of $\mathcal{D}$. Moreover, we recall the following fundamental facts.

Theorem A.1. *Let $\mathcal{D} \in \mathrm{Div}(K_g)$, $\omega \in \mathcal{M}^1(K_g)\backslash\{0\}$. Then*
(i).

$$i(\mathcal{D}) = r(\mathcal{D} - (\omega)), \quad g \in \mathbb{N}_0. \tag{A.33}$$

(ii). (Riemann-Roch theorem).

$$r(-\mathcal{D}) = \deg(\mathcal{D}) + i(\mathcal{D}) - g + 1, \quad g \in \mathbb{N}_0. \tag{A.34}$$

(iii). (Abel's theorem). $\mathcal{D} \in \mathrm{Div}(K_g)$, $g \in \mathbb{N}$ is principal if and only if

$$\deg(\mathcal{D}) = 0 \text{ and } \underline{\alpha}_{P_0}(\mathcal{D}) = \underline{0}. \tag{A.35}$$

(iv). (Jacobi's inversion theorem). Assume $g \in \mathbb{N}$, then $\underline{\alpha}_{P_0} : \mathrm{Div}(K_g) \to J(K_g)$ is surjective.

For notational convenience we agree to abbreviate

$$\mathcal{D}_Q : \begin{cases} K_g \to \{0,1\} \\ P \to \begin{cases} 1, & P = Q \\ 0, & P \neq Q \end{cases} \end{cases}, \quad \mathcal{D}_{\underline{Q}} : \begin{cases} K_g \to \{0,1,\dots,g\} \\ P \to \begin{cases} m & \text{if } P \text{ occurs } m\text{-times in } \{Q_1,\dots,Q_g\} \\ 0 & \text{if } P \notin \{Q_1,\dots,Q_g\} \end{cases} \end{cases}. \tag{A.36}$$

for $\underline{Q} = (Q_1, \ldots, Q_g) \in \sigma^g K_g$ ($\sigma^n K_g$ then n-th symmetric power of K_g). Moreover, $\sigma^n K_g$ can be identified with the set of positive divisors $0 < \mathcal{D} \in \operatorname{Div}(K_g)$ of degree n.

Lemma A.2. *Let $\mathcal{D}_{\underline{Q}} \in \sigma^g K_g$, $\underline{Q} = (Q_1, \ldots, Q_g)$. Then $1 \leq i(\mathcal{D}_Q) = s(\leq g/2)$ if and only if there are s pairs of the type $(P, P^*) \in \{Q_1, \ldots, Q_g\}$ (this includes, of course, branch points for which $P = P^*$).*

We emphasize that most results in this appendix immediately extend to the case where $\{E_n\}_{0 \leq n \leq 2g} \subset \mathbb{C}$. (In this case τ is no longer purely imaginary as stated in (A.18) but has a positive definite imaginary part.)

Appendix B. An Explicit Illustration of the Riemann-Roch Theorem

Finally, we give a brief illustration of the Riemann-Roch theorem in connection with KdV-type hyperelliptic curves, i.e., hyperelliptic curves branched at infinity, and explicitly determine a basis for the vector space $\mathcal{L}(-n\mathcal{D}_{P_\infty} - \mathcal{D}_{\underline{\hat{\mu}}(x_0)})$, $n \in \mathbb{N}_0$.

We freely use the notation introduced in Appendix A and refer, in particular, to the definition (A.31) of $\mathcal{L}(\mathcal{D})$ and the Riemann-Roch theorem stated in Theorem A.1 (ii). In addition, we use the short-hand notation

$$n\mathcal{D}_{P_\infty} + \mathcal{D}_{\underline{\hat{\mu}}(x_0)} = \sum_{m=1}^{n} \mathcal{D}_{P_\infty} + \sum_{j=1}^{g} \mathcal{D}_{\hat{\mu}_j(x_0)}, \quad n \in \mathbb{N}_0,\ \underline{\hat{\mu}}(x_0) = (\hat{\mu}_1(x_0), \ldots, \hat{\mu}_g(x_0)) \tag{B.1}$$

and recall that

$$\mathcal{L}(-n\mathcal{D}_{P_\infty} - \mathcal{D}_{\underline{\hat{\mu}}(x_0)}) = \{f \in \mathcal{M}(K_g) | f = 0 \text{ or } (f) + n\mathcal{D}_{P_\infty} + \mathcal{D}_{\underline{\hat{\mu}}(x_0)} \geq 0\}, \quad n \in \mathbb{N}_0. \tag{B.2}$$

With $\phi(P, x)$, $\psi(P, x, x_0)$ defined as in (3.15), (3.18) we obtain the following

Theorem B.1. *Assume $\mathcal{D}_{\underline{\hat{\mu}}(x_0)}$ to be nonspecial (i.e., $i(\mathcal{D}_{\underline{\hat{\mu}}(x_0)}) = 0$) and of degree $g \in \mathbb{N}$. For $n \in \mathbb{N}_0$, a basis for the vector space $\mathcal{L}(-n\mathcal{D}_{P_\infty} - \mathcal{D}_{\underline{\hat{\mu}}(x_0)})$ is given by*

$$\begin{cases} \{1\}, & n = 0 \\ \{\tilde{\pi}^j\}_{0 \leq j \leq (n-1)/2} \cup \{\tilde{\pi}^j \phi(., x_0)\}_{0 \leq j \leq (n-1)/2}, & n \text{ odd}, \\ \{\tilde{\pi}^j\}_{0 \leq j \leq n/2} \cup \{\tilde{\pi}^j \phi(., x_0)\}_{0 \leq j \leq (n-2)/2}, & n \text{ even} \end{cases} \tag{B.3}$$

or equivalently,

$$\mathcal{L}(-n\mathcal{D}_{P_\infty} - \mathcal{D}_{\underline{\hat{\mu}}(x_0)}) = \operatorname{span}\Big\{\frac{\partial^j}{\partial x^j}\psi(., x, x_0)\Big|_{x=x_0}\Big\}_{0 \leq j \leq n}. \tag{B.4}$$

Proof. The elements in (B.3) are easily seen to be linearly independent and belonging to $\mathcal{L}(-n\mathcal{D}_{P_\infty} - \mathcal{D}_{\hat{\underline{\mu}}(x_0)})$. It remains to be shown that they are maximal. From $0 = i(\mathcal{D}_{\hat{\underline{\mu}}(x_0)}) = i(\mathcal{D}_{nP_\infty} + \mathcal{D}_{\hat{\underline{\mu}}(x_0)})$ and the Riemann-Roch theorem (A.34) one obtains $r(-n\mathcal{D}_{P_\infty} - \mathcal{D}_{\hat{\underline{\mu}}(x_0)}) = n+1$ proving (B.3). In order to prove (B.4), one repeatedly uses the Schrödinger equation (3.21) to prove inductively that

$$\begin{aligned}
\frac{\partial^{2m+2}}{\partial x^{2m+2}}\psi(P,x,x_0) &= (-\tilde{\pi})^{m+1} + R_{2m+1}(P,x),\\
\frac{\partial^{2m+1}}{\partial x^{2m+1}}\psi(P,x,x_0) &= (-\tilde{\pi})^{m}\frac{\partial}{\partial x}\psi(P,x,x_0) + R_{2m}(P,x),
\end{aligned} \tag{B.5}$$

where $R_n(.,x_0) \in \mathcal{L}(-n\mathcal{D}_{P_\infty} - \mathcal{D}_{\hat{\underline{\mu}}(x_0)})$. □

Acknowledgments. F. G. would like to thank the organizers for their kind invitation to a most stimulating conference.

References

1. S. I. Al'ber, *Investigation of equations of Korteweg-de Vries type by the method of recurrence relations*, J. London Math. Soc. (2) **19**, 467–480 (1979) (Russian).
2. S. I. Al'ber, *On stationary problems for equations of Korteweg-de Vries type*, Commun. Pure Appl. Math. **34**, 259–272 (1981).
3. N. Aronszajn and W. F. Donoghue, *On exponential representations of analytic functions in the upper half-plane with positive imaginary part*, J. Anal. Math. **5**, 321–388 (1956–57).
4. E. D. Belokolos, A. I. Bobenko, V. Z. Enol'skii, A. R. Its, and V. B. Matveev, "*Algebro-Geometric Approach to Nonlinear Integrable Equations*", Springer, Berlin, 1994.
5. W. Bulla, F. Gesztesy, H. Holden, and G. Teschl, *Algebro-geometric quasi-periodic finite-gap solutions of the Toda and Kac-van Moerbeke hierarchy*, preprint, 1995.
6. J. L. Burchnall and T. W. Chaundy, *Commutative ordinary differential operators*, Proc. London Math. Soc. (2), **21**, 420–440 (1923).
7. J. L. Burchnall and T. W. Chaundy, *Commutative ordinary differential operators*, Proc. Roy. Soc. London **A118**, 557–583 (1928).
8. W. Craig, *The trace formula for Schrödinger operators on the line*, Commun. Math. Phys. **126**, 379–407 (1989).
9. L. A. Dickey, "*Soliton Equations and Hamiltonian Systems*", World Scientific, Singapore, 1991.
10. B. A. Dubrovin, *Periodic problems for the Korteweg-de Vries equation in the class of finite band potentials*, Theoret. Math. Phys. **9**, 215–223 (1975).
11. H. M. Farkas and I. Kra, "*Riemann Surfaces*", 2nd ed., Springer, New York, 1992.
12. J. D. Fay, "*Theta Functions on Riemann Surfaces*", Lecture Notes in Mathematics **352**, Springer, Berlin, 1973.
13. H. Flaschka, *On the inverse problem for Hill's operator*, Arch. Rat. Math. Anal. **59**, 293–309 (1975).
14. I. M. Gel'fand and L. A. Dikii, *Asymptotic behaviour of the resolvent of Sturm-Liouville equations and the algebra of the Korteweg-de Vries equations*, Russian Math. Surv. **30:5**, 77–113 (1975).
15. I. M. Gel'fand and L. A. Dikii, *Integrable nonlinear equations and the Liouville theorem*, Funct. Anal. Appl. **13**, 6–15 (1979).
16. F. Gesztesy, *New trace formulas for Schrödinger operators*, in "*Evolution Equations*", G. Ferreyra, G. R. Goldstein, and F. Neubrander (eds.), Marcel Dekker, New York, 1975, pp. 201–221.
17. F. Gesztesy and H. Holden, *Trace formulas and conservation laws for nonlinear evolution equations*, Rev. Math. Phys. **6**, 51–95 (1994).

18. F. Gesztesy and H. Holden, *On new trace formulae for Schrödinger operators*, Acta Applicandae Math. **39**, 315–333 (1995).
19. F. Gesztesy and B. Simon, *The xi function*, Acta Math. (to appear).
20. F. Gesztesy and R. Weikard, *Spectral deformations and Soliton equations*, in *"Differential Equations with Applications to Mathematical Physics"*, W. F. Ames, E. M. Harrell II, and J. V. Herod (eds.), Academic Press, Boston, 1993, pp. 101–139.
21. F. Gesztesy and R. Weikard, *Picard potentials and Hill's equation on a Torus*, Acta Math. (to appear).
22. F. Gesztesy, H. Holden, and B. Simon, *Absolute summability of the trace relation for certain Schrödinger operators*, Commun. Math. Phys. **168**, 137–161 (1995).
23. F. Gesztesy, H. Holden, B. Simon, and Z. Zhao, *Trace formulae and inverse spectral theory for Schrödinger operators*, Bull. Amer. Math. Soc. **29**, 250–255 (1993).
24. F. Gesztesy, H. Holden, B. Simon, and Z. Zhao, *Higher order trace relations for Schrödinger operators*, Rev. Math. Phys. (to appear).
25. H. Hochstadt, *On the determination of a Hill's equation from its spectrum*, Arch. Rat. Mech. Anal. **19**, 353–362 (1965).
26. A. R. Its and V. B. Matveev, *Schrödinger operators with finite-gap spectrum and N-soliton solutions of the Korteweg-de Vries equation*, Theoret. Math. Phys. **23**, 343–355 (1975).
27. C. G. T. Jacobi, *Über eine neue Methode zur Integration der hyperelliptischen Differentialgleichungen und über die rationale Form ihrer vollständigen algebraischen Integralgleichungen*, J. Reine Angew. Math. **32**, 220–226 (1846).
28. S. Kotani and M. Krishna, *Almost periodicity of some random potentials*, J. Funct. Anal. **78**, 390–405 (1988).
29. A. Krazer, *"Lehrbuch der Thetafunktionen"*, Chelsea, New York, 1970.
30. M. G. Krein, *Perturbation determinants and a formula for the traces of unitary and self-adjoint operators*, Sov. Math. Dokl. **3**, 707–710 (1962).
31. B. M. Levitan, *On the closure of the set of finite-zone potentials*, Math. USSR Sbornik **51**, 67–89 (1985).
32. B. M. Levitan, *"Inverse Sturm-Liouville Problems"*, VNU Science Press, Utrecht, 1987.
33. H. P. McKean, *Variation on a theme of Jacobi*, Commun. Pure Appl. Math. **38**, 669–678 (1985).
34. H. P. McKean and P. van Moerbeke, *The spectrum of Hill's equation*, Invent. Math. **30**, 217–274 (1975).
35. H. P. McKean and E. Trubowitz, *Hill's operator and hyperelliptic function theory in the presence of infinitely many branch points*, Commun. Pure Appl. Math. **29**, 143–226 (1976).
36. D. Mumford, *"Tata Lectures on Theta II"*, Birkhäuser, Boston, 1984.
37. R. Narasimhan, *"Compact Riemann Surfaces"*, Birkhäuser, Basel, 1992.
38. S. Novikov, S. V. Manakov, L. P. Pitaevskii, and V. E. Zakharov, *"Theory of Solitons"*, Consultants Bureau, New York, 1984.
39. B. Simon, *Spectral analysis of rank one perturbations and applications, proceedings, "Mathematical Quantum Theory II: Schrödinger Operators"*, J. Feldman, R. Froese, and L. M. Rosen (eds.), Amer. Math. Soc., Providence, RI, to appear.
40. E. Trubowitz, *The inverse problem for periodic potentials*, Commun. Pure Appl. Math. **30**, 321–337 (1977).

DEPARTMENT OF MATHEMATICS, UNIVERSITY OF MISSOURI, COLUMBIA, MO 65211, USA

E-mail address for F.G.: mathfg@mizzou1.missouri.edu

E-mail address for R.R.: mathgr29@mizzou1.missouri.edu

E-mail address for G.T.: mathgr42@mizzou1.missouri.edu

1991 Mathematics Subject Classification. Primary 35Q53, 34B24; Secondary 34L40, 58F07.

Operator Theory:
Advances and Applications, Vol. 87

ON SPECTRAL PROPERTIES OF SCHRÖDINGER-TYPE OPERATOR WITH COMPLEX POTENTIAL

EDWARD GRINSHPUN

Let T be the operator $(-\Delta)^m + q(x)$ on $L_2(\mathbb{R}^n)$. Assume that the "principal" part of $\Re q(x)$ is positive, "regular" and tends to infinity as $|x| \to \infty$, and $\Im q$ is relatively form-bounded with respect to $(-\Delta)^m + \Re q(x)$ with relative bound zero (therefore T has purely discrete spectrum). In the framework of the general perturbation approach, we study the spectral asymptotics and the Riesz basisness for the generalized eigenfunctions of T.

Consider the operator $T = (-\Delta)^m + q(x)$ on $L_2(\mathbb{R}^n)$ with complex potential q, where $1 \leq \Re q(x) \to +\infty$ as $|x| \to +\infty$, and $\Im q(x)$ is infinitesimally form bounded with respect to $\Re T = (-\Delta)^m + \Re q(x)$, so that T has purely discrete spectrum. We study the following natural questions: 1) the asymptotic behavior of the counting function of spectrum $N(r,T) = \#\{\text{eigenvalues of } T \text{ in the disk } \{|\lambda| \leq r\}\}$; 2) the basisness properties for the generalized eigenfunctions. The advanced and powerful "selfadjoint" methods (see [7,35,32,17,37,20,12]) seem to be not applicable in this situation. The excellent exposition of various "nonselfadjoint" approaches is given in [14] (see also [4,35,32,6]). S.Agmon [1,2] used the very natural Green's function approach, which however did not allow too strong singularities of the complex coefficients. We follow the Keldyš-Krein perturbation approach. This approach was successfully developed by A.S.Markus and V.I.Matsaev [26,28,27,25], who applied their method of the artificial gap in the spectrum and the entire functions technique. Combining this method with the "p-subordinated" or form "p-subordinated" perturbations [26,28,27,25,5] served the elliptic boundary value problems, where the counting function of spectrum grows polynomially. Application of these results to the Schrödinger operator is given in [35]. However in the Schrödinger operator case one needs more sensitive classes of perturbations. Our goal is that the admissible classes of the complex perturbations $\Im q$ can be described in the terms of the function $C(\varepsilon)$ in the subordination condition

$$|\langle \Im qu, u\rangle| \leq \varepsilon\langle \Re Tu, u\rangle + C(\varepsilon)\|u\|^2 \text{ for all } u \in D(A^{1/2}), \tag{1}$$

more precisely, in the terms of the Legendre transform of $-C(\varepsilon)$, which is determined by the rate of growth of $N(r, \Re T)$. For the Schrödinger-type operator this means that the "admissible" local singularities of $\Im q(x)$ are determined by the rate of growth of $\Re q(x)$ at infinity.

The rest of the paper is organized as follows. In Section 1 we present the general perturbation results on spectrum asymptotics (Theorems 1-5, Section 1.2) and Riesz basisness (Theorem 6, Section 1.3). The proofs are given in Appendices 1 and 2 respectively. In Section 2 we apply the results of Section 1 to the Schrödinger-type operator. For this purpose we describe (Appendix 3) the class of complex potentials, which satisfy (1) with the given function $C(\varepsilon)$.

The author is grateful to A.S.Markus, V.I.Matsaev, and M.I.Agranovich for valuable discussions.

1. GENERAL PERTURBATION RESULTS

1.1 Notations and definitions

Denote H to be a separable Hilbert space with the norm $\|\cdot\|$ and the scalar product $\langle\cdot,\cdot\rangle$. Let $A \geq 1$ be a selfadjoint operator on H with discrete spectrum and known asymptotics of the counting function of its spectrum $N(r, A)$ as $r \to +\infty$. Denote $Q(A)$ to be the domain of the sesquilinear form associated with A. Suppose the sesquilinear form $\mathbf{B}[u,v]$ satisfies $Q(\mathbf{B}) \supset Q(A)$, and given $\varepsilon \in (0,1]$ there exists $C(\varepsilon) > 0$ such that

$$|\mathbf{B}[u,u]| \leq \varepsilon\langle Au, u\rangle + C(\varepsilon)\|u\|^2 \qquad \text{for all } u \in Q(A). \tag{2}$$

The subject of our interest is the rate of growth of $C(\varepsilon)$ as $\varepsilon \to 0+$, therefore without loss of generality we impose the following assumptions:

$$C'(\varepsilon) < 0,\ C''(\varepsilon) > 0,\ \lim_{\varepsilon\to 0+} C(\varepsilon) = +\infty, \ \text{ and } C'(1) = -1 \tag{3}$$

(the last one is only for the sake of technical convenience). Here and further on we use the notations g' and g^{-1} for the derivative and the inverse function respectively. In the asymptotic relations $o(g(r))$ and $O(g(r))$ we always mean "as $r \to +\infty$". Denote $f(r)$ to be the Legendre transform of $-C(\varepsilon)$ (see [34]):

$$f(r) = \min_{\varepsilon\in(0,1]}\{\varepsilon r + C(\varepsilon)\}. \tag{4}$$

The obvious properties of $f(r)$ are collected in the following proposition.

Proposition 1. 1. $f(r) = r(-C')^{-1}(r) + C((-C')^{-1}(r))$ on $[1, +\infty)$;
2. $f' = (-C')^{-1} > 0$;
3. $f(r) \uparrow +\infty$, and $\dfrac{f(r)}{r} \downarrow 0$ as $r \to +\infty$.

Denote $A\dot{+}B$ to be the m-sectorial operator associated with the form sum $\mathbf{A}[u] + \mathbf{B}[u]$ [21, Section 6]. The operator $A\dot{+}B$ also has purely discrete spectrum [21, Theorem 6.3.4]). Denote $\sigma(A\dot{+}B)$ to be the spectrum of $A\dot{+}B$. For $\lambda \notin \sigma(A\dot{+}B)$ denote $R_\lambda(A\dot{+}B) = (A\dot{+}B - \lambda)^{-1}$. We write R_λ for $R_\lambda(A)$. Denote $N(r, A\dot{+}B)$ to be the number of the eigenvalues of the operator $A\dot{+}B$ in the disk $\{|\lambda| \leq r\}$. We recall some additional definitions (see [14]). A sequence $\mathbf{P}_j$ of subspaces of $\mathbf{H}$ is called a *basis of subspaces* if for any $u \in \mathbf{H}$ there exists the unique representation

$$u = \sum_{j=1}^{+\infty} u_j, \quad u_j \in \mathbf{P}_j. \tag{5}$$

A basis of subspaces is called *unconditional* if it remains a basis of $\mathbf{H}$ under any permutation of the subspaces appearing in it, i.e. if the series (5) converge unconditionally for each $u \in \mathbf{H}$. A linearly independent sequence $\{g_j\}$ of vectors in $\mathbf{H}$ is called *unconditional basis with brackets* if there exists a sequence of positive integers $\{m_k\}$, so that $\mathbf{P}_k = Span\{g_j,\ m_{k-1} \leq j \leq m_k - 1\}$ form the unconditional basis of subspaces.

We shall use the following notations. For $p \geq 1$

$L_{p,unif}(\mathbb{R}^n) = \{u : \sup_{x \in \mathbb{R}^n} \int_{|x-y|\leq 1} |u(y)|^p dy < \infty\}$, and

$L_{p,weak,unif} := \{u : \sup_{x \in \mathbb{R}^n} \sup_{t>0} t^p measure\{y : \ |y - x| \leq 1 \text{ and } |u(y)| > t\} < +\infty\}$.

1.2 Spectrum asymptotics

THEOREM 1 (see [26,28,25] in the case of p-subordinated and form p-subordinated perturbations). *Suppose the perturbation* $\mathbf{B}$ *satisfies* (2). *Then given* $\delta \in (0,1)$ *there exist* $r_0(\delta) > 0$ *and* $D(\delta) > 0$ *such that for all* $r > r_0(\delta)$

$$|N(r, A\dot{+}B) - N(r,A)| \leq D(\delta)[N(r + (2+\delta)f(r), A) - N(r - (2+\delta)f(r), A)]. \tag{6}$$

Theorem 1 implies the following Theorems 2, 3 and 4.

THEOREM 2 [15] (Preservation of the leading term for the polynomially growing $N(r,A)$). *Suppose*

$$\lim_{\substack{r \to \infty \\ \delta \to 0}} \frac{N(r + \delta r, A)}{N(r,A)} = 1. \tag{7}$$

If $\mathbf{B}$ *satisfies* (2) *then* $N(r, A\dot{+}B) = N(r,A)(1 + o(1))$.

THEOREM 3 (Preservation of the leading asymptotics term for the faster growth of $N(r,A)$). *Suppose the condition* (2) *holds with the corresponding function* f *satisfying*

$$\lim_{r \to +\infty} \frac{N(r + f(r), A)}{N(r,A)} = 1. \tag{8}$$

Then

$$N(r, A\dot{+}B) = N(r,A)(1 + o(1)). \tag{9}$$

THEOREM 4 (Preservation of the remainder). *Let* $N(r,A) = h(r) + o(g(r))$, *where* $h(r), g(r) \uparrow +\infty$, *and* $\frac{g(r)}{h(r)} \downarrow 0$ *as* $r \to +\infty$. *If the condition* (2) *holds with the corresponding function* $f(r)$ *satisfying*

$$h(r + f(r)) - h(r) = o(g(r)), \tag{10}$$

then also $N(r, A\dot{+}B) = h(r) + o(g(r))$. *The statement of the Theorem holds true with* $O(g(r))$ *in place of* $o(g(r))$.

THEOREM 5 (see [26] in the case of p-subordinated perturbations). *Theorems* 3 *and* 4 *are precise in the following sence.*

(a) If (9) *holds for all perturbations* $\mathbf{B}$ *satisfying* (2) *with the given function* $C(\varepsilon)$, *then* $f(r)$, *defined by* (4), *satisfies* (8).

(b) Let $N(r, A) = h(r) + o(g(r))$, as in Theorem 4. If $N(r, A\dot{+}B)$ has the same asymptotics for all perturbations **B** *satisfying* (2) *with the given function $C(\varepsilon)$, then $f(r)$, defined by* (4), *satisfies* (10).

1.3 The Riesz basisness

THEOREM 6 (see [27] in the case of the p-subordinated perturbations and [5] in the case of the form p-subordinated perturbations). *Suppose* $\liminf_{r\to+\infty} N(r, A)r^{-q} < +\infty$ *for some* $q > 0$, *and the condition (2) holds with the corresponding function* f *satisfying*

$$\liminf_{r\to+\infty}(N(r + (2+\delta)f(r), A) - N(r - (2+\delta)f(r), A)) < +\infty \text{ for some } \delta > 0, \text{ and} \quad (11)$$

$$\int_1^{+\infty} \frac{f(r)}{r^2} dr < \infty. \quad (12)$$

Then the generalized eigenvectors of the operator $A\dot{+}B$ form the unconditional basis with brackets.

Remark 1. The condition $\liminf_{r\to+\infty} N(r, A)r^{-q} < +\infty$ for some $q > 0$ is imposed for the "completeness purposes" (see lemma 7 in Appendix 2). For the "regularly" growing $N(r, A)$ the existence of the corresponding function $f(r)$ in (11) actually implies $N(r, A) = o(r)$.

2. APPLICATION TO THE SCHRÖDINGER-TYPE OPERATOR

Let $T = (-\Delta)^m + q(x)$ on $L_2(\mathbb{R}^n)$, where the conditions on the complex potential $q(x)$ will be imposed later (see [10,16] for the properties of the maximal Schrödinger operator with minimal apriori assumptions on the complex potential).

We call the auxillary real potential $Q(x)$ *regular* if $Q(x) \geq 1$, $\lim_{|x|\to\infty} Q(x) = +\infty$, and the selfadjoint operator $(-\Delta)^m + Q(x)$ possess the classical leading term of the asymptotics: $N(r, (-\Delta)^m + Q(x)) = W(r)(1 + o(1))$, where

$$W(r) = (2\pi)^{-n} v_n \int_{\{Q(x)<r\}} (r - Q(x))^{n/2m} dx, \quad (13)$$

v_n stands for the volume of the unit sphere in $\mathbb{R}^n$. The sufficient conditions for (13) can be found for example in [35,9,23,18]. Denote $\mu(t, Q) = measure\{x : Q(x) < t\}$. Applying Theorems 2-6 yields the following Theorems 7-10.

THEOREM 7 (The "principal part" of $\Re q$ grows polynomially). *Suppose there exists a real "regular" potential $Q(x)$ such that $\mu(t, Q)$ grows polynomially, more precisely,* $\lim_{\substack{r\to\infty\\ \delta\to 0}} \mu(r+\delta r, Q)(\mu(r, Q))^{-1} = 1$, *and $q - Q$ is infinitesimally form bounded with respect to $(-\Delta)^m + Q(x)$, that is*

$$|\langle (q(x) - Q(x))u, u\rangle| \leq \varepsilon\langle ((-\Delta)^m + Q(x))u, u\rangle + C(\varepsilon)\|u\|^2 \text{ for all } u \in C_0^\infty(\mathbb{R}^n). \quad (14)$$

Then $N(r, T) = W(r)(1 + o(1))$.

Example 1. The condition (14) holds if $q - Q = q_1 + q_2$, where $|q_2(x)| = o(Q(x))$, $|x| \to \infty$, and $q_1(x)$ satisfies one of the following conditions:

1. (see [33, Theorem 10.19])
   $m = 1$, $n = 3$ and $q_1 \in$ (Rolnik class) $= \{u : \int_{\mathbb{R}^3 \times \mathbb{R}^3} \frac{|u(x)||u(y)|}{|x-y|^2} dxdy < \infty\}$.

2. (see [8,31]) $q_1 \in L_{s,unif}(\mathbb{R}^n)$ where $s = \frac{n}{2m}$ for $n > 2m$, $s > 1$ for $n = 2m$, and $s = 1$ for $n < 2m$.

3. (see [11, Section 1.2]) $m = 1$ and
   $q_1 \in$ Kato class $= \{u : \lim_{a\downarrow 0}(\sup_{x\in\mathbb{R}^n} \int_{|x-y|\le a} |x-y|^{2-n}|u(y)|dy) = 0\}$ for $n \geq 3$, where $|x-y|^{2-n}$ should be replaced by $\ln|x-y|$ for $n = 2$ and by 1 for $n = 1$.

Example 2 (Perturbation by the "penetrable wall potential", see [19] for the definition). Let T be the operator on $L_2(\mathbb{R}^3)$ associated with the sesquilinear form

$$\langle Tu, u\rangle = \langle((-\Delta) + Q(x))u, u\rangle + \int_{S_a} p(\omega)|\hat{u}(\omega)|^2 d\omega,$$

(defined on $H^1(\mathbb{R}^3) \cap \{u : Q|u|^2 \in L_2(\mathbb{R}^3)\}$) where $Q(x)$ is "regular", $S_a = \{|x| = a\}$, $\hat{u}(\omega)$ is the trace of the H^1-function u on S_a, and $p(\omega)$ is a complex valued function on S_a. If

$$g(|p|) \in L_2(S_a) \tag{15}$$

for some positive increasing function g on $[0, +\infty)$ such that $\lim_{t\to+\infty} \frac{g(t)}{t} = +\infty$, then $N(r, T) = W(r)(1 + o(1))$.

In Theorems 8-10 below we assume the function $f(t)$ to satisfy the following condition:
Condition 1. *$f(t)$ is a positive differentiable function on $[1, +\infty)$ with $f(t) \uparrow +\infty$, and $f'(t) \downarrow 0$ as $t \to +\infty$ (in particular, $f(t)=o(t)$).*
THEOREM 8 (The "principal part" of $\Re q$ grows slower than the polynomial). *Suppose there exists real "regular" potential $Q(x)$ and function a $f(t)$ satisfying Condition 1 such that*

$$\lim_{t\to+\infty} \frac{\mu(t + f(t), Q) - \mu(t, Q)}{\mu(t, Q)} = 0, \tag{16}$$

and $q - Q = q_1 + q_2$, where

$$\langle f^{-1}(|q_1|)u, u\rangle \leq const\,\langle((-\Delta)^m + 1)u, u\rangle, \text{ for all } u \in C_0^\infty(\mathbb{R}^n), \tag{17}$$

$$q_2 \in L_{\infty,loc} \text{ with } |q_2(x)| = o(f(Q(x))) \text{ as } |x| \to \infty. \tag{18}$$

Then $N(r, (-\Delta)^m + q(x)) = W(r)(1 + o(1))$.

Example 3. The condition (17) holds in particular if $f^{-1}(|q_1|) \in L_{\frac{n}{2m},weak,unif}(\mathbb{R}^n)$ for $n > 2m$ (see [8]) and $f^{-1}(|q_1|) \in L_{1,unif}(\mathbb{R}^n)$ for $n < 2m$. The necessary and sufficient conditions in terms of capacities, including the case $n = 2m$, are given in [31]. The condition (16) means that $\mu(t, Q)$ grows slower than the exponent.

THEOREM 9 (Remainder in the asymptotic formula). *Let $m = 1$. Suppose there exists a real "regular" potential $Q(x) \sim |x|^k$, $k > 2$, such that [36]*

$$N(r, -\Delta + Q(x)) = W(r)(1 + O(r^{-(\frac{1}{2}+\frac{1}{k})}), \tag{19}$$

and $q - Q$ satisfies (17)-(18) with $f(r) = O(r^{\frac{1}{2}-\frac{1}{k}})$. Then the operator $-\Delta + q(x)$ has the same asymptotics (19). (It is easy to see that $W(r) \sim r^{n(\frac{1}{2}+\frac{1}{k})}$)

THEOREM 10 (The Riesz basisness). *Let $n < 2m$. Suppose $q = Q + q_1 + q_2$, where $Q(x)$ is a real "regular" potential with $\mu(t, Q) = o(t^{1-\frac{n}{2m}})$, and q_1, q_2 satisfy the following condition: q_2 satisfies (18) and $f^{-1}(|q_1|) \in L_{1,unif}(\mathbb{R}^n)$, where the function $f(t)$ (from Condition 1) is determined by*

$$\int_1^t \mu(s, Q)f'(s)ds + \int_t^{t+f(t)} \mu(s, Q)ds = O(t^{1-\frac{n}{2m}}) \text{ as } t \to +\infty. \tag{20}$$

Then the generalized eigenfunctions of the operator T form the unconditional (Riesz) basis with brackets.

Example 4.
a) Let $\mu(t, Q) = t^a$, $0 < a < 1 - \frac{n}{2m}$. Then $f(t) = O(t^{1-\frac{n}{2m}-a})$. This case is covered by the corresponding form p-subordination [5] results. In the following examples the form p-subordination condition does not work (b), or does not give the precise class of perturbations (c-d) (we also emphasize which one of the two integrals in (20) determines $f(r)$):
b) Let $\mu(t, Q) = t^{1-\frac{n}{2m}}(\log t)^{-1}$. Then $f(t) = O(\log t)$ (first integral).
c) Let $\mu(t, Q) = \log t$. Then $f(t) = O(t^{1-\frac{n}{2m}}(\log t)^{-1})$ (second integral).
d) Let $\mu(t, Q) = t^{1-\frac{n}{2m}-a} \log t$, $0 < a < 1 - \frac{n}{2m}$. Then $f(t) = O(t^a(\log t)^{-1})$ (both integrals).

APPENDIX 1. Proof of Theorem 1

Our proof is the modification of the p-subordination case proof [26,28,25]. It is based on the idea of the artificial gap in the spectrum due to A.S.Markus and V.I.Matsaev, and consists of the following 3 steps.
Step 1. The M.V.Keldyš-type [22] lemma (see also [26,3,4]).

Lemma 1. *Let $\mathbf{B}$ satisfy (2). Then for any $\delta > 0$ the spectrum of $A \dot{+} B$ (excluding possibly finitely many eigenvalues) is situated in the domain $\Phi \stackrel{def}{=} \{Re\lambda > 0 : |Im\lambda| < (2+\delta)f(|\lambda|)\}$, and there exists $r_1(\delta) > 0$ such that*

$$\|R_\lambda(A \dot{+} B)\| \leq \frac{1}{\delta f(|\lambda|)} \text{ for } \lambda \notin \Phi, |\lambda| > r_1(\delta). \tag{21}$$

Proof. Let $\lambda \notin \sigma(A)$.The identity

$$A \dot{+} B - \lambda = (A - \lambda)^{1/2}(I + R_\lambda^{1/2} B R_\lambda^{1/2})^{-1}(A - \lambda)^{1/2} \tag{22}$$

yields that it is enough to check the condition $\|R_\lambda^{1/2} B R_\lambda^{1/2}\| < 1$. We use the notations $(A - \lambda)^{\pm 1/2} \stackrel{def}{=} \int_1^\infty (t - \lambda)^{\pm 1/2} dE_t$, where E_t is the spectral family corresponding to A, $z^{1/2} = |z|^{1/2} \exp(i\frac{1}{2} \arg z)$, $-\pi < \arg z \leq \pi$, and B stands for the bounded operator associated with the sesquilinear form $\mathbf{B}[u]$ on the Hilbert space $Q(A)$ with the norm $\mathbf{A}[u]^{1/2}$. The following proposition concludes the proof.

Proposition 2. *Suppose the perturbation* **B** *satisfies* (2). *Let* $\Re\lambda > 1$, *and* $a(\lambda) \stackrel{def}{=} distance(\Re\lambda, \sigma(A)) < \Re\lambda$. *Then for* $\lambda \notin \sigma(A)$

$$\|R_\lambda^{1/2} B R_\lambda^{1/2}\| \le \begin{cases} \frac{2f(\Re\lambda + a(\lambda))}{(|\Im\lambda|^2 + a(\lambda)^2)^{1/2}} & for\ |\Im\lambda| < (a(\lambda)\Re\lambda)^{1/2}, \\ \frac{2f(|\lambda|)}{|\Im\lambda|} & for\ |\Im\lambda| \ge (a(\lambda)\Re\lambda)^{1/2}. \end{cases} \tag{23}$$

Proof of Proposition 2. Let $|\Im\lambda| < (a(\lambda)\Re\lambda)^{1/2}$. The second case is similar. Combining (2) with Polarization identity and the estimates

$$\|A^{1/2}R_\lambda^{1/2}\|^2 \le \sup_{t\in\sigma(A)} |\frac{t}{t-\lambda}| \le \frac{\Re\lambda + a(\lambda)}{(|\Im\lambda|^2 + a(\lambda)^2)^{1/2}}, \quad \|R_\lambda^{1/2}\|^2 \le \frac{1}{(|\Im\lambda|^2 + a(\lambda)^2)^{1/2}} \tag{24}$$

implies

$$|\langle R_\lambda^{1/2} B R_\lambda^{1/2} u, v\rangle| \le \frac{\varepsilon(\Re\lambda + a(\lambda)) + C(\varepsilon)}{(|\Im\lambda|^2 + a(\lambda)^2)^{1/2}} (\|u\|^2 + \|v\|)^2 \text{ for all } u, v \in \mathbf{H}.$$

Putting $\varepsilon = -(f')^{-1}(\Re\lambda + a(\lambda))$ yields (23). ∎

Step 2 (Hypothetical gap in the spectrum of A).

Lemma 2. *For each* $\delta \in (0,1)$ *there exists* $r_2(\delta)$ *such that if for* $r > r_2(\delta)$ $(r - (2+\delta)f(r),\ r + (2+\delta)f(r)) \cap \sigma(A) = \emptyset$ *for* $r > r_2(\delta)$, *then* $N(r, A) = N(r, A\dot{+}B)$.

Proof. Define for sufficiently large r and $a \in [0,1]$ the family of projections

$$P_{\Gamma_r}(A\dot{+}aB) = -\frac{1}{2\pi i}\int_{\Gamma_r} R_\lambda(A\dot{+}aB)d\lambda, \tag{25}$$

where $\Gamma_r = \{|\lambda| = r\}$. If $(r - (2+\delta)f(r),\ r + (2+\delta)f(r)) \cap \sigma(A) = \emptyset$ then for $\lambda \in \Gamma_r$ the operators $(A\dot{+}aB - \lambda)^{-1}$ are uniformly bounded and continuous with respect to $a \in [0,1]$. Therefore the dimensions of $P_{\Gamma_r}(A\dot{+}B)$ and $P_{\Gamma_r}(A)$ are equal. ∎

Step 3. The artificial gap in the spectrum.

Denote $E_{(a,b)}$ to be the spectral projection of the operator A corresponding to the interval (a,b). Denote $\gamma_{r,\delta} = (2+\delta)f(r)$, and consider the operators

$$K_r = (A - (r - \gamma_{r,\delta}))E_{(r-\gamma_{r,\delta},\, r]} + (A - (r + \gamma_{r,\delta}))E_{(r,\, r+\gamma_{r,\delta})} \text{ and } G_r = A - K_r.$$

We have $dim K_r = N(r + \gamma_{r,\delta}, A) - N(r - \gamma_{r,\delta}, A)$, $\|K_r\| \le \gamma_{r,\delta}$, $\sigma(G_r) \cap (r - \gamma_{r,\delta}, r + \gamma_{r,\delta}) = \emptyset$, and $\langle Au, u\rangle \le r(r - \gamma_{r,\delta})^{-1}\langle G_r u, u\rangle$ for $u \in Q(A)$. Suppose $\Gamma_r \cap \sigma(A\dot{+}B) = \emptyset$ (otherwise we replace r by $\hat{r} > r$, so that the disk $\{|\lambda| \in [r, \hat{r}]\}$ does not contain the eigenvalues of both A and $A\dot{+}B$). By Lemma 2, for $r > r_3(f,\delta) \stackrel{def}{=} max\ \{r_1(\frac{\delta}{2}), (f')^{-1}(\frac{\delta}{8(4+\delta)})\}$ we have $N(r, G_r\dot{+}B) = N(r, G_r)$. Since $N(r, A) = N(r, G_r)$, the argument principle [21, Section 4.6] implies $N(r, A\dot{+}B) - N(r, A) = N(r, A\dot{+}B) - N(r, G_r\dot{+}B) = (2\pi)^{-1}[argD(\lambda)]_{\Gamma_r}$, where $[argD(\lambda)]_{\Gamma_r}$ denotes the increment of the $argD(\lambda)$ along Γ_r,

$$D(\lambda) = det[(A\dot{+}B - \lambda)R_\lambda(G_r\dot{+}B)] = det[I + K_r R_\lambda(G_r\dot{+}B)]. \tag{26}$$

The operator K_r is finite dimensional therefore the Weinstein-Aronszajn determinant (26) is well defined (see [21, Section 3.4.3] and [14, Chapter 4]). Using the following Lemma with $U = \{\lambda :\ |Im\lambda| < 6\hat{f}(r) \text{ and } |Re\lambda - r| < \frac{\delta}{5}\hat{f}(r)\}$, and $F = \{\lambda :\ |Im\lambda| \le 5\hat{f}(r) \text{ and } |Re\lambda - r| \le \frac{\delta}{6}\hat{f}(r)\}$, where $\hat{f}(t) = f(\frac{r}{r-\gamma_{r,\delta}}t)$, concludes the proof.

Lemma 3 [26]. *Denote* $U = \{\lambda : |Im\lambda| < b_1\rho \text{ and } |Re\lambda - r| < a_1\rho\}$ *and* $F = \{\lambda : |Im\lambda| \le b_2\rho \text{ and } |Re\lambda - r| \le a_2\rho\}$ *where* $a_1 > a_2 > 0$, $b_1 > b_2 > 0$, *and* $\rho > 0$. *Suppose the operator valued function* $K(z)$ *bounded and continuous on* Γ_r *and holomorphic on* U; $K(z)$ *is finite dimensional for* $z \in U \cup \Gamma_r$ *with* $dim K(z) \le n$; $\|K(z)\| \le q < 1$ *for* $z \in \Gamma_r \setminus F$, *and* $\|K(z)\| \le M$ *for* $z \in U$. *If* Γ_r *does not contain the roots of* $det(I + K(z))$ *then* $|[arg\ det(I + K(z))]_{\Gamma_r}| \le D\,n$, *where the constant* D *depends on* a_1, a_2, b_1, b_2, M *and* q, *and is independent of* n, r, ρ *and* $K(z)$.

■

APPENDIX 2. Proof of Theorem 6

Our proof is based on Theorem 1. The key estimate is Lemma 8 below. In the case of p-subordinated perturbations [27] and form p-subordinated perturbations [5] the proof was based on the artificial gap method of Markus and Matsaev (which is also crucial for Theorem 1). Emphasizing the role of Theorem 1 clears up and simplifies the proof of Theorem 6. We start with several Lemmas.

Denote P_j to be orthogonal projection on $\mathbf{P}_j$, $P_j\mathbf{H} = \mathbf{P}_j$. The following lemma is well known (see [25, Chapter 1,Section 6]).

Lemma 4. *Let the sequence of subspaces* $\{\mathbf{P}_j\}_{j=1}^{+\infty}$ *be complete in* $\mathbf{H}$. *If there exists a sequence of disjoint orthogonal projections* $\{Q_j\}$ $(Q_jQ_k = \delta_{jk}Q_k)$ *such that* $\sum_{j=1}^{+\infty} |\langle (P_j - Q_j)u, v\rangle| < \infty$, *for all* $u, v \in \mathbf{H}$, *then* $\{\mathbf{P}_j\}_{j=1}^{+\infty}$ *forms the unconditional basis of subspaces for* $\mathbf{H}$.

Lemma 5. *Let* U *be a simply connected bounded domain,* F *be a compact subset of* U, $z_0 \in F$. *Then there exists a number* $C > 0$, *depending only on* U, F, *and* z_0 *such that any bounded holomorphic nonvanishing function g(z) on* U *satisfies the inequality:* $\log |g(z)(g(z_0))^{-1}| \ge -C \log(\sup\{|g(z)| : z \in U\}|g(z_0)|^{-1})$ *for* $z \in F$. *The constant* C *remains unchanged under the affine transformation (i.e. under simultanious replacing of* U, F, *and* z_0 *by* $aU+b$, $aF+b$ *and* $az_0 + b$ *correspondingly).*

Proof of Lemma 5. In the case of $U = \{|z| < r_1\}$, $F = \{|z| < r_2 < r_1\}$, and $z_0 = 0$ the lemma is proved in [24, Theorem 1.6.9]. The general case can be reduced to this one by means of the conformal mapping. ■

Lemma 6 (see [38] in the operator subordination case). *Let* $A \ge 1$ *be a selfadjoint operator with purely discrete spectrum* $\lambda_j(A) \to +\infty$, *as* $j \to \infty$ ($\lambda_j(A)$ *are numbered in the increasing order according to their multiplicities). If the sesquilinear form* $\mathbf{B}[u, v]$ *satisfies the subordination condition (2), then* $\mathbf{B}$ *is relatively form compact with respect to* A *(this means that the operator* $A^{-1/2}BA^{-1/2}$ *is compact on* H*).*

Proof of Lemma 6. Denote P_n to be the orthogonal projection on the span of the first n eigenvectors of A. Combining (2) with the Polarization identity implies

$$|\mathbf{B}[u, v]| \le (\varepsilon\mathbf{A}[u] + C(\varepsilon)\|u\|^2)^{1/2}(\varepsilon\mathbf{A}[v] + C(\varepsilon)\|v\|^2)^{1/2} \ \forall\varepsilon \in (0, 1] \text{ and } u, v \in \mathbf{H}. \quad (27)$$

Let $\|u\| = \|v\| = 1$, and $\delta > 0$. Putting $\varepsilon = 0.5\delta$ and choosing n_0 such that $C(0.5\delta)\|A^{-1/2}(I - P_{n_0})\| < 0.5\delta$, yields $|\langle (I - P_n)A^{-1/2}BA^{-1/2}(I - P_n)u, v\rangle| \le \delta$. Hence

$\|(I-P_n)A^{-1/2}BA^{-1/2}(I-P_n)\| \to 0$, as $n \to \infty$, and Lemma 6 follows from the identity $A^{-1/2}BA^{-1/2} = D_n + (I-P_n)A^{-1/2}BA^{-1/2}(I-P_n)$, where D_n stands for the finite dimensional operator $P_nA^{-1/2}BA^{-1/2}(I-P_n) + A^{-1/2}BA^{-1/2}P_n$. ∎

The following lemma seems to be known, excluding possibly its formulation in the terms of the infinitesimally form bounded perturbation (cf [30], and the p-subordination version in [5]). It is based on the Matsaev's resolvent estimate [29].

Lemma 7. *Let $A \geq 1$ be a selfadjoint operator with purely discrete spectrum. Suppose $N(r, A)$ satisfies the condition*

$$\liminf_{r\to+\infty} N(r,A)r^{-q} < +\infty \quad \textit{for some } q > 0. \tag{28}$$

If **B** *satisfies (2), then the generalized eigenvectors of the operator $A\dot{+}B$ are complete in* **H**.

Proof. Denote $Span(A\dot{+}B)$ to be the closure of the span of generalized eigenvectors of $A\dot{+}B$. We shall show that $Span(A\dot{+}B)^{\perp} = 0$. Indeed, let $v \in Span(A\dot{+}B)^{\perp}$. Without loss of generality assume that $0 \notin \sigma(A\dot{+}B)$. Consider for arbitrary $\|u\| = \|v\| = 1$ the function $F(\lambda) = \lambda\langle R_\lambda(A\dot{+}B)u, v\rangle = -\langle u, v\rangle - \lambda^{-1}\langle R_{\frac{1}{\lambda}}((A\dot{+}B)^{-1})u, v\rangle$. The function $F(\lambda)$ is analitic in **C** (see [13, Theorem 11.29]), and satisfies $F(0) = 0$. Since, for $\varepsilon > 0$ and $|\lambda| > const(\varepsilon)$, $\|R_\lambda(A\dot{+}B)\|$ is uniformly bounded outside the set $\Omega_\varepsilon = \{\Re\lambda > 0 : |\Im\lambda| \leq \varepsilon\Re\lambda\}$, it is sufficient to show that $F(\lambda)$ is bounded in Ω_ε. The Matsaev's resolvent estimate [29, Corollary 4] implies

$$\max_{|\lambda|=r} |\frac{1}{\lambda}\langle R_{\frac{1}{\lambda}}((A\dot{+}B)^{-1})u, v\rangle| \leq \exp(const(\delta)(1 + \int_0^{r(1+\delta)} \frac{\nu(t,(A\dot{+}B)^{-1})}{t}dt)), \tag{29}$$

where $\nu(t,(A\dot{+}B)^{-1}) = \#\{s_j((A\dot{+}B)^{-1}) \geq \frac{1}{t}\}$, $s_j(K)$ denotes the j-th eigenvalue of the operator $(K^*K)^{1/2}$. Since $(A\dot{+}B)^{-1} = A^{-1/2}(I + A^{-1/2}BA^{-1/2})^{-1}A^{-1/2}$, we have $\nu(t,(A\dot{+}B)^{-1}) \leq 2N(dt, A)$, where $d = \|(I + A^{-1/2}BA^{-1/2})^{-1}\|$. Therefore (29) implies $|F(\lambda)| \leq \exp(const_1 N(const_2|\lambda|, A))$ for $\lambda \in \Omega_\varepsilon$ with $|\lambda| > const(\varepsilon)$. Choosing ε to be sufficiently small and applying (28) with the Phragmén-Lindelöf Theorem concludes the proof. ∎

Lemma 8. *Under the conditions of Theorem 6 there exist a sequence $t_m \to +\infty$ and constants $a > 0, M > 0$, such that the strips $\{\Re\lambda \in \Phi_m \overset{def}{=} [t_m - af(t_m), t_m + af(t_m)]\}$ do not contain the eigenvalues of both A and $A\dot{+}B$, and*

$$\|(I + R_\lambda^{1/2}BR_\lambda^{1/2})^{-1}\| \leq M \quad \textit{for } \Re\lambda \in \Phi_m. \tag{30}$$

Proof. Consider the increasing sequence $s_m \to +\infty$ from the condition (11) such that $s_1 > r_4(\delta) \overset{def}{=} max\{r_0(\delta), r_1(\frac{\delta}{2}), (f')^{-1}(\frac{\delta}{4(2+\delta)^2})\}$, and for some constant $L > 0$

$$\#\{\lambda_j(A) \in (s_m - (2+\delta)f(s_m), s_m + (2+\delta)f(s_m))\} \leq L.$$

Denote $\gamma(r,\delta) = (2+\delta)f(r)$. For $r > r_4(\delta)$ and $t = r + \frac{\delta}{4}$ we have $(t - \gamma(t, 0.5\delta), t + \gamma(t, 0.5\delta)) \subset (r - \gamma(r,\delta), r + \gamma(r,\delta))$. Hence Theorem 1 yields

$$N(s_k + \frac{\delta}{4}f(s_k), A\dot{+}B) - N(s_k, A\dot{+}B) \leq L(2D(\frac{\delta}{2}) + 1).$$

The Dirichlet principle implies that there exist $a = a(\delta) \in (0, 0.25\delta)$ and $t_m \in (s_m, s_m + 0.25\delta f(s_m))$ such that the strips $\{\Re\lambda \in \Psi_m \stackrel{def}{=} (t_m - 2af(t_m), t_m + 2af(t_m))\}$ do not contain the eigenvalues of both A and $A\dot{+}B$. The identity (22) implies $Ker(I + R_\lambda^{1/2}BR_\lambda^{1/2}) = \emptyset$ for $\Re\lambda \in \Psi_m$. Therefore from Lemma 6 and the Fredholm Theorem it follows that the operators $(I + R_\lambda^{1/2}BR_\lambda^{1/2})^{-1}$ are bounded for $\Re\lambda \in \Psi_m$. In order to establish the uniform norm estimates, we use Lemma 3. Consider the sets $U_m := \{\lambda : |\Re\lambda - t_m| < 1.5af(t_m),\ |\Im\lambda| < (2+3\delta)f(t_m)\}$, $F_m := \{\lambda : |\Re\lambda - t_m| \le af(t_m),\ |\Im\lambda| \le (2+2.5\delta)f(t_m)\}$, and the points $z_m = t_m + i(2+\delta)f(t_m)$. For arbitrary $u \in H$ with $\|u\| = 1$ consider on U_m the analytic function $g(\lambda) = \langle (I + R_\lambda^{1/2}BR_\lambda^{1/2})u, u\rangle$. For $t_m > r_0$ Proposition 2 with Lemma 1 imply $\sup_{\lambda\in U_m} |g(\lambda| \le 1 + 4a^{-1} + 0.25\delta$, and $|g(z_m)| \ge \delta(4+2\delta)^{-1}$. Therefore Lemma 8 follows from Lemma 3. ∎

Proof of Theorem 6. Let t_m be from Lemma 8. Let $\gamma_m = \{\lambda : \Re\lambda = t_m, |\Im\lambda| \le t_m\}$, $\Gamma_\pm = \{\lambda : \Re\lambda > 0, \Im\lambda = \pm\Re\lambda\}$, and $\Gamma_m = \gamma_m \cup \gamma_{m+1} \cup (\Gamma_\pm \cap \{\Re\lambda \in [t_m, t_{m+1}]\})$. Consider the projections

$$P_m = -\frac{1}{2\pi i}\int_{\Gamma_m} R_\lambda(A\dot{+}B)d\lambda, \text{ and } Q_m = -\frac{1}{2\pi i}\int_{\Gamma_m} R_\lambda d\lambda. \tag{31}$$

Combining (21) and (12) with the identity

$$R_\lambda - R_\lambda(A\dot{+}B) = R_\lambda^{1/2}(R_\lambda^{1/2}BR_\lambda^{1/2})(I + R_\lambda^{1/2}BR_\lambda^{1/2})^{-1}R_\lambda^{1/2} \tag{32}$$

implies $\int_{\Gamma_\pm} |\langle (R_\lambda - R_\lambda(A\dot{+}B))u, v\rangle||d\lambda| < \infty$. Therefore, by Lemmas 4 and 7, it is sufficient to prove that there exists an infinite subsequence of $\{t_m\}$, $\{t_{m_k}\}$, such that

$$\sum_{k=1}^{+\infty}\int_{\gamma_{m_k}} |\langle (R_\lambda - R_\lambda(A\dot{+}B))u, v\rangle||d\lambda| < \infty, \quad \forall u, v \in \mathbf{H}. \tag{33}$$

Combining Lemmas 1 and 8 with (32) yields for $\Re\lambda = t_m$

$$|\langle (R_\lambda - R_\lambda(A\dot{+}B))u, v\rangle| \le \frac{const(\delta)f(t_m)}{(|\Im\lambda|^2 + a^2f(t_m)^2)^{1/2}}(\|R_\lambda^{1/2}u\|^2 + \|R_{\bar\lambda}^{1/2}v\|^2).$$

Let $u = \sum_{j=1}^{+\infty} u_j\phi_j$ be the expansion of u with respect to the basis of the eigenvectors of A, $\{\phi_j\}$. In order to prove (33) we estimate

$$\sum_{k=1}^{+\infty}\int_{\gamma_{m_k}} \frac{f(t_{m_k})}{(|\Im\lambda|^2 + a^2f(t_{m_k})^2)^{1/2}}\sum_{j=1}^{+\infty}\frac{1}{|\lambda_j(A) - \lambda|}|u_j|^2|d\lambda| \le$$
$$2\|u\|^2 \sup_{s\in\sigma(A)}\sum_{k=1}^{+\infty} f(t_{m_k})\int_0^{t_{m_k}} \frac{dy}{(y^2 + a^2f(t_{m_k})^2)^{1/2}(y^2 + |s - t_{m_k}|^2)^{1/2}}. \tag{34}$$

We write $\int_0^{t_{m_k}} = \int_0^{f(t_{m_k})} + \int_{f(t_{m_k})}^{t_{m_k}}$. Denote $t = t_{m_k}$. Since $|s - t_{m_k}| \ge af(t_{m_k})$, we have

$$\int_0^{f(t)} \le 2^{-1/2}\frac{1}{af(t)}\int_0^{f(t)}\frac{dy}{|s-t|^{1/2}y^{1/2}} \le \frac{2^{1/2}}{af(t)}\frac{f(t)^{1/2}}{|s-t|^{1/2}}, \text{ and}$$
$$\int_{f(t)}^{t} \le \frac{1}{2^{1/2}|s-t|^{1/2}}\int_{f(t)}^{t} y^{-3/2}dy \le \frac{2^{1/2}}{f(t)}\frac{f(t)^{1/2}}{|s-t|^{1/2}}.$$

Choosing t_{m_k} such that $\sum_{k=1}^{+\infty}(\frac{f(t_{m_k})}{t_{m_k}})^{1/2} < +\infty$ and $t_{m_{k+1}} \geq 4t_{m_k}$ yields for $s \in (t_j, t_{j+1})$

$$\sum_{k=1}^{+\infty}(\frac{f(t_{m_k})}{|s-t_{m_k}|})^{1/2} \leq \sum_{k=1}^{j-1} + \sum_{j+1}^{+\infty} + a^{-1/2} \leq \frac{1}{3}\sum_{k=1}^{j-1}(\frac{f(t_{m_k})}{t_{m_k}})^{1/2} + \frac{4}{3}\sum_{j+1}^{+\infty}(\frac{f(t_{m_k})}{t_{m_k}})^{1/2} + a^{-1/2}.$$

∎

APPENDIX 3. The subordination condition.

We study the class of potentials satisfying the subordination condition:

$$\langle |q|u, u\rangle \leq \; const\,(\varepsilon\langle(-\Delta)^m u, u\rangle + C(\varepsilon)\|u\|^2),\ \forall\varepsilon \in (0,1] \text{ and } u \in C_0^\infty(\mathbb{R}^n), \tag{35}$$

with the given function $C(\varepsilon)$.

Lemma 9. *Suppose*

$$\langle f^{-1}(|q|)u, u\rangle \leq const\,\langle((-\Delta)^m + 1)u, u\rangle, \text{ for all } u \in C_0^\infty(\mathbb{R}^n), \tag{36}$$

where f is the Legendre transform of $-C(\varepsilon)$, defined by (4). *Then q satisfies* (35). *In particular,* (35) *holds if*

$$f^{-1}(|q|) \in L_{\frac{n}{2m},weak,unif} \text{ for } n > 2m, \text{ and } f^{-1}(|q|) \in L_{1,unif} \text{ for } n < 2m. \tag{37}$$

Proof of Lemma 9. Notice that replacing in (35) $(-\Delta)^m$ with $(-\Delta)^m + 1$ leads to the equivalent inequality. Taking in (35) minimum with respect to ε and applying f^{-1} to both sides yields $f^{-1}(\langle\frac{|q|}{const}u, u\rangle\|u\|^{-2}) \leq \langle(-\Delta+1)u, u\rangle\|u\|^{-2}$. The convexity argument implies $f^{-1}(\langle|q|u, u\rangle\|u\|^{-2}) \leq \langle f^{-1}(|q|)u, u\rangle$, therefore (35) follows from (36). The last part of the statement of Lemma 9 follows from the corresponding embedding (see [8,31]). ∎

References

[1] S. Agmon. *Lectures on Elliptic Boundary Value Problems.* Van Nostrand Math. Studies, Prinston, 1965.

[2] S. Agmon. On kernels, eigenvalues, and eigenfunctions of operators related to elliptic problems. *Comm. Pure Appl. Math.*, 18:627–663, 1965.

[3] S. Agmon. On the eigenfunctions and on the eigenvalues of general elliptic boundary value problems. *Comm. Pure Appl. Math.*, 15:119–147, 1962.

[4] M.S. Agranovich. *Elliptic operators on closed manifolds.* Volume 63 of *Encyclopaedia of Mathematical Sciences, Partial Differential Equations*, Springer-Verlag, 1993. Transl. from *Itogi Nauki i Tekhniki: Sovrem. Probl. Mat., Fundament. Napr.*, v.63, VINITI, Moscow, 1989, pp. 5-129.

[5] M.S. Agranovich. On series with respect to root vectors of operators associated with forms having symmetric principal part. *Funct. Anal. Appl.*, 1994. Transl. from *Funkts. Analiz i Prilozh.*, vol. 28, N. 3, 1994, pp. 1-21.

[6] M.S. Agranovich and A.S. Markus. On spectral properties of elliptic pseudo-differential operators far from selfadjoint ones. *Zeitschrift für Analysis und ihre Anwendungen*, 3(8):237–260, 1989.

[7] M.Sh. Birman and M.Z. Solomyak. *Qualitative Analysis in Sobolev Imbedding Theorems and Applications to Spectral Theory.* Volume 114 of *AMS Translations, Ser. 2*, Providence, 1980.

[8] M.Sh. Birman and M.Z. Solomyak. Schrödinger operator. Estimates for number of bound states as function-theoretical problem. *AMS Translations, Ser. 2*, 150, 1992. Transl. from *Spectral Theory of Operators*, Novgorod, 1989.

[9] K.Kh. Boimatov. Asymptotic properties of the spectrum of Schrödinger operator. *Differential Equations*, 10(11):1492–1497, 1974.

[10] H. Brezis and T. Kato. Remarks on the Schrödinger operator with singular complex potentials. *J. Math. Pures Appl.*, 58(2):137–151, 1979.

[11] H.L. Cycon, R.G. Froese, W. Kirsch, and B. Simon. *Schrödinger Operators with Applications to Quantum Mechanics and Global Geometry.* Springer-Verlag, 1987.

[12] E.B. Davies. *Heat kernels and spectral theory.* Volume 92 of *Cambridge Tracts in Mathematics*, Cambridge University Press, 1980.

[13] N. Dunford and J.T. Schwartz. *Linear Operators. Part 2, Spectral Theory.* NewYork, London, 1963.

[14] I. Gohberg and M. Krein. *Introduction to the Theory of Linear Nonselfadjoint Operators.* Volume 18, AMS Transl. Math. Monographs, 1969.

[15] E. Grinshpun. Asymptotics of spectrum under infinitesimally form-bounded perturbation. *Integral Equations and Operator Theory*, 19(2):240–250, 1994.

[16] E. Grinshpun. Localization theorems for equality of minimal and maximal Schrödinger-type operators. *Journal of Functional Analysis*, 124:40–60, 1994.

[17] L. Hörmander. *The Analysis of Linear Partial Differential Operators.* Volume 4, Springer-Verlag, 1985.

[18] L. Hörmander. On the asymptotic distribution of the eigenvalues of pseudodifferential operators in $\mathbb{R}^n$. *Ark. Mat.*, 17:297–313, 1979.

[19] T. Ikebe and S.-I. Shimada. Spectral and scattering theory for the Schrödinger operators with penetrable wall potentials. *J. Math. Kyoto University*, 31(1), 1991.

[20] V.Ya. Ivrii. Semiclassical spectral asymptotics. *Asterisque*, (207):7–33, 1992.

[21] T. Kato. *Perturbation Theory for Linear Operators.* Springer Verlag, 1966.

[22] M.V. Keldyš. On the eigenvalues and eigenfunctions of certain classes of nonselfadjoint equations. *Doklady Akad. Nauk SSSR*, 77:11–14, 1951.

[23] S. Levendorskii. The approximate spectral projector method. *Acta Appl. Math*, (7):137–197, 1986.

[24] B.Ya. Levin. *Distribution of zeros of entire functions.* AMS, Providence, R.I., 1964.

[25] A.S. Markus. *Introduction to the Spectral Theory of Polynomial Operator Pencils.* Volume 71, AMS Transl. Math. Monographs, 1988.

[26] A.S. Markus and V.I. Matsaev. Comparison theorems for spectra of linear operators and spectral asymptotics. *Trudy. Moscov. Mat. Obsch.*, 45(1):133–181, 1982. English transl. in *Trans. Moscow Math. Soc.*, 45(1): 139-187, 1984.

[27] A.S. Markus and V.I. Matsaev. On the convergence of eigenvector expansions for an operator wich is close to being selfadjoint. *Math. Issled.*, (61):104–129, 1981. (Russian).

[28] A.S. Markus and V.I. Matsaev. Operators associated with sesquilinear forms and spectral asymptotics. *Math. Issled.*, (61):86–103, 1981. (Russian).

[29] V.I. Matsaev. A method of estimation of the resolvents of non-selfadjoint operators. *Soviet Math. Dokl.*, 5(1):236–240, 1964. Translated from *Dokl. Akad. Nauk. SSSR*, v.154, pp. 1034-1037, 1964.

[30] V.I. Matsaev. Some theorems on the completeness of root subspaces of the completely continuous operators. *Soviet Math. Dokl.*, 5(2):393–395, 1964. Translated from *Dokl. Akad. Nauk. SSSR*, v.155, pp. 273-276, 1964.

[31] V.G. Mazja. *Sobolev spaces.* Springer-Verlag, 1985.

[32] M. Reed and B. Simon. *Methods of Modern Mathematical Physics.* Volume 4. Analysis of Operators, Academic Press, 1978.

[33] M. Reed and B. Simon. *Methods of Modern Mathematical Physics.* Volume 2. Harmonic Analysis. Selfadjointness, Academic Press, 1975.

[34] R.T. Rockafellar. *Convex Analysis.* Princeton Univ. Press, 1970.

[35] G.V. Rosenbljum, M.Z. Solomyak, and M.A. Shubin. *Spectral Theory of Differential Operators.* Volume 64 of *Encyclopaedia of Mathematical Sciences, Partial Differential Equations*, Springer-Verlag, 1994. Transl. from *Itogi Nauki i Tekhniki: Sovrem. Probl. Mat., Fundament. Napr.*, v.64, VINITI, Moscow, 1989.

[36] H. Tamura. Asymptotic formulas with sharp remainder estimates for eigenvalues of elliptic operators of second order. *Duke Math. J.*, 49(1):87–119, 1982.

[37] D.G. Vassiliev. The distribution of eigenvalues of partial differential operators. In *Seminaire sur les Equations aux Derivees Partielles*, 1991-92. Exp. No. XVII, 19pp. Ecole Polytech., Palaiseau, 1992.

[38] S. Yakubov. *Completeness of root functions of regular differential operators.* Volume 71 of *Pitman Monographs and Surveys in Pure and Applied Mathematics*, 1994.

Department of Mathematics and Computer Science,
Bar - Ilan University,
Ramat - Gan 52900, Israel
E-mail: grinsh@bimacs.cs.biu.ac.il

AMS Subject Classification: Primary 35P10, 35P20; Secondary 35J10, 47A55

Operator Theory:
Advances and Applications, Vol. 87

COALGEBRAS AND SPECTRAL THEORY IN ONE AND SEVERAL PARAMETERS

LUZIUS GRUNENFELDER AND TOMAŽ KOŠIR

The coalgebraic versions of the primary decomposition theorem for a single linear map and for several commuting linear maps are proved. They lead to a description of the primary decomposition for multiparameter eigenvalue problems in terms of the underlying multiparameter system. Also the coalgebraic version of the primary decomposition theorem for a monic matrix polynomial is discussed.

1 Introduction

The main goal of our paper is to give an elementary and self-contained introduction to coalgebras and their use in spectral theory. We first introduce basic properties of coalgebras and then we prove the primary decomposition theorem for a linear map. When the underlying field is algebraically closed this result reduces to the theorem on the decomposition of a vector space into spectral subspaces. We also give versions of the primary decomposition theorem for a monic matrix polynomial and for several commuting linear maps. These results are well-known and coalgebraic techniques provide an alternative point of view. However, the coalgebraic versions of these results together with other coalgebraic techniques are essential for a solution of the root subspace problem for multiparameter systems. This problem was posed by Atkinson in [2, 3]. Several authors gave partial solutions of the problem using various methods in [4, 6, 8, 18, 20, 21]. A general solution using coalgebraic techniques was presented in [13]. We describe the problem and its solution in §7.

The structure of a coalgebra is dual to the structure of an algebra. Moreover, an algebra structure can be 'dualized' to give a structure of a coalgebra on a dual. The precise construction is discussed in §2. The particular case of the coalgebra structure on the 'dual' of a polynomial algebra is presented in §3. This case is used in the core of our paper, i.e. for §4-§7. There the primary decomposition theorem for a single linear map, for a matrix polynomial and for several commuting linear maps are proved. The result for several commuting linear maps is then applied to multiparameter systems. The 'dual' of a polynomial algebra carries, not only the structure of a coalgebra, but that of a (commutative and cocommutative) Hopf algebra [23] (see also [1, 22, 25]). However, for our paper the coalgebra structure is the important part.

2 Coalgebras and Comodules

The structure of a coalgebra is dual to the structure of an algebra. Let us first consider the structure of an algebra in a slightly different way than it is customary in linear algebra or operator theory. Suppose that $\mathcal{A}$ is an algebra over a field F. (In this paper F is the fixed underlying field for all vector spaces, algebras and coalgebras under consideration. We further assume that the characteristic of F is 0, although most of our discussion remains valid for more general fields.) The multiplication and the unit of $\mathcal{A}$ are linear maps $\mu : \mathcal{A}\otimes\mathcal{A} \to \mathcal{A}$ and $e : F \to \mathcal{A}$, respectively. They are given by $\mu(a \otimes b) = ab$ and $e(\alpha) = \alpha 1$, where 1 is the unit element in $\mathcal{A}$. The associativity is given by the equality

(i) $\mu(\mu \otimes I_{\mathcal{A}}) = \mu(I_{\mathcal{A}} \otimes \mu)$, i.e. by the commutative diagram

$$\begin{array}{ccc} \mathcal{A}\otimes\mathcal{A}\otimes\mathcal{A} & \stackrel{\mu\otimes I_{\mathcal{A}}}{\longrightarrow} & \mathcal{A}\otimes\mathcal{A} \\ \downarrow^{I_{\mathcal{A}}\otimes\mu} & & \downarrow^{\mu} \\ \mathcal{A}\otimes\mathcal{A} & \stackrel{\mu}{\longrightarrow} & \mathcal{A} \end{array},$$

and the unit law is given by the equality

(ii) $\mu(e \otimes I_{\mathcal{A}}) = I_{\mathcal{A}} = \mu(I_{\mathcal{A}} \otimes e)$, i.e. by the commutative diagram

$$\begin{array}{ccccc} F\otimes\mathcal{A} & \cong & \mathcal{A} & \cong & \mathcal{A}\otimes F \\ \downarrow^{e\otimes I_{\mathcal{A}}} & & \|^{I_{\mathcal{A}}} & & \downarrow^{I_{\mathcal{A}}\otimes e} \\ \mathcal{A}\otimes\mathcal{A} & \stackrel{\mu}{\longrightarrow} & \mathcal{A} & \stackrel{\mu}{\longleftarrow} & \mathcal{A}\otimes\mathcal{A} \end{array}.$$

Here $I_{\mathcal{A}}$ is the identity map on $\mathcal{A}$.

Dualizing the above structure, we say that a vector space $\mathcal{C}$ is a coalgebra if there exist linear maps $\delta : \mathcal{C} \to \mathcal{C}\otimes\mathcal{C}$ and $\varepsilon : \mathcal{C} \to F$ such that

(i') $(\delta \otimes I_{\mathcal{C}})\,\delta = (I_{\mathcal{C}} \otimes \delta)\,\delta$, i.e. the diagram

$$\begin{array}{ccc} \mathcal{C} & \stackrel{\delta}{\longrightarrow} & \mathcal{C}\otimes\mathcal{C} \\ \downarrow^{\delta} & & \downarrow^{I_{\mathcal{C}}\otimes\delta} \\ \mathcal{C}\otimes\mathcal{C} & \stackrel{\delta\otimes I_{\mathcal{C}}}{\longrightarrow} & \mathcal{C}\otimes\mathcal{C}\otimes\mathcal{C} \end{array}$$

commutes, and

(ii') $(\varepsilon \otimes I_{\mathcal{C}})\,\delta = I_{\mathcal{C}} = (I_{\mathcal{C}} \otimes \varepsilon)\,\delta$, i.e. the diagram

$$\begin{array}{ccccc} \mathcal{C}\otimes\mathcal{C} & \stackrel{\delta}{\longleftarrow} & \mathcal{C} & \stackrel{\delta}{\longrightarrow} & \mathcal{C}\otimes\mathcal{C} \\ \downarrow^{\varepsilon\otimes I_{\mathcal{C}}} & & \|^{I_{\mathcal{C}}} & & \downarrow^{I_{\mathcal{C}}\otimes\varepsilon} \\ F\otimes\mathcal{C} & \cong & \mathcal{C} & \cong & \mathcal{C}\otimes F \end{array}$$

commutes.

The maps δ and ε are called the *comultiplication* and the *counit*, respectively; and the properties (i') and (ii') are the *coassociativity* and the *counit law*, respectively.

If a vector subspace $\mathcal{B} \subset \mathcal{C}$ is a coalgebra for the restricted maps of δ and ε then we call $\mathcal{B}$ a *subcoalgebra* of $\mathcal{C}$. Note that for $\mathcal{B} \subset \mathcal{C}$ to be a subcoalgebra it suffices to require that $\delta(\mathcal{B}) \subset \mathcal{B} \otimes \mathcal{B}$.

If $\mathcal{A}$ is a finite-dimensional algebra then the canonical linear map $\varphi : \mathcal{A}^* \otimes \mathcal{A}^* \to (\mathcal{A} \otimes \mathcal{A})^*$, determined by $\varphi(f \otimes g)(a \otimes b) = f(a)g(b)$ for $f, g \in \mathcal{A}^*$ and $a, b \in \mathcal{A}$, is an isomorphism. We identify the vector spaces $\mathcal{A}^* \otimes \mathcal{A}^*$ and $(\mathcal{A} \otimes \mathcal{A})^*$ (via φ). Thus $\mathcal{A}^*$ becomes a coalgebra with comultiplication μ^* and counit e^*. If $\mathcal{A}$ is not finite-dimensional then $\varphi : \mathcal{A}^* \otimes \mathcal{A}^* \to (\mathcal{A} \otimes \mathcal{A})^*$ is not an isomorphism. In this case $\mathcal{A}^*$ is replaced by the subspace $\mathcal{A}^\circ$ of all representative functionals. A functional $f \in \mathcal{A}^*$ is called *representative* if its kernel contains a *cofinite* ideal (i.e. an ideal $\mathcal{I} \subset \mathcal{A}$ such that $\mathcal{A}/\mathcal{I}$ is a finite dimensional algebra). The map φ restricts to an isomorphism $\varphi : \mathcal{A}^\circ \otimes \mathcal{A}^\circ \to (\mathcal{A} \otimes \mathcal{A})^\circ$ (cf. [1, 22, 25] and also Proposition 2.1 below). If we identify $(\mathcal{A} \otimes \mathcal{A})^\circ$ and $\mathcal{A}^\circ \otimes \mathcal{A}^\circ$ via φ, then the vector space $\mathcal{A}^\circ$ is a coalgebra with comultiplication and counit

$$\delta = \mu^\circ : \mathcal{A}^\circ \to \mathcal{A}^\circ \otimes \mathcal{A}^\circ \text{ and } \varepsilon = e^\circ : \mathcal{A}^\circ \to F, \tag{1}$$

given by the restrictions of μ^* and e^* to $\mathcal{A}^\circ$. The coalgebra $\mathcal{A}^\circ$ is called the *coalgebra dual* of $\mathcal{A}$. Note that if $\mathcal{A}$ is finite-dimensional then $\mathcal{A}^\circ = \mathcal{A}^*$. The following result is found in [25, Lemma 6.0.1(b)].

PROPOSITION 2.1 *If $\mathcal{A}_1$ and $\mathcal{A}_2$ are two algebras then $(\mathcal{A}_1 \otimes \mathcal{A}_2)^\circ \cong \mathcal{A}_1^\circ \otimes \mathcal{A}_2^\circ$.*

PROOF. Suppose that $f \in (\mathcal{A}_1 \otimes \mathcal{A}_2)^\circ$ and that $\mathcal{K} \subset \ker f$ is a cofinite ideal. We imbed the algebra $\mathcal{A}_1$ in $\mathcal{A}_1 \otimes \mathcal{A}_2$ by mapping $a \mapsto a \otimes 1$ and the algebra $\mathcal{A}_2$ by $a \mapsto 1 \otimes a$. Since the inverse image of a cofinite ideal is a cofinite ideal it follows that

$$\mathcal{I}_1 = \{a \in \mathcal{A}_1;\ a \otimes 1 \in \mathcal{K}\} \text{ and } \mathcal{I}_2 = \{a \in \mathcal{A}_2;\ 1 \otimes a \in \mathcal{K}\}$$

are cofinite ideals in $\mathcal{A}_1$ and $\mathcal{A}_2$, respectively. Then $\mathcal{I} = \mathcal{A}_1 \otimes \mathcal{I}_2 + \mathcal{I}_1 \otimes \mathcal{A}_2$ is an ideal in $\mathcal{A}_1 \otimes \mathcal{A}_2$. Note that $\mathcal{I} \subset \mathcal{K}$. Let $\pi_i : \mathcal{A}_i \to \mathcal{A}_i/\mathcal{I}_i$ $(i = 1, 2)$ be the quotient map. Then $\mathcal{I}$ is the kernel of the quotient map $\pi_1 \otimes \pi_2$. Because $\mathcal{A}_1/\mathcal{I}_1 \otimes \mathcal{A}_2/\mathcal{I}_2$ is finite dimensional it follows that $\mathcal{I}$ is cofinite. Since $\mathcal{I} \subset \mathcal{K}$ there exists a unique functional $\overline{f} : \mathcal{A}_1/\mathcal{I}_1 \otimes \mathcal{A}_2/\mathcal{I}_2 \to F$ such that $f = \overline{f} \circ (\pi_1 \otimes \pi_2)$, and since the algebra $\mathcal{A}_1/\mathcal{I}_1 \otimes \mathcal{A}_2/\mathcal{I}_2$ is finite dimensional it follows that $(\mathcal{A}_1/\mathcal{I}_1 \otimes \mathcal{A}_2/\mathcal{I}_2)^* \cong (\mathcal{A}_1/\mathcal{I}_1)^* \otimes (\mathcal{A}_2/\mathcal{I}_2)^*$. In view of this identification the functional f may be written as a finite sum $\overline{f} = \sum_j \overline{f}_{1j} \otimes \overline{f}_{2j}$, where $\overline{f}_{ij} \in (\mathcal{A}_i/\mathcal{I}_i)^*$ $(i = 1, 2)$. Since $f = \sum_j \overline{f}_{1j}\pi_1 \otimes \overline{f}_{2j}\pi_2 \in \mathcal{A}_1^\circ \otimes \mathcal{A}_2^\circ$ it follows that $(\mathcal{A}_1 \otimes \mathcal{A}_2)^\circ \subset \mathcal{A}_1^\circ \otimes \mathcal{A}_2^\circ$.

To prove the opposite inclusion assume that $f_i \in \mathcal{A}_i^\circ$ $(i = 1, 2)$ and $\mathcal{I}_i \subset \ker f_i$ is a cofinite ideal. Then $f_1 \otimes f_2$ vanishes on the cofinite ideal $\mathcal{A}_1 \otimes \mathcal{I}_2 + \mathcal{I}_1 \otimes \mathcal{A}_2$ and so $f_1 \otimes f_2 \in (\mathcal{A}_1 \otimes \mathcal{A}_2)^\circ$. Hence also $\mathcal{A}_1^\circ \otimes \mathcal{A}_2^\circ \subset (\mathcal{A}_1 \otimes \mathcal{A}_2)^\circ$. □

The structure of a comodule over a coalgebra is dual to the structure of a module over an algebra. If $\mathcal{A}$ is an algebra and $\mathcal{M}$ a vector space then $\mathcal{M}$ is an $\mathcal{A}$-module if there is a linear map $\omega : \mathcal{A} \otimes \mathcal{M} \to \mathcal{M}$, called the *action* of $\mathcal{A}$ on $\mathcal{M}$, such that

(iii) $\omega\,(e \otimes I_{\mathcal{M}}) = I_{\mathcal{M}}$, i.e. the diagram

$$\begin{array}{ccc} F \otimes \mathcal{M} & \cong & \mathcal{M} \\ \downarrow^{e \otimes I_{\mathcal{M}}} & & \|^{I_{\mathcal{M}}} \\ \mathcal{A} \otimes \mathcal{M} & \xrightarrow{\omega} & \mathcal{M} \end{array}$$

commutes, and

(iv) $\omega\,(\mu \otimes I_{\mathcal{M}}) = \omega\,(I_{\mathcal{A}} \otimes \omega)$, i.e. the diagram

$$\begin{array}{ccc} \mathcal{A} \otimes \mathcal{A} \otimes \mathcal{M} & \xrightarrow{\mu \otimes I_{\mathcal{M}}} & \mathcal{A} \otimes \mathcal{M} \\ \downarrow^{I_{\mathcal{A}} \otimes \omega} & & \downarrow^{\omega} \\ \mathcal{A} \otimes \mathcal{M} & \xrightarrow{\omega} & \mathcal{M} \end{array}$$

commutes.

By duality, if $\mathcal{C}$ is a coalgebra and $\mathcal{R}$ a vector space then $\mathcal{R}$ is a $\mathcal{C}$-comodule if there is a linear map $\alpha : \mathcal{R} \to \mathcal{C} \otimes \mathcal{R}$, called the *coaction* of $\mathcal{C}$ on $\mathcal{R}$, such that

(iii') $(\varepsilon \otimes I_{\mathcal{R}})\,\alpha = I_{\mathcal{R}}$, i.e. the diagram

$$\begin{array}{ccc} \mathcal{R} & \xrightarrow{\alpha} & \mathcal{C} \otimes \mathcal{R} \\ \|^{I_{\mathcal{R}}} & & \downarrow^{\varepsilon \otimes I_{\mathcal{R}}} \\ \mathcal{R} & \cong & F \otimes \mathcal{R} \end{array}$$

commutes, and

(iv') $(\delta \otimes I_{\mathcal{R}})\,\alpha = (I_{\mathcal{C}} \otimes \alpha)\,\alpha$, i.e. the diagram

$$\begin{array}{ccc} \mathcal{R} & \xrightarrow{\alpha} & \mathcal{C} \otimes \mathcal{R} \\ \downarrow^{\alpha} & & \downarrow^{\delta \otimes I_{\mathcal{R}}} \\ \mathcal{C} \otimes \mathcal{R} & \xrightarrow{I_{\mathcal{C}} \otimes \alpha} & \mathcal{C} \otimes \mathcal{C} \otimes \mathcal{R} \end{array}$$

commutes.

If a vector subspace $\mathcal{S} \subset \mathcal{R}$ is a comodule for the restricted map of α, i.e. if $\alpha\,(\mathcal{S}) \subset \mathcal{C} \otimes \mathcal{S}$, then we call $\mathcal{S}$ a *subcomodule* of $\mathcal{R}$.

If V is a vector space then $\alpha = \delta \otimes I_V$ is a coaction on $\mathcal{C} \otimes V$. This follows directly from the properties (i') and (ii') for the comultiplication δ. The comodule $\mathcal{C} \otimes V$ is called *cofree.*

Suppose that $\mathcal{R}$ and $\mathcal{S}$ are $\mathcal{C}$-comodules with coactions $\alpha_{\mathcal{R}}$ and $\alpha_{\mathcal{S}}$, respectively. A linear map $\varphi : \mathcal{R} \to \mathcal{S}$ is called a *comodule map* if

(v') $\alpha_{\mathcal{S}}\varphi = (I_{\mathcal{C}} \otimes \varphi)\,\alpha_{\mathcal{R}}$, i.e. the diagram

$$\begin{array}{ccc} \mathcal{R} & \xrightarrow{\varphi} & \mathcal{S} \\ \downarrow^{\alpha_{\mathcal{R}}} & & \downarrow^{\alpha_{\mathcal{S}}} \\ \mathcal{C} \otimes \mathcal{R} & \xrightarrow{I_{\mathcal{C}} \otimes \varphi} & \mathcal{C} \otimes \mathcal{S} \end{array}$$

commutes.

Note that the kernel $\mathcal{K}$ of a comodule map φ is a subcomodule of $\mathcal{R}$ since the relation (v') implies that $\alpha_{\mathcal{R}}(\mathcal{K}) \subset \mathcal{C} \otimes \mathcal{K}$.

The following is usually called the *fundamental theorem for comodules* [1, 11, 25]. To keep our presentation elementary and complete we provide a simple proof. Note that the dimension of a coalgebra or a comodule is its dimension as a vector space over F.

LEMMA 2.2 *Suppose $\mathcal{R}$ is a $\mathcal{C}$-comodule. If V is a finite-dimensional (vector) subspace of $\mathcal{R}$ then V is contained in a finite-dimensional subcomodule of $\mathcal{R}$.*

PROOF. Since V is finite-dimensional there exists a finite-dimensional vector subspace $\mathcal{W} \subset \mathcal{R}$ such that

$$\alpha(V) \subseteq \mathcal{C} \otimes \mathcal{W}. \tag{2}$$

This is the case because $\alpha(v) = \sum_i c_i(v) \otimes u_i(v)$, where $c_i(v) \in \mathcal{C}$ and $u_i(v) \in \mathcal{R}$, is a finite sum for every $v \in V$. Then it suffices for $\mathcal{W}$ to be the span of all $u_i(v)$, where v ranges over a basis of V.

Next we want to show that $\mathcal{W}$ is a subcomodule. Since $(\varepsilon \otimes I_{\mathcal{R}})\alpha = I_{\mathcal{R}}$ (cf. (iii')) it follows that $\alpha^{-1}(\varepsilon \otimes I_{\mathcal{R}})^{-1}(\mathcal{W}) = \mathcal{W}$. We have also that $(\varepsilon \otimes I_{\mathcal{W}})^{-1}(\mathcal{W}) = \mathcal{C} \otimes \mathcal{W}$, and hence

$$\alpha^{-1}(\mathcal{C} \otimes \mathcal{W}) = \mathcal{W}. \tag{3}$$

Then the restriction of the coaction α to $\mathcal{W}$ maps into $\mathcal{C} \otimes \mathcal{W}$, and therefore $\mathcal{W}$ is a subcomodule. From the relations (2) and (3) it follows also that $V \subseteq \mathcal{W}$. □

Next we state the corresponding result for coalgebras.

LEMMA 2.3 *Suppose $\mathcal{C}$ is a coalgebra. If V is a finite-dimensional (vector) subspace of $\mathcal{C}$ then V is contained in a finite-dimensional subcoalgebra of $\mathcal{C}$.*

PROOF. By Lemma 2.2 applied to the cofree comodule $\mathcal{C} \otimes F \cong \mathcal{C}$ it follows that there is a finite-dimensional subcomodule $\mathcal{W}$ such that $V \subseteq \mathcal{W}$. Since the coaction on $\mathcal{C}$ is given by the comultiplication it follows that $\delta(\mathcal{W}) \subset \mathcal{C} \otimes \mathcal{W}$. The second equality in (ii') implies that also $\delta(\mathcal{W}) \subset \mathcal{W} \otimes \mathcal{C}$, and therefore $\delta(\mathcal{W}) \subset \mathcal{W} \otimes \mathcal{W}$, i.e. $\mathcal{W}$ is a subcoalgebra of $\mathcal{C}$. □

The *cotensor product* of two $\mathcal{C}$-comodules $\mathcal{R}$ and $\mathcal{S}$ is defined by the equalizer diagram

$$\mathcal{R} \otimes^{\mathcal{C}} \mathcal{S} \to \mathcal{R} \otimes \mathcal{S} \underset{\sigma_{12}(\alpha_{\mathcal{R}} \otimes I_{\mathcal{S}})}{\overset{I_{\mathcal{R}} \otimes \alpha_{\mathcal{S}}}{\rightrightarrows}} \mathcal{R} \otimes \mathcal{C} \otimes \mathcal{S},$$

i.e. $\mathcal{R} \otimes^{\mathcal{C}} \mathcal{S} = \ker(I_{\mathcal{R}} \otimes \alpha_{\mathcal{S}} - \sigma_{12}(\alpha_{\mathcal{R}} \otimes I_{\mathcal{S}}))$, where σ_{ij} switches the ith and jth tensor factor.

The following property of the cotensor product $\otimes^{\mathcal{C}}$ is used in §7. It is a special case of Proposition 2.1.1 of [16].

LEMMA 2.4 *If $\mathcal{K}$ is the kernel of a comodule map $\varphi : \mathcal{R} \to \mathcal{S}$ and if $\mathcal{L}$ is another comodule then $\mathcal{K} \otimes^{\mathcal{C}} \mathcal{L}$ is the kernel of the induced map $\varphi \otimes^{\mathcal{C}} I_{\mathcal{L}} : \mathcal{R} \otimes^{\mathcal{C}} \mathcal{L} \to \mathcal{S} \otimes^{\mathcal{C}} \mathcal{L}$.*

PROOF. Let $\tau_{\mathcal{PL}} = I_{\mathcal{P}} \otimes \alpha_{\mathcal{L}} - \sigma_{12}(\alpha_{\mathcal{P}} \otimes I_{\mathcal{L}})$, where $\mathcal{P}$ is either $\mathcal{K}$, $\mathcal{R}$ or $\mathcal{S}$. By definition we have that $\mathcal{P} \otimes^{\mathcal{C}} \mathcal{L} = \ker \tau_{\mathcal{PL}}$. Then the diagram

$$\begin{array}{ccccc} \mathcal{K} \otimes^{\mathcal{C}} \mathcal{L} & \rightarrow & \mathcal{R} \otimes^{\mathcal{C}} \mathcal{L} & \xrightarrow{\varphi \otimes^{\mathcal{C}} I_{\mathcal{L}}} & \mathcal{S} \otimes^{\mathcal{C}} \mathcal{L} \\ \downarrow & & \downarrow & & \downarrow \\ \mathcal{K} \otimes \mathcal{L} & \rightarrow & \mathcal{R} \otimes \mathcal{L} & \xrightarrow{\varphi \otimes I_{\mathcal{L}}} & \mathcal{S} \otimes \mathcal{L} \\ \downarrow{\tau_{\mathcal{KL}}} & & \downarrow{\tau_{\mathcal{RL}}} & & \downarrow{\tau_{\mathcal{SL}}} \\ \mathcal{K} \otimes \mathcal{C} \otimes \mathcal{L} & \rightarrow & \mathcal{R} \otimes \mathcal{C} \otimes \mathcal{L} & \xrightarrow{\varphi \otimes I_{\mathcal{C}} \otimes I_{\mathcal{L}}} & \mathcal{S} \otimes \mathcal{C} \otimes \mathcal{L} \end{array}$$

commutes. (Here an arrow without a label indicates an inclusion.) Since also $\mathcal{K} \otimes \mathcal{L} = \ker(\varphi \otimes I_{\mathcal{L}})$ and $\mathcal{K} \otimes \mathcal{C} \otimes \mathcal{L} = \ker(\varphi \otimes I_{\mathcal{C}} \otimes I_{\mathcal{L}})$ it follows that

$$\mathcal{K} \otimes^{\mathcal{C}} \mathcal{L} \subset \ker\left(\varphi \otimes^{\mathcal{C}} I_{\mathcal{L}}\right). \tag{4}$$

To prove the opposite inclusion suppose that $x \in \ker\left(\varphi \otimes^{\mathcal{C}} I_{\mathcal{L}}\right)$. Then it follows that $x \in \ker(\varphi \otimes I_{\mathcal{L}})$, and by definition of the cotensor that $x \in \ker \tau_{\mathcal{RL}}$. Hence $x \in \ker \tau_{\mathcal{KL}} = \mathcal{K} \otimes^{\mathcal{C}} \mathcal{L}$, and so $\ker\left(\varphi \otimes^{\mathcal{C}} I_{\mathcal{L}}\right) \subset \mathcal{K} \otimes^{\mathcal{C}} \mathcal{L}$. This relation together with (4) proves the lemma. □

3 The Coalgebra Dual of a Polynomial Algebra

The coalgebras we use in our paper are derived from the coalgebra dual of a polynomial algebra $F[x]$, where F is the underlying field of characteristic 0. A functional $f \in F[x]^*$ is determined by the values $f_k = f\left(x^k\right)$, $(k \geq 0)$. Thus we may identify $F[x]^*$ with the vector space of all infinite sequences $f = (f_k)_{k=0}^{\infty}$. If $f \in F[x]^{\circ}$ then $f(\mathcal{J}) = 0$ for some cofinite ideal $\mathcal{J}$. Since $F[x]$ is a principal ideal domain there exists a monic polynomial, say $p(x) = x^l - b_{l-1}x^{l-1} - \cdots - b_0$, such that $\mathcal{J}$ is the ideal generated by p. Then $f\left(x^k p(x)\right) = 0$ for all $k \geq 0$, i.e.

$$f_{k+l} = \sum_{i=0}^{l-1} b_i f_{k+i}, \ (k \geq 0). \tag{5}$$

Hence $(f_k)_{k=0}^{\infty}$ is a linearly recursive sequence. Conversely, if $f = (f_k)_{k=0}^{\infty}$ is a linearly recursive sequence then it satisfies (5) for some b_i and so $f(q) = 0$, for all q in $\mathcal{J}$, the ideal generated by $p(x) = x^l - b_{l-1}x^{l-1} - \cdots - b_0$. Thereafter we identify $F[x]^{\circ}$ with the space of all linearly recursive sequences.

If $m_x : F[x] \rightarrow F[x]$ is the map of multiplication by x then its dual $D : F[x]^{\circ} \rightarrow F[x]^{\circ}$ is given by

$$Df\left(x^k\right) = f \cdot \left(m_x\left(x^k\right)\right) = f\left(x^{k+1}\right). \tag{6}$$

From (6) it follows that $D(f_k)_{k=0}^{\infty} = (f_{k+1})_{k=0}^{\infty}$ and so D is the *shift operator* on $F[x]^{\circ}$.

The counit ε and the comultiplication δ of the coalgebra dual $\mathcal{A}^{\circ} = F[x]^{\circ}$ are given by (1). If $f = (f_k)_{k=0}^{\infty} \in F[x]^{\circ}$ then

$$\varepsilon(f)(1) = f(e(1)) = f_0 \tag{7}$$

and

$$\delta(f)(x^r \otimes x^s) = f(\mu(x^r \otimes x^s)) = f\left(x^{r+s}\right) = f_{r+s}. \tag{8}$$

The relation (7) implies that $\varepsilon(f) = f(1)$. To find a description of δ some additional notation is needed. Assume that $\mathcal{J}$ is an ideal, which is maximal among ideals contained in $\ker f$. Then we can assume that $\mathcal{J} = \langle p \rangle$ is a principal ideal generated by a monic polynomial p. Suppose $l = \deg p$ and consider the Hankel matrix $H = [f_{i+j-2}]_{i,j=1}^{l}$. Since $\mathcal{J}$ is maximal in $\ker f$ it follows that there is no linearly recursive relation for the sequence $(f_k)_{k=0}^{\infty}$ of degree less than l. A nonzero linear combination of columns of H would give a linear recursive relation for the sequence $(f_k)_{k=0}^{\infty}$ of degree strictly less than l. So, it follows that H is an invertible matrix. Let $H^{-1} = [g_{ij}]_{i,j=1}^{l}$. Then

$$\delta(f) = \sum_{i,j=1}^{l} g_{ij} D^{i-1} f \otimes D^{j-1} f. \tag{9}$$

To verify (9) observe that

$$\sum_{i,j=1}^{l} g_{ij} \left(D^{i-1} f \otimes D^{j-1} f \right) (x^r \otimes x^s) = \sum_{j=1}^{l} \sum_{i=1}^{l} g_{ij} f_{i-1+r} f_{j-1+s} = \sum_{j=1}^{l} \delta_{j-1,r} f_{j-1+s} = f_{r+s},$$

which coincides with (8).

For a monic polynomial $p \in F[x]$ we define $\mathcal{C}_p = \{f \in F[x]^\circ;\ f(p) = 0\}$. By the definition of D (see (6)) it follows that

$$\mathcal{C}_p = \{f \in F[x]^\circ;\ p(D) f = 0\}. \tag{10}$$

If we choose arbitrarily the values for $f \in \mathcal{C}_p$ at x^k for $k < \deg p$ then the values $f\left(x^k\right)$ for $k \geq \deg p$ are determined by (5). Thus it follows that

$$\dim \mathcal{C}_p = \deg p. \tag{11}$$

If p is a power of a monic irreducible polynomial q, i.e. $p(x) = q(x)^{l+1}$ then we write $\mathcal{C}_p = \mathcal{B}_q^{(l)}$ and

$$\mathcal{B}_q = \bigcup_{l=0}^{\infty} \mathcal{B}_q^{(l)}.$$

If q is of degree one, i.e. $q(x) = x - \lambda$ for some $\lambda \in F$, then we write $\mathcal{B}_q = \mathcal{B}_\lambda$. Note that a consequence of (10) is that D leaves each $\mathcal{C}_p$, and also each $\mathcal{B}_q$, invariant. By (9) it follows that $\delta(\mathcal{C}_p) \subset \mathcal{C}_p \otimes \mathcal{C}_p$ and $\delta(\mathcal{B}_q) \subset \mathcal{B}_q \otimes \mathcal{B}_q$, and so $\mathcal{C}_p$ and $\mathcal{B}_q$ are subcoalgebras of $F[x]^\circ$. If $\mathcal{J} = \langle p \rangle$ is maximal among the ideals contained in the kernel of f, then $\mathcal{C}_p$ is the minimal (finite-dimensional) coalgebra containing f (cf. Lemma 2.3).

LEMMA 3.1 *If $p(x) = q_1(x)^{l_1+1} q_2(x)^{l_2+1} \cdots q_s(x)^{l_s+1}$ is the factorization of the monic polynomial p into distinct monic irreducible factors q_i then $\mathcal{C}_p \cong \oplus_{i=1}^{s} \mathcal{B}_{q_i}^{(l_i)}$. Furthermore, as a coalgebra*

$$F[x]^\circ = \bigoplus_{q \in Q} \mathcal{B}_q, \tag{12}$$

where Q is the set of all monic irreducible polynomials in $F[x]$. If F is algebraically closed then $F[x]^\circ = \bigoplus_{\lambda \in F} \mathcal{B}_\lambda$.

PROOF. Let $\langle p\rangle$ denote the ideal generated by the polynomial p. Since $\left\langle\ p\ \right\rangle \subset \left\langle q_i^{l_i+1}\right\rangle$ for all i, it follows that

$$\mathcal{B}_{q_i}^{(l_i)} \subset \mathcal{C}_p. \tag{13}$$

Because q_i and q_j, $i \neq j$, are relatively prime there exist polynomials r_i and r_j such that $q_i^{l_i+1} r_i + q_j^{l_j+1} r_j = 1$. For $f \in \mathcal{B}_{q_i}^{(l_i)} \cap \mathcal{B}_{q_j}^{(l_j)}$ it follows then that

$$f\left(x^k\right) = f\left(x^k q_i^{l_i+1} r_i + x^k q_j^{l_j+1} r_j\right) = 0.$$

Therefore $f = 0$ and $\oplus_{i=1}^{s} \mathcal{B}_{q_i}^{(l_i)}$ is a direct sum. We use the equality (11) to obtain

$$\dim \mathcal{C}_p = \deg p = \sum_{i=1}^{s} (l_i + 1) \deg q_i = \sum_{i=1}^{s} \dim \mathcal{B}_{q_i}^{(l_i)}.$$

Because the relation (13) holds for all i it follows then that $\mathcal{C}_p = \oplus_{i=1}^{s} \mathcal{B}_{q_i}^{(l_i)}$.

We have seen above that $\mathcal{B}_{q_i}^{(l_i)} \cap \mathcal{B}_{q_j}^{(l_j)} = 0$ for two distinct irreducible polynomials q_i and q_j, and for all l_i and l_j. Hence $\mathcal{B}_{q_i} \cap \mathcal{B}_{q_j} = 0$. Then $\sum_{q \in Q} \mathcal{B}_q$, where Q is the set of all monic irreducible polynomials in $F[x]$, is a direct sum. For each $f \in F[x]^{\circ}$ there is a (nonzero) polynomial p such that $f \in \mathcal{C}_p$. Then the first part of this lemma implies that $f \in \bigoplus_{q \in Q} \mathcal{B}_q$, and hence the relation (12) follows. □

We refer to [23] for a more complete presentation of the structure of commutative and cocommutative Hopf algebra on $F[x]^{\circ}$.

4 The Primary Decomposition Theorem

The direct sum decomposition (12) of the coalgebra dual $F[x]^{\circ}$ is essential for the further discussion. It induces a decomposition of a cofree comodule

$$F[x]^{\circ} \otimes V = \bigoplus_{q \in Q} \mathcal{B}_q \otimes V. \tag{14}$$

Assume now that V is a finite-dimensional vector space. Let $A : V \to V$ be a linear map and let m be its (monic) minimal polynomial. As in §3 the map D is dual to the multiplication by x on $F[x]$. We denote the induced maps $I \otimes A$ and $D \otimes I$ on the comodule $F[x]^{\circ} \otimes V$ again by A and D, respectively. Clearly, A and D are comodule maps. Therefore the kernel $\mathcal{R}$ of $A - D : F[x]^{\circ} \otimes V \to F[x]^{\circ} \otimes V$ is a subcomodule of $F[x]^{\circ} \otimes V$. Then (14) induces a direct sum decomposition of $\mathcal{R}$, which is, as we show later in the section, the coalgebraic version of the primary decomposition theorem for A. First we prove an auxiliary result.

PROPOSITION 4.1 *If $\underline{v} \in F[x]^{\circ} \otimes V$ is such that $(\varepsilon \otimes I_V) D^i \underline{v} = 0$ for all i then $\underline{v} = 0$.*

PROOF. Suppose that $\underline{v} = \sum_{j=1}^{k} f_j \otimes v_j$, where $f_j \in F[x]^{\circ}$ and $v_j \in V$, and that v_j are linearly independent. Then $0 = (\varepsilon \otimes I_V) D^i \underline{v} = \sum_{j=1}^{k} f_j(x^i) v_j$, and so $f_j(x^i) = 0$ for all i and j. This implies that $f_j = 0$ and hence $\underline{v} = 0$. □

We denote by $\mathcal{R}_p$ the kernel of the restricted map $A - D : \mathcal{C}_p \otimes V \to \mathcal{C}_p \otimes V$.

LEMMA 4.2 *The kernel $\mathcal{R}_p$ is a subcomodule of $\mathcal{C}_p \otimes \ker p(A)$ and the composite $\mathcal{R}_p \hookrightarrow \mathcal{C}_p \otimes \ker p(A) \xrightarrow{\varepsilon \otimes I_V} \ker p(A)$ is bijective. In particular, if $p = m$ is the (monic) minimal polynomial of A then $\mathcal{R} \subset \mathcal{C}_m \otimes V$ and $\mathcal{R} \hookrightarrow \mathcal{C} \otimes V \xrightarrow{\varepsilon \otimes I_V} \ker m(A)$ is bijective.*

PROOF. Suppose that $p(x) = x^l + a_{l-1}x^{l-1} + \cdots + a_0$. Since A and D commute it follows that

$$p(A) - p(D) = r(A, D)(A - D),$$

where

$$r(A, D) = \sum_{i=0}^{l-1} a_{i+1} \sum_{j=0}^{i} A^j D^{i-j}$$

and $a_l = 1$. Because $p(D) f = 0$ for $f \in \mathcal{C}_p$ (see (10)) we have that $p(D) \underline{v} = 0$ for $\underline{v} \in \mathcal{C}_p \otimes V$. If also $\underline{v} \in \mathcal{R}_p$ then

$$p(A)\underline{v} = (p(A) - p(D))\underline{v} = r(A, D)(A - D)\underline{v} = 0.$$

This implies that

$$\mathcal{R}_p \subset \mathcal{C}_p \otimes \ker p(A),$$

and thus also that

$$(\varepsilon \otimes I_V)\mathcal{R}_p \subset \ker p(A). \tag{15}$$

Suppose that $g \in \mathcal{C}_p$ is given by

$$g\left(x^i\right) = \begin{cases} 0 & \text{if } i \neq l-1, \\ 1 & \text{if } i = l-1. \end{cases}$$

For $v \in \ker p(A)$ we define $\eta(v) = r(A, D)\, g \otimes v$. Then it follows that

$$(A - D)\eta(v) = (p(A) - p(D))\, g \otimes v = 0$$

and

$$(\varepsilon \otimes I_V)\eta(v) = \sum_{i=0}^{l-1} a_{i+1} \sum_{j=0}^{i} g\left(x^{i-j}\right) A^j v = v.$$

Hence $\ker p(A) \subset (\varepsilon \otimes I_V)\mathcal{R}_p$. The latter inclusion together with (15) implies that

$$(\varepsilon \otimes I_V)\mathcal{R}_p = \ker p(A).$$

It remains to show that $\varepsilon \otimes I_V : \mathcal{R}_p \to \ker p(A)$ is one-to-one. Suppose that $(\varepsilon \otimes I_V)\underline{v} = 0$ for $\underline{v} \in \mathcal{R}_p$. Then it follows that $A\underline{v} = D\underline{v}$ and therefore $(\varepsilon \otimes I_V) D^i \underline{v} = (\varepsilon \otimes I_V) A^i \underline{v} = A^i (\varepsilon \otimes I_V)\underline{v} = 0$ for all i. Then $\underline{v} = 0$ by Proposition 4.1, and hence $\varepsilon \otimes I_V$ is one-to-one.

The second part of the lemma follows from the first one if we set $p = m$. □

Note that if $p(x) = (x - \lambda)^k$ in Lemma 4.2 then $V_p = (\varepsilon \otimes I_V)\mathcal{R}_p = \ker (A - \lambda)^k$ is the kth root subspace of A at the eigenvalue λ.

Now we are ready to prove the coalgebraic version of the primary decomposition theorem [17, Thm. 12, p. 220].

THEOREM 4.3 *Suppose that* $m = q_1^{l_1+1} q_2^{l_2+1} \cdots q_s^{l_s+1}$ *is the factorization of the minimal polynomial* m *of* A *into distinct irreducible polynomials. Then*

$$\mathcal{R} = \bigoplus_{i=1}^{s} \mathcal{R}_i, \tag{16}$$

where $\mathcal{R}_i$ *is the kernel of* $A - D : \mathcal{B}_{q_i}^{(l_i)} \otimes V \to \mathcal{B}_{q_i}^{(l_i)} \otimes V$. *Moreover, if* $V_i = (\varepsilon \otimes I_V) \mathcal{R}_i$ *then*

$$V = \bigoplus_{i=1}^{s} V_i \tag{17}$$

and $V_i = \ker q_i^{l_i+1}(A)$.

PROOF. By Lemma 3.1 applied to the polynomial m we have that $\mathcal{C}_m = \bigoplus_{i=1}^{s} \mathcal{B}_{q_i}^{(l_i)}$. Since $\mathcal{B}_{q_i}^{(l_i)}$ is invariant for D it follows that $\mathcal{B}_{q_i}^{(l_i)} \otimes V$ is invariant for $A - D$. Then it follows that $\mathcal{R}_m = \bigoplus_{i=1}^{s} \mathcal{R}_i$. By Lemma 4.2 we have that $V_i = \ker q_i^{l_i+1}(A)$ and that $\varepsilon \otimes I_V : \mathcal{R} \to V$ is bijective. Thus $\varepsilon \otimes I_V$ carries the direct sum (16) to the direct sum (17). □

COROLLARY 4.4 *If* F *is algebraically closed and* $m = (x - \lambda_1)^{l_1+1} \cdots (x - \lambda_s)^{l_s+1}$ *then* $\mathcal{R} = \bigoplus_{i=1}^{s} \mathcal{R}_i$ *and* $V = \bigoplus_{i=1}^{s} V_i$, *where* $\mathcal{R}_i$ *is the kernel of* $A - D : \mathcal{B}_{\lambda_i}^{(l_i)} \otimes V \to \mathcal{B}_{\lambda_i}^{(l_i)} \otimes V$ *and* $V_i = (\varepsilon \otimes I_V) \mathcal{R}_i = \ker (A - \lambda_i I)^{l_i+1}$.

5 Monic Matrix Polynomials

We consider a monic matrix polynomial

$$L(x) = Ix^l + A_{l-1}x^{l-1} + \cdots + A_0,$$

where A_i are linear maps on a finite-dimensional vector space V. The matrix

$$C_L = \begin{bmatrix} 0 & I & \cdots & 0 \\ \vdots & \ddots & \ddots & \vdots \\ 0 & 0 & \cdots & I \\ -A_0 & -A_1 & \cdots & -A_{l-1} \end{bmatrix}$$

acting on V^l is called the *companion matrix of* L. Then there exist matrix polynomials $E(x)$ and $F(x)$ such that their inverses are again matrix polynomials and such that

$$E(x)(xI - C_L) F(x) = L(x) \oplus I_{l-1}. \tag{18}$$

Here I_{l-1} is the identity map on V^{l-1}. We refer to [10, pp. 13–14] for details.

With L we associate a comodule map $L(D) : F[x]^\circ \otimes V \to F[x]^\circ \otimes V$ by substituting D for x, i.e. $L(D) = D^l + A_{l-1}D^{l-1} + \cdots + A_1 D + A_0$.

LEMMA 5.1 *If* $\mathcal{R}$ *is the kernel of* $L(D)$ *then the composite map* $\mathcal{R} \hookrightarrow F[x]^\circ \otimes V \xrightarrow{\varepsilon^{(l)} \otimes I_V} V^l$, *where*

$$\varepsilon^{(l)} \otimes I_V = \begin{bmatrix} \varepsilon \otimes I_V & \varepsilon D \otimes I_V & \cdots & \varepsilon D^{l-1} \otimes I_V \end{bmatrix}^T : F[x]^\circ \otimes V \to V^l,$$

is bijective.

PROOF. Note that in (18) we can replace x by D since D commutes with maps acting on V^l. Because $E(D)$ and $F(D)$ are invertible linear maps it follows from (18) that

$$\dim \mathcal{R} = \dim \ker (D - C_L) = l \dim V. \tag{19}$$

Here the latter equality follows by Lemma 4.2. Suppose next that $\underline{v} \in \mathcal{R}$ is such that $\left(\varepsilon^{(l)} \otimes I_V\right) \underline{v} = 0$, i.e. $(\varepsilon D^i \otimes I_V) \underline{v} = 0$ for $i = 0, 1, \ldots, l-1$. Now assume that for some $k (\geq l)$ we have that $(\varepsilon D^i \otimes I_V) \underline{v} = 0$ for all $i < k$. Because $\underline{v} \in \mathcal{R}$ it follows that

$$\left(\varepsilon D^k \otimes I_V\right) \underline{v} = -\sum_{i=0}^{l-1} \left(A_i \varepsilon D^{k-l+i} \otimes I_V\right) \underline{v} = 0.$$

By induction we have then that $(\varepsilon D^i \otimes I_V) \underline{v} = 0$ for all i, and so Proposition 4.1 implies that $\underline{v} = 0$. Thus $\varepsilon^{(l)} \otimes I_V : \mathcal{R} \to V^l$ is injective and, since $\dim \mathcal{R} = \dim V^l$ by (19), it is bijective. □

The following is a version of the primary decomposition theorem for monic matrix polynomials. (See also [15, Thm. 5.3].) It follows directly using the decomposition (14) of $F[x]^{\circ} \otimes V$ and Lemma 5.1.

THEOREM 5.2 *Suppose that $\mathcal{R} = \ker L(D)$ and $m(x) = \det L(x)$. Then $\mathcal{R} = \oplus_q \mathcal{R}_q$, where the direct sum is over all monic irreducible divisors q of m and $\mathcal{R}_q$ is the kernel of $L(D) : \mathcal{B}_q \otimes V \to \mathcal{B}_q \otimes V$. Furthermore, $V^l = \oplus_q V_q$, where $V_q = \left(\varepsilon^{(l)} \otimes I_V\right) \mathcal{R}_q$.*

We remark that coalgebraic techniques are used to study regular matrix polynomials in [15], and for the general (including singular) one and several parameter matrix polynomials in [14].

6 Several Commuting Maps

In this section we generalize the results of §4 to the case of several commuting linear maps. The discussion follows steps similar to those in §4. The coalgebra used now is the coalgebra dual of a polynomial algebra in several variables.

If δ_i and ε_i $(i = 1, 2)$ are the comultiplication and the counit of the coalgebra dual $\mathcal{A}_i^{\circ}$ then $\delta = \sigma_{23}(\delta_1 \otimes \delta_2)$ and $\varepsilon = \varepsilon_1 \otimes \varepsilon_2$ are the comultiplication and the counit, respectively, for the coalgebra $\mathcal{A}_1^{\circ} \otimes \mathcal{A}_2^{\circ}$. Here we identify $(\mathcal{A}_1 \otimes \mathcal{A}_2)^{\circ}$ with $\mathcal{A}_1^{\circ} \otimes \mathcal{A}_2^{\circ}$ (see Proposition 2.1), and we recall that σ_{ij} switches the ith and jth tensor factor. If $F[\mathbf{x}]$ is the polynomial algebra in n variables $\mathbf{x} = (x_1, x_2, \ldots, x_n)$ then $F[\mathbf{x}] \cong F[x_1] \otimes F[x_2] \otimes \cdots \otimes F[x_n]$. So it follows that $F[\mathbf{x}]^{\circ} \cong F[x_1]^{\circ} \otimes F[x_2]^{\circ} \otimes \cdots \otimes F[x_n]^{\circ}$. The decomposition (12) applied to each tensor factor $F[x_i]^{\circ}$ yields the decomposition

$$F[\mathbf{x}]^{\circ} \cong \bigoplus_{\mathbf{q} \in \mathbf{Q}} \mathcal{B}_{\mathbf{q}}, \tag{20}$$

where $\mathbf{q} = (q_1, q_2, \ldots, q_n)$ is an n-tuple of monic irreducible polynomials (in one variable), $\mathbf{Q}$ the set of all such n-tuples, and $\mathcal{B}_{\mathbf{q}} = \mathcal{B}_{q_1} \otimes \mathcal{B}_{q_2} \otimes \cdots \otimes \mathcal{B}_{q_n}$. If q_i are linear, i.e. if $q_i(x) = x - \lambda_i$ for all i, then we write $\mathcal{B}_{\boldsymbol{\lambda}} = \mathcal{B}_{\lambda_1} \otimes \mathcal{B}_{\lambda_2} \otimes \cdots \otimes \mathcal{B}_{\lambda_n}$, where $\boldsymbol{\lambda} = (\lambda_1, \lambda_2, \ldots, \lambda_n) \in F^n$. We denote by D_i the map on $F[\mathbf{x}]^{\circ}$ induced by the map D on $F[x_i]^{\circ}$. If V is a finite-dimensional vector space then the map induced by D_i on a cofree comodule $F[\mathbf{x}]^\circ \otimes V$ is also denoted by D_i.

The following is an auxiliary result. It is an analogue of Proposition 4.1.

PROPOSITION 6.1 *If $\underline{v} \in F[\mathbf{x}]^\circ \otimes V$ is such that $(\varepsilon \otimes I_V) D_1^{i_1} D_2^{i_2} \cdots D_n^{i_n} \underline{v} = 0$ for all choices of indices $i_1, i_2, \ldots, i_n$ then $\underline{v} = 0$.*

PROOF. Suppose that $\underline{v} = \sum_{j=1}^k f_j \otimes v_j$, where $f_j \in F[\mathbf{x}]^\circ$ and $v_j \in V$, and that the v_j are linearly independent. Then $0 = (\varepsilon \otimes I_V) D_1^{i_1} D_2^{i_2} \cdots D_n^{i_n} \underline{v} = \sum_{j=1}^k f_j \left(x_1^{i_1} x_2^{i_2} \cdots x_n^{i_n}\right) v_j$, and so $f_j \left(x_1^{i_1} x_2^{i_2} \cdots x_n^{i_n}\right) = 0$ for all choices of indices $i_1, i_2, \ldots, i_n$ and j. This implies that $f_j = 0$ and hence $\underline{v} = 0$. □

Consider commuting linear maps $A_i : V \to V$ $(i = 1, 2, \ldots, n)$. They induce commuting comodule maps $I \otimes A_i$ on the cofree comodule $F[\mathbf{x}]^\circ \otimes V$, which we denote again by A_i. The comodule maps $A_i - D_i$ on $F[\mathbf{x}]^\circ \otimes V$ have a joint kernel

$$\mathcal{R} = \bigcap_{i=1}^n \ker(A_i - D_i),$$

which is a subcomodule of $F[\mathbf{x}]^\circ \otimes V$. Note that (20) induces the direct sum decomposition

$$F[\mathbf{x}]^\circ \otimes V = \bigoplus_{\mathbf{q} \in \mathbf{Q}} \mathcal{B}_{\mathbf{q}} \otimes V, \tag{21}$$

and then also a direct sum decomposition of $\mathcal{R}$. The latter gives the coalgebraic version of the primary decomposition theorem for several commuting maps (see Theorem 6.3).

If $\mathbf{p} = (p_1, p_2, \ldots, p_n)$ is an n-tuple of monic polynomials then we write $\mathcal{C}_{\mathbf{p}} = \mathcal{C}_{p_1} \otimes \mathcal{C}_{p_2} \otimes \cdots \otimes \mathcal{C}_{p_n}$. The restrictions of the comodule maps $A_i - D_i$ $(i = 1, 2, \ldots n)$ to $\mathcal{C}_{\mathbf{p}} \otimes V$ have a joint kernel $\mathcal{R}_{\mathbf{p}} = \mathcal{R} \bigcap (\mathcal{C}_{\mathbf{p}} \otimes V)$. Now we are set to prove the analogues of Lemma 4.2 and Theorem 4.3 for several commuting maps.

LEMMA 6.2 *The joint kernel $\mathcal{R}_{\mathbf{p}}$ is a subcomodule of $\mathcal{C}_{\mathbf{p}} \otimes \bigcap_{i=1}^n \ker p_i(A_i)$ and the composite $\mathcal{R}_{\mathbf{p}} \hookrightarrow \mathcal{C}_{\mathbf{p}} \otimes \bigcap_{i=1}^n \ker p_i(A_i) \xrightarrow{\varepsilon \otimes I_V} \bigcap_{i=1}^n \ker p_i(A_i)$ is bijective. In particular, if $p_i = m_i$ is the monic minimal polynomial of A_i then $\mathcal{R} \subset \mathcal{C}_{\mathbf{m}} \otimes V$, where $\mathbf{m} = (m_1, m_2, \ldots, m_n)$, and $\mathcal{R}_{\mathbf{p}} \hookrightarrow \mathcal{C}_{\mathbf{p}} \otimes \bigcap_{i=1}^n \ker p_i(A_i) \xrightarrow{\varepsilon \otimes I_V} V$ is bijective.*

PROOF. By Lemma 4.2 the kernel $\mathcal{R}_{ip_i}$ of $A_i - D_i : \mathcal{C}_{p_i} \otimes V \to \mathcal{C}_{p_i} \otimes V$ is a subcomodule of $\mathcal{C}_{p_i} \otimes \ker p_i(A_i)$. Then it follows that the kernel of $A_i - D_i : \mathcal{C}_{\mathbf{p}} \otimes V \to \mathcal{C}_{\mathbf{p}} \otimes V$ is a subcomodule of $\mathcal{C}_{\mathbf{p}} \otimes \ker p_i(A_i)$, and so $\mathcal{R}_{\mathbf{p}} \subset \mathcal{C}_{\mathbf{p}} \otimes \bigcap_{i=1}^n \ker p_i(A_i)$.

By Lemma 4.2 each of the maps $\varepsilon_i \otimes I_V : \mathcal{R}_i \to \ker p_i(A_i)$ is onto, so it follows that $\varepsilon \otimes I_V : \mathcal{R}_{\mathbf{p}} \to \bigcap_{i=1}^n \ker p_i(A_i)$ is also onto. To complete the proof we show that $\varepsilon \otimes I_V : \mathcal{R}_{\mathbf{p}} \to \bigcap_{i=1}^n \ker p_i(A_i)$ is one-to-one. Suppose that $\underline{v} \in \mathcal{R}_{\mathbf{p}}$ is such that $(\varepsilon \otimes I_V)\underline{v} = 0$. Then it follows that

$$(\varepsilon \otimes I_V) D_1^{i_1} D_2^{i_2} \cdots D_n^{i_n} \underline{v} = (\varepsilon \otimes I_V) A_1^{i_1} A_2^{i_2} \cdots A_n^{i_n} \underline{v} = A_1^{i_1} A_2^{i_2} \cdots A_n^{i_n} (\varepsilon \otimes I_V)\underline{v} = 0$$

for all choices of indices $i_1, i_2, \ldots, i_n$, and so $\underline{v} = 0$ by Proposition 6.1. □

THEOREM 6.3 *Suppose that* m_i *is the minimal polynomial of* A_i. *Then*

$$\mathcal{R} = \bigoplus_{\mathbf{q}} \mathcal{R}_{\mathbf{q}}, \tag{22}$$

where the sum is over all the n*-tuples of monic irreducible polynomials* $(q_1, q_2, \ldots, q_n)$ *such that*
$\bigcap_{i=1}^n \ker q_i(A_i) \neq 0$. *Moreover, if* $V_{\mathbf{q}} = (\varepsilon \otimes I_V)\mathcal{R}_{\mathbf{q}}$ *then*

$$V = \bigoplus_{\mathbf{q}} V_{\mathbf{q}} \tag{23}$$

and $V_{\mathbf{q}} = \bigcap_{i=1}^n \ker q_i^{l_i}(A_i)$, *where* l_i *is the multiplicity of* q_i *in* m_i.

PROOF. Since $\mathcal{B}_{\mathbf{q}}$ are invariant for all D_i it follows that $\mathcal{B}_{\mathbf{q}} \otimes V$ is invariant for all $A_i - D_i$. Then the direct sum $\mathcal{R} = \bigoplus_{\mathbf{q}} \mathcal{R}_{\mathbf{q}}$, where the sum is over all n-tuples of irreducible monic polynomials, is induced by (21). However, $\varepsilon \otimes I_V : \mathcal{R}_{\mathbf{q}} \to \bigcap_{i=1}^n \ker q_i(A_i)$ is bijective by Lemma 6.2, and hence $\mathcal{R}_{\mathbf{q}} = 0$ if $V_{\mathbf{q}} = \bigcap_{i=1}^n \ker q_i(A_i) = 0$. Now the result follows easily. □

COROLLARY 6.4 *If* F *is algebraically closed then*

$$\mathcal{R} = \bigoplus_{\boldsymbol{\lambda}} \mathcal{R}_{\boldsymbol{\lambda}} \quad \text{and} \quad V = \bigoplus_{\boldsymbol{\lambda}} V_{\boldsymbol{\lambda}},$$

where the sums are over all joint eigenvalues of $\{A_i\}_{i=1}^n$, $\mathcal{R}_{\boldsymbol{\lambda}}$ *is the joint kernel of the* $A_i - D_i : \mathcal{B}_{\boldsymbol{\lambda}} \otimes V \to \mathcal{B}_{\boldsymbol{\lambda}} \otimes V$, *and* $V_{\boldsymbol{\lambda}} = (\varepsilon \otimes I_V)\mathcal{R}_{\boldsymbol{\lambda}}$.

To conclude the section we remark that the problem posed by Davis [5] on minimal representations of commuting linear maps by tensor products was solved using the above coalgebraic construction in [12]. This shows that, in general, the maps D_i provide a minimal model for n-tuples of commuting maps (cf. [7]).

7 Multiparameter Systems

We consider a system of n-parameter linear pencils

$$W_i(\mathbf{x}) = \sum_{j=1}^n A_{ij}x_j - A_{i0}, \quad (i = 1, 2, \ldots, n). \tag{24}$$

For each i the A_{ij} are linear maps on a finite-dimensional vector space V_i. If $V = V_1 \otimes V_2 \otimes \cdots \otimes V_n$ then A_{ij} induces a linear map $A_{ij}^\dagger$ on V by acting on the ith tensor factor of V. The determinant $\Delta_0 : V \to V$ of the matrix

$$A = \begin{bmatrix} A_{11}^\dagger & A_{12}^\dagger & \cdots & A_{1n}^\dagger \\ A_{21}^\dagger & A_{22}^\dagger & \cdots & A_{2n}^\dagger \\ \vdots & \vdots & & \vdots \\ A_{n1}^\dagger & A_{n2}^\dagger & \cdots & A_{nn}^\dagger \end{bmatrix} \tag{25}$$

is a linear map. It is well-defined because any two entries from distinct rows commute. We define determinants Δ_j by replacing the jth column in (25) by the column $\left[A_{i0}^{\dagger}\right]_{i=1}^{n}$. We assume that the multiparameter system considered is *regular*, i.e. that Δ_0 is invertible. Then the linear maps $\Gamma_j = \Delta_0^{-1}\Delta_j$ are called the *associated maps* of the multiparameter system (24). Atkinson [3, Thm. 6.7.1–2] shows that the Γ_j commute, and that

$$\sum_{j=1}^{n} A_{ij}^{\dagger}\Gamma_j = A_{i0}^{\dagger}, \quad (i = 1, 2, \ldots, n). \tag{26}$$

The proof of the latter statement is similar to the scalar case via the adjoint matrix whose entries are the cofactors of determinants Δ_j. Note that the relations (26) are a generalization of *Cramer's Rule*. In matrix form they become $A\left[\Gamma_i\right]_{i=1}^{n} = \left[A_{i0}^{\dagger}\right]_{i=1}^{n}$. For further studies of multiparameter systems in the finite-dimensional and in the general Hilbert space setting we refer to [3, 6, 9, 19, 24, 26].

Atkinson [2, 3] asked for a description of the joint spectral subspace $\bigcap_{i=1}^{n} \ker\left(\Gamma_i - \lambda_i I\right)^N$, where N is large enough, e.g. $N \geq \dim V$, in terms of the original linear maps A_{ij} without constructing Γ_i explicitly. To answer this problem we now combine Atkinson's approach with the coalgebraic techniques outlined in the preceding sections.

For $i = 1, 2, \ldots, n$ we define comodule maps $W_i(\mathbf{D}) = A_{i0} - \sum_{j=1}^{n} A_{ij}D_j$ which act on the cofree comodules $F[\mathbf{x}]^{\circ} \otimes V_i$, and the induced maps $W_i(\mathbf{D})^{\dagger} = A_{i0}^{\dagger} - \sum_{j=1}^{n} A_{ij}^{\dagger}D_j$ which act on $F[\mathbf{x}]^{\circ} \otimes V$. From (26) it follows that

$$\sum_{j=1}^{n} A_{ij}^{\dagger}\left(\Gamma_j - D_j\right) = W_i(\mathbf{D})^{\dagger},$$

or, written in matrix form, that

$$A\left[\Gamma_j - D_j\right]_{j=1}^{n} = \left[W_i(\mathbf{D})^{\dagger}\right]_{i=1}^{n}. \tag{27}$$

Next consider the joint kernels $\mathcal{R}_{\Gamma} = \bigcap_{j=1}^{n} \ker\left(\Gamma_j - D_j\right)$ and $\mathcal{R}_{\mathbf{W}} = \bigcap_{i=1}^{n} \ker W_i(\mathbf{D})^{\dagger}$. Because A is invertible it follows by (27) that

$$\mathcal{R}_{\Gamma} = \mathcal{R}_{\mathbf{W}}, \tag{28}$$

and by Theorem 6.3 we have that $\mathcal{R}_{\Gamma} = \bigoplus_{\mathbf{q}} \mathcal{R}_{\mathbf{q}}$, where $\mathbf{q}$ ranges over all the n-tuples of irreducible polynomials such that $\bigcap_{i=1}^{n} \ker q_i\left(\Gamma_i\right) \neq 0$.

The following result answers Atkinson's question in full generality. It provides a remarkable example of coalgebraic techniques yielding new results in spectral theory. In the theorem we describe the comodule $\mathcal{R}_{\mathbf{q}}$ in terms of the kernels of $W_i(\mathbf{D})$. Then via $V_{\mathbf{q}} = \left(\varepsilon \otimes I_V\right)\mathcal{R}_{\mathbf{q}}$ we get a description of the joint spectral subspace for Γ_i in terms of the original linear maps A_{ij}. In particular, if all the components of $\mathbf{q}$ are linear (always the case if F is algebraically closed) then we answer to Atkinson's question. To simplify the notation we drop the index $\mathbf{q}$.

THEOREM 7.1 *If $\mathcal{R}_i$ is the kernel of $W_i(\mathbf{D}) : \mathcal{B} \otimes V_i \to \mathcal{B} \otimes V_i$ then*

$$\mathcal{R} = \mathcal{R}_1 \otimes^{\mathcal{B}} \mathcal{R}_2 \otimes^{\mathcal{B}} \cdots \otimes^{\mathcal{B}} \mathcal{R}_n.$$

PROOF. We write $\mathcal{V}_i = \mathcal{B} \otimes V_i$. Because the cotensor product preserves kernels by Lemma 2.4 it follows that $W_1(\mathbf{D})^\dagger = \mathcal{R}_1 \otimes^{\mathcal{B}} \mathcal{V}_2 \otimes^{\mathcal{B}} \cdots \otimes^{\mathcal{B}} \mathcal{V}_n$. Now suppose that

$$\bigcap_{i=1}^{k} \ker W_i(\mathbf{D})^\dagger = \mathcal{R}_1 \otimes^{\mathcal{B}} \cdots \otimes^{\mathcal{B}} \mathcal{R}_k \otimes^{\mathcal{B}} \mathcal{V}_{k+1} \otimes^{\mathcal{B}} \cdots \otimes^{\mathcal{B}} \mathcal{V}_n \tag{29}$$

for some $k < n$. We want to show that (29) holds when k is replaced by $k+1$ and thus prove the theorem by induction on k. If we apply Lemma 2.4 twice then we get that $\mathcal{R}_k \otimes^{\mathcal{B}} \mathcal{R}_{k+1}$ is the intersection of the kernels of

$$W_i(\mathbf{D})^\dagger : \mathcal{V}_k \otimes^{\mathcal{B}} \mathcal{V}_{k+1} \to \mathcal{V}_k \otimes^{\mathcal{B}} \mathcal{V}_{k+1} \quad (i = k, k+1).$$

Next we cotensor the comodule $\mathcal{V}_k \otimes^{\mathcal{B}} \mathcal{V}_{k+1}$ by $\mathcal{R}_1 \otimes^{\mathcal{B}} \cdots \otimes^{\mathcal{B}} \mathcal{R}_{k-1}$ on the left-hand side and by $\mathcal{V}_{k+2} \otimes^{\mathcal{B}} \cdots \otimes^{\mathcal{B}} \mathcal{V}_n$ on the right-hand side. Then it follows by applying Lemma 2.4 again that $\mathcal{R}_1 \otimes^{\mathcal{B}} \cdots \otimes^{\mathcal{B}} \mathcal{R}_{k+1} \otimes^{\mathcal{B}} \mathcal{V}_{k+2} \otimes^{\mathcal{B}} \cdots \otimes^{\mathcal{B}} \mathcal{V}_n$ is the intersection of the kernels of $W_i(\mathbf{D})^\dagger$ $(i = k, k+1)$ considered as comodule maps on $\mathcal{R}_1 \otimes^{\mathcal{B}} \cdots \otimes^{\mathcal{B}} \mathcal{R}_{k-1} \otimes^{\mathcal{B}} \mathcal{V}_k \otimes^{\mathcal{B}} \cdots \otimes^{\mathcal{B}} \mathcal{V}_n$. This, together with the inductive hypothesis, implies that

$$\bigcap_{i=1}^{k+1} \ker W_i(\mathbf{D})^\dagger = \mathcal{R}_1 \otimes^{\mathcal{B}} \cdots \otimes^{\mathcal{B}} \mathcal{R}_{k+1} \otimes^{\mathcal{B}} \mathcal{V}_{k+2} \otimes^{\mathcal{B}} \cdots \otimes^{\mathcal{B}} \mathcal{V}_n. \tag{30}$$

When $k+1 = n$ the relation (30) is $\mathcal{R}_{\mathbf{W}} = \mathcal{R}_1 \alpha^{\mathcal{B}} \mathcal{R}_2 \otimes^{\mathcal{B}} \cdots \otimes^{\mathcal{B}} \mathcal{R}_n$, and because $\mathcal{R}_{\mathbf{\Gamma}} = \mathcal{R}_{\mathbf{W}}$ it follows that $\mathcal{R}_{\mathbf{\Gamma}} = \mathcal{R}_1 \otimes^{\mathcal{B}} \mathcal{R}_2 \otimes^{\mathcal{B}} \cdots \otimes^{\mathcal{B}} \mathcal{R}_n$. □

Theorem 7.1 is proved in [13, Thm. 5.1]. There, the structure of elements of $\mathcal{R}$ is studied closely and an algorithm to construct a 'canonical' basis for V in the case $n = 2$ and F algebraically closed is discussed. A generalization to the case of Fredholm operators in a Hilbert space is considered as well. For further applications of coalgebraic techniques to multiparameter spectral theory see also [14].

References

[1] E. Abe. *Hopf Algebras.* Cambridge Univ. Press, 1980.

[2] F.V. Atkinson. Multiparameter Spectral Theory. *Bull. Amer. Math. Soc.*, 74:1–27, 1968.

[3] F.V. Atkinson. *Multiparameter Eigenvalue Problems.* Academic Press, 1972.

[4] P.A. Binding. Multiparameter Root Vectors. *Proc. Edin. Math. Soc.*, 32:19–29, 1989.

[5] C. Davis. Representing a Commuting Pair by Tensor Products. *Lin. Alg. Appl.*, 3:355–357, 1970.

[6] M. Faierman. *Two-parameter Eigenvalue Problems in Ordinary Differential Equations*, volume 205 of *Pitman Research Notes in Mathematics.* Longman Scientific and Technical, Harlow, U.K., 1991.

[7] C.K. Fong and A.R. Sourour. Renorming, Similarity and Numerical Ranges. *J. London Math. Soc.*, (2)18:511–518, 1978.

[8] G.A. Gadzhiev. On a Multitime Equation and its Reduction to a Multiparameter Spectral Problem. *Soviet. Math. Dokl.*, 32:710–713, 1985.

[9] G.A. Gadzhiev. *Introduction to Multiparameter Spectral Theory* (in Russian). Azerbaijan State University, Baku, 1987.

[10] I. Gohberg, P. Lancaster, and L. Rodman. *Matrix Polynomials.* Academic Press, 1982.

[11] L. Grunenfelder. Hopf-Algebren und Coradical. *Math. Z.*, 116:166–182, 1970.

[12] L. Grunenfelder and T. Košir. Representation of Commuting Maps by Tensor Products, to appear in *Lin. Alg. Appl.*

[13] L. Grunenfelder and T. Košir. An Algebraic Approach to Multiparameter Eigenvalue Problems, submitted.

[14] L. Grunenfelder and T. Košir. Koszul Cohomology for Finite Families of Comodule Maps and Applications, submitted.

[15] L. Grunenfelder and M. Omladič. Linearly Recursive Sequences and Operator Polynomials. *Lin. Alg. Appl.*, 182:127–145, 1993.

[16] L. Grunenfelder and R. Paré. Families Parametrized by Coalgebras. *J. Algebra*, 107:316–375, 1987.

[17] K. Hoffman and R. Kunze. *Linear Algebra.* Prentice Hall, second edition, 1971.

[18] G.A. Isaev. On Root Elements of Multiparameter Spectral Problems. *Soviet. Math. Dokl.*, 21:127–130, 1980.

[19] H.(G.A.) Isaev. *Lectures on Multiparameter Spectral Theory.* Dept. of Math. and Stats., University of Calgary, 1985.

[20] T. Košir. *Commuting Matrices and Multiparameter Eigenvalue Problems.* Ph.D. thesis, Dept. of Math. and Stats., University of Calgary, 1993.

[21] T. Košir. The Finite-dimensional Multiparameter Spectral Theory : The Nonderogatory Case. *Lin. Alg. Appl.*, 212/213:45–70, 1994.

[22] S. Montgomery. *Hopf Algebras and Their Actions on Rings.* CBMS Reg. Conf. Ser. in Math. 82, AMS, 1993.

[23] B. Peterson and E. Taft. The Hopf Algebra of Linearly Recursive Sequences. *Aequationes Math.*, 20:1–17, 1980.

[24] B.D. Sleeman. *Multiparameter Spectral Theory in Hilbert Space*, volume 22 of *Pitman Research Notes in Mathematics.* Pitman Publ. Ltd., London U.K., Belmont U.S.A., 1978.

[25] M.E. Sweedler. *Hopf Algebras.* Benjamin, New York, 1969.

[26] H. Volkmer. *Multiparameter Eigenvalue Problems and Expansion Theorems*, volume 1356 of *Lecture Notes in Mathematics.* Springer-Verlag, Berlin, New York, 1988.

Department of Mathematics, Statistics and Computing Science
Dalhousie University
Halifax, Nova Scotia
Canada, B3H 3J5

1991 Mathematics Subject Classification. 15A21, 16W30, 47A13.

Operator Theory:
Advances and Applications, Vol. 87

DESTABILIZATION OF INFINITE–DIMENSIONAL TIME–VARYING SYSTEMS VIA DYNAMICAL OUTPUT FEEDBACK

Birgit Jacob

It is shown that a time–varying system on a Banach space can be destabilized by linear output feedback with norm arbitrarily close to $\sup_{t\geq 0} \|\mathbb{L}_a S_t\|^{-1}$, where $\mathbb{L}_a$ denotes the input output operator and S_t is the operator of right shift by t on $L_p(a,\infty;Y)$.

1 INTRODUCTION

Over the last decade, problems of robust stability have been very popular in control theory. The problem of robust stability may be stated as follows. Given a nominal state–space system and a normed set of perturbations, which is the largest $r > 0$ which guarantees that for every perturbation with norm less than r the perturbed system is also stable?

In order to measure this largest bound r, Hinrichsen and Pritchard have introduced the concept of "stability radii" in [HP86b] and [HP86a]. There are a many of papers calculating the stability radii for different classes of perturbations in the finite–dimensional setting, see [HP90], [HP92], [HIP89] and the references therein. For extensions to infinite–dimensional systems we refer the reader to [PT89], [HP94] and [JDP95]. However, up till now in the time–varying situation only lower bounds for the stability radii were known. So it is the purpose of this paper to develop destabilization results for a wide class of time–varying infinite–dimensional state–space systems. Moreover, the results obtained in this paper are even new in a finite–dimensional setting.

We consider a nominal system of the form

$$x(t) = \Phi(t,t_1)x_1, \qquad t \geq t_1, \tag{1}$$

where $\Phi : \Gamma_a \to \mathcal{L}(X)$ is an exponentially stable mild evolution operator on X, $t_1 \in [a,\infty)$ and $x_1 \in X$. Supposing the mild evolution operator is subjected to structured perturbations, the perturbed trajectory is given by the solution of the following integral equation

$$x(t) = \Phi(t,t_1)x_1 + \int_{t_1}^{t} \Phi(t,\rho)B(\rho)P(C(\cdot)x(\cdot))(\rho)\,d\rho, \quad t \geq t_1, \tag{2}$$

where P is an unknown disturbance operator and B, C are given "scaling operators" defining the "structure" of the perturbation. Formally the integral equation (2) can also obtained by applying the feedback $u(t) = P(y)(t)$ to the time-varying system

$$\begin{aligned} x(t) &= \Phi(t,t_1)x_1 + \int_{t_1}^{t} \Phi(t,\rho)B(\rho)u(\rho)\,d\rho, \\ y(t) &= C(t)x(t). \end{aligned}$$

In this paper we address to the problem of finding a causal operator $P \in \mathcal{L}(L_p(a,\infty;Y), L_p(a,\infty;U))$ with minimal norm such that the origin of the perturbed equation (2) is not output stable. We treat this problem in a very general setup by assuming that the tuple (Φ, B, C) defines a stable tv–system (time–varying system). The concept of stable tv–systems was first introduced in [HP94] (see also [Jac95]), and it allows the same degree of unboundedness in the input– and output operators as regular systems [Wei89a] do in a time–invariant situation. If (Φ, B, C) defines a stable tv–system we prove in this paper that for every $\varepsilon > 0$ there exists a $P \in \mathcal{L}(L_p(a,\infty;Y), L_p(a,\infty;U))$, P causal, with $\|P\| < \sup_{t\geq 0} \|\mathbb{L}_a S_t\|^{-1} + \varepsilon$ such that the origin of the perturbed equation (2) is not output stable. Moreover, this result is sharp in the following sense. If (Φ, B, C) is a stable tv–system with Property (L) in [JDP95] it is shown that for every causal $P \in \mathcal{L}(L_p(a,\infty;Y), L_p(a,\infty;U))$ with $\|P\| < \sup_{t\geq 0} \|\mathbb{L}_a S_t\|^{-1}$ the origin of the corresponding perturbed equation (2) is output stable. Furthermore, it is illustrated by an example that such a destabilization result does not hold for linear memoryless perturbations.

In order to prove this destabilization result we follow an idea of Shamma and Zhao [SZ93] and approximate the input–output operator by causal operators with finite memory. They used this idea in order to prove results about robust stability for systems in an input–output setting. So the aim of this paper is to translate the result of Shamma and Zhao [SZ93] to infinite–dimensional time–varying state–space systems.

We proceeds as follows. In the next section we give some notations and present mild evolution operators. In Section 3 we introduce a fairly large class of infinite–dimensional time–varying state–space systems which allows for unbounded control and observation. Our set–up is in a time–invariant setting closely related to regular systems introduced by Weiss [Wei89b]. We prove a number of technical results which will be used in sequel. In Section 4 we formulate the problem precisely and recall some results which are proved in [JDP95]. Finally, in Section 5 we show the destabilization results.

2 NOTATION AND FUNDAMENTAL RESULTS

Let X and Y be arbitrary real or complex Banach spaces and $-\infty < s < t \leq \infty$. For every $1 \leq p < \infty$ we denote by $L_p(s,t;X)$ the space of measurable functions f with $\|f\|_p := \left(\int_s^t \|f(t)\|_X^p \,dt\right)^{1/p} < \infty$ and by $L_\infty(s,t;X)$ the space of measurable and essential bounded functions. Moreover, $L_{s,\infty}(s,t;\mathcal{L}(X,Y))$ is the space of strongly measurable functions f with $\|f\|_{L_{s,\infty}(s,t;\mathcal{L}(X,Y))} := \operatorname{ess\,sup}_{\rho\in[s,t]} \|f(\rho)\| < \infty$. We are also interested in the space $L_p^{loc}(s,\infty;X)$, $1 \le p \le \infty$, which contain all functions f with the property $f \in L_p(s,t;X)$ for every $s < t < \infty$. Moreover, by $H_p(s,t;X)$ we will denote the Banach space of functions $f : [s,t) \to X$ which satisfies $f, f' \in L_p(s,t;X)$ provided with the norm $\|f\|_{H_p(s,t;X)} := \left(\|f\|^p_{L_p(s,t;X)} + \|f'\|^p_{L_p(s,t;X)}\right)^{1/p}$. For a function $f \in L_p(s,t;X)$ we will denote by supp f the support of f. For our description of time–varying linear systems by mild evolution operators, the sets $\Gamma_a^b := \{(t,s) : a \le s \le t \le b\}$ and $\Gamma_a := \{(t,s) : a \le s \le t < \infty\}$, where $-\infty < a < b < \infty$, will be needed.

DEFINITION 2.1 *$\Phi : \Gamma_a \to \mathcal{L}(X)$ is said to be a* **mild evolution operator (on X)**, *if*

1) $\Phi(t,t) = I$ for each $t \in [a,\infty)$,

2) $\Phi(t,\sigma)\Phi(\sigma,s) = \Phi(t,s)$ for $a \le s \le \sigma \le t < \infty$,

3) The maps $\Phi(\cdot,s) : [s,\infty) \to \mathcal{L}(X)$ and $\Phi(t,\cdot) : [a,t] \to \mathcal{L}(X)$ are strongly continuous.

4) For every $t_1 \in (a,\infty)$ we have $M_\Phi^{t_1} := \sup_{(t,s)\in\Gamma_a^{t_1}} \|\Phi(t,s)\| < \infty$.

An example given by Gibson [Gib76] proves that in general 1)–3) do not imply 4).

DEFINITION 2.2 *We say that a mild evolution operator Φ is* **exponentially stable** *if there exist $\alpha, M > 0$ such that*

$$\|\Phi(t,s)\| \le Me^{-\alpha(t-s)}, \qquad \text{for all } (t,s) \in \Gamma_a.$$

For $\tau \ge 0$ and $k \in [a,\infty)$ S_τ will denote the operator of right shift by τ on $L_p(a,\infty;X)$ and π_k will denote the operator of truncation at k on $L_p(a,\infty;X)$. Supposing $\tau \in (a,\infty)$, $f \in L_q(a,\tau;X)$ and $g \in L_q(\tau,\infty;X)$ we denote by $f \underset{\tau}{\diamond} g \in L_q(a,\infty;X)$ the function defined by

$$\left(f \underset{\tau}{\diamond} g\right)(t) := \begin{cases} f(t) & , \ t \in [a,\tau) \\ g(t) & , \ t \in (\tau,\infty) \end{cases}.$$

Finally, we call an operator $P \in \mathcal{L}(L_p(a,\infty;X), L_p(a,\infty;Y))$ **causal** if and only if $\pi_t N \pi_t = \pi_t N$, for every $t \in [a,\infty)$. Causality is a fundamental property of physically realizable systems. It merely expresses that past and present output values do not depend on future input values.

3 SYSTEM DESCRIPTION

In order to define stable tv–systems we assume that $\Phi : \Gamma_a \to \mathcal{L}(X)$ is an exponentially stable mild evolution operator on X, $B : [a,\infty) \to \mathcal{L}(U,\overline{X})$, $C : [a,\infty) \to \mathcal{L}(\underline{X},Y)$ and we introduce the following hypotheses

(TV^{S}1) U and Y are Banach spaces. The Banach spaces $\underline{X}$, X and $\overline{X}$ satisfy $\underline{X} \hookrightarrow X \hookrightarrow \overline{X}$, i.e. the canonical injections are continuous with dense range.

(TV^{S}2) For all $(t,s) \in \Gamma_a$ the operator $\Phi(t,s)$ can be extended to a bounded linear operator on $\overline{X}$ (again denoted by $\Phi(t,s)$). Furthermore, for all $s \geq a$ and all $x \in \underline{X}$ we have $\Phi(t,s)x \in \underline{X}$ for a.e. $t \geq s$.

(TV^{S}3) For every $u \in L_p(a,T;U)$, $T \in (a,\infty)$, the map $\Phi(T,\cdot)B(\cdot)u(\cdot) : [a,T] \to \overline{X}$ is integrable.

An important role will be played by the input–state operator $\mathbb{M}_s$, $s \in [a,\infty)$, given by

$$(\mathbb{M}_s u)(t) = \int_s^t \Phi(t,\rho)B(\rho)u(\rho)\,d\rho, \qquad (t,s) \in \Gamma_a, u \in L_p(s,t;U),$$

where $p \geq 1$ is fixed. Note that because of (TV^{S}3) the integral is well–defined in $\overline{X}$.

(TV^{S}4) For every $(t,s) \in \Gamma_a$ and all $u \in H_p(s,t;U)$ we have $(\mathbb{M}_s u)(t) \in \underline{X}$.

(TV^{S}5) For $s \in [a,\infty)$ and $u \in L_p^{loc}(s,\infty;U)$, the map $\mathbb{M}_s u : [s,\infty) \to X$ is continuous.

(TV^{S}6) There exist positive numbers k_1 and κ_1 such that

$$\|(\mathbb{M}_s u)(t)\|_X \leq k_1 e^{\kappa_1(t-s)} \|u\|_{L_p(s,t;U)}, \qquad (t,s) \in \Gamma_a, u \in L_p(s,t;U).$$

(TV^{S}7) For all $s \in [a,\infty)$ $\mathbb{M}_s \in \mathcal{L}(L_p(s,\infty;U), L_p(s,\infty;X))$.

(TV^{S}8) There exists a positive number k_2 such that

$$\|C(\cdot)\Phi(\cdot,s)x\|_{L_p(s,\infty;Y)} \leq k_2\|x\|_X, \qquad x \in \underline{X}, s \in [a,\infty).$$

Moreover, for $x \in \underline{X}$ and $t \in (a,\infty)$ we have $\lim_{s \nearrow t} \|C(\cdot)\Phi(\cdot,s)x\|_{L_p(s,t;Y)} = 0$.

The input–output operator is given by

$$(\mathbb{L}_s u)(t) = C(t)\int_s^t \Phi(t,\rho)B(\rho)u(\rho)\,d\rho, \qquad (t,s) \in \Gamma_a \text{ and } u \in H_p(s,t;U).$$

(TV^{S}9) There exists a positive number k_3 such that

$$\|\mathbb{L}_s u\|_{L_p(s,t;Y)} \leq k_3\|u\|_{L_p(s,t;U)}, \qquad (t,s) \in \Gamma_a, u \in H_p(s,t;U).$$

REMARK 3.1 *1) Easy calculations show that the operator $\mathbb{C}_{(t,s)} \in \mathcal{L}(\underline{X}, L_p(s,t;Y))$ given by $(\mathbb{C}_{(t,s)}x)(\rho) := C(\rho)\Phi(\rho,s)x$ has a continuous extension $\overline{\mathbb{C}_{(t,s)}} \in \mathcal{L}(X, L_p(s,t;Y))$ and (TV^{S}8) holds for every $x \in X$. However, we also denote the extension of $\mathbb{C}_{(t,s)}$ to $\mathcal{L}(X, L_p(s,t;Y))$ by $\mathbb{C}_{(t,s)}$.*

2) $(TV^S 9)$ implies that the operator $\mathbb{L}_s \in \mathcal{L}(H_p(s,t;U), L_p(s,t;Y))$ has a continuous extension $\overline{\mathbb{L}_s} \in \mathcal{L}(L_p(s,t;U), L_p(s,t;Y))$. We also denote this extension by $\mathbb{L}_s$ and get $\mathbb{L}_s \in \mathcal{L}(L_p(s,\infty;U), L_p(s,\infty;Y))$ for all $s \in [a,\infty)$. Moreover, note that $\|\mathbb{L}_s\| = \|\mathbb{L}_a S_{s-a}\|$ holds for every $s \geq a$.

3) Assumption $(TV^S 6)$ may seem a little bit strange, as it does not immediately imply that the state trajectories are bounded. However, Lemma 3.3 shows that this is the case.

4) The assumptions $(TV^S 1)$-$(TV^S 9)$ are nearly the same as (TV1)-(TV4), $(TV5^+)$, $(TV6^+)$, (TV7) and (TV8) in [JDP95]. However, all results given in [JDP95] are also true if we assume that $(TV^S 1)$-$(TV^S 9)$ hold.

DEFINITION 3.2 *We call the tuple $(\Phi, B, C, \underline{X}, X, \overline{X}, U, Y, a)$ a* **stable tv–system (stable time–varying system)**, *if $a \in \mathbb{R}$, $\Phi : \Gamma_a \to \mathcal{L}(X)$ is an exponentially stable mild evolution operator on X, $B : [a,\infty) \to \mathcal{L}(U, \overline{X})$, $C : [a,\infty) \to \mathcal{L}(\underline{X}, Y)$ and $(TV^S 1)$-$(TV^S 9)$ hold.*

This definition of a stable tv–system is quoted from [Jac95] and for examples for stable tv–systems we refer the reader to [Jac95]. Moreover, in [Jac95] a concept of time–varying systems (tv–systems) is introduced which allow the same degree of unboundedness but do not assume that the system is stable. The conditions defining a tv–system are similar to hypothesis (TV1)-(TV6) in [JDP95]. However, in this paper we are only concerned with stable tv–systems. For simplicity we often write (Φ, B, C) instead of $(\Phi, B, C, \underline{X}, X, \overline{X}, U, Y, a)$.

If $(\Phi, B, C, \underline{X}, X, \overline{X}, U, Y, a)$ is a stable tv–system, then U presents the input space, X the state space and Y the output space. Moreover, the state trajectory generated by an initial condition $x(t_1) = x_1 \in X$ at initial time $t_1 \in [a,\infty)$ and a control function $u \in L_p^{loc}(t_1,\infty;U)$, is given by

$$x(t;t_1,x_1,u) = \Phi(t,t_1)x_1 + \int_{t_1}^{t} \Phi(t,\rho)B(\rho)u(\rho)\,d\rho, \quad t \in [t_1,\infty), \tag{3}$$

with associated output function

$$y(t;t_1,x_1,u) = C(t)\Phi(t,t_1)x_1 + C(t)\int_{t_1}^{t} \Phi(t,\rho)B(\rho)u(\rho)\,d\rho, \tag{4}$$

$t \in [t_1,\infty)$. By assumption the state trajectory $x(\cdot;t_1,x_1,u)$ is continuous on $[t_1,\infty)$ with values in X and $y(\cdot;t_1,x_1,u) \in L_p^{loc}(t_1,\infty;Y)$.

LEMMA 3.3 *For a stable tv–system (Φ, B, C) there exists a positive constant ρ such that $\|x(t;t_1,x_1,u)\|_X \leq \rho(\|x_1\|_X + \|u\|_{L_p(t_1,\infty;U)})$ for every $(t_1,x_1,u) \in [a,\infty) \times X \times L_p(t_1,\infty;U)$ and $t \geq t_1$.*

PROOF:
By Assumption (TV^{S}7), the exponential stability of Φ on X and the property $\|\mathbb{M}_s\| \leq \|\mathbb{M}_a\|$ for all $s \geq a$, we obtain the existence of a constant $\rho_1 > 0$ such that

$$\|x(\cdot;t_1,x_1,u)\|_{L_p(t_1,\infty;X)} \leq \rho_1(\|x_1\|_X + \|u\|_{L_p(t_1,\infty;U)})$$

for every $(t_1,x_1,u) \in [a,\infty) \times X \times L_p(t_1,\infty;U)$. Moreover, the exponential stability of Φ on X and Assumption (TV^{S}6) imply the existence of constants $M,\omega > 0$ such that

$$\|x(t;t_1,x_1,u)\|_X \leq Me^{\omega(t-t_1)}(\|x_1\|_X + \|u\|_{L_p(t_1,\infty;U)}) \tag{5}$$

for every $(t_1,x_1,u) \in [a,\infty) \times X \times L_p(t_1,\infty;U)$ and $t \geq t_1$. Choosing $(t_1,x_1,u) \in [a,\infty) \times X \times L_p(t_1,\infty;U)$ we note that for $t \geq s \geq t_1$ the following evolution property is satisfied $x(t;t_1,x_1,u) = x(t;s,x(s;t_1,x_1,u),u)$. Thus we find that for $t \geq t_1$

$$\begin{aligned}
&\left(\frac{1-e^{-2\omega p(t-t_1)}}{2\omega p}\|x(t;t_1,x_1,u)\|_X^p\right)^{1/p} \\
= \;& \left(\int_{t_1}^t \|x(t;t_1,x_1,u)\|_X^p e^{-2\omega p(t-\tau)}\,d\tau\right)^{1/p} \\
= \;& \left(\int_{t_1}^t \|x(t;\tau,x(\tau;t_1,x_1,u),u)\|_X^p e^{-2\omega p(t-\tau)}\,d\tau\right)^{1/p} \\
\leq \;& \left(\int_{t_1}^t M^p e^{-\omega p(t-\tau)}[\|x(\tau;t_1,x_1,u)\|_X + \|u\|_{L_p(t_1,\infty;U)}]^p\,d\tau\right)^{1/p} \qquad (6)\\
\leq \;& M\left(\int_{t_1}^t \|x(\tau;t_1,x_1,u)\|_X^p\,d\tau\right)^{1/p} + M\left(\int_{t_1}^t e^{-\omega p(t-\tau)}\,d\tau\right)^{1/p}\|u\|_{L_p(t_1,\infty;U)} \\
\leq \;& \rho[\|x_1\|_X + \|u\|_{L_p(t_1,\infty;U)}],
\end{aligned}$$

where $\rho > 0$ is independent of the initial data (t_1,x_1,f). Finally, (5) and (6) imply the statement. ∎

Finally, for $a \leq \tau_1 < \tau_2 \leq \infty$ and $s \in [a,\infty)$ we define

$$\begin{aligned}
d(\tau_1,\tau_2) &:= \|\mathbb{L}_{\tau_1}\|_{\mathcal{L}(L_p(\tau_1,\tau_2;U),L_p(\tau_1,\tau_2;Y))}, \\
d^+(s) &:= \inf_{\varepsilon>0} d(s,s+\varepsilon), \\
d^-(s) &:= \inf_{\varepsilon>0} d(s-\varepsilon,s).
\end{aligned}$$

(TV^{S}9) implies $d(\tau_1,\tau_2)$, $d^+(s)$, $d^-(s) < \infty$ for $a \leq \tau_1 < \tau_2 \leq \infty$ and $s \in [a,\infty)$. Moreover, $d(\tau_1,\tau_2) \leq d(\tilde{\tau}_1,\tilde{\tau}_2)$ for $a \leq \tilde{\tau}_1 \leq \tau_1 < \tau_2 \leq \tilde{\tau}_2 \leq \infty$.

DEFINITION 3.4 *We say a stable tv-system (Φ,B,C) has* **Property (L)** *if $d^+(s) = 0$ for all $s \in [a,\infty)$ and $d^-(s) = 0$ for all $s \in (a,\infty)$.*

Whether every stable tv–system has Property (L) is not known to the author. However, in [Jac95] sufficient conditions are given which guarantee that a stable tv–system has Property (L).

4 PROBLEM FORMULATION

Throughout this section we assume that (Φ, B, C) is a stable tv–system and we consider time–varying feedback system of the following kind:

$$x(t) = \Phi(t,t_1)x_1 + \int_{t_1}^{t} \Phi(t,\rho)B(\rho)P(f \underset{t_1}{\diamond} C(\cdot)x(\cdot))(\rho)\, d\rho, \tag{7}$$

where $t_1 \geq a$, $x_1 \in X$ and $f \in L_p(a,t_1;Y)$. Moreover the unknown time–varying feedback operator P satisfies $P \in \mathcal{L}(L_p(a,\infty;Y), L_p(a,\infty;U))$, P causal. We say $\boldsymbol{x} : [\boldsymbol{t_1}, \boldsymbol{\infty}) \to \boldsymbol{X}$ **is a solution of (7) on** $[\boldsymbol{t_1},\boldsymbol{\infty})$ **with initial data** $(\boldsymbol{t_1}, \boldsymbol{x_1}, \boldsymbol{f})$ if x is continuous on $[t_1,\infty)$, $C(\cdot)x(\cdot) \in L_p^{loc}(t_1,\infty;Y)$ and $x(t)$ satisfies (7) for all $t \in [t_1,\infty)$. In the following we will denote, if it exists, the solution of (7) on $[t_1,\infty)$ with initial data (t_1,x_1,f) by $x(\cdot;t_1,x_1,f)$. Due to the weak hypotheses (TV^S1)–(TV^S9) in general equation (7) does not have a solution on $[t_1,\infty)$ for every initial data $(t_1,x_1,f) \in [a,\infty) \times X \times L_p(a,t_1;Y)$. However, in [JDP95] sufficient conditions on (Φ, B, C) and P have been developed which guarantee that equation (7) has a unique solution on $[t_1,\infty)$ for every initial data (t_1,x_1,f). The following proposition borrowed from [JDP95] will be useful in sequel.

PROPOSITION 4.1 *If the stable tv–system (Φ, B, C) has Property (L) then equation (7) has for every $P \in \mathcal{L}(L_p(a,\infty;Y), L_p(a,\infty;U))$, P causal, and for every initial data $(t_1,x_1,f) \in [a,\infty) \times X \times L_p(a,t_1;Y)$ a unique solution on $[t_1,\infty)$.*

Since a basic requirement for stability is that solutions exist on $[t_1,\infty)$, $t_1 \geq a$, (no finite escape time) we introduce for $P \in \mathcal{L}(L_p(a,\infty;Y), L_p(a,\infty;U))$, P causal, the set

$$\begin{aligned}\mathcal{D}(P) \;:=\; \{&(t_1,x_1,f) \in [a,\infty) \times X \times L_p(a,t_1;Y) : \\ &(7) \text{ has a unique solution on } [t_1,\infty) \text{ with initial data } (t_1,x_1,f)\}.\end{aligned}$$

Moreover, we introduce the following stability concept for the closed loop system.

DEFINITION 4.2 – *For $P \in \mathcal{L}(L_p(a,\infty;Y), L_p(a,\infty;U))$, P causal, the origin of (7) is called* **globally L_p–stable**, *if there exist $\rho_1, \rho_2 > 0$ such that for all $(t_1,x_1,f) \in \mathcal{D}(P)$ we have*

$$\begin{aligned}\|x(t;t_1,x_1,f)\|_X &\leq \rho_1[\|x_1\|_X + \|f\|_{L_p(a,t_1;Y)}], \quad t \geq t_1, \\ \|x(\cdot;t_1,x_1,f)\|_{L_p(t_1,\infty;X)} &\leq \rho_2[\|x_1\|_X + \|f\|_{L_p(a,t_1;Y)}].\end{aligned}$$

– *For $P \in \mathcal{L}(L_p(a,\infty;Y), L_p(a,\infty;U))$, P causal, the origin of (7) is called* **output stable**, *if there exists a number $\rho > 0$ such that for all $(t_1,x_1,f) \in \mathcal{D}(P)$ we have*

$$\|C(\cdot)x(\cdot;t_1,x_1,f)\|_{L_p(t_1,\infty;Y)} \leq \rho[\|x_1\|_X + \|f\|_{L_p(a,t_1;Y)}].$$

In general output stability implies global L_p–stability (see [JDP95]). However, whether global L_p–stability implies output stability is not known to the author. The following proposition gives a sufficient condition which guarantees global L_p–stability of the origin of (7). For a proof see [JDP95].

PROPOSITION 4.3 *If the stable tv–system* (Φ, B, C) *has Property (L) and* $\|P\| < \sup_{t\geq 0} \|\mathbb{L}_a S_t\|^{-1}$ *then the origin of (7) is globally* L_p*-stable.*

In this paper we take an interest in the converse question:

> Does there exist a sequence $P_n \in \mathcal{L}(L_p(a,\infty;Y), L_p(a,\infty;U))$, P_n causal, with $\|P_n\| \searrow \sup_{t\geq 0} \|\mathbb{L}_a S_t\|^{-1}$ such that the origin of (7) is not globally L_p–stable?

However, this question will be discussed in the following section. We will see that under some additional weak conditions there exists such a sequence.

5 DESTABILIZATION RESULTS

Throughout this section we assume that (Φ, B, C) is a stable tv–system. In order to obtain destabilization results we follow Shamma and Zhao [SZ93] and approximate the operator $\mathbb{L}_a$ by causal operators with finite memory.

DEFINITION 5.1 *We say a causal operator* $Q \in \mathcal{L}(L_p(a,\infty;U), L_p(a,\infty;Y))$ *has* **finite memory** *if there exists a function* $\Psi : [a,\infty) \to [a,\infty)$ *such that* $\Psi(t) \geq t$ *and* $(I - \pi_{\Psi(t)})Q\pi_t = 0$ *for all* $t \geq a$*. The function* Ψ *is called the* **finite–memory function** *associated with* Q*.*

This definition of finite–memory states that inputs over a given finite duration are forgotten. Moreover, this finite–memory property need not be uniform in time, i.e. the difference $\Psi(t) - t$ need not be uniformly bounded. However, this definition is stronger than the definition of pointwise finite–memory in [SZ93]. The next lemma proves that the operator $\mathbb{L}_a$ can be approximated by operators with finite–memory.

LEMMA 5.2 *There exists a sequence of causal operators* $Q_n \in \mathcal{L}(L_p(a,\infty;U), L_p(a,\infty;Y))$ *with finite-memory such that* $\lim_{n\to\infty} \|\mathbb{L}_a - Q_n\|_{\mathcal{L}(L_p(a,\infty;U),L_p(a,\infty;Y))} = 0$.

PROOF:
First of all we choose $\delta > 0$ and $t \in [a,\infty)$. Then we obtain for $u \in L_p(a,\infty;U)$ and $T > t$

$$\begin{aligned}(\mathbb{L}_a \pi_t u)(T) &= C(T)\int_a^T \Phi(T,\rho)B(\rho)(\pi_t u)(\rho)\,d\rho = C(T)\int_a^t \Phi(T,\rho)B(\rho)u(\rho)\,d\rho \\ &= C(T)\Phi(T,t)\int_a^t \Phi(t,\rho)B(\rho)u(\rho)\,d\rho = \mathbb{C}_{(T,t)}[(\mathbb{M}_a u)(t)](T).\end{aligned}$$

Using the exponential stability of Φ there exists $\Psi_\delta(t) > t$ with

$$\|\Phi(\Psi_\delta(t), t)\|_{\mathcal{L}(X)} \leq (k_2 k_1 e^{\kappa_1(t-a)})^{-1}\delta.$$

Note, that k_1, k_2 and κ_1 are the constants in (TV^{S}6) and (TV^{S}8). The function $\Psi_\delta : [a,\infty) \to [a,\infty)$ defined in this way has the properties $\Psi_\delta(t) > t$ for $t \in [a,\infty)$ and

$$\begin{aligned}
\|[(I - \pi_{\Psi_\delta(t)})\mathbb{L}_a \pi_t](u)\|_{L_p(a,\infty;Y)} &= \|(\mathbb{L}_a \pi_t)(u)\|_{L_p(\Psi_\delta(t),\infty;Y)} \\
&= \|C(\cdot)\Phi(\cdot, \Psi_\delta(t))\Phi(\Psi_\delta(t), t)(\mathbb{M}_a u)(t)\|_{L_p(\Psi_\delta(t),\infty;Y)} \\
&\leq k_2 \|\Phi(\Psi_\delta(t), t)(\mathbb{M}_a u)(t)\|_X \quad (\text{use } (\mathrm{TV}^S 8)) \\
&\leq \delta \|u\|_{L_p(a,\infty;U)} \quad (\text{use } (\mathrm{TV}^S 6)).
\end{aligned}$$

Now we construct for $\varepsilon > 0$ a causal operator $Q \in \mathcal{L}(L_p(a,\infty;U), L_p(a,\infty;Y))$ with finite–memory such that $\|\mathbb{L}_a - Q\|_{\mathcal{L}(L_p(a,\infty;U),L_p(a,\infty;Y))} < \varepsilon$. Choosing $\varepsilon > 0$ we define $\alpha_n := \alpha_n(\varepsilon) := \varepsilon\left(\frac{1}{2}\right)^{n+1}$. Moreover, suppose Ψ_{α_n} is the function defined above. Then we choose a sequence $(s_n)_n \subset [a,\infty)$ in the following way $s_0 := a$, $s_{n+1} := \Psi_{\alpha_n}(s_n) + 1$ for $n \in \mathbb{N}_0$ and define the operator Q by

$$(Qu)(t) := \begin{cases} (\mathbb{L}_a u)(t) & , s_0 \leq t < s_1 \\ [\mathbb{L}_a(I - \pi_{s_{n-1}})u](t) & , s_n \leq t < s_{n+1} \quad \text{for} \quad n \in \mathbb{N} \end{cases},$$

where $u \in L_p(a,\infty;U)$ and $t \in [a,\infty)$. It is easy to see that $Q \in \mathcal{L}(L_p(a,\infty;U), L_p(a,\infty;Y))$ is causal and has finite–memory. Furthermore the lemma follows from

$$\begin{aligned}
\|\mathbb{L}_a u - Qu\|_{L_p(a,\infty;Y)} &\leq \sum_{j=0}^{\infty} \|\mathbb{L}_a u - Qu\|_{L_p(s_j,s_{j+1};Y)} \leq \sum_{j=1}^{\infty} \|\mathbb{L}_a \pi_{s_{j-1}} u\|_{L_p(s_j,s_{j+1};Y)} \\
&\leq \sum_{j=1}^{\infty} \|(I - \pi_{\Psi_{\alpha_{j-1}}(s_{j-1})})\mathbb{L}_a \pi_{s_{j-1}} u\|_{L_p(a,\infty;Y)} \\
&\leq \sum_{j=1}^{\infty} \alpha_{j-1} \|u\|_{L_p(a,\infty;U)} = \varepsilon \|u\|_{L_p(a,\infty;U)}
\end{aligned}$$

■

Thus the input–output operator $\mathbb{L}_a$ can be approximate by causal, finite–memory operators $Q \in \mathcal{L}(L_p(a,\infty;U), L_p(a,\infty;Y))$. Such operators are called operators with fading–memory (see [SZ93]). In order to get destabilization results the following two lemmas will be useful. Lemma 5.3 shows an interesting existence result for operators and Lemma 5.4 gives an useful result concerning causal, finite–memory operators (see also [SZ93]).

LEMMA 5.3 *Suppose $f_1 \in L_p(a,\infty;Y)$, $f_2 \in L_p(a,\infty;U)$ with* supp $f_1 \subseteq [T_1, T_2]$ *and* supp $f_2 \subseteq [T_3, T_4]$, *where $a \leq T_1 < T_2 < T_3 < T_4$. Then there exists a causal operator $P \in \mathcal{L}(L_p(a,\infty;Y), L_p(a,\infty;U))$ with*

- $Pf_1 = f_2$,
- supp $Pf \subseteq [T_3, T_4]$ *for all* $f \in L_p(a, \infty; Y)$,
- *If* $f \in L_p(a, \infty; Y)$ *with* supp $f \cap [T_1, T_2] = \emptyset$ *then* $Pf = 0$,
- $\|P\| = \|f_2\| / \|f_1\|$.

PROOF:
Using the Theorems of Hahn–Banach there exists $\Lambda \in L_p(T_1, T_2; Y)'$ with $\|\Lambda\| = 1$ and $\Lambda f_1 = \|f_1\|$. Then the operator $(Pf)(t) := \frac{1}{\|f_1\|} f_2(t) \Lambda(f|_{[T_1,T_2]})$ satisfies the required statements. ∎

LEMMA 5.4 *Suppose that* $Q \in \mathcal{L}(L_p(a, \infty; U), L_p(a, \infty; Y))$ *is causal and has finite-memory. Moreover, assume that* $\beta > \sup_{t \geq 0} \|QS_t\|^{-1}$. *Then there exist a causal operator* $P \in \mathcal{L}(L_p(a, \infty; Y), L_p(a, \infty; U))$ *with* $\|P\| < \beta$ *and functions* $f, g \in L_p^{loc}(a, \infty; U)$ *such that*
- *g has compact support,*
- $f \in L_p^{loc}(a, \infty; U) \backslash L_p(a, \infty; U)$,
- $Qf \in L_p^{loc}(a, \infty; Y) \backslash L_p(a, \infty; Y)$,
- $(I - PQ)f = g$.

PROOF:
Define $\alpha := \sup_{t \geq 0} \|QS_t\|^{-1}$ and suppose Ψ is the finite–memory function associated with Q. Thus $\|Q\| \geq \inf_{t \geq 0} \|QS_t\| = \frac{1}{\alpha} > \frac{1}{\beta}$ and so there exists a function $\widetilde{f_0} \in L_p(a, \infty; U)$ with $\|\widetilde{f_0}\| = 1$ and $\|Q\widetilde{f_0}\| > \frac{3}{4}\frac{1}{\alpha} + \frac{1}{4}\frac{1}{\beta}$. Choosing $t_1 \in (a, \infty)$ in such a way that $\|\pi_{t_1} Q\widetilde{f_0}\| \geq \frac{1}{2}\frac{1}{\alpha} + \frac{1}{2}\frac{1}{\beta}$, defining $f_0 := \frac{1}{\|\pi_{t_1}\widetilde{f_0}\|} \pi_{t_1} \widetilde{f_0}$ and $N_0 := \Psi(t_1)$ we obtain
- $\|f_0\| = 1$,
- supp $f_0 \subseteq [a, N_0]$,
- supp $Qf_0 \subseteq [a, N_0]$,
- $\|Qf_0\| \geq \frac{1}{2}\left(\frac{1}{\alpha} + \frac{1}{\beta}\right)$.

Using $\inf_{t \geq 0} \|QS_t\|^{-1} > \frac{1}{\beta}$, it is possible to find iterative sequences $(N_n)_{n \in \mathbb{N}_0} \subset [a, \infty)$ and $(f_n)_{n \in \mathbb{N}_0} \subset L_p(a, \infty; U)$ such that
- $\|f_n\| = 1$,
- $N_n > N_{n-1} + 1$,
- supp $f_n \subseteq [N_{n-1} + 1, N_n]$,
- supp $Qf_n \subseteq [N_{n-1} + 1, N_n]$,
- $\|Qf_n\| \geq \frac{1}{2}\left(\frac{1}{\alpha} + \frac{1}{\beta}\right)$,

where $N_{-1} := a - 1$. Now we define the function $f \in L_p^{loc}(a, \infty; U)$ by $f := \sum_{n=0}^{\infty} f_n$. It is easy to see that the function f defined above satisfies $f \in L_p^{loc}(a, \infty; U) \backslash L_p(a, \infty; U)$

and $Qf \in L_p^{loc}(a,\infty;Y)\backslash L_p(a,\infty;Y)$. By the previous lemma there exist causal operators $P_n \in \mathcal{L}(L_p(a,\infty;Y), L_p(a,\infty;U))$, $n \in \mathbb{N}_0$, with

- $P_n(Qf_n) = f_{n+1}$,
- $\|P_n\| \leq \left(\frac{1}{2}\left(\frac{1}{\alpha}+\frac{1}{\beta}\right)\right)^{-1}$,
- supp $P_n h \subseteq [N_n+1, N_{n+1}]$ for all $h \in L_p(a,\infty;Y)$,
- if $h \in L_p(a,\infty;Y)$ with supp $h \cap [N_{n-1}+1, N_n] = \emptyset$ then $P_n h = 0$.

Finally, we set

$$Ph := \sum_{n=0}^{\infty} P_n h \quad \text{for} \quad h \in L_p(a,\infty;Y)$$

and $g := f_0$, and so $P \in \mathcal{L}(L_p(a,\infty;Y), L_p(a,\infty;U))$, P causal, $\|P\| < \beta$ and

$$(I - PQ)f = \sum_{n=0}^{\infty}[f_n - PQf_n] = \sum_{n=0}^{\infty}[f_n - P_nQf_n] = f_0 = g.$$

This completes the proof. ■

Now we are ready to formulate the main theorem of this section.

THEOREM 5.5 *For every $\beta > \sup_{t\geq a}\|\mathbb{L}_t\|^{-1} = \sup_{t\geq 0}\|\mathbb{L}_a S_t\|^{-1}$ there exists a causal operator $P \in \mathcal{L}(L_p(a,\infty;Y), L_p(a,\infty;U))$ with $\|P\| < \beta$, such that the origin of (7) is not output stable.*

PROOF:
First of all, we choose a number α with $\beta > \alpha > \sup_{t\geq 0}\|\mathbb{L}_a S_t\|^{-1}$. By Lemma 5.2 there exists a sequence $(Q_n)_n \subset \mathcal{L}(L_p(a,\infty;U), L_p(a,\infty;Y))$, where every Q_n is causal and has finite–memory, such that $\|\mathbb{L}_a - Q_n\| \overset{n\to\infty}{\longrightarrow} 0$. Clearly, the sequence $(Q_nS_s)_n$ converges to $\mathbb{L}_a S_s$ for n tending to ∞ uniformly in s. Thus there exists a number $N_1 \in \mathbb{N}$ such that $\sup_{s\geq 0}\|Q_nS_s\|^{-1} < \alpha$ for $n \geq N_1$. Now the previous lemma implies the existence of causal operators $P_n \in \mathcal{L}(L_p(a,\infty;Y), L_p(a,\infty;U))$ with $\|P_n\| < \alpha$ and functions $f_n, g_n \in L_p^{loc}(a,\infty;U)$ with g_n has compact support, $f_n \in L_p^{loc}(a,\infty;U)\backslash L_p(a,\infty;U)$ and $Q_nf_n \in L_p^{loc}(a,\infty;Y)\backslash L_p(a,\infty;Y)$ such that

$$(I - P_nQ_n)f_n = g_n \quad \text{for} \quad n \geq N_1.$$

Moreover, we are able to find $N_2 \geq N_1$ such that $\|Q_n - \mathbb{L}_a\| < \frac{1}{\alpha}$ for every $n \geq N_2$. Thus

$$(I - (Q_n - \mathbb{L}_a)P_n)^{-1} = \sum_{k=0}^{\infty}((Q_n - \mathbb{L}_a)P_n)^k \in \mathcal{L}(L_p(a,\infty;Y))$$

and $(I - (Q_n - \mathbb{L}_a)P_n)^{-1}$ is causal for $n \geq N_2$. Furthermore, there is a number $N \geq N_2$ such that $P := P_N(I - (Q_N - \mathbb{L}_a)P_N)^{-1} \in \mathcal{L}(L_p(a,\infty;Y), L_p(a,\infty;U))$ is causal and satisfies

$\|P\| < \beta$. Defining $\tilde{y} := (I - (Q_N - \mathbb{L}_a)P_N)Q_N f_N$ we have $\tilde{y} \in L_p^{loc}(a,\infty;Y)\backslash L_p(a,\infty;Y)$ and $(I - \mathbb{L}_a P)\tilde{y} = (I - Q_N P_N)Q_N f_N = Q_N g_N$. Note that g_N has a compact support and that Q_N has finite–memory, and thus the function $Q_N g_N$ has also a compact support. Thus there exists a number $t_0 > a$ such that supp $Q_N g_N \subseteq [a, t_0)$. Note, that for every $u \in L_p^{loc}(a,\infty;U)$ and $t \geq t_0$ we have $(\mathbb{L}_{t_0}(u|_{[t_0,\infty)}))(t) = (\mathbb{L}_a u)(t) - (\mathbb{L}_a \pi_{t_0} u)(t)$. Thus for $t \geq t_0$ we get $\tilde{y}(t) = (\mathbb{L}_{t_0}(P\tilde{y})|_{[t_0,\infty)})(t) + (\mathbb{L}_a \pi_{t_0} P\tilde{y})(t)$. Finally, we define

$$\begin{aligned} x_0 &:= \int_a^{t_0} \Phi(t_0,\rho)B(\rho)(P\tilde{y})(\rho)\, d\rho \in X \\ f &:= \tilde{y}|_{[a,t_0)} \end{aligned}$$

and thus for $t \geq t_0$ we obtain

$$\begin{aligned} \tilde{y}(t) &= C(t)\int_{t_0}^{t} \Phi(t,\rho)B(\rho)P[f \underset{t_0}{\diamond} \tilde{y}|_{[t_0,\infty)}](\rho)\, d\rho + (\mathbb{L}_a \pi_{t_0} P\tilde{y})(t) \\ &= C(t)\Phi(t,t_0)x_0 + C(t)\int_{t_0}^{t} \Phi(t,\rho)B(\rho)P(f \underset{t_0}{\diamond} \tilde{y}|_{[t_0,\infty)})(\rho)\, d\rho, \end{aligned}$$

which proves the theorem. ■

In the special situation where the stable tv–system (Φ, B, C) has Property (L) we get a stronger result than Theorem 5.5. This result is formulated in the following corollary.

COROLLARY 5.6 *Suppose that the stable tv–system (Φ, B, C) has Property (L). Then there exists a sequence $(P_n)_n \subset \mathcal{L}(L_p(a,\infty;Y), L_p(a,\infty;U))$, P_n causal, with*

$$\|P_n\|_{\mathcal{L}(L_p(a,\infty;Y),L_p(a,\infty;U))} \searrow \sup_{s\geq 0} \|\mathbb{L}_a S_s\|^{-1}$$

such that the origin of the corresponding perturbed equation is not output stable.

PROOF: For every $\beta > \sup_{t\geq 0} \|\mathbb{L}_a S_t\|^{-1}$ the previous theorem shows that there exists a perturbation $P \in \mathcal{L}(L_p(a,\infty;Y), L_p(a,\infty;U))$, P causal, with $\|P\| < \beta$ such that the origin of the corresponding perturbed equation is not output stable. Assuming

$$\|P\| < \sup_{s\geq 0} \|\mathbb{L}_a S_s\|^{-1} = \sup_{t\geq a} \|\mathbb{L}_t\|^{-1},$$

Proposition 4.3 together with Property (L) of the system (Φ, B, C) imply that the origin of the corresponding perturbed equation is output stable, which is a contradiction. ■

Supposing that the stable tv–system (Φ, B, C) has Property (L) Proposition 4.3 and Corollary 5.6 classify the class of linear, dynamical perturbations in the following way:

- For $P \in \mathcal{L}(L_p(a,\infty;Y), L_p(a,\infty;U))$, P causal, with $\|P\| < \sup_{t\geq a} \|\mathbb{L}_t\|^{-1}$ the origin of the corresponding perturbed equation is output stable.

– There exists a sequence $(P_n)_n \subset \mathcal{L}(L_p(a,\infty;Y), L_p(a,\infty;U))$, P_n causal, with

$$\|P_n\|_{\mathcal{L}(L_p(a,\infty;Y),L_p(a,\infty;U))} \searrow \sup_{t\geq a} \|\mathbb{L}_t\|^{-1}$$

such that the origin of the corresponding perturbed equation is not output stable.

The following example which is quoted from [HIP89] shows that such a destabilization result is not true for linear, memoryless perturbations.

EXAMPLE 5.7 *Consider the scalar system*

$$\dot{x}(t) = a(t)x(t), \qquad t \geq 0$$

where $a(t) = -1 + k\alpha(t)$, $k \in \mathbb{R}$ *and* $\alpha \in PC([0,\infty),\mathbb{C})$ *is periodic with period* $3T$, $T = \ln 2$, *given by*

$$\alpha(t) := \begin{cases} 0 & , t \in [3iT, (3i+1)T) \\ 1 & , t \in [(3i+1)T, (3i+2)T), \\ -1 & , t \in [(3i+2)T, 3(i+1)T), \end{cases} \qquad i \in \mathbb{N}_0.$$

Then $a(\cdot)$ *is the generator of a strong evolution operator* Φ *on* $\mathbb{C}$ *and* $(\Phi, 1, 1, \mathbb{C}, \mathbb{C}, \mathbb{C}, \mathbb{C}, \mathbb{C}, 0)$ *is a stable tv–system. A definition of strong evolution operators can be found in [Jac95]. Moreover, in [HIP89] it has been shown that* $\|\mathbb{L}_0\|^{-1} = \sup_{s\geq 0} \|\mathbb{L}_s\|^{-1} < 1$, *but for every* $\Delta \in L_{s,\infty}(a,\infty;\mathcal{L}(Y,U))$ *with* $\|\Delta\| < 1$ *the origin of the corresponding perturbed equation is output stable.*

Due to the probable gap between output stability and global L_p–stability we cannot expect a similar result as Theorem 5.5 for global L_p–stability. However, under some additional assumptions we obtain the following two results which give sufficient conditions to guarantee that the origin of the perturbed equation is not globally L_p–stable. We omit the proofs since they follow immediately from Theorem 5.5 and Corollary 5.6, noting that for every stable tv–system with $\underline{X} = X$ and $C \in L_{s,\infty}(a,\infty;\mathcal{L}(X,Y))$, output stability and global L_p–stability are equivalent notions.

COROLLARY 5.8 *Suppose* (Φ, B, C) *is a stable tv–system with* $\underline{X} = X$ *and* $C \in L_{s,\infty}(a,\infty;\mathcal{L}(X,Y))$. *Then for every* $\beta > \sup_{t\geq 0} \|\mathbb{L}_a S_t\|^{-1}$ *there exists a perturbation* $P \in \mathcal{L}(L_p(a,\infty;Y), L_p(a,\infty;U))$, P *causal, with* $\|P\| < \beta$ *such that the origin of the corresponding perturbed equation is not globally* L_p*–stable.*

COROLLARY 5.9 *Let* (Φ, B, C) *a stable tv–system with Property (L),* $\underline{X} = X$ *and* $C \in L_{s,\infty}(a,\infty;\mathcal{L}(X,Y))$. *Then there exists a sequence* $(P_n)_n \subset \mathcal{L}(L_p(a,\infty;Y), L_p(a,\infty;U))$, P_n *causal, with*

$$\|P_n\|_{\mathcal{L}(L_p(a,\infty;Y),L_p(a,\infty;Y))} \searrow \sup_{s\geq 0} \|\mathbb{L}_a S_s\|^{-1}$$

such that the origin of the corresponding perturbed equation is not globally L_p*–stable.*

ACKNOWLEDGMENTS

My special thanks go to H. Logemann for suggesting the problem and for numerous helpful discussions which we had during the preparation of this paper.

REFERENCES

[Gib76] J. S. Gibson. The Riccati integral equations for optimal control problems in Hilbert spaces. *SIAM J. Control and Optimization*, 17(4):537–565, 1976.

[HIP89] D. Hinrichsen, A. Ilchmann, and A. J. Pritchard. Robustness of stability of time–varying linear systems. *Journal of Differential Equations*, 82(2):219–250, 1989.

[HP86a] D. Hinrichsen and A. J. Pritchard. Stability radii for structured perturbations and the algebraic Riccati equation. *System Control Lett.*, 8:105–113, 1986.

[HP86b] D. Hinrichsen and A. J. Pritchard. Stability radii of linear systems. *System Control Lett.*, 7:1–10, 1986.

[HP90] D. Hinrichsen and A. J. Pritchard. Real and complex stability radii: a survey. In *Proc. Workshop Control of Uncertain Systems, Bremen 1989*, Progress in System and Control Theory, pages 119–162. Birkhäuser, 1990.

[HP92] D. Hinrichsen and A. J. Pritchard. Destabilization by output feedback. *Differential and Integral Equations*, 5(2):357–386, 1992.

[HP94] D. Hinrichsen and A. J. Pritchard. Robust stability of linear evolution operators on Banach spaces. *Siam J. Control and Optimization*, 32(6):1503–1541, 1994.

[Jac95] B. Jacob. *Infinite dimensional time–varying state–space systems.* PhD thesis, University of Bremen, June 1995.

[JDP95] B. Jacob, D. Dragan, and A. J. Pritchard. Robust stability of infinite–dimensional time–varying systems with respect to nonlinear perturbations. *Integral Equations and Operator Theory*, 1995. to appear.

[PT89] A. J. Pritchard and S. Townley. Robustness of linear systems. *J. Differ. Equations*, 77(2):254–286, 1989.

[SZ93] J. S. Shamma and R. Zhao. Fading–memory feedback systems and robust stability. *Automatica*, 29(1):191–200, 1993.

[Wei89a] G. Weiss Admissibility of unbounded control operators. *SIAM Journal on Control and Optimization*, 27:527–545, 1989.

[Wei89b] G. Weiss. The representation of regular linear systems on Hilbert spaces. In W. Schappacher F. Kappel, K. Kunisch, editor, *"Distributed Parameter Systems", proc. of the conference in Vorau, Austria, July 1988*, pages 401–416. Birkhäuser Verlag Basel, 1989.

Department of Applied Mathematics,
University of Twente,
P. O. Box 217,
NL-7500 AE Enschede, The Netherlands

AMS CLASSIFICATIONS: 93C50, 93C22, 93C05, 93C60, 93D09

Operator Theory:
Advances and Applications, Vol. 87

PERTURBATIONS OF G-SELFADJOINT OPERATORS AND OPERATOR POLYNOMIALS WITH REAL SPECTRUM

P. LANCASTER,* A. MARKUS, and V. MATSAEV

Let G be a bounded and invertible selfadjoing operator on a Hilbert space $\mathcal{H}$ and consider bounded operators A satisfying $GA = A^*G$. Results are established concerning the spectrum of A when (A, G) are obtained by perturbation of a uniformly definitizable pair (A_0, G_0); the perturbations may be small, finite rank, or compact. These results are applied to similar perturbation problems for selfadjoint operator polynomials $\sum_{j=0}^{\ell} \lambda^j A_j$ $(A_j = A_j^*)$ (and include the case that A_ℓ is not invertible), as well as a factorization theorem for perturbed polynomials (in the case $A_\ell = I$). The appropriate connections with operator differential equations having stably bounded solutions are established.

1 Introduction

This work is a natural development of the theory of uniformly definitizable operators and quasihyperbolic operator polynomials initiated by the authors in reference [12]. To define these and other relevant concepts consider a Hilbert space $\mathcal{H}$ over $\mathbb{C}$ with inner product $(.,.)$, and let G be a bounded and invertible selfadjoint operator on $\mathcal{H}$. When G is indefinite, an indefinite inner product $[.,.]$ is defined on $\mathcal{H}$ by $[x, y] = (Gx, y)$ for all $x, y \in \mathcal{H}$. Then $(\mathcal{H}, [.,.])$ is known as a *Krein space*.

If A is a bounded linear operator on $\mathcal{H}$ its spectrum, $\sigma(A)$, and resolvent set $\rho(A)$, are defined in the usual way. Also $\lambda \in \mathbb{C}$ is in the *approximate point spectrum* of A, $\sigma_{\mathrm{ap}}(A)$, if there is a sequence $\{f_n\}_{n=1}^{\infty}$ in $\mathcal{H}$ such that

$$\|f_n\| = 1, \qquad \text{and} \qquad \|Af_n - \lambda f_n\| \to 0, \qquad (n \to \infty). \tag{1.1}$$

It is well known that $\partial\sigma(A) \subset \sigma_{\mathrm{ap}}(A)$, and it is clear that $\sigma_{\mathrm{ap}}(A) \subset \sigma(A)$.

The operator A is said to be *G-selfadjoint* if $[Ax, y] = [x, Ay]$ for all $x, y \in \mathcal{H}$ or, what is equivalent, $GA = A^*G$. If A is G-selfadjoint and $\lambda \in \sigma_{\mathrm{ap}}(A)$, λ is said to be of *plus type*, or of *minus type*, if for any sequence f_n satisfying (1.1) we have

$$\underline{\lim}[f_n, f_n] > 0, \qquad \text{or} \qquad \overline{\lim}[f_n, f_n] < 0,$$

*The work of this author was supported in part by a grant from the Natural Sciences and Engineering Research Council of Canada.

respectively. Points having either plus type or minus type are said to be of *determinate type.*

It is not difficult to see that points of $\sigma_{\mathrm{ap}}(A)$ which have determinate type are necessarily real. Those points of $\sigma_{\mathrm{ap}}(A) \cap \mathbf{R}$ which are not of determinate type are said to have *mixed type.* Finally we define $\sigma_+(A)$, $\sigma_-(A)$ to be the sets of all points in $\sigma_{\mathrm{ap}}(A)$ having plus, or minus, type, respectively.

Now a G-selfadjoint operator A is said to be definitizable, strongly definitizable, or uniformly definitizable (with respect to G) if there is a real polynomial p such that $Gp(A) \geq 0$, $Gp(A) > 0$, or $Gp(A) \gg 0$, respectively.

We remark that a definition of points of plus and minus types is well-known (p. 36 of [8]). However, the earlier definitions use the deeper notion of a spectral function. It is not difficult to show that, when the spectral function exists, the two definitions coincide. This is the case when the operator A is definitizable, for example (see [8]). In the recent paper [14] it is proved that, if λ_0 is a point of determinate type, then a spectral function exists on some neighbourhood of λ_0. This means that the two definitions of points of determinate type are always equivalent.

These concepts generalize ideas introduced in reference [11] in the case when $\mathcal{H}$ is finite dimensional. In that paper the finite dimensional theory is extended and applied to matrix polynomials and the effects of finite rank perturbations are investigated. A similar program is adopted here, but for the case in which $\mathcal{H}$ has infinite dimension. In particular, a selfadjoint operator polynomial has the form

$$L(\lambda) := \sum_{j=0}^{\ell} \lambda^j A_j$$

where $A_0, A_1, \ldots, A_\ell$ are bounded selfadjoint operators on $\mathcal{H}$. In the two papers mentioned above it is assumed that $A_\ell = I$ in which case the operator polynomial is said to be *monic.* Here, this condition is relaxed and, in the final section, the case that A_ℓ is not invertible (but $\sigma(L) \neq \mathbf{C}$) is investigated. In this respect, the study of finite dimensional problems is also extended. Throughout, the operator polynomial is required to be quasihyperbolic and it is shown that, in each case, this is naturally characterized by the condition that the spectrum of $L(\lambda)$ has determinate type (when suitably defined).

Sections 2 and 3 contain theorems concerning finite rank and compact perturbations of operator pairs A, G where A is uniformly definitizable with respect to G. In Section 4 quasihyperbolic operator polynomials (QHP) with invertible leading coefficient are introduced and preceding results are applied to investigate the effects of finite rank and compact perturbations of the coefficients of such a QHP. A well-known theorem of Gohberg implies that compact perturbations of an operator polynomial with real spectrum produce only countably many nonreal eigenvalues (provided the perturbed polynomial has at least one regular point). It is shown here that, for certain selfadjoint compact perturbations of QHP, the number of nonreal eigenvalues is finite.

A factorization theorem for perturbed monic QHP is presented in section 5. Section 6 contains a characterization of QHP (with invertible leading coefficient) in terms of the stable boundedness of solutions of a certain operator differential equation; a result in the tradition of earlier works by M.G. Krein and several others. Finally, in Section 7, the notion of QHP with noninvertible leading coefficient is developed and a theorem concerning the spectrum of compactly perturbed QHP of this kind is presented.

2 Perturbations of finite rank

Consider an operator A_0 which is strongly definitizable with respect to G_0 and then a pair A, G with $GA = A^*G$, and for which $A_0 - A$ and $G_0 - G$ have finite rank (and both G_0 and G are invertible). It is apparent that there is no nontrivial subspace $\mathcal{M}_0$ which is both A_0-invariant and G_0-neutral (i.e. for which $A\mathcal{M}_0 \subseteq \mathcal{M}_0$ and $(Gf, f) = 0$ for all $f \in \mathcal{M}_0$). We show first that if $\mathcal{M}$ is an A-invariant and G-neutral subspace, then an upper bound for the dimension of $\mathcal{M}$ (say $\dim \mathcal{M}$) can be obtained.

Recall that A is said to be strongly q-definitizable (with respect to G) if it is strongly definitizable and, among all definitizing polynomials, the least degree is $q - 1$.

Lemma 1 *Let A_0 be strongly q-definitizable with respect to G_0, let A be G-selfadjoint, and let*

$$\operatorname{rank}(A - A_0) = a, \quad \operatorname{rank}(G - G_0) = g.$$

Then if $\mathcal{M}$ is a subspace which is A-invariant and G-neutral then

$$\dim(\mathcal{M}) \leq (q-1)a + g \tag{2.1}$$

Proof. The proof is essentially that of [11] but, for completeness, we outline the argument again.

Let $\mathcal{M}$ be A-invariant and G-neutral and observe that, for any $x \in \mathcal{M}$ and any polynomial p, $p(A)x \in \mathcal{M}$ and hence

$$(Gp(A)x, x) = 0. \tag{2.2}$$

If p has degree $q - 1$ some algebraic manipulation leads to

$$\operatorname{rank}(Gp(A) - G_0p(A_0)) \leq (q-1)a + g. \tag{2.3}$$

Suppose that $G_0p(A_0) > 0$ and $\mathcal{Z} = \operatorname{Ker}(Gp(A) - G_0p(A_0))$. Then $\mathcal{Z}$ has codimension not less than $(q-1)a + g$ and if $\dim \mathcal{M} > (q-1)a + g$ then there is a nonzero $x \in \mathcal{M} \cap \mathcal{Z}$. So using (2.2),

$$0 = ((Gp(A) - G_0p(A_0))x, x) = (Gp(A)x, x) - (G_0p(A_0)x, x) < 0;$$

which is a contradiction. So (2.1) follows. □

Let $\lambda_1, \lambda_2, \ldots, \lambda_s$ be distinct real eigenvalues of A with corresponding G-neutral eigenvectors $\varphi_1, \varphi_2, \ldots, \varphi_s$. Let subspace $\mathcal{M}$ be the span of $\varphi_1, \ldots, \varphi_s$ together with all the root subspaces of A corresponding to eigenvalues in the halfplane $\operatorname{Im}\lambda > 0$. As all the components of this span are mutually G-orthogonal it follows that $\mathcal{M}$ is G-neutral and also A-invariant. Applying the lemma to this subspace we immediately obtain:

Theorem 1 *Let A_0 be strongly q-definitizable with respect to G_0, let A be G-selfadjoint, and let*

$$\operatorname{rank}(A - A_0) = a, \quad \operatorname{rank}(G - G_0) = g.$$

If m_+ denotes the sum of the algebraic multiplicities of eigenvalues of A with positive imaginary part, and if s is the number of distinct real eigenvalues of A with a corresponding neutral eigenvector, then

$$m_+ + s \leq (q-1)a + g. \tag{2.4}$$

Observe that the inequality (2.4) can be improved at the expense of some further complications. The theorem associates a one-dimensional G-neutral A-invariant subspace with each real eigenvalue λ_j, and there may be a larger G-neutral A-invariant subspace of λ_j. See Theorem 3.1 of [11] for the maximal dimension of a subspace of this kind, and compare with the notion of "rank of indefiniteness" introduced by Jonas and Langer (see [8]). In the case $g = 0$ Theorem 1 is included in results of the same paper of Jonas and Langer and is obtained by different methods.

Recall that, if A and G are obtained by finite rank perturbations of A_0 and G_0, respectively, then A is necessarily definitizable with respect to G (i.e. there is a polynomial p such that $Gp(A) \geq 0$), see [8] and [16].

3 Small and compact perturbations

The next theorem gives a "strong stability" property for uniformly definitizable operators, but let us first recall the effect of using a definitizing polynomial of minimal degree. Let A_0 be uniformly q-definitizable with respect to G_0 and p be a real polynomial of degree $q-1$ for which $G_0 p(A_0) \gg 0$. Then p has real and simple zeros $\mu_1 < \mu_2 < \ldots < \mu_{q-1}$ (see Lemma 1 of [12]). Writing $\mu_0 = -\infty$, $\mu_q = \infty$, $\sigma(A_0)$ lies in the union of q intervals

$$[a_k, b_k] \subset (\mu_{k-1}, \mu_k), \quad k = 1, 2, \ldots, q.$$

Furthermore, all points of $\sigma(A_0)$ in $[a_k, b_k]$ (for a fixed k) have the same type, and these types alternate as k increases from 1 through q (see the proof of Theorem 1 of [12]). The intervals $[a_k, b_k]$ are called "quasizones" of A_0.

Theorem 2 *Let A_0 be uniformly q-definitizable with respect to G_0. Then there is an $\varepsilon > 0$ such that, when A is G-selfadjoint and*

$$\|A - A_0\| < \varepsilon, \quad \|G - G_0\| < \varepsilon,$$

A is uniformly q-definitizable with respect to G.

Proof. By continuity, if $G_0 p(A_0) \gg 0$, ε can be chosen so small that $Gp(A) \gg 0$. To see that the degree of a definitizing polynomial cannot decrease with the perturbation, choose ε so that, after perturbation, the q quasizones remain separated. □

Theorem 3 *Let A_0 be uniformly q-definitizable with respect to G_0 and let A be G-selfadjoint with $A - A_0$ and $G - G_0$ compact operators. Then A has at most a finite number of nonreal eigenvalues and real points of $\sigma(A)$ which have mixed type (with respect to G).*

Proof. Since A_0 is uniformly q-definitizable with respect to G_0 it follows, as above, that $\sigma(A_0)$ is confined to q quasizones on the real line separated by gaps (or intervals) of positive length. As A is obtained from A_0 by a compact perturbation, that part of the spectrum of A appearing in the gaps of $\sigma(A_0)$ is discrete. So we may choose points $\alpha_1 < \alpha_2 < \ldots \alpha_{q-1}$, one in each gap, in such a way that $A - \alpha_k I$ is invertible for $k = 1, 2, \ldots, q-1$.

If we define $p(\lambda) = \prod_{k=1}^{q-1}(\lambda - \alpha_k)$ then p is a definitizing polynomial for A_0. Also,

$$Gp(A) = G\prod_{k=1}^{q-1}(A - \alpha_k I)$$

is invertible and, as $Gp(A)$ is a compact perturbation of $G_0 p(A_0)$, it defines a nondegenerate inner product on $\mathcal{H}$ with finite dimensional negative part, i.e. a Pontrjagin space. But, clearly, A is selfadjoint in the $Gp(A)$ inner product and points of $\sigma(A)$ of mixed type with respect to G one also of mixed type with respect to $Gp(A)$. Now the assertions of the theorem follow from well known results about selfadjoint operators on Pontrjagin spaces (see [9] and Section IX.4 of [2], for example). □

Remark 1 Another argument based on Theorems 1 and 2 leads immediately to a result very like Theorem 3.

Remark 2 Theorem 3 can be obtained from results of Jonas and Langer [8] in the case $G = G_0$ and more generally from a result of Jonas [7]. We prefer to give this direct and simple proof.

Remark 3 If A_0 is uniformly definitizable with respect to G_0 in Theorem 1, then it follows from (2.3) that the Pontrjagin space of the proof above has index $\kappa \leq (q-1)a + g$, then it follows from known results that the number s in (2.4) can be interpreted as the number of mixed points of spectrum of A.

4 Applications to operator polynomials

The preceding results can now be applied to the theory of selfadjoint operator polynomials. Let $A_0, A_1, \ldots, A_\ell$ be bounded selfadjoint operators on $\mathcal{H}$ with A_ℓ invertible. Then the function L defined on $\mathbb{C}$ by

$$L(\lambda) = \sum_{j=0}^{\ell} \lambda^j A_j \tag{4.1}$$

is a *selfadjoint operator polynomial*, and the spectrum of L, $\sigma(L)$, is the set of $\lambda \in \mathbb{C}$ for which $L(\lambda)$ is not invertible. The following definitions are immediate generalizations of those introduced in [12] for monic polynomials. A point $\lambda \in \sigma_{ap}(L)$ (see [12] for the natural definition) which is also real is said to be of *plus type* if for any sequence $\{f_n\} \subset \mathcal{H}$ satisfying $\|f_n\| = 1$ and $\|L(\lambda)f_n\| \to 0$ as $n \to \infty$ we also have

$$\underline{\lim}(L'(\lambda)f_n, f_n) > 0$$

as $n \to \infty$ (and $L'(\lambda) = \sum_{j=1}^{\ell} j\lambda^{j-1}A_j$). A similar definition applies for real points of $\sigma(L)$ of *minus type*. A point of *determinate type* has either plus or minus type and a point of $\sigma_{ap}(L)$ which is real and not of determinante type is said to be of *mixed type*. A polynomial of the form (4.1) is then said to be a *quasihyperbolic polynomial* (QHP) if $\sigma(L) \subseteq \mathbf{R}$ and all points of $\sigma(L)$ have determinate type. This is a natural generalization of the notion of "hyperbolic polynomial". See Section 7 of [12] for this connection.

Operators A and G on $\mathcal{H}^n$ are associated with $L(\lambda)$ and defined by the matrices

$$A = \begin{bmatrix} 0 & I & 0 & \dots & 0 \\ 0 & 0 & I & & \vdots \\ \vdots & \vdots & & \ddots & \\ 0 & 0 & \dots & & I \\ -\hat{A}_0 & -\hat{A}_1 & \dots & & -\hat{A}_{\ell-1} \end{bmatrix}, \quad G = \begin{bmatrix} A_1 & A_2 & \dots & A_{\ell-1} & A_\ell \\ A_2 & & \cdot^{\cdot^{\cdot}} & A_\ell & 0 \\ \vdots & \cdot^{\cdot^{\cdot}} & \cdot^{\cdot^{\cdot}} & & \vdots \\ A_{\ell-1} & A_\ell & \cdot^{\cdot^{\cdot}} & & \\ A_\ell & 0 & \dots & & 0 \end{bmatrix} \tag{4.2}$$

where $\hat{A}_j = A_\ell^{-1}A_j$, $j = 0, 1, \dots, \ell - 1$, and it is easily seen that A is G-selfadjoint. Also, $\sigma(L) = \sigma(A)$.

An important result of reference [12] states that a monic selfadjoint operator polynomial $L(\lambda)$ is a QHP if and only if the "linearization" A is uniformly definitizable with respect to G. The proof of this fact extends immediately to the case in which A_ℓ is merely invertible. Thus, as described at the beginning of Section 3, when $L(\lambda)$ is a QHP, its spectrum is contained in a finite number of quasizones, say q. With the obvious meaning, this situation is described by saying that "$L(\lambda)$ is a QHP with q quasizones".

Let us first observe that a QHP has a strong stability property derived from Lemma 2. Thus, if $L(\lambda)$ is a QHP with q quasizones and has the form (4.1), then all polynomials obtained from $L(\lambda)$ by sufficiently small selfadjoint perturbations of $A_0, A_1, \dots, A_\ell$ are also QHP with q quasizones. (This kind of result is well-known in finite dimensional spaces; see Chapters III.1, III.2 of [6], for example.)

Now consider finite rank and compact perturbations:

Theorem 4 *Let*

$$L_0(\lambda) = \sum_{j=0}^{\ell} \lambda^j A_j^0, \quad L(\lambda) = \sum_{j=0}^{\ell} \lambda^j A_j \tag{4.3}$$

be selfadjoint operator polynomials with A_ℓ and A_ℓ^0 invertible and let $L_0(\lambda)$ be a QHP *with q quasizones.*

(a) *If* $\operatorname{rank}(A_k - A_k^0) = a_k$, $k = 0, 1, \dots, \ell$, *m_+ is the total algebraic multiplicity of all eigenvalues of $L(\lambda)$ in the halfplane* $\operatorname{Im}\lambda > 0$, *and s is the number of distinct mixed real points of spectrum of $L(\lambda)$, then*

$$m_+ + s \le \sum_{j=0}^{\ell} (q + j - 1)a_j. \tag{4.4}$$

(b) *If the operators $A_k - A_k^0$ are compact ($k = 0, 1, \ldots, \ell$) then $m_+ + s$ is finite.*

Proof. For part (a) apply Remark 3 above to the linearizations A (for $L(\lambda)$) and A^0 (for $L_0(\lambda)$). Simple rank estimates are used as in the proof of Theorem 8.1 of [11].

For part (b) apply Theorem 3 to the linearizations. □

Note that in the special case of $n = 2$ and a "strongly hyperbolic" hypothesis, statement (b) follows from results of Shkalikov and Pliev (see [17]).

5 A factorization theorem in the monic case

The next result generalizes a factorization theorem of Langer (Theorem 7 of [13]). A QHP is the unperturbed system here, rather than the hyperbolic system implied by Langer's hypotheses. Nevertheless, our generalization uses other basic results on factorization, some of which are collected in the following statement. (See Theorems 1 and 3 of [13]). We remark that the assumption of a definite leading coefficient seems to be essential for the classical polynomial factorization theorems. See Krupnik et al. (reference [10]) for results of a different kind for quadratic matrix QHP.

Theorem 5 (Langer). *Let $L(\lambda)$ be a monic selfadjoint operator polynomial of degree ℓ and let A be its companion operator (put $A_\ell = I$ in (4.2)). If A has an invariant subspace $\mathcal{M}_+$ which is maximal G-nonnegative, then $L(\lambda)$ admits a factorization $L(\lambda) = L_-(\lambda)L_+(\lambda)$ where $L_+(\lambda)$, $L_-(\lambda)$ are monic operator polynomials of degrees $\left[\frac{\ell+1}{2}\right]$, $\left[\frac{\ell}{2}\right]$, respectively, the spectrum of $L_+(\lambda)$ coincides with the spectrum of $A|\mathcal{M}_+$ and, if $\mathcal{M}_-$ is the G-orthogonal complement of $\mathcal{M}_+$, the spectrum of $L_-(\lambda)$ coincides with the spectrum of $A|\mathcal{M}_-$.*

The *essential spectrum* of an operator A (or of an operator polynomial $L(\lambda)$) is the subset of its spectrum consisting of points which do not have finite type (see [4]), and is written $\sigma_{\text{ess}}(A)$ (or as $\sigma_{\text{ess}}(L)$). We also write σ_{ess}^+, σ_{ess}^- for points of σ_{ess} of plus type and minus type, respectively. Note the following simple fact, which can be deduced from Lemma 12.1 of [15], for example.

Lemma 2 *If $L(\lambda)$ is a selfadjoint operator polynomial with invertible leading coefficient and linearization A (as in (4.2)), then $\sigma_{\text{ess}}(L) = \sigma_{\text{ess}}(A)$.*

Finally, let us partition the nonreal spectrum of $L(\lambda)$, say $\sigma_0(L)$ in the form

$$\sigma_0(L) = \sigma_1 \cup \sigma_2 \tag{5.1}$$

where $\sigma_1 \cap \sigma_2 = \emptyset$ and $\sigma_2 = \overline{\sigma_1}$ (thus, no conjugate pairs are contained in σ_1, or in σ_2). Our factorization theorem now takes the form:

Theorem 6 *Let*

$$L_0(\lambda) = \lambda^\ell I + \sum_{j=0}^{\ell-1} \lambda^j A_j^0, \quad L(\lambda) = \lambda^\ell I + \sum_{j=0}^{\ell-1} \lambda^j A_j$$

be monic selfadjoint operator polynomials with $L_0(\lambda)$ *a* QHP *and* $A_k - A_k^0$ *compact for* $k = 0, 1, \ldots, \ell-1$. *When* $\sigma_0(L)$ *is partitioned as in* (5.1), $L(\lambda)$ *admits a factorization* $L(\lambda) = L_-(\lambda)L_+(\lambda)$ *with the following properties:*

(i) $L_+(\lambda)$ *is monic with degree* $\left[\frac{\ell+1}{2}\right]$, $\sigma_0(L_+) = \sigma_2$, *and* $\sigma_{\text{ess}}(L_+) = \sigma_{\text{ess}}^+(L_0)$.

(ii) $L_-(\lambda)$ *is monic with degree* $\left[\frac{\ell}{2}\right]$, $\sigma_0(L_-) = \sigma_1$, *and* $\sigma_{\text{ess}}(L_-) = \sigma_{\text{ess}}^-(L_0)$.

Proof. Define A, G for $L(\lambda)$ and A_0, G_0 for $L_0(\lambda)$ as in (4.2) (with $A_\ell = A_\ell^0 = I$), and note that A is G-selfadjoint, A_0 is G_0-selfadjoint. Since $L_0(\lambda)$ is a QHP it follows from Theorem 2 of [12], that A_0 has an invariant subspace $\mathcal{M}_+$ which is maximal uniformly G_0-positive. Now it follows from Theorem 6 of [13] that A has an invariant subspace $\mathcal{M}$ which is maximal G-nonnegative and $\sigma(A|\mathcal{M}) \cap \sigma_0 = \sigma_1$, $\sigma_{\text{ess}}(A|\mathcal{M}) = \sigma_{\text{ess}}(A_0|\mathcal{M}_+)$. A similar statement applies with "G-nonnegative" replaced by "G-nonpositive".

It remains to apply Lemma 2 and Theorem 5. □

6 Differential equations with stably bounded solutions

The main objective of this section is a result in which a QHP $L(\lambda) := \sum_{j=0}^{\ell} \lambda^j A_j$ is characterized through properties of solutions of the time-invariant operator differential equation

$$\sum_{j=0}^{\ell} A_j \left(i\frac{d}{dt}\right)^j x(t) = 0. \tag{6.1}$$

This equation is said to have *stably bounded solutions* if all solutions are bounded for all t and remain bounded under small selfadjoint perturbations of the coefficients $A_0, A_1, \ldots, A_\ell$. We are to prove that equation (6.1) has stably bounded solutions if and only if $L(\lambda)$ is a QHP. Results of this kind have a history including contributions by Krein, Gelfand and Lidskii, Yakubovic, and others. Our result is the operator version of a theorem of Gohberg et al. (see [5] and [6]). It is proved there for the case when $\mathcal{H}$ is finite dimensional. More historical details can be found in those sources.

First we need two technical lemmas.

Lemma 3 *Let* A_0, A_1 *be bounded selfadjoint operators on* $\mathcal{H}$ *and let there exist two sequences* $\{f_n\}$, $\{g_n\}$ *in* $\mathcal{H}$ *such that, as* $n \to \infty$,

$$\|f_n\| = \|g_n\| = 1, \quad \|A_0 f_n\| \to 0, \quad \|A_0 g_n\| \to 0, \tag{6.2}$$

and

$$\lim(A_1 f_n, f_n) \geq 0, \quad \lim(A_1 g_n, g_n) \leq 0. \tag{6.3}$$

Then there exists a sequence of vectors $\{h_m\}$ such that

$$\|h_n\| = 1, \quad \|A_0 h_n\| \to 0, \quad (A_1 h_n, h_n) \to 0 \tag{6.4}$$

Proof. Define $d_n = \inf_{z \in \mathbf{C}} \|g_n - z f_n\|$. If $d_n \to 0$ as $n \to \infty$ then it follows from (6.3) that both $(A_1 f_n, f_n) \to 0$ and $(A_1 g_n, g_n) \to 0$. Hence conditions (6.4) can be obtained by taking $h_n = f_n$ for each n.

If $d_n \not\to 0$ we can assume that $d_n \geq \delta > 0$. Let P_n be the orthogonal projection on the two dimensional subspace span$\{f_n, g_n\}$. Then P_n can be represented in the form

$$P_n = (\cdot, f_n) f_n + (\cdot, u_n) u_n$$

where

$$u_n = \frac{g_n - (g_n, f_n) f_n}{\|g_n - (g_n, f_n) f_n\|}.$$

Hence

$$\|A_o P_n\| \leq \|A_0 f_n\| + \|A_0 u_n\| \tag{6.5}$$

and

$$\|A_0 u_n\| \leq \frac{\|A_0 g_n\| + \|A_0 f_n\|}{\|g_n - (g_n, f_n) f_n\|} \leq \delta^{-1}(\|A_0 g_n\| + \|A_0 f_n\|). \tag{6.6}$$

It follows from (6.2), (6.5) and (6.6) that $\|A_0 P_n\| \to 0$ as $n \to \infty$. On the other hand, it follows from (6.3) and the connectedness of the unit sphere in $\mathbf{C}^2$ that there exist vectors h_n in span$\{f_n, g_n\}$ such that $\|h_n\| = 1$ and $\lim(A_1 h_n, h_n) = 0$. Furthermore, since $\|A_0 P_n\| \to 0$, it is seen that, also, $\|A_0 h_n\| \to 0$. □

Lemma 4 *Let A_0, A_1 be bounded selfadjoint operators on $\mathcal{H}$ and assume that, for some $\varepsilon > 0$, there is an $h \in \mathcal{H}$ such that $\|h\| = 1$, $\|A_0 h\| < \varepsilon$ and $|(A_1 h, h)| < \varepsilon$. Then there exists a pair of selfadjoint operators A_0', A_1' such that $A_0' h = 0$, $(A_1' h, h) = 0$ and*

$$\|A_0' - A_0\| < 2\varepsilon, \quad \|A_1' - A_1\| < \varepsilon \tag{6.7}$$

Proof. Let P be the orthogonal projection onto span$\{h\}$. Thus, $P = (\cdot, h)h$. Define $A_0' = (I - P)A_0, (I - P)$ so that $A_0' h = 0$. It is easy to see that

$$\|PA_0\| = \|A_0 h\|, \quad \|(I - P)A_0 P\| \leq \|A_0 P\| = \|A_0 h\|$$

and hence that

$$\|A_0' - A_0\| \leq 2\|A_0 h\| < 2\varepsilon.$$

If we define $A_1' = A_1 - (A_1 h, h)$ then we obtain $(A_1' h, h) = 0$ and $\|A_1' - A_1\| = |(A_1 h, h)| < \varepsilon$. □

The next theorem is also required for the main result of this section, but is of independent interest. It generalizes a construction of Gohberg et al. (pages 254–5 of [6]).

Theorem 7 *Let $L(\lambda)$ be a selfadjoint operator polynomial of degree ℓ. If there is at least one point $\lambda_0 \in \sigma(L) \cap \mathbf{R}$ which is not of determinate type, then there is a selfadjoint and arbitrarily small polynomial perturbation of $L(\lambda)$ such that the perturbed polynomial has nonreal points of spectrum.*

It will be clear from our construction that there is a perturbing polynomial whose degree does not exceed $\max(\ell - 2, 1)$ which has the desired effect. Furthermore, all coefficients can have finite rank.

Proof. If $\lambda_0 \notin \sigma_{\rm ap}(L)$ then $\lambda_0 \notin \partial\sigma(L)$, i.e. there is a complex neighbourhood of λ_0 contained in $\sigma(L)$ and the result follows trivially. So we may suppose that $\lambda_0 \in \sigma_{\rm ap}(L)$. Write $L(\lambda) = \sum_{j=0}^{\ell} \lambda^j A_j$. We can assume that $\lambda_0 = 0$ and then choose sequences $\{f_n\}$ and $\{g_n\}$ so that A_0 and A_1 satisfy the conditions of Lemma 3. Then, at the expense of small selfadjoint perturbations, Lemma 4 can be used to pass to the case when there exists an $h \in \mathcal{H}$ such that $\|h\| = 1$, $A_0 h = 0$, and $(A_1 h, h) = 0$. Now let $\mathcal{N} = \operatorname{span}\{h\}$, $\mathcal{H} = \mathcal{N} \oplus \mathcal{M}$ and, with respect to this decomposition write

$$\begin{aligned} L(\lambda) &= \begin{bmatrix} \lambda^2 e_1(\lambda) & \lambda A_{21}^* + \lambda^2 E_2(\bar{\lambda})^* \\ \lambda A_{21} + \lambda^2 E_2(\lambda) & E_3(\lambda) \end{bmatrix}, \\ &= \lambda \begin{bmatrix} 0 & A_{21}^* \\ A_{21} & 0 \end{bmatrix} + \lambda^2 F(\lambda) + \begin{bmatrix} 0 & 0 \\ 0 & E_3(\lambda) \end{bmatrix}, \end{aligned}$$

where the lower right entry of $F(\lambda)$ is zero.

For $\varepsilon > 0$ consider

$$L_\varepsilon(\lambda) := L(\lambda) + \varepsilon \begin{bmatrix} 0 & iA_{21}^* \\ -iA_{21} & 0 \end{bmatrix} + \varepsilon^2 F(\lambda).$$

Then for any $\varepsilon > 0$ we have $L_\varepsilon(i\varepsilon)h = 0$, and this proves the theorem. □

Now the main result of the section can be established. It is an infinite dimensional generalization of Theorem III.2.2 of reference [6].

Theorem 8 *The following statements are equivalent for a selfadjoint polynomial, $L(\lambda) = \sum_{j=0}^{\ell} \lambda^j A_j$, with A_ℓ invertible.*

(i) *$L(\lambda)$ is a* QHP *(i.e. $\sigma(L) \subseteq \mathbf{R}$ and all points of $\sigma(L)$ are of determinate type).*

(ii) *The equation* (6.1) *has stably bounded solutions.*

(iii) *$\sigma(L)$ is real, and this property is preserved under small selfadjoint perturbations of the coefficients $A_0, A_1, \ldots, A_\ell$.*

Proof. (i) $\Rightarrow$ (ii). Define operators A and G on $\mathcal{H}^\ell$ as in equations (4.2). As noted in Section 4, $L(\lambda)$ a QHP implies that A is uniformly definitizable with respect to G. Then, by Theorem 3 of [12], A is similar to a selfadjoint operator. Consequently, all solutions of the differential equation

$$\frac{dy}{dt} = iAy(t) \tag{6.8}$$

are bounded. (Note that, for this equation, boundedness on $\mathbf{R}$ is equivalent to boundedness on $[0, \infty)$. See Chapter 4 of [3].) If $x(t)$ is any solution of equation (6.1) then $y(t) = \langle x(t), \frac{1}{i}\frac{dx}{dt}, \ldots, (\frac{1}{i}\frac{d}{dt})^{\ell-1}x\rangle$ is a solution of (6.8) and, consequently, $x(t)$ is bounded. But all selfadjoint polynomials close enough to $L(\lambda)$ are QHP as well (see Section 4). Hence the solutions of (6.1) are stably bounded.

(ii) $\Rightarrow$ (iii). From (ii) it follows that the solutions of (6.8) are also bounded. This implies that A is similar to a selfadjoint operator (see Theorem II.3.1 of [3], for example) and, consequently $\sigma(A) = \sigma(L) \subseteq \mathbf{R}$. But equation (6.1) has *stably* bounded solutions, and this implies property (iii).

(iii) $\Rightarrow$ (i). This follows immediately from Theorem 7. □

7 The case of noninvertible leading coefficient

In this section it will be shown that Theorem 4 can be extended to include selfadjoint polynomials (as in equation (4.3)) for which the unperturbed polynomial $L_0(\lambda)$ has a noninvertible leading coefficient. This requires suitable extensions of our basic definitions followed by simple transformation of the eigenvalue parameter which produces an unperturbed polynomial with invertible leading coefficient.

Let $L(\lambda) = \sum_{j=0}^{\ell} \lambda^j A_j$ with $A_0, A_1, \ldots, A_\ell$ selfadjoint and A_ℓ not invertible. In this case we say that $\infty \in \sigma(L)$. Note that the case $A_\ell = 0$ is not excluded. It is clear that there is a sequence $\{f_n\}_{n=1}^{\infty}$ such that, as $n \to \infty$,

$$\|f_n\| = 1, \quad \|A_\ell f_n\| \to 0, \tag{7.1}$$

(i.e. $\infty \in \sigma_{\text{ap}}(L)$). We shall say that ∞ *is a point of determinate type* if it follows from (7.1) that either $\underline{\lim}(A_{\ell-1}f_n, f_n) > 0$ or $\overline{\lim}(A_{\ell-1}f_n, f_n) < 0$. More precisely, we say that ∞ has *plus* or *minus type* in the first or second case, respectively.

Now, when A_ℓ is not invertible, $L(\lambda)$ is said to be a *quasihyperbolic polynomial* (QHP) if $\sigma(L) \subseteq \mathbf{R}$ and all points of $\sigma(L)$ (including ∞) are of determinate type. Also, $L(\lambda)$ is said to be *regular* if there is an $\alpha \in \mathbf{R}$ such that $\alpha \notin \sigma(L)$.

For a regular QHP we transform the eigenvalue parameter as follows. Choose an $\alpha \in \mathbf{R}$, $\alpha \notin \sigma(L)$ and set $\mu = (\lambda - \alpha)^{-1}$, $\lambda = \alpha + \mu^{-1}$. Then define

$$M(\mu) := \mu^\ell L(\alpha + \mu^{-1}) = \sum_{j=0}^{\ell} \frac{\mu^{\ell-j}}{j!} L^{(j)}(\alpha). \tag{7.2}$$

It is apparent that $M(\mu)$ is a selfadjoint operator polynomial with invertible leading coefficient.

Lemma 5 *Let $L(\lambda)$ be a regular selfadjoint polynomial and suppose $\mu \neq 0$. Then $\lambda \in \sigma(L)$ if and only if $\mu \in \sigma(M)$. Furthermore, λ is a point of determinate type for $L(\lambda)$ if and only if μ $(= (\lambda - \alpha)^{-1})$ is a point of determinate type for $M(\mu)$. If ℓ is even then the types of λ and μ are different (i.e. one is plus and the other is minus). If ℓ is odd and $\mu > 0$ then the types of λ and μ are different. If ℓ is odd and $\mu < 0$ then the types of λ and μ are the same.*

Proof. We have

$$M'(\mu) = \ell\mu^{\ell-1}L(\lambda) - \mu^{\ell-2}L'(\lambda). \tag{7.3}$$

Also, for a normalized sequence $\{f_n\}$ we have $\|L(\lambda)f_n\| \to 0$ as $n \to \infty$ if and only if $\|M(\mu)f_n\| \to 0$ as $n \to \infty$. In this case (7.3) yields

$$\lim_{n\to\infty}\{(M'(\mu)f_n, f_n) + \mu^{\ell-2}(L'(\lambda)f_n, f_n)\} = 0$$

and the results follow from this relation. □

Lemma 6 *The point $\mu = 0$ is a point of plus (minus) type for $M(\mu)$ if and only if ∞ is a point of plus(minus) type for $L(\lambda)$.*

Proof. Let ∞ be a point of plus type for $L(\lambda)$. Then, by definition, there is a normalized sequence $\{f_n\}$ such that, as $n \to \infty$, $\|A_\ell f_n\| \to 0$ and

$$\underline{\lim}(A_{\ell-1}f_n, f_n) > 0.$$

Then, using (7.2), $\|M(0)f_n\| \to 0$ as well. Furthermore,

$$M'(0) = A_{\ell-1} + \ell\alpha A_\ell \tag{7.4}$$

and so

$$\underline{\lim}(M'(0)f_n, f_n) = \underline{\lim}(A_{\ell-1}f_n, f_n) > 0.$$

Thus $\mu = 0$ is a point of plus type for $M(\mu)$. The converse statement also follows from equation (7.4). The case of points of minus type is similar. □

Now the next result follows immediately:

Lemma 7 *If $L(\lambda)$ is a* QHP *then $M(\mu)$ (of equation* (7.2)*) is a* QHP *with invertible leading coefficient.*

Note that this parameter transformation can produce one less quasi-zone for $M(\mu)$ than $L(\lambda)$. If α is chosen in a gap between quasi-zones of $L(\lambda)$ then $M(\mu)$ will have no more quasi-zones than $L(\lambda)$.

Now we can obtain a perturbation theorem like Theorem 4 which applies where A_ℓ is not invertible.

Theorem 9 *Let*

$$L_0(\lambda) = \sum_{j=0}^{\ell} \lambda^j A_j^0, \quad L(\lambda) = \sum_{j=0}^{\ell} \lambda^j A_j$$

be selfadjoint operator polynomials with A_ℓ^0 not invertible and $L_0(\lambda)$ a QHP with q quasi-zones. Assume also that there is an $\alpha \in \mathbf{R}$ such that $L_0(\alpha)$ and $L(\alpha)$ are invertible.

(a) *If rank $(A_k - A_k^0) = a_k$, $k = 0, 1, \ldots, \ell$ and m_+, s are defined as in Theorem 4, then*

$$m_+ + s \leq \sum_{k=0}^{\ell} (q + \ell - 1 - \frac{1}{2}k) a_k.$$

(b) *If the operators $A_k - A_k^0$ are compact $(k = 0, 1, \ldots, \ell)$ then $m_+ + s$ is finite.*

Proof. Define

$$M_0(\mu) = \mu^\ell L_0(\alpha + \mu^{-1}), \quad M(\mu) = \mu^\ell L(\alpha + \mu^{-1}).$$

Then, by Lemma 7, $\mathcal{M}_0(\mu)$ is a QHP with invertible leading coefficient. Also, using (7.2), we have

$$M_0(\mu) - M(\mu) = \sum_{j=0}^{\ell} \frac{\mu^{\ell-j}}{j!} (L_0^{(j)}(\alpha) - L^{(j)}(\alpha)) \tag{7.5}$$

(a) It suffices to estimate m_+ and s for $M(\mu)$ and, for this purpose, we may use inequality (4.4). From (7.5) we obtain

$$\begin{aligned} b_j := \operatorname{rank}(M_j^0 - M_j) &= \operatorname{rank}(L_0^{(n-j)}(\alpha) - L^{(n-j)}(\alpha)) \\ &\leq \sum_{k=\ell-j}^{\ell} \operatorname{rank}(A_k^0 - A_k) = \sum_{k=\ell-j}^{\ell} a_k. \end{aligned} \tag{7.6}$$

Then (4.4) gives

$$\begin{aligned} m_+ + s &\leq \sum_{j=0}^{\ell} (q + j - 1) b_j \leq \sum_{k=0}^{\ell} a_k \sum_{j=\ell-k}^{\ell} (q + j - 1) \\ &= \sum_{k=0}^{\ell} (q + \ell - 1 - \frac{1}{2}k) a_k. \end{aligned} \tag{7.7}$$

(b) When $A_k - A_k^0$ are compact, $k = 0, 1, \ldots, \ell$, it follows from (7.5) that $M(\mu)$ is a compact perturbation of $M_0(\mu)$. Now apply part (b) of Theorem 4 to show that $M(\mu)$, and hence $L(\lambda)$ has finitely many nonreal eigenvalues and mixed real eigenvalues. □

Remark. It is apparent that the estimates leading to (7.7) are crude, and may be significantly improved in special cases. For example, if A_0^0 and A_0 are invertible, then we may put $\alpha = 0$ and, in (7.6), we obtain $b_j = \operatorname{rank}(A_{\ell-j}^0 - A_{\ell-j}) = a_{\ell-j}$. Then (7.7) becomes

$$m_+ + s \leq \sum_{j=0}^{\ell} (q + j - 1) a_{\ell-j}.$$

References

[1] L. Barkwell, P. Lancaster, and A.S. Markus. Gyroscopically stabilized systems: a class of quadratic eigenvalue problems with real spectrum. *Canadian Jour. of Math.*, (1992), **44**, 42–53.

[2] J. Bognar. *Indefinite Inner Product Spaces.* Springer–Verlag, New York, 1974.

[3] J.L. Daleckii and M.G. Krein. *Stability of Solutions of Differential Equations in Banach Space.* American Math. Society, Providence, 1974. (Translation of Russian edition of 1970.)

[4] I. Gohberg, S. Goldberg, and M.A. Kaashoek. *Classes of Linear Operators.* Vol. 1, Birhaüser Verlag, 1990.

[5] I. Gohberg, P. Lancaster and L. Rodman. Perturbations of H-selfadjoint matrices, with applications to differential equations. *Int. Eq. and Operator Theory,* (1982) **5**, 718–757.

[6] I. Gohberg, P. Lancaster, and L. Rodman. *Matrices and Indefinite Scalar Products.* Birkhäuser, Basel, 1983.

[7] P. Jonas. On a class of selfadjoint operators in Krein space and their compact perturbations. *Int. Eq. and Operator Theory,* (1988) **11**, 351–384.

[8] P. Jonas and H. Langer. Compact perturbations of definitizable operators. *Jour. of Operator Theory,* (1979) **2**, 63–77.

[9] M.G. Krein and H. Langer. The spectral function of a selfadjoint operator in a space with indefinite metric. *Soviet Math. Doklady,* (1963), **4**, 1236–1239.

[10] I. Krupnik, P. Lancaster, and A. Markus. Factorization of selfadjoint quadratic matrix polynomials with real spectrum. *Jour. Lin. Multilinear Algebra,* (to appear).

[11] P. Lancaster, A.S. Markus, and Q. Ye. Low rank perturbations of strongly definitizable transformations and matrix polynomials. *Lin. Alg. and its Applications,* (1994) **197**, **198**, 3-29.

[12] P. Lancaster, A.S. Markus, and V.I. Matsaev. Definitizable operators and quasihyperbolic polynomials. *Jour. Functional Anal.* (1995) **131**, 1-28.

[13] H. Langer. Factorization of operator pencils. *Acta. Sci. Math.,* (1976) **38**, 83-96.

[14] H. Langer, A.S. Markus, and V.I. Matsaev. *Locally definite operators in indefinite inner product spaces.* (Submitted.)

[15] A.S. Markus. *Introduction to the Spectral Theory of Polynomial Operator Pencils.* American Math. Society, Providence, 1988. (Translation of Russian edition of 1986.)

[16] L. Rodman. On factorization of selfadjoint operator polynomials. *Proc. Symposium Pure Math.*, Vol. 51, Part 2, 295–306. (American Math. Society, Providence, 1990).

[17] A.A. Shkalikov and V.T. Pliev. Compact perturbations of strongly damped operator pencils. *Math. Notes*, (1989) **45**, 167–174. (Translation of Math. Zametki.)

P. LANCASTER
Department of Mathematics and Statistics
University of Calgary
Calgary, Alberta $T2N$ $1N4$
Canada

A. MARKUS
Department of Mathematics and Computer Science
Ben Gurion University of the Negev
Beer Sheva
Israel

V. MATSAEV
School of Mathematical Sciences
Tel Aviv University
Ramat Aviv 69978
Israel

AMS Classifications: 47A05, 47A55, 47A56

Operator Theory:
Advances and Applications, Vol. 87

DEFINITIZABLE G-UNITARY OPERATORS AND THEIR APPLICATIONS TO OPERATOR POLYNOMIALS

P. LANCASTER*, A. MARKUS, and V. MATSAEV

Let G be a bounded and invertible selfadjoint operator on a Hilbert space $\mathcal{H}$ and consider linear operators U on $\mathcal{H}$ satisfying $U^*GU = G$. Then U is said to be G-unitary and, if there is a trigonometric polynomial q such that $Gq(U) \gg 0$ then U is said to be uniformly definitizable. Such operators are characterized by properties of their spectrum. Results are also established concerning the spectrum of U when (U, G) is obtained by perturbation of a uniformly definitizable pair (U_0, G_0); the perturbations may be small, finite rank, or compact. Similar results are obtained for certain operator polynomials whose spectrum is symmetric with respect to the unit circle.

1 Introduction

Let G be an invertible selfadjoint operator on a Hilbert space $\mathcal{H}$ with inner product $(\cdot,\cdot)$. In a familiar way G can be used to define a (generally) indefinite scalar product $[\cdot,\cdot]$ on $\mathcal{H}$ by writing $[x,y] = (Gx,y)$ for all $x, y \in \mathcal{H}$. A bounded operator U on $\mathcal{H}$ is said to be *G-unitary* if it satisfies $U^*GU = G$. Also, a bounded operator A on $\mathcal{H}$ is said to be G-selfadjoint if $GA = A^*G$.

The notions of definitizable G-unitary and G-selfadjoint operators were introduced by Langer in 1964 (see [10]). We make definitions of these concepts which differ only in details from those of Langer. First, a rational function q is called a *trigonometric polynomial* if it has the form

$$q(z) = a_0 + \sum_{j=1}^{k} (a_j z^j + \bar{a}_j z^{-j}) \tag{1.1}$$

where $a_0 \in \mathbb{R}$, and we note that q is real-valued when $z \in \mathbb{T}$, the unit circle. A G-unitary operator U is said to be *definitizable* if there is a trigonometric polyomial q such that $Gq(U) \geq 0$. Similarly, a G-selfadjoint operator A is said to be definitizable if there is a real polynomial p such that $Gp(A) \geq 0$. The term definitizable is prefixed by *strongly-* or *uniformly-* if the

*The work of this author was supported in part by a grant from the Natural Sciences and Engineering Research Council of Canada.

symbol $\geq$ is replaced by $>$ or $\gg$, respectively. We note that all G-unitary and G-selfadjoint operators are definitizable when $\mathcal{H}$ with the scalar product $[\cdot,\cdot]$ is a Pontrjagin space (i.e. when either $G_+ = \frac{1}{2}(|G| + G)$ or $G_- = \frac{1}{2}(|G| - G)$ has finite rank (see Section 18 of [8] and [11])).

Stimulated by applications in mechanics and boundary value problems (see [3] and [14]) the authors have developed a theory of uniformly definitizable G-selfadjoint operators in [12] and [13] with applications to selfadjoint operator polynomials. Briefly, it is our objective in this paper to follow a similar program for G-unitary operators. The strong forms of definitizability admit a more transparent characterization of the spectrum of an operator than is possible with the mere definitizable assumption (see Theorem 1 below), and this facilitates the development of perturbation arguments and applications to operator polynomials (which generalize results developed in [7] in the case that $\mathcal{H}$ is finite dimensional).

In Section 2 the basic characterization of uniformly definitizable G-unitary operators in terms of simple spectral properties is established. Taking advantage of well-known results recorded by Daleckii and Krein [4], (but originating with Phillips [15], Derguzov [5], [6], and Jonas [9]) a number of other characterizations follow readily. Section 3 contains discussion of the spectral properties of G-unitary operators U which is obtained after compact perturbations of both G_0 and U_0, where U_0 is uniformly definitizable with respect to G_0. Sections 4 and 5 contain analysis of similar properties of some operator polynomials $L(\lambda) = \sum_{j=0}^{2k} \lambda^j A_j$ (with A_{2k} invertible) with the property that the spectrum of L is contained in $\mathbf{T}$.

2 Preliminary definitions and results

As indicated above, a G-unitary operator U on $\mathcal{H}$ is said to be uniformly definitizable (with respect to G) if there is a Fourier polynomial q such that $Gq(U) \gg 0$.

Lemma 1 *A uniformly definitizable G-unitary operator is similar to a unitary operator on $\mathcal{H}$.*

Proof. Let U be G-unitary and $Gq(U) \gg 0$ for a trigonometric polynomial q. Then it is easily verified that U is unitary in the definite scalar product $\{\cdot,\cdot\}$ defined on $\mathcal{H}$ by $\{f,g\} = (Gq(U)f,g)$. But then the scalar products $\{\cdot,\cdot\}$ and $(\cdot,\cdot)$ are equivalent. □

Let us recall a basic definition of [12]. If A is a bounded operator on $\mathcal{H}$ then a point λ of the approximate point spectrum of A ($\sigma_{\text{ap}}(A)$) is a point of *plus type*, or of *minus type*, if for any sequence $\{f_n\} \subseteq \mathcal{H}$ satisfying $\|f_n\| = 1$ and $\|Af_n - \lambda f_n\| \to 0$ as $n \to \infty$, we have either

$$\underline{\lim}_{n\to\infty}(Gf_n, f_n) > 0, \quad \text{or} \quad \overline{\lim}_{n\to\infty}(Gf_n, f_n) < 0,$$

respectively. In either of these cases λ is said to have *determinate type*. It is easily seen that, if $\lambda \in \sigma_{\text{ap}}(A)$ then

$$[Af_n, Af_n] - |\lambda|^2[f_n, f_n] \to 0.$$

Consequently, for a G-unitary operator U with $\lambda \in \sigma_{\text{ap}}(U)$ we have $(1-|\lambda|^2)[f_n, f_n] \to 0$ and, if λ has determinate type, it follows that $|\lambda| = 1$. Those points of $\sigma_{\text{ap}}(U) \cap \mathbf{T}$ which are not of determinate type are said to have *mixed type.*

The next lemma is crucial and is an analogue of Lemma 5 of [12]. The method of proof is adapted from that source.

Lemma 2 *Let U be G-unitary and assume that $\sigma(U) \subseteq \mathbf{T}$ and all points of $\mathbf{T}$ are either regular points of U or points of $\sigma(U)$ of plus type. Then $G \gg 0$.*

Proof. Note first of all that, as in Lemma 4 of [12], there exist $\varepsilon > 0$ and $\delta > 0$ such that if $\lambda \in \mathbf{T}$ then, for $f \in \mathcal{H}$,

$$\|(U-\lambda I)f\| < \varepsilon\|f\| \Rightarrow (Gf, f) > \delta\|f\|^2. \tag{2.1}$$

Consider a number $r > 1$ and the circles $\Gamma := \{z : |z| = r\}$, $\gamma := \{z : |z| = r^{-1}\}$. Then, because $\sigma(U) \subseteq \mathbf{T}$ we have

$$\begin{aligned}
2\pi i I &= \left(\int_\Gamma - \int_\delta\right)(zI-U)^{-1}\,dz,\\
&= i\int_0^{2\pi}\{r(re^{i\theta}I-U)^{-1} - r^{-1}(r^{-1}e^{i\theta}I-U)^{-1}\}e^{i\theta}\,d\theta,\\
&= i\int_0^{2\pi} r^{-1}(r^{-1}e^{i\theta}I-U)^{-1}\{r(r^{-1}e^{i\theta}I-U) - r^{-1}(re^{i\theta}I-U)\}r(re^{i\theta}I-U)^{-1}e^{i\theta}\,d\theta,\\
&= i(-r+r^{-1})\int_0^{2\pi}(r^{-1}e^{i\theta}I-U)^{-1}U(re^{i\theta}I-U)^{-1}e^{i\theta}\,d\theta.
\end{aligned}$$

Now write $re^{-i\theta}(r^{-1}e^{i\theta}I-U)U^{-1} = U^{-1} - re^{-i\theta}I$ and we obtain

$$I = \frac{(r^2-1)}{2\pi}\int_0^{2\pi}(re^{-i\theta}I-U^{-1})^{-1}(re^{i\theta}I-U)^{-1}\,d\theta).$$

Since $U^*GU = G$ it follows that $G(re^{-i\theta}I-U^{-1})^{-1} = (re^{-i\theta}I-U^*)^{-1}G$ and hence that

$$[f,f] = \frac{r^2-1}{2\pi}\int_0^{2\pi}[(re^{i\theta}I-U)^{-1}f, (re^{i\theta}I-U)^{-1}f]\,d\theta. \tag{2.2}$$

Let f be a fixed unit vector and define the subsets of $[0, 2\pi]$:

$$E_1(r) = \left\{\theta \in [0,2\pi] : \|(re^{i\theta}I-U)^{-1}f\| \le \frac{2}{\varepsilon}\right\}, \quad E_2(r) = [0,2\pi]\backslash E_1(r).$$

Let $\theta \in E_2(r)$ and $h(\theta) = (re^{i\theta}I-U)^{-1}f$. Then

$$(e^{i\theta}I-U)h(\theta) = f - (r-1)e^{i\theta}h(\theta)$$

and hence

$$\begin{aligned}
\|(e^{i\theta}I-U)h(\theta)\| &\le 1 + (r-1)\|h(\theta)\|\\
&= (\|h(\theta)\|^{-1} + r - 1)\|h(\theta)\| \le \left(\frac{\varepsilon}{2} + r - 1\right)\|h(\theta)\|
\end{aligned}$$

since $\theta \in E_2(r)$. If $r < 1 + \frac{1}{2}\varepsilon$, then we obtain

$$\|(e^{i\theta} I - U)h(\theta)\| < \varepsilon \|h(\theta)\|$$

and applying (2.1) to $h(\theta)$,

$$[h(\theta), h(\theta)] > \delta \|h(\theta)\|^2 > 0 \tag{2.3}$$

for any $\theta \in E_2(r)$.

It follows from (2.2) and (2.3) that

$$[f, f] \geq \frac{r^2 - 1}{2\pi} \int_{E_1(r)} [h(\theta), h(\theta)]\, d\theta. \tag{2.4}$$

But

$$\left| \int_{E_1(r)} [h(\theta), h(\theta)]\, d\theta \right| \leq \int_{E_1(r)} \|G\|\, \|h(\theta)\|^2\, d\theta \leq \|G\| \left(\frac{2}{\varepsilon}\right)^2 2\pi$$

since, by definition of $E_1(r)$, we have $\|h(\theta)\| \leq 2/\varepsilon$. Thus (2.4) gives

$$[f, f] \geq -\frac{4(r^2 - 1)}{\varepsilon^2} \|G\|.$$

Since $r > 1$ and $r - 1$ may be arbitrarily small we have, in fact, $[f, f] \geq 0$ for any vector f. Since G is invertible this now implies that $G \gg 0$. □

Theorem 1 *Let U be a G-unitary operator. Then the following statements are equivalent:*

(a) *U is uniformly definitizable.*

(b) *$\sigma(U) \subseteq \mathbf{T}$ and all points of $\sigma(U)$ have determinate type.*

Proof. (a) $\Rightarrow$ (b). It follows from Lemma 1 that $\sigma(U) \subseteq \mathbf{T}$. The argument that all points of $\sigma(U)$ have determinate type is now just that of Lemma 3 of [12] (with the real definitizing polynomial replaced by a trigonometric polynomial).

(b) $\Rightarrow$ (a). If $\sigma(U)$ consists entirely of points of one type (either plus or minus) then, by Lemma 2, $G \gg 0$ or $G \ll 0$, and U is definitized by the polynomial $q(z) = 1$ or by $q(z) = -1$.

If $\sigma(U)$ contains points of both plus and minus types then for some integer $k \geq 1$ there are $2k$ distinct closed arcs $\{\Delta_j\}_{j=1}^{2k}$ on $\mathbf{T}$ such that all points of $\sigma(U)$ on Δ_{2j-1} (on Δ_{2j}) are of plus (of minus) type, $\sigma(U) \subseteq \bigcup_{j=1}^{2j} \Delta_j$, and the types alternate with j. Choose one point on $\mathbf{T}$ between each pair of arcs and construct a trigonometric polynomial $q(z)$ of the form (1.1) with zeros at the $2k$ chosen points, and at no others. In addition we may construct q so that $q(z) > 0$ on the arcs Δ_{2j-1} and $q(z) < 0$ on the arcs Δ_{2j}, $j = 1, 2, \ldots, k$.

Now let $\mathcal{H}_j$ be the spectral subspace of U corresponding to Δ_j $j = 1, 2, \ldots 2k$ and we have a G-orthogonal direct sum $\mathcal{H} = \mathcal{H}_1 \dot{+} \cdots \dot{+} \mathcal{H}_{2k}$. By Lemma 2

$$(-1)^{j-1}[f_j, f_j] \geq \delta_j \|f_j\|^2$$

for all $f_j \in \mathcal{H}_j$ and some $\delta_j > 0$. But, by Lemma 1, $U|H_j$ is a unitary operator in the positive definite scalar product $(-1)^{j-1}[f, f]$ and hence $[q(U)f_j, f_j] \geq \gamma_j \|f_j\|^2$ for all $f_j \in \mathcal{H}_j$ and some $\gamma_j > 0$. finally, we obtain

$$[q(U)f, f] \geq \gamma \|f\|^2$$

for all $f \in \mathcal{H}$ and some $\gamma > 0$. □

Remark 1 It seems natural to apply a Cayley transform technique to establish this result from Theorem 1 of [12]. However, this line or argument holds only if $\sigma(U) \neq \mathbf{T}$ and the argument of Lemma 2 is still necessary.

Remark 2 Daleckii and Krein (see p. 48 of [4]) introduce the following definition: a bounded operator A on $\mathcal{H}$ is said to be *normally G-decomposable* if there is a decomposition $\sigma(A) = \sigma_1 \cup \sigma_2$, $\sigma_1 \cap \sigma_2 = \emptyset$ and, in the corresponding spectral decomposition $\mathcal{H} = \mathcal{H}_1 \dot{+} \mathcal{H}_2$, the A-invariant subspaces $\mathcal{H}_1$ and $\mathcal{H}_2$ are uniformly G-definite. It follows from the proof of Theorem 1 that conditions (a) and (b) are equivalent to:

(c) *U is normally G-decomposable.*

Let us pursue another line of thought developed by Daleckii and Krein (p. 49 of [4]). A G-unitary operator U is said to be *stable* if

$$\sup_{n>0} \|U^n\| < \infty, \quad (\sup_{n\in\mathbf{Z}} \|U^n\| < \infty). \tag{2.5}$$

(These conditions are, in fact, equivalent because $U^{-n} = G^{-1}(U^n)^*G$). Further, a stable G-unitary operator U is said to be *strongly stable* if there is a number $\delta > 0$ such that any G-unitary U' satisfying $\|U' - U\| < \delta$ is also stable.

Remark 3 Theorem I.8.3 of [4] shows that (c) is equivalent to:

(d) *U is strongly stable,*

and hence that (a), (b), (c) and (d) are equivalent. Furthermore it is shown in Section I.8 of [4] that (a)–(d) are also equivalent to the following statements:

(e) *There is a number $\delta > 0$ such that all G-unitary operators U' satisfying $\|U' - U\| < \delta$ are similar to unitary operators on $\mathcal{H}$.*

(f) *There is a number $\delta > 0$ such that any G-unitary operator U' for which $\|U' - U\| < \delta$ has the property $\sigma(U') \subseteq \mathbf{T}$.*

Remark 4 It is apparent that the property of uniform definitizability (property (a)) is stable under small perturbations of (G, U), to (G', U'), say, where U' is G'-unitary. Consequently, properties (d), (e), and (f) are also preserved under such simultaneous perturbations of G and U.

3 Compact perturbations

As in Remark 4, consider simultaneous perturbations of a pair U_0, G_0 where U_0 is G_0-unitary and uniformly definitizable, but now the perturbations are compact rather than "small".

Theorem 2 *Let U_0 be G_0-unitary and uniformly definitizable. Let U be G-unitary with $U - U_0$ and $G - G_0$ compact operators. Then U has at most a finite number of eigenvalues which are not on $\mathbf{T}$ and points of $\sigma(U)$ on $\mathbf{T}$ with mixed type.*

Proof. If $\sigma(U_0)$ contains points of both plus and minus types the proof of Theorem 2 of [13] is easily adapted to this situation.

If $\sigma(U_0)$ consists of points of only one type (plus or minus) then Lemma 2 implies $G_0 \gg 0$ or $G_0 \ll 0$. Since $G - G_0$ is compact the space $\mathcal{H}$ with the indefinite scalar product generated by G is a Pontrjagin space. The result now follows from those of Section IX.4 of [2], for example. □

Remark 5 In the special case of Theorem 2 when $U - U_0$ and $G - G_0$ have finite rank, it is not difficult to find estimates for the number of eigenvalues not on $\mathbf{T}$ plus the number of eigenvalues on $\mathbf{T}$ of mixed type. See Theorem 1 of [13].

4 Operator polynomials quasihyperbolic on T

Let $A_0, A_1, \dots, A_{2k}$ be bounded linear operators on $\mathcal{H}$ with A_{2k} invertible and consider the difference equation

$$A_0 x_m + A_1 x_{m+1} + \cdots + A_{2k} x_{m+2k} = 0, \quad m = 0, 1, 2, \dots, \tag{4.1}$$

where $\{x_m\}_{m=0}^{\infty} \subset \mathcal{H}$. Introduce a "companion operator", C_L for the polynomial $L(\lambda) := \sum_{j=0}^{2k} \lambda^j A_j$ by writing

$$C_L = \begin{bmatrix} 0 & I & 0 & \dots & 0 \\ 0 & 0 & I & & 0 \\ \vdots & & & \ddots & \vdots \\ & & & & I \\ -\tilde{A}_0 & -\tilde{A}_1 & \dots & & -\tilde{A}_{2k-1} \end{bmatrix} \tag{4.2}$$

where $\tilde{A}_j = A_{2k}^{-1} A_j$, $j = 0, 1, \dots, 2k-1$. Observe first of all that all solutions of (4.1) are bounded if and only if C_L is stable in the sense that $\sup_{n>0} \|C_L^n\| < \infty$. (See Section 2.11 of [16])

In order to apply the preceding ideas it is necessary to find an (indefinite) scalar product in which C_L is unitary. Following Section II.2.4 of [7] this can be done in the following case: Assume also that

$$A_j^* = A_{2k-j}, \quad j = 0, 1, \dots, 2k, \tag{4.3}$$

(an assumption that is natural for some finite difference methods). Then the rational function $\hat{L}(\lambda) := \lambda^{-k}L(\lambda)$ is selfadjoint on T, i.e. $(\hat{L}(\lambda))^* = \hat{L}(\bar{\lambda}^{-1}))$ and so $(\hat{L}(\lambda))^* = \hat{L}(\lambda)$ when $|\lambda| = 1$.

Now consider the indefinite scalar product on $\mathcal{H}^{2k}$ generated by the invertible selfadjoint operator

$$\hat{B}_L = i \begin{bmatrix} & & & A_{2k} & \dots & 0 \\ & 0 & & \vdots & \ddots & \vdots \\ & & & A_{k+1} & \dots & A_{2k} \\ -A_0 & \dots & -A_{k-1} & & & \\ \vdots & \ddots & \vdots & & 0 & \\ 0 & \dots & -A_0 & & & \end{bmatrix}.$$

Then it is easily verified (see p. 161 of [7]) that C_L is $\hat{B}_L$-unitary.

Let $\lambda_0 \in \mathsf{T}$ be in the approximate spectrum of $L(\lambda)$, i.e. there is a normalized sequence $\{f_n\}_{n=1}^{\infty}$ such that $\|L(\lambda_0)f_n\| \to 0$ as $n \to \infty$. Such a point is said to have *plus type* (or *minus type*) if, for all sequences $\{f_n\}$ satisfying $\|f_n\| = 1$ and $\|L(\lambda)f_n\| \to 0$ as $n \to \infty$, we have either $\underline{\lim}(i\lambda_0\hat{L}'(\lambda_0)f_n, f_n) > 0$ (or $\overline{\lim}(i\lambda_0\hat{L}'(\lambda_0)f_n, f_n) < 0$), respectively. Points of plus or minus type are said to have *determinate type.*

Now we make an important definition. An operator polynomial $L(\lambda) = \sum_{j=0}^{2k} \lambda^j A_j$ with A_{2k} invertible, and satisfying (4.3) is said to be *quasihyperbolic on* T (is a QHP on T) if $\sigma(L) \subseteq \mathsf{T}$ and consists of points of determinate type.

Lemma 3 *A point $\lambda_0 \in \mathsf{T}$ has plus or minus type for the polynomial $L(\lambda)$ (satisfying* (4.2)*) if and only if it has plus or minus type, respectively, as a point of the approximate spectrum of C_L (with respect to $\hat{B}_L$).*

Proof. The proof is readily adapted from that of Lemma 7 of [12]. If $\{f^{(m)}\}$ is a normalized approximate eigenvector sequence for C_L at $\lambda_0 \in \mathsf{T}$, and $f^{(m)} = (f_0^{(m)}, \dots f_{2k-1}^{(m)})$ where $f_j^m \in \mathcal{H}$, then a pivotal role is played by the relation

$$\lim_{m\to\infty}(f_0^{(m)}, i\lambda_0\hat{L}_0'(\lambda_0)f_0^{(m)}) = \lim_{m\to\infty}(f^{(m)}, \hat{B}_L f^{(m)}).$$

□

It then follows immediately from Theorem 1 that:

Theorem 3 *$L(\lambda)$ is a* QHP *on* T *if and only if C_L is uniformly definitizable with respect to $\hat{B}_L$.*

5 Other characterizations of QHP on T

The main theorem of this section gives characterizations of QHP on T in terms of "strong stability" properties of $L(\lambda)$ itself and of the solutions of the difference equation (4.1). Analogous results are obtained in [13] for operator polynomials which are selfadjoint in the sense that $A_j^* = A_j$ for each j and the corresponding differential equations. We can take advantage of those ideas by using a Möbius transformation.

Lemma 4 *Let* $L(\lambda) = \sum_{j=0}^{2k} \lambda^j A_j$, $\hat{L}(\lambda) = \lambda^{-k} L(\lambda)$, $\mu = i(\frac{1-\lambda}{1+\lambda})$ *and*

$$S(\mu) = (\lambda+1)^{-2k} L(\lambda) = (\lambda+1)^{-2k} \lambda^k \hat{L}(\lambda) = \left(\frac{1+\mu^2}{4}\right)^k \hat{L}(\lambda). \tag{5.1}$$

Then:

1°) *the rational function* $\hat{L}(\lambda)$ *is selfadjoint on* $\mathbf{T}$ *(i.e. satisfies* (4.3)*) if and only if the polynomial* $S(\mu)$ *is selfadjoint on* $\mathbf{R}$ *(i.e. the coefficients of powers of* μ *are selfadjoint),*

2°) $\sigma(L) \subseteq \mathbf{T}$ *if and only if* $\sigma(S) \subseteq \mathbf{R}$,

3°) *a point* $\lambda_0 \in \mathbf{T}$ *is of plus (minus) type for* $L(\lambda)$ *if and only if the point* $\mu_0 = i(\frac{1-\lambda_0}{1+\lambda_0})$ *is a point of the same type for* $S(\mu)$.

Proof. Statements 1°) and 2°) follow from the relation

$$S(\mu) = \left(\frac{1+\mu^2}{4}\right)^k \hat{L}(\lambda).$$

¿From the same equation it is easy to see that

$$S'(\mu) = \frac{(1+\mu^2)^{k-1}}{2^{2k-1}} i\lambda \hat{L}'(\lambda) + \frac{k(1+\mu^2)^{k-1}\mu}{2^{2k-1}} \hat{L}(\lambda). \tag{5.2}$$

But the conditions $S(\mu)f_n \to 0$ and $L(\lambda)f_n \to 0$ for a normalized sequence $\{f_n\}$ are equivalent. Hence statement 3°) follows from (5.2). □

Lemma 5 *Let* $L(\lambda)$ *satisfy* (4.3). *If there is at least one point* $\lambda_0 \in \sigma(L) \cap \mathbf{T}$ *which is not of determinate type then, for any* $\varepsilon > 0$, *there are operators* $B_0, B_1, \ldots B_{2k}$ *on* $\mathcal{H}$ *such that* $B_j^* = B_{2k-j}$, $\|B_j - A_j\| < \varepsilon$, $j = 0, 1, \ldots, 2k$, *and the polynomial* $M(\lambda) := \sum_{j=0}^{2k} \lambda^j B_j$ *has points of spectrum which do not belong to* $\mathbf{T}$.

Proof. Define the selfadjoint polynomial $S(\mu)$ as in Lemma 4. Then $\mu_0 = i(\frac{1-\lambda_0}{1+\lambda_0})$ is not of determinate type (for $S(\mu)$). By Theorem 6 of [10] there exists a selfadjoint operator polynomial $Q(\mu)$ such that not all points of $\sigma(Q)$ are real and the degree of Q does not exceed $2k$, while corresponding coefficients of $S(\mu)$ and $Q(\mu)$ are arbitrarily close. Then the polynomial

$$M(\lambda) := (\lambda+1)^{2k} Q\left(i\frac{1-\lambda}{1+\lambda}\right)$$

has all the required properties. □

Returning to the difference equation (4.1) (with A_{2k} invertible and conditions (4.3)), we say that solutions of (4.1) are *stably bounded* if all solutions of every difference equation

$$B_0 x_m + B_1 x_{m+1} + \cdots + B_{2k} x_{m+2k} = 0, \quad m = 0, 1, 2, \ldots, \tag{5.3}$$

with $B_j^* = B_{2k-j}$ and $\|B_j - A_j\|$ small enough (for $j = 0, 1, \ldots, 2k$) are bounded.

Lemma 6 *If all solutions of* (4.1) *are bounded then* $\sigma(L) \subseteq \mathbf{T}$.

Proof. Since A_{2k} is invertible $L(\lambda)$ is invertible for all $\lambda \geq \rho$ with ρ sufficiently large. For arbitrary $f \in \mathcal{H}$ write

$$x_m := \frac{1}{2\pi i}\int_{|\lambda|=\rho} \lambda^m L^{-1}(\lambda) f \, d\lambda$$

for $m = 0, 1, 2, \ldots,$. Then for $m \geq 0$

$$\begin{aligned}\sum_{j=0}^{2k} A_j x_{m+j} &= \frac{1}{2\pi i}\int_{|\lambda|=\rho} \sum_{j=0}^{2k} A_j \lambda^{m+j} L^{-1}(\lambda) f \, d\lambda \\ &= \frac{1}{2\pi i}\int_{|\lambda|=\rho} \lambda^m L(\lambda) L^{-1}(\lambda) f \, d\lambda \\ &= \frac{1}{2\pi i}\int_{|\lambda|=\rho} \lambda^m \, d\lambda f = 0.\end{aligned}$$

Thus $\{x_m\}_{m=0}^{\infty}$ is a solution of (4.1) and, by hypothesis, $\sup \|x_m\| < \infty$. Consequently the series $\sum_{m=0}^{\infty} \lambda^{-m} x^m$ converges for all $|\lambda| > 1$. But for $|\lambda| > \rho$ we have

$$\begin{aligned}\sum_{m=0}^{\infty} \lambda^{-m} x^m &= \frac{1}{2\pi i}\int_{|\lambda|=\rho} \sum_{m=0}^{\infty} \left(\frac{\mu}{\lambda}\right)^m L^{-1}(\mu) d\mu f \\ &= \frac{\lambda}{2\pi i}\int_{|\lambda|=\rho} \frac{L^{-1}(\mu)}{\lambda - \mu} d\mu f = \lambda L^{-1}(\lambda) f,\end{aligned}$$

and hence $L^{-1}(\lambda) f$ is analytic for $|\lambda| > 1$. But in this argument f is arbitrary, and it follows that $L(\lambda)$ is invertible for $|\lambda| > 1$.

¿From the equality $L(\lambda^{-1}) = \lambda^{-2k}(L(\bar{\lambda}))^*$ it follows that $L(\lambda)$ is also invertible for $|\lambda| < 1$. Hence $\sigma(L) \subseteq \mathbf{T}$. □

Theorem 4 *Let* $L(\lambda) = \sum_{j=0}^{2k} \lambda^j A_j$ *satisfy* (4.3) *with* A_{2k} *invertible. Then the following statements are equivalent:*

(i) $L(\lambda)$ *is a* QHP *on* $\mathbf{T}$;

(ii) *the solutions of the difference equation* (4.1) *are stably bounded;*

(iii) *there exists a number* $\delta > 0$ *such that* $\sigma(M) \subseteq \mathbf{T}$ *for any polynomial* $M(\lambda) = \sum_{j=0}^{2k} \lambda^j B_j$ *such that* $B_j^* = B_{2k-j}$ *and* $\|B_j - A_j\| < \delta$ *for* $j = 0, 1, \ldots, 2k$.

Proof. (iii) $\Rightarrow$ (i) is just Lemma 5.

(ii) $\Rightarrow$ (iii) follows from Lemma 6.

(i) $\Rightarrow$ (ii). By Lemma 3, statement (i) implies that $\sigma(C_L) \subseteq \mathbf{T}$ and all points of $\sigma(C_L)$ are of determinate type with respect to $\hat{B}_L$. Also, as in Remark 3, C_L is strongly stable in the sense that there is a number $\delta > 0$ such that, if G' is selfadjoint and invertible on $\mathcal{H}^{2k}$ and U' is a G'-unitary operator for which $\|U' - C_L\| + \|G' - B_L\| < \delta$ then U' is stable with respect to G'. Choose $G' = \hat{B}_M$ and $U' = C_M$ where $M(\lambda) = \sum_{j=0}^{2k} \lambda^j B_j$, $B_j^* = B_{2k-j}$ and $\|B_j - A_j\|$ is sufficiently small ($j = 0, 1, \ldots, 2k$). Then it is found that C_M is stable and all solutions of (5.3) are bounded. □

Note that the equivalence of statements (i) and (ii) is a generalization of Theorem III.2.6 of [7] from finite to infinite dimensional space $\mathcal{H}$.

The following analogue of Theorem 3 (part (b)) of [13] follows immediately from Theorems 2 and 3.

Theorem 5 *Let $L(\lambda)$ be a* QHP *on* $\mathbf{T}$. *Let operators $T_0, T_1, \ldots, T_{2k}$ on $\mathcal{H}$ be compact and satisfy $T_j^* = T_{2k-j}$, $j = 0, 1, \ldots, 2k$ with $A_{2k} + T_{2k}$ invertible. Then the operator polynomial*

$$M(\lambda) := L(\lambda) + \sum_{j=0}^{2k} \lambda^j T_j$$

has at most finitely many eigenvalues which are not on $\mathbf{T}$ *and points of $\sigma(U)$ on* $\mathbf{T}$ *of mixed type.*

Note that a natural extension of Remark 6 holds in this context as well.

References

[1] T. Ya. Azizov and I.S. Iokhvidov. *Linear Operators in Spaces with an Indefinite Metric.* Wiley, Chichester, 1989.

[2] J. Bognar. *Indefinite Inner Product Spaces.* Springer–Verlag, New York, 1974.

[3] L. Barkwell, P. Lancaster, A.S. Markus. Gyroscopically stabilized systems: a class of quadratic eigenvalue problems with real spectrum. *Canadian J. Math.* **44** (1992), 42–53.

[4] Ju. L. Daleckii and M.G. Krein. *Stability of Solutions of Differential Equations in Banach Space.* American Math. Soc., Providence, 1974. (Translations of Math. Monographs, Vol. **43**.)

[5] V.I. Derguzov. On the stability of the solutions of the Hamiltonian equations with unbounded periodic operator coefficients. *Mat. Sb.* **63** (1964), 591–619. (Russian.)

[6] V.I. Derguzov. Sufficient conditions for the stability of Hamiltonian equations with unbounded periodic coefficients, *Mat. Sb.* **64** (1964), 419–435. (Russian.)

[7] I. Gohberg, P. Lancaster, and L. Rodman. *Matrices and Indefinite Scalar Products.* Birkhäuser, Basel, 1983 (OT 8).

[8] I.S. Iokhvidov and M.G. Krein. Spectral theory of operators in spaces with an indefinite metric. *American Math. Soc. Translations* **34** (1969), 283–373. (Translation of Russian original of 1959.)

[9] P. Jonas. *Einige Betrachtungen zur Stabilität kanonischer Differentialgleichungen in Hilbertraum.* Diplomarbeit, Techn. Univ. Dresden, 1964.

[10] H. Langer. *Spektraltheorie linearer Operatoren in J-Räumen und einige Anwendungen auf die Schar $L(\lambda) = \lambda^2 I + \lambda B + C$.* Habilitationschrift, Tech. Univ. Dresden, 1964.

[11] H. Langer. *Spectral Functions of Definitizable Operators in Krein Spaces.* Springer–Verlag, Lecture Notes in Math., Vol. **948** (1982), 1–46.

[12] P. Lancaster, A.S. Markus, and V.I. Matsaev. Definitizable operators and quasihyperbolic operator polynomials. *J. Functional. Anal.* **131** (1995), 1–28.

[13] P. Lancaster, A. Markus, and V. Matsaev. *Perturbations of G-selfadjoint Operators and Operator Polynomials with Real Spectrum*, (the preceding paper in this volume).

[14] P. Lancaster, A. Shkalikov and Q. Ye. Strongly definitizable linear pencils in Hilbert space. *Integral Eq. and Operator Theory* **17** (1993), 338–360.

[15] R.S. Phillips. *The extension of dual subspaces invariant under an algebra.* Proc. Internat. Sympos. Linear Spaces (Jerusalem, 1960), Jerusalem Academic Press, Jerusalem; Pergamon Press, Oxford, 1961, pp. 366–398.

[16] L. Rodman. *An Introduction to Operator Polynomials.* Birkhaüser, Basel, 1989 (OT 38).

P. LANCASTER
Department of Mathematics and Statistics
University of Calgary
Calgary, Alberta *T2N 1N4*
Canada

A. MARKUS
Department of Mathematics and Computer Science
Ben Gurion University of the Negev
Beer Sheva
Israel

V. MATSAEV
School of Mathematical Sciences
Tel Aviv University
Ramat Aviv 69978
Israel

AMS Classification Numbers: 47A05, 47A55, 47A56

Operator Theory:
Advances and Applications, Vol. 87

SYSTEM THEORETIC ASPECTS OF COMPLETELY SYMMETRIC SYSTEMS

RAIMUND J. OBER

System theoretic aspects of completely symmetric systems will be discussed both for discrete time and continuous time systems. Realization theoretic results are presented. Necessary and sufficient conditions are given for the boundedness of the observability and reachability operators. The asymptotic, exponential/power stability of a completely symmetric system is characterized through the support of its defining measure. For continuous time systems the boundedness of the system operators is analyzed.

1 Introduction

In this paper we consider completely symmetric systems. Finite dimensional completely symmetric systems or relaxation systems have received a considerable amount of attention (see e.g. [17]). The primary aim of this paper is to examine this class of systems in the infinite dimensional case. In the Russian literature such systems have been investigated in their connection to operator nodes and operator extension problems ([4], [16]). Transfer functions of completely symmetric systems are Stieltjes functions for which there is a rich literature, see e.g. [1] for their role in operator theory and [7] for their role in the theory of differential equations. System theoretic investigations of subclasses of this class of systems can be found for example in [6], [3], [9].

In this paper we present a system theoretic study of this class of systems without additional assumptions such as the boundedness of the input and output operators for continuous time systems. We investigate both discrete time and continuous time systems. Particular emphasis is placed on the analysis of system theoretic properties through properties of the transfer function. The reachability and observability of the systems is characterized through the so-called defining measure of the system. It will be shown that a completely symmetric system has bounded reachability and observability operator if and only if the defining measure is a Carleson measure. The exponential stability of a completely symmetric system is characterized through the support of the defining measure. A realization result is also given.

Discrete time systems are analyzed first in Section 2. Continuous time systems are then investigated in the subsequent section. The bilinear transform that was studied in [10] will be used to translate several of the discrete time results to a continuous time setting.

1.1 Notation

The set of all real numbers is denoted by $\mathcal{R}$ and the set of all complex numbers is denoted by $\mathcal{C}$. If $A \subseteq \mathcal{R}$ then χ_A denotes the characteristic function of the set A, i.e. $\chi_A(\lambda) = 1$ for $\lambda \in A$ and $\chi_A(\lambda) = 0$ for $\lambda \in \mathcal{R} \setminus A$.

All Banach spaces considered in this paper are spaces over the complex field $\mathcal{C}$. Given a Hilbert space our convention is that the scalar product is linear in the first component and anti-linear in the second component. For H_1, H_2 Hilbert spaces, $L(H_1, H_2)$ denotes the space of bounded linear operators $T : H_1 \to H_2$. For an operator T on a Hilbert space the spectrum is denoted by $\sigma(T)$. The point spectrum is denoted by $\sigma_p(T)$. For an operator T the Hilbert space adjoint is denoted by T^*. The open unit disk is denoted by $\mathcal{D}$, i.e. $\mathcal{D} = \{z \in \mathcal{C} \mid |z| < 1\}$. The exterior of the closed unit disk is denoted by $\mathcal{D}_e$, i.e. $\mathcal{D}_e = \{z \in \mathcal{C} \mid |z| > 1\}$. We denote by RHP the open right half plane, i.e. $RHP = \{s \in \mathcal{C} \mid Re(s) > 0\}$.

For a measurable function $F : \Omega \to \mathcal{C}$ we say that the integral $\int_\Omega f d\nu$ exists if $\int_\Omega |f| d\nu < \infty$. For a regular positive Borel measure ν on a subset A of $\mathcal{R}$, the Hilbert space of functions on A that are square integrable with respect to ν is denoted by $L^2(A, \nu)$.

The Hardy space $H^\infty(RHP)$ is the Banach space of functions analytic in RHP and uniformly bounded in RHP with norm $\|f\|_\infty := \sup_{s \in RHP} |f(s)|$, for $f \in H^\infty(RHP)$. The Hardy space $H^2(RHP)$ is the Hilbert space of analytic functions in RHP, such that $\sup_{\substack{x \in \mathcal{R} \\ x>0}} \int_{\mathcal{R}} |f(x+iy)|^2 dy < \infty$, with norm $\|f\|_2 = \left(\sup_{\substack{x \in \mathcal{R} \\ x>0}} \int_{\mathcal{R}} |f(x+iy)|^2 dy\right)^{\frac{1}{2}}$, for $f \in H^2(RHP)$.

2 Discrete time systems

A quadruple (A_d, B_d, C_d, D_d) of operators is called a discrete-time system with input space U, output space Y and state space X with U, Y, X being separable Hilbert spaces if A_d is a contraction on X, $B_d \in L(U, X)$, $C_d \in L(X, Y)$, $D_d \in L(U, Y)$. The system is called admissible if $-1 \notin \sigma_p(A_d)$ and $\lim_{\lambda > 1, \lambda \to 1} C_d(\lambda I + A_d)^{-1} B_d$ exists in the norm topology. We denote by $D_X^{U,Y}$ the set of admissible systems with state space X, input space U and output space Y.

We now define what we mean by completely symmetric discrete time systems. In this paper we only consider single input single output systems, i.e. systems such that U and Y are one-dimensional.

Definition 2.1 *A discrete-time (admissible) single input single output system* (A_d, B_d, C_d, D_d) *is called* completely symmetric *if it coincides with its dual system, i.e. if*

$$A_d = A_d^*, \quad B_d = C_d^*, D_d = D_d^*,$$

and $\pm 1 \notin \sigma_p(A_d)$.

The following proposition gives a characterization of the transfer function of a completely symmetric system.

Proposition 2.1 *Let (A_d, B_d, C_d, D_d) be a completely symmetric discrete time system with transfer function $G_d(z) = C_d(zI - A_d)^{-1}B_d + D_d$, $z \in \mathcal{D}_e$. Set $G_d^{\perp}(z) := \frac{1}{z}[G_d(\frac{1}{z}) - G_d(\infty)]$, $z \in \mathcal{D}$. Then there exists a unique positive finite Borel measure ν on $[-1,1]$ such that for $z \in \mathcal{D}$,*

$$G_d^{\perp}(z) = \sum_{n=0}^{\infty} a_n z^n = \int_{[-1,1]} \frac{1}{1-tz} d\nu(t),$$

and for $z \in \mathcal{D}_e$,

$$G_d(z) = \int_{[-1,1]} \frac{1}{z-t} d\nu(t) + D_d$$

with $a_n := \int_{[-1,1]} t^n d\nu(t)$, $n \geq 0$.
Moreover,

1. $$\nu(\{-1,1\}) = 0,$$

2. $$\lim_{n\to\infty} a_n = 0.$$

In particular, if $A_d = \int_{[-1,1]} t dE(t)$, is the spectral decomposition of A_d with Borel σ-algebra Ω then ν is the Borel measure given by

$$\nu(\omega) = \langle E(\omega)B_d, B_d \rangle, \quad \omega \in \Omega.$$

Proof: We have for $z \in \mathcal{D}$,

$$\begin{aligned} G_d^{\perp}(z) &= \frac{1}{z}C_d(\frac{1}{z}I - A_d)^{-1}B_d = C_d(I - zA_d)^{-1}B_d \\ &= C_d \sum_{n=0}^{\infty}(zA_d)^n B_d = \sum_{n=0}^{\infty} C_d z^n A_d^n B_d = \sum_{n=0}^{\infty} B_d^* z^n A_d^n B_d \\ &= \sum_{n=0}^{\infty} B_d^* z^n \int_{[-1,1]} t^n dE(t) B_d = \sum_{n=0}^{\infty} z^n \int_{[-1,1]} t^n d \langle E(t)B_d, B_d \rangle \\ &= \sum_{n=0}^{\infty} z^n \int_{[-1,1]} t^n d\nu(t) = \int_{[-1,1]} \sum_{n=0}^{\infty} z^n t^n d\nu(t) \\ &= \int_{[-1,1]} \frac{1}{1-tz} d\nu(t), \end{aligned}$$

where $A_d = \int_{[-1,1]} t dE(t)$, is the spectral decomposition of A_d with Borel σ-algebra Ω and ν is the Borel measure given by $\nu(\omega) = \langle E(\omega)B_d, B_d \rangle$, $\omega \in \Omega$. This measure is finite and positive since for $\omega \in \Omega$

$$\nu(\omega) = \langle E(\omega)B_d, B_d \rangle = \langle E(\omega)B_d, E(\omega)B_d \rangle = \|E(\omega)B_d\|^2 \leq \|B_d\|^2$$

and

$$\nu(\omega) = \langle E(\omega)B_d, B_d \rangle \geq 0.$$

Assume that there is another positive finite regular Borel measure μ such that

$$a_n = \int_{[-1,1]} t^n d\nu(t) = \int_{[-1,1]} t^n d\mu(t), \qquad n \geq 0.$$

Let f be a continuous function on $[-1,1]$. By Weierstrass's theorem for $\epsilon > 0$ there exists a polynomial $p_n(t) = \sum_{k=0}^{n} \lambda_n t^n$ such that $\sup_{t\in[-1,1]} |f(t) - p_n(t)| < \epsilon$. Then since $\int_{[-1,1]} p_n d\mu = \int_{[-1,1]} p_n d\nu$,

$$\begin{aligned} |\int_{[-1,1]} f d\mu - \int_{[-1,1]} f d\nu| &= |\int_{[-1,1]} f d\mu - \int_{[-1,1]} p_n d\mu - (\int_{[-1,1]} f d\nu - \int_{[-1,1]} p_n d\nu)| \\ &\leq \int_{[-1,1]} |f - p_n| d\mu + \int_{[-1,1]} |f - p_n| d\nu \\ &\leq \epsilon\mu([-1,1]) + \epsilon\nu([-1,1]). \end{aligned}$$

Hence $\int_{[-1,1]} f d\mu = \int_{[-1,1]} f d\nu$ for all continuous functions f on $[-1,1]$. Therefore, by the Riesz representation theorem ([13], p. 40) $\mu = \nu$.

1.) Note that since by assumption $\pm 1 \notin \sigma_p(A_d)$, it follows ([14], Theorem 12.29) that $E(\{-1,+1\}) = 0$. Hence $\nu(\{-1,+1\}) =< E(\{-1,+1\}B_d, B_d >= 0$.

2.) Clearly, $|t^n| < 1$ and $\lim_{n\to\infty} t^n = 0$ for $t \in]-1,1[$. As $\nu(\{-1,1\}) = 0$ we have for $n \geq 0$ that

$$a_n = \int_{[-1,1]} t^n d\nu(t) = \int_{[-1,1]} t^n d\nu(t) = \int_{]-1,1[} t^n d\nu(t).$$

Since the measure ν is finite we have by the Lebesgue dominated convergence theorem that

$$a_n = \int_{]-1,1[} t^n d\nu(t) \to 0$$

as $n \to \infty$. □

Given a completely symmetric discrete-time system or its transfer function we call the measure ν constructed in the previous Proposition the *defining measure* of the system or transfer function.

We now show that functions with the above integral representation are analytic outside the support $supp(\nu)$ of the measure ν, where $supp(\nu)$ is the complement of the largest open set A with $\nu(A) = 0$.

Lemma 2.1 *Let ν be a finite positive regular Borel measure on $[-1,1]$ such that $\nu(\{-1,+1\}) = 0$. Then the function f given by*

$$z \mapsto f(z) := \int_{[-1,1]} \frac{1}{z-t} d\nu(t)$$

is analytic on $C \setminus supp(\nu)$.

A consequence of this Lemma is that if $G_d(z)$, $z \in \mathcal{D}_e$, is the transfer function of a completely symmetric system (A_d, B_d, C_d, D_d), G_d can be extended analytically to $C \setminus supp(\nu)$ where ν is the defining measure of the system. The continuation has the same integral representation

$$G_d(z) = \int_{[-1,1]} \frac{1}{z-t} d\nu(t) + D_d,$$

$z \in C \setminus supp(\nu)$.

2.1 Stability

A discrete-time system (A_d, B_d, C_d, D_d) is called *asymptotically stable* if $\lim_{n\to\infty} A_d^n x = 0$ for $x \in X$ and *power stable* if there exists $0 \le r < 1$ and $0 \le M < \infty$ such that $\|A_d^n\| \le Mr^n$ for $n = 0, 1, 2, \ldots$.

Lemma 2.2 *Let (A_d, B_d, C_d, D_d) be a completely symmetric discrete-time system. Then*

1. *(A_d, B_d, C_d, D_d) is asymptotically stable.*
2. *(A_d, B_d, C_d, D_d) is power stable if and only if $\sigma(A_d) \subseteq [-\alpha, \alpha]$ for some $0 \le \alpha < 1$.*

Proof: Let $A_d = \int_{[-1,1]} \lambda dE(\lambda)$ be the spectral representation of A_d. Note that as $\pm 1 \notin \sigma_p(A_d)$, we have that $E(\{-1, 1\}) = 0$. Let x be a vector in the state space X. Then for $n \ge 0$,

$$\|A_d^n x\|^2 = \int_{[-1,1]} |\lambda^n|^2 dE_{x,x}(\lambda) = \int_{]-1,1[} |\lambda^{2n}| dE_{x,x}(\lambda),$$

as $E(\{-1, 1\}) = 0$. Since $|\lambda^{2n}| \to 0$, as $n \to \infty$, it follows by the Lebesgue dominated convergence theorem that $\|A_d^n x\|^2 \to 0$. Hence the system is asymptotically stable.

It follows from the formula for the spectral radius, i.e. $\rho(A_d) = \inf_{n\ge1} \|A_d^n\|^{1/n}$ that the system is power stable if and only if the spectral radius is strictly less than 1. This is of course the case here if and only if $\sigma(A_d) \subseteq [-\alpha, \alpha]$ for some $0 \le \alpha < 1$. □

2.2 Observability and reachability

For a discrete-time system (A_d, B_d, C_d, D_d) the *observability operator* $\mathcal{O}_d$ is defined by $\mathcal{O}_d : D(\mathcal{O}_d) \to l_Y^2$; $x \mapsto (C_d A_d^n x)_{n\ge0}$, where $D(\mathcal{O}_d) = \{x \in X \mid (C_d A_d^n x)_{n\ge0} \in l_Y^2\}$. The system is said to have bounded observability operator if $D(\mathcal{O}_d) = X$ in which case $\mathcal{O}_d$ is a bounded operator. The system is called *observable* if it has a bounded observability operator with zero kernel. The *reachability operator* $\mathcal{R}_d$ of the system is defined by $\mathcal{R}_d : D(\mathcal{R}_d) \to X$; $\mathcal{R}_d((u_i)_{1\le i\le k}) = \sum_{i=0}^{\infty} A^i B u_i$, where $D(\mathcal{R}_d)$ is the set of finite sequences in l_U^2. The system is said to have bounded reachability operator if $\mathcal{R}_d$ extends to a bounded operator on l_U^2. If such an extension exists, the extension will also be called the reachability operator and will be also be denoted by $\mathcal{R}_d$. The system is called *reachable* if it has a bounded reachability operator with dense range.

For power stable systems it is easily seen that they have bounded observability respectively reachability operators.

Lemma 2.3 *A power stable completely symmetric discrete-time system has bounded reachability and observability operator.*

Proof: The result is easily verified. □

In order to give a characterization of the boundedness of the observability and reachability operators for general completely symmetric systems we need to introduce the notion of a Hankel operator. Let $h_n \in \mathcal{C}$ for $n = 0, 1, \ldots$ and consider the operator $H : l_U^2 \to l_Y^2$ given by the matrix $H = (h_{n+m})_{n,m\ge0}$.

Lemma 2.4 *Let (A_d, B_d, C_d, D_d) be a completely symmetric discrete-time system. Then $D(\mathcal{O}_d)$ is dense in X, $\mathcal{O}_d = \mathcal{R}_d^*$ and the following statements are equivalent.*

1. *The system has bounded reachability operator.*
2. *The system has bounded observability operator.*
3. *The Hankel operator H is bounded where H is given by the matrix $H = (C_d A_d^{i+j} B_d)_{i,j\geq 0}$.*

Moreover, the system is observable if and only if it is reachable.

Proof: The proof follows from duality arguments and the fact that $H = \mathcal{O}\mathcal{R}$. □

The following theorem is now an immediate consequence of Widom's theorem that characterizes the boundedness of positive Hankel operators (see [11]).

Theorem 2.1 *Let (A_d, B_d, C_d, D_d) be a completely symmetric discrete-time system with defining measure ν. The system has bounded reachability respectively observability operator if and only if ν is a Carleson measure on $[-1,1]$ which is the case if and only if*

$$\nu([\alpha,1]) + \nu([-1,-\alpha]) = O(1-\alpha)$$

as $\alpha \to 1$.

Proof: The theorem follows by combining the previous Lemma with Widom's theorem ([11], Theorem 1.6). □

We can now address the problem of the observability and reachability of completely symmetric discrete time systems.

Theorem 2.2 *Let (A_d, B_d, C_d, D_d) be a completely symmetric discrete-time system with bounded reachability and observability operator. Let $A_d = \int_{[-1,1]} \lambda dE(\lambda)$, be the spectral decomposition of A_d with spectral decomposition E defined on the Borel σ-algebra Ω on $[-1,1]$. Then the system is reachable/observable if and only if*

$$\cap_{\omega\in\Omega} \ker(C_d E(\omega)) = \{0\}.$$

Proof: We show that $\cap_{\omega\in\Omega} \ker(C_d E(\omega)) = ker(\mathcal{O}_d)$, where $\mathcal{O}_d$ is the observability operator of the system. Let $x \in ker(\mathcal{O}_d)$, then for each $n \geq 0$, $y \in Y$,

$$0 =< y, C_d A_d^n x >=< C_d^* y, A_d^n x >=< C_d^* y, \int_{[-1,1]} \lambda^n dE(\lambda) x >= \int_{[-1,1]} \lambda^n dE_{x,C_d^* y}(\lambda).$$

By Weierstrass's theorem and the Riesz representation theorem ([13], Theorem 6.19) this implies that the complex Borel measure $\omega \mapsto E_{x,C_d^* y}(\omega) =< C_d^* y, E(\omega)x >$ on $[-1,1]$ is the zero measure, i.e. $< y, C_d E(\omega)x >= 0$ for all $\omega \in \Omega$ and therefore $x \in \ker(C_d E(\omega))$ for all $\omega \in \Omega$. Hence $x \in \cap_{\omega\in\Omega} \ker(C_d E(\omega))$.

Let now $x \in \bigcap_{\omega\in\Omega} \ker(C_d E(\omega))$. Then for $y \in Y$, $\omega \in \Omega$,

$$E_{x,C_d^* y}(\omega) = < C_d^* y, E(\omega)x > = < y, C_d E(\omega)x > = 0.$$

Hence $E_{x,C_d^* y}(\omega) = 0$ for all $\omega \in \Omega$ and $n \geq 0$

$$0 = \int_{[-1,1]} \lambda^n dE_{x,C_d^* y}(\lambda) = < C_d^* y, \int_{[-1,1]} \lambda^n dE(\lambda)x > = < C_d^* y, A_d^n x > = < y, C_d A_d^n x > .$$

This implies that $C_d A_d^n x = 0$, $n \geq 0$, and therefore $\mathcal{O}_d x = 0$, i.e. $x \in Ker(\mathcal{O}_d)$. □

2.3 Realization theory

In Proposition 2.1 we showed that the transfer function of a discrete time completely symmetric system has a particular integral representation that is determined by the defining measure ν. The defining measure was shown to be a positive finite Borel measure on $[-1,1]$ such that $\nu(\{-1,1\}) = 0$. In the following realization result we are going to show that the converse is also true. Given a positive finite Borel measure ν on $[-1,1]$ such that $\nu(\{-1,1\}) = 0$ we establish the existence of a completely symmetric discrete time system whose defining measure is ν.

Theorem 2.3 *Let ν be a positive finite Borel measure on $[-1,1]$, such that $\nu(\{-1,1\}) = 0$. Let $c \in \mathcal{C}$ and let*

$$G_d(z) := \int_{[-1,1]} \frac{1}{z-t} d\nu(t) + c$$

for $z \in \mathcal{C} \setminus supp(\nu)$. Let $X = L^2([-1,1],\nu)$ and define

$$B_d : \mathcal{C} \to X, \qquad u \mapsto \chi_{[-1,1]} u;$$

$$A_d : X \to X, \qquad x \mapsto Mx;$$

$$C_d := B_d^*,$$

$$D_d := c,$$

where $(Mx)(t) = tx(t)$, $t \in [-1,1]$.
Then

1. *(A_d, B_d, C_d, D_d) is a completely symmetric discrete time system whose transfer function is G_d.*

2. *The system (A_d, B_d, C_d, D_d) has bounded reachability or observability operators if and only if ν is a Carleson measure.*

3. *If ν is a Carleson measure then the system is observable and reachable.*

Proof: 1.) Clearly B_d and A_d are bounded operators and A_d is self-adjoint. Since $\sigma(A_d) \subseteq [-1,1]$ and A_d is self-adjoint, A_d is a contraction. As $\nu(\{-1,1\}) = 0$, we have that $\sigma_p(A_d) \subseteq]-1,1[$. Hence the system has a transfer function G_d^1 which is analytic on $\mathcal{D}_e$, where for $z \in \mathcal{D}_e$, and $u, y \in \mathcal{C}$,

$$< y, G_d^1(z)u >=< y, (C_d(zI - A_d)^{-1}B_d + D_d)u >=< B_d y, (zI - A_d)^{-1}B_d u > + < y, D_d u >$$

$$=< \chi_{[-1,1]}y, (zI - M)^{-1}\chi_{[-1,1]}u > +yc\bar{u} = y\int_{[-1,1]} \frac{1}{z-t} d\nu(t)\bar{u} + yc\bar{u} = yG_d(z)\bar{u}.$$

Hence the system is a realization of G_d. Clearly the system is completely symmetric.
2.) This is a consequence of Theorem 2.1 .
3.) Let ν be a Carleson measure. The system is reachable and therefore also observable if $range(\mathcal{R})$ is dense in $L^2(\nu)$. Let $u = (u_0, u_1, \ldots, u_n, 0, 0, \ldots)$ then for $t \in [-1,1]$,

$$(\mathcal{R}u)(t) = (\sum_{i=0}^{n} A^i B u_i)(t) = (\sum_{i=0}^{n} M^i \chi_{[-1,1]} u_i)(t) = \sum_{i=0}^{n} t^i u_i.$$

Hence $range(\mathcal{R})$ is dense in $L^2(\nu)$ if the polynomial functions $(t^i)_{i\geq 0}$ span $L^2(\nu)$. But this is the case by ([13], p.69) and Weierstrass's theorem. That the system is observable follows by duality. □

An observable and reachable discrete-time system (A_d, B_d, C_d, D_d) with reachability operator $\mathcal{R}$ and observability operator $\mathcal{O}$ is called *par-balanced* if $\mathcal{O}^*\mathcal{O} = \mathcal{R}\mathcal{R}^*$. The duality properties of a completely symmetric observable and reachable system imply that such a system is par-balanced.

The following proposition is due to N. Young ([18]) and shows that a par-balanced realization is unique up to a unitary state-space transformation.

Lemma 2.5 *Let (A_d, B_d, C_d, D_d) be a reachable and observable par-balanced realization of a transfer function G. Then all reachable and observable par-balanced realizations of the transfer function G are given by $(UA_dU^*, UB_d, C_dU^*, D_d)$, where U is unitary.*

Hence we have the following Lemma.

Lemma 2.6 *Let G_d be the transfer function of a completely symmetric discrete time system. Then (A_d, B_d, C_d, D_d) is a completely symmetric realization of G_d if and only if (A_d, B_d, C_d, D_d) is a par-balanced realization of G_d.*

In the following Lemma the spectral minimality of a completely symmetric system is established.

Corollary 2.1 *Let ν be a positive finite measure on $[-1,1]$ such that $\nu(\{-1,1\}) = 0$ and assume that ν is a Carleson measure. Let*

$$G_d(z) = \int_{[-1,1]} \frac{1}{z-t} d\nu(t),$$

for $z \notin supp(\nu)$. If (A_d, B_d, C_d, D_d) is a par-balanced realization of G_d, then

$$\sigma(G_d) = \sigma(A_d) = supp(\nu),$$

where $\sigma(G_d)$ denotes the set of singularities of G_d, i.e. those points in the complex plane at which G_d has no analytic extension. Moreover, the spectrum of A_d has only simple multiplicity.

Proof: The realization of Theorem 2.3 is par-balanced. Since by Lemma 2.5 all par-balanced realizations are related to this realization by a unitary transformation, we can assume without loss of generality that (A_d, B_d, C_d, D_d) is the realization of Theorem 2.3. This realization is reachable, observable and A_d is self-adjoint. Therefore, it is spectrally minimal (see [3],[5]), i.e. $\sigma(G_d) = \sigma(A_d)$ and by ([12], p.229), $\sigma(A_d) = supp(\nu)$. Moreover, by ([12], p.232), A_d only has simple spectrum. □

In the following corollary the stability question is addressed again.

Corollary 2.2 *Let ν be a positive finite measure on $[-1, 1]$ such that $\nu(\{-1, 1\}) = 0$ and assume that ν is a Carleson measure. Let*

$$G_d(z) = \int_{[-1,1]} \frac{1}{z-t} d\nu(t),$$

for $z \notin supp(\nu)$. If (A_d, B_d, C_d, D_d) is a par-balanced realization of G_d, then the system is asymptotically stable. It is power stable if and only if

$$supp(\nu) \subseteq [-\alpha, \alpha]$$

for some $0 \leq \alpha < 1$.

Proof: This follows immediately from the previous corollary and Lemma 2.2. □

3 Continuous-time systems

In this section we will consider continuous time completely symmetric systems. To study these systems in appropriate generality we need to deal with systems with unbounded operators. Such systems are now defined.

If A is the generator of a strongly continuous semigroup of contractions on the Hilbert space X then $D(A)$ is a Hilbert space with inner product induced by the graph norm $||x||_A^2 := ||x||^2 + ||Ax||_X^2$, $x \in D(A)$. Denote by $D(A)^{(')}$ the Hilbert space of antilinear functionals on $(D(A), ||\cdot||_A)$ with norm $||f||' := \sup_{||x||_A \leq 1} |f(x)|$, $f \in D(A)^{(')}$. We then have the rigged structure

$$D(A) \subseteq X \subseteq D(A)^{(')}.$$

For the adjoint $(A^*, D(A^*))$ we have similarly $D(A^*) \subseteq X \subseteq D(A^*)^{(')}$. We can now define admissible continuous-time systems (see [10]).

Definition 3.1 *A quadruple of operators* (A_c, B_c, C_c, D_c) *is called an* admissible *continuous time system with state space* X*, input space* U *and output space* Y*, where* X*,* U*,* Y *are separable Hilbert spaces, if*

1. $(A_c, D(A_c))$ *is the generator of a strongly continuous semigroup of contractions on* X*.*
2. $B_c : U \to (D(A_c^*)^{(')}, \|\cdot\|')$ *is a bounded linear operator.*
3. $C_c : D(C_c) \to Y$ *is linear with* $D(C_c) = D(A_c) + (I - A_c)^{-1}B_cU$ *and* $C_{c|D(A_c)} : (D(A_c), \|\cdot\|_{A_c}) \to Y$ *is bounded.*
4. $C_c(I - A_c)^{-1}B_c \in L(U, Y)$*.*
5. A_c*,* B_c*,* C_c *are such that* $\lim_{\substack{s\in\Re \\ s\to+\infty}} C_c(sI - A_c)^{-1}B_c = 0$ *in the norm topology.*

We write $C_X^{U,Y}$ for the set of admissible continuous time systems with input space U, output space Y and state space X.

In order to define what we mean by a completely symmetric continuous time system we need to recall the definition of the dual of an admissible continuous time system (see [10]).

Definition 3.2 *Let* $(A_c, B_c, C_c, D_c) \in C_X^{U,Y}$*. Then the* dual system $(\tilde{A}_c, \tilde{B}_c, \tilde{C}_c, \tilde{D}_c)$ *of* $(A_c, B_c, C_c$ *is given by*

1. $(\tilde{A}_c, D(\tilde{A}_c)) := (A_c^*, D(A_c^*))$*, the adjoint operator of* $(A_c, D(A_c))$*.*
2. $\tilde{B}_c : Y \to D(A_c)^{(')}$*;* $y \mapsto \tilde{B}_c(y)[\cdot] :=< y, C_c(\cdot) >$*.*
3. $\tilde{C}_c : D(\tilde{C}_c) \to U$*,* $D(\tilde{C}_c) = D(\tilde{A}_c) + (I - \tilde{A}_c)^{-1}\tilde{B}_cY$*, where* $\tilde{C}_c x_0$ *is defined by*

$$< u, \tilde{C}_c x_0 >= B_c(u)[x_0]$$

for $x_0 \in D(A_c^*)$*,* $u \in U$*, and*

$$< \tilde{C}_c x_0, u >=< y_0, C_c(I - A_c)^{-1}B_c u >$$

for $x_0 = (I - \tilde{A}_c)^{-1}\tilde{B}_c y_0$*,* $y_0 \in Y$*,* $u \in U$*.*

4. $\tilde{D}_c := D_c^* : Y \to U$*.*

The dual system of an admissible system is admissible. If the continuous time transfer function $G(s) : RHP \to L(U, Y)$ has an admissible realization (A_c, B_c, C_c, D_c), then the dual system $(\tilde{A}_c, \tilde{B}_c, \tilde{C}_c, \tilde{D}_c)$ is a realization of the transfer function $\tilde{G}(s) := (G(\bar{s}))^*$, $s \in RHP$.

We now define a completely symmetric continuous-time system. As in the discrete-time case we restrict ourselves to systems with one dimensional input and output spaces.

Definition 3.3 *An admissible system* (A_c, B_c, C_c, D_c) *with one dimensional input and output space is called* completely symmetric *if*

$$A_c = \tilde{A}_c, \quad B_c = \tilde{C}_c, \quad D_c = \tilde{D}_c$$

and $0 \notin \sigma_p(A_c)$*, where* $(\tilde{A}_c, \tilde{B}_c, \tilde{C}_c, \tilde{D}_c)$ *is the dual system.*

Our method of analysis of continuous time completely symmetric systems is mainly based on relating these systems to discrete time completely symmetric systems. This will be done by the bilinear transform between continuous time and discrete time admissible systems. For a discussion of the background of this technique and the particular formulation which we will need see [10].

In the following theorem (see [10]) we introduce the map $T : D_X^{U,Y} \to C_X^{U,Y}$ that transforms discrete time systems to continuous time systems.

Theorem 3.1 *Let* $(A_d, B_d, C_d, D_d) \in D_X^{U,Y}$, *then* $T((A_d, B_d, C_d, D_d)) =: (A_c, B_c, C_c, D_c) \in C_X^{U,Y}$, *where the operators* A_c, B_c, C_c, D_c *are defined as follows:*

1. $A_c := (I + A_d)^{-1}(A_d - I) = (A_d - I)(I + A_d)^{-1}$, $D(A_c) := D((I + A_d)^{-1})$. *This operator generates a strongly continuous semigroup of contractions on* X.

2. *The operator* B_c *is given by:*

$$B_c := \sqrt{2}(I + A_d)^{-1}B_d : U \to D(A_c^*)^{(')};$$

$$u \mapsto \sqrt{2}(I + A_d)^{-1}B_d(u)[x] :=< B_d(u), (I + A_d^*)^{-1}(x) >_X .$$

3. *The operator* C_c *is given by:*

$$C_c : D(C_c) \to Y; \qquad x \mapsto \lim_{\substack{\lambda \to 1 \\ \lambda > 1}} \sqrt{2}C_d(\lambda I + A_d)^{-1}x,$$

where $D(C_c) = D(A_c) + (I - A_c)^{-1}B_cU$. *On* $D(A_c)$ *we have* $C_{c|D(A_c)} = \sqrt{2}C_d(I + A_d)^{-1}$.

4. $D_c := D_d - \lim_{\substack{\lambda \to 1 \\ \lambda > 1}} C_d(\lambda I + A_d)^{-1}B_d$.

Moreover, let the admissible discrete time system (A_d, B_d, C_d, D_d) *be a realization of the transfer function*

$$G_d(z) : \mathcal{D}_e \to L(U, Y),$$

i.e. $G_d(z) = C_d(zI - A_d)^{-1}B_d + D_d$ *for* $z \in \mathcal{D}_e$. *Then*

$$(A_c, B_c, C_c, D_c) = T((A_d, B_d, C_d, D_d))$$

is an admissible continuous time realization of the transfer function

$$G_c(s) := G_d\left(\frac{1+s}{1-s}\right) : RHP \to L(U, Y),$$

$s \in RHP$.

The inverse map is considered in the next theorem ([10]).

Theorem 3.2 *Let* $(A_c, B_c, C_c, D_c) \in C_X^{U,Y}$, *then* $T^{-1}((A_c, B_c, C_c, D_c)) := (A_d, B_d, C_d, D_d) \in D_X^{U,Y}$, *where the operators* A_d, B_d, C_d, D_d *are defined as*

1. $A_d := (I + A_c)(I - A_c)^{-1}$, *and for* $x \in D(A_c)$ *we have that* $A_d x = (I - A_c)^{-1}(I + A_c)x$.

2. $B_d := \sqrt{2}(I - A_c)^{-1}B_c$.

3. $C_d := \sqrt{2}C_c(I - A_c)^{-1}$.

4. $D_d := C_c(I - A_c)^{-1}B_c + D_c$.

Moreover, let the admissible continuous time system (A_c, B_c, C_c, D_c) *be a realization of the transfer function* $G_c : RHP \to L(U, Y)$, *i.e.* $G_c(s) = C_c(sI - A_c)^{-1}B_c + D_c$, $s \in RHP$. *Then*

$$(A_d, B_d, C_d, D_d) = T^{-1}((A_c, B_c, C_c, D_c))$$

is an admissible discrete time realization of the transfer function

$$G_d(z) := G_c\left(\frac{z-1}{z+1}\right), \qquad z \in \mathcal{D}_e.$$

The following Lemma shows that T maps completely symmetric discrete-time systems to completely symmetric continuous-time systems.

Lemma 3.1 *Let* (A_d, B_d, C_d, D_d) *be an admissible discrete-time system and let* $(A_c, B_c, C_c, D_c) := T((A_d, B_d, C_d, D_d))$. *Then* (A_c, B_c, C_c, D_c) *is completely symmetric if and only if* (A_d, B_d, C_d, D_d) *is completely symmetric.*

Proof: This follows immediately from the fact that the map T maps the dual system of (A_d, B_d, C_d, D_d) to the dual system of (A_c, B_c, C_c, D_c) (see [10]). Moreover, $+1 \notin \sigma_p(A_d)$ if and only if $0 \notin \sigma_p(A_c)$. Note that $-1 \notin \sigma_p(A_d)$ by the definition of admissibility. □

In order to be able to define the bilinear transform for a discrete-time completely symmetric system (A_d, B_d, C_d, D_d) the following admissibility condition (Section 2). has to be satisfied. It is required that the limit $\lim_{\substack{\lambda>1 \\ \lambda\to 1}} C_d(\lambda I + A_d)^{-1}B_d$ exists. If

$$G_d(z) = C_d(zI - A_d)^{-1}B_d + D_d = \int_{[-1,1]} \frac{1}{z-t} d\nu(t) + D_d,$$

$z \notin C \setminus supp(\nu)$, is the transfer function of the system this is equivalent to requiring that

$$\lim_{\substack{\lambda>1 \\ \lambda\to 1}} C_d(\lambda I + A_d)^{-1}B_d = \lim_{\substack{\lambda>1 \\ \lambda\to 1}} \int_{[-1,1]} \frac{1}{\lambda+t} d\nu(t)$$

exists.

The following Lemma gives a necessary and sufficient condition for a discrete time completely symmetric system to be admissible.

Lemma 3.2 *Let* (A_d, B_d, C_d, D_d) *be a completely symmetric discrete-time system with transfer function*

$$G_d(z) = C_d(zI - A_d)^{-1}B_d + D_d = \int_{[-1,1]} \frac{1}{z-t} d\nu(t) + D_d,$$

$z \notin C \setminus supp(\nu)$, where ν is the defining measure. Then the system is admissible, i.e. $\lim_{\substack{\lambda>1\\ \lambda\to 1}} C_d(\lambda I + A_d)^{-1}B_d$ exists if and only if the integral

$$\int_{[-1,1]} \frac{1}{1+t} d\nu(t)$$

exists. Moreover, if $\int_{[-1,1]} \frac{1}{1+t} d\nu(t)$ exists then

$$\lim_{\substack{\lambda>1\\ \lambda\to 1}} C_d(\lambda I + A_d)^{-1}B_d = \int_{[-1,1]} \frac{1}{1+t} d\nu(t).$$

Proof: Let for $\lambda \geq 1$, $t \in [-1,1]$

$$h_\lambda(t) = \frac{1}{\lambda + t}$$

Clearly $h_\lambda(t) > 0$ for $t \in [-1,1]$. Let $\lambda_1 > \lambda_2 > 1$. Then for $t \in [-1,1]$

$$h_{\lambda_1}(t) - h_{\lambda_2}(t) = \frac{1}{\lambda_1 + t} - \frac{1}{\lambda_2 + t} = \frac{\lambda_2 - \lambda_1}{(\lambda_1 + t)(\lambda_2 + t)} < 0.$$

Hence as $\lambda \to 1$, $\lambda > 1$, h_λ monotonically increases to h_1. Assume that

$$\int_{[-1,1]} \frac{1}{1+t} d\nu(t) = \int_{[-1,1]} h_1(t) d\nu(t)$$

exists. Then by Lebesgue's monotone convergence theorem

$$\lim_{\substack{\lambda>1\\ \lambda\to 1}} C_d(\lambda I + A_d)^{-1}B_d = \lim_{\substack{\lambda>1\\ \lambda\to 1}} \int_{[-1,1]} h_\lambda(t) d\nu(t) = \int_{[-1,1]} h_1(t) d\nu(t)$$

and the system is admissible.

Now assume that the system is admissible, i.e. $\lim_{\substack{\lambda>1\\ \lambda\to 1}} C_d(\lambda I + A_d)^{-1}B_d$ exists and is finite, then by Fatou's Lemma

$$0 \leq \int_{[-1,1]} \frac{1}{1+t} d\nu(t) = \int_{[-1,1]} h_1(t) d\nu(t) = \int_{[-1,1]} \liminf_{\substack{\lambda>1\\ \lambda\to 1}} h_\lambda(t) d\nu(t)$$

$$\leq \liminf_{\substack{\lambda>1\\ \lambda\to 1}} \int_{[-1,1]} h_\lambda(t) d\nu(t) = \lim_{\substack{\lambda>1\\ \lambda\to 1}} C_d(\lambda I + A_d)^{-1}B_d.$$

Therefore $\int_{[-1,1]} \frac{1}{1+t} d\nu(t)$ exists. This completes the proof. □

In the context of the boundedness condition of Theorem 2.1 the following result is of interest for admissible discrete time systems.

Lemma 3.3 *Let ν be a positive finite Borel measure on $[-1,1]$ such that $\nu(\{-1\}) = 0$ and such that $\int_{[-1,1]} \frac{1}{1+t} d\nu(t)$ exists. Then*

$$\nu([-1,-\alpha]) = O(1-\alpha)$$

as $\alpha \to 1$.

Proof: We have that

$$\int_{[-1,1]} \frac{1}{1+t} d\nu(t) \geq \int_{[-1,-\alpha]} \frac{1}{1+t} d\nu(t) \geq \frac{1}{1-\alpha} \nu([-1,-\alpha])$$

which implies the claim. □

In the following Lemma many of the technical details are worked out that are necessary to translate the results on the transfer functions of discrete time completely symmetric systems to the continuous time case.

Lemma 3.4 *Let ν be a finite positive Borel measure on $[-1,1]$, such that $\nu(\{-1,1\}) = 0$. Let*

$$\rho : [-1,1] \to [-\infty, 0]; \quad t \mapsto \frac{t-1}{t+1},$$

where we take $\rho(-1) = -\infty$. Then

$$\mu(A) := \int_A \frac{1}{2}(1-r)^2 d(\nu\rho^{-1})(r) = \int_{\rho^{-1}(A)} \frac{2}{(1+t)^2} d\nu(t)$$

for all Borel sets A in $[-\infty, 0]$, defines a, not necessarily finite, positive regular Borel measure on $[-\infty, 0]$, such that $\mu(\{-\infty, 0\}) = 0$. We therefore consider μ as a positive regular Borel measure on $]-\infty, 0]$.

Moreover,

1.
$$\nu(A) = \int_{\rho(A)} d(\nu\rho^{-1})(r) = \int_{\rho(A)} \frac{2}{(1-r)^2} d\mu(r),$$

for all Borel sets A in $[-1,1]$.

2. *for f a measurable function on $[-1,1]$, $\int_{[-1,1]} f(t)d\nu(t)$ exists if and only if $\int_{]\infty,0]}(f \circ \rho^{-1})(r)\frac{2}{(1-r)^2}d\mu(r)$ exists. If one of the integral exists, both integrals exist.*

   For g a measurable function on $]-\infty, 0]$, $\int_{]-\infty,0]} g(r)d\mu(r)$ exists if and only if $\int_{[-1,1]}(g \circ \rho)(t)\frac{2}{(1+t)^2}d\nu(t)$ exists. If one of the integrals exists, both integrals are equal.

3. *The map $V : L^2([-1,1],\nu) \to L^2(]\infty,0],\mu)$ with*

$$(V(f))(r) = \left(\frac{\sqrt{2}}{1-r} f\left(\frac{1+r}{1-r}\right)\right), \qquad -\infty < r \leq 0,$$

is unitary with inverse $V^{-1} : L^2(]-\infty,0],\mu) \to L^2([-1,1],\nu)$, where

$$(V^{-1}(g))(t) = \left(\frac{\sqrt{2}}{1+t} g\left(\frac{t-1}{t+1}\right)\right), \qquad -1 < t \leq 1,$$

and $(V^{-1}(g))(-1)$ arbitrary.

$$supp(\mu) = \rho(supp(\nu) \setminus \{-1\}),$$
$$supp(\nu) \setminus \{-1\} = \rho^{-1}(supp(\mu)).$$

5. *μ is such that $\int_{]-\infty,0]} \frac{1}{1-r} d\mu(r)$ exists if and only if ν is such that $\int_{[-1,1]} \frac{1}{1+t} d\nu(t)$ exists. If one of the integrals exists then both are equal.*

6. *if μ is such that $\int_{]-\infty,0]} \frac{1}{1-r} d\mu(r) < \infty$ then $\mu([r,0]) = O(r)$ as $r \to 0-$ if and only if $\nu([-1,-\alpha]) + \nu([\alpha,1]) = O(1-\alpha)$ as $\alpha \to 1$.*

7. *if μ is such that $\int_{]-\infty,0]} \frac{1}{1-r} d\mu(r)$ exists then*
$$G_c(s) = \int_{]-\infty,0]} \frac{1}{s-r} d\mu(r)$$
is an analytic function on $\mathcal{C} \setminus supp(\mu) = \mathcal{C} \setminus \rho^{-1}(supp(\nu) \setminus \{-1\})$.
Moreover, if
$$G_d(z) = \int_{[-1,1]} \frac{1}{z-t} d\nu(t) + \int_{[-1,1]} \frac{1}{1+t} d\nu(t)$$
for $z \in \mathcal{C} \setminus supp(\nu)$ then
$$G_c(s) = G_d\left(\frac{1+s}{1-s}\right),$$
for $s \in \mathcal{C} \setminus supp(\mu)$. Also
$$\lim_{\substack{s \to +\infty \\ s \in \Re}} \int_{]-\infty,0]} \frac{1}{s-r} d\mu(r) = 0.$$

Proof: Let $\tilde{\mu}$ be the finite positive regular Borel measure on $[-\infty, 0]$ defined by
$$\tilde{\mu}(A) := (\nu\rho^{-1})(A) := \nu(\rho^{-1}(A))$$
for each Borel set A in $[-\infty, 0]$. Note (see e.g. [2], Theorem 6.12, p. 213) that for each measurable function f on $[-1,1]$ we have that $\int_{[-1,1]} f(t) d\nu(t)$ exists if and only if $\int_{[-\infty,0]} (f \circ \rho^{-1})(r) d(\nu\rho^{-1})(r)$ exists. If one of the integrals exists both are equal. For the measure μ defined by
$$\mu(A) := \int_A \frac{1}{2}(1-r)^2 d(\nu\rho^{-1})(r) = \int_A \frac{1}{2}(1-r)^2 d\tilde{\mu}(r) = \int_{\rho^{-1}(A)} \frac{2}{(1+t)^2} d\nu(t)$$
we therefore have
$$\nu(A) = \int_{\rho(A)} d(\nu\rho^{-1})(r) = \int_{\rho(A)} \frac{2}{(1-r)^2} d\mu(r)$$
for each Borel set A in $[-1,1]$. We have used that if $r = \frac{t-1}{t+1}$, for $t \in [-1,1]$, then $t = \frac{1+r}{1-r}$ and $\frac{2}{(1+t)^2} = \frac{1}{2}(1-r)^2$. We have that
$$\mu(\{-\infty, 0\}) = \int_{\nu(\{-1,1\})} \frac{2}{(1+t)^2} d\nu(t) = 0,$$

since $\nu(\{-1,1\}) = 0$. Hence we can consider ν as a positive Borel measure on $]-\infty, 0]$. We have also shown 1.).
2.) Follows immediately from the proof of 1.).
3.) Follows immediately from 2.).
4.) This is verified easily.
5.) We have

$$\int_{]-\infty,0]} \frac{1}{1-r} d\mu(r) = \int_{]-\infty,0]} \frac{1}{2}(1-r)d(\nu\rho^{-1})(r) = \int_{[-1,1]} \frac{1}{2}(1-\frac{t-1}{t+1})d\nu(t)$$
$$= \int_{[-1,1]} \frac{1}{1+t} d\nu(t),$$

which implies the claim.
6.) Note that by Lemma 3.3 $\nu([-1,-\alpha]) = O(1-\alpha)$ as $\alpha \to 1$ since $\int_{]-\infty,0]} \frac{1}{1-r} d\mu(r) = \int_{[-1,1]} \frac{1}{1+t} d\nu(t)$ exists. For $0 < \alpha < 1$, we have that $-1 < \rho(\alpha) < 0$. Then

$$\nu([\alpha,1]) = \int_{\rho([\alpha,1])} \frac{2}{(1-r)^2} d\mu(r) = \int_{[\rho(\alpha),0]} \frac{2}{(1-r)^2} d\mu(r).$$

This identity implies that

$$\frac{1}{2}\mu([\rho(\alpha),0]) \le \frac{2}{(1-\rho(\alpha))^2}\mu([\rho(\alpha),0]) \le \int_{[\rho(\alpha),0]} \frac{2}{(1-r)^2} d\mu(r)$$
$$= \nu([\alpha,1]) \le \frac{2}{(1-0)^2}\mu([\rho(\alpha,0]) = 2\mu([\rho(\alpha),0]).$$

Since $0 \le -\rho(\alpha) \le \frac{-2\rho(\alpha)}{1-\rho(\alpha)} = 1-\alpha \le -2\rho(\alpha)$ we therefore have

$$\frac{\mu([\rho(\alpha),0])}{-4\rho(\alpha)} = \frac{1}{2}\frac{\mu([\rho(\alpha),0])}{-2\rho(\alpha)} \le \frac{\nu([\alpha,1])}{1-\alpha} \le 2\mu([\rho(\alpha),0]) \le -\rho(\alpha).$$

As $\alpha \to 1$ if and only if $\rho(\alpha) \to 0-$, these inequalities imply that $\mu([r,0]) = O(r)$ as $r \to 0-$ if and only if $\nu([\alpha,1[) = O(1-\alpha)$ as $\alpha \to 1$.
7.) Assume that the measure μ is such that $\int_{]-\infty,0]} \frac{1}{1-r} d\mu(r)$ exists. Then by 5.) $\int_{[-1,1]} \frac{1}{1+t} d\nu(t)$ exists. Hence by Lemma 2.1

$$G_d(z) = \int_{[-1,1]} \frac{1}{z-t} d\nu(t) + \int_{[-1,1]} \frac{1}{1+t} d\nu(t)$$

defines an analytic function on $\mathcal{C} \setminus supp(\nu)$. For $s \in \mathcal{C} \setminus supp(\mu)$ let

$$G_c(s) = G_d\left(\frac{1+s}{1-s}\right) = \int_{[-1,1]} \frac{1}{\frac{1+s}{1-s} - t} d\nu(t) + \int_{[-1,1]} \frac{1}{1+t} d\nu(t)$$
$$= \int_{[-1,1]} \frac{1}{s - \frac{t-1}{t+1}} \frac{2}{(1+t)^2} d\nu(t)$$
$$= \int_{]-\infty,0]} \frac{1}{s-r} \frac{1}{2}(1-r)^2 d(\nu\rho^{-1})(r)$$

$$= \int_{]-\infty,0]} \frac{1}{s-r} d\mu(r).$$

Note that for $s \in \Re$, $s > 0$, the function

$$f_s : \ ]-\infty, 0] \to \Re; \ \ r \mapsto \frac{1}{s-r}$$

is positive. Also for $s > 1$, $f_s \leq f_1$ and by assumption $f_1 = |f_1|$ is integrable. By Lebesgue's dominated convergence theorem we have as $\lim_{s\to\infty} f_s = 0$ that

$$\lim_{\substack{s\to\infty \\ s\in\Re}} \int_{]-\infty,0]} \frac{1}{s-r} d\mu(r) = \lim_{s\to\infty} \int_{]-\infty,0]} f_s(r) d\mu(r) = \int_{]-\infty,0]} \lim_{s\to\infty} f_s(r) d\mu(r)$$

$$= \int_{]-\infty,0]} 0 d\mu(r) = 0.$$

□

Remark 3.1 *In [11] a transform technique similar but not identical to the one in the previous Lemma was used to analyze unitarily equivalent Hankel operators.*

We now show that the transfer function of a completely symmetric continuous time system has an integral representation similar to discrete time completely symmetric systems.

Proposition 3.1 *Let (A_c, B_c, C_c, D_c) be a completely symmetric continuous time system with transfer function $G_c(s) = C_c(sI - A_c)^{-1}B_c + D_c$, $s \in RHP$. Then there exists a unique positive regular Borel measure μ on $]-\infty, 0]$ such that*

$$G_c(s) = \int_{]-\infty,0]} \frac{1}{s-r} d\mu(r) + D_c, \qquad s \in RHP.$$

Moreover,

1. *G_c can be extended analytically to $C \setminus supp(\mu)$ where the extension is given by*

$$G_c(s) = \int_{]-\infty,0]} \frac{1}{s-t} d\mu(t) + D_c, \qquad s \in C \setminus supp(\mu).$$

2. *the integrals*

$$\int_{]-\infty,0]} \frac{1}{1-r} d\mu(r)$$

and

$$\int_{]-\infty,0]} \frac{1}{(1-r)^2} d\mu(r)$$

exist.

Proof: Let $(A_d, B_d, C_d, D_d) = T^{-1}((A_c, B_c, C_c, D_c))$ be the corresponding discrete time admissible system. Since the bilinear transform preserves duality, the discrete time system is an admissible completely symmetric system. Let G_d be the transfer function of (A_d, B_d, C_d, D_d). By Proposition 2.1 there exists a unique positive finite Borel measure ν such that

$$G_d(z) = \int_{[-1,1]} \frac{1}{z-t} d\nu(t) + D_d, \qquad z \in \mathcal{D}_e.$$

Let G_c be the transfer function of the continuous time system. Then by Theorem 3.1

$$G_c(s) = G_d\left(\frac{1+s}{1-s}\right), \qquad s \in RHP.$$

Let μ be the positive Borel measure on $]-\infty, 0]$ constructed in Lemma 3.4. Since the discrete time system is admissible we have by Lemma 3.4 that $\int_{]-\infty,0]} \frac{1}{1-r} d\mu(r) = \int_{[-1,1]} \frac{1}{1+t} d\nu(t)$ exist. Using Lemma 3.4 part 7, for $s \in RHP$,

$$G_c(s) = \int_{[-1,1]} \frac{1}{\frac{1+s}{1-s} - t} d\nu(t) + D_d = \int_{]-\infty,0]} \frac{1}{s-r} d\mu(r) + D_d - \int_{[-1,1]} \frac{1}{1+t} d\nu(t)$$

Since $\lim_{\substack{s \in \mathcal{R} \\ s \to \infty}} \int_{]-\infty,0]} \frac{1}{s-r} d\mu(r) = 0$ we have that $D_c = D_d - \int_{[-1,1]} \frac{1}{1+t} d\nu(t)$. Hence

$$G_c(s) = \int_{]-\infty,0]} \frac{1}{s-r} d\mu(r) + D_c.$$

Also by Lemma 3.4 part 7 G_c is analytic on $\mathcal{C} \setminus supp(\mu)$. The uniqueness of μ follows from the fact that G_c and G_d are bilinearly related and that ν is unique. This shows 1.)
2.) That $\int_{]-\infty,0]} \frac{1}{1-r} d\mu(r)$ exists has already been established. To complete the proof note that

$$\int_{]-\infty,0]} \frac{1}{(1-r)^2} d\mu(r) = \int_{]-\infty,0]} \frac{1}{2} \frac{1}{(1-r)^2} (1-r)^2 d(\nu\rho^{-1})(r) = \int_{[-1,1]} \frac{1}{2} d\nu(t) < \infty.$$

□

As in the discrete-time case we refer to the measure μ as the *defining measure* of the continuous time completely symmetric system or its transfer function.

3.1 Stability

A continuous-time system (A_c, B_c, C_c, D_c) is *asymptotically stable* if $\lim_{t\to\infty} e^{tA_c}x = 0$ for all $x \in X$ and *exponentially stable* if there exists $0 \leq M < \infty$ and $\omega < 0$ such that $\|e^{tA_c}\| \leq Me^{\omega t}$ for all $t \geq 0$.

Proposition 3.2 *Let (A_c, B_c, C_c, D_c) be a completely symmetric continuous-time system. Then*

1. *(A_c, B_c, C_c, D_c) is asymptotically stable.*

2. (A_c, B_c, C_c, D_c) *is exponentially stable if and only if* $\sigma(A_c) \subseteq]-\infty, \beta]$ *for some* $\beta < 0$.

Proof: 1.) The asymptotic stability follows from the discrete-time result by applying the fact ([15]) that a semigroup is asymptotically stable if and only if the co-generator is also asymptotically stable.
2.) Let $(e^{tA_c})_{t\geq 0}$ be the semigroup of contractions with generator A_c. Let $A_c = \int_{]-\infty,0]} \lambda dE_c(\lambda)$ be the spectral decomposition of A_c. Then by the functional calculus for unbounded selfadjoint operators ([14]), for $t \geq 0$

$$e^{tA_c} = \int_{]-\infty,0]} e^{t\lambda} dE_c(\lambda).$$

It follows by ([8], Proposition A-III, 2.1) that the semigroup is exponentially stable if and only if $r(e^{A_c}) < 1$, where $r(T)$ is the spectral radius of the operator T. But by the spectral mapping theorem for selfadjoint operators

$$\sigma(e^{A_c}) = \overline{e^{\sigma(A_c)}}.$$

This implies that the semigroup is exponentially stable if and only if $\sigma(A_c) \subseteq]-\infty, \beta]$ for some $\beta < 0$. □

3.2 Observability and Reachability

The definition of observability and reachability of admissible continuous time systems is now given.

Definition 3.4 *Let* $(A_c, B_c, C_c, D_c) \in C_X^{U,Y}$, *then the operator*

$$\mathcal{O}_c : D(\mathcal{O}_c) \to L_Y^2([0,\infty[); x \mapsto (C_c e^{tA_c} x)_{t\geq 0}$$

is called the observability operator *of the system* (A_c, B_c, C_c, D_c), *where* $D(\mathcal{O}_c) =$

$$\{x \in X \mid C_c e^{tA_c} x \text{ exists for almost all } t \in [0,\infty[\text{ and } (C_c e^{tA_c} x)_{t\geq 0} \in L_Y^2([0,\infty[)\}.$$

We say that (A_c, B_c, C_c, D_c) *has a bounded observability operator if* $D(A_c) \subseteq D(\mathcal{O}_c)$ *and* $\mathcal{O}_c$ *extends to a bounded operator on* X. *This extension will also be denoted by* $\mathcal{O}_c$.

If (A_c, B_c, C_c, D_c) *has a bounded observability operator* $\mathcal{O}_c$ *such that* $ker(\mathcal{O}_c) = \{0\}$, *then the system* (A_c, B_c, C_c, D_c) *is called* observable.

Let $(\tilde{A}_c, \tilde{B}_c, \tilde{C}_c, \tilde{D}_c)$ *be the dual of* (A_c, B_c, C_c, D_c). *If the observability operator* $\tilde{\mathcal{O}}_c$ *of* $(\tilde{A}_c, \tilde{B}_c, \tilde{C}_c, \tilde{D}_c)$ *is a bounded operator on* X, *the adjoint of* $\tilde{\mathcal{O}}_c$ *is called the* reachability operator, *denoted* $\mathcal{R}_c$ *of* (A_c, B_c, C_c, D_c), *i.e.* $\mathcal{R}_c = \tilde{\mathcal{O}}_c^*$. *If* $\mathcal{R}_c$ *exists and* $range(\mathcal{R}_c)$ *is dense in* X, *the system* (A_c, B_c, C_c, D_c) *is said be* reachable.

We need to define Hankel operators for continuous time transfer functions.

Definition 3.5 *If G_c is a $L(U,Y)$ valued function analytic on RHP, then the operator*

$$H_{G_c,RHP} : D(H_{G_c,RHP}) \to H_Y^2(RHP); f \mapsto P_+ M_{G_c} Rf$$

where

$$Rf(s) = f(-s)$$

$$M_{G_c} \quad \text{multiplication operator by } G_c$$

$$P_+ \quad \text{projection on } H_Y^2(RHP)$$

with $D(H_{G_c,RHP}) = \{f \in H_U^2(RHP) : \ f \text{ rational, } G_c Rf \text{ has non-tangential limit a.e. on } i\Re \text{ that is in } L_Y^2(i\Re)\}$ *is called the* Hankel operator $H_{G_c,RHP}$ *with symbol G_c. If $H_{G_c,RHP}$ extends to a bounded operator on $H_U^2(RHP)$, this extension is also called the Hankel operator $H_{G_c,RHP}$.*

If it is clear from the context that the Hankel operator is defined with respect to RHP we will drop the subscript RHP and write H_G instead of $H_{G,RHP}$.

Lemma 3.5 *Let (A_c, B_c, C_c, D_c) be a completely symmetric continuous time system and let G_c be its transfer function. If $\mathcal{O}_c$ is the observability operator and $\mathcal{R}_c$ the reachability operator, then $D(\mathcal{O}_c)$ is dense in X, $\mathcal{O}_c = \mathcal{R}_c^*$ and the following statements are equivalent,*

1. *the system has bounded reachability operator.*
2. *the system has bounded observability operator.*
3. *the Hankel operator $H_{G_c,RHP}$ is bounded.*

Moreover, the system is observable if and only if it is reachable.

Proof: The proof follows from the discrete time result and the fact (Theorem 7.7 in [10]) that under the bilinear transform the discrete time observability (reachability/Hankel) operator and the continuous time observability (reachability/Hankel) operator are unitarily equivalent. □

We can now characterize the boundedness of the observability/reachability operator of a continuous time completely symmetric system.

Theorem 3.3 *Let (A_c, B_c, C_c, D_c) be a completely symmetric continuous time system with defining measure μ. The system has bounded reachability/observability operator if and only if*

$$\mu([r, 0]) = O(r)$$

as $r \to 0-$.

Proof: Let $(A_d, B_d, C_d, D_d) := T^{-1}((A_c, B_c, C_c, D_c))$ be the corresponding discrete time system with defining measure ν. By Proposition 3.1 $\int_{]-\infty,0]} \frac{1}{1-r} d\mu(r)$ exists. Hence by Lemma 3.4 part 6; $\nu([-1,-\alpha]) + \nu([\alpha,1]) = O(1-\alpha)$ as $\alpha \to 1$ if and only if $\mu([r,0]) = O(r)$ as $r \to 0-$. Since by Theorem 7.7 in [10] the observability operator of the continuous time and discrete time system are unitarily equivalent, the result now follows from the discrete time result (Theorem 2.1). □

We now establish a reachability/observability criterion for continuous time completely symmetric systems.

Corollary 3.1 *Let (A_c, B_c, C_c, D_c) be a completely symmetric continuous-time system with bounded reachability and observability operator. Let $A_c = \int_{-\infty}^{0} \lambda dE(\lambda)$ be the spectral decomposition of A_c, with spectral family E defined on the Borel σ-algebra on $]-\infty, 0]$. Then the system is observable/reachable if and only if*

$$\cap_{\omega\in\Omega}\{x \in D(A_c) \mid x \in ker(C_c E(\omega))\} = \{0\}.$$

Proof: Let $(A_d, B_d, C_d, D_d) := T^{-1}((A_c, B_c, C_c, D_c))$ be the corresponding discrete time system. The proof is based on the fact that (A_d, B_d, C_d, D_d) is observable if and only if (A_c, B_c, C_c, D_c) is observable ([10]). Also since A_d and A_c are related by a Cayley transformation, E is also the spectral family associated with the spectral decomposition of A_d. Note that by ([14], p.365)

$$E(\omega)(I + A_d)^{-1} \subseteq (I + A_d)^{-1} E(\omega)$$

for $\omega \in \Omega$. We first show that $\cap_{\omega\in\Omega}\{x \in D(A_c) \mid x \in ker(C_c E(\omega))\} \neq \{0\}$ implies that $\cap_{\omega\in\Omega} ker C_d E(\omega) \neq \{0\}$. Let $\omega \in \Omega$ and $x \in D(A_c) = D((I + A_d)^{-1})$ such that $C_c E(\omega)x = 0$. Let $y = (I + A_d)^{-1}x$. Then by the above

$$E(\omega)(I + A_d)^{-1}x = (I + A_d)^{-1}E(\omega)x$$

and therefore $E(\omega)x \in D((I + A_d)^{-1})$. Therefore

$$C_d E(\omega)y = C_d E(\omega)(I + A_d)^{-1}x = C_d(I + A_d)^{-1}E(\omega)x = \frac{1}{\sqrt{2}} C_c E(\omega)x = 0.$$

Hence $\cap_{\omega\in\Omega} ker C_d E(\omega) \neq \{0\}$. We now show the other implication. Let $\omega \in \Omega$ and $x \neq 0$ such that $C_d E(\omega)x = 0$. Set $y = (I + A_d)x$. Note that $y \neq 0$ since $-1 \notin \sigma_p(A_d)$. Then $E(\omega)x = E(\omega)(I+A_d)^{-1}y = (I+A_d)^{-1}E(\omega)y$ and therefore $E(\omega)y \in D((I+A_d)^{-1}) = D(A_c)$. Also

$$0 = C_d E(\omega)x = C_d E(\omega)(I + A_d)^{-1}y = C_d(I + A_d)^{-1}E(\omega)y = \frac{1}{\sqrt{2}} C_c E(\omega)y.$$

Hence $y \in \cap_{\omega\in\Omega}\{x \in D(A_c) \mid C_c E(\omega)x = 0\}$. □

3.3 Realization theory

To most efficiently study realization theory for continuous time completely symmetric systems we first apply the bilinear transform to the discrete time realization of Theorem 2.3

Lemma 3.6 *Let (A_d, B_d, C_d, D_d) be the discrete time realization of Theorem 2.3 of the transfer function*

$$G_d(z) = \int_{[-1,1]} \frac{1}{z-t} d\nu(t) + c, \qquad z \in C \setminus supp(\nu),$$

where ν is a positive finite Borel measure on $[-1,1]$ such that $\int_{[-1,1]} \frac{1}{1+t} d\nu(t) < \infty$ and c is a constant. Let

$$(A_c, B_c, C_c, D_c) := T((A_d, B_d, C_d, D_d))$$

be the corresponding continuous time system. Then the state space of the continuous time system is $X = L^2([-1,1], \nu)$. The operators of the continuous time system are given by

1.

$$A_c : D(A_c) \to X;$$

where

$$(A_c x)(t) = \begin{cases} \left(\frac{t-1}{t+1}\right) x(t), & -1 < t \leq 1 \\ 0, & t = -1 \end{cases}$$

and

$$D(A_c) = \{x \in L^2([-1,1], \nu) \mid \int_{[-1,1]} \frac{|x(t)|^2}{(1+t)^2} d\nu(t) < \infty\}.$$

2.

$$B_c : U \to (D(A_c))^{(')},$$

where for $u \in U$

$$B_c(u) : D(A_c) \to C; \quad x \mapsto \sqrt{2} \int_{[-1,1]} \frac{1}{1+t} \overline{x(t)} d\nu(t) u.$$

3. *For $x \in D(A_c)$,*

$$C_c x = \sqrt{2} \int_{[-1,1]} \frac{1}{1+t} x(t) d\nu(t).$$

If $x \in (I - A_c)^{-1} B_c u$, then $x = \chi_{[-1,1]} u$ for some $u \in U$, and

$$C_c x = \int_{[-1,1]} \frac{1}{1+t} d\nu(t) u.$$

4.

$$D_c = c - \int_{[-1,1]} \frac{1}{1+t} d\nu(t).$$

Proof: We use Theorem 2.3.
1.) We know that $D(A_c) = D((I + A_d)^{-1})$. But

$$D((I + A_d)^{-1}) = \{x \in L^2([-1,1],\nu) \mid \left(\frac{1}{1+t}x(t)\right)_{-1\leq t\leq 1} \in L^2([-1,1],\nu)\}.$$

For $x \in D(A_c)$

$$A_c x = (A_d - I)(I + A_d)^{-1}x = \left(\left(\frac{t-1}{t+1}\right)x(t)\right)_{-1\leq t\leq 1}.$$

2.) Since A_c is self adjoint, $(D(A_c^*))^{(\prime)} = (D(A_c))^{(\prime)}$. For $u \in U$, $x \in D(A_c)$

$$B_c(u)[x] = \sqrt{2} < B_d u, (I + A_d^*)^{-1}x >= \sqrt{2}\int_{[-1,1]} \chi_{[-1,1]}(t)u\frac{1}{1+t}\overline{x(t)}d\nu(t)$$

$$= \sqrt{2}\int_{[-1,1]} \frac{1}{1+t}\overline{x(t)}d\nu(t)u.$$

3.) For $x \in D(A_c)$,

$$C_c x = \sqrt{2}C_d(I + A_d)^{-1}x.$$

For $y \in Y$,

$$< C_c x, y >_Y = < \sqrt{2}C_d(I + A_d)^{-1}x, y >_Y = \sqrt{2} < (I + A_d)^{-1}x, C_d^* y >_Y$$

$$= \sqrt{2}\int_{[-1,1]} \frac{1}{1+t}x(t)\chi_{[-1,1]}(t)\bar{y}d\nu(t)$$

$$= \sqrt{2}\int_{[-1,1]} \frac{1}{1+t}x(t)d\nu(t)\bar{y}.$$

Hence

$$C_c x = \sqrt{2}\int_{[-1,1]} \frac{1}{1+t}x(t)d\nu(t).$$

If $x \in (I - A_c)^{-1}B_c U$, then $x = \frac{1}{\sqrt{2}}B_d u = \frac{1}{\sqrt{2}}\chi_{[-1,1]}u$ for some $u \in U$ (see p.448 in [10]). Hence

$$C_c x = \frac{1}{\sqrt{2}}C_c B_d u = \lim_{\substack{\lambda\to 1\\ \lambda>1}} C_d(\lambda I + A_d)^{-1}B_d u = \int_{[-1,1]} \frac{1}{1+t}d\nu(t)u$$

by the admissibility of (A_d, B_d, C_d, D_d) and Lemma 3.2.
4.) The expression for D_c follows since

$$D_c = D_d - \lim_{\substack{\lambda\to 1\\ \lambda>1}} C_d(\lambda I + A_d)^{-1}B_d = c - \int_{[-1,1]} \frac{1}{1+t}d\nu(t).$$

□

We are now in a position to prove the realization theorem for continuous time completely symmetric systems.

Theorem 3.4 *Let μ be a positive regular Borel measure on $]-\infty,0]$ such that $\mu(\{0\}) = 0$ and $\int_{]-\infty,0]} \frac{1}{1-r} d\mu(r) < \infty$ and let*

$$G_c(s) = \int_{]-\infty,0]} \frac{1}{s-r} d\mu(r)$$

for $s \in \mathcal{C} \setminus supp(\mu)$. Let $X = L^2(]-\infty,0],\mu)$. Define

1.

$$A_c : D(A_c) \to X; \quad x \mapsto (rx(r))_{-\infty<r\leq 0} ,$$

with $D(A_c) = \{x \in L^2(]-\infty,0],\mu) \mid \int_{]-\infty,0]} |x(r)r|^2 d\mu(r) < \infty\}$.

2.

$$B_c : U \to D(A_c)^{(')};$$

$$u \mapsto \left(x \mapsto \int_{]-\infty,0]} \overline{x(r)} d\mu(r) u \right).$$

3. *For $x \in D(A_c)$,*

$$C_c x = \int_{]-\infty,0]} x(r) d\mu(r).$$

For $x \in (I - A_c)^{-1} B_c U$ we have $x(r) = \frac{\sqrt{2}}{1-r} \chi_{]-\infty,0]}(r) u$, $\infty < r \leq 0$, for some $u \in U$, then set

$$C_c x = \int_{]-\infty,0]} \frac{1}{1-r} d\mu(r) u.$$

4. *$D_c = 0$.*

Then

1. *the system (A_c, B_c, C_c, D_c) is an admissible completely symmetric system with transfer function G_c.*
2. *the system (A_c, B_c, C_c, D_c) has bounded reachability and observability operator if and only if $\mu([r,0]) = O(r)$ as $r \to 0-$.*
3. *if the system (A_c, B_c, C_c, D_c) has bounded reachability and observability operator then the system is reachable and observable.*

Proof: Note that by Lemma 3.4 G_c is an analytic function on $\mathcal{C} \setminus supp(\mu)$ such that $\lim_{s\to\infty, s\in\Re} G_c(s) = 0$. Let ν be the measure on $[-1,1]$ as in Lemma 3.4 then by Lemma 3.4part 4

$$G_c(s) = G_d\left(\frac{1+s}{1-s}\right)$$

for $s \in \mathcal{C} \setminus supp(\mu)$, where

$$G_d(z) = \int_{[-1,1]} \frac{1}{z-t} d\nu(t) + \int_{[-1,1]} \frac{1}{1+t} d\nu(t),$$

$z \in C \setminus supp(\nu)$. Let (A_d, B_d, C_d, D_d) be the realization of G_d given in Theorem 2.3. Then by Lemma 3.6 and Theorem 3.1

$$(A_c', B_c', C_c', D_c') := T((A_d, B_d, C_d, D_d))$$

is an admissible completely symmetric continuous time realization of G_c with state space $L^2([-1,1],\nu)$. To obtain a realization with state space $L^2(]-\infty,0],\mu)$ we perform a state space transformation with the unitary operator V of Lemma 3.4part 3. We will show that the resulting system (A_c, B_c, C_c, D_c) is as defined in the statement of the Theorem.

By Lemma 3.6

$$D(A_c') = \{x \in L^2([-1,1],\nu) \mid \int_{[-1,1]} \frac{|x(t)|^2}{(1+t)^2} d\nu(t) < \infty\}.$$

Let $x \in D(A_c')$ then by Lemma 3.4

$$\int_{[-1,1]} \frac{|x(t)|^2}{(1+t)^2} d\nu(t) = \int_{]-\infty,0]} \frac{\left|x\left(\frac{1+r}{1-r}\right)\right|}{\left(1+\frac{1+r}{1-r}\right)^2} \frac{2}{(1-r)^2} d\mu(r)$$

$$= \int_{]-\infty,0]} \left|\frac{\sqrt{2}}{1-r} x\left(\frac{1+r}{1-r}\right)\right|^2 \frac{(1-r)^2}{2} d\mu(r) = \int_{]-\infty,0]} |(V(x))(r)|^2 \frac{(1-r)^2}{2} d\mu(r).$$

Hence

$$V(D(A_c')) = \{g \in L^2(]-\infty,0],\mu) \mid \int_{]-\infty,0]} \left|g(r)\frac{(1-r)^2}{\sqrt{2}}\right|^2 d\mu(r) < \infty\}$$

$$= \{g \in L^2(]-\infty,0],\mu) \mid \int_{]-\infty,0]} |g(r)r|^2 d\mu(r) < \infty\}.$$

To determine A_c let $g \in V(D(A_c))$. Then

$$(V^{-1}g)(t) = \frac{\sqrt{2}}{1+t} g\left(\frac{t-1}{t+1}\right)$$

and

$$(A_c V^{-1} g)(t) = \sqrt{2} \frac{t-1}{(t+1)^2} g\left(\frac{t-1}{t+1}\right),$$

$-1 < t \leq 1$. Hence

$$(V A_c V^{-1} g)(r) = \frac{\sqrt{2}}{1-r} \sqrt{2} \frac{\frac{1+r}{1-r} - 1}{\left(\frac{1+r}{1-r}+1\right)^2} g\left(\frac{\frac{1+r}{1-r}-1}{\frac{1+r}{1-r}+1}\right) = r g(r)$$

for $-\infty < r \leq 0$.

To determine B_c let $u \in U$, $x \in D(A_c')$, then

$$B_c'(u)[x] = \sqrt{2} \int_{[-1,1]} \frac{1}{1+t} \overline{x(t)} d\nu(t) u = \sqrt{2} \int_{]-\infty,0]} \frac{1}{1+\frac{1+r}{1-r}} \overline{x\left(\frac{1+r}{1-r}\right)} \frac{2}{(1-r)^2} d\mu(r) u$$

$$= \int_{]-\infty,0]} \overline{\frac{\sqrt{2}}{1-r} x\left(\frac{1+r}{1-r}\right)} d\mu(r) u$$
$$= \int_{]-\infty,0]} \overline{(Vx)(r)} d\mu(r) u.$$

Hence for $g \in D(A_c)$, $u \in U$,

$$B_c(u)[g] = \int_{]-\infty,0]} \overline{g(r)} d\mu(r) u.$$

To determine C_c recall that for $x \in D(A_c')$

$$C_c' x = \sqrt{2} \int_{[-1,1]} \frac{1}{1+t} x(t) d\nu(t) = \sqrt{2} \int_{]-\infty,0]} \frac{1}{1+\frac{1+r}{1-r}} x\left(\frac{1+r}{1-r}\right) \frac{2}{(1-r)^2} d\mu(r)$$
$$= \int_{]-\infty,0]} \frac{\sqrt{2}}{1-r} x\left(\frac{1+r}{1-r}\right) d\mu(r)$$
$$= \int_{]-\infty,0]} (Vx)(r) d\mu(r).$$

Hence for $g \in D(A_c)$,

$$C_c g = \int_{]-\infty,0]} g(r) d\mu(r).$$

If $x \in (I - A_c')^{-1} B_c' U$, then $x = \chi_{]-\infty,0]} u$ for some $u \in U$. Therefore

$$C_c' x = \int_{[-1,1]} \frac{1}{1+t} d\nu(t) u = \int_{]-\infty,0]} \frac{1}{1+\frac{1+r}{1-r}} \frac{2}{(1-r)^2} d\mu(r) u$$
$$= \int_{]-\infty,0]} \frac{1}{1-r} d\mu(r) u = \int_{]-\infty,0]} \frac{1}{\sqrt{2}} (Vx)(r) d\mu(r)$$
$$= C_c V(x).$$

Since by Lemma 3.4

$$\lim_{\substack{s \to \infty \\ s \in \Re}} G_c(s) = 0,$$

we have that $D_c = 0$.

Since (A_c', B_c', C_c', D_c') is a completely symmetric admissible realization of G_c. Since V is unitary (A_c, B_c, C_c, D_c) is a completely symmetric admissible system whose transfer function is G_c. This shows 1.)

2.) Follows from Theorem 3.3.

3.) The bilinear transform and unitary map V preserve the boundedness of the observability and reachability operators. They also preserve observability and reachability of the system ([10]). Since under the assumption (A_d, B_d, C_d, D_d) is reachable and observable this implies the reachability and observability of (A_c, B_c, C_c, D_c). □

An admissible observable and reachable continuous time system (A_c, B_c, C_c, D_c) is called par-balanced if $\mathcal{O}_c^* \mathcal{O}_c = \mathcal{R}_c \mathcal{R}_c^*$, where $\mathcal{O}_c$ is the observability operator and $\mathcal{R}_c$ is the reachability operator. As in the discrete time case the duality properties of a completely symmetric observable/reachable system imply that such a system is par-balanced. We have the following result on the uniqueness of completely symmetric or equivalently par-balanced realizations.

Lemma 3.7 *let G_c be the transfer function of a completely symmetric system. Then*

1. *(A_c, B_c, C_c, D_c) is an observable/reachable completely symmetric realization of G_c if and only if (A_c, B_c, C_c, D_c) is a par-balanced realization of G_c.*
2. *if (A_c, B_c, C_c, D_c) is an observable/reachable completely symmetric (par-balanced) realization of G_c, then all other observable/reachable completely symmetric (par-balanced) realization are given by $(UA_cU^*, UB_c, C_cU^*, D_c)$, where U is a unitary operator.*

Proof: The proof follows from Lemma 2.5 and Lemma 3.1 since the bilinear transform preserves complete symmetry, par-balancedness and unitary equivalence of systems ([10]). □

In the following corollary the boundedness of the input and output operators is investigated.

Corollary 3.2 *Let (A_c, B_c, C_c, D_c) be a completely symmetric reachable/observable continuous time system with defining measure μ. Then*

1. *the input operator $B_c : U \to X$ is bounded if and only if $C_c : X \to Y$ is bounded.*
2. *the input operator B_c /the output operator C_c is bounded if and only if μ is a finite measure.*
3. *if (A_c, B_c, C_c, D_c) is the realization given in Theorem 3.4 and if B_c is bounded then B_c can be represented as*

$$B_c : U \to X; \quad u \mapsto \chi_{]-\infty,0]}u.$$

Proof: By the previous Lemma we can assume that the system (A_c, B_c, C_c, D_c) is the realization given in Theorem 3.4.
1.) The statement is a consequence of the duality between B_c and C_c.
2.) Assume that μ is finite then using the Cauchy-Schwarz inequality, for $x \in D(A_c)$,

$$|C_c x| = |\int_{]-\infty,0]} x(t)d\mu(t)| \leq \left(\int_{]-\infty,0]} d\mu(t)\right)^{1/2} \|x\|_{L^2(]-\infty,0],\mu)}.$$

Hence C_c is bounded on $(D(A_c), \|\cdot\|_X)$. Since $D(A_c)$ is dense in $(X, \|\cdot\|_X)$, C_c can be extended to a bounded operator on X.

Assume that C_c extends to a bounded operator on $(X, \|\cdot\|_X)$ but μ is not finite. Then there exists $x \in X = L^2(]-\infty,0],\mu)$ such that $x \notin L^1(]-\infty,0],\mu)$. The integral representation of C_c implies that C_c is not bounded.
3.) If B_c acts as a bounded operator then for $u \in U$, $x \in D(A_c)$,

$$B_c(u)[x] = \int_{]-\infty,0]} \overline{x(t)}d\mu(r)u = < \chi_{]-\infty,0]}u, x >$$

which implies the claim. □

We can now establish the spectral minimality of observable/reachable completely symmetric systems.

Corollary 3.3 *Let μ be a positive regular Borel measure on $]-\infty, 0]$ such that $\int_{]-\infty,0]} \frac{1}{1-r} d\mu(r)$ exists and $\mu([r,0]) = O(r)$ as $r \to 0-$. Let*

$$G_c(s) = \int_{]-\infty,0]} \frac{1}{s-r} d\mu(r)$$

for $s \in C \setminus supp(\mu)$. If (A_c, B_c, C_c, D_c) is a par-balanced respectively observable/reachable completely symmetric realization of G_c, then

$$\sigma(G_c) = \sigma(A_c) = supp(\mu),$$

where $\sigma(G_c)$ denotes the set of singularities of G_c. Moreover, the spectrum of A_c has only simple multiplicity.

Proof: The follows from the discrete time result, the spectral mapping theorem for selfadjoint operators, Lemma 3.4 part 4 and Lemma 3.4 part 7. □

The stability properties of completely symmetric systems are now considered again.

Corollary 3.4 *Let (A_c, B_c, C_c, D_c) be a completely symmetric observable/reachable continuous time system with defining measure μ. Then*

1. *the system is asymptotically stable.*
2. *the system is exponentially stable if and only if*
$$supp(\mu) \subseteq]-\infty, -\alpha]$$
for some $\alpha > 0$.
3. *if the system is exponentially stable, the Hankel operator $H_{G_c,RHP}$ is compact, where G_c is the transfer function of the system.*

Proof: 1.) Follows from Proposition 3.2 .
2.) Follows from Proposition 3.2 and the spectral minimality of the realization (Corollary 3.3).
3.) Under this condition the transfer function is analytic and bounded in the right half plane and is continuous on the extended imaginary axis. The result therefore follows by Hartmann's theorem (see e.g. [11]). □

We now consider conditions for the boundedness of A_c.

Corollary 3.5 *Let (A_c, B_c, C_c, D_c) be a completely symmetric continuous time reachable/observab system with defining measure μ. Then*

1. *A_c is bounded if and only if*
$$supp(\mu) \subseteq [-\alpha, 0]$$
for some $\alpha > 0$.

2. if A_c is bounded then B_c and C_c is bounded.

Proof: Since the system is observable and reachable we have by Corollary 3.3 that $\sigma(\mu) = \sigma(A_c)$. If A_c is bounded then $\sigma(A_c)$ is compact. 1.) now follows since a selfadjoint operator is bounded if and only if the spectrum is bounded.
2.) Follows from 1.) and Corollary 3.2) part2. □

4 Acknowledgements

This research was supported by NSF grants: DMS-9304696, DMS-9501223.

References

[1] N. Akhiezer and I. Glazman. *Theory of linear operators in Hilbert space.* Dover, 1993.

[2] P. Billingsley. *Probability and Measure.* Wiley, 2nd edition, 1986.

[3] R. Brockett and P. Fuhrmann. Normal symmetric dynamical systems. *SIAM Journal Control and Optimization*, 14:107–119, 1976.

[4] M. Brodskii. *Triangular and Jordan Representations of Linear Operators*, volume 32. Translations of the American Mathematical Society, 1971.

[5] A. Feintuch. Spectral minimality for infinite dimensional linear systems. *SIAM Journal Control and Optimization*, 14(5):945–950, 1976.

[6] P. Fuhrmann. *Linear systems and operators in Hilbert space.* McGraw-Hill, 1981.

[7] I. Kac and M. Krein. On the spectral functions of the string. *American Mathematical Society Translations*, 103(2), 1974.

[8] R. Nagel. *One parameter semigroups of positive operators*, volume 1184 of *Lecture Notes in Mathematics.* Springer Verlag, 1986.

[9] R. J. Ober. A parametrization approach to infinite dimensional balanced systems and their approximation. *IMA Journal of Mathematical Control and Information*, 4(1):263–279, 1987.

[10] R. J. Ober and S. Montgomery-Smith. Bilinear transformation of infinite dimensional state-space systems and balanced realizations of nonrational transfer functions. *SIAM Journal Control and Optimization*, 28(2):438–465, 1990.

[11] S. Power. *Hankel operators on Hilbert space*, volume 64 of *Research Notes in Mathematics.* Pitman Advanced Publishing Program, 1982.

[12] M. Reed and B. Simon. *Methods of mathematical physics I: Functional Analysis.* Academic Press, 1980.

[13] W. Rudin. *Real and Complex Analysis.* McGraw Hill, 3rd edition, 1987.

[14] W. Rudin. *Functional Analysis.* McGraw Hill, 2nd edition, 1991.

[15] B. Sz-Nagy and C. Foias. *Harmonic Analysis of Operators on Hilbert space.* North Holland, Amsterdam, 1970.

[16] E. Tsekanovskii. Accretive extensions and problems on the Stieltjes operator-valued functions relations. In *Operator Theory: Advances and Applications*, volume 59. Birkäuser Verlag Basel, 1992.

[17] J. Willems. Realization of systems with internal passivity and symmetrical constraints. *Journal of the Franklin Institute*, 301:605–621, 1976.

[18] N. Young. Balanced realizations in infinite dimensions. In *Operator theory, Advances and Applications*, volume 19, pages 449–470. Birkhäuser Verlag, 1986.

Center for Engineering Mathematics EC 35
University of Texas at Dallas
Richardson, TX 75083
USA
email: ober@utdallas.edu

93B15, 93B20, 93B28, 47N70, 93D20

Operator Theory:
Advances and Applications, Vol. 87

CONTRACTIVE COMPLETION OF BLOCK MATRICES AND ITS APPLICATION TO $\mathcal{H}_\infty$ CONTROL OF PERIODIC SYSTEMS *

Li Qiu Tongwen Chen

Design of $\mathcal{H}_\infty$-optimal controllers for discrete-time periodic systems requires proper handling of a causality constraint, which in turn is related to factorization and contractive completion problems associated with block lower-triangular matrices. For a given block upper-triangular matrix, this paper gives a parametrization of all possible contractive completions. The unique contractive completion which minimizes an entropy function is also given. This is then applied to $\mathcal{H}_\infty$ control of periodic systems: We explicitly characterize the set of all periodic, causal controllers which achieve a certain closed-loop $\mathcal{H}_\infty$ norm bound and also give the unique controller which further minimizes a linear-exponential-quadratic-Gaussian cost functional.

1 Introduction

$\mathcal{H}_\infty$-optimal control of linear time-invariant (LTI) systems has been thoroughly studied; its importance in robust control is widely recognized, see the books [8, 12, 24] and the references therein.

$\mathcal{H}_\infty$-optimal control of discrete-time linear *periodic* systems was first studied in [7, 10] in the one-block case and was later extended to the general case in [22]. The common technique used in the study of periodic systems is lifting [17], which amounts to extending input and output spaces of periodic systems and obtaining equivalent LTI systems. This process of lifting has the norm-preserving property and therefore allows an equivalent $\mathcal{H}_\infty$ control problem to be posed for the lifted LTI systems. However, lifting also introduces a

*This research was supported by the Hong Kong Research Grants Council and the Natural Sciences and Engineering Research Council of Canada.

design constraint due to causality requirement of the controllers; this causality constraint requires that the feedthrough terms in the lifted controllers be block lower-triangular.

In the general case in [22], this constraint was treated using a convex search over a finite-dimensional space. However, it is possible to give explicit solutions to the $\mathcal{H}_\infty$ design problem using factorization involving block triangular matrices, as is discussed in [5] for multirate sampled-data systems. In [22, 5], only *one* solution was computed for the associated $\mathcal{H}_\infty$ problems. In this paper we shall characterize all possible solutions achieving a certain $\mathcal{H}_\infty$ performance and also give the unique solution which further minimizes an auxiliary cost functional. Similar techniques have also been applied to multirate sample-data systems [21].

The new result for $\mathcal{H}_\infty$ periodic control is based on the study of completing partial (block upper-triangular) matrices to contractions, or equivalently, the matrix distance problem stated as follows: Given a full matrix and a certain associated block lower-triangular structure, find all possible block lower-triangular matrices which are within a pre-specified distance, measured by the spectral norm, from the given matrix. Such problems were studied before [3, 4, 23]: In [3, 4] the solutions are derived based on J-spectral factorizations using operator theory and the Krein-space geometry; in [23] the solution may be considered as a finite-dimensional analogue of Schur's algorithm and is derived by using elementary matrix algebra. In this paper, we shall present another solution which keeps the flavor of J-spectral factorization of [3, 4] but uses only elementary linear algebra.

J-spectral factorization roots deeply in $\mathcal{H}_\infty$ control theory, which was studied before primarily via matrix interpolations [8, 11]. For an overview on the role of J-spectral factorization in matrix interpolations, see [2].

We remark that it is possible to develop a complete theory for the $\mathcal{H}_\infty$ control problem with the causality constraint if, as suggested in [2, Section 3] and [13, Chapter 7], one redefines the $\mathcal{H}_\infty$ space to be the set of bounded analytic matrices on the unit disk which are block lower-triangular when evaluated at the origin. However, we feel that our approach in this paper is advantageous because it connects the existing standard discrete time $\mathcal{H}_\infty$ solution in [14] with the solution to the matrix completion problem and the new results are obtained with relatively less effort.

The organization of this paper is as follows. In the next section, we state precisely our $\mathcal{H}_\infty$ control problem for periodic systems, convert the problem via lifting into an equivalent problem for LTI systems with a causality constraint on the controllers, and relate the causality condition to a certain block lower-triangular structure.

In Section 3, we solve the matrix distance problem stated earlier using elementary linear algebra. The proofs of the results are also given.

In Section 4, the results in Section 3 are applied to our $\mathcal{H}_\infty$ control problem for periodic systems. For a given $\mathcal{H}_\infty$ norm bound (normalized to 1), we characterize the set of all periodic controllers satisfying the causality constraint and achieving the $\mathcal{H}_\infty$ norm bound for the closed-loop system. Furthermore, we give the controller which minimizes a linear-exponential-quadratic-Gaussian functional, or equivalently, an entropy cost.

Section 5 offers some concluding remarks.

Now we introduce some notation. Given an operator K and two operator matrices

$$P = \begin{bmatrix} P_{11} & P_{12} \\ P_{21} & P_{22} \end{bmatrix}, \quad Q = \begin{bmatrix} Q_{11} & Q_{12} \\ Q_{21} & Q_{22} \end{bmatrix},$$

the linear fractional transformation associated with P and K is denoted

$$\mathcal{F}(P,K) = P_{11} + P_{12}K(I - P_{22}K)^{-1}P_{21}$$

and the star product of P and Q is

$$P \star Q = \begin{bmatrix} P_{11} + P_{12}Q_{11}(I - P_{22}Q_{11})^{-1}P_{21} & P_{12}(I - Q_{11}P_{22})^{-1}Q_{12} \\ Q_{21}(I - P_{22}Q_{11})^{-1}P_{21} & Q_{21}(I - P_{22}Q_{11})^{-1}P_{22}Q_{12} + Q_{22} \end{bmatrix}.$$

Here, we assume that the domains and co-domains of the operators are compatible and the inverses exist. With these definitions, we have

$$\mathcal{F}(P, \mathcal{F}(Q,K)) = \mathcal{F}(P \star Q, K).$$

2 $\mathcal{H}_\infty$ Periodic Control and Lifting

Let ℓ be the space of discrete-time signals, possibly vector-valued, defined on the time set $\{0,1,2,\cdots\}$. Let U be the unit delay operator on ℓ and U^* the unit advance operator. For a positive integer l, a linear, causal discrete-time system $\tilde{G}$ is *l-periodic* if $(U^*)^l\tilde{G}U^l = \tilde{G}$. A 1-periodic system is normally known as time-invariant.

A linear l-periodic system can be viewed as an LTI system via lifting [17]. Define the *lifting operator* L_l via $v = L_l\tilde{v}$:

$$\{v(0), v(1), \cdots\} \mapsto \left\{ \begin{bmatrix} \tilde{v}(0) \\ \tilde{v}(1) \\ \vdots \\ \tilde{v}(l-1) \end{bmatrix}, \begin{bmatrix} \tilde{v}(l) \\ \tilde{v}(l+1) \\ \vdots \\ \tilde{v}(2l-1) \end{bmatrix}, \cdots \right\}. \quad (1)$$

L_n maps ℓ to ℓ^l, the external direct sum of l copies of ℓ. The inverse L_l^{-1}, mapping ℓ^l to ℓ, amounts to reversing the operation in (1). The lifted system is defined as

$$G = L_l\tilde{G}L_l^{-1}.$$

It is a fact that G is LTI iff $\tilde{G}$ is linear and l-periodic. Moreover, since the lifting preserves the norm on ℓ_2, G is ℓ_2-bounded iff $\tilde{G}$ is and in this case they have the same ℓ_2-induced norm: $\|G\| = \|\tilde{G}\|$.

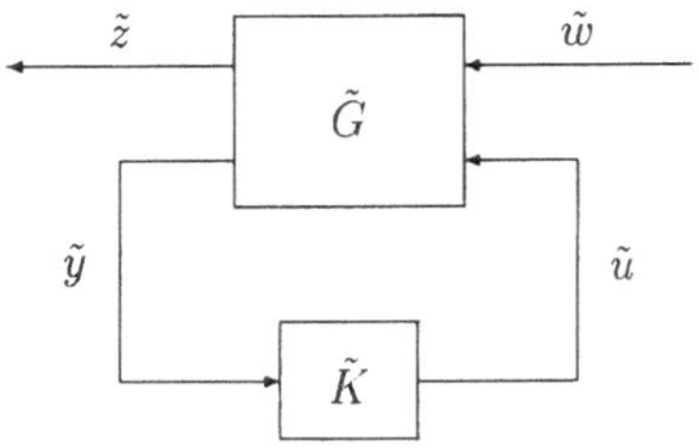

Figure 1: The discrete-time periodic system

The linear periodic control system to be studied is shown in Figure 1. Here $\tilde{G}$, the system to be controlled, is l-periodic and causal with two input vectors, the exogenous input $\tilde{w}$ and the control input $\tilde{u}$, and two output vectors, the output to be controlled $\tilde{z}$ and the measured output $\tilde{y}$; $\tilde{K}$, which processes $\tilde{y}$ and generates $\tilde{u}$, is the controller to be designed. Since $\tilde{G}$ is l-periodic, we shall require that $\tilde{K}$ be l-periodic and causal. (The causality is for implementability of the controller.) We shall be interested in only finite-dimensional $\tilde{G}$ and $\tilde{K}$, i.e., those $\tilde{G}$ and $\tilde{K}$ which have finite-dimensional state space realizations. Let us take any minimal state space realization of $\tilde{G}$ and $\tilde{K}$ in Figure 1; the closed system is said to be internally stable if the state vectors of $\tilde{G}$ and $\tilde{K}$ tend to zero from every initial condition.

The $\mathcal{H}_\infty$ control problem is as follows: Given $\tilde{G}$, design $\tilde{K}$ so that the closed-loop system is internally stable and the map $\tilde{w} \mapsto \tilde{z}$, denoted $T_{\tilde{z}\tilde{w}}$, has ℓ_2-induced norm less than a pre-specified number γ, or, $\|T_{\tilde{z}\tilde{w}}\| < \gamma$. By normalization, we can take $\gamma = 1$. Clearly, the solutions to this $\mathcal{H}_\infty$ control problem, if they exist, are not unique. We first seek a characterization of all solutions, and then find among those solutions the unique one which minimizes the following linear-exponential-quadratic-Gaussian (LEQG) cost:

$$\Omega(T_{\tilde{z}\tilde{w}}) = \lim_{T\to\infty} \frac{2}{T} \ln \mathbf{E}\left\{\exp\left[\frac{1}{2}\sum_{k=0}^{T-1} \tilde{z}'(k)\tilde{z}(k)\right]\right\},$$

where $\tilde{z}$ is the response of the closed-loop system when the input $\tilde{w}$ is a Gaussian white noise with zero mean and unit covariance, and $\mathbf{E}$ is the expectation operator.

Previous work on such an $\mathcal{H}_\infty$ problem are [7, 10], which studied a special case (the one-block problem), and [22]; none of them contained the characterization of all possible solutions or derived the particular solution minimizing the LEQG cost.

Now we lift the system in Figure 1 to get Figure 2, where all signals are lifted, e.g., $w = L_n\tilde{w}$, and the two lifted systems

$$G = L_l\tilde{G}L_l^{-1}, \quad K = L_l\tilde{K}L_l^{-1},$$

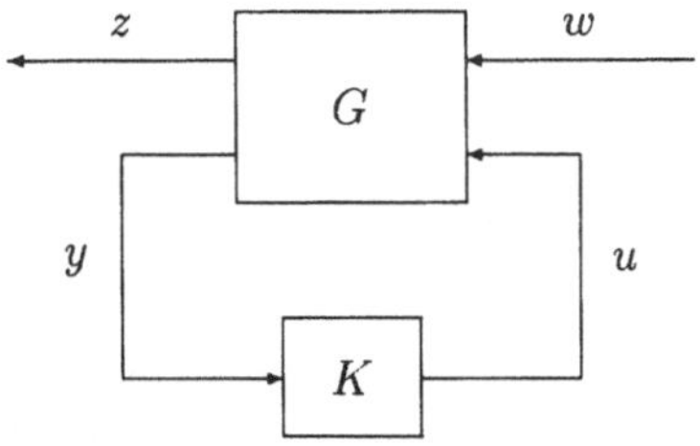

Figure 2: The lifted LTI system

are both LTI and causal with transfer functions

$$\hat{G}(\lambda) = \begin{bmatrix} \hat{G}_{11}(\lambda) & \hat{G}_{12}(\lambda) \\ \hat{G}_{21}(\lambda) & \hat{G}_{22}(\lambda) \end{bmatrix}, \quad \hat{K}(\lambda)$$

respectively. Here $\hat{G}(\lambda)$ is partitioned according to the dimensions of its two inputs and two outputs. (In the transfer functions, we used λ-transforms instead of the more traditional z-transforms, where $\lambda = z^{-1}$.)

To proceed, we need to introduce some notation. Let F be either $\mathbf{R}$ or $\mathbf{C}$. Define the set of block matrices:

$$\mathcal{M}(\mathsf{F}^{m\times n}) := \left\{ \begin{bmatrix} M_{11} & \cdots & M_{1l} \\ \vdots & & \vdots \\ M_{l1} & \cdots & M_{ll} \end{bmatrix} : \; M_{ij} \in \mathsf{F}^{m\times n} \right\}.$$

The integer l is not reflected in the notation since we will assume it is fixed. The block lower-triangular subset of $\mathcal{M}(\mathsf{F}^{m\times n})$, denoted by $\mathcal{T}(\mathsf{F}^{m\times n})$, consists of all matrices with $M_{ij} = 0$, $i < j$, and the strictly block lower-triangular subset, $\mathcal{T}_s(\mathsf{F}^{m\times n})$, consists of matrices with $M_{ij} = 0$, $i \leq j$.

Let the dimensions of $\tilde{w}$, $\tilde{z}$, $\tilde{u}$, $\tilde{y}$ be p, q, m, n, respectively. Due to causality of $\tilde{G}$ and $\tilde{K}$, the transfer functions of the lifted systems satisfy $\hat{G}_{11}(0) \in \mathcal{T}(\mathbf{R}^{q\times p})$, $\hat{G}_{12}(0) \in \mathcal{T}(\mathbf{R}^{q\times m})$, $\hat{G}_{21}(0) \in \mathcal{T}(\mathbf{R}^{n\times p})$, $\hat{G}_{22}(0) \in \mathcal{T}(\mathbf{R}^{n\times m})$, and $\hat{K}(0) \in \mathcal{T}(\mathbf{R}^{m\times n})$. It follows from [17] that each LTI causal nl-input ml-output controller K satisfying $\hat{K}(0) \in \mathcal{T}(\mathbf{R}^{m\times n})$ corresponds to an l-periodic, causal, n-input m-output controller.

Letting T_{zw} be the closed-loop map $w \mapsto z$ in Figure 2, we have

$$\|T_{\tilde{z}\tilde{w}}\| = \|T_{zw}\| = \|\hat{T}_{zw}\|_\infty,$$

the last quantity being the $\mathcal{H}_\infty$ norm of the transfer function for T_{zw}. By using the techniques in [9], we can show that when $\|\hat{T}_{zw}\|_\infty < 1$,

$$\Omega(T_{\tilde{z}\tilde{w}}) = \frac{1}{l}\mathcal{I}(\hat{T}_{zw}),$$

where $\mathcal{I}(\hat{T}_{zw})$, the entropy of $\hat{T}_{zw}$, is defined as

$$\mathcal{I}(\hat{T}_{zw}) = -\frac{1}{2\pi}\int_{-\pi}^{\pi} \ln\det\left[I - \hat{T}_{zw}^{*}(e^{j\omega})\hat{T}_{zw}(e^{j\omega})\right] d\omega.$$

Hence the equivalent LTI $\mathcal{H}_\infty$ problem is: Given the LTI system G from the lifting of $\tilde{G}$, characterize all LTI and causal controllers K satisfying $\hat{K}(0) \in \mathcal{T}(\mathsf{R}^{m\times n})$ such that the closed-loop system is internally stable and $\|\hat{T}_{zw}\|_\infty < 1$; furthermore, find the unique such controller which minimizes $\mathcal{I}(\hat{T}_{zw})$.

One cannot apply standard $\mathcal{H}_\infty$ techniques [14, 18, 15, 16] to this problem directly because the causality constraint on $\hat{K}(0)$: $\hat{K}(0) \in \mathcal{T}(\mathsf{R}^{m\times n})$. How to handle this constraint is the main concern of this paper.

3 Matrix Contractive Completion

In this section we consider the following matrix completion problem: Given $M \in \mathcal{M}(\mathsf{F}^{m\times n})$, characterize all $T \in \mathcal{T}(\mathsf{F}^{m\times n})$ such that $\|M+T\| < 1$; and then find the one, among those T characterized, which minimizes

$$\mathcal{I}(M+T) := -\ln\det[I - (M+T)^*(M+T)].$$

We shall need the following notation: For a block matrix

$$M = \begin{bmatrix} M_{11} & \cdots & M_{1q} \\ \vdots & & \vdots \\ M_{p1} & \cdots & M_{pq} \end{bmatrix}$$

and integers i_1, i_2, j_1, j_2 with $1 \le i_1 \le i_2 \le p$ and $1 \le j_1 \le j_2 \le q$, we write the submatrix

$$[M_{ij}]_{i=i_1,j=j_1}^{i_2,j_2} := \begin{bmatrix} M_{i_1j_1} & \cdots & M_{i_1j_2} \\ \vdots & & \vdots \\ M_{i_2j_1} & \cdots & M_{i_2j_2} \end{bmatrix}.$$

The following theorem gives several equivalent conditions for the solvability of the matrix completion problem.

Theorem 1 *Let $M \in \mathcal{M}(\mathsf{F}^{m\times n})$. The following statements are equivalent:*

(a) For $k = 1, 2, \ldots, l-1$, $\|[M_{ij}]_{i=1,j=k+1}^{k,l}\| < 1$.

(b) There exists $T \in \mathcal{T}(\mathsf{F}^{m\times n})$ such that $\|M+T\| < 1$.

(c) *There exists*

$$W = \begin{bmatrix} W_{11} & W_{12} \\ W_{21} & W_{22} \end{bmatrix}$$

with $W_{11} \in \mathcal{T}(\mathsf{F}^{m\times m})$, $W_{12} \in \mathcal{T}(\mathsf{F}^{m\times n})$, $W_{21} \in \mathcal{T}_s(\mathsf{F}^{n\times m})$, *and* $W_{22} \in \mathcal{T}(\mathsf{F}^{n\times n})$ *such that* $W^*JW = G^*JG$, *where*

$$G = \begin{bmatrix} I & M \\ 0 & I \end{bmatrix}, \quad J = \begin{bmatrix} I & 0 \\ 0 & -I \end{bmatrix}.$$

(d) *There exists*

$$P = \begin{bmatrix} P_{11} & P_{12} \\ P_{21} & P_{22} \end{bmatrix}$$

with $P_{11} \in \mathcal{T}(\mathsf{F}^{m\times n})$, $P_{12} \in \mathcal{T}(\mathsf{F}^{m\times m})$, $P_{21} \in \mathcal{T}(\mathsf{F}^{n\times n})$, $P_{22} \in \mathcal{T}_s(\mathsf{F}^{n\times m})$, *and* P_{12}, P_{21} *both invertible such that*

$$\begin{bmatrix} M + P_{11} & P_{12} \\ P_{21} & P_{22} \end{bmatrix}$$

is unitary.

About the proof of this theorem, (a) $\Leftrightarrow$ (b) follows from the Arveson's distance formula [6]; (d) $\Rightarrow$ (b) is obvious since $\|M + P_{11}\| < 1$ and $P_{11} \in \mathcal{T}(\mathsf{F}^{m\times n})$; the rest is rather involved and hence is left in the appendix. Matrices W and P in conditions (c) and (d) are essential to the solution of the matrix problem; their existence is proven constructively and hence W and P can be computed if condition (a) in Theorem 1 holds, which is easily verifiable.

If condition (c) in Theorem 1 holds, we have $W^*JW = G^*JG$. Because G and J are invertible, so is W. Furthermore, the $(1,1)$ block of the equation $W^*JW = G^*JG$ reads

$$W_{11}^* W_{11} - W_{21}^* W_{21} = I,$$

which implies that W_{11} is invertible too. Using this condition, we can parametrize all solutions to our matrix problem.

Theorem 2 *Let* $M \in \mathcal{M}(\mathsf{F}^{m\times n})$ *and assume condition (c) in Theorem 1 is satisfied. Then the set of all* $T \in \mathcal{T}(\mathsf{F}^{m\times n})$ *such that* $\|M + T\| < 1$ *is given by*

$$\left\{ T = Q_1 Q_2^{-1} : \begin{bmatrix} Q_1 \\ Q_2 \end{bmatrix} = W^{-1} \begin{bmatrix} U \\ I \end{bmatrix}, \ U \in \mathcal{T}(\mathsf{F}^{m\times n}), \text{ and } \|U\| < 1 \right\}. \tag{2}$$

Proof: Let $V = W^{-1}$ and partition V compatibly. Then $V_{11} \in \mathcal{T}(\mathsf{F}^{m\times m})$, $V_{12} \in \mathcal{T}(\mathsf{F}^{m\times n})$, $V_{21} \in \mathcal{T}_s(\mathsf{F}^{n\times m})$, and $V_{22} \in \mathcal{T}(\mathsf{F}^{n\times n})$. Since W_{11} is invertible, so is V_{22} by some calculation. Now inverting both sides of $W^*JW = G^*JG$ and noting $J^{-1} = J$, we get $G^{-1}JG^{*-1} = VJV^*$. The $(2,2)$ block of the latter equation gives

$$V_{21}V_{21}^* - V_{22}V_{22}^* = -I,$$

which implies $\|V_{22}^{-1}V_{21}\| < 1$. Since

$$Q_2 = V_{22} + V_{21}U = V_{22}(I + V_{22}^{-1}V_{21}U),$$

it follows that for every $U \in \mathcal{T}(\mathsf{F}^{m\times n})$ with $\|U\| < 1$, Q_2^{-1} exists and belongs to $\mathcal{T}(\mathsf{F}^{n\times n})$. Letting T be given by (2), we have

$$\begin{aligned}
(M+T)^*(M+T) - I &= \begin{bmatrix} M+T \\ I \end{bmatrix}^* J \begin{bmatrix} M+T \\ I \end{bmatrix} \\
&= \begin{bmatrix} T \\ I \end{bmatrix}^* G^*JG \begin{bmatrix} T \\ I \end{bmatrix} \\
&= Q_2^{*-1} \begin{bmatrix} U \\ I \end{bmatrix}^* V^*G^*JGV \begin{bmatrix} U \\ I \end{bmatrix} Q_2^{-1} \\
&= Q_2^{*-1} \begin{bmatrix} U \\ I \end{bmatrix}^* J \begin{bmatrix} U \\ I \end{bmatrix} Q_2^{-1} \\
&= Q_2^{*-1}(U^*U - I)Q_2^{-1} \\
&< 0.
\end{aligned}$$

This implies $\|M + T\| < 1$.

Conversely, suppose $T \in \mathcal{T}(\mathsf{F}^{m\times n})$ with $\|M + T\| < 1$. Define

$$\begin{aligned}
\begin{bmatrix} U_1 \\ U_2 \end{bmatrix} &:= W \begin{bmatrix} T \\ I \end{bmatrix} \\
&= WG^{-1} \begin{bmatrix} M+T \\ I \end{bmatrix}.
\end{aligned} \tag{3}$$

Then we have $U_1 \in \mathcal{T}(\mathsf{F}^{m\times n})$ and $U_2 \in \mathcal{T}(\mathsf{F}^{n\times n})$. Since

$$\begin{aligned}
U_1^*U_1 - U_2^*U_2 &= \begin{bmatrix} U_1 \\ U_2 \end{bmatrix}^* J \begin{bmatrix} U_1 \\ U_2 \end{bmatrix} \\
&= \begin{bmatrix} M+T \\ I \end{bmatrix}^* G^{*-1}W^*JWG^{-1} \begin{bmatrix} M+T \\ I \end{bmatrix} \\
&= \begin{bmatrix} M+T \\ I \end{bmatrix}^* J \begin{bmatrix} M+T \\ I \end{bmatrix} \\
&= (M+T)^*(M+T) - I \\
&< 0,
\end{aligned}$$

it follows that U_2 is invertible and $\|U_1U_2^{-1}\| < 1$. Define $U := U_1U_2^{-1}$ and

$$\begin{bmatrix} Q_1 \\ Q_2 \end{bmatrix} := W^{-1} \begin{bmatrix} U \\ I \end{bmatrix},$$

it follows from (3) that $Q_1 = TU_2^{-1}$ and $Q_2 = U_2^{-1}$. Hence $T = Q_1Q_2^{-1}$, which satisfies the parametrization in (2). □

Alternatively, we can use condition (d) in Theorem 1 to find all solutions to the same matrix problem.

Theorem 3 *Let $M \in \mathcal{M}(\mathsf{F}^{m\times n})$ and assume condition (d) in Theorem 1 is satisfied. Then the set of all $M \in \mathcal{T}(\mathsf{F}^{m\times n})$ such that $\|M+T\|<1$ is given by*

$$\{T = \mathcal{F}(P,U):\ U \in \mathcal{T}(\mathsf{F}^{m\times n}) \text{ and } \|U\|<1\}. \tag{4}$$

Proof: Since the matrix

$$\begin{bmatrix} M+P_{11} & P_{12} \\ P_{21} & P_{22} \end{bmatrix}$$

is unitary and P_{12}, P_{21} are invertible, it follows from [20] that the map

$$U \mapsto \mathcal{F}\left(\begin{bmatrix} M+P_{11} & P_{12} \\ P_{21} & P_{22} \end{bmatrix}, U\right) = M + \mathcal{F}(P,U)$$

is a bijection from the open unit ball of $\mathcal{M}(\mathsf{F}^{m\times n})$ onto itself. What is left to show is that $\mathcal{F}(P,U) \in \mathcal{T}(\mathsf{F}^{m\times n})$ iff $U \in \mathcal{T}(\mathsf{F}^{m\times n})$. The "if" part follows from simple matrix manipulation. For the "only if" part, assume $T := \mathcal{F}(P,U) \in \mathcal{T}(\mathsf{F}^{m\times n})$ for some $U \in \mathcal{M}(\mathsf{F}^{m\times n})$; we need to show that U too belongs to $\mathcal{T}(\mathsf{F}^{m\times n})$. From

$$T = P_{11} + P_{12}U(I-P_{22}U)^{-1}P_{21},$$

we obtain after some algebra

$$P_{12}^{-1}(T-P_{11})P_{21}^{-1} = [I + P_{12}^{-1}(T-P_{11})P_{21}^{-1}P_{22}]U. \tag{5}$$

Since

$$\begin{aligned} I + P_{12}^{-1}(T-P_{11})P_{21}^{-1}P_{22} &= I + P_{12}^{-1}P_{12}U(I-P_{22}U)^{-1}P_{21}P_{21}^{-1}P_{22}] \\ &= I + U(I-P_{22}U)^{-1}P_{22} \\ &= (I-UP_{22})^{-1}, \end{aligned}$$

it follows that $I + P_{12}^{-1}(T-P_{11})P_{21}^{-1}P_{22}$ is invertible. Hence from (5)

$$U = [I + P_{12}^{-1}(T-P_{11})P_{21}^{-1}P_{22}]^{-1}P_{12}^{-1}(T-P_{11})P_{21}^{-1}$$

Therefore U belongs to $\mathcal{T}(\mathsf{F}^{m\times n})$. □

The characterizations in Theorems 2 and 3 also give easy expression to the T which minimizes $\mathcal{I}(M+T)$.

Theorem 4 *Let $M \in \mathcal{M}(\mathsf{F}^{m\times n})$ and assume condition (c) or (d) in Theorem 1 is satisfied. Then the unique T satisfying $\|M+T\|<1$ which minimizes $\mathcal{I}(M+T)$ is given by $T = P_{11}$ or $T = -W_{11}^{-1}W_{12}$.*

Proof: We will show that $T = P_{11}$ minimizes the entropy, the proof that $T = -W_{11}^{-1}W_{12}$ also minimizes the entropy is omitted. According to Theorem 3, all T satisfying $\|M+T\| < 1$ are characterized by (4), Consequently, all resulting $M + T$ are given by

$$\left\{\mathcal{F}\left(\begin{bmatrix} M+P_{11} & P_{12} \\ P_{21} & P_{22} \end{bmatrix}, U\right) : \ U \in \mathcal{T}(\mathsf{F}^{m\times n}) \text{ and } \|U\| < 1\right\}.$$

By Lemma 2 i) in [16], we obtain

$$\mathcal{I}(M+T) = \mathcal{I}(U) + \mathcal{I}(M+P_{11}) + 2\ln|\det(I - P_{22}U)|.$$

Notice that the second term is independent of U and $P_{22}U \in \mathcal{T}_s(\mathsf{F}^{n\times m})$, which implies that the third term is zero. Therefore the U which minimizes $\mathcal{I}(M+T)$ is given by $U = 0$. □

One implication of Theorem 4 is that although W given in condition (c) and P in condition (d) of Theorem 1 are not unique, P_{11} and $-W_{11}^{-1}W_{12}$ are uniquely determined and they are equal.

4 All $\mathcal{H}_\infty$ Suboptimal Periodic Controllers

Now we return to the $\mathcal{H}_\infty$ periodic control problem stated in Section 2: Given LTI G resulted from the lifting of $\tilde{G}$, characterize all LTI, causal K with $\hat{K}(0) \in \mathcal{T}(\mathsf{R}^{m\times n})$ that stabilize G and achieve $\||\mathcal{F}(\hat{G}, \hat{K})\||_\infty < 1$. This problem is called the constrained $\mathcal{H}_\infty$ problem. To make use of the standard results in the literature and to simplify our solution, we make the following assumptions:

1. $\hat{G}_{12}(0)$ has full column rank, $\hat{G}_{21}(0)$ has full row rank.
2. $\hat{G}_{12}$ and $\hat{G}_{21}$ have no transmission zeros on the unit circle.
3. $\hat{G}_{22}(0) \in \mathcal{T}_s(\mathsf{R}^{n\times m})$ or, equivalently, $\tilde{G}$ is strictly causal.

For this constrained $\mathcal{H}_\infty$ problem to be solvable, it is necessary that the unconstrained problem be solvable. Hence first we drop the causality constraint temporarily and consider the corresponding unconstrained $\mathcal{H}_\infty$ problem: Find all LTI, causal K to stabilize G and achieve $\||\mathcal{F}(\hat{G}, \hat{K})\||_\infty < 1$. This is a standard problem and has been extensively studied in, e.g., [14]. Several solutions to the standard $\mathcal{H}_\infty$ problem exist in the literature. Here we adopt the solution in [14]. Assume the solvability conditions are satisfied, then all stabilizing controllers K satisfyinging $\|\mathcal{F}(\hat{G}, \hat{K})\|_\infty < 1$ are characterized by

$$\left\{\hat{K} = \mathcal{F}\left(\begin{bmatrix} 0 & I \\ I & -\hat{G}_{22}(0) \end{bmatrix} \star \hat{M}, \hat{\Phi}\right) : \quad \hat{\Phi} \in \mathcal{RH}_\infty, \ \|\hat{\Phi}\|_\infty < 1, \right.$$
$$\left. \text{and } I + \hat{G}_{22}(0)\mathcal{F}[\hat{M}(0), \hat{\Phi}(0)] \text{ is invertible.}\right\} \quad (6)$$

Here the $\mathcal{RH}_\infty$ matrix

$$\hat{M} = \begin{bmatrix} \hat{M}_{11} & \hat{M}_{12} \\ \hat{M}_{21} & \hat{M}_{22} \end{bmatrix}$$

is not uniquely given in [14] and by using Cholesky factorizations we can always choose $\hat{M}$ so that

$$\begin{aligned} \hat{M}_{12}(0) &\in \mathcal{T}(\mathbf{R}^{m\times m}) \\ \hat{M}_{21}(0) &\in \mathcal{T}(\mathbf{R}^{n\times n}) \\ \hat{M}_{22}(0) &= 0. \end{aligned}$$

Furthermore, $\hat{M}_{12}(0)$ and $\hat{M}_{21}(0)$ are invertible.

Now let us return to the constrained $\mathcal{H}_\infty$ problem.

Theorem 5 *The constrained $\mathcal{H}_\infty$ problem is solvable iff the corresponding unconstrained problem is solvable and there exists $T \in \mathcal{T}(\mathbf{R}^{m\times n})$ such that*

$$\| - \hat{M}_{12}(0)^{-1}\hat{M}_{11}(0)\hat{M}_{21}(0)^{-1} + T\| < 1. \tag{7}$$

Proof: Obviously, the corresponding unconstrained problem has to be solvable in order for the constrained problem to be solvable. Assume that the unconstrained problem is solvable. Since $\hat{G}_{22}(0) \in \mathcal{T}_s(\mathbf{R}^{n\times m})$, it follows that $\hat{K}(0) \in \mathcal{T}(\mathbf{R}^{m\times n})$ iff

$$\mathcal{F}[\hat{M}(0), \hat{\Phi}(0)] = \hat{M}_{11}(0) + \hat{M}_{12}(0)\hat{\Phi}(0)\hat{M}_{21}(0) \in \mathcal{T}(\mathbf{F}^{m\times n}).$$

Pre- and post-multiply this by $\hat{M}_{12}(0)^{-1}$ and $\hat{M}_{21}(0)^{-1}$ respectively to get

$$\hat{M}_{12}(0)^{-1}\mathcal{F}[\hat{M}(0), \hat{\Phi}(0)]\hat{M}_{21}(0)^{-1} = \hat{M}_{12}(0)^{-1}\hat{M}_{11}(0)\hat{M}_{21}(0)^{-1} + \hat{\Phi}(0).$$

If there exists $\mathcal{F}[\hat{M}(0), \hat{\Phi}(0)] \in \mathcal{T}(\mathbf{F}^{m\times n})$ such that $\|\hat{\Phi}(0)\| < 1$, then $\hat{M}_{12}(0)^{-1}\mathcal{F}[\hat{M}(0), \hat{\Phi}(0)]$ $\hat{M}_{21}(0)^{-1} \in \mathcal{T}(\mathbf{R}^{m\times n})$ and

$$\| - \hat{M}_{12}(0)^{-1}\hat{M}_{11}(0)\hat{M}_{21}(0)^{-1} + \hat{M}_{12}(0)^{-1}\mathcal{F}[\hat{M}(0), \hat{\Phi}(0)]\hat{M}_{21}(0)^{-1}\| < 1.$$

Conversely, if $T \in \mathcal{T}(\mathbf{R}^{m\times n})$ such that (7) is true, let

$$\Phi = -\hat{M}_{12}(0)^{-1}\hat{M}_{11}(0)\hat{M}_{21}(0)^{-1} + T$$

Then

$$\hat{K} = \mathcal{F}\left(\begin{bmatrix} 0 & I \\ I & -\hat{G}_{22}(0) \end{bmatrix} \star \hat{M}, \Phi\right)$$

achieves $\hat{K}(0) \in \mathcal{T}(\mathbf{R}^{m\times n})$. □

The solvability condition for the corresponding unconstrained problem is given in [14]; the existence of $T \in \mathcal{T}(\mathbf{R}^{m\times n})$ such that (7) is satisfied can be checked easily by using Theorem 1. If the conditions in Theorem 5 are satisfied, then there exists

$$P = \begin{bmatrix} P_{11} & P_{12} \\ P_{21} & P_{22} \end{bmatrix}$$

with $P_{11} \in \mathcal{T}(\mathbf{R}^{m\times n})$, $P_{12} \in \mathcal{T}(\mathbf{R}^{m\times m})$, $P_{21} \in \mathcal{T}(\mathbf{R}^{n\times n})$, $P_{22} \in \mathcal{T}_s(\mathbf{R}^{n\times m})$, and P_{12}, P_{21} both invertible such that

$$U = \begin{bmatrix} -\hat{M}_{12}^{-1}(0)\hat{M}_{11}(0)\hat{M}_{21}^{-1}(0) + P_{11} & P_{12} \\ P_{21} & P_{22} \end{bmatrix}$$

is orthogonal (Theorem 1). Define

$$\hat{N} = \begin{bmatrix} 0 & I \\ I & -D_{22} \end{bmatrix} \star \hat{M} \star U.$$

It is easy to check that $\hat{N}_{11}(0) \in \mathcal{T}(\mathbf{R}^{m\times n})$, $\hat{N}_{12}(0) \in \mathcal{T}(\mathbf{R}^{m\times m})$, $\hat{N}_{21}(0) \in \mathcal{T}(\mathbf{R}^{n\times n})$, $\hat{N}_{22}(0) \in \mathcal{T}_s(\mathbf{R}^{n\times m})$, and $\hat{N}_{12}(0)$, $\hat{N}_{21}(0)$ are both invertible. The set in (6) can be rewritten as

$$\{\hat{K} = \mathcal{F}(\hat{N}, \hat{\Phi}) : \hat{\Phi} \in \mathcal{RH}_\infty, \|\hat{\Phi}\|_\infty < 1, I - \hat{N}_{22}(0)\hat{\Phi}(0) \text{ is invertible}\}.$$

Theorem 6 *Assume solvability of the constrained $\mathcal{H}_\infty$ problem. Then the set of all controllers solving the problem is given by*

$$\{\hat{K} = \mathcal{F}(\hat{N}, \hat{\Phi}) : \hat{\Phi} \in \mathcal{RH}_\infty, \|\hat{\Phi}\|_\infty < 1, \hat{\Phi}(0) \in \mathcal{T}(\mathbf{R}^{m\times n})\ \}.$$

Proof: First notice that $I - \hat{N}_{22}(0)\hat{\Phi}(0)$ is always invertible if $\hat{\Phi}(0) \in \mathcal{T}(\mathbf{R}^{m\times n})$. the special properties of $\hat{N}(0)$ guarantees that $\hat{K}(0) \in \mathcal{T}(\mathbf{R}^{m\times n})$ iff $\hat{\Phi}(0) \in \mathcal{T}(\mathbf{R}^{m\times n})$. Then the result follows immediately. □

It follows from Theorem 6 that all $\mathcal{H}_\infty$ suboptimal closed-loop transfer functions are

$$\{\mathcal{F}(\hat{G}, \hat{K}) = \mathcal{F}(\hat{J}, \hat{\Phi}) :\ \hat{\Phi} \in \mathcal{RH}_\infty,\ \|\hat{\Phi}\|_\infty < 1,\ \hat{\Phi}(0) \in \mathcal{T}(\mathbf{R}^{m\times n})\}$$

where

$$\hat{J} = \begin{bmatrix} \hat{J}_{11} & \hat{J}_{12} \\ \hat{J}_{21} & \hat{J}_{22} \end{bmatrix} = \hat{G} \star \hat{N}.$$

It follows from the internal stability requirement that $\hat{J} \in \mathcal{RH}_\infty$. Also we have $\hat{J}_{11}(0) \in \mathcal{T}(\mathbf{R}^{m\times n})$, $\hat{J}_{12}(0) \in \mathcal{T}(\mathbf{R}^{m\times m})$, $\hat{J}_{21}(0) \in \mathcal{T}(\mathbf{R}^{n\times n})$, and $\hat{J}_{22}(0) \in \mathcal{T}_s(\mathbf{R}^{n\times m})$. Since $\|\mathcal{F}(\hat{J}, \hat{\Phi})\| < 1$ for all $\hat{\Phi} \in \mathcal{RH}_\infty$ with $\|\hat{\Phi}\|_\infty < 1$, it can be shown by using an idea in [19] that there exists a scalar transfer function $\hat{d} \in \mathcal{RH}_\infty$ with $\hat{d}^{-1} \in \mathcal{RH}_\infty$ such that

$$\left\| \begin{bmatrix} \hat{J}_{11} & \hat{d}\hat{J}_{12} \\ \hat{d}^{-1}\hat{J}_{21} & \hat{J}_{22} \end{bmatrix} \right\|_\infty < 1.$$

Consequently, we can find $\hat{J}_{13}$, $\hat{J}_{23}$, $\hat{J}_{31}$, $\hat{J}_{32}$, $\hat{J}_{33}$, all belonging to $\mathcal{RH}_\infty$, such that

$$\hat{J}_{aug} = \begin{bmatrix} \hat{J}_{11} & \hat{d}\hat{J}_{12} & \hat{J}_{13} \\ \hat{d}^{-1}\hat{J}_{21} & \hat{J}_{22} & \hat{J}_{23} \\ \hat{J}_{31} & \hat{J}_{32} & \hat{J}_{33} \end{bmatrix}$$

is para-unitary. Then another way to characterize the $\mathcal{H}_\infty$ suboptimal closed-loop transfer functions is

$$\left\{\mathcal{F}\left(\hat{J}_{aug}, \begin{bmatrix} \hat{\Phi} & 0 \\ 0 & 0 \end{bmatrix}\right) : \quad \hat{\Phi} \in \mathcal{RH}_\infty, \quad \|\hat{\Phi}\|_\infty < 1, \quad \hat{\Phi}(0) \in \mathcal{T}(\mathsf{R}^{m\times n})\right\}.$$

By Lemma 2 i) in [16],

$$\begin{aligned} \mathcal{I}[\mathcal{F}(\hat{G},\hat{K})] &= \mathcal{I}\left(\begin{bmatrix} \hat{\Phi} & 0 \\ 0 & 0 \end{bmatrix}\right) + \mathcal{I}(\hat{J}_{11}) + 2\ln\left|\det\left(I - \begin{bmatrix} \hat{J}_{22}(0) & \hat{J}_{23}(0) \\ \hat{J}_{32}(0) & \hat{J}_{33}(0) \end{bmatrix}\begin{bmatrix} \hat{\Phi}(0) & 0 \\ 0 & 0 \end{bmatrix}\right)\right| \\ &= \mathcal{I}(\hat{\Phi}) + \mathcal{I}(\hat{J}_{11}) + 2\ln|\det[I - \hat{J}_{22}(0)\hat{\Phi}(0)]| \\ &= \mathcal{I}(\hat{\Phi}) + \mathcal{I}(\hat{J}_{11}). \end{aligned}$$

The last equality is due to $\hat{J}_{22}(0)\hat{\Phi}(0) \in \mathcal{T}_s(\mathsf{R}^{m\times m})$. Therefore, the minimum of $\mathcal{I}[\mathcal{F}(\hat{G},\hat{K})]$ is achieved at $\hat{\Phi} = 0$. The following theorem is thus obtained.

Theorem 7 *The minimum entropy controller is given by $\hat{K}_{ME} = \hat{N}_{11}$. The minimum entropy value of the closed-loop transfer function is $\mathcal{I}(\hat{J}_{11})$.*

5 Concluding Remarks

Using a parametrization of all solutions for a certain matrix completion problem, we obtained a simple characterization of all $\mathcal{H}_\infty$ suboptimal periodic controllers satisfying a causality constraint. This characterization also gives a simple expression of the unique $\mathcal{H}_\infty$ suboptimal periodic controller which further minimizes an LEQG cost. The results obtained are explicit and require no numerical optimization. The ideas used in the paper can be applied to study similar problems involving multirate sampled-data controllers as in [5, 21].

Appendix: Proof of Theorem 1

To prove (c) $\Rightarrow$ (d), let

$$W = \begin{bmatrix} W_{11} & W_{12} \\ W_{21} & W_{22} \end{bmatrix}$$

with $W_{11} \in \mathcal{T}(\mathsf{F}^{m\times m})$, $W_{12} \in \mathcal{T}(\mathsf{F}^{m\times n})$, $W_{21} \in \mathcal{T}_s(\mathsf{F}^{n\times m})$, and $W_{22} \in \mathcal{T}(\mathsf{F}^{n\times n})$ satisfy $W^*JW = G^*JG$. Define

$$P = \begin{bmatrix} P_{11} & P_{12} \\ P_{21} & P_{22} \end{bmatrix} := \begin{bmatrix} -W_{11}^{-1}W_{12} & W_{11}^{-1} \\ W_{22} - W_{21}W_{11}^{-1}W_{12} & W_{21}W_{11}^{-1} \end{bmatrix}.$$

It is easy to see that $P_{11} \in \mathcal{T}(\mathsf{F}^{m\times n})$, $P_{12} \in \mathcal{T}(\mathsf{F}^{m\times m})$, $P_{21} \in \mathcal{T}(\mathsf{F}^{n\times n})$, $P_{22} \in \mathcal{T}_s(\mathsf{F}^{n\times m})$. Since W_{11} and W are invertible (see the remarks below Theorem 1), so are P_{12} and P_{21}. It can be verified then that

$$\begin{bmatrix} M + P_{11} & P_{12} \\ P_{21} & P_{22} \end{bmatrix}$$

is a unitary matrix.

It remains to show (a) $\Rightarrow$ (c). As it is well-known, the inertia of a Hermitian matrix H is an ordered triple $\{\pi_+(H), \pi_-(H), \pi_0(H)\}$ of positive numbers, where $\pi_+(H), \pi_-(H), \pi_0(H)$ are numbers of positive, negative, and zero eigenvalues of H respectively, all counting multiplicities. In the following, we denote G^*JG by H and prove two claims related to H.

Claim 1 *Matrices*

$$\begin{bmatrix} [H_{ij}]_{i=k,j=k}^{l,l} & [H_{ij}]_{i=k,j=l+k}^{l,2l} \\ [H_{ij}]_{i=l+k,j=k}^{2l,l} & [H_{ij}]_{i=l+k,j=l+k}^{2l,2l} \end{bmatrix}, \quad k = 1, 2, \ldots, l,$$

are invertible and their inertias are $\{(l-k+1)m, (l-k+1)n, 0\}$ *respectively.*

Proof: The claim is obviously true when $k = 1$. Now assume $2 \le k \le l$. Since

$$H = G^*JG = \begin{bmatrix} I & M \\ M^* & M^*M - I \end{bmatrix},$$

we have

$$\begin{aligned} &\begin{bmatrix} [H_{ij}]_{i=k,j=k}^{l,l} & [H_{ij}]_{i=k,j=l+k}^{l,2l} \\ [H_{ij}]_{i=l+k,j=k}^{2l,l} & [H_{ij}]_{i=l+k,j=l+k}^{2l,2l} \end{bmatrix} \\ &= \begin{bmatrix} I & [M_{ij}]_{i=k,j=k}^{l,l} \\ ([M_{ij}]_{i=k,j=k}^{l,l})^* & ([M_{ij}]_{i=1,j=k}^{l,l})^*[M_{ij}]_{i=1,j=k}^{l,l} - I \end{bmatrix} \\ &= \begin{bmatrix} I & [M_{ij}]_{i=k,j=k}^{l,l} \\ ([M_{ij}]_{i=k,j=k}^{l,l})^* & ([M_{ij}]_{i=1,j=k}^{k-1,l})^*[M_{ij}]_{i=1,j=k}^{k-1,l} + ([M_{ij}]_{i=k,j=k}^{l,l})^*[M_{ij}]_{i=k,j=k}^{l,l} - I \end{bmatrix} \\ &= \begin{bmatrix} I & 0 \\ ([M_{ij}]_{i=k,j=k}^{l,l})^* & I \end{bmatrix} \begin{bmatrix} I & 0 \\ 0 & ([M_{ij}]_{i=1,j=k}^{k-1,l})^*[M_{ij}]_{i=1,j=k}^{k-1,l} - I \end{bmatrix} \begin{bmatrix} I & [M_{ij}]_{i=k,j=k}^{l,l} \\ 0 & I \end{bmatrix}. \end{aligned}$$

Since $\|[M_{ij}]_{i=1,j=k}^{k-1,l}\| < 1$, the claim follows immediately. □

Claim 2 *Matrices*

$$\begin{bmatrix} [H_{ij}]_{i=k,j=k}^{l,l} & [H_{ij}]_{i=k,j=l+k+1}^{l,2l} \\ [H_{ij}]_{i=l+k+1,j=k}^{2l,l} & [H_{ij}]_{i=l+k+1,j=l+k+1}^{2l,2l} \end{bmatrix}, \quad k = 1, 2, \ldots, l-1,$$

are invertible and their inertias are $\{(l-k+1)m, (l-k)n, 0\}$.

Proof: Following the same argument as in the proof of Claim 1, we can show that

$$\begin{bmatrix} [H_{ij}]_{i=k,j=k}^{l,l} & [H_{ij}]_{i=k,j=l+k+1}^{l,2l} \\ [H_{ij}]_{i=l+k+1,j=k}^{2l,l} & [H_{ij}]_{i=l+k+1,j=l+k+1}^{2l,2l} \end{bmatrix}$$

$$= \begin{bmatrix} I & 0 \\ ([M_{ij}]_{i=k,j=k+1}^{l,l})^* & I \end{bmatrix} \begin{bmatrix} I & 0 \\ 0 & ([M_{ij}]_{i=1,j=k+1}^{k-1,l})^*[M_{ij}]_{i=1,j=k+1}^{k-1,l} - I \end{bmatrix} \begin{bmatrix} I & [M_{ij}]_{i=k,j=k+1}^{l,l} \\ 0 & I \end{bmatrix}.$$

Since $[M_{ij}]_{i=1,j=k+1}^{k-1,l}$ is a submatrix of $[M_{ij}]_{i=1,j=k+1}^{k,l}$, we also have $\|[T_{ij}]_{i=1,j=k+1}^{k-1,l}\| < 1$. The claim thus follows. □

Now let us permute the rows and columns of H and J to form

$$\tilde{H} = \begin{bmatrix} \tilde{H}_{11} & \tilde{H}_{21}^* & \cdots & \tilde{H}_{l1}^* \\ \tilde{H}_{21} & \tilde{H}_{22} & \cdots & \tilde{H}_{l2}^* \\ \vdots & \vdots & & \vdots \\ \tilde{H}_{l1} & \tilde{H}_{l2} & \cdots & \tilde{H}_{ll} \end{bmatrix}, \quad \tilde{J} = \begin{bmatrix} \tilde{J}_1 & 0 & \cdots & 0 \\ 0 & \tilde{J}_2 & \cdots & 0 \\ \vdots & \vdots & \ddots & \vdots \\ 0 & 0 & \cdots & \tilde{J}_l \end{bmatrix}$$

where

$$\tilde{H}_{ij} = \begin{bmatrix} H_{ij} & H_{i(l+j)} \\ H_{(l+i)j} & H_{(l+i)(l+j)} \end{bmatrix}, \quad \tilde{J}_i = \begin{bmatrix} I & 0 \\ 0 & -I \end{bmatrix}.$$

With this permutation, the desired factorization becomes $\tilde{W}^*\tilde{J}\tilde{W} = \tilde{H}$ where $\tilde{W}$ belongs to $\mathcal{T}(\mathsf{F}^{(m+n)\times(m+n)})$:

$$\tilde{W} = \begin{bmatrix} \tilde{W}_{11} & 0 & \cdots & 0 \\ \tilde{W}_{21} & \tilde{W}_{22} & \cdots & 0 \\ \vdots & \vdots & \ddots & \vdots \\ \tilde{W}_{l1} & \tilde{W}_{l2} & \cdots & \tilde{W}_{ll} \end{bmatrix}.$$

Further partition gives

$$\tilde{W}_{ij} = \begin{bmatrix} (\tilde{W}_{ij})_{11} & (\tilde{W}_{ij})_{12} \\ (\tilde{W}_{ij})_{21} & (\tilde{W}_{ij})_{22} \end{bmatrix}.$$

Hence $W_{21} \in \mathcal{T}_s(\mathsf{F}^{n\times m})$ implies $(\tilde{W}_{ii})_{21} = 0$.

Claim 3 *Matrices*

$$\tilde{H}_{kk} - [\tilde{H}_{kj}]_{j=k+1}^{l}([\tilde{H}_{ij}]_{i=k+1,j=k+1}^{l,l})^{-1}[\tilde{H}_{ik}]_{i=k+1}^{l}, \quad k = 1, 2, \ldots, l-1,$$

are invertible and their inertias are $\{m, n, 0\}$ *respectively. If these matrices are further partitioned into* 2×2 *block matrices with* $m \times m$ *(1,1) blocks, then their (1,1) blocks are positive definite.*

Proof: By Claim 1, $[\tilde{H}_{ij}]_{i=k,j=k}^{l,l}$ and $[\tilde{H}_{ij}]_{i=k+1,j=k+1}^{l,l}$ are invertible and their inertias are $\{(l-k+1)m, (l-k+1)n, 0\}$ and $\{(l-k)m, (l-k)n, 0\}$ respectively. Write

$$[\tilde{H}_{ij}]_{i=k,j=k}^{l,l} = \begin{bmatrix} \tilde{H}_{kk} & [\tilde{H}_{kj}]_{j=k+1}^{l} \\ [\tilde{H}_{ik}]_{i=k+1}^{l} & [\tilde{H}_{ij}]_{i=k+1,j=k+1}^{l,l} \end{bmatrix} =: \begin{bmatrix} A & B^* \\ B & C \end{bmatrix}.$$

Note that

$$\begin{bmatrix} A & B^* \\ B & C \end{bmatrix} = \begin{bmatrix} I & B^*C^{-1} \\ 0 & I \end{bmatrix} \begin{bmatrix} A - B^*C^{-1}B & 0 \\ 0 & C \end{bmatrix} \begin{bmatrix} I & 0 \\ C^{-1}B & I \end{bmatrix}.$$

Hence $\tilde{H}_{kk} - [\tilde{H}_{kj}]_{j=k+1}^{l}([\tilde{H}_{ij}]_{i=k+1,j=k+1}^{l,l})^{-1}[\tilde{H}_{ik}]_{i=k+1}^{l}$ is invertible and its inertia is $\{m, n, 0\}$.

If we apply the same argument to matrix

$$\begin{bmatrix} H_{kk} & [H_{kj}]_{j=k+1}^{l} \\ [H_{ik}]_{i=k+1}^{l} & [\tilde{H}_{ij}]_{i=k+1,j=k+1}^{l,l} \end{bmatrix}$$

and notice that its inertia is $\{(l-k+1)m, (l-k)n, 0\}$ (Claim 2), then we see that the inertia of $H_{kk} - [H_{kj}]_{j=k+1}^{l}([\tilde{H}_{ij}]_{i=k+1,j=k+1}^{l,l})^{-1}[H_{ik}]_{i=k+1}^{l}$ is $\{m, 0, 0\}$. This matrix is exactly the (1,1) block of $\tilde{H}_{kk} - [\tilde{H}_{kj}]_{j=k+1}^{l}([\tilde{H}_{ij}]_{i=k+1,j=k+1}^{l,l})^{-1}[\tilde{H}_{ik}]_{i=k+1}^{l}$.

Suppose now that we can carry out the following computation:

For i from l to 1,

find $\tilde{W}_{ii}$ with $(W_{ii})_{21} = 0$ such that

$$\tilde{W}_{ii}^*\tilde{J}_i\tilde{W}_{ii} = \begin{cases} \tilde{H}_{ll} & \text{if } i = l \\ \tilde{H}_{ii} - \begin{bmatrix} \tilde{W}_{(i+1)i}^* & \cdots & \tilde{W}_{li}^* \end{bmatrix} \begin{bmatrix} \tilde{J}_{i+1} & \cdots & 0 \\ \vdots & \ddots & \vdots \\ 0 & \cdots & \tilde{J}_l \end{bmatrix} \begin{bmatrix} \tilde{W}_{(i+1)i} \\ \vdots \\ \tilde{W}_{li} \end{bmatrix} & \text{if } i < l; \end{cases} \tag{8}$$

for $j = 1, \ldots, i-1$, let

$$\tilde{W}_{ij} = \tilde{J}_i\tilde{W}_{ii}^{*-1}\left(\tilde{H}_{ij} - \sum_{k=i+1}^{l} \tilde{W}_{ki}^*\tilde{J}_k\tilde{W}_{kj}\right). \tag{9}$$

end

end

Then we obtain all $\tilde{W}_{ij}$ for $i = 1, 2, \ldots, l$, $j = 1, 2, \ldots, i$, and it is straightforward to check that we have $\tilde{W}^*\tilde{J}\tilde{W} = \tilde{H}$. In order to show that the above computation can be carried out, we need to show that the factorization in (8) can be done and $\tilde{W}_{ii}$, $i = 1, 2, \ldots, l$, are invertible. For this purpose, a technical lemma is needed.

Lemma 1 *Given a nonsingular Hermitian matrix* $A = \begin{bmatrix} A_{11} & A_{12} \\ A_{12}^* & A_{22} \end{bmatrix} \in \mathsf{F}^{(m+n)\times(m+n)}$ *and* $J = \begin{bmatrix} I & 0 \\ 0 & -I \end{bmatrix} \in \mathsf{F}^{(m+n)\times(m+n)}$, *there exists* $B = \begin{bmatrix} B_{11} & B_{12} \\ 0 & B_{22} \end{bmatrix} \in \mathsf{F}^{(m+n)\times(m+n)}$ *such that* $B^*JB = A$ *if and only if* $A_{11} > 0$ *and the inertia of* A *is* $\{m, n, 0\}$.

Proof: The necessity is obvious since $B_{11}^*B_{11} = A_{11}$ and the inertia is invariant under congruence. To see the sufficiency, notice that if $A_{11} > 0$, then

$$\begin{bmatrix} A_{11} & A_{12} \\ A_{12}^* & A_{22} \end{bmatrix} = \begin{bmatrix} I & 0 \\ A_{12}^*A_{11}^{-1} & I \end{bmatrix} \begin{bmatrix} A_{11} & 0 \\ 0 & A_{22} - A_{12}^*A_{11}^{-1}A_{12} \end{bmatrix} \begin{bmatrix} I & A_{11}^{-1} \\ 0 & I \end{bmatrix}.$$

The inertia of A being $\{m, n, 0\}$ implies that $A_{22} - A_{12}^*A_{11}^{-1}A_{12} < 0$. Hence we can define

$$B_{11} = A_{11}^{\frac{1}{2}}, \quad B_{12} = A_{11}^{-\frac{1}{2}}A_{12}, \quad B_{22} = (A_{12}^*A_{11}^{-1}A_{12} - A_{22})^{\frac{1}{2}}.$$

With this definition, $B^*JB = A$ is satisfied. □

By this lemma, it becomes obvious that when $i = l$ the factorization in (8) can be done and $\tilde{W}_{ll}$ is invertible. When $i < l$, we have

$$\begin{bmatrix} \tilde{W}_{(i+1)(i+1)}^* & \cdots & \tilde{W}_{l(i+1)}^* \\ \vdots & \ddots & \vdots \\ 0 & \cdots & \tilde{W}_{ll}^* \end{bmatrix} \begin{bmatrix} \tilde{J}_{i+1} & \cdots & 0 \\ \vdots & \ddots & \vdots \\ 0 & \cdots & \tilde{J}_l \end{bmatrix} \begin{bmatrix} \tilde{W}_{(i+1)i} \\ \vdots \\ \tilde{W}_{li} \end{bmatrix} = \begin{bmatrix} \tilde{H}_{(i+1)i} \\ \vdots \\ \tilde{H}_{li} \end{bmatrix}.$$

Then

$$\begin{aligned} &\tilde{H}_{ii} - \begin{bmatrix} \tilde{W}_{(i+1)i}^* & \cdots & \tilde{W}_{li}^* \end{bmatrix} \begin{bmatrix} \tilde{J}_{i+1} & \cdots & 0 \\ \vdots & \ddots & \vdots \\ 0 & \cdots & \tilde{J}_l \end{bmatrix} \begin{bmatrix} \tilde{W}_{(i+1)i} \\ \vdots \\ \tilde{W}_{li} \end{bmatrix} \\ = \; & \tilde{H}_{ii} - \begin{bmatrix} \tilde{H}_{(i+1)i}^* & \cdots & \tilde{H}_{li}^* \end{bmatrix} \begin{bmatrix} \tilde{W}_{(i+1)(i+1)} & \cdots & 0 \\ \vdots & \ddots & \vdots \\ \tilde{W}_{l(i+1)} & \cdots & \tilde{W}_{ll} \end{bmatrix}^{-1} \\ & \begin{bmatrix} \tilde{J}_{i+1} & \cdots & 0 \\ \vdots & \ddots & \vdots \\ 0 & \cdots & \tilde{J}_l \end{bmatrix} \begin{bmatrix} \tilde{W}_{(i+1)(i+1)}^* & \cdots & \tilde{W}_{l(i+1)}^* \\ \vdots & \ddots & \vdots \\ 0 & \cdots & \tilde{W}_{ll}^* \end{bmatrix}^{-1} \begin{bmatrix} \tilde{H}_{(i+1)i} \\ \vdots \\ \tilde{H}_{li} \end{bmatrix} \\ = \; & \tilde{H}_{ii} - \begin{bmatrix} \tilde{H}_{(i+1)i}^* & \cdots & \tilde{H}_{li}^* \end{bmatrix} \begin{bmatrix} \tilde{H}_{(i+1)(i+1)} & \cdots & \tilde{H}_{l(i+1)}^* \\ \vdots & & \vdots \\ \tilde{H}_{l(i+1)} & \cdots & \tilde{H}_{ll} \end{bmatrix}^{-1} \begin{bmatrix} \tilde{H}_{(i+1)i} \\ \vdots \\ \tilde{H}_{li} \end{bmatrix}. \end{aligned}$$

It then follows from Claim 3 that the factorization in (8) can be carried out and the resulting $\tilde{W}_{ii}$ are invertible.

After we get $\tilde{W}$, certain row and column permutations will give us W which satisfies $W^*JW = G^*JG$ in the standard matrix representation.

References

[1] W. Arveson, "Interpolation problems in nest algebras," *J. Functional Analysis*, vol. 20, pp. 208–233, 1975.

[2] J.A. Ball, "Nevanlinna-Pick interpolation: generalizations and applications," in *Recent Results in Operator Theory, Vol. I*, (Ed. J.B. Conway and B.B. Morrel), Longman Scientific & Technical, Essex, pp. 51–94, 1988.

[3] J.A. Ball and I. Gohberg, "A commutant lifting theory for triangular matrices with diverse applications," *Integral Equations and Operator Theory*, vol. 8, pp. 205-267, 1985.

[4] J.A. Ball and I. Gohberg, "Shift invariant subspaces, factorization, and interpolation for matrices I: The canonical case," *Linear Algebra and Its Applications*, vol. 74, pp. 87–150, 1986.

[5] T. Chen and L. Qiu, "$\mathcal{H}_\infty$ design of general multirate sampled-data control systems," *Automatica*, vol. 30, pp. 1139-1152, 1994.

[6] K.R. Davidson, *Nest Algebras*, Longman Scientific & Technical, Essex, England, 1988.

[7] A. Feintuch, P. P. Khargonekar, and A. Tannenbaum, "On the sensitivity minimization problem for linear time-varying periodic systems," *SIAM J. Control and Optimization*, vol. 24, pp. 1076-1085, 1986.

[8] B. A. Francis, *A Course in $\mathcal{H}_\infty$ Control Theory*, Springer-Verlag, New York, 1987.

[9] K. Glover, J. C. Doyle, "State-space formulae for all stabilizing controllers that satisfy an $\mathcal{H}_\infty$-norm bound and relations to risk sensitivity," *Systems & Control Letters*, vol. 11, pp. 167-172, 1988.

[10] T. T. Georgiou and P. P. Khargonekar, "A constructive algorithm for sensitivity optimization of periodic systems," *SIAM J. Control and Optimization*, vol. 25, pp. 334-340, 1987.

[11] M. Green, K. Glover, D. Limebeer, and J. Doyle, "A J-spectral factorization approach to $\mathcal{H}_\infty$ control," *SIAM J. Control and Optimization*, vol. 28, pp. 1350-1371, 1990.

[12] M. Green and D. Limebeer, *Linear Robust Control*, Prentice-Hall, Englewood Cliffs, 1994.

[13] J.W. Helton, *Operator Theory, Analytic Functions, Matrices, and Electrical Engineering*, C.B.M.S. Regional Conference Series in Mathematics, no. 68, American Mathematical Society, Providence, 1987.

[14] P. A. Iglesias and G. Glover, "Sate-space approach to discrete-time $\mathcal{H}_\infty$ control," *Int. J. Control*, vol. 54, pp. 1031–1073, 1991.

[15] P. A. Iglesias, D. Mustafa, and G. Glover, "Discrete time $\mathcal{H}_\infty$ controllers satisfying a minimum entropy criterion," *Systems & Control Letters*, vol. 14, pp. 275-286, 1990.

[16] P. A. Iglesias, and D. Mustafa, "State-space solution of the discrete-time minimum entropy control problem via separation," *IEEE Trans. Automat. Control*, vol. 38, pp. 1525-1530, 1993.

[17] P. P. Khargonekar, K. Poolla, and A. Tannenbaum, "Robust control of linear time-invariant plants using periodic compensation," *IEEE Trans. Automat. Control*, vol. 30, pp. 1088-1096, 1985.

[18] D. Mustafa and K. Glover, *Minimum Entropy $\mathcal{H}_\infty$ Control, Lecture Notes in Control and Information Sciences*, vol. 146, Springer-Verlag, 1991.

[19] R. Redheffer, "Inequalities for a matrix Riccati equation", *J. Math. Mech.*, vol. 8, pp. 349-367, 1959.

[20] R. M. Redheffer, "On a certain linear fractional transformation," *J. Math. Phys.*, vol. 39, pp. 269-286, 1960.

[21] L. Qiu and T. Chen, "Multirate sampled-data systems: all $\mathcal{H}_\infty$ suboptimal controllers and the minimum entropy controller," *Proc. 33th IEEE Conf. on Decision and Control*, pp. 3707–3712, 1994.

[22] P. G. Voulgaris, M. A. Dahleh, and L. S. Valavani, "$\mathcal{H}_\infty$ and $\mathcal{H}_2$ optimal controllers for periodic and multi-rate systems," *Automatica*, vol. 30, pp. 251-263, 1994.

[23] H. J. Woerdeman, "Strictly contractive and positive completions for block matrices," *Linear Algebra and Its Applications*, vol. 136, pp. 105, 1990.

[24] K. Zhou, J.C. Doyle, and K. Glover, *Robust and Optimal Control*, Prentice-Hall, Englewood Cliffs, 1995.

Li Qiu
Dept of Electrical and Electronic Engineering
Hong Kong University of Science and Technology
Clear Water Bay, Kowloon, Hong Kong

Tongwen Chen
Dept of Electrical and Computer Engineering
University of Calgary
Calgary, Alberta, Canada T2N 1N4

AMS Classification: 47A20, 93B35

Operator Theory:
Advances and Applications, Vol. 87

Spline approximation methods for Wiener-Hopf operators

Steffen Roch[1]

In the present paper, we introduce an algebra of approximation sequences both for singular integral operators with piecewise continuous coefficients and for Wiener-Hopf operators with piecewise continuous generating function. By means of localization techniques and of the two-projections-theorem, necessary and sufficient conditions for the stability of sequences in this algebra are derived.

1 Introduction

Let F denote the Fourier transform acting on the Schwartz space by

$$(Ff)(t) = \int_{\mathbb{R}} e^{-2\pi ist} f(s)\, ds\,, \quad t \in \mathbb{R}.$$

The Fourier transform has an inverse given by

$$(F^{-1}f)(t) = \int_{\mathbb{R}} e^{2\pi ist} f(s)\, ds\,, \quad t \in \mathbb{R},$$

and F and F^{-1} extend continuously to bounded and unitary operators on the Hilbert space L^2 which are denoted by F and F^{-1} again. As usual, the inner product of two functions $f, g \in L^2$ is

$$(f, g) = \int_{\mathbb{R}} f(s)\overline{g(s)}\, ds\,.$$

Given a bounded measurable function a on the real line $\mathbb{R}$, the operator $W^0(a)$ of *Fourier convolution* by a,

$$W^0(a) : L^2 \to L^2\,, \quad f \mapsto F^{-1}aFf$$

is bounded on L^2, and the restriction $W(a)$ of $W^0(a)$ onto the subspace of L^2 consisting of all functions which vanish on the negative semi-axis is called the *Wiener-Hopf operator with generating function a*.

[1]Supported by a DFG Heisenberg grant

In this paper we shall be concerned with *piecewise continuous* functions, i.e. with functions a having finite one-sided limits $a(t \pm 0)$ at each point $t \in \mathbb{R}$ as well as finite limits $a(\pm\infty)$ at infinity. Examples for Wiener-Hopf operators with piecewise continuous generating function are the classical Wiener-Hopf operators

$$(Af)(t) = f(t) + \int_0^\infty k(t-s)f(s)\,ds\,, \quad t \geq 0,$$

with $k \in L^1$ which can be rewritten as $A = W(1+a)$ with $a = Fk$ (observe that a is continuous on $\mathbb{R}$ and $a(\pm\infty) = 0$ by the Riemann-Lebesgue theorem), and the singular integral operator

$$(Sf)(t) = \frac{1}{\pi i}\int_0^\infty \frac{f(s)}{s-t}\,ds\,, \quad t \geq 0,$$

(existing in sense of a Cauchy principal value integral) in which case $S = W(\mathrm{sgn})$ with $\mathrm{sgn}(s)$ denoting the sign of s.

For the approximate solution of the Wiener-Hopf equation

$$W(a)u = f \tag{1}$$

we employ spline projection methods. The spline spaces considered here are supposed to be of a special (but natural) structure, namely, we start with a *mother spline* φ, that is, with a bounded, measurable, and compactly supported function φ satisfying the following conditions:

$$\sum_{k\in\mathbb{Z}} \varphi(x-k) \equiv 1 \quad \text{for} \quad x \in \mathbb{R}, \tag{2}$$

$$\sum_{k\in\mathbb{Z}} \int_{\mathbb{R}} \varphi(t+k)\overline{\varphi(t)}dt \cdot z^k \neq 0 \quad \text{for} \quad z \in \mathbf{T} \tag{3}$$

where $\mathbf{T}$ refers to the unit circle $|z| = 1$ (the sums in (2) and (3) are actually finite due to the boundedness of the support of φ.) Then we set $\varphi_{kn}(t) := \varphi(nt-k)$ and define the spline space S_n as the smallest closed subspace of L^2 containing all functions φ_{kn} with $k \in \mathbb{Z}$. For example one can take $\varphi = \chi_{[0,1]}$, the characteristic function of the interval $[0,1]$, or $\varphi = \chi_{[0,1]} * \ldots * \chi_{[0,1]}$, the d–fold convolution of $\chi_{[0,1]}$ by itself. Then (2) and (3) are satisfied, and S_n is just the space of all L^2-functions which are polynomials of degree d over each interval $[k, k+1]$, and which are $d-1$ times continuously differentiable on $\mathbb{R}$.

The *Galerkin projection* L_n is the operator mapping L^2 onto S_n such that $(L_n f, \varphi_{kn}) = (f, \varphi_{kn})$ for all $f \in L^2$ and $k \in \mathbb{Z}$. Condition (3) ensures the existence of L_n, whereas (2) involves the strong convergence of L_n to the identity operator I as $n \to \infty$ (compare Sections 2.7 and 2.8 in [3]).

For the Galerkin method for solving the Wiener-Hopf equation we replace (1) by the sequence of equations

$$L_n W(a) u_n = L_n f\,, \quad n = 1, 2, \ldots \tag{4}$$

with the solutions u_n being sought in the spline space S_n. Our concern is the *applicability* of method (4) to equation (1), i.e. the problem whether the equations (4) are uniquely solvable for all $n \geq n_o$ and for all right sides $f \in L^2$, and whether the sequence $(u_n)_{n \geq n_0}$ converges to a solution u of (1). Since L_n converges strongly to I, the applicability of (2) is equivalent to the *stability* of the sequence $(L_n W(a)|_{S_n})$. (A sequence (A_n) of operators is *stable* if A_n is invertible for all sufficiently large n, and if $\sup ||A^{-1}|| < \infty$.)

Our stability criterion for the Galerkin method for Wiener-Hopf operators will be (partially) given in terms of Toeplitz operators. So let l^2 denote the Hilbert space of all sequences $(x_n)_{n\in\mathbb{Z}}$ of complex numbers with inner product $((x_n), (y_n)) = \sum_{n\in\mathbb{Z}} x_n \overline{y_n}$.

Given a bounded measurable function a on the unit circle define its kth Fourier coefficient a_k by $a_k = \int_0^1 a(e^{2\pi is}) e^{-2\pi iks} ds$. The *Laurent operator* $T^0(a)$ acts on finitely supported sequences $(x_n) \in l^2$ as

$$T^0(a)(x_n) = (y_n) \quad \text{with} \quad y_n = \sum_{k\in\mathbb{Z}} a_{n-k} x_k ,$$

and this operator extends by continuity to a bounded operator on all of l^2 which is denoted by $T^0(a)$ again. The *Toeplitz operator* $T(a)$ *with generating function* a is defined as the restriction of $T^0(a)$ onto the subspace of l^2 consisting of all sequences (x_n) with $x_n = 0$ whenever $n < 0$.

We finally introduce functions λ and σ on the unit circle by

$$\lambda(z) = \sum_{k\in\mathbb{Z}} \left(\int_{\mathbf{R}} \varphi(s+k)\overline{\varphi(s)}\, ds \right) \cdot z^k ,$$

$$\sigma(z) = \sum_{k\in\mathbb{Z}} \left(\int_{\mathbf{R}} (S_{\mathbf{R}}\varphi)(s+k)\overline{\varphi(s)}\, ds \right) \cdot z^k$$

where $S_{\mathbf{R}} = W^0(\mathrm{sgn})$ is the singular integral operator on the real axis. The function λ is actually a polynomial in z, hence bounded and measurable, and one can show that σ is bounded and measurable, too (see [3], Section 2.11.1, and compare also 2.11.3 where it is shown that σ is even piecewise continuous on $\mathbf{T}$ and has its only discontinuity at the point 1).

Now our stability theorem can be formulated as follows.

Theorem 1 *The Galerkin method (4) applies to the Wiener-Hopf equation (1) if and only if the operators* $W(a)$ *and*

$$T(a(+\infty)(\lambda^{-1}\sigma + 1)/2 + a(-\infty)(\lambda^{-1}\sigma - 1)/2) + K,$$

where K *is a certain compact operator on* l^2*, are invertible.*

(The invertibility of λ is a consequence of (3).)

Of course, the appearance of the (undetermined) compact perturbation K is an unpleasant effect which - nevertheless - lies in the nature of the matter. It results simply from overlapping basis spline functions φ_{kn}. Thus, if $\varphi = \chi_{[0,1]}$, then $K = 0$. Moreover, as in the case of the Galerkin method for singular integral operators or Mellin operators (see [6] and

[3], Chapter 4), there seems to be a canonical method to modify the spline spaces in such a way that the appearance of perturbations of this type can be completely avoided and that the resulting stability conditions become much weaker. The analogue of these techniques for Wiener-Hopf equations will be considered in a forthcoming paper. For spline spaces of another type, this programme has already been carried out by Elschner [2]. See also Prössdorf and Silbermann [5].

The paper is organized as follows. In Section 2 we summarize and prove some technical lemmata. Then, in Section 3, we introduce an algebra of approximation sequences which, besides the sequence of the Galerkin method for Wiener-Hopf equations, contains a bulk of other interesting approximation sequences, e.g. for operators of the form $\sum a_k W^0(b_k)$ where a_k and b_k are certain piecewise continuous functions. This algebra will be completely analyzed, i.e. we shall derive necessary and sufficient stability conditions for each sequence in it. Finally, in the forth section, we are going to specialize these general results to concrete classes of operators and methods.

2 Technical preliminaries

Let $\mathcal{A}$ denote the set of all sequences (A_n) of operators $A_n : S_n \to S_n$ having the property that there is an operator $W(A_n)$ such that $A_k L_k \to W(A_n)$ and $(A_k L_k)^* \to W(A_n)^*$ strongly as $k \to \infty$. Provided with operations

$$(A_n) + (B_n) := (A_n + B_n), \quad (A_n)(B_n) := (A_n B_n), \quad \alpha(A_n) := (\alpha A_n),$$

with involution $(A_n)^* := (A_n^*)$ and with norm $\|(A_n)\| := \sup \|A_n\|$, the set $\mathcal{A}$ becomes a C^*-algebra.

The following proposition is a very special case of the so-called Lifting theorem (see [3], Theorem 1.8 and Proposition 3.8 for the general situation and for a proof).

Proposition 1 *(a) The set*

$$\mathcal{J} = \{(L_n K|_{S_n}) + (C_n) \text{ where } K \text{ is compact and } \|C_n\| \to 0\}$$

is a closed two-sided **-ideal of* $\mathcal{J}$.
(b) If $(A_n) \in \mathcal{A}$ *and* $A = W(A_n)$ *then the approximation method* (A_n) *applies to* A *if and only if the operator* A *is invertible and if the coset* $(A_n) + \mathcal{J}$ *is invertible in the quotient algebra* $\mathcal{A}/\mathcal{J}$.

In order to mention some concrete sequences of operators belonging to the algebra $\mathcal{A}$ we recall from [3], Theorems 2.5 and 2.6, that the spline space S_n is isomorphic to l^2 with the isomorphism given by

$$E_n : l^2 \to S_n, \quad (x_k) \mapsto \sum x_k \varphi_{kn}$$

and

$$E_{-n} : S_n \to l^2, \quad \sum x_k \varphi_{kn} \mapsto (x_k).$$

Moreover, $\|E_n\| \leq Cn^{-1/2}$ and $\|E_{-n}\| \leq Cn^{1/2}$. Further, we denote by P the projection operator

$$P : l^2 \to l^2, \quad (x_k) \mapsto (\ldots 0, 0, x_0, x_1, \ldots).$$

Proposition 2 *(a) Let a be a piecewise continuous function on the unit circle. Then the sequence $(E_n T^0(a) E_{-n})$ belongs to $\mathcal{A}$, and*

$$W(E_n T^0(a) E_{-n}) = \frac{a(1+0) + a(1-0)}{2} I - \frac{a(1+0) - a(1-0)}{2} S_{\mathbb{R}}$$

where $a(1 \pm 0)$ denote the one-sided limits of a at $1 \in \mathbf{T}$ and the $+$ sign is related with the clockwisely taken limit.
(b) The sequence $(E_n P E_{-n})$ is in $\mathcal{A}$ and

$$W(E_n P E_{-n}) = \chi_{[0,\infty)} I$$

with χ_M referring here and hereafter to the characteristic function of the set M.

For a proof see Proposition 3.13 of [3].

In order to construct some more examples of sequences in $\mathcal{A}$ we introduce the shift operators U_s and D_s for $s \in \mathbb{R}$ by

$$U_s : L^2 \to L^2, \quad (U_s f)(t) = f(t-s)$$

$$D_s : L^2 \to L^2, \quad (D_s f)(t) = e^{-2\pi i t s} f(t),$$

the discretized shifts $U_{s,n}$ and $D_{s,n}$ by

$$U_{s,n} : l^2 \to l^2, \quad U_{s,n}(x_k) = (x_{k-\{sn\}})$$

where $\{x\}$ refers to the smallest integer which is greater than or equal to x, and

$$D_{s,n} : l^2 \to l^2, \quad D_{s,n}(x_k) = (e^{-2\pi i s k/n} x_k),$$

and we finally set

$$\hat{U}_{s,n} : S_n \to S_n, \quad \hat{U}_{s,n} = E_n U_{s,n} E_{-n},$$

$$\hat{D}_{s,n} : S_n \to S_n, \quad \hat{D}_{s,n} = E_n D_{s,n} E_{-n}.$$

Proposition 3 *If (A_n) is a sequence in $\mathcal{A}$ then the sequences $(\hat{D}_{s,n}^{-1} A_n \hat{D}_{s,n})$ and $(\hat{U}_{s,n}^{-1} A_n \hat{U}_{s,n})$ belong to $\mathcal{A}$, too, and*

$$W(\hat{D}_{s,n}^{-1} A_n \hat{D}_{s,n}) = D_s^{-1} W(A_n) D_s,$$

$$W(\hat{U}_{s,n}^{-1} A_n \hat{U}_{s,n}) = U_s^{-1} W(A_n) U_s.$$

Proof. We claim that

$$\|\hat{D}_{s,n} L_n - D_s L_n\| \to 0 \quad \text{as} \quad n \to \infty. \tag{5}$$

Let $f = \sum_{k=-l}^{l} a_k \varphi_{kn} \in S_n$. Then

$$\begin{aligned} \|(\hat{D}_{s,n} - D_s)f\|^2 &= \int_{-\infty}^{\infty} \left| \sum_{k=-l}^{l} \left(e^{-2\pi i s k/n} - e^{-2\pi i s x}\right) a_k \varphi_{kn}(x) \right|^2 dx \\ &= \int_{-\infty}^{\infty} \left| \sum_{k=-l}^{l} \left(e^{-2\pi i s k/n} - e^{-2\pi i s x}\right) a_k \varphi(nx - k) \right|^2 dx \\ &= \frac{1}{n} \int_{-\infty}^{\infty} \left| \sum_{k=-l}^{l} \left(e^{-2\pi i s k/n} - e^{-2\pi i s x/n}\right) a_k \varphi(x - k) \right|^2 dx. \end{aligned}$$

Suppose for definiteness that the support of φ in contained in the interval $[-a, a]$ with $a \in \mathbb{Z}$ and $a > 0$. Then the latter integral is equal to

$$\frac{1}{n} \int_{-\infty}^{\infty} \left| \sum_{k=-l}^{l} \sum_{m=-a}^{a-1} \left(e^{-2\pi i s k/n} - e^{-2\pi i s x/n}\right) a_k \chi_{[m,m+1]}(x-k) \varphi(x-k) \right|^2 dx$$

$$= \frac{1}{n} \int_{-\infty}^{\infty} \left| \sum_{r=-l-a}^{l+a+1} \sum_{m=m_r}^{M_r} \left(e^{-2\pi i s \frac{r-m}{n}} - e^{-2\pi i s \frac{x}{n}}\right) a_{r-m} \chi_{[r,r+1]}(x) \varphi(x - r - m) \right|^2 dx \qquad (6)$$

with certain integers $m_r \le M_r$ where $M_r - m_r \le 2a$. The integral (6) is the same as

$$\frac{1}{n} \sum_{r=-l-a}^{l+a+1} \int_{r}^{r+1} \left| \sum_{m=m_r}^{M_r} \left(e^{-2\pi i s \frac{r-m}{n}} - e^{-2\pi i s \frac{x}{n}}\right) a_{r-m} \varphi(x - r - m) \right|^2 dx$$

$$\le \frac{1}{n} \sum_{r=-l-a}^{l+a+1} \int_{r}^{r+1} \left(\sum_{m=m_r}^{M_r} \left| e^{-2\pi i s \frac{r-m}{n}} - e^{-2\pi i s \frac{x}{n}} \right| |a_{r-m}| \, |\varphi(x - r - m)| \right)^2 dx. \qquad (7)$$

Set $C := \sup_x |\varphi(x)|$ and denote the modul of continuity of the function $g(t) = \exp(2\pi i s t)$ by $\omega(g, y)$, i.e.

$$\omega(g, y) = \sup\{|g(t_1) - g(t_2)| \quad \text{with} \quad |t_1 - t_2| < y\}.$$

Then (7) can be estimated by

$$C^2 \omega(g, a/n)^2 \frac{1}{n} \sum_{r=-l-a}^{l+a+1} \left(\sum_{m=m_r}^{M_r} |a_{r-m}| \right)^2$$

$$\le C^2 C_1 \omega(g, a/n)^2 \frac{1}{n} \sum_{r=-l-a}^{l+a+1} \left(\sum_{m=m_r}^{M_r} |a_{r-m}|^2 \right)$$

$$\le 2a C^2 C_1 \omega(g, a/n)^2 \frac{1}{n} \sum_{r=-l}^{l} |a_k|^2.$$

Hence,

$$\begin{aligned}\|(\hat{D}_{s,n} - D_s)f\| &\leq C_2\omega(g, a/n)n^{-1/2}\|E_{-n}f\| \\ &\leq C_2\omega(g, a/n)\|f\|\end{aligned}$$

for all $f \in S_n$ which are finite sums of the basis splines, and with a constant C_2 being independent of f. Since these functions are dense in S_n, and since $\omega(g, a/n) \to 0$ as $n \to \infty$, we get our claim (5).

Now it is easy to see that the first assertion of the proposition holds:

$$\begin{aligned}\hat{D}_{s,n}^{-1}A_n\hat{D}_{s,n}L_n &= \hat{D}_{s,n}^{-1}L_n \cdot A_nL_n \cdot \hat{D}_{s,n}L_n \\ &\to D_{-s}W(A_n)D_s = D_s^{-1}W(A_n)D_s\end{aligned}$$

strongly as $n \to \infty$.

The second assertion,

$$\text{s-lim}_{n\to\infty}\hat{U}_{s,n}^{-1}A_n\hat{U}_{s,n}L_n = U_s^{-1}W(A_n)U_s\,, \tag{8}$$

can be verified easily since $\hat{U}_{s,n} = U_{\{sn\}/n}|_{S_n}$, and since the operators $U_{\{sn\}/n}$ are defined on all of L^2. The strong continuity of the function $t \mapsto U_t$ is evident, hence, $U_{\{sn\}/n} \to U_s$ strongly as $n \to \infty$ which yields (8). ■

Clearly, if $A \in L(L^2)$, then the sequence $(L_nA|_{S_n})$ belongs to $\mathcal{A}$, and $W(L_nA|_{S_n}) = A$. In particular, the sequences $(L_nfI|_{S_n})$ and $(L_nW^0(f)|_{S_n})$ belong to $\mathcal{A}$ for all bounded functions f. In what follows we are going to examine some commutator relations of these sequences modulo the ideal $\mathcal{J}$. For, we abbreviate the commutator $xy - yx$ of two elements of an algebra by $[x, y]$. Further we write $\dot{\mathbb{R}}$ for the compactification of the real axis by one point ∞ (thus, $\dot{\mathbb{R}}$ can be thought of as a circle) and $\bar{\mathbb{R}}$ for the compactification of $\mathbb{R}$ by the two points $+\infty$ and $-\infty$ (which can be viewed as a closed interval then).

Proposition 4 *(a) If $f \in C(\dot{\mathbb{R}})$ then the commutators*

$$[(L_nfI|_{S_n}), (E_nT^0(a)E_{-n})] \quad \textit{and} \quad [(L_nfI|_{S_n}), (E_nPE_{-n})]$$

belong to $\mathcal{J}$ for all piecewise continuous functions a having their only discontinuity at the point $1 \in \mathbf{T}$.
(b) Let $(A_n) \in \mathcal{A}$. If the commutators $[(L_nfI|_{S_n}), (A_n)]$ belong to $\mathcal{J}$ for all $f \in C(\dot{\mathbb{R}})$ then the commutators

$$[(L_nfI|_{S_n}), (\hat{D}_{s,n}^{-1}A_n\hat{D}_{s,n})], \tag{9}$$

$$[(L_nfI|_{S_n}), (\hat{U}_{s,n}^{-1}A_n\hat{U}_{s,n})] \tag{10}$$

belong to $\mathcal{J}$ for all $f \in C(\dot{\mathbb{R}})$, too.

Proof. Assertion (a) can be found in [3], Proposition 3.18. In order to verify (b) we are going to show that the commutators (9) and (10) coincide modulo $\mathcal{J}$ with

$$(\hat{D}_{s,n}^{-1})[(L_nfI|_{S_n}), (A_n)](\hat{D}_{s,n}), \tag{11}$$

$$(\hat{U}_{s,n}^{-1})[(L_n f_{-s} I|_{S_n}), (A_n)](\hat{U}_{s,n}) \tag{12}$$

where $f_{-s}(t) = f(t-s)$, respectively, and that

$$(\hat{D}_{s,n}^{-1})(L_n K|_{S_n} + C_n)(\hat{D}_{s,n}) \in \mathcal{J} \tag{13}$$

$$(\hat{U}_{s,n}^{-1})(L_n K|_{S_n} + C_n)(\hat{U}_{s,n}) \in \mathcal{J} \tag{14}$$

whenever K is compact and $\|C_n\| \to 0$. Write (9) as

$$(L_n fI|_{S_n})(\hat{D}_{s,n}^{-1} A_n \hat{D}_{s,n}) - (\hat{D}_{s,n}^{-1} A_n \hat{D}_{s,n})(L_n fI|_{S_n})$$

$$= (\hat{D}_{s,n}^{-1})[(\hat{D}_{s,n} L_n f \hat{D}_{s,n}^{-1}), (A_n)](\hat{D}_{s,n}).$$

The sequence $(\hat{D}_{s,n} L_n f \hat{D}_{s,n}^{-1} - D_s L_n f L_n D_s^{-1}|_{S_n})$ tends to zero by (5) and belongs to $\mathcal{J}$. Furthermore one has the well-known commutator relation

$$\|L_n D_s - D_s L_n\| \to 0 \tag{15}$$

(in [3], Theorem 2.8, it is shown that $\|L_n fI - fL_n\| \to 0$ for all $f \in C(\dot{\mathbb{R}})$, but the proof given there makes only use of the continuity of f on $\mathbb{R}$ and of the fact that the modul of continuity $\omega(f, 1/n)$ tends to zero as $n \to \infty$. Both facts hold for the function $f(x) = \exp(2\pi isx)$, too.) Hence, the sequence $(D_s L_n f L_n D_s^{-1}|_{S_n} - L_n D_s f D_s^{-1}|_{S_n})$ goes to zero, too, and since $D_s f D_s^{-1} = f$ we arrive at (11).

Analogously, (10) is equal to

$$(\hat{U}_{s,n}^{-1})[(\hat{U}_{s,n} L_n f \hat{U}_{s,n}^{-1}), (A_n)](\hat{U}_{s,n}).$$

Since $\hat{U}_{s,n} L_n = L_n \hat{U}_{s,n}$ and $\hat{U}_{s,n} = U_{\{sn\}/n}|_{S_n}$ we obtain

$$(\hat{U}_{s,n} L_n f \hat{U}_{s,n}^{-1}) = (L_n U_{\{sn\}/n} f U_{-\{sn\}/n}|_{S_n}),$$

and the uniform continuity of f on $\mathbb{R}$ entails that

$$\|U_{\{sn\}/n} f U_{-\{sn\}/n} - U_s f U_{-s}\|_\infty \to 0$$

as $n \to \infty$ which yields (12). The same arguments show that modulo $\mathcal{J}$

$$(\hat{D}_{s,n}^{-1})(L_n K|_{S_n} + C_n)(\hat{D}_{s,n}) = (L_n D_s^{-1} K D_s|_{S_n}),$$

$$(\hat{U}_{s,n}^{-1})(L_n K|_{S_n} + C_n)(\hat{U}_{s,n}) = (L_n U_s^{-1} K U_s|_{S_n})$$

which gives (13) and (14), respectively. ∎

In order to establish the analogue of Proposition 4 for the sequences $(L_n W^0(f)|_{S_n})$ in place of $(L_n fI|_{S_n})$ we describe the structure of the approximation operators $L_n W^0(f)|_{S_n}$. For this goal, define operators $P_n : L^2 \to L^2$ by

$$P_n f = (x_k) \quad \text{with} \quad x_k = \int_{\mathbb{R}} f(\frac{s+k}{n}) \overline{\varphi(s)}\, ds.$$

These operators are bounded, and $\|P_n\| \leq Cn^{1/2}$ (see [3], Proposition 2.13). Thus, the operators $P_n E_n : l^2 \to l^2$ are correctly defined, and it is not hard to see that they are independent of n and that $P_n E_n$ is just the Laurent operator $T^0(\lambda)$. This operator is invertible (what is a consequence of (3)), and $L_n = E_n T^0(\lambda)^{-1} P_n$ (compare [3], Propositions 2.13 and 2.14).

Proposition 5 *(a) Let a be a piecewise continuous function on $\mathbb{R}$. For each n, there is a function $a_n^\dagger$ in $L^\infty(\mathbf{T})$ such that*

$$P_n W^0(a) E_n = T^0(a_n^\dagger) \quad and \quad E_{-n} L_n W^0(a) E_n = T^0(\lambda^{-1} a_n^\dagger).$$

(b) If φ is sufficiently smooth (say, piecewise C^1) then

$$a_n^\dagger(e^{2\pi i t}) = \sum_{k \in \mathbb{Z}} a(-n(k+t)) \, |(F\bar{\varphi})(k+t)|^2, \quad t \in (0,\, 1).$$

The proof proceeds exactly as that one in [3], Sections 2.7.6 and 2.11.3. The condition imposed on φ in part (b) is needed to guarantee the convergence of the series. Here are some special choices of the generating function a.

Example 1 If $a \equiv 1$ then $W^0(a)$ is the identity operator and $P_n W^0(a) E_n = P_n E_n = T^0(\lambda)$. Thus, $1_n^\dagger = \lambda$ and

$$\lambda(e^{2\pi i t}) = \sum_{k \in \mathbb{Z}} |(F\bar{\varphi})(k+t)|^2. \tag{16}$$

Example 2 If $a = \text{sgn}$ then $W^0(a) = S_{\mathbb{R}}$, hence $\text{sgn}_n^\dagger = \sigma$ and

$$\sigma(e^{2\pi i t}) = -\sum_{k \in \mathbb{Z}} \text{sgn}(k + 1/2) |(F\bar{\varphi})(k+t)|^2 \tag{17}$$

(notice that $\text{sgn}(-n(k+t)) = -\text{sgn}(k+1/2)$ for all $n \in \mathbb{Z}^+$, $k \in \mathbb{Z}$ and $t \in (0,\, 1)$). In both examples, $a_n^\dagger$ is independent of n.

Example 3 Let a be a piecewise continuous function with compact support, say $\text{supp}\, a \subseteq [-m,\, m]$ with $m \in \mathbb{Z}^+$. Then $a(-n(k+t))$ does not vanish if and only if

$$-m \le -n(k+t) \le m \quad \text{resp.} \quad -m/n \le k+t \le m/n,$$

and in case $n > m$ this involves (since $t \in (0,\, 1)$) that $k \in \{0, -1\}$. Further, if $k = 0$ then the inequalities $-m/n \le t \le m/n$ and $0 < t$ imply that $0 < t \le m/n$, whereas $1 - m/n \le t < 1$ in case $k = -1$. Summarizing this, we get

Proposition 6 *If $\text{supp}\, a \subseteq [-m,\, m]$ and $n > 2m$ then*

$$a_n^\dagger(e^{2\pi i t}) = \begin{cases} a(-nt)|(F\bar{\varphi})(t)|^2 & \text{if } t \in (0,\, m/n], \\ 0 & \text{if } t \in (m/n,\, 1-m/n), \\ a(-n(t-1))|(F\bar{\varphi})(t-1)|^2 & \text{if } t \in [1-m/n,\, 1), \end{cases}$$

and this representation holds for arbitrary mother splines φ (no convergence problems occur).

Now we have the following analogue of Proposition 4.

Proposition 7 *(a) If $f \in C(\dot{\mathbb{R}})$ then the commutators*

$$[(L_n W^0(f)|_{S_n}),\, (E_n T^0(a) E_{-n})] \quad and \quad [(L_n W^0(f)|_{S_n}),\, (E_n P E_{-n})]$$

belong to the ideal $\mathcal{J}$ for all bounded functions a.
(b) Let $(A_n) \in \mathcal{A}$. If the commutators $[(L_n W^0(f)|_{S_n}), (A_n)]$ belong to $\mathcal{J}$ for all $f \in C(\dot{\mathbb{R}})$ then the commutators

$$[(L_n W^0(f)|_{S_n}), (\hat{D}_{s,n}^{-1} A_n \hat{D}_{s,n})], \tag{18}$$

$$[(L_n W^0(f)|_{S_n}), (\hat{U}_{s,n}^{-1} A_n \hat{U}_{s,n})] \tag{19}$$

belong to $\mathcal{J}$ for all $f \in C(\dot{\mathbb{R}})$, too.

Proof. (a) The first assertion is obvious since

$$L_n W^0(f) E_n T^0(a) E_{-n} - E_n T^0(a) E_{-n} L_n W^0(f)|_{S_n}$$

$$= E_n T^0(\lambda^{-1} f_n^\dagger) T^0(a) E_{-n} - E_n T^0(a) T^0(\lambda^{-1} f_n^\dagger) E_{-n} = 0$$

by Proposition 5. For a proof of the second part of assertion (a), let K denote the operator $P - E_{-n} L_n \chi_{\mathbb{R}^+} E_n$. This operator is compact (see [3], Section 2.11.4). Further set

$$a(x) = \begin{cases} 0 & \text{if} \quad x < -1, \\ (x+1)/2 & \text{if} \quad -1 \le x \le 1, \\ 1 & \text{if} \quad 1 < x, \end{cases}$$

and let $b := \chi_{\mathbb{R}^+} - a$. Then

$$E_n P E_{-n} = L_n a I|_{S_n} + L_n b I|_{S_n} + E_n K E_{-n},$$

and it remains to show that

$$[(L_n W^0(f)|_{S_n}), (L_n a I|_{S_n})] \in \mathcal{J}, \tag{20}$$

$$[(L_n W^0(f)|_{S_n}), (L_n b I|_{S_n})] \in \mathcal{J}, \tag{21}$$

$$[(L_n W^0(f)|_{S_n}), (E_n K E_{-n})] \in \mathcal{J} \tag{22}$$

for all functions $a \in C(\bar{\mathbb{R}})$ and all compactly supported functions b and for all compact operators K.

Assertion (20) is again a consequence of the commutator relation

$$\|L_n a I - a L_n\| \to 0 \quad \text{as} \quad n \to \infty \tag{23}$$

which holds for all functions $a \in C(\bar{\mathbb{R}})$ (observe that $\omega(a, 1/n) \to 0$ for all of these functions), and of the compactness of the operator$W^0(f) a I - a W^0(f)$ (see, e.g., [7], Proposition 12.6 (b)).

For a proof of (21), choose a compactly supported continuous function c on $\mathbb{R}$ such that $c(x) = 1$ for $x \in [-1, 1]$. Then

$$(L_n W^0(f) L_n b I|_{S_n}) = (L_n W^0(f) L_n c b I|_{S_n}) = (L_n W^0(f) c I|_{S_n})(L_n b I|_{S_n}) + (G_n)$$

with $\|G_n\| \to 0$, again by the commutator relation (23). The operator $W^0(f)cI$ can be written as $f(\infty)cI + W^0(f')cI$ where $f'(\infty) = 0$ and, hence, $W^0(f')cI$ is compact (see [7], Proposition 12.6 (a)). Thus,

$$(L_n W^0(f) L_n bI|_{S_n}) - f(\infty)(L_n bI|_{S_n}) \in \mathcal{J}$$

and, analogously,

$$(L_n b L_n W^0(f)|_{S_n}) - f(\infty)(L_n bI|_{S_n}) \in \mathcal{J}$$

whence (21) follows.

For (22), introduce projection operators R_k $(k \geq 1)$ by

$$R_k : l^2 \to l^2, \quad (x_n) \mapsto (\ldots, 0, 0, x_{-k}, \ldots, x_{k-1}, 0, 0, \ldots).$$

Because of $R_k \to I$ as $k \to \infty$, the compact operator K can be approximated as closely as desired by operators of the form $R_k K R_k$, and so it is sufficient to replace the K in (22) by $R_k K R_k$. Let further $[-a, a]$ be an interval containing the support of the mother spline φ. Then, clearly,

$$E_n R_k K R_k E_{-n} = \chi_{[(-k-a)/n,\,(k+a)/n]} E_n R_k K R_k E_{-n},$$

and if c denotes a compactly supported continuous function which is equal to 1 on the interval $[(-k-a)/n, (k+a)/n]$ then

$$(L_n W^0(f)|_{S_n})(E_n R_k K R_k E_{-n}) =$$

$$= f(\infty)(E_n R_k K R_k E_{-n}) + (L_n W^0(f') c E_n R_k K R_k E_{-n})$$

$$= f(\infty)(E_n R_k K R_k E_{-n}) + (L_n W^0(f') cI|_{S_n})(E_n R_k K R_k E_{-n}).$$

The sequence $(E_n R_k K R_k E_{-n})$ is in $\mathcal{A}$, and its strong limit is equal to 0 (see [3], Proposition 3.13], and the sequence $(L_n W^0(f') cI|_{S_n})$ is even in the ideal $\mathcal{J}$ (recall that $W^0(f')cI$ is compact). Hence,

$$(L_n W^0(f)|_{S_n})(E_n R_k K R_k E_{-n}) - f(\infty)(E_n R_k K R_k E_{-n}) \in \mathcal{J}$$

and, analogously,

$$(E_n R_k K R_k E_{-n})(L_n W^0(f)|_{S_n}) - f(\infty)(E_n R_k K R_k E_{-n}) \in \mathcal{J}$$

which gives (22).

(b) We claim that the commutators (18) and (19) coincide with

$$(\hat{D}_{s,n}^{-1})[(L_n W^0(f_s)|_{S_n}),\, (A_n)](\hat{D}_{s,n}), \tag{24}$$

$$(\hat{U}_{s,n}^{-1})[(L_n W^0(f)|_{S_n}),\, (A_n)](\hat{U}_{s,n}) \tag{25}$$

where $f_s(t) = f(s+t)$, respectively. The remaining parts of the proof proceed as in Proposition 4 then.

As in the proof of that proposition we get that (18) is nothing else than

$$(\hat{D}_{s,n}^{-1})[(\hat{D}_{s,n}L_nW^0(f)\hat{D}_{s,n}^{-1}),\,(A_n)](\hat{D}_{s,n}),$$

and employing relations (5) and (15) we find that

$$(\hat{D}_{s,n}L_nW^0(f)\hat{D}_{s,n}^{-1}) - (L_nD_sW^0(f)D_s^{-1}|_{S_n}) \in \mathcal{J}.$$

But, as one easily checks, $D_sW^0(f)D_s^{-1} = W^0(f_s)$. Similarly, (19) is the same as

$$(\hat{U}_{s,n}^{-1})[(\hat{U}_{s,n}L_nW^0(f)\hat{U}_{s,n}^{-1}),\,(A_n)](\hat{U}_{s,n}),$$

and $\hat{U}_{s,n}L_nW^0(f)\hat{U}_{s,n}^{-1} = L_nW^0(f)|_{S_n}$ since $\hat{U}_{s,n}L_n = L_nU_{\{sn\}/n}$ and $W^0(f)$ is invariant with respect to the shift $W^0(f) \mapsto U_xW^0(f)U_x^{-1}$. ∎

Finally, the following quasicommutator relations hold.

Proposition 8 *(a) If $f \in C(\dot{\mathbb{R}})$ and $g \in PC$ (= the algebra of the piecewise continuous functions) then*

$$(L_nfI|_{S_n})(L_ngI|_{S_n}) - (L_nfgI|_{S_n}) \in \mathcal{J}.$$

(b) If $f, g \in PC$ and $f(\pm\infty) = g(\pm\infty) = 0$ then

$$(L_nfI|_{S_n})(L_nW^0(g)|_{S_n}) \in \mathcal{J} \quad \textit{and} \quad (L_nW^0(g)|_{S_n})(L_nfI|_{S_n}) \in \mathcal{J}.$$

(c) If $f, g \in PC$ and $f(+\infty) = f(-\infty)$ or $g(+\infty) = g(-\infty)$ then

$$(L_nW^0(f)|_{S_n})(L_nW^0(g)|_{S_n}) - (L_nW^0(fg)|_{S_n}) \in \mathcal{J}.$$

Proof. (a) This is a well-known fact which follows immediately from the commutator relation (23).

(b) The hypothesis $f(\pm\infty) = g(\pm\infty) = 0$ ensures that both f and g can be approximated (in the supremum norm) by compactly supported functions, f' and g' say. Let c be a continuous function which is compactly supported, too, and which takes the value 1 on the support of f'. Then

$$(L_nf'I|_{S_n})(L_nW^0(g')|_{S_n}) = (L_nf'cL_nW^0(g')|_{S_n})$$

$$= (L_nf'I|_{S_n})(L_ncW^0(g')|_{S_n}) + (G_n)$$

where $\|G_n\| \to 0$ by (23) and, thus, $(G_n) \in \mathcal{J}$. The operator $cW^0(g')$ is compact (Proposition 12.6 (a) in [7]), hence, $(L_ncW^0(g')|_{S_n}) \in \mathcal{J}$. The second assertion can be verified analogously.

(c) Let, for definiteness, $f(+\infty) = f(-\infty)$, and write f as $f(+\infty) + (f - f(+\infty))$. The assertion is obviously correct for the constant function $f(+\infty)$ in place of f. Further, the function $f - f(+\infty)$ is continuous at infinity and equal to 0 there. Hence, this function can be approximated by a compactly supported and piecewise continuous one.

Similarly, we decompose the function g into

$$\frac{g(+\infty) + g(-\infty)}{2} + \frac{g(+\infty) - g(-\infty)}{2}\mathrm{sgn} + g'$$

where $g'(\pm\infty) = 0$, and g' can be approximated by compactly supported functions again. So it remains to consider the quasicommutators

$$(L_n W^0(f)|_{S_n})(L_n W^0(g)|_{S_n}) - (L_n W^0(fg)|_{S_n}), \tag{26}$$

$$(L_n W^0(f)|_{S_n})(L_n S_{\mathbf{R}}|_{S_n}) - (L_n W^0(f) S_{\mathbf{R}}|_{S_n}) \tag{27}$$

with compactly supported and piecewise continuous functions f and g (recall that $S_{\mathbf{R}} = W^0(\mathrm{sgn})$).

Let $[-k, k]$ be an interval containing the supports of f and g. Then, for all $n \geq 2k$,

$$L_n W^0(f)|_{S_n} = E_n T^0(\lambda^{-1}) T^0(f_n^\dagger) E_{-n}$$

with $f_n^\dagger$ given as in Proposition 6, and the uniform boundedness of the norms $\|E_n\|\,\|E_{-n}\|$ with respect to n entails that

$$\|L_n W^0(f) L_n W^0(g)|_{S_n} - L_n W^0(fg)|_{S_n}\|$$

$$\|E_n T^0(\lambda^{-1} f_n^\dagger) T^0(\lambda^{-1} g_n^\dagger) E_{-n} - E_n T^0(\lambda^{-1}(fg)_n^\dagger) E_{-n}\|$$

$$\leq C \|T^0(\lambda^{-1} f_n^\dagger) T^0(\lambda^{-1} g_n^\dagger) - T^0(\lambda^{-1}(fg)_n^\dagger)\|$$

$$= C \sup_{z \in \mathbf{T}} |\lambda(z)^{-1} f_n^\dagger(z) \cdot \lambda(z)^{-1} g_n^\dagger(z) - \lambda(z)^{-1}(fg)_n^\dagger(z)| \tag{28}$$

(see [1], Proposition 2.2, for a proof that $\|T^0(a)\| = \sup_{z\in\mathbf{T}} |a(z)| =: \|a\|_\infty$ for all functions $a \in L^\infty(\mathbf{T})$). By Proposition 6, the supremum in (28) has to be taken only for those $z = e^{2\pi i y}$ with $y \in (0, k/n]$ or $y \in [1 - k/n, 1)$. Consider, e.g., the first case. Then (28) reads as

$$C \sup_{y \in (0,\, k/n]} \big|\lambda(e^{2\pi i y})^{-1} f(-ny) |(F\bar{\varphi})(y)|^2 \cdot \lambda(e^{2\pi i y})^{-1} g(-ny) |(F\bar{\varphi})(y)|^2$$

$$-\lambda(e^{2\pi i y})^{-1} f(-ny) g(-ny) |(F\bar{\varphi})(y)|^2\big|$$

$$\leq C \|f\|_\infty \|g\|_\infty \sup_{y \in (0,\, k/n]} \big|\lambda(e^{2\pi i y})^{-2} |(F\bar{\varphi})(y)|^4 - \lambda(e^{2\pi i y})^{-1} |(F\bar{\varphi})(y)|^2\big|. \tag{29}$$

Both functions $y \mapsto \lambda(e^{2\pi i y})^{-1}$ and $y \mapsto |(F\bar{\varphi})(y)|^2$ are continuous (since λ is a polynomial and $\bar{\varphi}$ is in L^1) and take the value 1 at $y = 0$ (compare [3], Proposition 2.19). Hence, the mapping

$$y \mapsto \lambda(e^{2\pi i y})^{-2} |(F\bar{\varphi})(y)|^4 - \lambda(e^{2\pi i y})^{-1} |(F\bar{\varphi})(y)|^2$$

is continuous and has a zero at $y = 0$ whence follows that the supremum in (29) tends to 0 as $n \to \infty$. This shows that

$$\|L_n W^0(f) L_n W^0(g)|_{S_n} - L_n W^0(fg)|_{S_n}\| \to 0 \quad \text{as} \quad n \to \infty$$

which, in particular, implies that the quasicommutator (26) is in $\mathcal{J}$.

Analogously, we find for (27)

$$\|L_n W^0(f) L_n S_{\mathbf{R}}|_{S_n} - L_n W^0(f) S_{\mathbf{R}}|_{S_n})\|$$

$$\leq C \sup_{z \in \mathbb{T}} \left| \lambda(z)^{-1} f_n^\dagger(z) \cdot \lambda(z)^{-1} \sigma(z) - \lambda(z)^{-1} (f \cdot \mathrm{sgn})_n^\dagger(z) \right|$$

and if the supremum is again considered over $z = e^{2\pi i y}$ with $y \in (0,\, k/n]$ only, then the latter item is equal to

$$C \sup_{y \in (0,\, k/n]} \left| \lambda(e^{2\pi i y})^{-1} f(-ny) |(F\bar{\varphi})(y)|^2 \cdot \lambda(e^{2\pi i y})^{-1} \sigma(e^{2\pi i y}) \right.$$

$$\left. - \lambda(e^{2\pi i y})^{-1} f(-ny) \mathrm{sgn}(-ny) |(F\bar{\varphi})(y)|^2 \right|$$

$$\leq C \|\lambda^{-1}\|_\infty \|f\|_\infty \|F\bar{\varphi}\|_\infty^2 \sup_{y \in (0,\, k/n]} \left| \lambda(e^{2\pi i y})^{-1} \sigma(e^{2\pi i y}) - \mathrm{sgn}(-ny) \right| .$$

For $y \in (0,\, k/n]$ we have $\mathrm{sgn}(-ny) = -1$, further the function $y \mapsto \lambda(e^{2\pi i y})^{-1} \sigma(e^{2\pi i y})$ is continuous on $(0,\, k/n]$ and possesses the one-sided limit -1 as $y \searrow 0$ (see [3], Theorem 2.15). These facts combine to give that

$$\|L_n W^0(f) L_n S_{\mathbb{R}}|_{S_n} - L_n W^0(f) S_{\mathbb{R}}|_{S_n})\| \to 0 \quad \text{as} \quad n \to \infty$$

and, hence, the quasicommutator (27) belongs to $\mathcal{J}$. ∎

3 An algebra of approximation sequences for Wiener-Hopf operators

We let PC_1 stand for the algebra of all piecewise continuous functions on the unit circle $\mathbf{T}$ which are continuous on $\mathbb{T} \setminus \{1\}$, and we consider the smallest closed subalgebra $\mathcal{B}$ of the algebra $\mathcal{A}$ which contains all sequences (G_n) with $\|G_n\| \to 0$ as $n \to \infty$, all sequences $(E_n T^0(a) E_{-n})$ with $a \in PC_1$, as well as the sequence $(E_n P E_{-n})$, and which has the property that, whenever a sequence (A_n) is in $\mathcal{B}$ then the sequences $(\hat{D}_{s,n}^{-1} A_n \hat{D}_{s,n})$ and $(\hat{U}_{s,n}^{-1} A_n \hat{U}_{s,n})$ belong for all s to $\mathcal{B}$, too. The correctness of this definition is a consequence of Propositions 2 and 3.

To motivate this definition, observe that the smallest closed subalgebra $\mathcal{D}$ of $L(L^2)$ which contains all operators aI of multiplication and all operators $W^0(b)$ of Fourier convolution by piecewise continuous functions a and b also admits the following equivalent characterization: it is the smallest closed subalgebra of $L(L^2)$ which contains the operators $\chi_{\mathbb{R}^+}$ and $S_{\mathbb{R}}$ and which has the property that, whenever an operator A belongs to $\mathcal{D}$ then the shifted operators $D_s^{-1} A D_s$ and $U_s^{-1} A U_s$ belong to this algebra, too. Thus, an algebra of approximation sequences for operators in $\mathcal{D}$ should reflect these properties, namely it should contain approximation sequences for the operators $\chi_{\mathbb{R}^+}$ and $S_{\mathbb{R}}$ (these are actually the sequences $(E_n P E_{-n})$ and $(E_n T^0(a) E_{-n})$, respectively), and it should be invariant with respect to the "shifts" $(A_n) \mapsto (\hat{D}_{s,n}^{-1} A_n \hat{D}_{s,n})$ and $(A_n) \mapsto (\hat{U}_{s,n}^{-1} A_n \hat{U}_{s,n})$. Here are some further sequences which belong to $\mathcal{B}$.

Proposition 9 *(a) If $a \in PC(\mathbb{R}) \cap C(\mathbb{R} \setminus \mathbb{Z}$ then the sequence $(L_n aI|_{S_n})$ is in $\mathcal{B}$.*
(b) If $a \in PC$ then the sequences $(L_n W^0(a)|_{S_n})$ and $(L_n W(a)|_{S_n})$ belong to $\mathcal{B}$.

Proof. For part (a) see [3], Propositions 3.15 and 3.24. For part (b) we first remark that the sequence

$$(L_n W^0(\mathrm{sgn})|_{S_n}) = (L_n S_{\mathbb{R}}|_{S_n}) = (E_n T^0(\lambda^{-1}\sigma)E_{-n})$$

belongs to $\mathcal{B}$ (see [3], Theorem 2.15 for the inclusion $\lambda^{-1}\sigma \in PC_1$). Thus, the sequence $(\hat{D}_{s,n}^{-1} L_n W^0(\mathrm{sgn})\hat{D}_{s,n})$ belongs to $\mathcal{B}$, too, and since

$$\|\hat{D}_{s,n}^{-1} L_n W^0(\mathrm{sgn})\hat{D}_{s,n} - L_n D_s^{-1} W^0(\mathrm{sgn}) D_s|_{S_n})\| \to 0$$

by the commutator relations (5) and (15), and since $D_s^{-1} W^0(\mathrm{sgn}) D_s = W^0(\mathrm{sgn}_s)$ with $\mathrm{sgn}_s(t) = -1$ if $t < -s$ and $\mathrm{sgn}_s(t) = 1$ if $t > -s$, one concludes that $(L_n W^0(a)|_{S_n}) \in \mathcal{B}$ for each piecewise constant function a. But the piecewise constant functions lie dense in the piecewise continuous ones which yields our claim for the sequence $(L_n W^0(a)|_{S_n})$. The proof for the sequences $(L_n W(a)|_{S_n})$ is similar and makes use of the fact that $(L_n S|_{S_n}) \in \mathcal{B}$ (see [3], Section 2.11.4 and Proposition 2.9 (a)). ∎

Further one can show that $\mathcal{J} \subseteq \mathcal{B}$ and, thus, $\mathcal{J}$ is a closed two-sided *−ideal of $\mathcal{B}$ (compare Proposition 3.17 in [3]). Since C^*-subalgebras of C^*-algebras are inverse closed we conclude from the Lifting Proposition 1 that a sequence (A_n) in $\mathcal{B}$ is stable if and only if the operator $A = W(A_n)$ is invertible and if the coset $(A_n) + \mathcal{J}$ is invertible in the quotient algebra $\mathcal{B}/\mathcal{J}$. So we are left with studying invertibility in $\mathcal{B}/\mathcal{J}$. We shall do this by a two-and-a-half-fold localization via the local principle of Allan and Douglas which reads as follows.

Local principle *(See, e.g., [1], 1.31.) Let* **B** *be a C^*- algebra and* **C** *be a *-subalgebra in the center of* **B** *(= the set of all elements which commute with each other element of* **B***). For each maximal ideal x of* **C** *let I_x denote the smallest closed *- ideal of* **B** *which contains x. Then an element $b \in$* **B** *is invertible if and only if the cosets $b + I_x$ are invertible for all maximal ideals x.*

Proposition 4 entails that the cosets $(L_n fI|_{S_n}) + \mathcal{J}$ belong to the center of $\mathcal{B}/\mathcal{J}$ for all $f \in C(\dot{\mathbb{R}})$. So it makes sense to reify the local principle with $\mathcal{B}/\mathcal{J}$ and

$$\mathcal{C} = \left\{(L_n fI|_{S_n}) + \mathcal{J}, \quad f \in C(\dot{\mathbb{R}})\right\}$$

in place of **B** and **C**, respectively. The maximal ideal space of **C** has been determined in [3], Propositions 3.21 and 3.9: it is homeomorphic to $\dot{\mathbb{R}}$, and the maximal ideal corresponding to $x \in \dot{\mathbb{R}}$ is

$$\left\{(L_n fI|_{S_n}) + \mathcal{J}, \quad f \in C(\dot{\mathbb{R}}) \quad \text{with} \quad f(x) = 0\right\}.$$

The outcome of this first localization step is certain *local algebras* $\mathcal{B}_x := (\mathcal{B}/\mathcal{J})/I_x$ for $x \in \dot{\mathbb{R}}$ with canonical homomorphisms $\Phi_x : \mathcal{B} \to \mathcal{B}_x$, $b \mapsto (b + \mathcal{J}) + I_x$. In case x is finite these algebras can be studied by constructing *locally equivalent representations.*

Proposition 10 *Let $(A_n) \in \mathcal{B}$ and $s \in \mathbb{R}$. Then the strong limit*

$$W_s(A_n) := \text{s-lim}_{k\to\infty} U_{s,k}^{-1} E_{-k} A_k E_k U_{s,k}$$

exists, and the mapping $W_s : \mathcal{B} \to L(l^2)$ *is a* **- homomorphism. In particular,*

$$W_s(E_n P E_{-n}) = \begin{cases} 0 & \textit{if} \quad s < 0 \\ P & \textit{if} \quad s = 0 \\ I & \textit{if} \quad s > 0, \end{cases} \tag{30}$$

$$W_s(E_n T^0(a) E_{-n}) = T^0(a), \tag{31}$$

$$W_s(L_n K|_{S_n}) = 0 \quad \textit{for } K \textit{ compact}, \tag{32}$$

$$W_s(L_n f I|_{S_n}) = f(s) I \quad \textit{for } f \in C(\dot{\mathbb{R}}), \tag{33}$$

and, moreover, for each sequence $(A_n) \in \mathcal{B}$ *and* $t \in \mathbb{R}$,

$$W_s(\hat{D}_{t,n}^{-1} A_n \hat{D}_{t,n}) = W_s(A_n), \tag{34}$$

$$W_s(\hat{U}_{t,n}^{-1} A_n \hat{U}_{t,n}) = W_{s+t}(A_n). \tag{35}$$

Proof. The identities (30) - (33) are shown in [3], Propositions 3.14, 3.16 and 3.17. For (34), we have by definition,

$$W_s(\hat{D}_{t,n}^{-1} A_n \hat{D}_{t,n}) = \text{s-lim } U_{s,n}^{-1} E_{-n} \hat{D}_{t,n}^{-1} A_n \hat{D}_{t,n} E_n U_{s,n}$$

$$= \text{s-lim } U_{s,n}^{-1} D_{t,n}^{-1} U_{s,n} \cdot U_{s,n}^{-1} E_{-n} A_n E_n U_{s,n} \cdot U_{s,n}^{-1} D_{t,n} U_{s,n}, \tag{36}$$

and the strong limit s-lim $U_{s,n}^{-1} D_{t,n} U_{s,n}$ exists. Indeed,

$$U_{s,n}^{-1} D_{t,n} U_{s,n} = e^{2\pi i t \{sn\}/n} D_{t,n}, \tag{37}$$

where the numbers $e^{2\pi i t\{sn\}/n}$ converge to $e^{2\pi i s t}$, and the operators $D_{t,n} = \text{diag}\,(e^{-2\pi i t k/n})$ converge strongly to the identity operator (since they converge entry-wise to this operator). Hence, s-lim $U_{s,n}^{-1} D_{t,n} U_{s,n} = e^{2\pi i s t} I$ and, analogously, s-lim $U_{s,n}^{-1} D_{t,n}^{-1} U_{s,n} = e^{-2\pi i s t} I$ whence (34) follows via (36).

Finally, (35) is also a consequence of Propositions 3.14, 3.16 and 3.17 of [3] in case (A_n) is a sequence of the form $(\hat{U}_{v,n}^{-1} B_n \hat{U}_{v,n})$ with (B_n) referring to one of the sequences mentioned in (30) - (33). The appearance of additional shifts $\hat{D}_{s,n}$ in (A_n) cannot change the matter due to (37) and the strong convergences established above. ■

Let $\mathcal{T}$ stand for the smallest closed subalgebra of $L(l^2)$ which contains the projection P and all Laurent operators $T^0(a)$ with $a \in PC_1$. The preceding proposition shows that the W_s are actually homomorphisms from $\mathcal{B}$ into $\mathcal{T}$. Moreover, we conclude from (32) that the ideal $\mathcal{J}$ lies in the kernel of W_s. So the quotient homomorphisms

$$\mathcal{B}/\mathcal{J} \to \mathcal{T}, \quad (A_n) + \mathcal{J} \mapsto W_s(A_n)$$

are correctly defined, and we denote them by W_s again. Furthermore, identity (33) entails that the local ideal I_s belongs to the kernel of the quotient W_s, hence, one has a natural mapping

$$\mathcal{B}_s \to \mathcal{T}, \quad \Phi_s(A_n) \mapsto W_s(A_n) \tag{38}$$

which will again be denoted by W_s for brevity.

Proposition 11 *The mapping (38) is an isometrical isomorphism between $\mathcal{B}_s$ and $\mathcal{T}$. In particular, the local coset $\Phi_s(A_n)$ is invertible if and only if the operator $W_s(A_n)$ is invertible. (In this sense, W_s is locally equivalent.)*

Proof. A surjective *-homomorphism between C^*-algebras which preserves spectra is automatically an isometry. So it remains to prove the second assertion of the proposition.

It is evident that invertibility of $\Phi_s(A_n)$ involves that of $W_s(A_n)$. For the reverse implication we verify that

$$\Phi_s(E_n W_s(A_n) E_{-n}) = \Phi_s(A_n) \quad \text{for all} \quad (A_n) \in \mathcal{B}. \tag{39}$$

Indeed, once (39) is established, one can argue as follows: If $W_s(A_n)$ is invertible (in $L(l^2)$) then it is invertible in $\mathcal{T}$ (inverse closedness), then the sequence $(E_n W_s(A_n) E_{-n})$ is invertible in $\mathcal{B}$ (its inverse is $(E_n W_s(A_n)^{-1} E_{-n}) \in \mathcal{B}$), then the coset $\Phi_s(E_n W_s(A_n) E_{-n})$ is invertible in $\mathcal{B}_s$ which, on its hand, coincides with $\Phi_s(A_n)$ by (39).

The mappings $(A_n) \mapsto \Phi_s(A_n)$ and $(A_n) \mapsto \Phi_s(E_n W_s(A_n) E_{-n})$ are continuous homomorphisms. So it suffices to verify (39) for the generating sequences of the algebra $\mathcal{B}$ in place of (A_n). Thus, since $U_{t,n}^{-1} T^0(a) U_{t,n} = T^0(a)$ and $D_{t,n}^{-1} P D_{t,n} = P$ and since (37) holds, it remains to prove (39) for the sequences $(E_n D_{t,n}^{-1} T^0(a) D_{t,n} E_{-n})$ with $a \in PC_1$ and $(E_n U_{t,n}^{-1} P U_{t,n} E_{-n})$ in place of (A_n). For the latter one this has been done in [3], Proposition 3.22. For the first one, write a as $\alpha\lambda^{-1}\sigma + c$ with a number α being chosen in such a manner that c becomes continuous on all of $\mathbf{T}$. Then

$$\Phi_s(E_n D_{t,n}^{-1} T^0(\lambda^{-1}\sigma) D_{t,n} E_{-n}) = \Phi_s(L_n W^0(\mathrm{sgn}_t)|_{S_n})$$

with $\mathrm{sgn}_t(x) = \mathrm{sgn}(x+t)$, whereas

$$\Phi_s(E_n W_s(E_n D_{t,n}^{-1} T^0(\lambda^{-1}\sigma) D_{t,n} E_{-n}) E_{-n})$$

$$= \Phi_s(E_n T^0(\lambda^{-1}\sigma) E_{-n}) = \Phi_s(L_n S_{\mathbf{R}}|_{S_n}) = \Phi_s(L_n W^0(\mathrm{sgn})|_{S_n})$$

by (31) and (34). Now

$$\Phi_s(L_n W^0(\mathrm{sgn}_t)|_{S_n}) - \Phi_s(L_n W^0(\mathrm{sgn})|_{S_n}) = \Phi_s(L_n W^0(\mathrm{sgn}_t - \mathrm{sgn})|_{S_n}).$$

The function $\mathrm{sgn}_t - \mathrm{sgn}$ is compactly supported and, hence,

$$(L_n W^0(\mathrm{sgn}_t - \mathrm{sgn})|_{S_n})(L_n f I|_{S_n}) \in \mathcal{J} \tag{40}$$

for every compactly supported continuous function f by Proposition 8 (b). Let, in addition, $f(s) = 1$ then $\Phi_s(L_n f I|_{S_n})$ is the identity element in $\mathcal{B}_s$. Thus, application of the local homomorphisms Φ_s to (40) yields the assertion in case $a = \lambda^{-1}\sigma$.

For $a = c$, approximate c by a polynomial, and so suppose without loss that $a = \tau_k$ with $\tau_k(z) = z^k$. Then

$$\Phi_s(E_n D_{t,n}^{-1} T^0(\tau_k) D_{t,n} E_{-n}) = \Phi_s(e^{2\pi i t k/n} E_n T^0(\tau_k) E_{-n})$$

as in (37) since $T^0(\tau_k)$ is nothing but a shift operator on l^2, and

$$\Phi_s(E_n W_s(E_n D_{t,n}^{-1} T^0(\tau_k) D_{t,n} E_{-n}) E_{-n}) = \Phi_s(E_n T^0(\tau_k) E_{-n})$$

by Proposition 10, (31) and (34). But $e^{2\pi itk/n} - 1 \to 0$ as $n \to \infty$, hence

$$\Phi_s(e^{2\pi itk/n} E_n T^0(\tau_k) E_{-n}) = \Phi_s(E_n T^0(\tau_k) E_{-n}). \quad \blacksquare$$

Our next goal is the local algebra $\mathcal{B}_\infty$. We examine this algebra by a second application of the local principle. By Proposition 7, all cosets $\Phi_\infty(L_n W^0(f)|_{S_n})$ with $f \in C(\dot{\mathbb{R}})$ lie in the center of $\mathcal{B}_\infty$, and it makes sense to localize $\mathcal{B}_\infty$ over its central subalgebra

$$\mathcal{C}_\infty := \left\{ \Phi_\infty(L_n W^0(f)|_{S_n}) \, , \quad f \in C(\dot{\mathbb{R}}) \right\}.$$

Proposition 12 *The maximal ideal space of $\mathcal{C}_\infty$ is homeomorphic to $\dot{\mathbb{R}}$, and the maximal ideal corresponding to $x \in \dot{\mathbb{R}}$ is*

$$\left\{ \Phi_\infty(L_n W^0(f)|_{S_n}) \, , \quad f \in C(\dot{\mathbb{R}}) \quad \textit{with} \quad f(s) = 0 \right\}.$$

Proof. Suppose the coset $\Phi_\infty(L_n W^0(f)|_{S_n})$ to be invertible. Then there are sequences $(B_n) \in \mathcal{B}$ and $(C_n) \in \mathcal{B}$ with $(C_n) + \mathcal{J} \in I_\infty$ such that $L_n W^0(f) B_n = L_n + C_n$. Letting n go to infinity we obtain $W^0(f)B = I + C$ where B belongs to the algebra $\mathcal{D}$ and C lies in the smallest closed ideal of this algebra which contains the compact operators and all operators gI of multiplication by continuous functions g vanishing at infinity. It is not hard to see that, for each operator A in $\mathcal{D}$, the strong limit s-lim $U_{-n} A U_n$ exists and that, in particular, $U_{-n} W^0(f) U_n \to W^0(f)$, $U_{-n} g U_n \to 0$ whenever g is continuous at infinity and equal to zero there, and $U_{-n} K U_n \to 0$ for each compact operator K. Thus, $W^0(f) \cdot \text{s-lim}\, U_{-n} B U_n = I$ which implies invertibility of $W^0(f)$ and, hence, that of f.

If, conversely, f is invertible, then Proposition 8 (c) yields

$$\Phi_\infty(L_n W^0(f)|_{S_n}) \cdot \Phi_\infty(L_n W^0(f^{-1})|_{S_n}) = \Phi_\infty(I|_{S_n}),$$

i.e. the coset $\Phi_\infty(L_n W^0(f)|_{S_n})$ is invertible. This shows that the algebras $\mathcal{C}_\infty$ and $C(\dot{\mathbb{R}})$ are isomorphic to each other which proves our claim. $\blacksquare$

The result of this second localization is local algebras $\mathcal{B}_{\infty,x} := \mathcal{B}_\infty / I_{\infty,x}$ for $x \in \dot{\mathbb{R}}$ with canonical homomorphisms $\Phi_{\infty,x} : \mathcal{B} \to \mathcal{B}_{\infty,x}$. Let us start with the local algebras $\mathcal{B}_{\infty,x}$ where x is finite.

Proposition 13 *If x is finite, then the local algebra $\mathcal{B}_{\infty,x}$ is generated by the identity element and by the cosets $p := \Phi_{\infty,x}(E_n P E_{-n})$ and $q := \Phi_{\infty,x}(L_n W^0(\chi_{[x,\infty)})|_{S_n})$, both elements are projections (= self-adjoint and idempotent), and the spectrum of pqp is the interval* [0, 1].

Proof. We have already seen that $\mathcal{B}_{\infty,x}$ is generated by the set of all elements of the form $\Phi_{\infty,x}(E_n U_{t,n}^{-1} P U_{t,n} E_{-n})$ and $\Phi_{\infty,x}(E_n D_{t,n}^{-1} T^0(a) D_{t,n} E_{-n})$ with $t \in \mathbb{R}$ and $a \in PC_1$. The identity

$$\Phi_{\infty,x}(E_n U_{t,n}^{-1} P U_{t,n} E_{-n}) = \Phi_{\infty,x}(E_n P E_{-n})$$

for all t is shown in the proof of Proposition 3.22 in [3] (and holds even for Φ_∞ in place of $\Phi_{\infty,x}$). Further, as in the proof of Proposition 11, write a as $\alpha\lambda^{-1}\sigma + c$ with c continuous and find

$$\Phi_{\infty,x}(E_n D_{t,n}^{-1} T^0(\lambda^{-1}\sigma) D_{t,n} E_{-n}) = \Phi_{\infty,x}(L_n W^0(\text{sgn}_t)|_{S_n}),$$

and it is easy to see that the latter coset coincides with $\Phi_{\infty,x}(I|_{S_n})$ if $t < x$ and with $\Phi_{\infty,x}(-I|_{S_n})$ if $t > x$. Finally, if we consider $T^0(\tau_k)$ in place of $T^0(c)$ then we get (as in the proof of Proposition 11 again)

$$\Phi_{\infty,x}(E_n D_{t,n}^{-1} T^0(\tau_k) D_{t,n} E_{-n}) = \Phi_{\infty,x}(E_n T^0(\tau_k) E_{-n}).$$

Choose a compactly supported continuous function g on $\mathbb{R}$ with $g(x) = 1$. Then the coset $\Phi_{\infty,x}(L_n W^0(g)|_{S_n})$ is the identity element in $\mathcal{B}_{\infty,x}$, and thus,

$$\Phi_{\infty,x}(E_n T^0(\tau_k) E_{-n}) - \Phi_{\infty,x}(E_n I E_{-n}) = \Phi_{\infty,x}(E_n T^0(\tau_k - 1) E_{-n} L_n W^0(g)|_{S_n}),$$

$$= \Phi_{\infty,x}(E_n T^0(\tau_k - 1) T^0(\lambda^{-1} g_n^\dagger) E_{-n})$$

and $\|T^0((\tau_k - 1)\lambda^{-1} g_n^\dagger)\| = \sup_z |(z^k - 1)\lambda^{-1}(z) g_n^\dagger(z)|$ goes to zero as $n \to \infty$ by the same arguments as used in the proof of proposition 8. Thus, $\mathcal{B}_{\infty,x}$ is generated by the identity and by the cosets $\Phi_{\infty,x}(E_n P E_{-n})$ and $\Phi_{\infty,x}(L_n W^0(\mathrm{sgn}_x)|_{S_n})$, and this proves the first part of the assertion since $(1 + \mathrm{sgn}_x)/2 = \chi_{[x,\infty)}$. Further, both elements are projections. This is evident for p and can be seen for q as follows (here we take $x = 0$ for simplicity, the proof for $x \neq 0$ is analogous). Choose a function g as before. Then

$$q^2 - q = \Phi_{\infty,x}\left(((L_n W^0(\chi_{\mathbb{R}^+})|_{S_n})^2 - (L_n W^0(\chi_{\mathbb{R}^+})|_{S_n}))(L_n W^0(g)|_{S_n})\right)$$

$$= \Phi_{\infty,x}\left(E_n T^0(((\frac{1+\lambda^{-1}\sigma}{2})^2 - (\frac{1+\lambda^{-1}\sigma}{2}))\lambda^{-1} g_n^\dagger) E_{-n}\right),$$

and $\|((\frac{1+\lambda^{-1}\sigma}{2})^2 - (\frac{1+\lambda^{-1}\sigma}{2}))\lambda^{-1} g_n^\dagger\|_\infty$ goes to 0 as $n \to \infty$ (remember that the function $((\frac{1+\lambda^{-1}\sigma}{2})^2 - (\frac{1+\lambda^{-1}\sigma}{2}))$ is continuous at $1 \in \mathbf{T}$ and has a zero there). Thus, $q^2 = q$.

It remains to compute the spectrum of pqp; for, we need some further knowledge about the local algebra $\mathcal{B}_{\infty,\infty}$, and that's why we defer the computation of the spectrum to Corollary 1 below. ∎

Now two-projections-theorems apply to give a precise description of the local algebra $\mathcal{B}_{\infty,x}$ for $x \neq \infty$ (compare, e.g., [4], [8], [3], Theorem 1.10).

Proposition 14 *Let $x \in \mathbb{R}$. Then the local algebra $\mathcal{B}_{\infty,x}$ is isomorphic to the subalgebra of the algebra $C([0,1], rm\, \mathbb{C}^{2\times 2})$ of all continuous 2×2-matrix-valued functions on $[0,1]$ which are diagonal at 0 and 1. The isomorphism sends*

$$\Phi_{\infty,x}(I|_{S_n}) \quad \text{into} \quad y \mapsto \begin{pmatrix} 1 & 0 \\ 0 & 1 \end{pmatrix},$$

$$\Phi_{\infty,x}(E_n P E_{-n}) \quad \text{into} \quad y \mapsto \begin{pmatrix} 1 & 0 \\ 0 & 0 \end{pmatrix},$$

$$\Phi_{\infty,x}(L_n W^0(\chi_{[x,\infty)})|_{S_n}) \quad \text{into} \quad y \mapsto \begin{pmatrix} y & \sqrt{y(1-y)} \\ \sqrt{y(1-y)} & 1-y \end{pmatrix}.$$

In particular, an element of $\mathcal{B}_{\infty,x}$ is invertible if and only if the associated matrix function is invertible.

We agree upon denoting the matrix function associated with the sequence (A_n) locally at ∞, x by $W_{\infty,x}(A_n)$.

Our final goal is the local algebra $\mathcal{B}_{\infty,\infty}$.

Proposition 15 *The algebra $\mathcal{B}_{\infty,\infty}$ is generated by all cosets $\Phi_{\infty,\infty}(E_nT^0(a)E_{-n})$ with $a \in PC_1$ and $\Phi_{\infty,\infty}(E_nPE_{-n})$. This algebra is commutative.*

Proof. We have already seen that the local algebra at ∞,∞ is generated by all cosets $\Phi_{\infty,\infty}(E_nU_{t,n}^{-1}PU_{t,n}E_{-n})$ and $\Phi_{\infty,\infty}(E_nD_{t,n}^{-1}T^0(a)D_{t,n}E_{-n})$. As in the proof of Proposition 13 one gets $\Phi_{\infty,\infty}(E_nU_{t,n}^{-1}PU_{t,n}E_{-n}) = \Phi_{\infty,\infty}(E_nPE_{-n})$ and, decomposing a into $\alpha\lambda^{-1}\sigma + c$, one obtains

$$\Phi_{\infty,\infty}(E_nD_{t,n}^{-1}T^0(\lambda^{-1}\sigma)D_{t,n}E_{-n}) = \Phi_{\infty,\infty}(L_nW^0(\text{sgn}_t)|_{S_n})$$

$$= \Phi_{\infty,\infty}(L_nW^0(\text{sgn})|_{S_n}) = \Phi_{\infty,\infty}(E_nT^0(\lambda^{-1}\sigma)E_{-n})$$

(clearly, $\text{sgn}_t - \text{sgn}$ is continuous at infinity and equal to zero there). Moreover, for $c = \tau_k$,

$$\Phi_{\infty,\infty}(E_nD_{t,n}^{-1}T^0(\tau_k)D_{t,n}E_{-n}) = \Phi_{\infty,\infty}(e^{2\pi itk/n}E_nT^0(\tau_k)E_{-n})$$

$$= \Phi_{\infty,\infty}(E_nT^0(\tau_k)E_{-n}).$$

Hence,

$$\Phi_{\infty,\infty}(E_nD_{t,n}^{-1}T^0(a)D_{t,n}E_{-n}) = \Phi_{\infty,\infty}(E_nT^0(a)E_{-n}).$$

for all functions $a \in PC_1$ and all $t \in \mathbb{R}$. For the commutativity, consider

$$\left[\Phi_{\infty,\infty}(E_nPE_{-n}),\, \Phi_{\infty,\infty}(E_nT^0(\lambda^{-1}\sigma)E_{-n})\right]$$

$$= \left[\Phi_{\infty,\infty}(L_n\chi_{\mathbb{R}^+}I|_{S_n}),\, \Phi_{\infty,\infty}(L_nW^0(\text{sgn})|_{S_n})\right]. \tag{41}$$

Locally (at ∞,∞) seen, one can replace $\chi_{\mathbb{R}^+}$ by a function g which is continuous on $\bar{\mathbb{R}}$ and satisfies $g(-\infty) = 0$, $g(+\infty) = 1$. Analogously, replace sgn by $2g - 1$. Thus, instead of (41) it remains to deal with the commutator

$$\left[\Phi_{\infty,\infty}(L_ngI|_{S_n}),\, \Phi_{\infty,\infty}(L_nW^0(g)|_{S_n})\right] = \Phi_{\infty,\infty}([(L_ngI|_{S_n}),\, (L_nW^0(g)|_{S_n})])$$

$$= \Phi_{\infty,\infty}(L_n(gW^0(g) - W^0(g)gI)|_{S_n})$$

by the commutator relation (23). But the operator $gW^0(g) - W^0(g)gI$ is compact (see, e.g., [7], Proposition 12.6 (b)) which shows that (41) actually vanishes. Finally,

$$\left[\Phi_{\infty,\infty}(E_nPE_{-n}),\, \Phi_{\infty,\infty}(E_nT^0(\tau_k)E_{-n})\right] = \Phi_{\infty,\infty}(E_n(PT^0(\tau_k) - T^0(\tau_k)P)E_{-n}) = 0$$

since $PT^0(\tau_k) - T^0(\tau_k)P$ is compact as one easily checks. ∎

Proposition 16 *If* $(A_n) \in \mathcal{B}$ *then the strong limits*

$$W_\pm(A_n) = \text{s-lim}_{n\to\infty} E_{-n} U_{\mp n} A_n U_{\pm n} E_n$$

exist, and the mappings $W_\pm$ *are* **-homomorphisms form* $\mathcal{B}$ *into the algebra* $T^0(PC_1)$ *of all Laurent operators with generating function in* PC_1. *In particular,*

$$W_\pm(E_n U_{t,n}^{-1} P U_{t,n} E_{-n}) = \begin{cases} I & \text{for } W_+ \\ 0 & \text{for } W_-, \end{cases} \tag{42}$$

$$W_\pm(E_n D_{t,n}^{-1} T^0(a) D_{t,n} E_{-n}) = T^0(a), \tag{43}$$

$$W_\pm(L_n aI|_{S_n}) = a(\pm\infty) I \quad \text{for} \quad a \in PC(\mathbb{R}) \cap C(\mathbb{R} \setminus \mathbb{Z}), \tag{44}$$

$$W_\pm(L_n W^0(a)|_{S_n}) = T^0(a(+\infty)\frac{1+\lambda^{-1}\sigma}{2} + a(-\infty)\frac{1-\lambda^{-1}\sigma}{2}) \quad \text{for} \quad a \in PC, \tag{45}$$

$$W_\pm(L_n K|_{S_n}) = 0 \quad \text{for } K \text{ compact.} \tag{46}$$

Proof. The sequences $(E_n U_{t,n}^{-1} P U_{t,n} E_{-n})$ and $(E_n D_{t,n}^{-1} T^0(a) D_{t,n} E_{-n})$ with $t \in \mathbb{R}$ and $a \in PC_1$ generate $\mathcal{B}$. Thus, the first assertion of the proposition follows immediately from (42) and (43). In order to verify these identities, abbreviate the shift operator $T^0(\tau_k)$ to V_k. Using $U_n E_n = E_n V_{n^2}$ and $E_{-n} U_{-n} = V_{-n^2} E_{-n}$ one gets

$$W_\pm(E_n U_{t,n}^{-1} P U_{t,n} E_{-n}) = \text{s-lim} V_{\mp n^2 + \{tn\}} P V_{\pm n^2 - \{tn\}}$$

which yields (42). Analogously,

$$W_\pm(E_n D_{t,n}^{-1} T^0(a) D_{t,n} E_{-n}) = \text{s-lim} V_{\mp n^2} D_{t,n}^{-1} T^0(a) D_{t,n} V_{\pm n^2}$$

$$= \text{s-lim} D_{t,n}^{-1} V_{\mp n^2} T^0(a) V_{\pm n^2} D_{t,n} \quad (\text{use } (37))$$

$$= \text{s-lim} D_{t,n}^{-1} T^0(a) D_{t,n} = T^0(a)$$

(recall that $D_{t,n} \to I$ strongly as $n \to \infty$). This proves (42) and (43), and having these two strong limits in mind, one easily derives (44) and (45). Finally, (46) holds for the compact operators $fS_{\mathbb{R}} - S_{\mathbb{R}} fI$ with $f \in C(\dot{\mathbb{R}})$ in place of K (use the commutator relation (23) again as well as (44) and (45)), and since each compact operator can be approximated in the operator norm by operators of the form $\sum a_i(f_i S_{\mathbb{R}} - S_{\mathbb{R}} f_i) b_i I$ with a_i, b_i, $f_i \in C(\dot{\mathbb{R}})$ (compare Exercise E 3.1 in [3]), (46) holds for arbitrary K. ∎

By (46), the ideal $\mathcal{J}$ lies in the kernel of $W_\pm$, and one can introduce the quotient homomorphisms

$$\mathcal{B}/\mathcal{J} \to T^0(PC_1), \quad (A_n) + \mathcal{J} \mapsto W_\pm(A_n),$$

denoted by $W_\pm$ again. Further, if f is continuous on $\dot{\mathbb{R}}$ and $f(\infty) = 0$ then $W_\pm(L_n fI|_{S_n}) = 0$, hence the local ideal generated by these cosets lies in the kernel of $W_\pm$, and one has natural homomorphisms

$$\mathcal{B}_\infty \to T^0(PC_1), \quad \Phi_\infty(A_n) \mapsto W_\pm(A_n),$$

again denoted by $W_\pm$. Finally, by (45), if f is as above then also $W_\pm(L_n W^0(f)|_{S_n}) = 0$, whence result canonical homomorphisms

$$\mathcal{B}_{\infty,\infty} \to T^0(PC_1), \quad \Phi_{\infty,\infty}(A_n) \mapsto W_\pm(A_n),$$

again denoted by $W_\pm$.

Now, (42) entails that

$$W_\pm(\Phi_{\infty,\infty}(E_n P E_{-n})) = \begin{cases} I & \text{for } W_+ \\ 0 & \text{for } W_-, \end{cases}$$

hence, the cosets $\Phi_{\infty,\infty}(E_n P E_{-n})$ and $\Phi_{\infty,\infty}(E_n(I-P)E_{-n})$ are nontrivial projections in $\mathcal{B}_{\infty,\infty}$ which, moreover, belong to the center of this algebra by Proposition 15. So we can localize $\mathcal{B}_{\infty,\infty}$ with respect to

$$\mathcal{C}_{\infty,\infty} = \text{alg}\,\{\Phi_{\infty,\infty}(E_n P E_{-n}),\, \Phi_{\infty,\infty}(E_n(I-P)E_{-n})\},$$

being a commutative C^*-algebra with maximal ideal space $\{-1,\, 1\}$ and with maximal ideals

$$\text{rm } \mathbb{C}\Phi_{\infty,\infty}(E_n P E_{-n}) \quad \text{and} \quad \text{rm } \mathbb{C}\Phi_{\infty,\infty}(E_n(I-P)E_{-n})$$

corresponding to -1 and 1, respectively. The result of this "half" localization is local algebras $\mathcal{B}_{\infty,\infty,\pm} = \mathcal{B}_{\infty,\infty}/I_{\infty,\infty,\pm}$ with local homomorphisms $\Phi_{\infty,\infty,\pm} : \mathcal{B} \to \mathcal{B}_{\infty,\infty,\pm}$.

The local ideals $I_{\infty,\infty,\pm}$ belong to the kernels of $W_\pm$, respectively, and this finally yields canonical quotient homomorphisms

$$\mathcal{B}_{\infty,\infty,\pm} \to T^0(PC_1), \quad \Phi_{\infty,\infty,\pm}(A_n) \mapsto W_\pm(A_n). \tag{47}$$

Proposition 17 *The homomorphisms (47) are *-isomorphisms between the algebras $\mathcal{B}_{\infty,\infty,\pm}$ and $T^0(PC_1)$.*

Proof. As in the proof of Proposition 11, it suffices to show that the homomorphisms (47) preserve spectra. We shall do this by verifying that

$$\Phi_{\infty,\infty,\pm}(E_n W_\pm(A_n) E_{-n}) = \Phi_{\infty,\infty,\pm}(A_n) \quad \text{for all } (A_n) \in \mathcal{B}. \tag{48}$$

Let $A_n = E_n U_{t,n}^{-1} P U_{t,n} E_{-n}$. Then $\Phi_{\infty,\infty,+}(E_n W_+(A_n) E_{-n}) = \Phi_{\infty,\infty,+}(E_n P E_{-n})$, and the cosets $\Phi_{\infty,\infty,+}(E_n P E_{-n})$ and $\Phi_{\infty,\infty,+}(E_n U_{t,n}^{-1} P U_{t,n} E_{-n})$ are equal to $\Phi_{\infty,\infty,+}(L_n \chi_{\mathbb{R}^+} I|_{S_n})$ and $\Phi_{\infty,\infty,+}(L_n \chi_{[\{tn\}/n,\infty)}|_{S_n})$, respectively (see the proof of Proposition 3.22 in [3]). Clearly, both cosets coincide, and the proof for the "-"sign is analogous.

If $A_n = E_n D_{t,n}^{-1} T^0(a) D_{t,n} E_{-n}$, then $\Phi_{\infty,\infty,+}(E_n W_+(A_n) E_{-n}) = \Phi_{\infty,\infty,+}(E_n T^0(a) E_{-n})$, and it remains to show that

$$\Phi_{\infty,\infty,+}(E_n T^0(a) E_{-n}) = \Phi_{\infty,\infty,+}(E_n D_{t,n}^{-1} T^0(a) D_{t,n} E_{-n}). \tag{49}$$

In case $a = \lambda^{-1}\sigma$, the sequences $(E_n T^0(a) E_{-n})$ and $(E_n D_{t,n}^{-1} T^0(a) D_{t,n} E_{-n})$ are (modulo $\mathcal{J}$) equal to the sequences $(L_n W^0(\text{sgn})|_{S_n})$ and $(L_n W^0(\text{sgn}_t)|_{S_n})$, and these sequences coincide at $\infty, \infty, +$. In case $a = c$ with c continuous,we have already remarked that (49) holds (see the proof of Proposition 11). So, (48) is established for all generating sequences of the algebra $\mathcal{B}$ and, since $W_\pm$ and $\Phi_{\infty,\infty,+}$ are continuous homomorphisms, for all sequences in $\mathcal{B}$. ■

Corollary 1 *The spectrum of*

$$pqp = \Phi_{\infty,x}(E_n P E_{-n})\Phi_{\infty,x}(L_n W^0(\chi_{[x,\infty)})|_{S_n})\Phi_{\infty,x}(E_n P E_{-n})$$

is equal to $[0, 1]$.

Proof. Let, for definiteness, $x = 0$. Then

$$pqp = \Phi_{\infty,x}(E_n P T^0(\frac{1+\lambda^{-1}\sigma}{2}) P E_{-n}) = \Phi_{\infty,x}(E_n T(\frac{1+\lambda^{-1}\sigma}{2}) E_{-n}).$$

Let a be an arbitrary function in PC_1. Then, as shown in [3], Proposition 3.22, the spectrum of $\Phi_\infty(E_n T(a) E_{-n})$ coincides with the Fredholm spectrum of the Toeplitz operator $T(a)$, i.e. with the union of the range $a(\mathbb{T})$ of a with the line segment $[a(1-0),\, a(1+0)]$ (see, e.g. [1], Theorem 4.86, or [3], Section 2.5.2). Further, by the local principle,

$$[a(1-0),\, a(1+0)] \cup a(\mathbb{T}) = \bigcup_{y \in \dot{\mathbb{R}}} \operatorname{spec}(\Phi_{\infty,y}(E_n T(a) E_{-n})).$$

The spectrum of $\Phi_{\infty,\infty}(E_n T(a) E_{-n})$ is the same as that of $T^0(a)$, i.e. the range of a, by the preceding proposition. If $y \in \mathbb{R} \setminus \{0\}$ then (here we make use of the decomposition $a = \alpha\lambda^{-1}\sigma + c$ again)

$$\begin{aligned}\Phi_{\infty,y}(E_n T(a) E_{-n}) &= \alpha\Phi_{\infty,y}(L_n S|_{S_n}) + \Phi_{\infty,y}(E_n T(c) E_{-n}) \\ &= \begin{cases} \alpha + c(1) = a(1-0) & \text{if } \; y > 0 \\ \alpha - c(1) = a(1+0) & \text{if } \; y < 0 \end{cases}\end{aligned}$$

(compare the proof of Proposition 13). Hence, $\operatorname{spec}(\Phi_{\infty,y}(E_n T(a) E_{-n})) \subseteq a(\mathbb{T})$ for all $y \neq 0$, and this yields the inclusion

$$[a(1-0),\, a(1+0)] \subseteq \operatorname{spec}(\Phi_{\infty,0}(E_n T(a) E_{-n}))$$

which finishes the proof. ∎

Summarizing these results we get the following.

Theorem 2 *A sequence* (A_n) *belonging to the algebra* $\mathcal{B}$ *is stable if and only if the operators* $W(A_n)$, $W_s(A_n)$ *and* $W_\pm(A_n)$ *as well as the matrix functions* $W_{\infty,s}(A_n)$ *are invertible for all* $s \in \mathbb{R}$.

4 Approximation methods for composed operators

Let a_i and c_i be piecewise continuous on $\mathbb{R}$ and continuous on $\mathbb{R} \setminus \mathbb{Z}$, and let b_i be arbitrary piecewise continuous functions. Consider the operator

$$A = \sum_{i=1}^{r} a_i W^0(b_i) c_i I + K$$

which is composed by multiplication operators and Fourier convolution operators and by a compact perturbation K. Special choices of the functions a_i, b_i, c_i lead to the singular integral operator $a_1 I + a_2 S_{\mathbb{R}}$ and to the Wiener-Hopf operator $W(b)$.

Proposition 18 *The sequence $(L_n A|_{S_n})$ belongs to $\mathcal{B}$.*

Proof. As in [3], Section 3.8.1, one can show (using KMS-techniques) that the sequence $(L_n a S_{\mathbb{R}} cI|_{S_n})$ is in $\mathcal{B}$ for all functions $a, c \in PC \cap C(\mathbb{R} \setminus \mathbb{Z})$. The shift argument employed in the proof of Proposition 9 (b) yields that then $(L_n a W^0(b) cI|_{S_n}) \in \mathcal{B}$ for all piecewise continuous functions b (recall that the operators D_s and aI commute). ∎

Thus, the stability analysis from Section 3 applies, and we have to compute some strong limits as well as 2×2-matrix functions.

Clearly, $W(L_n A|_{S_n}) = A$. For the computation of $W_s(L_n A|_{S_n})$ with $s \in \mathbb{R}$ we make use of the fact that W_s is locally equivalent. Namely, we can replace A by an operator A_s which behaves at s in the very same way as A, and then we compute $W_s(L_n A_s|_{S_n})$. For A as above, one can choose

$$A_s = \sum_i \left(a_i(s-0)\chi_{(-\infty,s)} + a_i(s+0)\chi_{[s,\infty)}\right)\left(b_i(+\infty)W^0(\chi_{\mathbb{R}^+}) + b_i(-\infty)W^0(\chi_{\mathbb{R}^-})\right) \cdot$$

$$\left(c_i(s-0)\chi_{(-\infty,s)} + c_i(s+0)\chi_{[s,\infty)}\right) I$$

if $s \in \mathbb{Z}$ and

$$A_s = \sum_i a_i(s)c_i(s)\left(b_i(+\infty)W^0(\chi_{\mathbb{R}^+}) + b_i(-\infty)W^0(\chi_{\mathbb{R}^-})\right)$$

if $s \notin \mathbb{Z}$. Computing the strong limits yield for $s \in \mathbb{Z}$

$$W_s(L_n A_s|_{S_n}) =$$

$$= \sum_i (a_i(s-0)Q + a_i(s+0)P)T^0(b_i(+\infty)\frac{1+\lambda^{-1}\sigma}{2} + b_i(-\infty)\frac{1-\lambda^{-1}\sigma}{2})$$

$$(c_i(s-0)Q + c_i(s+0)P) + R \tag{50}$$

where we set $Q = I - P$, and where R refers to a certain compact (even finite rank) operator which is zero if all a_i and c_i are continuous at s, and for $s \notin \mathbb{Z}$

$$W_s(L_n A_s|_{S_n}) = \sum_i a_i(s)c_i(s)T^0(b_i(+\infty)\frac{1+\lambda^{-1}\sigma}{2} + b_i(-\infty)\frac{1-\lambda^{-1}\sigma}{2}). \tag{51}$$

Let us emphasize once more that the compact operator R appearing in (50) results from the structure of the spline space under consideration which, for example, involves that $E_{-n}L_n S_{\mathbb{R}} E_n$ is the Laurent operator $T^0(\lambda^{-1}\sigma)$ whereas $E_{-n}L_n S E_n$ differs from the Toeplitz operator $T(\lambda^{-1}\sigma)$ by a compact term.

For the computation of the matrix function $W_{\infty,s}(L_n A|_{S_n}))$ we again replace A by an operator $A_{\infty,s}$ which behaves at ∞, s in the same manner as A. One can choose

$$A_{\infty,s} = \sum_i \left(a_i(+\infty)\chi_{\mathbb{R}^+} + a_i(-\infty)\chi_{\mathbb{R}^-}\right)\left(b_i(s-0)W^0(\chi_{(-\infty,s)}) + b_i(s+0)W^0(\chi_{[s,\infty)})\right)$$

$$\left(c_i(+\infty)\chi_{\mathbb{R}^+} + c_i(-\infty)\chi_{\mathbb{R}^-}\right) I,$$

and the matrix function $W_{\infty,s}(L_nA_{\infty,s}|_{S_n}))$ is equal to

$$y \mapsto \sum_i \begin{pmatrix} a_i(+\infty) & 0 \\ 0 & a_i(-\infty) \end{pmatrix} \cdot$$

$$\begin{pmatrix} b_i(s+0)y + b_i(s-0)(1-y) & (b_i(s+0) - b_i(s-0))\sqrt{y(1-y)} \\ (b_i(s+0) - b_i(s-0))\sqrt{y(1-y)} & b_i(s+0)(1-y) + b_i(s-0)y \end{pmatrix} \cdot$$

$$\begin{pmatrix} c_i(+\infty) & 0 \\ 0 & c_i(-\infty) \end{pmatrix}.$$

One can show that the Fredholmness of the operator A already implies the invertibility of this function (compare [7], Section 15, where it is explained that Fredholmness in the algebra $\mathcal{D}$ can be studied by the same procedure as that one used in Section 3: a twofold localization which, at ∞, s, exactly yields local algebras generated by two idempotents, say p and q with $\operatorname{spec}(pqp) = [0, 1]$ again).

Finally, for the local algebra at ∞, ∞ we replace A by

$$A_{\infty,\infty} = \sum_i (a_i(+\infty)\chi_{\mathbb{R}^+} + a_i(-\infty)\chi_{\mathbb{R}^-})\,(b_i(+\infty)W^0(\chi_{\mathbb{R}^+}) + b_i(-\infty)W^0(\chi_{\mathbb{R}^-}))$$

$$(c_i(+\infty)\chi_{\mathbb{R}^+} + c_i(-\infty)\chi_{\mathbb{R}^-})\,I,$$

and obtain

$$W_{\infty,\infty,\pm}(L_nA|_{S_n}) = W_{\infty,\infty,\pm}(L_nA_{\infty,\infty}|_{S_n})$$

$$= \begin{cases} T^0(\sum_i a_i(+\infty)c_i(+\infty)(b_i(+\infty)\frac{1+\lambda^{-1}\sigma}{2} + b(-\infty)\frac{1-\lambda^{-1}\sigma}{2})) & \text{if the sign is } + \\ T^0(\sum_i a_i(-\infty)c_i(-\infty)(b_i(+\infty)\frac{1+\lambda^{-1}\sigma}{2} + b(-\infty)\frac{1-\lambda^{-1}\sigma}{2})) & \text{if the sign is } -. \end{cases}$$

Summarizing we find

Theorem 3 *The Galerkin method* $(L_nA|_{S_n})$ *applies to the operator* $A = \sum_{i=1}^r a_iW^0(b_i)c_iI + K$ *if and only if*

- *the operator A is invertible,*
- *the operators in (50) are invertible for all $s \in \mathbb{Z}$ where the a_i and c_i are not continuous,*
- *the functions*

$$\sum_i a_i(s)c_i(s)(b_i(+\infty)\frac{1+\lambda^{-1}\sigma}{2} + b(-\infty)\frac{1-\lambda^{-1}\sigma}{2})$$

are invertible for all $s \in (\mathbb{R} \setminus \mathbb{Z}) \cup \{\pm\infty\}$ and for all $s \in \mathbb{R}$ where all functions a_i and c_i are continuous.

Specifying this result to the Wiener-Hopf operator $A = \chi_{\mathbf{R}^+} W^0(a)\chi_{\mathbf{R}^+} I + \chi_{\mathbf{R}^-} I$ yields Theorem 1, and specification to the singular integral operator $a_1 I + a_2 S_{\mathbf{R}}$ gives Theorem 3.11 in [3].

Moreover, it is possible to investigate other spline approximation methods. If, e.g., one has besides φ another mother spline, say ψ then one can consider the generalized Galerkin method $L_n^{\varphi,\psi} A u_n = L_n^{\varphi,\psi} f$ where the solutions u_n are seeked in S_n^{φ}, but the test space is S_n^{ψ} now. Under these conditions, the sequence $(L_n^{\varphi,\psi} A|_{S_n^{\varphi}})$ belongs to $\mathcal{B}$ again, and Theorem 3 remains valid with the functions λ and σ replaced by

$$\lambda^{\varphi,\psi}(z) = \sum_{k\in\mathbb{Z}} \left(\int_{\mathbf{R}} \varphi(s+k)\,\overline{\psi(s)}\,ds \right) \cdot z^k ,$$

$$\sigma^{\varphi,\psi}(z) = \sum_{k\in\mathbb{Z}} \left(\int_{\mathbf{R}} (S_{\mathbf{R}}\varphi)(s+k)\,\overline{\psi(s)}\,ds \right) \cdot z^k .$$

In the same way one can include collocation and qualocation methods (see [3], Sections 3.8.2 - 3.8.4, where this is done for singular integral operators, ans Section 3.4.2 for a possible (slight) generalization of this algebra $\mathcal{B}$ by introducing an additional parameter).

Let us finally remark that the homomorphisms W, W_s, $W_{\infty,s}$ and $W_{\pm}$ are *fractal* in sense of [9]. Thus, the machinery developed in [9] applies, and it allows to determine the asymptotic behaviour of the norms, the condition numbers, the s-numbers and the ε- pseudospectra for all sequences in $\mathcal{B}$.

References

[1] A. Böttcher, B. Silbermann, Analysis of Toeplitz operators. – Akademie–Verlag, Berlin, 1990, and Springer Verlag, Berlin, 1990.

[2] J. Elschner, On spline approximation for a class of non-compact integral equations. – Report R-MATH-09/88 des Karl-Weierstrass-Instituts für Mathematik, Berlin, 1988.

[3] R. Hagen, S. Roch, B. Silbermann, Spectral Theory of Approximation Methods for Convolution Equations. – Birkhuser Verlag Basel - Boston - Berlin, 1995.

[4] P. R. Halmos, Two subspaces. – Transactions Am. Math. Soc. **144**(1969), 381 – 389.

[5] S. Prössdorf, B. Silbermann, Numerical analysis for integral and related operator equations. – Akademie-Verlag, Berlin, 1991, and Birkhäuser Verlag, Basel – Boston – Stuttgart, 1991.

[6] S. Roch, Spline approximation methods cutting off singularities. – ZfAA **13**(1994), 2, 329-345.

[7] S. Roch, B. Silbermann, Algebras of convolution operators and their image in the Calkin algebra. – Report R-MATH-05/90 des Karl-Weierstrass-Instituts fr Mathematik, Berlin, 1990.

[8] S. Roch, B. Silbermann, Algebras generated by idempotents and the symbol calculus for singular integral operators. - IEOT **11**(1988), 385 - 419.

[9] S. Roch, B. Silbermann, C^*-algebra techniques in numerical analysis. - Preprint TU Chemnitz, 1994.

Universität Leipzig
Mathematisches Institut
Augustusplatz 10 - 11
D 04109 Leipzig
Germany
e-mail: roch@mathematik.uni-leipzig.de

AMS subject classification: 45E10, 47B35, 65R20

Operator Theory:
Advances and Applications, Vol. 87

INERTIA CONDITIONS FOR THE MINIMIZATION OF QUADRATIC FORMS IN INDEFINITE METRIC SPACES

ALI H. SAYED, BABAK HASSIBI, AND THOMAS KAILATH

We study the relation between the solutions of two minimization problems with indefinite quadratic forms. We show that a complete link between both solutions can be established by invoking a fundamental set of inertia conditions. While these inertia conditions are automatically satisfied in a standard Hilbert space setting, which is the case of classical least-squares problems in both the deterministic and stochastic frameworks, they nevertheless turn out to mark the differences between the two optimization problems in indefinite metric spaces. Applications to H^∞-filtering, robust adaptive filtering, and approximate total-least-squares methods are included.

1 INTRODUCTION

Given two invertible Hermitian matrices $\{\Pi, W\}$, a column vector y, and an arbitrary matrix A of appropriate dimensions, we study the relation between the following two minimization problems:

$$\min_z \left[z^*\Pi^{-1}z + (y - Az)^*W^{-1}(y - Az)\right], \tag{1}$$

where z is a column vector of unknowns, and

$$\min_K \{\Pi - KA\Pi - \Pi A^*K^* + K[A\Pi A^* + W]K^*\}, \tag{2}$$

where K is a matrix. The symbol "$*$" stands for Hermitian conjugation (complex conjugation for scalars). If we denote the cost function that appears in (2) by $J(K)$,

$$J(K) \triangleq \Pi - KA\Pi - \Pi A^*K^* + K[A\Pi A^* + W]K^*. \tag{3}$$

then by the minimization in (2) we mean finding a K^o such that for any complex column vector a, and for all K, we have $a^*J(K^o)a \leq a^*J(K)a$.

An interpretation of both optimization criteria (1) and (2) in terms of estimation problems in indefinite metric spaces is provided in the next sections. Here we only wish to emphasize that both cost functions in (1) and (2) are quadratic in the respective independent variables z and K, and that they can also be rewritten in the following revealing forms:

$$\min_z \begin{bmatrix} z^* & y^* \end{bmatrix} \begin{bmatrix} \Pi^{-1} + A^*W^{-1}A & -A^*W^{-1} \\ -W^{-1}A & W^{-1} \end{bmatrix} \begin{bmatrix} z \\ y \end{bmatrix}, \tag{4}$$

and

$$\min_K \begin{bmatrix} I & -K \end{bmatrix} \begin{bmatrix} \Pi & \Pi A^* \\ A\Pi & A\Pi A^* + W \end{bmatrix} \begin{bmatrix} I \\ -K^* \end{bmatrix}, \tag{5}$$

where the central matrices

$$\begin{bmatrix} \Pi^{-1} + A^*W^{-1}A & -A^*W^{-1} \\ -W^{-1}A & W^{-1} \end{bmatrix} \text{ and } \begin{bmatrix} \Pi & \Pi A^* \\ A\Pi & A\Pi A^* + W \end{bmatrix}, \tag{6}$$

are in fact the inverses of each other, as detailed below.

Moreover, and contrary to standard quadratic minimization problems, the weighting matrices $\{\Pi, W\}$ in (1) and (2) are allowed to be indefinite (i.e., they are not restricted to being positive-definite). Consequently, the central matrices in (4) and (5) are generally indefinite. For this reason, solutions to (1) and (2) are not always guaranteed to exist. However, when they exist, we shall show that the expressions for the solutions, and the conditions for their existence, are very closely related. This relation will be established via a fundamental set of inertia conditions. Here, by the inertia of an invertible Hermitian matrix X, we mean a pair of integers, denoted by $I_+(X)$ and $I_-(X)$, where

$$\begin{aligned} I_+(X) &\triangleq \text{number of strictly positive eigenvalues of } X, \\ I_-(X) &\triangleq \text{number of strictly negative eigenvalues of } X. \end{aligned}$$

Note also that since X is assumed invertible, it has no zero eigenvalues and, consequently,

$$I_+(X) + I_-(X) = \text{number of columns (or rows) of the matrix } X.$$

The significance of the relations to be established between problems (1) and (2) is the following. It often happens in applications that one is interested in solving quadratic problems of the form (1), with indefinite weighting matrices. A particular example that has received increasing attention in the last decade is the class of so-called H^∞-filtering and control problems, as suggested by several of the references at the end of this paper – see, e.g., the recent books [GL95, ZDG96] for more details and references on the topic. In this context, the Π matrix in (1) is further restricted to be positive-definite and the W matrix is indefinite but of the special form $W = \text{diag.}\{I, -\gamma^2 I\}$, for a given positive constant γ^2. Here we shall treat the general class of optimization problems suggested by (1) where both $\{\Pi, W\}$ are allowed to be arbitrary indefinite matrices. For example, the special case $\Pi = -\rho^2 I$ and $W = I$ turns out to be useful in approximate solutions of the so-called total least-squares (TLS) or errors-in-variables methods.

On the other hand, problems of the form (2) are characteristic of state-space estimation formulations, where a so-called Kalman filter procedure is available as an efficient computational scheme for determining the solution in the presence of state-space structure, as pointed out in [HSK93]. By relating the solutions of (1) and (2) we shall then be able to apply Kalman-type algorithms to the solution of (1), as well as obtain a complete set of inertia conditions that will automatically test for the existence of solutions to (1), without discarding the available information from the solution of (2).

In the sequel, we shall use capital letters to denote matrices (e.g., A) and small letters to denote vectors.

2 An Inertia Result for Linear Transformations

We first establish a preliminary inertia result that tells us how the inertia of the matrices Π and W is affected by transformations of the form

$$(A\Pi A^* + W) \quad \text{and} \quad (\Pi^{-1} + A^*W^{-1}A), \tag{7}$$

for arbitrary matrices A of appropriate dimensions. The reason for choosing these transformations is because the positivity of the matrices in (7) will be shown to be equivalent to necessary and sufficient conditions for the solvability of the problems (1) and (2). Hence, by studying how their inertia depends on $\{\Pi, W\}$, we shall be able to conclude how the choice of $\{\Pi, W\}$ affects the solvability of problems (1) and (2) – see Theorem 2.1 below. Also, a justification for the name *linear* transformations that appears in the title of this section will become clear further ahead, where it will be shown that the matrix A can be interpreted as the coefficient matrix of a linear model.

We start by noting that the matrices in (6) are indeed the inverses of each other and, consequently, that their inertia coincide. For this purpose, we form the square Hermitian matrix

$$G \triangleq \begin{bmatrix} \Pi & \Pi A^* \\ A\Pi & A\Pi A^* + W \end{bmatrix}, \tag{8}$$

and note that the Schur decomposition of G into a (block) *lower-diagonal-upper* triangular form leads to

$$G = \begin{bmatrix} I & 0 \\ A & I \end{bmatrix} \begin{bmatrix} \Pi & 0 \\ 0 & W \end{bmatrix} \begin{bmatrix} I & A^* \\ 0 & I \end{bmatrix}. \tag{9}$$

This establishes, in view of the assumptions on Π and W, that G is invertible. Its inverse is given by

$$G^{-1} = \begin{bmatrix} I & -A^* \\ 0 & I \end{bmatrix} \begin{bmatrix} \Pi^{-1} & 0 \\ 0 & W^{-1} \end{bmatrix} \begin{bmatrix} I & 0 \\ -A & I \end{bmatrix} = \begin{bmatrix} \Pi^{-1} + A^*W^{-1}A & -A^*W^{-1} \\ -W^{-1}A & W^{-1} \end{bmatrix}, \tag{10}$$

which thus establishes our earlier claim that the matrices in (6) are the inverses of each other.

Note also that the Schur decompositions in (9) and (10) are in fact congruence relations. This shows, in view of Sylvester's law of inertia [Gan59], that G and G^{-1} have the same positive and negative inertia as the block diagonal matrix $(\Pi \oplus W)$,

$$I_+(G) = I_+(G^{-1}) = I_+(\Pi \oplus W), \quad I_-(G) = I_-(G^{-1}) = I_-(\Pi \oplus W).$$

Here, the notation $A \oplus B$ stands for a block diagonal matrix,

$$(A \oplus B) \triangleq \begin{bmatrix} A & 0 \\ 0 & B \end{bmatrix}.$$

We state this preliminary result in the following lemma.

Lemma 2.1 (Inertia of G) *Given $\{\Pi, W\}$ Hermitian and invertible. Then, for any matrix A of appropriate dimensions, the block matrix*

$$G \triangleq \begin{bmatrix} \Pi & \Pi A^* \\ A\Pi & A\Pi A^* + W \end{bmatrix},$$

has the same positive and negative inertia as the block diagonal matrix $(\Pi \oplus W)$,

$$I_+(G) = I_+(\Pi \oplus W), \quad I_-(G) = I_-(\Pi \oplus W). \tag{11}$$

Proof: The proof is immediate from the congruence relation (9) and from Sylvester's law of inertia. ■

A less immediate inertia result follows if we instead perform a (block) *upper-diagonal-lower* triangular factorization of G. In this case, we need to further assume that the lower-right corner element of G is also invertible, viz.,

$$(A\Pi A^* + W) \text{ is invertible.} \tag{12}$$

It then follows that the matrix $(\Pi^{-1} + A^*W^{-1}A)$ will be invertible, as is immediate from the matrix inversion formula

$$(\Pi^{-1} + A^*W^{-1}A)^{-1} = \Pi - \Pi A^*(A\Pi A^* + W)^{-1}A\Pi. \tag{13}$$

This is in fact a useful preliminary result for our later analysis and a stronger statement is given below.

Lemma 2.2 (Invertibility Conditions) *Assume* $\{\Pi, W\}$ *are invertible. Then, for any matrix* A *of appropriate dimensions,* $(A\Pi A^* + W)$ *is invertible if, and only if,* $(\Pi^{-1} + A^*W^{-1}A)$ *is invertible.*

Proof: The result follows from the matrix inversion formulas

$$(\Pi^{-1} + A^*W^{-1}A)^{-1} = \Pi - \Pi A^*(A\Pi A^* + W)^{-1}A\Pi,$$

and

$$(A\Pi A^* + W)^{-1} = W^{-1} - W^{-1}A\left[\Pi^{-1} + A^*W^{-1}A\right]^{-1}A^*W^{-1}.$$

■

The right-hand side of (13) is simply the Schur complement of G with respect to its lower right block entry. We can therefore write the alternative Schur decomposition

$$G = \tag{14}$$

$$\begin{bmatrix} I & \Pi A^*(A\Pi A^* + W)^{-1} \\ 0 & I \end{bmatrix} \begin{bmatrix} (\Pi^{-1} + A^*W^{-1}A)^{-1} & 0 \\ 0 & A\Pi A^* + W \end{bmatrix} \begin{bmatrix} I & 0 \\ (A\Pi A^* + W)^{-1}A\Pi & I \end{bmatrix}.$$

It again follows from Sylvester's law of inertia, and under the additional assumption (12), that G has the same inertia as the block diagonal matrix $[(\Pi^{-1} + A^*W^{-1}A) \oplus (A\Pi A^* + W)]$. We establish a stronger statement in the following theorem.

Theorem 2.1 (Fundamental Inertia Result) *Given* $\{\Pi, W\}$ *Hermitian and invertible. Then, for any matrix* A *of appropriate dimensions, the following inertia equalities hold,*

$$\begin{aligned} I_+(\Pi \oplus W) &= I_+[(\Pi^{-1} + A^*W^{-1}A) \oplus (A\Pi A^* + W)], & (15) \\ I_-(\Pi \oplus W) &= I_-[(\Pi^{-1} + A^*W^{-1}A) \oplus (A\Pi A^* + W)], & (16) \end{aligned}$$

if, and only if, $(A\Pi A^* + W)$ *is invertible.*

Proof: If $(A\Pi A^* + W)$ is invertible then the triangular decomposition (14) is applicable, thus leading to a congruence relation. This shows that G has the same inertia as

$$\left[(\Pi^{-1} + A^*W^{-1}A) \oplus (A\Pi A^* + W)\right].$$

The inertia equalities of the theorem then follow from (11).

Conversely, assume the inertia conditions (15) and (16) hold. Then the total number of nonzero eigenvalues of the block diagonal matrix $[(\Pi^{-1}+A^*W^{-1}A)\oplus(A\Pi A^*+W)]$ is equal to $(n+N)$, which is also the size of this block matrix. Here, n is the size of Π and N is the size of W. Consequently, none of the eigenvalues of either $(\Pi^{-1}+A^*W^{-1}A)$ or $(A\Pi A^*+W)$ can be zero. This implies that we must necessarily have an invertible matrix $(A\Pi A^*+W)$.

■

The inertia conditions (15) and (16) will play an important role in our analysis. In simple terms, they show how the inertia of the matrices $\{\Pi, W\}$ affects the inertia of the matrices $\{(A\Pi A^*+W), (\Pi^{-1}+A^*W^{-1}A)\}$, and vice-versa. In the special case of positive-definite matrices $\{\Pi, W\}$, we see that relation (16) becomes unnecessary and relation (15) is trivialized.

3 The Indefinite-Weighted Least-Squares Problem

We now return to the optimization problems (1) and (2) and proceed to a closer study of both criteria. We shall also motivate both problems by arguing that they can be related to estimation problems in indefinite metric spaces. We start with the first problem (1), which we shall refer to, for reasons to be clarified soon, as the indefinite-weighted least-squares problem (IWLS, for short).

Problem 3.1 (IWLS Problem) *Given invertible Hermitian matrices $\{\Pi, W\}$, a column vector y, and a matrix A of appropriate dimensions, we are interested in determining, if possible, the optimal $\hat{z}$ that solves the optimization problem:*

$$\min_z \left[z^*\Pi^{-1}z + (y-Az)^*W^{-1}(y-Az)\right]. \tag{17}$$

3.1 Interpretation as an Estimation Problem with an Indefinite Metric

The problem (17) admits an interpretation in terms of an estimation problem as follows. We may regard z as a column vector of n unknown parameters that is related to the vector y via a linear relation of the form

$$y = Az + v, \tag{18}$$

where v denotes the mismatch between the value of y and the value of Az. In signal processing literature, the y is called the *observation* vector, the v is called the *noise* vector, and the objective is to use the available data y in order to come up with an estimate for the unknown vector z. The problem is posed as one of minimizing a quadratic cost function of the same form as in (17) but with *positive-definite* matrices $\{\Pi, W\}$ [Hay91, PRLN92, SK94]. It is well known in such cases that for any positive-definite matrix W, and for any complex-valued column vectors a and b in C^n, the scalar quantity $a^*W^{-1}b$ is a well-defined inner product, denoted by $< b, a >$, and, consequently, least-squares solutions can be found by orthogonally projecting onto appropriate linear subspaces – see, e.g., [SK94] for a recent survey on the topic in the positive-definite case and along the lines of this paper.

Here, however, we allow for indefinite matrices $\{\Pi, W\}$, thus leading to a least-squares problem with indefinite weighting matrices. Now a bilinear form $a^*W^{-1}b$ is not guaranteed to satisfy the positivity condition $a^*W^{-1}a > 0$ for all nonzero a. We thus say that C^n, coupled with a bilinear form $a^*W^{-1}b$ with W indefinite, is an indefinite metric space. More generally, an indefinite metric space $\{\mathcal{K}, <.,.>_{\mathcal{K}}\}$ is defined as a vector space that satisfies two simple requirements (see, e.g., [Bog74, GLR83] for more details):

(i) $\mathcal{K}$ is linear over the field of complex numbers $\mathcal{C}$, and

(ii) $\mathcal{K}$ possesses a bilinear form, $<.,.>_{\mathcal{K}}$, such that for any $a, b, c \in \mathcal{K}$, and for any $\alpha, \beta \in \mathcal{C}$, we have

$$\begin{aligned} <\alpha a + \beta b, c>_{\mathcal{K}} &= \alpha <a, c>_{\mathcal{K}} + \beta <b, c>_{\mathcal{K}}, \\ <b, a>_{\mathcal{K}} &= <a, b>_{\mathcal{K}}^{*} . \end{aligned}$$

In particular, the quantity $<a, a>_{\mathcal{K}}$ is in general indefinite. This is in contrast to a Hilbert space setting, $\{\mathcal{H}, <.,.>_{\mathcal{H}}\}$, where for any $a \in \mathcal{H}$ the quantity $<a, a>_{\mathcal{H}}$ is necessarily nonnegative.

In the formulation (17), each of the terms $z^*\Pi^{-1}z$ and $(y - Az)^*W^{-1}(y - Az)$ may be indefinite. Note also that we can rewrite the cost function in (17) in the form:

$$\min_{z} \left(\begin{bmatrix} 0 \\ y \end{bmatrix} - \begin{bmatrix} I \\ A \end{bmatrix} z \right)^{*} \begin{bmatrix} \Pi & 0 \\ 0 & W \end{bmatrix}^{-1} \left(\begin{bmatrix} 0 \\ y \end{bmatrix} - \begin{bmatrix} I \\ A \end{bmatrix} z \right),$$

where the central matrix $(\Pi \oplus W)^{-1}$ is indefinite. This further highlights that the cost function in (17) is an *indefinite* quadratic cost function.

Also, in estimation problems it often happens that the linear model (18) arises as a consequence of repeated experiments. That is, one collects several observation vectors $\{y_i\}$ that are also linearly related to the same unknown z, say via

$$y_i = A_i z + v_i,$$

where the A_i are given matrices of appropriate dimensions, and the v_i are the corresponding noise components. If we collect several such observations into matrix form and write

$$\underbrace{\begin{bmatrix} y_0 \\ y_1 \\ \vdots \\ y_N \end{bmatrix}}_{y} = \underbrace{\begin{bmatrix} A_0 \\ A_1 \\ \vdots \\ A_N \end{bmatrix}}_{A} z + \underbrace{\begin{bmatrix} v_0 \\ v_1 \\ \vdots \\ v_N \end{bmatrix}}_{v},$$

we again obtain the linear model (18) and we are back to the problem of estimating z from the y by solving (17).

3.2 Solution of the IWLS Problem

Let $J(z)$ denote the quadratic cost function that appears in (17),

$$\begin{aligned} J(z) &\triangleq z^*\Pi^{-1}z + (y - Az)^*W^{-1}(y - Az), \qquad (19) \\ &= z^*[\Pi^{-1} + A^*W^{-1}A]z - y^*W^{-1}Az - z^*A^*W^{-1}y + y^*W^{-1}y. \end{aligned}$$

Every $\hat{z}$ at which the gradient of $J(z)$ with respect to z vanishes is called a ***stationary*** point of $J(z)$. A stationary point $\hat{z}$ may or may not be a minimum of $J(z)$ as clarified by the following statement.

Theorem 3.1 (Solution of the IWLS Problem) *The stationary points $\hat{z}$ of $J(z)$ in (19), if they exist, are solutions of the linear system of equations*

$$[\Pi^{-1} + A^*W^{-1}A]\hat{z} = A^*W^{-1}y. \qquad (20)$$

There exists a unique stationary point if, and only if, $[\Pi^{-1} + A^*W^{-1}A]$ is invertible. In this case, it is given by

$$\hat{z} = \left[\Pi^{-1} + A^*W^{-1}A\right]^{-1} A^*W^{-1}y, \tag{21}$$

and the corresponding value of the cost function is

$$J(\hat{z}) = y^* \left[W + A\Pi A^*\right]^{-1} y. \tag{22}$$

Moreover, this unique point is a minimun if, and only if, the coefficient matrix is positive-definite,

$$(\Pi^{-1} + A^*W^{-1}A) > 0. \tag{23}$$

Proof: It is straightforward to verify, by differentiation, that the gradient of $J(z)$ with respect to z^* is equal to $([\Pi^{-1} + A^*W^{-1}A]z - A^*W^{-1}y)$. Therefore, the stationary points of $J(z)$, when they exist, must satisfy the linear system of equations

$$[\Pi^{-1} + A^*W^{-1}A]\hat{z} = A^*W^{-1}y.$$

This has a unique solution $\hat{z}$ if, and only if, the coefficient matrix is invertible. Also, the Hessian matrix is equal to $[\Pi^{-1} + A^*W^{-1}A]$, which thus needs to be positive-definite for a unique minimum solution with respect to z.

■

Note that in contrast to positive-definite least-squares problems (i.e., when $\Pi > 0$ and $W > 0$) where $[\Pi^{-1} + A^*W^{-1}A]$ is always guaranteed to be positive for any A and, consequently, a unique minimizing solution of $J(z)$ always exists, the IWLS problem may or may not have a minimum, and actually may not even have a stationary point if a solution to (20) does not exist.

4 The Equivalent Estimation Problem

We now study the second optimization criterion (2) and also present an interpretation for it in terms of an estimation problem in an indefinite metric space. We shall refer to this problem as the equivalent estimation problem (or EE, for short).

Problem 4.1 (The EE Problem) *Given invertible Hermitian matrices $\{\Pi, W\}$, and a matrix A of appropriate dimensions, we are interested in determining, if possible, the optimal K^o that solves the optimization problem (in the sense explained after (3)):*

$$\min_K \{\Pi - KA\Pi - \Pi A^*K^* + K[A\Pi A^* + W]K^*\}. \tag{24}$$

4.1 Interpretation as an Estimation Problem with an Indefinite Metric

An interpretation for this problem is the following. We consider column vectors $\{\mathbf{y}, \mathbf{v}, \mathbf{z}\}$ that are linearly related via the expression

$$\mathbf{y} = A\mathbf{z} + \mathbf{v}, \tag{25}$$

and where the individual entries $\{\mathbf{y}_i, \mathbf{v}_i, \mathbf{z}_i\}$ of the vectors $\{\mathbf{y}, \mathbf{v}, \mathbf{z}\}$ are all elements of an indefinite metric space, say $\mathcal{K}'$.

For two vectors $\{\mathbf{a}, \mathbf{b}\}$, with entries $\{\mathbf{a}_i, \mathbf{b}_j\}$ in $\mathcal{K}'$, we write $< \mathbf{a}, \mathbf{b} >_{\mathcal{K}'}$ to denote a matrix whose entries are the individual $< \mathbf{a}_i, \mathbf{b}_j >_{\mathcal{K}'}$. In a Hilbert setting, an analogy arises with the space of scalar-valued zero-mean random variables, say $\mathcal{E}$: for two column vectors $\mathbf{p}$ and $\mathbf{q}$ of random variables, the bilinear form $E\mathbf{p}\mathbf{q}^*$ is a matrix whose individual entries are $E\mathbf{p}_i\mathbf{q}_j^*$ (see, e.g., [AM79, Kai81]). Note that to distinguish between the elements in $\mathcal{K}$ and $\mathcal{K}'$, we are using boldface letters to denote the variables of the equivalent problem.

The variables $\{\mathbf{v}, \mathbf{z}\}$ can be regarded as having Gramian matrices $\{W, \Pi\}$ and cross Gramian zero, namely

$$W \triangleq <\mathbf{v}, \mathbf{v}>_{\mathcal{K}'}, \quad \Pi \triangleq <\mathbf{z}, \mathbf{z}>_{\mathcal{K}'}, \quad <\mathbf{z}, \mathbf{v}>_{\mathcal{K}'} = 0.$$

Under these conditions, it follows from the linear model (25) that the Gramian matrix of $\mathbf{y}$ is equal to

$$<\mathbf{y}, \mathbf{y}>_{\mathcal{K}'} = A\Pi A^* + W.$$

Let $J(K)$ denote the quadratic cost function that appears in (24),

$$J(K) \triangleq \Pi - KA\Pi - \Pi A^* K^* + K[A\Pi A^* + W]K^*. \tag{26}$$

It is then immediate to see that $J(K)$ can be interpreted as the Gramian matrix of the vector difference $(\mathbf{z} - K\mathbf{y})$, viz.,

$$J(K) = <\mathbf{z} - K\mathbf{y}, \mathbf{z} - K\mathbf{y}>_{\mathcal{K}'}.$$

Every K^o at which the gradient of $a^*J(K)a$ with respect to a^*K vanishes for all a is called a *stationary* solution of $J(K)$ [Note from (26) that $a^*J(K)a$ is a function of a^*K]. A stationary point K^o may or may not be a minimum as clarified further ahead.

Hence, solving for the stationary solutions K^o can also be interpreted as solving the problem of linearly estimating $\mathbf{z}$ from $\mathbf{y}$.

Definition 4.1 (Linear Estimates) *A linear estimate of* $\mathbf{z}$ *given* $\mathbf{y}$ *is defined by*

$$\hat{\mathbf{z}} \triangleq K^o\mathbf{y}, \tag{27}$$

where K^o *is a stationary solution of (24). This estimate is uniquely defined if* K^o *is unique. It is further said to be the optimal linear estimate if* K^o *is the unique minimizing solution of (24).*

4.2 Solution of the EE Problem

We now state and prove the solution of (24).

Theorem 4.1 (Solution of the EE Problem) *The stationary points* K^o *of* $J(K)$*, if they exist, are solutions of the linear system of equations*

$$\Pi A^* = K^o[A\Pi A^* + W]. \tag{28}$$

There exists a unique stationary point K^o *if, and only if,* $(A\Pi A^* + W)$ *is invertible. In this case, it is given by*

$$K^o = \left[\Pi^{-1} + A^*W^{-1}A\right]^{-1} A^*W^{-1}, \tag{29}$$

and the corresponding value of the cost function is

$$J(K^o) = \left[\Pi^{-1} + A^*W^{-1}A\right]^{-1}. \tag{30}$$

The unique linear estimate of the corresponding $\mathbf{z}$ *in (27) is*

$$\hat{\mathbf{z}} = \left[\Pi^{-1} + A^*W^{-1}A\right]^{-1} A^*W^{-1}\mathbf{y}. \tag{31}$$

Moreover, this unique point K^o *is a minimum (and, correspondingly,* $\hat{\mathbf{z}}$ *is optimal) if, and only if, the coefficient matrix is positive-definite,*

$$(A\Pi A^* + W) > 0. \tag{32}$$

Proof: The proof follows the same lines of Theorem 3.1 when applied to the now scalar-valued cost function $a^*J(K)a$, where a is any column vector (recall the explanation below (3)). In particular, it is immediate to see that any stationary solution K^o, if it exists, must satisfy the orthogonality condition $< \mathbf{z} - K^o\mathbf{y}, \mathbf{y} >_{\mathcal{K}'} = 0$, which leads to the linear system of equations

$$\Pi A^* = K^o[A\Pi A^* + W].$$

A unique stationary point K^o then exists as long as $[A\Pi A^* + W]$ is invertible, thus leading to the expression

$$K^o = \Pi A^*[A\Pi A^* + W]^{-1}. \tag{33}$$

But in view of the matrix inversion formula, and Lemma 2.2,

$$[A\Pi A^* + W]^{-1} = W^{-1} - W^{-1}A\left[\Pi^{-1} + A^*W^{-1}A\right]^{-1} A^*W^{-1},$$

we can also write

$$K^o = \left[\Pi^{-1} + A^*W^{-1}A\right]^{-1} A^*W^{-1}.$$

The necessary and sufficient condition for this solution to correspond to a minimum is $(A\Pi A^* + W) > 0$, as follows if we evaluate the Hessian matrix of $a^*J(K)a$.

■

The matrices that appear in (33) can be interpreted as follows:

$$< \mathbf{z}, \mathbf{y} >_{\mathcal{K}'} = \Pi A^*, \quad < \mathbf{y}, \mathbf{y} >_{\mathcal{K}'} = A\Pi A^* + W.$$

We therefore conclude that the following equivalent equalities also hold:

$$K^o = < \mathbf{z}, \mathbf{y} >_{\mathcal{K}'} < \mathbf{y}, \mathbf{y} >_{\mathcal{K}'}^{-1}, \tag{34}$$

$$\hat{\mathbf{z}} = < \mathbf{z}, \mathbf{y} >_{\mathcal{K}'} < \mathbf{y}, \mathbf{y} >_{\mathcal{K}'}^{-1} \mathbf{y}. \tag{35}$$

5 Relations between the IWLS and EE Problems

We now compare expressions (31) and (21). We see that if we make the identifications: $\hat{\mathbf{z}} \leftarrow \hat{z}$ and $\mathbf{y} \leftarrow y$, then both expressions coincide. This means that the IWLS problem and the equivalent problem have the same expressions for the stationary points, $\hat{z}$ and $\hat{\mathbf{z}}$. But while a minimum for the IWLS problem (17) exists as long as $(\Pi^{-1} + A^*W^{-1}A) > 0$, the equivalent problem (24), on the other hand, has a minimum at K^o if, and only if, $(W + A\Pi A^*) > 0$.

This indicates that both problems are not generally guaranteed to have simultaneous minima. In the special case of positive-definite matrices $\{\Pi, W\}$, both conditions

$$(\Pi^{-1} + A^*W^{-1}A) > 0 \quad \text{and} \quad (W + A\Pi A^*) > 0,$$

are simultaneously met. But this situation does not hold for general indefinite matrices Π and W. A question of interest then is the following: given that one problem has a unique stationary solution, say the EE problem (24), and given that this solution has been computed, is it possible to verify whether the other problem, say the IWLS problem (17) admits a minimizing solution without explicitly checking for its positivity condition $(\Pi^{-1} + A^*W^{-1}A) > 0$?

The relevance of this question is that, as we shall see in a later section, when state-space structure is further imposed on the data, an efficient recursive procedure can be derived for the solution of the equivalent problem (24). Hence, once a connection is established with the IWLS problem (17), the solution of the latter should follow immediately.

We shall see that this is indeed possible by invoking the inertia results of Sec. 2. To begin with, the following result is a consequence of Lemma 2.2.

Lemma 5.1 (Simultaneous Stationary Points) *The IWLS problem (17) has a unique stationary point $\hat{z}$ if, and only if, the equivalent problem (24) has a unique stationary point K^o.*

Proof: The IWLS problem (17) has a unique stationary point $\hat{z}$ iff $(\Pi^{-1}+A^*W^{-1}A)$ is nonsingular. Likewise, the equivalent problem (24) has a unique stationary point K^o iff $(W + A\Pi A^*)$ is nonsingular. But, according to Lemma 2.2, the nonsingularity of one matrix implies the nonsingularity of the other, which thus establishes the desired result. ∎

This means that both optimization problems are always guaranteed to simultaneously have unique stationary solutions $\hat{z}$ and K^o, regardless of the invertible matrices $\{\Pi, W\}$ and for any A. That is, once we find a unique stationary solution K^o for the equivalent problem (24), we are at least guaranteed a unique stationary solution $\hat{z}$ for the IWLS problem. But we are in fact interested in a stronger result. We would like to verify whether this stationary solution $\hat{z}$ is a minimum or not. We would also like to be able to settle this question by exploiting the solution of the equivalent problem (24), and without explicitly checking the positivity condition that is required on $(\Pi^{-1} + A^*W^{-1}A)$ in the IWLS case (17).

The next statement is one of the main conclusions of this paper since it provides a set of inertia conditions that allows us to check the solvability of the IWLS problem (17) in terms of the inertia properties of the Gramian matrix $(A\Pi A^* + W)$ associated with the equivalent problem (24).

Theorem 5.1 (Fundamental Inertia Conditions) *Given invertible and Hermitian matrices Π and W, and an arbitrary matrix A of appropriate dimensions, the optimization problem (1) (i.e., the IWLS problem (17)) has a unique minimizing solution $\hat{z}$ if, and only if,*

$$\begin{aligned} I_-[W + A\Pi A^*] &= I_-[\Pi \oplus W], \\ I_+[W + A\Pi A^*] &= I_+[\Pi \oplus W] - n, \end{aligned}$$

where $n \times n$ is the size of Π.

Proof: Assume the IWLS problem has a unique minimizing solution. This means that we necessarily have

$$(\Pi^{-1} + A^*W^{-1}A) > 0.$$

We then obtain from Lemma 5.1 that $(W + A\Pi A^*)$ is also invertible.

In view of Theorem 2.1 we conclude that we must have

$$\begin{aligned} I_+(\Pi \oplus W) &= I_+[(\Pi^{-1} + A^*W^{-1}A) \oplus (A\Pi A^* + W)], \\ I_-(\Pi \oplus W) &= I_-[(\Pi^{-1} + A^*W^{-1}A) \oplus (A\Pi A^* + W)]. \end{aligned}$$

But $I_-[(\Pi^{-1} + A^*W^{-1}A)] = 0$ and $I_+[(\Pi^{-1} + A^*W^{-1}A)] = n$. Hence,

$$\begin{aligned} I_-[W + A\Pi A^*] &= I_-[\Pi \oplus W], \\ I_+[W + A\Pi A^*] &= I_+[\Pi \oplus W] - n. \end{aligned}$$

Conversely, assume the above inertia relations hold. It follows that the number of (strictly positive and strictly negative) eigenvalues of $(W + A\Pi A^*)$ is equal to the size of W. Therefore, $(W + A\Pi A^*)$ has no zero eigenvalues and is thus invertible. It follows from Lemma 2.2 that $(\Pi^{-1} + A^*W^{-1}A)$ is also invertible. We further invoke Theorem 2.1 to conclude that

$$I_-[(\Pi^{-1} + A^*W^{-1}A)] = I_-(\Pi \oplus W) - I_-[(W + A\Pi A^*)],$$

which thus establishes that we necessarily have

$$I_-[(\Pi^{-1} + A^*W^{-1}A)] = 0.$$

Therefore, $(\Pi^{-1} + A^*W^{-1}A) > 0$ and the IWLS problem (1) has a unique minimum.

■

The importance of the above theorem is that it allows us to check whether a minimizing solution exists to the IWLS problem (17) by comparing the inertia of the Gramian matrix of the equivalent problem, viz., $(W + A\Pi A^*)$, with the inertia of $(\Pi \oplus W)$. This is relevant because, as we shall see in the next section, when state-space structure is further imposed, we can derive an efficient procedure that allows us to keep track of the inertia of $(W + A\Pi A^*)$. In particular, the procedure will produce a sequence of matrices $\{R_{e,i}\}$ such that

$$\text{Inertia}(W + A\Pi A^*) = \text{Inertia}\ (R_{e,0} \oplus R_{e,1} \oplus R_{e,2} \ldots).$$

The theorem then shows that "all" we need to do is compare the inertia of the given matrices Π and W with that of the matrices $\{R_{e,i}\}$ that are made available via the recursive procedure.

Equally important is that this procedure will further allow us to compute the quantity $\hat{\mathbf{z}}$ in (27). But since we argued above that $\hat{\mathbf{z}}$ has the same expression as $\hat{z}$, the stationary solution of (17), then the procedure will also provide us with $\hat{z}$.

In summary, by establishing an explicit relation between both problems (17) and (24), we shall be capable of solving either problem via the solution of the other. In the special case of positive-definite quadratic cost functions, this point of view was fully exploited in [SK94] in order to establish a close link between known results in Kalman filtering theory and more recent results in adaptive filtering theory. In particular, it was shown in [SK94] that once such an equivalence relation is established, the varied forms of adaptive filtering algorithms can be obtained by writing down different variants of the so-called Kalman filter.

The discussion in this paper, while it provides a similar connection for indefinite quadratic cost functions, it shows that a satisfactory link can be established via an additional set of inertia conditions. These conditions are necessary because, contrary to the case of positive-definite quadratic cost functions, minimizing solutions are not always guaranteed to exist in the indefinite case. Note that in the positive case (i.e., Π and W positive), the inertia conditions of Theorem 5.1 are automatically satisfied.

We may finally remark that the above inertia conditions include, as special cases, the well-known conditions for the existence of H^∞-controllers and filters, as will be clarified in later sections.

6 Incorporating State-Space Structure

Now that we have established the exact relationship between the two basic optimization problems (1) and (2), we shall proceed to study an important special case of the equivalent problem (2).

More specifically, we shall pose an optimization problem that will be of the same form as (2) except that the associated A matrix will have considerable structure in it. In particular, the A matrix will be block-lower triangular and its individual entries will be further parameterized in terms of matrices $\{F_i, G_i, H_i\}$ that arise from an underlying state-space assumption. This will allow us to derive an efficient computational scheme for the solution of the corresponding optimization problem (2). The scheme is an extension to the indefinite case of a well-known Kalman filtering algorithm [HSK93].

6.1 Statement of the State-Space Problem

We consider an indefinite metric space $\mathcal{K}'$ and continue to employ the notation $<\mathbf{a},\mathbf{b}>_{\mathcal{K}'}$ to denote a matrix with entries $<\mathbf{a}_i,\mathbf{b}_j>_{\mathcal{K}'}$, where $\{\mathbf{a}_i,\mathbf{b}_j\}\in\mathcal{K}'$ are the individual entries of the columns $\mathbf{a}$ and $\mathbf{b}$.

We further consider vectors $\{\mathbf{y}_i,\mathbf{x}_i,\mathbf{u}_i,\mathbf{v}_i\}$, all with entries in $\mathcal{K}'$, and assume that they are related via state-space equations of the form

$$\begin{aligned}\mathbf{x}_{i+1} &= F_i\mathbf{x}_i + G_i\mathbf{u}_i,\\ \mathbf{y}_i &= H_i\mathbf{x}_i + \mathbf{v}_i\,,\quad i\geq 0,\end{aligned}\tag{36}$$

where F_i, H_i, and G_i are known $n\times n$, $p\times n$, and $n\times m$ matrices, respectively. It is further assumed that the Gramian matrices of $\{\mathbf{u}_i,\mathbf{v}_i,\mathbf{x}_0\}$ are known, say

$$<\mathbf{v}_i,\mathbf{v}_i>_{\mathcal{K}'} = R_i,\quad <\mathbf{u}_i,\mathbf{u}_i>_{\mathcal{K}'} = Q_i,\quad <\mathbf{x}_0,\mathbf{x}_0>_{\mathcal{K}'} = \Pi_0.$$

We also assume that the following relations hold for all $i\neq j$,

$$<\mathbf{v}_i,\mathbf{v}_j>_{\mathcal{K}'} = 0,\quad <\mathbf{u}_i,\mathbf{u}_j>_{\mathcal{K}'} = 0,\quad <\mathbf{v}_i,\mathbf{x}_0>_{\mathcal{K}'} = 0,\quad <\mathbf{u}_i,\mathbf{x}_0>_{\mathcal{K}'} = 0,$$

as well as $<\mathbf{v}_i,\mathbf{u}_j>_{\mathcal{K}'}=0$ for all i,j. More compactly, we may write the above requirements in the following form

$$<\begin{bmatrix}\mathbf{u}_i\\ \mathbf{v}_i\\ \mathbf{x}_0\end{bmatrix},\begin{bmatrix}\mathbf{u}_j\\ \mathbf{v}_j\\ \mathbf{x}_0\end{bmatrix}>_{\mathcal{K}'} = \begin{bmatrix}Q_i\delta_{ij} & 0 & 0\\ 0 & R_i\delta_{ij} & 0\\ 0 & 0 & \Pi_0\end{bmatrix},\tag{37}$$

where δ_{ij} is the Kronecker delta function that is equal to unity when $i=j$ and zero otherwise. The matrices $\{Q_i, R_i, \Pi_0\}$ are possibly indefinite.

The quantities $\{\mathbf{u}_i,\mathbf{v}_i,\mathbf{x}_0\}$ are assumed *unknown* and only the $\{\mathbf{y}_i\}$ are *known*. In other words, we assume that we have a collection of vectors $\{\mathbf{y}_i\}$ that we know arose from a state-space model of the form (36), with known $\{F_i, G_i, H_i\}$, but with no further access to the $\{\mathbf{u}_i,\mathbf{v}_i,\mathbf{x}_0\}$, except for the knowledge of their Gramian matrices as in (37).

The state-space structure (36) leads to a linear relation between the vectors $\{\mathbf{y}_i\}$ and the vectors $\{\mathbf{x}_0,\mathbf{u}_i\}_{i=0}^{N-1}$. Indeed, if we collect the $\{\mathbf{y}_i\}_{i=0}^{N}$ and the $\{\mathbf{v}_i\}_{i=0}^{N}$ into two column vectors, $\{\mathbf{y},\mathbf{v}\}$, respectively,

$$\mathbf{y}\triangleq\begin{bmatrix}\mathbf{y}_0\\ \mathbf{y}_1\\ \vdots\\ \mathbf{y}_N\end{bmatrix},\quad \mathbf{v}\triangleq\begin{bmatrix}\mathbf{v}_0\\ \mathbf{v}_1\\ \vdots\\ \mathbf{v}_N\end{bmatrix},\tag{38}$$

and define the column vector,

$$\mathbf{z} \triangleq \begin{bmatrix} \mathbf{x}_0 \\ \mathbf{u}_0 \\ \mathbf{u}_1 \\ \vdots \\ \mathbf{u}_{N-1} \end{bmatrix} \triangleq \begin{bmatrix} \mathbf{x}_0 \\ \mathbf{u} \end{bmatrix}, \tag{39}$$

it then follows from the state-space equations that

$$\mathbf{y} = A\mathbf{z} + \mathbf{v},$$

where A is the block-lower triangular matrix

$$A \triangleq \begin{bmatrix} H_0 & & & \\ H_1 F^{[0,0]} & H_1 G_0 & & \\ H_2 F^{[1,0]} & H_2 F^{[1,1]} G_0 & & \\ \vdots & \vdots & \ddots & \\ H_N F^{[N-1,0]} & H_N F^{[N-1,1]} G_0 & \dots & H_N G_{N-1} \end{bmatrix}. \tag{40}$$

Here, the notation $F^{[i,j]}$, $i \geq j$, stands for

$$F^{[i,j]} \triangleq F_i F_{i-1} \dots F_j.$$

Moreover, the Gramian matrices of the variables $\{\mathbf{z}, \mathbf{v}, \mathbf{y}\}$ so defined are easily seen to be, in view of the assumptions (37),

$$< \mathbf{z}, \mathbf{z} >_{\mathcal{K}'} = (\Pi_0 \oplus Q_0 \dots \oplus Q_{N-1}), \tag{41}$$

$$< \mathbf{v}, \mathbf{v} >_{\mathcal{K}'} = (R_0 \oplus R_1 \oplus \dots \oplus R_N). \tag{42}$$

More compactly, we shall write

$$< \mathbf{z}, \mathbf{z} >_{\mathcal{K}'} \triangleq \Pi, \quad < \mathbf{v}, \mathbf{v} >_{\mathcal{K}'} \triangleq W, \tag{43}$$

where the $\{\Pi, W\}$ are block diagonal matrices as defined in (41) and (42). We can now pose the following problem.

Problem 6.1 (State-Space Estimation Problem) *Consider the state-space model (36) and given the $\{\mathbf{y}, A, \Pi, W\}$ as above, determine a matrix K, and conditions on $\{A, \Pi, W\}$, so as to minimize the Gramian matrix*

$$\min_K < \mathbf{z} - K\mathbf{y}, \mathbf{z} - K\mathbf{y} >_{\mathcal{K}'} . \tag{44}$$

The optimal solution K^o, when it exists, can be used to define $K^o\mathbf{y}$ as the optimal linear estimate for $\mathbf{z}$. We denote this by

$$\hat{\mathbf{z}} \triangleq K^o \mathbf{y}.$$

In other words, we have posed the problem of linearly estimating $\mathbf{z}$ from $\mathbf{y}$ so as to minimize the Gramian matrix of the error signal, $\mathbf{z} - K\mathbf{y}$. This Gramian matrix can be expanded and the problem is easily seen to be equivalent to

$$\min_K \{\Pi - KA\Pi - \Pi A^* K^* + K[A\Pi A^* + W]K^*\},$$

where we have used (41) and (42).

We thus see that, given a state-space model of the form (36) and (37), the problem of linearly estimating the variables $\{\mathbf{x}_0, \mathbf{u}_0, \ldots, \mathbf{u}_{N-1}\}$ from the variables $\{\mathbf{y}_0, \mathbf{y}_1, \ldots, \mathbf{y}_N\}$ leads to an optimization problem of the same form as in (2): it requires that we determine a coefficient matrix K that minimizes $J(K)$. The optimal K^o is then used to define the optimal linear estimate of the desired variables via $\hat{\mathbf{z}} = K^o\mathbf{y}$. In case K^o is simply a unique stationary solution of $J(K)$, but not necessarily the minimum solution, we shall refer to $\hat{\mathbf{z}}$ as simply the linear estimate of $\mathbf{z}$ given $\mathbf{y}$, instead of the *optimal* linear estimate.

Using the result of Theorem 4.1, a unique linear estimate $\hat{\mathbf{z}}$ exists as long as $(A\Pi A^* + W)$ is invertible, where the matrices $\{A, \Pi, W\}$ are now as defined above. Moreover, when this happens the estimate $\hat{\mathbf{z}}$ is given by the expression

$$\hat{\mathbf{z}} = \left[\Pi^{-1} + A^*W^{-1}A\right]^{-1} A^*W^{-1}\mathbf{y}. \tag{45}$$

Alternatively, and using (35), we also write for later reference,

$$\hat{\mathbf{z}} = <\mathbf{z}, \mathbf{y}>_{\mathcal{K}'} <\mathbf{y}, \mathbf{y}>_{\mathcal{K}'}^{-1} \mathbf{y}. \tag{46}$$

While the expression (45) is analytically satisfactory, it however does not exploit two important facts that occur under the assumption of the state-space structure, namely that the matrices $\{\Pi, W\}$ are block diagonal and, more importantly, that the matrix A is now block-lower triangular. The entries of A are also completely parameterized by the matrices $\{F_i, G_i, H_i\}$ that describe the state-space model (36).

We shall see in the sequel that these two facts can be exploited in order to provide an alternative method for computing the solution $\hat{\mathbf{z}}$. While (45) provides a global expression for $\hat{\mathbf{z}}$, we shall argue that it will be more convenient to introduce a recursive procedure for computing $\hat{\mathbf{z}}$.

Remark on Notation. We shall from now on write $\mathbf{z}_N$ instead of $\mathbf{z}$ to indicate that it includes $\mathbf{x}_0$ and the vectors $\{\mathbf{u}_j\}$ up to time $N-1$, as defined in (39). That is, the subindex N indicates which vectors $\{\mathbf{u}_j\}$ are included in the definition of $\mathbf{z}$. We shall then write $\hat{\mathbf{z}}_{N|N}$ instead of simply $\hat{\mathbf{z}}$ to indicate that it is the estimate of $\mathbf{z}_N$ that is obtained by using the vectors $\{\mathbf{y}_i\}$ up to time N. That is, the $\{\mathbf{y}_0, \mathbf{y}_1, \ldots, \mathbf{y}_N\}$ are used in (45),

$$\hat{\mathbf{z}}_{N|N} = \left[\Pi^{-1} + A^*W^{-1}A\right]^{-1} A^*W^{-1}\mathbf{y}. \tag{47}$$

More generally, the estimate of $\mathbf{z}_N$ that is based on a different number of vectors $\{\mathbf{y}_j\}$, say up to time k, will be correspondingly indicated by $\hat{\mathbf{z}}_{N|k}$. In other words, the first subindex indicates which vectors $\{\mathbf{u}_j\}$ are included in the definition of the variable $\mathbf{z}$ and the second subindex indicates which vectors $\{\mathbf{y}_j\}$ are used in the estimation of $\mathbf{z}$.

These notational changes are necessary because we shall find it useful later to also define, for each i, the vector $\mathbf{z}_i$,

$$\mathbf{z}_i \triangleq \begin{bmatrix} \mathbf{x}_0 \\ \mathbf{u}_0 \\ \mathbf{u}_1 \\ \vdots \\ \mathbf{u}_{i-1} \end{bmatrix}, \tag{48}$$

which contains $\mathbf{x}_0$ and the vectors $\{\mathbf{u}_j\}$ up to time $(i-1)$. Correspondingly, the estimate of $\mathbf{z}_i$ that is based on vectors $\{\mathbf{y}_j\}$ up to a time k will be indicated by $\hat{\mathbf{z}}_{i|k}$.

6.2 A Strong Regularity Condition on the Gramian Matrix

Let $\hat{\mathbf{z}}_{N|i}$ denote the unique linear estimate of $\mathbf{z}_N$ that is based on the vectors $\{\mathbf{y}_0, \mathbf{y}_1, \ldots, \mathbf{y}_i\}$. That is, only the output vectors up to time i are used. By definition, this means that we should determine a coefficient matrix, say K_i^o, such that

$$\hat{\mathbf{z}}_{N|i} = K_i^o \begin{bmatrix} \mathbf{y}_0 \\ \mathbf{y}_1 \\ \vdots \\ \mathbf{y}_i \end{bmatrix}, \tag{49}$$

and K_i^o is the unique stationary solution of

$$J(K_i) \triangleq \langle \mathbf{z}_N - K_i \begin{bmatrix} \mathbf{y}_0 \\ \mathbf{y}_1 \\ \vdots \\ \mathbf{y}_i \end{bmatrix}, \mathbf{z}_N - K_i \begin{bmatrix} \mathbf{y}_0 \\ \mathbf{y}_1 \\ \vdots \\ \mathbf{y}_i \end{bmatrix} \rangle_{\mathcal{K}'} . \tag{50}$$

If we define

$$W_i \triangleq (R_0 \oplus R_1 \oplus \ldots \oplus R_i), \quad \Pi_i \triangleq (\Pi_0 \oplus Q_0 \oplus \ldots Q_{i-1}), \tag{51}$$

and

$$A_i \triangleq \begin{bmatrix} H_0 & & & \\ H_1 F^{[0,0]} & H_1 G_0 & & \\ H_2 F^{[1,0]} & H_2 F^{[1,1]} G_0 & & \\ \vdots & \vdots & \ddots & \\ H_i F^{[i-1,0]} & H_i F^{[i-1,1]} G_0 & \ldots & H_i G_{i-1} \end{bmatrix}, \tag{52}$$

then, as before, the problem (50) has a unique stationary solution K_i^o if, and only if,

$$W_i + \begin{bmatrix} A_i & 0 \end{bmatrix} \Pi \begin{bmatrix} A_i^* \\ 0 \end{bmatrix} = W_i + A_i \Pi_i A_i^* \text{ is invertible.}$$

A minimizing solution requires the positivity of this matrix. In any case, due to the block diagonal structure of $\{W, \Pi\}$ and due to the block lower-triangular structure of A, it is immediate to see that $(W_i + A_i \Pi_i A_i^*)$ is in fact a leading submatrix of $(W + A\Pi A^*)$.

To further clarify the implications of this observation, let R_y denote the Gramian matrix of the vector $\mathbf{y}$ in (27), i.e.,

$$R_y \triangleq \langle \mathbf{y}, \mathbf{y} \rangle_{\mathcal{K}'} = W + A\Pi A^*. \tag{53}$$

The existence of a unique stationary solution K^o to $J(K)$ in (44) then requires the invertibility of R_y. Likewise, the existence of unique stationary solutions K_i^o in (50), for $0 \leq i < N$, requires the invertibility of the leading (block) submatrices of R_y. We shall therefore assume here that all the leading (block) submatrices of R_y are invertible in order to guarantee the existence of unique stationary solutions K_i^o to the estimation problems (50) for $0 \leq i \leq N$. In this case, we say that R_y is (block) strongly regular.

Under this assumption, we can introduce the unique (block) lower-diagonal-upper triangular factorization

$$R_y \triangleq LDL^*, \tag{54}$$

where L is chosen to have unit diagonal entries and D is a block diagonal matrix whose entries are denoted by

$$D \triangleq \{R_{e,0}, R_{e,1}, \ldots, R_{e,N}\}.$$

The sizes of the blocks $R_{e,i}$ are $p \times p$, in accordance with the $p \times 1$ dimension of each $\mathbf{y}_i$. Also, the (block) strong regularity of R_y guarantees the invertibility of the $\{R_{e,i}\}$.

6.3 Orthogonalization via the Gram-Schmidt Procedure

In this section we shall argue that, under the strong regularity condition on the Gramian matrix R_y, a recursive procedure that allows us to directly update $\hat{\mathbf{z}}_{N|i}$ to $\hat{\mathbf{z}}_{N|i+1}$ is possible without explicitly computing K_{i+1}^o. This will be first achieved by "orthogonalizing" the output vectors $\{\mathbf{y}_j\}$, as we now explain.

Introduce the variables $\{\mathbf{e}_i\}$ defined by (these variables are often known as the *innovation* variables in the signal processing literature)

$$\mathbf{e} \triangleq L^{-1}\mathbf{y} \quad \text{or} \quad L\mathbf{e} = \mathbf{y}, \tag{55}$$

where $\mathbf{e}$ denotes the collection of the $\mathbf{e}_i$,

$$\mathbf{e} \triangleq \begin{bmatrix} \mathbf{e}_0 \\ \mathbf{e}_1 \\ \vdots \\ \mathbf{e}_N \end{bmatrix}.$$

It is immediate to conclude that the Gramian matrix of $\mathbf{e}$ is block diagonal since

$$< \mathbf{e}, \mathbf{e} >_{\mathcal{K}'} = < L^{-1}\mathbf{y}, L^{-1}\mathbf{y} >_{\mathcal{K}'} = L^{-1}R_y L^{-*} = D = (R_{e,0} \oplus R_{e,1} \oplus \ldots \oplus R_{e,N}).$$

Note that the vectors $\mathbf{e}$ and $\mathbf{y}$ are linearly related via an invertible transformation. They, therefore, span the same linear space. Also, and more importantly, the estimate of a variable $\mathbf{z}$ given the $\mathbf{y}$ is equal to the estimate of $\mathbf{z}$ given the $\mathbf{e}$. We prove this fact below and then discuss its ramifications.

Lemma 6.1 (Estimation Based on the $\{\mathbf{e}_i\}$) *Let $\hat{\mathbf{z}}$ denote the unique linear estimate of $\mathbf{z}$ given $\mathbf{y}$. That is, $\hat{\mathbf{z}} = K^o\mathbf{y}$, where K^o is the unique stationary solution of $< \mathbf{z} - K\mathbf{y}, \mathbf{z} - K\mathbf{y} >_{\mathcal{K}'}$. Let also $\hat{\mathbf{z}}_e$ denote the unique linear estimate of $\mathbf{z}$ given $\mathbf{e}$. That is, $\hat{\mathbf{z}}_e = K^{o,e}\mathbf{e}$, where $K^{o,e}$ is the unique stationary solution of $< \mathbf{z} - K^e\mathbf{e}, \mathbf{z} - K^e\mathbf{e} >_{\mathcal{K}'}$. Then $\hat{\mathbf{z}} = \hat{\mathbf{z}}_e$ and $K^o = K^{o,e}L^{-1}$.*

Proof: We know from (35) that estimating a variable $\mathbf{z}$ from $\mathbf{y}$ amounts to

$$\begin{aligned} \hat{\mathbf{z}} &= < \mathbf{z}, \mathbf{y} >_{\mathcal{K}'} < \mathbf{y}, \mathbf{y} >_{\mathcal{K}'}^{-1} \mathbf{y}, \\ &= < \mathbf{z}, L\mathbf{e} >_{\mathcal{K}'} < L\mathbf{e}, L\mathbf{e} >_{\mathcal{K}'}^{-1} L\mathbf{e}, \\ &= < \mathbf{z}, \mathbf{e} >_{\mathcal{K}'} < \mathbf{e}, \mathbf{e} >_{\mathcal{K}'}^{-1} \mathbf{e}, \\ &= \hat{\mathbf{z}}_e. \end{aligned}$$

■

The result also clearly holds for estimating $\mathbf{z}$ from a subcollection $\{\mathbf{y}_0, \ldots, \mathbf{y}_i\}$. In other words, we can work with the $\{\mathbf{e}_i\}$ instead of the $\{\mathbf{y}_i\}$. This corresponds to a change of basis and its main advantage is that the $\{\mathbf{e}_i\}$ are orthogonal in $\mathcal{K}'$, i.e.,

$$< \mathbf{e}_i, \mathbf{e}_j >_{\mathcal{K}'} = R_{e,i}\delta_{ij}.$$

Lemma 6.2 (Recursive Computation) *Let $\hat{\mathbf{z}}_{N|N}$ denote the unique linear estimate of $\mathbf{z}_N$ that is based on the vectors $\{\mathbf{y}_i\}$ up to time N. Then it can be recursively updated as follows:*

$$\hat{\mathbf{z}}_{N|N} = \hat{\mathbf{z}}_{N|N-1} + <\mathbf{z}_N, \mathbf{e}_N>_{\mathcal{K}'} R_{e,N}^{-1}\mathbf{e}_N. \tag{56}$$

Proof: It follows from Lemma 6.1 that

$$\begin{aligned}
\hat{\mathbf{z}} = \hat{\mathbf{z}}_{N|N} &= <\mathbf{z}_N, \mathbf{e}>_{\mathcal{K}'} <\mathbf{e}, \mathbf{e}>_{\mathcal{K}'}^{-1} \mathbf{e}, \\
&= \sum_{j=0}^{N} <\mathbf{z}_N, \mathbf{e}_j>_{\mathcal{K}'} <\mathbf{e}_j, \mathbf{e}_j>_{\mathcal{K}'}^{-1} \mathbf{e}_j, \\
&= \sum_{j=0}^{N-1} <\mathbf{z}_N, \mathbf{e}_j>_{\mathcal{K}'} <\mathbf{e}_j, \mathbf{e}_j>_{\mathcal{K}'}^{-1} \mathbf{e}_j + <\mathbf{z}_N, \mathbf{e}_N>_{\mathcal{K}'} <\mathbf{e}_N, \mathbf{e}_N>_{\mathcal{K}'}^{-1} \mathbf{e}_N, \\
&= \hat{\mathbf{z}}_{N|N-1} + <\mathbf{z}_N, \mathbf{e}_N>_{\mathcal{K}'} R_{e,N}^{-1}\mathbf{e}_N.
\end{aligned}$$

■

For this recursive scheme to be complete, we still need to show the following. Given the state-space model (36),

(i) How to compute the $\{\mathbf{e}_i\}$?

(ii) How to compute the $\{R_{e,i}\}$?

(iii) How to compute the $\{<\mathbf{z}_N, \mathbf{e}_i>_{\mathcal{K}'}\}$?

6.4 Computation of the $\{\mathbf{e}_i\}$ via a Kalman-Type Procedure

The computation of the variables $\{\mathbf{e}_i\}$ can be achieved via a standard Gram-Schmidt procedure:

- Let $\mathbf{e}_0 = \mathbf{y}_0$.
- Then form $\mathbf{e}_1$ by subtracting from $\mathbf{y}_1$ its linear estimate that is based on $\mathbf{y}_0$, written as $\hat{\mathbf{y}}_{1|0}$,

$$\mathbf{e}_1 = \mathbf{y}_1 - \hat{\mathbf{y}}_{1|0} = \mathbf{y}_1 - <\mathbf{y}_1, \mathbf{y}_0>_{\mathcal{K}'}<\mathbf{y}_0, \mathbf{y}_0>_{\mathcal{K}'}^{-1} \mathbf{y}_0 = \mathbf{y}_1 - <\mathbf{y}_1, \mathbf{e}_0>_{\mathcal{K}'}<\mathbf{e}_0, \mathbf{e}_0>_{\mathcal{K}'}^{-1} \mathbf{e}_0.$$

- Then form $\mathbf{e}_2$ by subtracting from $\mathbf{y}_2$ its linear estimate that is based on $\{\mathbf{y}_0, \mathbf{y}_1\}$, written as $\hat{\mathbf{y}}_{2|1}$,

$$\mathbf{e}_1 = \mathbf{y}_2 - \hat{\mathbf{y}}_{2|1} = \mathbf{y}_2 - <\mathbf{y}_2, \mathbf{e}_0>_{\mathcal{K}'}<\mathbf{e}_0, \mathbf{e}_0>_{\mathcal{K}'}^{-1} \mathbf{e}_0 - <\mathbf{y}_2, \mathbf{e}_1>_{\mathcal{K}'}<\mathbf{e}_1, \mathbf{e}_1>_{\mathcal{K}'}^{-1} \mathbf{e}_1.$$

More generally, we have

$$\mathbf{e}_i = \mathbf{y}_i - \hat{\mathbf{y}}_{i|i-1}, \tag{57}$$

where $\hat{\mathbf{y}}_{i|i-1}$ denotes the linear estimate of $\mathbf{y}_i$ that is based on $\{\mathbf{y}_0, \mathbf{y}_1, \ldots, \mathbf{y}_{i-1}\}$. It is immediate to conclude from the second line of the state-equations (36), by linearity and by the fact that $<\mathbf{v}_i, \mathbf{y}_j>_{\mathcal{K}'}= 0$ for $j < i$, that

$$\hat{\mathbf{y}}_{i|i-1} = H_i\hat{\mathbf{x}}_{i|i-1},$$

where $\hat{\mathbf{x}}_{i|i-1}$ now denotes the linear estimate of $\mathbf{x}_i$ that is based on $\{\mathbf{y}_0, \mathbf{y}_1, \ldots, \mathbf{y}_{i-1}\}$. We thus see that

$$\mathbf{e}_i = \mathbf{y}_i - H_i\hat{\mathbf{x}}_{i|i-1}, \tag{58}$$

and the computation of $\mathbf{e}_i$ is reduced to that of $\hat{\mathbf{x}}_{i|i-1}$.

Theorem 6.1 (Recursive Kalman Algorithm) *Consider the state-space model (36) and assume the Gramian matrix, $R_y = W + A\Pi A^*$, of the vector $\mathbf{y}$, defined in (38), is (block) strongly regular. The variables $\{\mathbf{e}_i\}$ defined via (55) or (57) can be recursively computed as follows. Start with $\hat{\mathbf{x}}_{0|-1} = 0$, $P_0 = \Pi_0$, and repeat for $i \geq 0$:*

$$\mathbf{e}_i = \mathbf{y}_i - H_i\hat{\mathbf{x}}_{i|i-1}, \tag{59}$$

$$\hat{\mathbf{x}}_{i+1|i} = F_i\hat{\mathbf{x}}_{i|i-1} + K_{p,i}\mathbf{e}_i, \tag{60}$$

$$K_{p,i} = F_iP_iH_i^*R_{e,i}^{-1}, \tag{61}$$

$$R_{e,i} = R_i + H_iP_iH_i^*, \tag{62}$$

$$P_{i+1} = F_iP_iF_i^* + G_iQ_iG_i^* - K_{p,i}R_{e,i}K_{p,i}^*. \tag{63}$$

Proof: In view of the recursive formula (56) (taking $\mathbf{x}_{i+1}$ as the variable $\mathbf{z}$) we have

$$\hat{\mathbf{x}}_{i+1|i} = \hat{\mathbf{x}}_{i+1|i-1} + < \mathbf{x}_{i+1}, \mathbf{e}_i >_{\mathcal{K}'} R_{e,i}^{-1}\mathbf{e}_i = \hat{\mathbf{x}}_{i+1|i-1} + K_{p,i}\mathbf{e}_i, \tag{64}$$

where we have defined $K_{p,i} \triangleq < \mathbf{x}_{i+1}, \mathbf{e}_i >_{\mathcal{K}'} < \mathbf{e}_i, \mathbf{e}_i >_{\mathcal{K}'}^{-1}$. It also follows from the first line of (36), and from the fact that $< \mathbf{u}_i, \mathbf{y}_j >_{\mathcal{K}'} = 0$ for $j < i$, that

$$\hat{\mathbf{x}}_{i+1|i-1} = F_i\hat{\mathbf{x}}_{i|i-1} + G_i\hat{\mathbf{u}}_{i|i-1} = F_i\hat{\mathbf{x}}_{i|i-1} + 0 = F_i\hat{\mathbf{x}}_{i|i-1}.$$

Substituting into (64) we obtain (60). To complete the argument we still need to show how to compute the $K_{p,i}$. Define the error quantity $\tilde{\mathbf{x}}_{i|i-1} \triangleq \mathbf{x}_i - \hat{\mathbf{x}}_{i|i-1}$, and let P_i denote its Gramian matrix, $P_i \triangleq < \tilde{\mathbf{x}}_{i|i-1}, \tilde{\mathbf{x}}_{i|i-1} >_{\mathcal{K}'}$. Then

$$\mathbf{e}_i = \mathbf{y}_i - H_i\hat{\mathbf{x}}_{i|i-1} = H_i\mathbf{x}_i - H_i\hat{\mathbf{y}}_{i|i-1} + \mathbf{v}_i = H_i\tilde{\mathbf{x}}_{i|i-1} + \mathbf{v}_i. \tag{65}$$

But it is immediate to note that $< \mathbf{v}_i, \tilde{\mathbf{x}}_{i|i-1} >_{\mathcal{K}'} = 0$ and, hence, (62) follows. Moreover,

$$< \mathbf{x}_{i+1}, \mathbf{e}_i >_{\mathcal{K}'} = F_i < \mathbf{x}_i, \mathbf{e}_i >_{\mathcal{K}'} + G_i < \mathbf{u}_i, \mathbf{e}_i >_{\mathcal{K}'} . \tag{66}$$

Now

$$< \mathbf{x}_i, \mathbf{e}_i >_{\mathcal{K}'} = < \mathbf{x}_i, \tilde{\mathbf{x}}_{i|i-1} >_{\mathcal{K}'} H_i^* + < \mathbf{x}_i, \mathbf{v}_i >_{\mathcal{K}'} = P_iH_i^* + 0,$$

while

$$< \mathbf{u}_i, \mathbf{e}_i >_{\mathcal{K}'} = < \mathbf{u}_i, \tilde{\mathbf{x}}_{i|i-1} >_{\mathcal{K}'} H_i^* + < \mathbf{u}_i, \mathbf{v}_i >_{\mathcal{K}'} = 0,$$

so that we can write

$$K_{p,i} \triangleq < \mathbf{x}_{i+1}, \mathbf{e}_i >_{\mathcal{K}'} R_{e,i}^{-1} = F_iP_iH_i^*R_{e,i}^{-1}. \tag{67}$$

Therefore $\{K_{p,i}, R_{e,i}\}$ can be determined once we have the Gramian matrices $\{P_i\}$.

The most direct method for computing the $\{P_i\}$ is to seek a recursion for $\tilde{\mathbf{x}}_{i+1|i}$ and then form P_{i+1}. In fact, from the model equations (36) and the estimator equation (60) we obtain

$$\tilde{\mathbf{x}}_{i+1|i} = F_{p,i}\tilde{\mathbf{x}}_{i|i-1} + \begin{bmatrix} G_i & -K_{p,i} \end{bmatrix} \begin{bmatrix} \mathbf{u}_i \\ \mathbf{v}_i \end{bmatrix},$$

where we have defined $F_{p,i} = F_i - K_{p,i}H_i$. Now it follows that P_i obeys the recursion (63).

■

We should remark here that the above recursive formulas extend the so-called Kalman filter to an indefinite metric space [HSK93]. The recursions have exactly the same form as those of the Kalman filter, except for the fact that the Gramian matrices $\{\Pi_0, R_i, Q_i\}$ are allowed to be indefinite. Also, the recursion (63) for P_i (with (61) and (62) inserted in (63)) is known as the Riccati difference equation.

An important fall out of the above algorithm is that the inertia of the Gramian matrix $<\mathbf{y}, \mathbf{y}>_{\mathcal{K}'}$ is completely determined by the inertia of the $\{R_{e,i}\}$.

Corollary 6.1 (Inertia of the Gramian Matrix) *Consider the state-space model (36) and let R_y denote the Gramian matrix of the vector $\mathbf{y}$ defined in (38), viz.,*

$$R_y = W + A\Pi A^*,$$

where $\{W, \Pi, A\}$ are as defined in (40), (41), and (42). The R_y is further assumed (block) strongly regular. Then

$$\textit{Inertia of}\,(W + A\Pi A^*) = \textit{Inertia of}\,(R_{e,0} \oplus R_{e,1} \oplus \ldots \oplus R_{e,N}). \tag{68}$$

Proof: This follows from the congruence relation $R_y = LDL^*$, where $D = (R_{e,0} \oplus R_{e,1} \oplus \ldots \oplus R_{e,N})$. ∎

6.5 Recursive Estimation of $\{\mathbf{x}_0, \mathbf{u}_0, \ldots, \mathbf{u}_{N-1}\}$

We already know how to recursively evaluate the $\{\mathbf{e}_i\}$ and the corresponding Gramian matrices $\{R_{e,i}\}$. We now return to (56), viz.,

$$\hat{\mathbf{z}}_{N|N} = \hat{\mathbf{z}}_{N|N-1} + <\mathbf{z}_N, \mathbf{e}_N>_{\mathcal{K}'} R_{e,N}^{-1} \mathbf{e}_N, \tag{69}$$

and show how to evaluate the terms $\{<\mathbf{z}_N, \mathbf{e}_i>_{\mathcal{K}'}\}$. Once this is done, we shall have an algorithm for the recursive update of the estimates $\{\hat{\mathbf{z}}_{N|i}\}$. Recall that $\hat{\mathbf{z}}_{N|i}$ was defined as the unique linear estimate of $\mathbf{z}_N$ based on the $\{\mathbf{y}_0, \mathbf{y}_1, \ldots, \mathbf{y}_i\}$.

Theorem 6.2 (Recursive Smoothing Solution) *Assume R_y is (block) strongly regular. Then the stationary solution $\hat{\mathbf{z}}$ is equal to $\hat{\mathbf{z}}_{N|N}$, where $\hat{\mathbf{z}}_{N|N}$ can be recursively computed as follows: start with $\hat{\mathbf{z}}_{N|-1} = 0$ and repeat for $i = 0, 1, \ldots, N$:*

$$\hat{\mathbf{z}}_{N|i} = \hat{\mathbf{z}}_{N|i-1} + K_{z,i} H_i^* R_{e,i}^{-1} \mathbf{e}_i,$$

where

$$K_{z,i+1} = K_{z,i}\left[F_i - K_{p,i}H_i\right]^* + \begin{bmatrix} 0 \\ I \\ 0 \end{bmatrix} Q_i G_i^*, \quad K_{z,0} = \begin{bmatrix} \Pi_0 \\ 0 \end{bmatrix}.$$

The identity matrix in the recursion for $K_{z,i+1}$ occurs at the position that corresponds to the entry $\mathbf{u}_i$.

Proof: Recall that $\mathbf{e}_i = H_i\tilde{\mathbf{x}}_{i|i-1} + \mathbf{v}_i$. Therefore,

$$\begin{aligned}\hat{\mathbf{z}}_{N|i} &= \hat{\mathbf{z}}_{N|i-1} + < \mathbf{z}_N, \mathbf{e}_i >_{\mathcal{K}'} R_{e,i}^{-1}\, \mathbf{e}_i\,, \\ &= \hat{\mathbf{z}}_{N|i-1} + < \mathbf{z}_N, \tilde{\mathbf{x}}_{i|i-1} >_{\mathcal{K}'} H_i^* R_{e,i}^{-1}\, \mathbf{e}_i.\end{aligned}$$

We now define $K_{z,i} \stackrel{\Delta}{=} < \mathbf{z}_N, \tilde{\mathbf{x}}_{i|i-1} >_{\mathcal{K}'}$, and note that

$$\begin{aligned}K_{z,i+1} = < \mathbf{z}_N, \tilde{\mathbf{x}}_{i+1|i} >_{\mathcal{K}'} &= < \mathbf{z}_N, \left[F_i\tilde{\mathbf{x}}_{i|i-1} - K_{p,i}\mathbf{e}_i + G_i\mathbf{u}_i\right] >_{\mathcal{K}'}\,, \\ &= K_{z,i}\left[F_i - K_{p,i}H_i\right]^* + \begin{bmatrix} 0 \\ I \\ 0 \end{bmatrix} Q_i G_i^*.\end{aligned}$$

∎

A remark is due here. Recall that we have defined $\hat{\mathbf{z}}_{N|i}$ in (50) as the unique linear estimator of $\mathbf{z}_N$ that is based on the vectors $\{\mathbf{y}_0, \mathbf{y}_1, \ldots, \mathbf{y}_i\}$. Now $\mathbf{z}_N$ is a vector containing the $\{\mathbf{x}_0, \mathbf{u}_0, \mathbf{u}_1, \ldots, \mathbf{u}_{N-1}\}$. By linearity, it follows that the entries of $\hat{\mathbf{z}}_{N|i}$ can be interpreted as the linear estimates of the corresponding entries of $\mathbf{z}_N$ given the $\{\mathbf{y}_0, \mathbf{y}_1, \ldots, \mathbf{y}_i\}$. That is, we have

$$\hat{\mathbf{z}}_{N|i} = \begin{bmatrix} \hat{\mathbf{x}}_{0|i} \\ \hat{\mathbf{u}}_{0|i} \\ \hat{\mathbf{u}}_{1|i} \\ \vdots \\ \hat{\mathbf{u}}_{N-1|i} \end{bmatrix},$$

where the notation $\hat{\mathbf{x}}_{0|i}$ denotes the linear estimate of $\mathbf{x}_0$ that is based on $\{\mathbf{y}_0, \ldots, \mathbf{y}_i\}$. Likewise, $\hat{\mathbf{u}}_{j|i}$ denotes the linear estimate of $\mathbf{u}_j$ that is based on the same vectors $\{\mathbf{y}_0, \ldots, \mathbf{y}_i\}$. But it follows from (37) that $< \mathbf{u}_j, \mathbf{y}_k >_{\mathcal{K}'} = 0$ for all $j \geq k$. This implies that

$$\hat{\mathbf{u}}_{i|i} = \hat{\mathbf{u}}_{i+1|i} = \ldots = \hat{\mathbf{u}}_{N-1|i} = 0.$$

Consequently, the last entries of $\hat{\mathbf{z}}_{N|i}$ are in fact zero,

$$\hat{\mathbf{z}}_{N|i} = \begin{bmatrix} \hat{\mathbf{x}}_{0|i} \\ \hat{\mathbf{u}}_{0|i} \\ \vdots \\ \hat{\mathbf{u}}_{i-1|i} \\ 0 \\ \vdots \\ 0 \end{bmatrix}. \tag{70}$$

If we introduce the definition of $\mathbf{z}_i$ as in (48), i.e., a vector composed of $\mathbf{x}_0$ and the $\{\mathbf{u}_j\}$ up to time $(i-1)$, then we can rewrite (70) more compactly as follows:

$$\hat{\mathbf{z}}_{N|i} = \begin{bmatrix} \hat{\mathbf{z}}_{i|i} \\ 0 \end{bmatrix}. \tag{71}$$

That is, the leading nonzero entries of the successive $\hat{\mathbf{z}}_{N|i}$ are precisely the entries of $\hat{\mathbf{z}}_{i|i}$.

7 A Recursive IWLS Problem in the Presence of State-Space Structure

In order to further appreciate the results of the earlier sections, let us first summarize what has been concluded in the state-space context.

Starting with a state-space model (36), with entries in an indefinite metric space $\mathcal{K}'$, we defined two vectors $\mathbf{z}$ and $\mathbf{y}$ as in (38) and (39). The vector $\mathbf{y}$ contained the output vectors $\{\mathbf{y}_i\}$ and the vector $\mathbf{z}$ contained the vectors $\{\mathbf{x}_0, \mathbf{u}_0, \ldots, \mathbf{u}_{N-1}\}$. We then used $\mathbf{z}$ and $\mathbf{y}$ as a motivation to introduce a quadratic minimization problem. This was achieved by defining the linear estimate of $\mathbf{z}$ given $\mathbf{y}$ as the vector $\hat{\mathbf{z}}$ obtained via $\hat{\mathbf{z}} = K^o\mathbf{y}$, where K^o was defined as the unique stationary solution of the cost function

$$J(K) = <\mathbf{z} - K\mathbf{y}, \mathbf{z} - K\mathbf{y}>_{\mathcal{K}'} = \Pi - KA\Pi - \Pi A^*K^* + K[A\Pi A^* + W]K^*. \tag{72}$$

We then observed that $J(K)$ is a special case of the optimization problem (2) introduced earlier in the paper, and hence the solution $\hat{\mathbf{z}}$, also denoted by $\hat{\mathbf{z}}_{N|N}$, could be obtained via the global expression (45),

$$\hat{\mathbf{z}} = \left[\Pi^{-1} + A^*W^{-1}A\right]^{-1} A^*W^{-1}\mathbf{y}.$$

But we further showed that in this case, and due to the state-space assumptions (36) and (37), the matrices $\{\Pi, W, A\}$ have extra structure in them. In particular, the $\{\Pi, W\}$ were shown to be diagonal matrices in (41) and (42), and the A matrix was shown to be block lower triangular in (40). As a result, we then argued that this structure can in fact be exploited in order to derive a recursive scheme that would allow us to directly update the estimate $\hat{\mathbf{z}}_{N|i}$ to $\hat{\mathbf{z}}_{N|i+1}$, starting with $\hat{\mathbf{z}}_{N|-1} = 0$ and ending with the desired solution $\hat{\mathbf{z}}_{N|N}$. This was achieved by the recursions of Theorem 6.2, which in turn rely on the recursions of Theorem 6.1. These recursions assume that the Gramian matrix R_y is (block) strongly regular so that the stationary solutions K_i^o that correspond to each estimate $\hat{\mathbf{z}}_{N|i}$ are uniquely defined.

Now, in view of the discussion at the beginning of Sec. 5, the above solution $\hat{\mathbf{z}}_{N|N}$ has the same expression as the solution $\hat{z}$ of a related minimization problem of the form (1). Indeed, it is rather immediate to write down the IWLS problem whose stationary point matches the above $\hat{\mathbf{z}}$ (or $\hat{\mathbf{z}}_{N|N}$). We simply use (72) to conclude that the related problem of the form (1) is the following:

$$\min_{z=\begin{bmatrix} x_0 \\ u \end{bmatrix}} \left\{\begin{bmatrix} x_0 \\ u \end{bmatrix}^* \Pi^{-1} \begin{bmatrix} x_0 \\ u \end{bmatrix} + \left(y - A\begin{bmatrix} x_0 \\ u \end{bmatrix}\right)^* W^{-1} \left(y - A\begin{bmatrix} x_0 \\ u \end{bmatrix}\right)\right\}. \tag{73}$$

Equivalently, using (42), (41), and (40), this can also be written as

$$\min_{\{x_0, u_0, \ldots, u_{N-1}\}} \left[x_0^*\Pi_0^{-1}x_0 + \sum_{j=0}^{N}(y_j - H_jx_j)^*R_j^{-1}(y_j - H_jx_j) + \sum_{j=0}^{N-1} u_j^*Q_j^{-1}u_j\right], \tag{74}$$

subject to

$$x_{j+1} = F_jx_j + G_ju_j. \tag{75}$$

Likewise, the IWLS problem whose stationary solution $\hat{z}_i$ matches the $\hat{\mathbf{z}}_{i|i}$ is

$$\min_{\{x_0, u_0, \ldots, u_{i-1}\}} \left[x_0^*\Pi_0^{-1}x_0 + \sum_{j=0}^{i}(y_j - H_jx_j)^*R_j^{-1}(y_j - H_jx_j) + \sum_{j=0}^{i-1} u_j^*Q_j^{-1}u_j\right], \tag{76}$$

subject to $x_{j+1} = F_j x_j + G_j u_j$. That is, only vectors $\{y_j\}$ up to time i are included. The stationary solution $\hat{z}_{i|i}$ exists and is unique if, and only if, using (51) and (52),

$$\Pi_i^{-1} + A_i^* W_i^{-1} A_i \text{ is invertible.}$$

This implies, in view of Lemma 2.2, that $(W_i + A_i \Pi A_i^*)$ is also invertible. We thus have the following preliminary conclusion, which shows that the strong regularity assumption that we imposed earlier on the Gramian matrix R_y is not a restriction. It is in fact necessary if we are interested in all the stationary solutions $\{\hat{z}_{i|i}\}$.

Lemma 7.1 (Strong Regularity) *The stationary solutions $\hat{z}_{i|i}$ are uniquely defined for all $0 \leq i \leq N$ if, and only if, the matrix $(W + A\Pi A^*)$ is (block) strongly regular.*

Proof: Since $\{W, \Pi\}$ are block diagonal and A is block lower triangular, the (block) leading submatrices of $(W + A\Pi A^*)$ are of the form $(W_i + A_i \Pi A_i^*)$. But we argued above that $\hat{z}_{i|i}$ is uniquely defined iff $(W_i + A_i \Pi A_i^*)$ is invertible. Since this holds for all $0 \leq i \leq N$, we conclude that $(W + A\Pi A^*)$ is necessarily (block) strongly regular.

■

In other words, recall that we have established earlier in Lemma 5.1 that the standard optimization problems (1) and (2) are always guaranteed to simultaneously have unique stationary solutions $\hat{z}$ and K^o (and also $\hat{\mathbf{z}}$). The above result then extends this conclusion to the successive solutions $\{\hat{z}_{i|i}, \hat{\mathbf{z}}_{i|i}\}$ of (50) and (76). That is, when state-space structure is incorporated into both optimization criteria, and recursive stationarization is employed, it also holds that the criteria have simultaneous stationary points.

Problem 7.1 (The IWLS State-Space Problem) *For each i, define the quadratic cost function*

$$J_i(x_0, u_0, \ldots, u_{i-1}) \triangleq \left[x_0^* \Pi_0^{-1} x_0 + \sum_{j=0}^{i} (y_j - H_j x_j)^* R_j^{-1} (y_j - H_j x_j) + \sum_{j=0}^{i-1} u_j^* Q_j^{-1} u_j \right]. \tag{77}$$

We are interested in minimizing, when possible, the J_i over $(x_0, u_0, \ldots, u_{i-1})$, for all $0 \leq i \leq N$, and subject to the state-space constraint $x_{i+1} = F_i x_i + G_i u_i$.

Before stating the conditions that would allow us to check whether the existence of minima for all J_i exist, we shall first consider the following:

(i) We shall show how to recursively compute the unique stationary points $\{\hat{z}_{i|i}\}$ when they exist.

(ii) We shall then derive conditions for these points to be minima.

In order to highlight the possibilities that may occur in the indefinite case, let us assume for now that the $\{J_i\}$ have unique stationary points $\{\hat{z}_{i|i}\}$, so that $(W + A\Pi A^*)$ is guaranteed to be (block) strongly regular, as proven in Lemma 7.1.

Now, each one of the stationary points $\hat{z}_{i|i}$ may or may not be a minimum in its own right, and this is independent of whether among the earlier solutions $\{\hat{z}_{j|j}\}_{j<i}$ we have minima or not. This is in contrast to the recursive minimization of quadratic cost functions with positive-definite weighting matrices, where all the solutions $\hat{z}_{i|i}$ are guaranteed to be minima. In the indefinite case however, it may happen that at a particular time instant, say the ith instant, the $\hat{z}_{i|i}$ is a minimum

of J_i, while in the next time instant, the $\hat{z}_{i+1|i+1}$ is not a minimum of J_{i+1}. This is because, the minimality of one requires the positivity of $(\Pi_i^{-1} + A_i^* W_i^{-1} A_i)$, while the minimality of the other requires the positivity of $(\Pi_{i+1}^{-1} + A_{i+1}^* W_{i+1}^{-1} A_{i+1})$, and the positivity of these two matrices do not imply each other. In particular, the second matrix contains new entries, such as Q_i, R_{i+1}, and an extra row in A_{i+1}. These entries can destroy the positivity of $(\Pi_{i+1}^{-1} + A_{i+1}^* W_{i+1}^{-1} A_{i+1})$. This situation does not occur with positive-definite quadratic forms because, in this case, the weighting matrices $\{\Pi, W\}$ are positive-definite and hence, $(\Pi_i^{-1} + A_i^* W_i^{-1} A_i)$ is positive-definite for all i.

7.1 Fundamental Inertia Conditions

The following result, for example, establishes under what condition J_N has a minimum at $\hat{z}_{N|N}$.

Lemma 7.2 (Minimization of J_N) *Consider a quadratic cost function as in (77) and subject to $x_{i+1} = F_i x_i + G_i u_i$. The quantities $\{x_0, u_0, \ldots, u_{N-1}\}$ are the unknowns. Let $m \times m$ denote the size of each Q_i. Likewise, let $n \times n$ denote the size of Π_0. Define*

$$\Pi \triangleq (\Pi_0 \oplus Q_0 \ldots \oplus Q_{N-1}), \quad W \triangleq (R_0 \oplus R_1 \oplus \ldots \oplus R_N).$$

Assume $(W + A\Pi A^*)$ is (block) strongly regular (i.e., the J_i are guaranteed to have unique stationary points $\hat{z}_{i|i}$ for all $0 \leq i \leq N$). Then J_N has a minimum with respect to these unknowns (i.e., the last stationary point $\hat{z}_{N|N}$ is a minimum) if, and only if,

$$\begin{aligned} I_-[\Pi \oplus W] &= I_-\{R_{e,0} \oplus \ldots \oplus R_{e,N}\}, \\ I_+[\Pi \oplus W] &= I_+\{R_{e,0} \oplus \ldots \oplus R_{e,N}\} + n + mN, \end{aligned}$$

where the matrices $\{R_{e,i}\}$ are recursively computed as follows:

$$R_{e,i} = H_i P_i H_i^* + R_i,$$

$$P_{i+1} = F_i P_i F_i^* + G_i Q_i G_i^* - K_{p,i} R_{e,i} K_{p,i}^*, \quad P_0 = \Pi_0,$$
$$K_{p,i} = F_i P_i H_i^* R_{e,i}^{-1}.$$

Proof: Note here that the size of Π is $(n + mN) \times (n + mN)$. It then follows from Theorem 5.1 that problem (74) (or, equivalently, (73)) has a minimum if, and only if,

$$\begin{aligned} I_-[W + A\Pi A^*] &= I_-[\Pi \oplus W], \\ I_+[W + A\Pi A^*] &= I_+[\Pi \oplus W] - n - mN. \end{aligned}$$

The result of the lemma now follows by invoking Corollary 6.1, which states that the matrix $(W + A\Pi A^*)$ has the same inertia as $\{R_{e,0} \oplus \ldots \oplus R_{e,N}\}$. This last statement holds as a result of the strong regularity of $(W + A\Pi A^*)$.

■

An immediate conclusion is the following special case where the Π matrix is itself positive-definite and, hence, its negative inertia is zero while its positive-inertia is equal to the number of its columns (or rows), $n + mN$.

Corollary 7.1 (A Special Case) *Consider the same setting of Lemma 7.2. Assume further that $\Pi_0 > 0$ and the $\{Q_i\}_{i=0}^{N-1}$ are positive-definite. Then J_N has a minimum with respect to z_N if, and only if,*

$$\begin{aligned} I_-\{R_0 \oplus \ldots \oplus R_N\} &= I_-\{R_{e,0} \oplus \ldots \oplus R_{e,N}\}, \\ I_+\{R_0 \oplus \ldots \oplus R_N\} &= I_+\{R_{e,0} \oplus \ldots \oplus R_{e,N}\}. \end{aligned}$$

The above results were concerned with the existence of a minimum for the last cost function J_N. More generally, we are interested in checking whether each $\hat{z}_{i|i}$ is a minimum of the corresponding J_i. This is addressed in the following statement.

Theorem 7.1 (Recursive Minimization of $\{J_i\}$) *Consider a quadratic cost function as in (77) and subject to $x_{i+1} = F_i x_i + G_i u_i$. The quantities $\{x_0, u_0, \ldots, u_{N-1}\}$ are the unknowns. Let $m \times m$ denote the size of each Q_i. Likewise, let $n \times n$ denote the size of Π_0. Define*

$$\Pi \triangleq (\Pi_0 \oplus Q_0 \ldots \oplus Q_{N-1}), \quad W \triangleq (R_0 \oplus R_1 \oplus \ldots \oplus R_N).$$

Then each J_i has a minimum with respect to $\{x_0, u_0, \ldots, u_{i-1}\}$ if, and only if,

$$I_-\left[\Pi_0 \oplus R_0\right] = I_-\{R_{e,0}\}, \tag{78}$$

$$I_+\left[\Pi_0 \oplus R_0\right] = I_+\{R_{e,0}\} + n, \tag{79}$$

and, for $i = 1, 2, \ldots, N$,

$$I_-\{Q_{i-1} \oplus R_i\} = I_-\{R_{e,i}\}, \tag{80}$$

$$I_+\{Q_{i-1} \oplus R_i\} = I_+\{R_{e,i}\} + m. \tag{81}$$

Moreover, when the stationary solutions (or minima) of the J_i are uniquely defined, the value of each J_i at its unique stationary solution (or minimum) $\hat{z}_{i|i}$ is given by

$$J_i(\hat{z}_{i|i}) = \sum_{j=0}^{i} e_i^* R_{e,i}^{-1} e_i, \tag{82}$$

where $e_i = (y_i - H_i \hat{x}_{i|i-1})$.

Proof: The proof is by induction. Minimizing J_0 over x_0 requires the inertia conditions (78) and (79), as is obvious for example from Lemma 7.2 specialized to $N = 0$. Likewise, the minimization of J_1 requires

$$\begin{aligned} I_-\left[\Pi_0 \oplus Q_0 \oplus R_0 \oplus R_1\right] &= I_-\{R_{e,0} \oplus R_{e,1}\}, \\ I_+\left[\Pi_0 \oplus Q_0 \oplus R_0 \oplus R_1\right] &= I_+\{R_{e,0} \oplus R_{e,1}\} + n + m, \end{aligned}$$

which by virtue of (78) and (79) yield (80) and (81) for $i = 1$. Continuing in this fashion we establish the result for $i > 1$.

To establish (82) we recall that the value of a quadratic cost function of the form (1) at its stationary solution is given by (22), which in the present context translates to

$$J_i(\hat{z}_{i|i}) = \begin{bmatrix} y_0^* & y_1^* & \ldots & y_i^* \end{bmatrix} [W_i + A_i \Pi_i A_i^*]^{-1} \begin{bmatrix} y_0 \\ y_1 \\ \vdots \\ y_i \end{bmatrix}.$$

But we know from the discussion in the earlier section (viz., (53) and (55)) that if we introduce the triangular factorization of the matrix $(W_i + A_i\Pi_i A_i^*)$, say

$$(W_i + A_i\Pi_i A_i^*) = L_i D_i L_i^*,$$

then

$$L_i \begin{bmatrix} e_0 \\ e_1 \\ \vdots \\ e_i \end{bmatrix} = \begin{bmatrix} y_0 \\ y_1 \\ \vdots \\ y_i \end{bmatrix},$$

and $D_i = (R_{e,0} \oplus \ldots \oplus R_{e,i})$. Consequently,

$$J_i(\hat{z}_{i|i}) = \begin{bmatrix} e_0^* & e_1^* & \ldots & e_i^* \end{bmatrix} D_i^{-1} \begin{bmatrix} e_0 \\ e_1 \\ \vdots \\ e_i \end{bmatrix} = \sum_{j=0}^{i} e_i^* R_{e,i}^{-1} e_i.$$

■

It is also clear from the discussions in Sec. 5 that the recursions of Theorem 6.2, with the proper identifications $\hat{\mathbf{z}}_{N|i} \leftarrow \hat{z}_{N|i}$, $\mathbf{y}_i \leftarrow y_i$, $\hat{\mathbf{x}}_{i|i-1} \leftarrow \hat{x}_{i|i-1}$, $\mathbf{u}_i \leftarrow u_i$, can be used to compute the stationary solutions $\hat{z}_{i|i}$ of (77). In particular, and according to the discussions that led to (70), we also have that the stationary solutions $\hat{z}_{i|i}$ are related to the $\hat{z}_{N|i}$, given below in the statement of the theorem, as follows:

$$\hat{z}_{N|i} = \begin{bmatrix} \hat{x}_{0|i} \\ \hat{u}_{0|i} \\ \vdots \\ \hat{u}_{i-1|i} \\ 0 \\ \vdots \\ 0 \end{bmatrix} = \begin{bmatrix} \hat{z}_{i|i} \\ 0 \end{bmatrix}. \tag{83}$$

That is, the leading entries of $\hat{z}_{Ni|i}$ denote the stationary solution of J_i with respect to $\{x_0, u_0, \ldots, u_{i-1}\}$

Theorem 7.2 (Recursive Solution of (77)) *Consider a quadratic cost function as in (77) and subject to $x_{i+1} = F_i x_i + G_i u_i$. The quantities $\{x_0, u_0, \ldots, u_{N-1}\}$ are the unknowns. Assume $(W + A\Pi A^*)$ is (block) strongly regular with $\{W, A, \Pi\}$ defined as in (40), (41), and (42). Let*

$$z_N \triangleq \begin{bmatrix} x_0 \\ u_0 \\ \vdots \\ u_{N-1} \end{bmatrix}.$$

The stationary solution, $\hat{z}_{i|i}$, of

$$\min_z \left[x_0^* \Pi_0^{-1} x_0 + \sum_{j=0}^{i} (y_j - H_j x_j)^* R_j^{-1} (y_j - H_j x_j) + \sum_{j=0}^{i-1} u_j^* Q_j^{-1} u_j \right], \tag{84}$$

can be recursively computed as follows: start with $\hat{z}_{N|-1} = 0$ *and repeat for* $i = 0, 1, \ldots, N$:

$$\hat{z}_{N|i} = \hat{z}_{N|i-1} + K_{z,i} H_i^* R_{e,i}^{-1} (y_i - H_i \hat{x}_{i|i-1}),$$

where

$$K_{z,i+1} = K_{z,i} \left[F_i - K_i R_{e,i}^{-1} H_i \right]^* + \begin{bmatrix} 0 \\ I \\ 0 \end{bmatrix} Q_i G_i^*, \quad K_{z,0} = \begin{bmatrix} \Pi_0 \\ 0 \end{bmatrix},$$

and

$$\hat{x}_{i+1|i} = F_i \hat{x}_{i|i-1} + K_{p,i}(y_i - H_i \hat{x}_{i|i-1}), \quad \hat{x}_{0|-1} = 0.$$

Remark. It may happen that the last term in the definition of the quadratic cost function J_i in (77) also includes the extra tem $u_i^* Q_i^{-1} u_i$, say

$$J_i(x_0, u_0, \ldots, u_{i-1}) \triangleq \left[x_0^* \Pi_0^{-1} x_0 + \sum_{j=0}^{i} (y_j - H_j x_j)^* R_j^{-1} (y_j - H_j x_j) + \sum_{j=0}^{i} u_j^* Q_j^{-1} u_j \right]. \quad (85)$$

In this case, the unknown variable u_i only appears in the quadratic term $u_i^* Q_i^{-1} u_i$, and it thus follows that minimization with respect to the u_i requires the positivity of Q_i. Hence, successive minimization of the J_i would additionally require that the $\{Q_i\}$ be positive-definite, which is a special case that often arises in the context of H^∞-problems, with the additional constraint $\Pi_0 > 0$. It is thus rather immediate to handle this case. All we need to do is to simply impose a positivity condition on the $\{Q_i\}$. This motivates us to consider the following two corollaries.

Corollary 7.2 (Some Positive Weighting Matrices) *Consider the same setting as in Theorem 7.1 and further assume that the* $\{Q_i\}_{i=0}^{N-1}$ *are positive-definite. Assume also that* $\Pi_0 > 0$. *Then each* J_i *has a minimum with respect to* $\{x_0, u_0, \ldots, u_{i-1}\}$ *if, and only if, for all* i,

$$\textit{Inertia}\{R_i\} = \textit{Inertia}\{R_{e,i}\}. \quad (86)$$

In this case, it follows that

$$P_i \geq 0 \quad \textit{for} \quad 0 \leq i \leq N. \quad (87)$$

[In fact, P_0 *is strictly positive since it is equal to* Π_0*].*

Proof: The inertia conditions (86) follow immediately as a special case of Theorem 7.1. We now establish the nonnegativity of the Riccati variables $\{P_i\}$. This is achieved by induction. Assume the result is valid up to time j, i.e., $\{P_0, P_1, \ldots, P_j\}$ are nonnegative-definite and let us prove that P_{j+1} is also nonnegative-definite.

It follows from (86) that $R_{e,j} = (R_j + H_j P_j H_j^*)$ and R_j must have the same inertia and, consequently, that $(R_j + H_j P_j H_j^*)$ is invertible.

Since P_j is nonnegative-definite, we can factor it into $P_j = M_j M_j^*$, where the number of columns of M_j is equal to the rank of P_j. Defining $\bar{H}_j \triangleq H_j M_j$ we can write $(R_j + H_j P_j H_j^*) = (R_j + \bar{H}_j \bar{H}_j^*)$.

The invertibility of $(R_j + \bar{H}_j \bar{H}_j^*)$ now implies, by virtue of Lemma 2.2, that $(I + \bar{H}_j^* R_j^{-1} H_j)$ is also invertible. Using the result of Theorem 2.1 we have that

$$\begin{aligned} I_+(I \oplus R_j) &= I_+[(I + \bar{H}_j^* R_j^{-1} \bar{H}_j) \oplus (R_j + \bar{H}_j \bar{H}_j^*)], \\ I_-(I \oplus R_j) &= I_-[(I + \bar{H}_j^* R_j^{-1} \bar{H}_j) \oplus (R_j + \bar{H}_j \bar{H}_j^*)]. \end{aligned}$$

But since

$$\text{Inertia}\{R_j + \bar{H}_j\bar{H}_j^*\} = \text{Inertia}\{R_j\},$$

we conclude that I and $(I+\bar{H}_j^* R_j^{-1}\bar{H}_j)$ must have the same inertia and, hence, $(I+\bar{H}_j^* R_j^{-1}H_j) > 0$. Now the Riccati recursion (63) implies that

$$\begin{aligned} P_{j+1} &= F_j\left[P_j - P_jH_j^*(R_j + H_jP_jH_j^*)^{-1}H_jP_j\right]F_j^* + G_jQ_jG_j^*, \\ &= F_jM_j\left[I - \bar{H}_j^*(R_j + \bar{H}_j\bar{H}_j^*)^{-1}\bar{H}_j\right]M_j^*F_j^* + G_jQ_jG_j^*, \\ &= F_jM_j\left[I + \bar{H}_j^*R_j^{-1}\bar{H}_j\right]^{-1}M_j^*F_j^* + G_jQ_jG_j^*. \end{aligned}$$

But since $(I + \bar{H}_j^*R_j^{-1}\bar{H}_j) > 0$ and $G_jQ_jG_j^* \geq 0$, we conclude that $P_{j+1} \geq 0$.

■

The next statement further assumes that the $\{F_i\}$ are invertible.

Corollary 7.3 (Positive Weights and Invertible $\{F_i\}$) *Consider the same setting as in Theorem 7.1 and further assume that the $\{Q_i\}_{i=0}^{N-1}$ are positive-definite. Assume also that $\Pi_0 > 0$ and that the $\{F_i\}$ are invertible. Then the following two statements provide equivalent necessary and sufficient conditions for each J_i to have a minimum with respect to $\{x_0, u_0, \ldots, u_{i-1}\}$.*

(i) All $\{J_i\}$ have minima iff, for $0 \leq i \leq N$,

$$P_i^{-1} + H_i^*R_i^{-1}H_i > 0. \tag{88}$$

(ii) All $\{J_i\}$ have minima iff, for $0 \leq i \leq N$,

$$P_{i+1} - G_iQ_iG_i^* > 0\ . \tag{89}$$

It further follows in the minimum case that, for all i,

$$P_{i+1} > 0. \tag{90}$$

Proof: A simple inductive argument establishes the result. It follows from Corollary 7.2 that $R_{e,0} = (R_0 + H_0\Pi_0H_0^*)$ and R_0 must have the same inertia and, consequently, that $(R_0 + H_0\Pi_0H_0^*)$ is invertible. Lemma 2.2 then implies that $(\Pi_0^{-1} + H_0^*R_0^{-1}H_0)$ is also invertible. Using the result of Theorem 2.1 we have that

$$\begin{aligned} I_+(\Pi_0 \oplus R_0) &= I_+[(\Pi_0^{-1} + H_0^*R_0^{-1}H_0) \oplus (H_0\Pi_0H_0^* + R_0)], \\ I_-(\Pi_0 \oplus R_0) &= I_-[(\Pi_0^{-1} + H_0^*R_0^{-1}H_0) \oplus (H_0\Pi_0H_0^* + R_0)]. \end{aligned}$$

But since

$$\text{Inertia}\{R_0 + H_0\Pi_0H_0^*\} = \text{Inertia}\{R_0\},$$

we conclude that Π_0 and $(\Pi_0^{-1} + H_0^*R_0^{-1}H_0)$ must have the same inertia and, hence, $(\Pi_0^{-1} + H_0^*R_0^{-1}H_0) > 0$ since $\Pi_0 > 0$. Now the Riccati recursion (63) implies that

$$\begin{aligned} P_1 &= F_0\left[\Pi_0 - \Pi_0H_0^*(R_0 + H_0\Pi_0H_0^*)^{-1}H_0\Pi_0\right]F_0^* + G_0Q_0G_0^*, \\ &= F_0\left[\Pi_0^{-1} + H_0^*R_0^{-1}H_0\right]^{-1}F_0^* + G_0Q_0G_0^*. \end{aligned}$$

The invertibility of F_0 guarantees the positive-definiteness of $F_0\left[\Pi_0^{-1}+H_0^*R_0^{-1}H_0\right]^{-1}F_0^*$. But since $Q_0>0$ we also have that $G_0Q_0G_0^*\geq 0$. Consequently, $P_1>0$. We can now repeat the argument to conclude that the conditions (88) hold for all i.

The equivalence of conditions (88) and (89) follow from the fact that for all i we have

$$P_{i+1}-G_iQ_iG_i^* \;=\; F_i\left[P_i^{-1}+H_i^*R_i^{-1}H_i\right]^{-1}F_i^*.$$

■

Conditions of the form (88) are the ones most cited in H^∞–applications (e.g., [YS91]). Here we see that they are related to the inertia conditions (86). These inertia conditions also arise in the H^∞–context (see, e.g., [GL95][p. 495] and Lemma 8.1 further ahead), where R_i has the additional structure $R_i=(-\gamma^2 I\oplus I)$. Here, we have derived these conditions as special cases of the general statement of Theorem 7.1, which holds for arbitrary indefinite matrices $\{\Pi_0,Q_i,R_i\}$, while the H^∞–results hold only for positive-definite matrices $\{\Pi_0,Q_i\}$ and for matrices R_i of the above form. Note also that testing for (88) not only requires that we compute the P_i (via a Riccati recursion (63)), but also that we invert P_i and R_i at each step and then check for the positivity of $P_i^{-1}+H_i^*R_i^{-1}H_i$. The inertia tests given by (86), on the other hand, employ the quantities $R_{e,i}$ and R_i, which are $p\times p$ matrices (as opposed to P_i which is $n\times n$). These tests can be used as the basis for alternative computational variants that are based on square-root ideas, as pursued in [HSK94].

8 An Application to H$^\infty$-Filtering

We now illustrate the applicability of the earlier results to a problem in H^∞-filtering. For this purpose, we consider a state-space model of the form

$$x_{i+1}=F_ix_i+G_iu_i\,,\quad y_i=H_ix_i+v_i\,, \tag{91}$$

where $\{x_0,u_i,v_i\}$ are unknown deterministic signals and $\{y_i\}_{i=0}^N$ are known (or measured) signals. Let $s_j=L_jx_j$ be a linear transformation of the state-vector x_j, where L_j is a known matrix.

Let $\hat{s}_{j|j}$ denote a function of the $\{y_k\}$ up to and including time j. For every time instant i we define the quadratic cost function

$$J_i(x_0,u_0,\ldots,u_i)\triangleq x_0^*\Pi_0^{-1}x_0+\sum_{j=0}^{i}u_j^*Q_j^{-1}u_j+\sum_{j=0}^{i}v_j^*v_j-\gamma^{-2}\sum_{j=0}^{i}(\hat{s}_{j|j}-L_jx_j)^*(\hat{s}_{j|j}-L_jx_j), \tag{92}$$

where $\{\Pi_0,Q_j\}$ are given positive-definite matrices, and γ is a given positive real number.

Problem 8.1 (An H$^\infty$–Filtering Problem) *Determine, if possible, functions*

$$\{\hat{s}_{0|0},\hat{s}_{1|1},\ldots,\hat{s}_{N|N}\},$$

in order to guarantee that

$$J_i>0 \;\; \textit{for} \;\; i=0,1,\ldots,N. \tag{93}$$

The positivity requirement (93) can be interpreted as imposing an upper bound on the following ratios (for nonzero denominators)

$$\frac{\sum_{j=0}^{i}(\hat{s}_{j|j}-L_jx_j)^*(\hat{s}_{j|j}-L_jx_j)}{x_0^*\Pi_0^{-1}x_0+\sum_{j=0}^{i}u_j^*Q_j^{-1}u_j+\sum_{j=0}^{i}v_j^*v_j} < \gamma^2, \text{ for } 0\le i\le N.$$

Using $v_j = y_j - H_jx_j$, we can rewrite the expression for J_i in the equivalent form

$$J_i = x_0^*\Pi_0^{-1}x_0+\sum_{j=0}^{i}\left(\begin{bmatrix}\hat{s}_{j|j}\\ y_j\end{bmatrix}-\begin{bmatrix}L_j\\ H_j\end{bmatrix}x_j\right)^*\begin{bmatrix}-\gamma^{-2}I & 0\\ 0 & I\end{bmatrix}\left(\begin{bmatrix}\hat{s}_{j|j}\\ y_j\end{bmatrix}-\begin{bmatrix}L_j\\ H_j\end{bmatrix}x_j\right)+\sum_{j=0}^{i}u_j^*Q_j^{-1}u_j,$$

which is a quadratic cost function in the unknowns $\{x_0, u_0, \ldots, u_i\}$ since the $\{y_j, \hat{s}_{j|j}\}_{j=0}^{i}$ can be expressed in terms of $\{x_0, u_0, \ldots, u_i\}$. Therefore, each J_i will be positive if, and only if, it has a minimum with respect to $\{x_0, u_0, \ldots, u_i\}$ and, moreover, the value of J_i at its minimum is positive.

8.1 Solvability Conditions

We thus see that we are faced with the problem of minimizing a quadratic cost function of the same general form as in (85), and also (84), where the column vector

$$\begin{bmatrix}\hat{s}_{j|j}\\ y_j\end{bmatrix},$$

and the block matrices

$$\begin{bmatrix}-\gamma^2I & 0\\ 0 & I\end{bmatrix} \text{ and } \begin{bmatrix}L_j\\ H_j\end{bmatrix},$$

now play the roles of $\{y_j, R_j, H_j\}$ in (85). That is, the auxiliary state-space model that we may invoke here, with variables in an indefinite space $\mathcal{K}'$, takes the form

$$\begin{aligned}\mathbf{x}_{i+1} &= F_i\mathbf{x}_i + G_i\mathbf{u}_i,\\ \begin{bmatrix}\hat{\mathbf{s}}_{j|j}\\ \mathbf{y}_j\end{bmatrix} &= \begin{bmatrix}L_j\\ H_j\end{bmatrix}\mathbf{x}_i+\bar{\mathbf{v}}_i,\end{aligned}$$

with

$$<\begin{bmatrix}\mathbf{u}_i\\ \bar{\mathbf{v}}_i\\ \mathbf{x}_0\end{bmatrix},\begin{bmatrix}\mathbf{u}_j\\ \bar{\mathbf{v}}_j\\ \mathbf{x}_0\end{bmatrix}>_{\mathcal{K}'}=\begin{bmatrix}Q_i\delta_{ij} & 0 & 0\\ 0 & (-\gamma^2I\oplus I)\delta_{ij} & 0\\ 0 & 0 & \Pi_0\end{bmatrix}.$$

We then conclude from Corollary 7.2, and according to the remark after Theorem 7.2, that each J_i will admit a minimizing solution if, and only if, the corresponding $R_{e,i}$ and R_i have the same inertia. In the present context, we have

$$R_i \triangleq \begin{bmatrix}-\gamma^2I & 0\\ 0 & I\end{bmatrix} \text{ and } R_{e,i}\triangleq\begin{bmatrix}-\gamma^2I & 0\\ 0 & I\end{bmatrix}+\begin{bmatrix}L_i\\ H_i\end{bmatrix}P_i\begin{bmatrix}L_i\\ H_i\end{bmatrix}^*,$$

where P_i satisfies the Riccati difference equation

$$\begin{aligned}P_{i+1} &= F_i\left[P_i-P_i\begin{bmatrix}L_i\\ H_i\end{bmatrix}^*\left\{\begin{bmatrix}L_i\\ H_i\end{bmatrix}P_i\begin{bmatrix}L_i\\ H_i\end{bmatrix}^*+\begin{bmatrix}-\gamma^2I & 0\\ 0 & I\end{bmatrix}\right\}^{-1}\begin{bmatrix}L_i\\ H_i\end{bmatrix}P_i\right]F_i^*+G_iQ_iG_i^*,\\ &= F_i\left[P_i^{-1}+\begin{bmatrix}L_i^* & H_i^*\end{bmatrix}\begin{bmatrix}-\gamma^2I & 0\\ 0 & I\end{bmatrix}^{-1}\begin{bmatrix}L_i\\ H_i\end{bmatrix}\right]^{-1}F_i^*+G_iQ_iG_i^*,\\ &= F_i\left[P_i^{-1}+H_i^*H_i-\gamma^{-2}L_i^*L_i\right]^{-1}F_i^*+G_iQ_iG_i^*.\end{aligned}$$

Lemma 8.1 (Inertia Conditions) *The J_i in (92) admit unique minima with respect to the quantities $\{x_0, u_0, \ldots, u_i\}$ if, and only if, the matrices*

$$\begin{bmatrix} -\gamma^2 I & 0 \\ 0 & I \end{bmatrix} \quad and \quad \begin{bmatrix} -\gamma^2 I + L_i P_i L_i^* & L_i P_i H_i^* \\ H_i P_i L_i^* & I + H_i P_i H_i^* \end{bmatrix}, \tag{94}$$

have the same inertia for all i. In this case, it also follows that all the leading submatrices of the above two matrices have the same inertia, i.e.,

$$\begin{aligned} I + H_i P_i H_i^* &> 0, \\ (-\gamma^2 I + L_i P_i L_i^*) - L_i P_i H_i^* (I + H_i P_i H_i^*)^{-1} H_i P_i L_i^* &< 0. \end{aligned}$$

Proof: The first part of the Lemma follows from Corollary 7.2. But recall also from the statement of the Corollary that the resulting P_i are further guaranteed to be nonnegative-definite, i.e., $P_i \geq 0$. It thus follows that $(I + H_i P_i H_i^*) > 0$. That is, the lower-right corner elements of both matrices in (94) have the same positive inertia. Consequently, it also holds that all the leading submatrices of the two matrices in (94) have the same inertia.

■

If the F_i are further assumed invertible, then we also conclude from Corollary 7.3 that the following alternative conditions can be used to guarantee the existence of minima for the J_i in (92),

$$P_i^{-1} + H_i^* H_i - \gamma^{-2} L_i^* L_i > 0, \quad \text{for } 0 \leq i \leq N. \tag{95}$$

8.2 Construction of a Solution

To end our discussion, we still need to show how to determine the estimates $\hat{s}_{j|j}$ once the existence of minima for the J_i are guaranteed. These estimates have to be chosen so as to guarantee that the values of the successive J_i at their minima are positive.

We shall illustrate the construction by induction. Assume that the $\{\hat{s}_{0|0}, \ldots, \hat{s}_{i-1|i-1}\}$ have already been chosen and that the values of the $\{J_0, J_1, \ldots, J_{i-1}\}$ are positive at their respective minima (recall expression (82)). In particular,

$$\sum_{j=0}^{i-1} e_j^* R_{e,j}^{-1} e_j > 0.$$

In order to guarantee $J_i > 0$ we need to choose $\hat{s}_{i|i}$ so as to result in

$$e_i^* R_{e,i}^{-1} e_i + \sum_{j=0}^{i-1} e_j^* R_{e,j}^{-1} e_j > 0.$$

This can be achieved in many ways and the choice is nonunique. One possibility is to choose $\hat{s}_{i|i}$ so as to meet the condition

$$e_i^* R_{e,i}^{-1} e_i > 0, \tag{96}$$

or, equivalently,

$$\begin{bmatrix} e_{i,s}^* & e_{i,y}^* \end{bmatrix} \begin{bmatrix} -\gamma^2 I + L_i P_i L_i^* & L_i P_i H_i^* \\ H_i P_i L_i^* & I + H_i P_i H_i^* \end{bmatrix}^{-1} \begin{bmatrix} e_{i,s} \\ e_{i,y} \end{bmatrix} > 0, \tag{97}$$

where we have partitioned the e_i accordingly, viz.,

$$e_i \triangleq \begin{bmatrix} \hat{s}_{i|i} \\ y_i \end{bmatrix} - \begin{bmatrix} L_i \\ H_i \end{bmatrix} \hat{x}_{i|i-1} \triangleq \begin{bmatrix} e_{i,s} \\ e_{i,y} \end{bmatrix}.$$

Here $\hat{x}_{i|i-1}$ is constructed recursively as indicated in Theorem 7.2,

$$\hat{x}_{i+1|i} = F_i \hat{x}_{i|i-1} + K_{p,i} \left(\begin{bmatrix} \hat{s}_{i|i} \\ y_i \end{bmatrix} - \begin{bmatrix} L_i \\ H_i \end{bmatrix} \hat{x}_{i|i-1} \right), \quad \hat{x}_{0|-1} = 0, \tag{98}$$

with

$$R_{e,i} = \begin{bmatrix} -\gamma^2 I & 0 \\ 0 & I \end{bmatrix} + \begin{bmatrix} L_i \\ H_i \end{bmatrix} P_i \begin{bmatrix} L_i \\ H_i \end{bmatrix}^*, \quad K_{p,i} = F_i P_i \begin{bmatrix} L_i^* & H_i^* \end{bmatrix} R_{e,i}^{-1}.$$

We may now introduce the lower-diagonal-upper factorization of the central matrix in (97), viz.,

$$\begin{bmatrix} -\gamma^2 I + L_i P_i L_i^* & L_i P_i H_i^* \\ H_i P_i L_i^* & I + H_i P_i H_i^* \end{bmatrix}^{-1} = \tag{99}$$

$$\begin{bmatrix} I & 0 \\ -(I + H_i P_i H_i^*)^{-1} H_i P_i L_i^* & I \end{bmatrix} \begin{bmatrix} \Delta^{-1} & 0 \\ 0 & (I + H_i P_i H_i^*)^{-1} \end{bmatrix} \begin{bmatrix} I & 0 \\ -(I + H_i P_i H_i^*)^{-1} H_i P_i L_i^* & I \end{bmatrix}^*,$$

where we have defined, for compactness of notation,

$$\Delta \triangleq (-\gamma^2 I + L_i P_i L_i^*) - L_i P_i H_i^* (I + H_i P_i H_i^*)^{-1} H_i P_i L_i^*,$$

which we know, from Lemma 8.1, to be a negative definite matrix.

We can then rewrite (97) in the form

$$\begin{bmatrix} e_{i,s}^* - e_{i,y}^* (I + H_i P_i H_i^*)^{-1} H_i P_i L_i^* & e_{i,y}^* \end{bmatrix} \begin{bmatrix} \Delta^{-1} & 0 \\ 0 & (I + H_i P_i H_i^*)^{-1} \end{bmatrix} \begin{bmatrix} e_{i,s} - L_i P_i H_i^* (I + H_i P_i H_i^*)^{-1} e_{i,y} \\ e_{i,y} \end{bmatrix}.$$

This is a quadratic expression in the variable $e_{i,s} = \hat{s}_{i|i} - L_i \hat{x}_{i|i-1}$, and since $\Delta < 0$ and $(I + H_i P_i H_i^*) > 0$, the positivity condition (96) can be met by setting

$$e_{i,s} - L_i P_i H_i^* (I + H_i P_i H_i^*)^{-1} e_{i,y} = 0,$$

or, equivalently,

$$\hat{s}_{i|i} - L_i \hat{x}_{i|i-1} = L_i P_i H_i^* (I + H_i P_i H_i^*)^{-1} [y_i - H_i \hat{x}_{i|i-1}].$$

Therefore, a possible choice for $\hat{s}_{i|i}$ is the following

$$\hat{s}_{i|i} = L_i \left[\hat{x}_{i|i-1} + P_i H_i^* (I + H_i P_i H_i^*)^{-1} (y_i - H_i \hat{x}_{i|i-1}) \right].$$

This choice simplifies (98) to the following (using the factorization (99) for $R_{e,i}^{-1}$ in the expression for $K_{p,i}$)

$$\hat{x}_{i+1|i} = F_i \left[\hat{x}_{i|i-1} + P_i H_i^* (I + H_i P_i H_i^*)^{-1} (y_i - H_i \hat{x}_{i|i-1}) \right].$$

We summarize the results in the following statement.

Lemma 8.2 (A Solution of the H$^\infty$–Problem) *Problem 8.1 has a solution if, and only if, for all $0 \le i \le N$, the matrices*

$$\begin{bmatrix} -\gamma^2 I & 0 \\ 0 & I \end{bmatrix} \quad \text{and} \quad \begin{bmatrix} -\gamma^2 I + L_i P_i L_i^* & L_i P_i H_i^* \\ H_i P_i L_i^* & I + H_i P_i H_i^* \end{bmatrix}, \tag{100}$$

have the same inertia. In this case, one possible construction for the estimates $\{\hat{s}_{i|i}\}$ is the following:

$$\hat{s}_{i|i} \;=\; L_i \left[\hat{x}_{i|i-1} + P_i H_i^* (I + H_i P_i H_i^*)^{-1}(y_i - H_i \hat{x}_{i|i-1})\right], \tag{101}$$

where the $\hat{x}_{i|i-1}$ is constructed recursively via

$$\hat{x}_{i+1|i} = F_i \left[\hat{x}_{i|i-1} + P_i H_i^* (I + H_i P_i H_i^*)^{-1}(y_i - H_i \hat{x}_{i|i-1})\right], \;\; \hat{x}_{0|-1} = 0, \tag{102}$$

and

$$P_{i+1} = F_i \left[P_i - P_i \begin{bmatrix} L_i \\ H_i \end{bmatrix}^* \left\{ \begin{bmatrix} L_i \\ H_i \end{bmatrix} P_i \begin{bmatrix} L_i \\ H_i \end{bmatrix}^* + \begin{bmatrix} -\gamma^2 I & 0 \\ 0 & I \end{bmatrix} \right\}^{-1} \begin{bmatrix} L_i \\ H_i \end{bmatrix} P_i \right] F_i^* + G_i Q_i G_i^*, \tag{103}$$

with the initial condition $P_0 = \Pi_0$.

9 An Application to Robust Adaptive Filters

We now consider another example that can, in effect, be regarded as a special case of the H^∞–problem studied in Sec. 8. Here, however, some simplifications occur that are worth considering separately.

We therefore assume that we have the following special state-space model

$$x_{i+1} = x_i\,, \;\; y_i = H_i x_i + v_i\,, \tag{104}$$

where $\{x_0, v_i\}$ are unknown deterministic signals and $\{y_i\}_{i=0}^N$ are known (or measured) signals. Compared with the model (91) we see that we are now assuming $u_i = 0$ and $F_i = I$. In fact, the arguments that follow can also be applied to any invertible matrix F_i (especially the arguments after Lemma 9.1).

The equations (104) show that the vector x_i does not change with time and is therefore equal to the initial unknown vector x_0. That is, we can as well regard the equations (104) as representing a collection of measured vectors $\{y_i\}$ that are linearly related to an unknown vector x_0,

$$y_i = H_i x_0 + v_i,$$

and the objective is to estimate the x_0 in a certain sense. A classical criterion is to solve a positive-definite least-squares problem of the form (see, e.g., [SK94])

$$\min_{x_0} \left[x_0^* \Pi_0^{-1} x_0 + \sum_{i=0}^{N} (y_i - H_i x_0)^* W_i^{-1} (y_i - H_i x_0) \right], \tag{105}$$

where $\{\Pi_0, W_i\}$ are given positive-definite weighting matrices. In this case, a minimizing solution is always guaranteed to exist and, under some extra conditions on the matrices $\{\Pi_0, H_i\}$, a recursive scheme is in fact possible, thus leading to the famed Recursive-Least-Squares (RLS) algorithm.

Here, however, we allow for indefinite weighting matrices $\{\Pi_0, W_i\}$, along the same lines studied in Sec. 8. More specifically, we let $\hat{x}_{j|j}$ denote a function of the $\{y_k\}$ up to and including time j. Since $x_j = x_0$, we shall also write $\hat{x}_{0|j}$ instead of $\hat{x}_{j|j}$.

For every time instant i we also define the quadratic cost function

$$J_i(x_0) \triangleq x_0^*\Pi_0^{-1}x_0 + \sum_{j=0}^{i} v_j^* v_j - \gamma^{-2}\sum_{j=0}^{i}(\hat{x}_{0|j} - x_0)^*(\hat{x}_{0|j} - x_0), \tag{106}$$

where $\{\Pi_0\}$ is a given positive-definite matrix, and γ is a given positive number.

Problem 9.1 (A Robust Adaptive Filter) *Determine, if possible, functions*

$$\{\hat{x}_{0|0}, \hat{x}_{0|1}, \ldots, \hat{x}_{0|N}\},$$

in order to guarantee that

$$J_i > 0 \quad \textit{for} \quad i = 0, 1, \ldots, N. \tag{107}$$

The positivity requirement (107) can be interpreted as imposing an upper bound on the following ratios (for nonzero denominators)

$$\frac{\sum_{j=0}^{i}(\hat{x}_{0|j} - x_0)^*(\hat{x}_{0|j} - x_0)}{x_0^*\Pi_0^{-1}x_0 + \sum_{j=0}^{i} v_j^* v_j} \; < \; \gamma^2, \text{ for } 0 \le i \le N.$$

Using $v_j = y_j - H_j x_0$, we can also write the above ratios in the form

$$\frac{\sum_{j=0}^{i} \|\hat{x}_{0|j} - x_0\|^2}{x_0^*\Pi_0^{-1}x_0 + \sum_{j=0}^{i} \|y_j - H_j x_0\|^2} \; < \; \gamma^2, \text{ for } 0 \le i \le N. \tag{108}$$

Comparing with (105), we see that the cost function of (105) now appears in the denominator of (108) (with $W_i = I$). Hence, instead of minimizing (105) over x_0, we are now interested in determining estimates for x_0 in order to guarantee that the energy in the error due to estimating x_0 is upper-bounded by γ^2 times the energy of the uncertainties, viz., the denominator in (108).

We can again rewrite the expression for J_i in the equivalent form

$$J_i \;=\; x_0^*\Pi_0^{-1}x_0 + \sum_{j=0}^{i}\left(\begin{bmatrix}\hat{x}_{0|j}\\ y_j\end{bmatrix} - \begin{bmatrix} I \\ H_j\end{bmatrix}x_0\right)^*\begin{bmatrix}-\gamma^{-2}I & 0\\ 0 & I\end{bmatrix}\left(\begin{bmatrix}\hat{x}_{0|j}\\ y_j\end{bmatrix} - \begin{bmatrix} I \\ H_j\end{bmatrix}x_0\right),$$

which is a quadratic cost function in the unknown $\{x_0\}$. We can now use Lemma 8.2 to conclude the following (by setting $L_i = I, F_i = I, G_i = 0, Q_i = 0, x_i = x_0$).

Lemma 9.1 (Solution of the Adaptive Problem) *Problem 9.1 has a solution if, and only if, for all* $0 \le i \le N$, *the matrices*

$$\begin{bmatrix}-\gamma^2 I & 0\\ 0 & I\end{bmatrix} \quad \textit{and} \quad \begin{bmatrix}-\gamma^2 I + P_i & P_i H_i^*\\ H_i P_i & I + H_i P_i H_i^*\end{bmatrix}, \tag{109}$$

have the same inertia. In this case, one possible construction for the estimates $\{\hat{x}_{0|i}\}$ *is the following:*

$$\hat{x}_{0|i} \;=\; \hat{x}_{0|i-1} + P_i H_i^*(I + H_i P_i H_i^*)^{-1}(y_i - H_i\hat{x}_{0|i-1}), \;\; \hat{x}_{0|-1} = 0, \tag{110}$$

where

$$P_{i+1} = \left[P_i - P_i\begin{bmatrix} I \\ H_i \end{bmatrix}^* \left\{\begin{bmatrix} I \\ H_i \end{bmatrix} P_i \begin{bmatrix} I \\ H_i \end{bmatrix}^* + \begin{bmatrix} -\gamma^2 I & 0 \\ 0 & I \end{bmatrix}\right\}^{-1} \begin{bmatrix} I \\ H_i \end{bmatrix} P_i\right], \tag{111}$$

with the initial condition $P_0 = \Pi_0$.

We now argue that the solvability condition can in fact be simplified in the adaptive case. For this purpose, we shall invoke the conclusions of Corollary 7.3. Indeed, it follows from the statement of the corollary that Problem 9.1 has a solution if, and only if, for all $0 \leq i \leq N$,

$$P_i^{-1} + \begin{bmatrix} I & H_i^* \end{bmatrix} \begin{bmatrix} -\gamma^2 I & 0 \\ 0 & I \end{bmatrix} \begin{bmatrix} I \\ H_i \end{bmatrix} > 0,$$

or, equivalently,

$$P_i^{-1} + H_i^* H_i - \gamma^2 I > 0. \tag{112}$$

A simpler statement is the following.

Lemma 9.2 (A Solvability Condition for the Adaptive Problem) *Problem 9.1 has a solution if, and only if,*

$$P_{i+1} > 0 \ \textit{for} \ 0 \leq i \leq N. \tag{113}$$

Proof: This follows from second condition of Corollary 7.3, using $G_i = 0$.

■

The condition (113) is indeed natural in the adaptive context. To clarify this, we note that it follows from the Riccati recursion (111) that

$$P_{i+1}^{-1} = P_i^{-1} + \begin{bmatrix} I & H_i^* \end{bmatrix} \begin{bmatrix} -\gamma^2 I & 0 \\ 0 & I \end{bmatrix} \begin{bmatrix} I \\ H_i \end{bmatrix}, \tag{114}$$

with initial condition $P_0^{-1} = \Pi_0^{-1}$. This implies, by recurrence, that

$$P_{i+1}^{-1} = \Pi_0^{-1} + \sum_{j=0}^{i} \begin{bmatrix} I & H_j^* \end{bmatrix} \begin{bmatrix} -\gamma^2 I & 0 \\ 0 & I \end{bmatrix} \begin{bmatrix} I \\ H_j \end{bmatrix}, \tag{115}$$

which, in view of expression (20) in Theorem 3.1, is precisely the coefficient matrix of the linear system of equations that provides us with $\hat{x}_{0|j}$. The conclusion (113) is then immediate once we also recall from the statement of Theorem 3.1 that a minimum is guaranteed as long as the coefficient matrix is positive-definite.

10 An Application to Total Least-Squares Methods

We now consider a third application that deals with the so-called total-least-squares (or errors-in-variables) method for the solution of linear systems of equations, $Ax \approx b$ (e.g., [LS83, HV91]). The notation $Ax \approx b$ means that due to possible errors (measurement errors, modelling errors, etc) the vector b does not necessarily lie in the range space of the matrix A, denoted by $\mathcal{R}(A)$. If indeed we had $b \in \mathcal{R}(A)$, then a solution x would exist to the equations $Ax = b$. In general, however, one has to settle for an approximate solution $\hat{x}$. In least-squares methods, it is often assumed that the vector b is possibly erroneous, while the matrix A is known and one proceeds to solve for the vector $\hat{x}$ that minimizes the Euclidean distance between $A\hat{x}$ and b, say

$$\min_{x} \|Ax - b\|^2. \tag{116}$$

This is clearly a special case of the quadratic cost function (1) with $\Pi \to \infty I$, $W = I$, and the notational changes $y \leftarrow b$, $z \leftarrow x$. All solutions $\hat{x}$ are well-known to satisfy the so-called normal system of equations

$$(A^*A)\hat{x} = A^*b. \tag{117}$$

Total least-squares (TLS, for short) methods, on the other hand, allow us to also handle possible errors in the matrix A itself. For this reason, they have been receiving increasing attention, especially in the signal processing community. The TLS problem seeks a matrix $\hat{M}$ and a vector $\hat{x}$ that minimize the following Frobenius norm:

$$\min_{M,x} \left\| \begin{bmatrix} M - A & Mx - b \end{bmatrix} \right\|_F^2. \tag{118}$$

Here, M is regarded as an approximation for A, which in its turn is used to determine an $\hat{x}$ that guarantees $b \in \mathcal{R}(\hat{M})$.

The solution of the above TLS problem is well-known and is given by the following construction [HV91][p. 36]. Assume A is $(N+1) \times n$ with $N \geq n$, as is often the case. Let $\{\sigma_1, \ldots, \sigma_n\}$ denote the singular values of A, with $\sigma_1 \geq \sigma_2 \geq \ldots \geq \sigma_n \geq 0$. Let also $\{\bar{\sigma}_1, \ldots, \bar{\sigma}_n, \bar{\sigma}_{n+1}\}$ denote the singular values of the extended matrix $\begin{bmatrix} A & b \end{bmatrix}$, with $\bar{\sigma}_i \geq 0$. If $\bar{\sigma}_{n+1} < \sigma_n$, then the unique solution $\hat{x}$ of (118) is given by

$$\hat{x} = (A^*A - \bar{\sigma}_{n+1}^2 I)^{-1} A^*b. \tag{119}$$

Moreover, the matrix $\hat{M}$ is constructed from the SVD of $\begin{bmatrix} A & b \end{bmatrix}$. In fact, a similar construction for $\hat{x}$ also exists in terms of the data available from the SVD. But here we shall instead focus on the representation (119) of the solution $\hat{x}$. Note also that the condition $\bar{\sigma}_{n+1} < \sigma_n$ assures that $(A^*A - \bar{\sigma}_{n+1}^2 I)$ is a positive-definite matrix, since σ_n^2 is the smallest eigenvalue of A^*A.

Comparing (119) with the solution of the indefinite quadratic problem (1), as given in Theorem 3.1, expression (20), we see that we can make the identifications

$$\Pi \leftarrow -\bar{\sigma}_{n+1}^{-2} I \quad \text{and} \quad W \leftarrow I,$$

along with $y \leftarrow b$ and $z \leftarrow x$. That is, we can regard (119) as the solution of the following indefinite problem

$$\min_{x} \left[-\bar{\sigma}_{n+1}^2 x^*x + (b - Ax)^*(b - Ax) \right], \tag{120}$$

which is clearly a special case of (1) in two respects: the Π matrix is negative-definite and a multiple of the identity, and the W matrix is simply the identity. Indeed, the minimum of (120) exists as long as $(-\bar{\sigma}_{n+1}^2 I + A^*A)$ is positive-definite, which is guaranteed by the assumption $\bar{\sigma}_{n+1} < \sigma_n$.

Note though that the solution $\hat{x}$ of the TLS problem (119) requires a singular value decomposition (SVD), which may be computationally expensive. But more important perhaps, is that this may hinder the possibility of recursive updates of the solution $\hat{x}$. More specifically, if an extra row is added to the matrix A and, correspondingly, if an extra entry is added to the vector b, then the SVD of the new extended matrix $\begin{bmatrix} A & b \end{bmatrix}$ will need to be computed again in order to evaluate the new solution $\hat{x}$.

An examination of expression (119), however, shows that the SVD step only affects the choice of the Π matrix. This suggests that a recursive scheme should be possible if one relaxes the criterion (118) and allows for other choices of the Π matrix in (120), say

$$\Pi^{-1} = -\rho^2 I,$$

for a nonnegative real number ρ^2 that is chosen by the user. In particular, any choice that satisfies $\rho^2 < \sigma_n$ will still result in a positive-definite matrix $[-\rho^2 I + A^* A]$. We may also employ a diagonal matrix of the form

$$\Pi^{-1} = -\text{diagonal}\ \{\rho_0^2, \rho_1^2, \ldots, \rho_{n-1}^2\},$$

with several nonnegative entries $\{\rho_i^2\}$. This would allow us to give different weights to the different entries of x and will also give us more freedom in controlling the existence of solutions to the recursive procedure described below.

We may also remark that the idea of replacing an optimal problem by a suboptimal one is frequent in many areas, including for example H^∞–problems, and this is often due to the computational burden that may be required by an optimal formulation.

Problem 10.1 (Approximate TLS Problem) *Consider a matrix A, with rows $\{a_i\}_{i=0}^N$, a vector b with entries $\{b(i)\}_{i=0}^N$, and a diagonal matrix $\Pi^{-1} = -diag\{\rho_i^2\}$. Define, for each i, the quadratic cost function*

$$J_i \triangleq \left[x^* \Pi^{-1} x + \sum_{j=0}^{i} |b(j) - a_j x|^2 \right].$$

Let $\hat{x}_i$ denote a stationary solution of J_i. We are interested in the following:

(i) A recursive update that relates $\hat{x}_i$ to $\hat{x}_{i+1}$. For this purpose, we shall assume that (recall Lemma 7.2) $[I + A\Pi A^*]$ is strongly regular. This suggests a criterion for choosing the Π matrix.

(ii) A condition that guarantees that the last estimate $\hat{x}_N$ is indeed a minimum of J_N.

The answers to the above questions are rather immediate if we invoke the results of Sec. 7 and, in particular, Lemma 7.2 and its corollary, and Theorems 7.1 and 7.2.

Lemma 10.1 (Solution of the Approximate TLS Problem) *A recursive construction of the solution can be obtained as follows, assuming $[I + A\Pi A^*]$ is strongly regular:*

(i) The successive stationary solutions are related via

$$\hat{x}_i = \hat{x}_{i-1} + \frac{P_i a_i^*}{1 + a_i P_i a_i^*} [b(i) - a_i \hat{x}_{i-1}], \quad \hat{x}_{-1} = 0, \tag{121}$$

$$P_{i+1} = P_i - \frac{P_i a_i^* a_i P_i}{1 + a_i P_i a_i^*}, \quad P_0 = \Pi = -diag\{\rho_i^2\}. \tag{122}$$

(ii) J_N has a minimum at $\hat{x}_N$ if, and only if, the matrix $[\Pi^{-1} + A^* A]$ is positive-definite. Under the assumption of strong regularity of $[I + A\Pi A^*]$, this positivity condition is also equivalent to $P_{N+1} > 0$ since, as argued after the proof of Lemma 9.1, we can also verify here that P_{N+1} is the inverse of $[\Pi^{-1} + A^* A]$. Indeed, from (122) we obtain

$$P_{i+1}^{-1} = P_i^{-1} + a_i^* a_i, \quad P_0 = \Pi^{-1}. \tag{123}$$

We emphasize, however, that the above is only a special case of the quadratic forms studied in this paper. For example, one may choose other forms for the diagonal matrices Π and W, such as allowing for positive entries in Π and for negative entries in W, or other convenient combinations.

11 Concluding Remarks

We have posed two minimization problems in indefinite metric spaces and established a link between their solutions via a fundamental set of inertia conditions. These conditions were derived under very general assumptions and later specialized to important special cases that arise in H^∞–filtering, robust adaptive filtering, and approximate TLS methods. In the H^∞–context, for instance, the inertia results of Corollary 7.2 can be used as the basis for alternative computational variants that are based on square-root ideas. This point of view is detailed in [HSK94]. More generally, the inertia conditions of Theorem 7.1 can also form the basis for general square-root algorithms and this will be discussed elsewhere.

Further connections with system theory and recent applications to problems in linear and non-linear robust adaptive filtering can be found in [SR95a, SR95b, SR95c, RS95, RS96].

Acknowledgment

This work was supported in part by a grant from the National Science Foundation under award no. MIP-9409319, and by the Army Research Office under contract DAAL03-89-K-0109.

The first author would also like to thank Prof. Karl Aström for suggesting possible connections between robust filtering and TLS methods.

References

[AM79] B. D. O. Anderson and J. B. Moore. *Optimal Filtering.* Prentice-Hall Inc., NJ, 1979.

[Bog74] J. Bognar. *Indefinite Inner Product Spaces.* Springer-Verlag, New York, 1974.

[DGKF89] J. C. Doyle, K. Glover, P. Khargonekar, and B. Francis. State-space solutions to standard H_2 and H_∞ control problems. *IEEE Transactions on Automatic Control,* 34(8):831–847, August 1989.

[Gan59] F. R. Gantmacher. *The Theory of Matrices.* Chelsea Publishing Company, NY, 1959.

[GL95] M. Green and D. J. N. Limebeer. *Linear Robust Control.* Prentice Hall, NJ, 1995.

[GLR83] I. Gohberg, P. Lancaster, and L. Rodman. *Matrices and Indefinite Scalar Products.* Birkhäuser Verlag, Basel, 1983.

[Gri93] M. J. Grimble. Polynomial matrix solution of the H^∞ filtering problem and the relationship to Riccati equation state-space results. *IEEE Trans. on Signal Processing,* 41(1):67–81, January 1993.

[Hay91] S. Haykin. *Adaptive Filter Theory.* Prentice Hall, Englewood Cliffs, NJ, second edition, 1991.

[HSK93] B. Hassibi, A. H. Sayed, and T. Kailath. Linear estimation in Krein spaces - Part I: Theory. *IEEE Transactions on Automatic Control,* 41(1), pp. 18–33, 1996.

[HSK94] B. Hassibi, A. H. Sayed, and T. Kailath. Square-root arrays and Chandrasekhar recursions for H^∞ problems. In *Proc. Conference on Decision and Control,* Orlando, FL, December 1994.

[HV91] S. Van Huffel and J. Vandewalle. *The Total Least Squares Problem: Computational Aspects and Analysis.* SIAM, Philadelphia, 1991.

[Kai81] T. Kailath. *Lectures on Wiener and Kalman Filtering.* Springer-Verlag, NY, second edition, 1981.

[KN91] P.P. Khargonekar and K. M. Nagpal. Filtering and smoothing in an $H^\infty-$ setting. *IEEE Trans. on Automatic Control*, AC-36:151–166, 1991.

[LS83] L. Ljung and T. Söderström. *Theory and Practice of Recursive Identification.* MIT Press, Cambridge, MA, 1983.

[LS91] D. J. Limebeer and U. Shaked. New results in H^∞-filtering. In *Proc. Int. Symp. on MTNS*, pages 317–322, June 1991.

[PRLN92] J. G. Proakis, C. M. Rader, F. Ling, and C. L. Nikias. *Advanced Digital Signal Processing.* Macmillan Publishing Co., New York, NY, 1992.

[SK94] A. H. Sayed and T. Kailath. A state-space approach to adaptive RLS filtering. *IEEE Signal Processing Magazine*, 11(3):18–60, July 1994.

[YS91] I. Yaesh and U. Shaked. H^∞–optimal estimation: The discrete time case. In *Proc. Inter. Symp. on MTNS*, pages 261–267, Kobe, Japan, June 1991.

[SR95a] A. H. Sayed and M. Rupp. A time-domain feedback analysis of adaptive gradient algorithms via the Small Gain Theorem. *Proc. SPIE Conference on Advanced Signal Processing: Algorithms, Architectures, and Implementations* , F.T. Luk, ed., vol. 2563, pp. 458–469, San Diego, CA, July 1995.

[SR95b] A. H. Sayed and M. Rupp. A class of adaptive nonlinear H^∞-filters with guaranteed l_2-stability. *Proc. IFAC Symposium on Nonlinear Control System Design*, vol. 1, pp. 455–460, Tahoe City, CA, June 1995.

[SR95c] A. H. Sayed and M. Rupp. A feedback analysis of Perceptron learning for neural networks. In *Proc. 29th Asilomar Conference on Signals, Systems, and Computers*, Pacific Grove, CA, Oct. 1995.

[RS95] M. Rupp and A. H. Sayed. A robustness analysis of Gauss-Newton recursive methods. In *Proc. Conference on Decision and Control*, vol. 1, pp. 210–215, New Orleans, LA, Dec. 1995. Also to appear in *Signal Processing*, 1996.

[RS96] M. Rupp and A. H. Sayed. A time-domain feedback analysis of filtered-error adaptive gradient algorithms. To appear in *IEEE Transactions on Signal Processing*, 1996.

[ZDG96] K. Zhou, J. C. Doyle, and K. Glover. *Robust and Optimal Control*, Prentice Hall, NJ, 1996.

Ali H. Sayed
Center for Control Engineering and Computation
Dept. of Electrical and Computer Engineering
University of California
Santa Barbara, CA 93106
USA

Thomas Kailath
Information Systems Laboratory
Dept. of Electrical Engineering
Stanford University
Stanford, CA 94305
USA

Babak Hassibi
Information Systems Laboratory
Dept. of Electrical Engineering
Stanford University
Stanford, CA 94305
USA

AMS Classification. 93E24, 93E11, 93E10, 93B36, 93B40, 60G35.

Operator Theory:
Advances and Applications, Vol. 87

BOUNDS FOR THE WIDTH OF THE INSTABILITY INTERVALS IN THE MATHIEU EQUATION

P. N. SHIVAKUMAR * and QIANG YE†

In this paper, we give upper and lower bounds for the width of instability intervals $a_n - b_n$ (for $n \geq max\{\frac{h^2+1}{2}, 3\}$ and for any n if $h^2 < 1$) of the Mathieu equation

$$y'' + (\lambda - 2h^2 \cos 2\theta)y = 0$$

using techniques of infinite matrices. The results are in agreement with the asymptotic approximations given by Avran and Simon [3] and with those of Hochstadt [5] who used continued fraction techniques.

1. Introduction

We consider the Mathieu equation

$$y''(\theta) + (\lambda - 2h^2 \cos 2\theta)y(\theta) = 0 \tag{1}$$

where λ is an eigenvalue parameter. Let $\{a_{2n}\}_0^\infty$, $\{a_{2n+1}\}_0^\infty$, $\{b_{2n}\}_1^\infty$, $\{b_{2n+1}\}_0^\infty$ denote the eigenvalues of (1) corresponding respectively to the following boundary conditions

$(B1)$ $\quad y'(0) = y'(\frac{\pi}{2}) = 0;$

$(B2)$ $\quad y'(0) = y(\frac{\pi}{2}) = 0;$

$(B3)$ $\quad y(0) = y'(\frac{\pi}{2}) = 0;$

$(B4)$ $\quad y(0) = y(\frac{\pi}{2}) = 0.$

*Research supported in part by the Natural Sciences and Engineering Research Council of Canada under Grant OGP0007899

†Research supported in part by the Natural Sciences and Engineering Research Council of Canada under Grant OGP0137369

It is well-known that if $h^2 > 0$ (cf [6, p. 119]),

$$a_0 < b_1 < a_1 < b_2 < a_2 < b_3 < a_3 < \cdots .$$

The Mathieu equation arises in a variety of applications ranging from classical mechanics to electromagnetics, see chapter 4 of [6]. Of particular interest are, for example, the instability interval $[a_n, b_n]$ and its widths $a_n - b_n$, which arise in the theory of parametric resonance in classical mechanics (see [1, p. 261]). Other physical significance of the instability intervals includes forbidden regions of the spectrum in the one electron theory of solids, where it is called "gap" in [3].

Asymptotic expansions have been the main tool in the study of the widths of instability intervals and early results can be found in [6, p. 121] for example. More precise asymptotic expansions for $a_n - b_n$ were given by Harrell [4] and Avran and Simon [3]. Hochstadt [5] gave a refined formula using the methods of continued fractions. In [8], Shivakumar, Williams and Rudraiah discuss the eigenvalues of (1) as a particular case of a more general theory of eigenvalues of diagonally dominant infinite matrices using estimates for the inverse elements. Subsequent works have mainly concentrated on Hill's equation, which has a form similar to (1) but with a more general periodic potential, see [2, 9]. In this paper, we use the techniques of infinite matrices [8] to derive both upper and lower bounds for the above widths $a_n - b_n$. The bounds lead to the asymptotic expansion given by Hochstadt [5] and are therefore tight at least asymptotically.

We point out that there are some similarities between the approaches of infinite matrices and continued fractions (see the remarks before lemma 5); however, the infinite matrix method seems to have some advantage owing to its simpler form. We also mention that our bounds hold for all $n \geq max\{\frac{h^2+1}{2}, 3\}$ (and all n if $h^2 < 1$) and are thus more widely applicable than asymptotic expansions.

Throughout the paper, we will assume $h^2 > 0$ and we shall present our results for $a_{2n} - b_{2n}$ only. The bounds for $a_{2n+1} - b_{2n+1}$ can be derived similarly and we omit the details.

2. Preliminaries

In this section, we outline the infinite matrix approach and present some preliminary results. Most of our results are stated under the condition $n \geq max\{\frac{h^2+1}{2}, 3\}$, but if $h^2 < 1$, they also hold in the same or a similar form for $n = 1,\ 2$. For the ease of presentation, the case $h^2 < 1$ will be discussed by some remarks after relevant lemmas.

If $y(\theta)$ is a solution to (1) with (B1) corresponding to the eigenvalue a_{2n}, it can be expanded as

$$y(\theta) = \sum_{k=1}^{\infty} x_k \cos 2(k-1)\theta.$$

Substituting it into (1), it is easy to check that $x = (x_1, x_2, \cdots, x_k, \cdots)^t \in l_2$ satisfies

$$A_{ce} x = a_{2n} x$$

where A_{ce} is the infinite tridiagonal matrix

$$A_{ce} = \begin{bmatrix} 0 & h^2 & & & & & \\ 2h^2 & 4\cdot 1^2 & h^2 & & & & \\ & h^2 & 4\cdot 2^2 & \ddots & & & \\ & & \ddots & \ddots & h^2 & & \\ & & & h^2 & 4k^2 & \ddots & \\ & & & & \ddots & \ddots & \ddots \\ & & & & & \ddots & \ddots \end{bmatrix}$$

and $\{a_{2n}\}$ are eigenvalues of A_{ce} on l_2. It is easy to see that A_{ce} is similar to a symmetric matrix, which is indeed self-adjoint with compact inverse (see the proof of Lemma 3). Similarly, it can be derived that $\{b_{2n}\}$ are eigenvalues of

$$B_{ce} = \begin{bmatrix} 4\cdot 1^2 & h^2 & & & & & \\ h^2 & 4\cdot 2^2 & h^2 & & & & \\ & h^2 & 4\cdot 3^2 & \ddots & & & \\ & & \ddots & \ddots & h^2 & & \\ & & & h^2 & 4k^2 & \ddots & \\ & & & & \ddots & \ddots & \ddots \\ & & & & & \ddots & \ddots \end{bmatrix}.$$

Note that B_{ce} is a submatrix of A_{ce}. We shall use the infinite matrix approach of [8] to study the eigenvalues a_{2n}, b_{2n}. We shall use the self-adjoint property and therefore we consider infinite matrices acting on l_2. Although the discussions in [8] are based on l_1 or l_∞, most of them can be easily adapted to the l_2 case. We begin by quoting the following lemma, which is an l_2 version of the Gershgorin theorem in [8, Theorem 1].

LEMMA 1. *Let $T = (t_{ij})_{i,j=1}^{\infty}$ be a matrix operator in l_2 and let λ be an eigenvalue of T. Then $\lambda \in U_{i=1}^{\infty} R_i$, where $R_i = \{z \in C : |z - t_{ii}| \leq \sum_{j=1, j\neq i}^{\infty} |t_{ij}|\}$.*

LEMMA 2. *For $n \geq max\{\frac{h^2+1}{2}, 3\}$,*

$$4n^2 - 2h^2 \leq b_{2n} \leq a_{2n} \leq 4n^2 + 2h^2.$$

PROOF: Consider the Gershgorin discs of $A_{ce} + 4h^2 I$, which are defined to be

$$R_i = \{z \in C : |z - 4i^2 - 4h^2| \leq 2h^2\}, \quad \text{for } i \geq 2$$

and $R_0 = \{z \in C : |z - 4h^2| \leq h^2\}$, $R_1 = \{z \in C : |z - 4 - 4h^2| \leq 3h^2\}$. Then for any $n \geq \max\{\frac{h^2+1}{2}, 3\}$ and any $i \leq n-1$, it is easy to check that

$$4i^2 + 2h^2 \leq 4n^2 - 2h^2, \text{ and } 4\cdot 1^2 + 3h^2 \leq 4n^2 - 2h^2.$$

This shows that R_n is disjoint from the rest Gershgorin discs R_i $(i \neq n)$. Now let D be the diagonals of $A_{ce} + 4h^2 I$ and $F = A_{ce} + 4h^2 I - D$. Since D^{-1} is compact and $\|D^{-1}F\|_2 < 1$ by calculation, $A_{ce} + 4h^2 I = D(I + D^{-1}F)$ has compact inverse. So using Lemma 1 and the argument in Theorem 2 of [8] and noting that a_{2n} is a function of h^2, the eigenvalue $a_{2n} + 4h^2$ of $A_{ce} + 4h^2 I$ is an analytic function of h^2 and thus varies continuously within the Gershgorin disc R_n as h^2 increases from 0. Therefore $|a_{2n} - 4n^2| \leq 2h^2$. The bound for b_{2n} is proved similarly and $b_{2n} \leq a_{2n}$ is well-known. □

If $h^2 < 1$, then similar inequalities hold for $n = 1, 2$, i.e.,

$$4 - 3h^2 \leq b_2 \leq a_2 \leq 4 + 3h^2 \quad \text{and} \quad 4 \cdot 2^2 - 2h^2 \leq b_4 \leq a_4 \leq 4 \cdot 2^2 + 2h^2.$$

Let $A^{i,j}$ denote the $(j - i + 1) \times (j - i + 1)$ submatrix of A_{ce} consisting of the rows and columns from $(i + 1)$ to $(j + 1)$, i.e.

$$A^{0,j} = \begin{bmatrix} 0 & h^2 & & & & \\ 2h^2 & 4\cdot 1^2 & h^2 & & & \\ & h^2 & 4\cdot 2^2 & h^2 & & \\ & & \ddots & \ddots & \ddots & \\ & & & \ddots & \ddots & h^2 \\ & & & & h^2 & 4j^2 \end{bmatrix}$$

and

$$A^{i,j} = \begin{bmatrix} 4i^2 & h^2 & & & \\ h^2 & 4(i+1)^2 & h^2 & & \\ & \ddots & \ddots & \ddots & \\ & & \ddots & \ddots & h^2 \\ & & & h^2 & 4j^2 \end{bmatrix} \quad \text{for } i \geq 1.$$

In this notation, j is allowed to extend to ∞.

LEMMA 3. *For $n \geq max\{\frac{h^2+1}{2}, 3\}$, $A^{n+1,\infty} - a_{2n}I$ is invertible and $(A^{n+1,\infty} - a_{2n}I)^{-1}$ is compact and self-adjoint on l_2. The same holds with a_{2n} replaced by b_{2n}.*

PROOF: Let D be the diagonals of $A^{n+1,\infty} - a_{2n}I$ i.e. $D = \text{diag}[4k^2 - a_{2n}]_{k=n+1}^{\infty}$, and let $F = A^{n+1,\infty} - a_{2n}I - D$. D^{-1} is compact since $4k^2 - a_{2n} \to \infty$ as $k \to \infty$. By Lemma 2, for $k \geq n + 1$,

$$4k^2 - a_{2n} \geq 4(n+1)^2 - a_{2n} \geq 8n + 4 - 2h^2.$$

Thus $\|D^{-1}F\|_2 < 1$. Hence $A^{n+1,\infty} - a_{2n}I = D(I + D^{-1}F)$ has compact inverse, which by the symmetry is self-adjoint. □

Again, the lemma will be true for all $n = 1, 2$ provided $h^2 < 1$.

We now define $D^{i,j}(\lambda) \equiv \det(A^{i,j} - \lambda I)$, if $j < \infty$. The following lemma is easily checked and will be valid for $n = 1$, 2 as well if $h^2 < 1$.

LEMMA 4. *For* $n \geq max\{\frac{h^2+1}{2}, 3\}$, $A^{0,n-1} - a_{2n}I$ *and* $A^{1,n-1} - b_{2n}I$ *are invertible and*

$$e_n^*(A^{0,n-1} - a_{2n}I)^{-1}e_n = \frac{D^{0,n-2}(a_{2n})}{D^{0,n-1}(a_{2n})}$$

$$e_{n-1}^*(A^{1,n-1} - b_{2n}I)^{-1}e_{n-1} = \frac{D^{1,n-2}(b_{2n})}{D^{1,n-1}(b_{2n})}$$

where $e_{n-1} = (0, \cdots, 0, 1)^t \in R^{n-1}$, $e_n = (0, \cdots, 0, 1)^t \in R^n$.

The following lemma gives the equations that a_{2n} and b_{2n} satisfy in the infinite matrix form. They were originally derived in [8, Theorem 8]. We remark that these equations seem to be equivalent to the classical ones based on continued fractions (see [6, p.118], for example), which were used by Hochstadt to derive the asymptotic result in [5]. However, the simple form that the infinite matrix approach has makes it easy in bounding several quantities involved.

LEMMA 5. *For* $n \geq max\{\frac{h^2+1}{2}, 3\}$, a_{2n}, b_{2n} *satisfy*

$$-a_{2n} + 4n^2 = h^4 e_n^*(A^{0,n-1} - a_{2n}I)^{-1}e_n + h^4 e_1^*(A^{n+1,\infty} - a_{2n}I)^{-1}e_1,$$

$$-b_{2n} + 4n^2 = h^4 e_{n-1}^*(A^{1,n-1} - b_{2n}I)^{-1}e_{n-1} + h^4 e_1^*(A^{n+1,\infty} - b_{2n}I)^{-1}e_1$$

where $e_{n-1} = (0, \cdots, 0, 1)^t \in R^{n-1}$, $e_n = (0, \cdots, 0, 1)^t \in R^n$ *and* $e_1 = (1, 0, \cdots\cdots)^t \in l_2$.

PROOF: The proof is similar to the one in Theorem 8 of [8] and here we give a brief outline only. First, by Lemmas 3 and 4, $A^{0,n-1} - a_{2n}I$ and $A^{n+1,\infty} - a_{2n}I$ are invertible. Then by expanding $(A_{ce} - a_{2n}I)x = 0$ and using the tridiagonal structure, we obtain $x_{n-1} = -h^2 x_n e_n^*(A^{0,n-1} - a_{2n}I)^{-1}e_n$ and $x_{n+1} = -h^2 x_n e_1^*(A^{n+1,\infty} - a_{2n}I)^{-1}e_1$. Substituting x_{n-1}, x_n into the equation obtained from the nth entry of $(A_{ce} - a_{2n}I)x = 0$ and noting that $x_n \neq 0$, we obtain the first equation. The second equation is obtained similarly by noting that $B_{ce} = A^{1,\infty}$. □

If $h^2 < 1$, the above lemma holds for a_4 and b_4. However, for a_2, b_2, the equations are slightly different and they are as follows

$$-a_2 + 4 = -2h^4 a_2^{-1} + h^4 e_1^*(A^{2,\infty} - a_2 I)^{-1}e_1, \tag{2}$$

$$-b_2 + 4 = h^4 e_1^*(A^{2,\infty} - b_{2n}I)^{-1}e_1.$$

LEMMA 6. *For* $n \geq max\{\frac{h^2+1}{2}, 3\}$,

$$0 < e_{n-1}^*(A^{1,n-1} - a_{2n}I)^{-1}e_{n-1} - e_{n-1}^*(A^{1,n-1} - b_{2n}I)^{-1}e_{n-1} \leq \frac{a_{2n} - b_{2n}}{(8n - 4 - 3h^2)^2},$$

$$0 < e_1^*(A^{n+1,\infty} - a_{2n}I)^{-1}e_1 - e_1^*(A^{n+1,\infty} - b_{2n}I)^{-1}e_1 \leq \frac{a_{2n} - b_{2n}}{(8n + 4 - 3h^2)^2}, \tag{3}$$

PROOF: We denote by $\lambda_k(M)$ the kth eigenvalue of a matrix M. First note that for $k = 1, ..., n-1$

$$\begin{aligned}\lambda_k(A^{1,n-1} - b_{2n}I) &\leq 4(n-1)^2 + h^2 - b_{2n} \leq 4(n-1)^2 + h^2 - 4n^2 + 2h^2 \\ &= -8n + 4 + 3h^2 < 0\end{aligned}$$

where we have used Lemmas 1 and 2. So

$$\| (A^{1,n-1} - b_{2n}I)^{-1} \|_2 \leq \frac{1}{8n - 4 - 3h^2}.$$

Similarly

$$\| (A^{1,n-1} - a_{2n}I)^{-1} \|_2 \leq \frac{1}{8n - 4 - 3h^2}.$$

Also from $a_{2n} > b_{2n}$, we have $0 > \lambda_k((A^{1,n-1} - a_{2n}I)^{-1}) > \lambda_k((A^{1,n-1} - b_{2n}I)^{-1})$. Then

$$e^*_{n-1}(A^{1,n-1} - a_{2n}I)^{-1}e_{n-1} > e^*_{n-1}(A^{1,n-1} - b_{2n}I)^{-1}e_{n-1}.$$

Furthermore,

$$\begin{aligned}&e^*_{n-1}(A^{1,n-1} - a_{2n}I)^{-1}e_{n-1} - e^*_{n-1}(A^{1,n-1} - b_{2n}I)^{-1}e_{n-1} \\ &= e^*_{n-1}[(A^{1,n-1} - a_{2n}I)^{-1} - (A^{1,n-1} - b_{2n}I)^{-1}]e_{n-1} \\ &= (a_{2n} - b_{2n})e^*_{n-1}[(A^{1,n-1} - b_{2n}I)^{-1}(A^{1,n-1} - a_{2n}I)^{-1}]e_{n-1} \\ &\leq (a_{2n} - b_{2n}) \| (A^{1,n-1} - b_{2n}I)^{-1} \|_2 \cdot \| (A^{1,n-1} - a_{2n}I)^{-1} \|_2 \\ &\leq \frac{a_{2n} - b_{2n}}{(8n - 4 - 3h^2)^2}.\end{aligned}$$

For the second inequality, we note that

$$\begin{aligned}\lambda_k(A^{n+1,\infty} - a_{2n}I) &\geq 4(n+1)^2 - h^2 - a_{2n} \geq 4(n+1)^2 - h^2 - 4n^2 - 2h^2 \\ &= 8n + 4 - 3h^2 > 0.\end{aligned}$$

The rest is proved similarly using the fact that $(A^{n+1,\infty} - a_{2n}I)^{-1}$ and $(A^{n+1,\infty} - b_{2n}I)^{-1}$ are compact and self-adjoint (Lemma 3). □

If $h^2 < 1$, the lemma is true for $n = 2$ based on the same proof. For $n = 1$, the second inequality (3) becomes

$$0 < e^*_1(A^{2,\infty} - a_2I)^{-1}e_1 - e^*_1(A^{2,\infty} - b_2I)^{-1}e_1 \leq \frac{a_2 - b_2}{16(3 - h^2)^2}. \tag{4}$$

LEMMA 7. *For any λ and $n \geq 3$,*

$$D^{0,n-2}(\lambda)D^{1,n-1}(\lambda) - D^{1,n-2}(\lambda)D^{0,n-1}(\lambda) = 2(h^4)^{n-1}$$

PROOF: By expanding along the last rows of $D^{0,n-1}(\lambda)$, $D^{1,n-1}(\lambda)$, we obtain

$$\begin{aligned}&D^{0,n-2}(\lambda)D^{1,n-1}(\lambda) - D^{1,n-2}(\lambda)D^{0,n-1}(\lambda) \\ &= D^{0,n-2}(\lambda)[(4(n-1)^2 - \lambda)D^{1,n-2}(\lambda) - h^4D^{1,n-3}(\lambda)] \\ &- D^{1,n-2}(\lambda)[(4(n-1)^2 - \lambda)D^{0,n-2}(\lambda) - h^4D^{0,n-3}(\lambda)] \\ &= h^4[D^{0,n-3}(\lambda)D^{1,n-2}(\lambda) - D^{1,n-3}(\lambda)D^{0,n-2}(\lambda)].\end{aligned}$$

Now, the lemma follows from applying the above recursively and noting that

$$D^{0,1}(\lambda)D^{1,2}(\lambda) - D^{1,1}(\lambda)D^{0,2}(\lambda) = 2(h^4)^2.$$

□

Our last lemma gives upper and lower bounds on the determinants of tridiagonal matrices, which are an application of a more general result of Ostrowski's [7].

LEMMA 8. *For* $n \geq max\{\frac{h^2+1}{2}, 3\}$ *and* $\lambda \geq 4n^2 - 2h^2$,

$$(\lambda - 4(n-1)^2)\prod_{k=0}^{n-2}(\lambda - 4k^2 - \frac{h^4}{4n-2-h^2}) \leq D^{0,n-1}(\lambda) \leq (\lambda - 4(n-1)^2)\prod_{k=0}^{n-2}(\lambda - 4k^2 + \frac{h^4}{4n-2-h^2})$$

$$(\lambda - 4(n-1)^2)\prod_{k=1}^{n-2}(\lambda - 4k^2 - \frac{h^4}{4n-2-h^2}) \leq D^{1,n-1}(\lambda) \leq (\lambda - 4(n-1)^2)\prod_{k=1}^{n-2}(\lambda - 4k^2 + \frac{h^4}{4n-2-h^2}).$$

PROOF: For $A^{0,n-1} - \lambda I = (a_{ij})_{i,j=1}^n$, it is easy to see that

$$\sigma \equiv \max_{1\leq i\leq n}\{\frac{\sum_{j\neq i}|a_{ij}|}{|a_{ii}|}\} < \frac{2h^2}{\lambda - 4(n-1)^2} \leq \frac{h^2}{4n-2-h^2}.$$

Then the lemma follows from (8) of [7, p.27] by taking $k = n-1$. □

3. Bounds on $a_{2n} - b_{2n}$

In this section, we present our results on bounds of $a_{2n} - b_{2n}$. The bound also leads to the asymptotic expansion in [5]. We first present a general bound in the following theorem, which depends on determinants and is less convenient in computation.

THEOREM 1. *For* $n \geq max\{\frac{h^2+1}{2}, 3\}$, *we have*

$$\begin{aligned}
& \frac{2h^{4n}}{|D^{0,n-1}(4n^2+2h^2)D^{1,n-1}(4n^2+2h^2)|}(1 - \frac{2h^4}{(8n-4-3h^2)^2}) \\
< \; & \frac{2h^{4n}}{|D^{0,n-1}(a_{2n})D^{1,n-1}(a_{2n})|}(1 - \frac{2h^4}{(8n-4-3h^2)^2}) \\
< \; & a_{2n} - b_{2n} \\
< \; & \frac{2h^{4n}}{|D^{0,n-1}(a_{2n})D^{1,n-1}(a_{2n})|} < \frac{2h^{4n}}{|D^{0,n-1}(4n^2-2h^2)D^{1,n-1}(4n^2-2h^2)|}.
\end{aligned}$$

PROOF: Using Lemma 5 and Lemma 6, we have

$$\begin{aligned}
a_{2n} - b_{2n} &= h^4e_{n-1}^*(A^{1,n-1} - b_{2n})^{-1}e_{n-1} - h^4e_n^*(A^{0,n-1} - a_{2n})^{-1}e_n \\
&\quad + h^4e_1^*(A^{n+1,\infty} - b_{2n})^{-1}e_1 - h^4e_1^*(A^{n+1,\infty} - a_{2n})^{-1}e_1 \\
&< h^4e_{n-1}^*(A^{1,n-1} - a_{2n}I)^{-1}e_{n-1} - h^4e_n^*(A^{0,n-1} - a_{2n})^{-1}e_n \\
&= h^4\frac{D^{1,n-2}(a_{2n})}{D^{1,n-1}(a_{2n})} - h^4\frac{D^{0,n-2}(a_{2n})}{D^{0,n-1}(a_{2n})} \qquad \text{(by Lemma 4)}
\end{aligned}$$

$$
\begin{aligned}
&= -h^4 \frac{D^{0,n-2}(a_{2n})D^{1,n-1}(a_{2n}) - D^{0,n-1}(a_{2n})D^{1,n-2}(a_{2n})}{D^{0,n-1}(a_{2n})D^{1,n-1}(a_{2n})} \\
&= -h^4 \frac{2(h^4)^{n-1}}{D^{0,n-1}(a_{2n})D^{1,n-1}(a_{2n})} \qquad \text{(by Lemma 7)} \\
&= \frac{2h^{4n}}{|D^{0,n-1}(a_{2n})| \cdot |D^{1,n-1}(a_{2n})|} \\
&< \frac{2h^{4n}}{|D^{0,n-1}(4n^2-2h^2)D^{1,n-1}(4n^2-2h^2)|}
\end{aligned}
$$

where we note that $|D^{0,n-1}(\lambda)|$ and $|D^{1,n-1}(\lambda)|$ are monotonic increasing for $4n^2 - 2h^2 < \lambda < 4n^2 + 2h^2$. On the other hand, by lemma 6 again,

$$
\begin{aligned}
a_{2n} - b_{2n} &= h^4 e_{n-1}^*(A^{1,n-1} - a_{2n})^{-1} e_{n-1} - h^4 e_n^*(A^{0,n-1} - a_{2n})^{-1} e_n \\
&\quad + h^4 e_{n-1}^*(A^{1,n-1} - b_{2n})^{-1} e_{n-1} - h^4 e_{n-1}^*(A^{1,n-1} - a_{2n} I)^{-1} e_{n-1} \\
&\quad + h^4 e_1^*(A^{n+1,\infty} - b_{2n} I)^{-1} e_1 - h^4 e_1^*(A^{n+1,\infty} - a_{2n})^{-1} e_1 \\
&\geq h^4 e_{n-1}^*(A^{1,n-1} - a_{2n})^{-1} e_{n-1} - h^4 e_n^*(A^{0,n-1} - a_{2n})^{-1} e_n \\
&\quad - h^4 \frac{a_{2n} - b_{2n}}{(8n-4-3h^2)^2} - h^4 \frac{a_{2n} - b_{2n}}{(8n+4-3h^2)^2}
\end{aligned}
$$

Thus

$$
\begin{aligned}
a_{2n} - b_{2n} &\geq \frac{h^4 e_{n-1}^*(A^{1,n-1} - a_{2n})^{-1} e_{n-1} - h^4 e_n^*(A^{0,n-1} - a_{2n})^{-1} e_n}{1 + \frac{2h^4}{(8n-4-3h^2)^2}} \\
&\geq \frac{2h^{4n}}{|D^{0,n-1}(a_{2n})| \cdot |D^{1,n-1}(a_{2n})|} \cdot \left(1 - \frac{2h^4}{(8n-4-3h^2)^2}\right) \\
&\geq \frac{2h^{4n}}{|D^{0,n-1}(4n^2+2h^2)D^{1,n-1}(4n^2+2h^2)|} \cdot \left(1 - \frac{h^4}{(8n-4-3h^2)^2}\right).
\end{aligned}
$$

□

Note that the expression in the lower bound is positive. The determinants in the above bounds can be replaced by the estimates of Lemma 8, leading to the following more transparent bound.

THEOREM **2**. *For $n \geq max\{\frac{h^2+1}{2}, 3\}$, we have*

$$
\frac{8h^{4n}}{4^{2n}[(2n-1)!]^2 K_+} \left(1 - \frac{h^4}{8(2n-1-h^2)^2}\right) \leq a_{2n} - b_{2n} \leq \frac{8h^{4n}}{4^{2n}[(2n-1)!]^2 K_-}
$$

where

$$
K_\pm = \left(1 \pm \frac{3h^2}{4n^2}\right) \prod_{k=1}^{n-1} \left(1 \pm \frac{3h^2}{4n^2 - 4k^2}\right)^2.
$$

PROOF: Clearly $h^4/(4n-2-h^2) \leq h^2$ for $n \geq max\{\frac{h^2+1}{2}, 3\}$. Then applying Lemma 8 to $|D^{0,n-1}(4n^2+2h^2)|$ and $|D^{1,n-1}(4n^2+2h^2)|$,

$$\begin{aligned}
&|D^{0,n-1}(4n^2+2h^2)D^{1,n-1}(4n^2+2h^2)| \\
\geq\ & \left(4n^2-2h^2-\frac{h^4}{4n-2-h^2}\right)\prod_{k=1}^{n-2}\left(4n^2-2h^2-4k^2-\frac{h^4}{4n-2-h^2}\right)^2 \\
\geq\ & \left(4n^2-3h^2\right)\prod_{k=1}^{n-1}\left(4n^2-4k^2-3h^2\right)^2 \\
=\ & 4^{2n}[(2n-1)!]^2K_-.
\end{aligned}$$

The upper bound is derived similarly and the theorem follows from Theorem 1. □

In section 2, we discussed the cases $n = 1,\ 2$ if $h^2 < 1$. In particular, Theorem 1 and Theorem 2 hold for $n = 2$ if $h^2 < 1$ by the same proofs. For $n = 1$, however, we use (2) and (4) to obtain

$$\begin{aligned}
a_2 - b_2 &= 2h^4a_2^{-1} + h^4e_1^*(A^{2,\infty}-b_2)^{-1}e_1 - h^4e_1^*(A^{2,\infty}-a_2)^{-1}e_1 \\
&\leq 2h^4a_2^{-1} < 2h^4/(4-3h^2),
\end{aligned}$$

and

$$a_2 - b_2 > 2h^4a_2^{-1} - h^4\frac{a_2-b_2}{16(3-h^2)^2}.$$

Thus

$$\frac{8h^4}{4^2(1-\frac{3}{4}h^2)}\left(1-\frac{h^4}{16(3-h^2)^2}\right) < a_2 - b_2 < \frac{8h^4}{4^2(1-\frac{3}{4}h^2)}.$$

In deriving Theorem 2 from Theorem 1, there are numerous places where the bounds can be made tighter. However, the bound in Theorem 2 is simple in form and is close enough to the asymptotic result. The bound in Theorem 1 contains a_{2n}, which we replaced by the rough bound $4n^2 \pm 2h^2$ of Lemma 2. Of course, if a more accurate estimate is available for a particular a_{2n}, better bounds would be obtained from Theorem 1. In particular, asymptotically, we can replace a_{2n} by its asymptotic expansion, namely $a_{2n} = 4n^2 + O(\frac{h^4}{n^2})$. Then by expanding the determinants, it can be reduced to

$$D^{0,n-1}(a_{2n}) = 4^n\Pi_{k=0}^{n-1}(4n^2-4k^2)\left[1+O(\frac{h^4}{n^2})\right],$$

and a similar one for $D^{1,n-1}(a_{2n})$. Thus Theorem 1 leads to the following asymptotic expansion

$$a_{2n} - b_{2n} \sim \frac{8h^{4n}}{4^{2n}[(2n-1)!]^2}\left[1+O(\frac{h^4}{n^2})\right].$$

which was obtained in [5].

References

[1] V. I. Arnol'd, Ordinary Differential Equations, Springer-Verlag, Berlin Heidelberg 1992.

[2] V. I. Arnol'd, Remarks on perturbation theory for problems of Mathieu type, Russian Math. Surveys 38(1983):215-233.

[3] J. Avran and B. Simon, The asymptotics of the gap in the Mathieu equation, Annals Phys. 134(1981):76-84.

[4] M. E. Harrell, On the effect of the boundary conditions on the eigenvalues of ordinary differential equations, Amer. J. Math. supplement(1981):139-150.

[5] H. Hochstadt, On the width of the instability intervals of the Mathieu equation. SIAM J. Math. Anal. 15(1984):105-107.

[6] J. Meixner and F.W. Schaefke, Mathieuische Funktionen und Spharoidfunktionen, Springer, Berlin-Gottingen-Heildelberg, 1954.

[7] A. M. Ostrowski, Note on bounds for determinants with dominant principal diagonal, Proc. Amer. Math. Soc. 3(1952):26-30.

[8] P.N. Shivakumar, J.J. Williams and N. Rudraiah, Eigenvalues for infinite matrices, Linear Alge. Appl. 96(1987):35-63.

[9] M. I. Weinstein and J. B. Keller, Asymptotic behavior of stability regions for Hill's equation, SIAM J. Appl. Math. 47(1987):941-958.

P. N. Shivakumar
Department of Applied Mathematics and
Institute of Industrial Mathematical Sciences
University of Manitoba
Winnipeg, Manitoba, Canada R3T 2N2

Qiang Ye
Department of Applied Mathematics
University of Manitoba
Winnipeg, Manitoba, Canada R3T 2N2

AMS Subject Classification: 47B37, 34B30, 34L15

Operator Theory:
Advances and Applications, Vol. 87

OPERATOR PENCILS ARISING IN ELASTICITY AND HYDRODYNAMICS: THE INSTABILITY INDEX FORMULA.

A.A.SHKALIKOV

Dedicated to Professor Peter Lancaster on the occasion of his 65th birthday

The main object of the paper is a quadratic operator pencil of the form

$$A(\lambda) = F\lambda^2 + (D + iG)\lambda + T,$$

with unbounded operator coefficients acting in Hilbert space. It is assumed that F, T are selfadjoint and boundedly invertible and $D \geqslant 0, G$ are symmetric and T-bounded. Pencils of this form arise as abstract models for concrete problems in elastisity and hydrodynamics. We investigate the relations between the classical and generalized spectra and under additional hypotheses prove the formula for the number of eigenvalues of $A(\lambda)$ in the right-half plane. The proof of this formula is based on the preliminary investigation of maximal semidefinite invariant subspaces in the root subspaces corresponding to the pure imaginary eigenvalues of a dissipative operator in Krein or Pontrjagin spaces.

INTRODUCTION

The plan of the present paper is the following. In Section 1 we consider some concrete problems arising in elasticity and hydrodynamics. Further we prefer to work with abstract formulations of physical problems under consideration. For this purpose we provide general classes of operator pencils with unbounded operator coefficients related to problems of origin. The main object of the paper is an operator pencil of the form

$$A(\lambda) = \lambda^2 F + (D + iG)\lambda + T,$$

The support of this work by a grant from the International Soros Foundation and the Russian Fundamental Research Foundation is greatfully acknowledged.

where F and T are selfadjoint and boundedly invertible operators, while $D \geqslant 0$ and G are symmetric and T-bounded. The study of the pencil $A(\lambda)$ is realized in Section 3. In particular, we introduce the concepts of the classical and the generelized spectra and investigate the relations between them. We associate the linear pencil

$$\mathbf{A}(\lambda) := \mathbf{T} - \lambda \mathbf{W} := -\begin{pmatrix} D+iG & T \\ -J & 0 \end{pmatrix} - \lambda \begin{pmatrix} F & 0 \\ 0 & J \end{pmatrix}, \qquad J = T|T|^{-1},$$

with the quadratic pencil $A(\lambda)$. It turns out that the operator $\mathbf{T}$ is dissipative in the space $\mathbf{H} = H \times H_1$, where H_1 coincides with the domain of the operator $|T|^{1/2}$ and equipped with the norm $(\cdot,\cdot)_1 = (|T|^{1/2}\cdot, |T|^{1/2}\cdot)$. Generally, the spectrum $\sigma(\mathbf{A})$ of the linearization $\mathbf{A}(\lambda)$ coincides with neither the classical nor the generalized spectrum of $A(\lambda)$. However, we prove that $\sigma(\mathbf{A})$ coincides with the generalized spectrum of $A(\lambda)$ in the open right half plane if the operator $\mathbf{W}$ generates a Pontrjagin space metric. In this case $\sigma(\mathbf{A})$ in the right half plane consists of finitely many eigenvalues, say $\kappa(A)$, and the number $\kappa(A)$ characterizes the index of instability of the equation

$$A\left(\frac{du}{dt}\right) = F\frac{d^2u}{dt^2} + (D+iG)\frac{du}{dt} + Tu = 0, \quad u = u(t).$$

The problem on stability for such kind of equations has a long background and apparently was originated by Kelvin and Tait [KT] (in the end of Section 3 we present a short historical review related to this problem). The main result of the paper is the instability index formula

$$\kappa(A) = \nu(F) + \nu(T) - \varepsilon^+(A)$$

where $\nu(F)$ and $\nu(T)$ are the numbers of the negative eigenvalues of the operators F and T respectively, while $\varepsilon^+(A)$ is expressed in terms of the lengths and the sign characteristics of Jordan chains corresponding to the pure imaginary eigenvalues of $A(\lambda)$. In particular, if all the pure imaginary eigenvalues of $A(\lambda)$ are of definite type then $\varepsilon^+(A)$ coincides with the number of the first type eigenvalues of $A(\lambda)$ (see the definitions in Section 3).

The results of Section 2 on root subspaces of linear dissipative pencils seem at the first sight to be isolated from the main subject of the paper. However, these results form a theoretical base to prove the index formula in Section 3. In our opinion, they have also an independent interest.

In Section 4 we return to the physical problems of origin and present the corollories of our abstract results. Here we also demonstrate how the index formula can be applied to estimate the number of the nonreal eigenvalues of a selfadjoint operator pencil.

1. Classes of unbounded operator pencils.

Small oscillations of an elastic thin beam of unit length with external and internal damping (so called Kelvin-Voigt material) are described by the equation

$$\frac{\partial^4 u}{\partial x^4} + \frac{\partial}{\partial t}\frac{\partial^2}{\partial x^2}\left(\alpha(x)\frac{\partial^2 u}{\partial x^2}\right) + \frac{\partial}{\partial x}\left(g(x)\frac{\partial u}{\partial x}\right) + \beta(x)\frac{\partial u}{\partial t} + \rho(x)\frac{\partial^2 u}{\partial t^2} = 0. \tag{1.1}$$

Here $x \in [0,1]$, $t \in \mathbb{R}^+$, and $u(x,t)$ is the transverse displacement at position x and time t. The function $\alpha(x)$ determines the internal damping and takes generally small values. The function $\beta(x) \geqslant 0$ determines the distribution of viscous damping, $\rho(x) > 0$ defines the mass distribution and $g(x)$ is responsible for the forces of contraction or tension (see more details in [PI], for example).

As the equation is considered on the finite interval, we have to submit solutions of (1.1) to some boundary conditions. For the sake of definitness we consider the case when both ends of the beam are clamped, i.e.

$$u(0,t) = \frac{\partial u(x,t)}{\partial x}\Big/_{x=0} = u(1,t) = \frac{\partial u(x,t)}{\partial x}\Big/_{x=1} = 0. \tag{1.2}$$

Separating variables $u(x,t) = y(x)\, e^{\lambda t}$, we obtain the following spectral problem

$$\begin{aligned} &\rho^{-1}(x)[y^{(4)}(x) + (g(x)y'(x))'] \\ &\qquad + \lambda\rho^{-1}(x)[(\alpha(x)y''(x))'' + \beta(x)y(x)] + \lambda^2 y(x) = 0, \end{aligned} \tag{1.3}$$

$$y(0) = y'(0) = y(1) = y'(1) = 0. \tag{1.4}$$

Suppose that $\rho(x),\ \beta(x) \in C[0,1], g(x) \in C^1[0,1]$ and $\alpha(x) \in C^2[0,1]$. According to the physical sense we have $\rho(x) > 0,\ \alpha(x) \geqslant 0,\ \beta(x) \geqslant 0$ and either $g(x) \geqslant 0$ or $g(x) \leqslant 0$. Then quadratic eigenvalue problem (1.3), (1.4) is represented in the form

$$[\lambda^2 I + \lambda(D_\alpha + D_\beta) + A + C]\, y(x) = 0, \tag{1.5}$$

where operators $D_\alpha,\ D_\beta,\ A,\ C$ act in Hilbert space $H = L_2([0,1],\ \rho(x))$ with the scalar product

$$(y,z) = \int_0^1 \rho(x)\, y(x)\, \overline{z(x)}\, dx,$$

and are defined by the equalities

$$(1.6)\qquad \begin{aligned} &(Ay)(x) = \rho^{-1}(x)y^{(4)}(x), \quad (D_\alpha y)(x) = \rho^{-1}(x)(\alpha(x)y''(x))'',\\ &(D_\beta y)(x) = \rho^{-1}(x)\beta(x)y(x), \quad (Cy)(x) = \rho^{-1}(x)(g(x)y'(x))' \end{aligned}$$

on the domains

$$\mathcal{D}(A) = \mathcal{D}(D_\alpha) = \mathcal{D}(C) = \{y|\ y \in W_2^4[0,1],\ y(0) = y'(0) = y(1) = y'(1) = 0\},$$
$$\mathcal{D}(D_\beta) = L_2([0,1],\ \rho(x)) = H.$$

We denote by I the identity operator and by $W_2^k[0,1]$ ($k \in \mathbb{N}^+$) the Sobolev spaces.

Naturally, it is more fruitful to study an abstract operator pencil of the form (1.5) rather then problem (1.3), (1.4). We have only to extract the most essential properties of the operators (1.6). We observe that these operators satisfy the following conditions (the terminology of unbounded operator theory we borrow from the book [Ka]):

i) $A = A^* \gg 0$ (i.e. A is selfadjoint and uniformly positive), and $T := A + C$ is selfadjoint and bounded below;

ii) D_α and D_β are nonnegative symmetric A-bounded operators.

iii) the identity operator I and the operator C are A-compact (or T-compact) and hence T has finitely many negative eigenvalues.

Some results on spectrum of problem (1.3), (1.4) in the case $\alpha(x) = const$ were reported by Pivovarchik [P1]. The comprehensive study of abstract pencil (1.5) with $D_\alpha = \alpha A$, $C = 0$ was carried out by Lancaster and Shkalikov [LS]. Additional results in the case $D_\alpha = \alpha A$, $C \neq 0$ were obtained in a recent paper by Shkalikov and Griniv [SG]. New problems appear in the case $\alpha \neq const$, as pencil (1.5) in this situation has nontrivial essential spectrum. However, we leave an interesting problem on the spectrum localization of pencil (1.5) with $D_\alpha \neq \alpha A$ for another occasion. We will deal with pencil (1.5) (and more general ones) mainly in view of the application of our index formula.

A more interesting example for the application of the index formula comes from hydrodynamics. Namely, small transverse oscillations of ideal incompressible fluid in a pipe of finite length are described by the equation which is obtained from (1.1) if we add in the left hand side of (1.1) the "gyroscopic" term

$$2sv\partial^2 u/\partial x\partial t.$$

Here v is the velocity of the fluid and s depends on the mass of the pipe and the fluid (see [ZKM], for example). The physical meanings of the functions in (1.1) are subject to change in this situation. In particular, $g(x) = v^2$. Assuming $sv = const$ and repeating the previous arguments we come to the following quadratic spectral problem

$$[\lambda^2 I + \lambda(D_\alpha + D_\beta + iG) + A + C]y = 0, \tag{1.7}$$

where

$$Gy = -2sviy', \qquad \mathcal{D}(G) = \mathcal{D}(A),$$

and D_α, D_β, A, C are defined as in (1.6). The last operators retain the properties *i)-iii)*. The most essential properties of the operator G are the following:

iv) G is a symmetric T-bounded operator;

v) G is a T-compact operator.

It is also of interest to consider equation (1.1) on the semiaxis $x \in \mathbb{R}^+$ (see the papers of Pivovarchik [P2] and Griniv [Gr]). Assuming that the left end of a beam is clamped we define the operator coefficients in (1.5) by equalities (1.6) on the domains

$$\mathcal{D}(A) = \mathcal{D}(D_\alpha) = \mathcal{D}(D_\beta) = \mathcal{D}(C) = \{y|\ y \in W_2^4[0,\infty],\ y(0) = y'(0) = 0\}.$$

Obviously, this definition is correct if we assume in addition that all the functions $\rho(x)$, $\rho^{-1}(x)$, $\alpha(x)$, $\beta(x)$, $g(x)$ are bounded on $\mathbb{R}^+$. In this case, the properties *i)-ii)* are retained, however, the property *iii)* is not true any more. This makes the problem much more complicated. Nevertheless, under some additional assumptions on the behaviour of the function $g(x)$ at ∞ (see [Gr]) the important property

vi) $T = A + C$ has finitely many negative eigenvalues

remains valid.

Analogously, equation (1.1) with the additional "gyroscopic" term can be considered on the semiaxis $\mathbb{R}^+$ with respect to the variable x. In this case we obtain a pencil of the form (1.7) whose coefficients satisfy the properties *i)-ii), iv)* and also the property *vi)* under additional assumptions on the behaviour of the function $g(x)$ found in the paper [Gr].

2. Root subspaces of linear dissipative pencils and their properties

In this section we deal with a linear dissipative operator pencil

$$A(\lambda) = T - \lambda W,$$

where W is a bounded selfadjoint operator, while T is a closed dissipative operator in Hilbert space H. This means that T is closed and

$$Im\,(Tx, x) \geqslant 0 \qquad \text{for all} \;\; x \in \mathcal{D}(T),$$

and $\mathcal{D}(T)$ is the domain of T. Through all the section we also assume that there exists at least one point μ_0 belonging to the open upper half plane $\mathbb{C}^+$ such that $A(\mu_0)$ has a bounded inverse, i.e. $\mu_0 \in \rho(A)$.

If the operator W has a bounded inverse then the spectrum $\sigma(A)$ and the root subspaces $\mathcal{L}_\mu(A)$ of the pencil $A(\lambda)$ coincide with those of the operator $A = W^{-1}T$. Hence, in this case spectral problems for the pencil $A(\lambda)$ are equivalent to those for dissipative operators in Krein or Pontrjagin spaces (see [AI], ch.II, §2). In the sequel we prefer to deal with the linear pencil $A(\lambda)$. The motivation for this becomes clear when considering the corresponding operator differential equations. Moreover, at least formally, we obtain more general results, as we do not always assume that W generates a regular indefinite metric.

The basic goal of this section is to prove formula (2.17). This formula is based on the well-known fundamental result on the existence of a maximal W-nonnegative A-invariant subspace in Pontrjagin space and on the explicit construction of maximal W-nonnegative subspaces corresponding to real normal eigenvalues of the pencil $A(\lambda)$ or of the operator $A = W^{-1}T$. In the paper [S1] the author considered dissipative operator pencils of an arbitrary oder $n \geqslant 1$ and constructed for such pencils *regular* canonical systems corresponding to real normal eigenvalues. This construction allows us to define the *sign characteristics* for Jordan chains and to realize the construction of a maximal W-nonnegative subspace $\mathcal{L}_\mu^+$ in the root subspace $\mathcal{L}_\mu$ corresponding to a real eigenvalue μ. The additional details for linear pencils were given in the unpublished manuscript [S2]. We note also the papers of Kostyuchenko and Orazov [KO] (devoted to the case of a selfadjoint operator T) and Gomilko [G] related to this topic. However, our construction is new and, perhaps simpler, even for selfadjoint pencils. In addition we obtain the information on the connection of the middle elements of mutually adjoint canonical systems. This information is essentially used when considering half range completeness and minimality problems (see [S1]). Recently Ran and Temme [RT] investigated an analogous problem from another point of view. Here we present some results of [S2] concerning this subject.

Let μ be an eigenvalue of the pencil $A(\lambda) = T - \lambda W$ and

$$y_j^0, y_j^1, \dots, y_j^{p_j}, \qquad j = 1, \dots, N, \tag{2.1}$$

be a canonical system of eigen and associated elements (or Jordan chains) corresponding to μ (see [Ke]). The linear span of all elements (2.1) is denoted $\mathcal{L}_\mu(A)$ or simply $\mathcal{L}_\mu$ and is called the root subspace corresponding to the eigenvalue μ. An eigenvalue μ is said to be *normal* if $A(\lambda)$ is invertible in a punctured neighbourhood of μ and the number $N = Ker(T - \mu W)$ as well as the lengths $p_j + 1$ of Jordan chains (2.1) are finite. It is known [Ke] that the principal part of the Laurent expansion of the function $A^{-1}(\lambda)$ at the pole μ has the representation

$$\sum_{j=1}^{N} \sum_{s=o}^{p_j} \frac{(\cdot, x_j^s) y_j^0 + \dots + (\cdot, x_j^0) y_j^s}{(\lambda - \mu)^{p_j + 1 - s}}, \tag{2.2}$$

where the *adjoint* system

$$x_j^0, x_j^1, \dots, x_j^{p_j}, \quad j = 1, \dots, N, \tag{2.3}$$

is uniquely determined by the choice of system (2.1). It turns out that the adjoint system (2.3) is a canonical system of Jordan chains corresponding to the eigenvalue $\bar{\mu}$ of the pencil $A^*(\lambda) = T^* - \lambda W$.

Further the upper index is always used for numeration of associated elements while the the lower one numerates eigenvalues and canonical chains simultaneously, i.e. each eigenvalue is counted as many times as its geometric multiplicity. The set of all eigenvalues of the pencil $A(\lambda)$ is denoted $\sigma_p(A)$. For the subset in $\sigma_p(A)$ consisting of the normal eigenvalues we reserve the notation $\sigma_d(A)$ (the discrete spectrum). Notice that canonical system of Jordan chains (2.1) is well defined for any $\mu \in \sigma_p(A)$ (possibly, consisting of infinitely many elements), however, adjoint system (2.3) is well defined only for $\mu \in \sigma_d(A)$. As usually the indefinite scalar product (Wx, x) is denoted $[x, x]$.

Although some of the subsequent propositions are essentially known, we present their proofs here for the reader's convenience. New constructions are started from Proposition 2.6.

Proposition 2.1. *Let $\mathcal{L}_+^0$ be the minimal subspace containing the root subspaces corresponding to all $\mu \in \mathbb{C}^+ \cap \sigma_p(A)$. Then $\mathcal{L}_+^0$ is a W-nonnegative subspace.*

Proof. (Cf. [AI], Ch.2, Corollary 2.22). We present here another, shorter proof. Suppose eigenvalues are numerated as many times as their geometric multiplicity. Let us consider the functions

$$u_j^h(t) = e^{i\mu_j t}\left(y_j^h + \frac{it}{1!}y_j^{h-1} + \ldots + \frac{(it)^h}{h!}y_j^0\right), \qquad h = 0, 1, \ldots, p_j,$$

where $y_j^0, \ldots, y_j^{p_j}$ are Jordan chains corresponding to the eigenvalues $\mu_j \in \mathbb{C}^+$. It is easily seen that the functions $u_j^h(t)$ satisfy the equation

$$iWu'(t) + Tu(t) = 0.$$

Any linear combination $u(t) = \sum c_{j,h}u_j^h(t)$ also satisfies this equation, therefore

$$\begin{aligned}[u(\xi), u(\xi)]' &= (Wu'(\xi), u(\xi)) + (u(\xi), Wu'(\xi)) \\ &= (iTu(\xi), u(\xi)) + (u(\xi), iTu(\xi)) = -2Im\,(Tu(\xi), u(\xi)).\end{aligned}$$

As all the functions $u_j^h(t)$ vanish at ∞, so does $u(t)$. Integrating the last equality from t to ∞ we obtain

$$[u(t), u(t)] = 2\int_t^\infty Im\,(Tu(\xi), u(\xi)) \geqslant 0.$$

In particular $[u(0), u(0)] \geqslant 0$ for all $u(0) = \sum c_{j,h}y_j^h$. By the definition the set of these elements is dense in $\mathcal{L}_+^0$, hence, $\mathcal{L}_+^0$ is a W-nonnegative subspace. □

Proposition 2.2. *Let (2.1) be a canonical system corresponding to a real eigenvalue μ. If $[\gamma]$ is the integer part of a number γ then the elements*

$$y_k^0, y_k^1, \ldots, y_k^{\alpha_k}, \qquad k = 1, \ldots, N, \qquad \alpha_k = \left[\frac{p_k}{2}\right], \tag{2.4}$$

belong to $\mathcal{D}(T^*)$ and $T^*y_k^h = Ty_k^h$ for all $1 \leqslant k \leqslant N$, $0 \leqslant h \leqslant \alpha_k$.

Proof. First we notice that T^* is well defined, as the operator T is closed by assumption (see [Ka], Ch.3, §5.5). Now, let us prove the following: *If $x \in \mathcal{D}(T)$ and $Im\,(Tx, x) = 0$ then $x \in \mathcal{D}(T^*)$ and $T^*x = Tx$.* (Cf. [AI], Ch.2, Theorem 2.15). To prove this fact, we introduce an indefinite product in the space $\mathbf{H} = H \times H$ as follows

$$\langle\{x_1, x_2\}, \{y_1, y_2\}\rangle = i(x_1, y_2) - i(x_2, y_1).$$

As T is dissipative, we have

$$\langle \mathbf{x}, \mathbf{x}\rangle = 2Im(Tx, x) \geqslant 0 \qquad \text{for all } \mathbf{x} = \{x, Tx\} \in \Gamma(T),$$

where $\Gamma(T)$ is the graph of T. If $x \in \mathcal{D}(T)$ and $Im\,(Tx,x)=0$ then by virtue of Cauchy-Schwarz-Bunyakovskii inequality we obtain

$$|(x,Tz)-(Tx,z)| = |\langle \mathbf{x},\mathbf{z}\rangle| \leqslant \langle \mathbf{x},\mathbf{x}\rangle^{1/2}\langle \mathbf{z},\mathbf{z}\rangle^{1/2} = 0 \qquad \text{for all } \mathbf{z}=\{z,Tz\}\in\Gamma(T).$$

Hence, $(Tz,x)=(z,Tx)$ for all $z\in\mathcal{D}(T)$. From the definition of the adjoint operator we obtain $x\in\mathcal{D}(T^*)$ and $T^*x=Tx$.

Now let us prove the assertion of Proposition 2.2. As the elements of system (2.1) are Jordan chains, we have

$$(2.5) \qquad (T-\mu W)y_k^h = Wy_k^{h-1}, \qquad 0\leqslant h\leqslant p_k \quad (y_k^{-1}:=0).$$

In particular,

$$Im\,((T-\mu W)\,y_k^0,\,y_k^0) = Im\,(Ty_k^0,y_k^0)=0.$$

Therefore, $y_k^0\in D(T^*)$ and $Ty_k^0=T^*y_k^0$. Now we can end the proof by induction. Suppose that for some $h\leqslant\alpha_k$ we have proved that

$$y_k^s\in D(T^*) \quad \text{and} \quad Ty_k^s=T^*y_k^s \qquad \text{for } s=0,1,\dots,h-1.$$

As $2h\leqslant p_k$, we find

$$(Wy_k^{h-1},y_k^h)=(y_k^{h-1},(T-\mu W)y_k^{h+1})=(Wy_k^{h-2},y_k^{h+1})=\dots$$
$$=(y_k^0,(T-\mu W)y_k^{2h})=0.$$

Hence, $Im\,(Wy_k^{h-1}+\mu Wy_k^h,\,y_k^h)=0$ and

$$Im\,(Ty_k^h,\,y_k^h)=Im\,((T-\mu W)y_k^h-Wy_k^{h-1},\,y_k^h)=0.$$

As before we deduce that $y_k^h\in D(T^*)$ and $Ty_k^h=T^*y_k^h$. □

Proposition 2.3. *Let (2.1) be a canonical system corresponding to a real eigenvalue* μ *. Then*

$$(2.6) \qquad [y_k^h,y_j^s]=0, \qquad j=1,\dots,N, \qquad h\leqslant[(p_k-1)/2], \quad s\leqslant[(p_j-1)/2].$$

If $p_j\neq p_k$ *then (2.6) hold for all* $s\leqslant[p_j/2], h\leqslant[p_k/2]$.

Proof. Suppose $p_j\leqslant p_k$. Then it follows from our assumptions that $h+s+1\leqslant p_k$. Taking into account (2.5) and the equalities $Ty_j^s=T^*y_j^s$ (Proposition 2.2) we find

$$(y_k^h,Wy_j^s)=((T-\mu W)y_k^{h+1},y_j^s)=(y_k^{h+1},Wy_j^{s-1})=\dots$$
$$=(y_k^{h+s+1},(T^*-\mu W)y_j^0)=0.$$

and the equalities (2.6) follow. □

Proposition 2.4. *Let ν and μ be eigenvalues of the pencils $A(\lambda)$ and $A^*(\lambda)$ respectively. If $\nu \neq \overline{\mu}$ then the root subspaces $\mathcal{L}_\nu(A)$ and $\mathcal{L}_\mu(A^*)$ are W-orthogonal. In particular, truncated Jordan chains (2.4) corresponding to a real eigenvalue μ of the pencil $A(\lambda)$ are W-orthogonal to any root subspace $\mathcal{L}_\nu(A)$ if $\nu \neq \mu$.*

Proof. Let $y^0, \dots, y^p \in \mathcal{L}_\nu(A)$, $x^0, \dots, x^q \in \mathcal{L}_\mu(A^*)$ be Jordan chains and $\nu \neq \bar{\mu}$. Using (2.5) we obtain

$$\begin{aligned}(Ty^s, x^l) =&\nu[y^s, x^l] + [y^{s-1}, x^l]\\ =&(y^s, T^* x^l) = \bar{\mu}[y^s, x^l] + [y^s, x^{l-1}], \qquad (y^{-1} := x^{-1} := 0).\end{aligned}$$

In particular, from these equalities we have $[y^0, x^0] = 0$. Now, the proof of the first assertion is ended by induction with respect to the index $s + l$. The second assertion follows from Proposition 2.2. □

Proposition 2.5. *Let (2.1) and (2.3) be mutually adjoint canonical systems corresponding to normal eigenvalues μ_j which are enumerated according to their geometric multiplicity. Then the following biorthogonality relations hold:*

$$[y_k^h, x_j^s] = -\delta_{k,j}\delta_{h,p_j-s}, \tag{2.7}$$

where $\delta_{m,n}$ is the Kronecker symbol.

Proof. (Cf.[Ke]). We have

$$A(\lambda)y_k^h = [A(\mu_k) - (\lambda - \mu_k)W]\, y_k^h = W y_k^{h-1} - (\lambda - \mu_k) W y_k^h, \qquad 0 \leqslant h \leqslant p_k,$$

where as before it is assumed that $y_k^{-1} := 0$. Using the representation (2.2) we obtain

$$\begin{aligned}y_k^h &= A^{-1}(\lambda)A(\lambda)y_k^h\\ &= \sum_{j=N_1}^{N_2}\sum_{s=0}^{p_j}\left[\frac{(\cdot, x_j^s)\,y_j^0 + \dots + (\cdot, x_j^0)\,y_j^s}{(\lambda - \mu_j)^{p_j+1-s}} + R(\lambda)\right]\left[-(\lambda - \mu_k)W y_k^h + W y_k^{h-1}\right],\end{aligned} \tag{2.8}$$

where $R(\lambda)$ is a holomorphic operator function at the point $\mu = \mu_j$ and $N_2 - N_1 + 1$ is the geometric multiplicity of the eigenvalue μ. We may assume that $N_1 = 1$, $N_2 = N$, $p_1 \geqslant p_2 \geqslant \dots \geqslant p_N$.

Suppose that $\mu_k \neq \mu_j$. If we take h=0 and compare the coefficients of the powers $(\lambda - \mu_j)^{-p_j-1+s}$, $0 \leqslant s \leqslant p_j$, we find

$$-\sum_{p_j=p_1} [y_k^0, x_j^0] y_j^0 = 0, \tag{2.9}$$

$$-\sum_{p_j=p_1} [y_k^0, x_j^1] y_j^0 - \sum_{p_j=p_1} [y_k^0, x_j^0] y_j^1 - \sum_{p_j=p_1-1} [y_k^0, x_j^0] y_j^0 = 0. \tag{2.10}$$

We do not write out the other coefficients corresponding to the indices $s \geqslant 2$. We also notice that the third term in (2.10) should be omitted if there are no Jordan chains of length $(p_1 + 1) - 1$. It follows from the definition of a canonical system that the elements $\{y_j^0\}_1^N$ are linearly independent. Hence, from (2.9) we have

$$[y_k^0, x_j^0] = 0, \qquad \text{for all indices } j \text{ such that } p_j = p_1. \tag{2.11}$$

Now, it follows from (2.10) and (2.11) that

$$[y_k^0, x_j^0] = 0 \quad \text{if} \quad p_j = p_1 - 1; \qquad [y_k^0, x_j^1] = 0 \quad \text{if} \quad p_j = p_1.$$

Repeating the argument we find $[y_k^0, x_j^s] = 0$ for all indices $0 \leqslant s \leqslant p_j$. Using the last equalities and taking $h = 1, 2, \dots, p_k$, we find subsequently

$$[y_k^1, x_j^s] = 0, \dots, [y_k^{p_k}, x_j^s] = 0 \qquad \text{for all } 0 \leqslant s \leqslant p_j.$$

The same arguments can be applied in the case $\mu_k = \mu_j$. Comparing the coefficients of the powers $(\lambda - \mu_j)^{\nu}$ in (2.10) it is found that, for $h = 0, 1, \dots, p_k$,

$$-[y_j^h, x_j^s] = \delta_{h, p_j - s},$$

and relations (2.7) follow. □

Let a canonical system (2.1) correspond to a real normal eigenvalue μ. Denote by $\mathcal{S}_\mu^0$ the span of elements

$$y_k^0, y_k^1, \dots, y_k^{\beta_k}, \qquad k = 1, \dots, N, \quad \beta_k = [(p_k - 1)/2] \tag{2.12}$$

(if $p_k = 0$, we assume that $\beta_k = -1$ and the element y_k^0 does not belong to $\mathcal{S}_\mu^0$). Let us fix an index k, $1 \leqslant k \leqslant N$. If the number $p_k + 1$ is even we set $\mathcal{S}_{\mu_k} := \mathcal{S}_\mu^0$. If $p_k + 1$ is odd we denote by $\mathcal{S}_{\mu_k}$ the span of elements (2.12) combined with the elements $y_j^{\alpha_j}, \alpha_j = [p_j/2]$, where index j runs through all the values such that $p_j = p_k$. Similary, by replacing chains (2.1) with adjoint chains (2.3) we construct subspaces $(\mathcal{S}_\mu^0)^*$ and $\mathcal{S}_{\mu_k}^*$. We emphasize that, according to our agreement about the enumeration of eigenvalues, the subspaces $\mathcal{S}_{\mu_k}$ are generally different although $\mu_k = \mu$.

Proposition 2.6. *For all nonzero real normal eigenvalues μ the following equalities hold*

$$\mathcal{S}_\mu^0 = (\mathcal{S}_\mu^0)^*, \quad \mathcal{S}_{\mu_k} = \mathcal{S}_{\mu_k}^* \qquad \text{for all } 1 \leqslant k \leqslant N.$$

Proof. Suppose that $y_k^h \in \mathcal{S}_{\mu_k}$ and $x_k^h \notin \mathcal{S}_{\mu_k}$. It follows from Proposition 2.2 that

$$x_k^0, x_k^1, \dots, x_k^{\alpha_k}, \qquad k = 1, \dots, N, \quad \alpha_k = [p_k/2],$$

are chains of EAE of the pencil $A(\lambda)$ as well as of $A^*(\lambda)$. Since (2.1) is a canonical system, we have the representation

$$x_k^h = \sum_{j=1}^{N} \sum_{s=0}^{h} c_{j,s} y_j^s, \qquad if\ 0 \leqslant h \leqslant \alpha_j = [p_j/2]. \tag{2.13}$$

We have assumed that $x_k^h \notin \mathcal{S}_{\mu_k}$, therefore, at least one of the numbers $c_{j,s}$ in (2.12) is not equal to zero for $s > \beta_j = [(p_j - 1)/2]$, $p_j < p_k$. In this case, however, $x_j^{p_j - s} \in \mathcal{S}_{\mu_j}^*$ i.e. $p_j - s \leqslant [p_j/2]$. Applying Proposition 2.3 with respect to the pencil $A^*(\lambda)$ we find

$$[x_k^h, x_j^{p_j - s}] = 0.$$

On the other hand it follows from Proposition 2.5 and representation (2.13) that

$$[x_k^h, x_j^{p_j - s}] = -c_{j,s}.$$

Hence, the assumption $x_k^h \notin \mathcal{S}_{\mu_k}$ is not valid. The equality $\mathcal{S}^0 = (\mathcal{S}^0)^*$ is proved in a similar way. □

Proposition 2.7. *A canonical system (2.1) corresponding to a real normal eigenvalue μ of the pencil $A(\lambda)$ can be chosen in such a way that*

$$[y_j^{\alpha_j}, y_l^{\alpha_l}] = \varepsilon_j \delta_{j,l}, \qquad \alpha_j = [p_j/2], \qquad \varepsilon_j = \begin{cases} 0 & \text{if } p_j + 1 \text{ is even} \\ \pm 1 & \text{if } p_j + 1 \text{ is odd} \end{cases} \tag{2.14}$$

for all indices $1 \leqslant j, l \leqslant N$.

Proof. Fix an index k such that $p_k + 1$ is odd. Assume that there are q chains of the length $p_k + 1$, i.e. $p_j = p_k$ for $j = k, k+1, \dots, k+q-1$. According to the definition of $\mathcal{S}_{\mu_k}$ we have $\dim \mathcal{S}_{\mu_k} \ominus \mathcal{S}_\mu^0 = q$. Let P_k be the orthoprojector onto the subspace $\mathcal{S}_{\mu_k}$. It follows from the biorthogonality relations (2.7) that the selfadjoint operator $P_k W P_k$ has exactly q nonzero eigenvalues which correspond to an orthogonal basis $\{\varphi_s\}_1^q$. We can replace, if necessary, chains (2.1) corresponding to indeces $l = k, k+1, \dots, k+q-1$, by their linear combinations and obtain a new canonical system such that the system $\{\varphi_l\}_1^q$ coincides with $\{y_s^{\alpha_k}\}_k^{k+q-1}$. Then, after a proper norming, the relations (2.14) hold for all indices $l, j = k, k+1, \dots, k+q-1$. We can repeat the same arguments for any other index r such that $\mathcal{S}_{\mu_r} \neq \mathcal{S}_{\mu_k}$. Taking into account that the subspaces $\mathcal{S}_{\mu_r}$ and $\mathcal{S}_{\mu_k}$ are W-orthogonal (Proposition 2.3), we obtain relations (2.14) for all indices such that $1 \leqslant j, l \leqslant N$. □

Proposition 2.8. *Let a canonical system (2.1) correspond to a real normal eigenvalue μ and satisfy relations (2.14). Then for all indices j such that $p_j = 2\alpha_j$ the elements $x_j^{\alpha_j}$ of the adjoint system (2.3) have the representation*

$$x_j^{\alpha_j} = -\varepsilon_j y_j^{\alpha_j} + y, \qquad \varepsilon_j = \pm 1, \quad \text{where } y \in \mathcal{S}_\mu^0. \tag{2.15}$$

In other words: there exists a canonical system (2.1) such that for Jordan chains of odd legth the middle elements $x_j^{\alpha_j}$ of its adjoint system have representation (2.15).

Proof. As $x_j^{\alpha_j} \in \mathcal{S}_{\mu_j}^* = \mathcal{S}_{\mu_j}$, we have

$$x_j^{\alpha_j} = \sum_{p_l = p_j} c_l y_l^{\alpha_k} + y, \qquad \text{where } y \in \mathcal{S}_\mu^0.$$

Now, if canonical system (2.1) satisfies relations (2.14) then $c_l = -\varepsilon_j \delta_{j,l}$, and relation (2.15) follow. □

A canonical system (2.1) which satisfies relations (2.14) or (2.15) is said to be *regular*. The numbers ε_j in (2.15) are said to be *sign characteristics*. We note that for linear selfadjoint pencils the sign characteristics are determined in a different way, namely, $\varepsilon_j = \pm 1$ for Jordan chains of any length (see [GLR], Ch.3, and [KS], Lemma 2). Simple examples show that for dissipative pencils the definite sign characteristics can not be well defined for Jordan chains of even length. In this situation it is convenient to assume that the *sign characteristics $\varepsilon_j = 0$ for all chains of even length $p_j + 1$*. It is supposed that this agreement holds through the rest of the paper.

Let (2.1) be a regular canonical system corresponding to a normal real eigenvalue μ. Denote by $\mathcal{L}_\mu^+$ ($\mathcal{L}_\mu^-$) the span of elements (2.12) combined with $y_j^{\alpha_j}$ satisfying relations (2.14) with $\varepsilon_j = +1$ ($\varepsilon_j = -1$). Then according to the definition of the sign charactheristics we have

$$\dim \mathcal{L}_\mu^+ = \sum_{k=1}^N (\varepsilon_k^+ + [(p_k - 1)/2]), \qquad \text{where} \quad \varepsilon_k^+ = \max(0, \varepsilon_k). \tag{2.16}$$

Proposition 2.9. *Let μ be a real normal eigenvalue of the pencil $A(\lambda) = T - \lambda W$. Then $\mathcal{L}_\mu^+$ is a maximal W-nonpositive subspace in the root subspace $\mathcal{L}_\mu$.*

Proof. It follows from Propositions 2.2 and 2.7 that $\mathcal{L}_\mu^+$ is a W-nonnegative subspace. Assume that $\mathcal{L}_\mu^+ \subset \mathcal{L}' \subset \mathcal{L}_\mu$, where $\mathcal{L}'$ is also W-nonnegative subspace, and there exists an element $y \in \mathcal{L}'$ such that $y \notin \mathcal{L}_\mu^+$. Obviously, $y \notin \mathcal{L}_\mu^-$, as the assumptions $y \in \mathcal{L}_\mu^-, y \notin \mathcal{L}_\mu^+$ imply $[y, y] < 0$. Therefore, $y \notin \mathcal{L}_\mu^+ \cup \mathcal{L}_\mu^-$. Now, using (2.7) we can find an element $y_k^h \in \mathcal{S}_\mu^0$ such that $[y_k^h, y] = \gamma \neq 0$. Denote $z = ay_k^h + \gamma y$. Then $[z, z] = |\gamma|^2(a + [y, y]) \to -\infty$ if $a \to -\infty$. On the other hand $[z, z] \geqslant 0$, as $z \in \mathcal{L}'$ and $\mathcal{L}'$ is by assumption W-nonnegative. This contradiction ends the proof. □

Denote by $\mathcal{L}$ the minimal subspace containing the root subspaces $\mathcal{L}_\mu(A)$ corresponding to all the eigenvalues $\mu \in \mathbb{C}^+$ and all the root subspaces $\mathcal{L}_\mu(A)$ corresponding to normal real eigenvalues. Analogously, let $\mathcal{L}^+$ be the minimal subspace containing $\mathcal{L}_\mu$ for all $\mu \in \mathbb{C}^+ \cap \sigma_p(A)$ and all the subspaces $\mathcal{L}_\mu^+$ corresponding to the normal real eigenvalues. For a selfadjoint operator C we introduce the (well-known) notations

$$\pi(C) = rank\, C^+, \quad \text{where } C^+ = (|C| + C)/2, \qquad \nu(C) = \pi(-C).$$

Further, we use the following fundamental result.

Theorem on a maximal nonnegative invariant subspace. *Suppose W generates a Pontrjagin space, i.e. W is boundedly invertible and $\nu(W) < \infty$. If $A = W^{-1}T$ and $\rho(A) \cap \mathbb{C}^+ \neq \varnothing$ then there exists a maximal A-invariant W-nonnegative subspace $H^+ \subset H$, $\dim H^+ = \nu(W)$, such that the spectrum of the restriction $A/_{H^+}$ lie in $\overline{\mathbb{C}}^+$, and in $\mathbb{C}^+$ coincides with the spectrum of A.*

Proof. In the case $T = T^*$ this is a well-known Pontrjagin theorem [P]. For a maximal W-dissipative operator A in Pontrjagin space the theorem was proved by Krein and Langer [KL], and by Azizov [A] (see [AI] and references therein). □

Theorem 2.10. *The subspace $\mathcal{L}^+$ defined above is a maximal W-nonnegative subspace in $\mathcal{L}$. If W generates a Pontrjagin space and all the real eigenvalues of the pencil $A(\lambda)$ are normal then $\mathcal{L}^+$ is a maximal W-nonnegative subspace in the whole space H.*

Proof. It follows from Proposition 2.4 and the definition that $\mathcal{L}^+$ is a W-nonnegative subspace. As $\mathcal{L}_\mu^+$ is a maximal W-nonnegative subspace in $\mathcal{L}_\mu$ for any $\mu \in \sigma_d(A) \cap \mathbb{R}$ (Proposition 2.9), we have that $\mathcal{L}^+$ possesses the same property in $\mathcal{L}$.

Now, let W generate a Pontrjagin space and all the real eigenvalues of the pencil $A(\lambda)$

are normal. According to the generalized Pontrjagin theorem there exists a maximal W-nonnegative subspace H^+ in H, $\dim H^+ = \nu(W)$, such that $\mathcal{L}_+^0 \subset H^+ \subset \mathcal{L}$, where $\mathcal{L}_+^0$ is defined in Proposition 2.1. As the subspace $H^+ \cap \mathcal{L}_\mu$ is W-nonnegative in $\mathcal{L}_\mu$ and $\mathcal{L}_\mu^+$ is a maximal nonnegative subspace in $\mathcal{L}_\mu$ (Proposition 2.9), we have: $\dim(H^+ \cap \mathcal{L}_\mu) \leqslant \dim \mathcal{L}_\mu^+$ (see, for example, [AI], Ch.I, §4). Then it follows that

$$\dim H^+ = \dim \mathcal{L}_+^0 + \sum_{\mu \in \mathbb{R} \cap \sigma_d} \dim(H^+ \cap \mathcal{L}_\mu) \leqslant \dim \mathcal{L}^+, \qquad \sigma_d := \sigma_d(A).$$

On the other hand, it is known ([AI], Ch.I, §4) that $\dim \mathcal{L}^+ \leqslant \nu(W) = \dim H^+$. Hence, $\dim \mathcal{L}^+ = \dim H^+$ and from this it follows that $\mathcal{L}^+$ is a maximal W-nonnegative subspace in the whole H. □

Corollary 2.11. *Let W be boundedly invertible, $\nu(W) < \infty$, and all the real eigenvalues of $A(\lambda)$ be normal. Then the following formula is valid*

$$\kappa(A) + \sum_{\mu_k \in \mathbb{R} \cap \sigma_d} (\varepsilon_k^+ + [(p_k - 1)/2]) = \nu(W), \qquad \varepsilon_k^+ = \max(0, \varepsilon_k). \tag{2.17}$$

Here $\kappa(A)$ is the total algebraic multiplicity of all eigenvalues in $\mathbb{C}^+$ and ε_k $(p_k + 1)$ are the sign characteristics (the lengths) of Jordan chains of regular canonical systems corresponding to real normal eigenvalues μ_k

Proof. It follows from formula (2.16) and Theorem 2.10. □

Remark 2.12. Formula (2.17) is not applicable if the pencil $A(\lambda)$ has real eigenvalues which are embedded into the essential spectrum. In this case we do not know how to determine the sign characteristics and how to realize the explicit construction of a maximal W-nonnegative subspace in the the root subspace $\mathcal{L}_\mu(A)$. However, the following inequality is always valid (cf. [AI], Ch.2, Theorem 2.26)

$$\kappa(A) + \sum_{\mu_k \in \mathbb{R} \cap \sigma_p} [(p_k - 1)/2] \leqslant \nu(W), \qquad \sigma_p := \sigma_p(A). \tag{2.18}$$

This inequality is much more simple and follows directly from Propositions 2.1, 2.3 and 2.4. It expresses the fact that the linear span of all root subspaces $\mathcal{L}_\mu$ corresponding to $\mu \in \sigma_p(A) \cap \mathbb{C}^+$ and all the truncated root subspaces $\mathcal{S}_\mu^0$ corresponding to $\mu \in \sigma_p(A) \cap \mathbb{R}$ forms a W-nonnegative subspace (not necessarily a maximal one). Indeed, using (2.17) we can improve (2.18) and write the following inequality

$$\kappa(A) + \sum_{\mu_k \in \mathbb{R} \cap \sigma_p} (\varepsilon_k^+ + [(p_k - 1)/2]) \leqslant \nu(W), \qquad \sigma_p := \sigma_p(A), \tag{2.19}$$

where $\varepsilon_k^+ = \max(0, \varepsilon_k)$ if $\mu_k \in \sigma_d$ and $\varepsilon_k^+ = 0$ if $\mu_k \in \sigma_p \setminus \sigma_d$.

3. Quadratic dissipative pencils and the instability index formula.

In this section we study a quadratic operator pencil of the form

$$A(\lambda) = \lambda^2 F + (D + iG)\lambda + T. \tag{3.1}$$

Further it is always assumed that the coefficients in (3.1) are operators in Hilbert space H satisfying the following conditions:

i) F is a selfadjoint bounded and boundedly invertible operator;

ii) T is defined on the domain $\mathcal{D}(T), T = T^*$ and T is boundedly invertible;

iii) D and G are symmetric T-bounded operators (i.e. D and G are symmetric, $\mathcal{D}(D) \subset \mathcal{D}(T)$ and $\mathcal{D}(G) \subset \mathcal{D}(T)$). Moreover, $D \geqslant 0$.

These assumptions imply that $A(\lambda)$ is a quadratic dissipative pencil with respect to the imaginary axis in the following sense (see [S1])

$$Im\,(\zeta A(i\zeta)x, x) = \zeta^2(Dx, x) \geqslant 0 \quad \text{for all } x \in \mathcal{D}(T) \text{ and } \zeta \in \mathbb{R}.$$

One may expect that the quadratic dissipative pencil (3.1) can be transformed into a linear dissipative pencil. Indeed, such a linearization will be realized below. However, working with unbounded pencils we come to some new problems which do not arise when considering pencils with bounded coefficients. In particular, the spectrum of a linearization may not coincide with the spectrum of the original pencil.

According to our assumptions $A(\lambda)$ is well defined for each $\lambda \in \mathbb{C}$ on the domain $\mathcal{D}(T)$. Hence, the first natural definition of the resolvent set $\rho(A)$ is the following: $\zeta \in \rho(A)$ if $A(\zeta)$ with the domain $\mathcal{D}(T)$ has a bounded inverse. To give another definition, we consider the scale of Hilbert spaces H_θ, $\theta \in \mathbb{R}\,(H_0 = H)$ generated by the selfadjoint operator $S^2 := |T| := (T^2)^{1/2}$. Namely, if $\theta > 0$ we set $H_\theta = \{x | x \in \mathcal{D}(S^\theta)\}$ with the norm $\|x\|_\theta = \|S^\theta x\|$. If $\theta < 0$, the space H_θ is defined as the closure of H with respect to the norm $\|x\|_\theta = \|S^\theta x\|$.

Let us associate the pencil

$$\hat{A}(\lambda) = \lambda^2 \hat{F} + \lambda(\hat{D} + i\hat{G}) + J$$

with the pencil $A(\lambda)$. Here

$$\hat{F} = S^{-1}FS^{-1}, \hat{D} = S^{-1}DS^{-1}, \hat{G} = S^{-1}GS^{-1}, J = T^{-1}|T|.$$

Obviously $\hat{F}$ and J are bounded. From the next Proposition it follows that $\hat{D}$ and $\hat{G}$ are also bounded in H.

Proposition 3.1. *Let S be an uniformly positive selfadjoint operator and B be a symmetric operator such that $D(B) \supset D(S^2)$. Then the operator $S^{\theta-2}BS^{-\theta}$ defined on the domain $D(S^{\theta-2})$ is bounded in H for all $0 \leqslant \theta \leqslant 2$. Equivalently, B is bounded as an operator acting from H_θ into $H_{\theta-2}$.*

Proof. As B is closable, the assumption $D(B) \supset D(S^2)$ implies that $B: \ H_2 \to H$ is a bounded operator (this follows immediately from the closed graph theorem). Hence, the adjoint operator $B^*: \ H \to H_{-2}$ is also bounded. As $B^* \supset B$, we have that $B: \ H \to H_{-2}$ is bounded. Now, applying the interpolation theorem (see [LM], Ch.1, for example) we find that $B: \ H_\theta \to H_{\theta-2}$ is bounded for all $0 \leqslant \theta \leqslant 2$. □

Let $\sigma(\hat{A})$ be the spectrum of the pencil $\hat{A}(\lambda)$ with bounded operator coefficients in the space H. It is easily seen that $\sigma(\hat{A})$ coincides with the spectrum of $A(\lambda)$ considered as the operator function in the space H_{-1} on the domain $\mathcal{D}(A) = H_1$. Both our definitions of the spectra are better understood (especially for the specialists working with partial differential operators) if we say the following: $\sigma(A)$ is the spectrum of the pencil $A(\lambda)$ considered in the "classical" space H while $\sigma(\hat{A})$ is its spectrum in the generalised space H_{-1}.

Generally, $\sigma(A) \neq \sigma(\hat{A})$. What is the connection between the classical and the generalized spectra? Some light is cast on this problem by the next propositons. It will be convenient to define in the complex plane the open set $\rho_m(A) := \rho(A) \cup \sigma_d(A)$. The set $\rho_m(\hat{A})$ is defined analogously. In the other words $\rho_m(A)$ and $\rho_m(\hat{A})$ are the domains where the operator functions $A^{-1}(\lambda)$ is finite meromorphic in the spaces H and H_{-1}, respectively.

Proposition 3.2. *In the domain $\rho_m(A) \cap \rho_m(\hat{A})$ all the eigenvalues and Jordan chains of $A(\lambda)$ in the spaces H and H_{-1} coincide.*

Proof. Let us consider the pencils

$$\tilde{A}(\lambda) := A(\lambda)S^{-2} = \lambda^2 FS^{-2} + \lambda(DS^{-2} + iGS^{-2}) + J,$$
$$\tilde{A}^*(\lambda) := \lambda^2 S^{-2}F + \lambda(S^{-2}D - iS^{-2}G) + J.$$

We have already noticed (Proposition 3.1) that all the operator coefficients of these pencils are bounded operators in H. Moreover, $\tilde{A}(\lambda)$ and $\tilde{A}^*(\bar{\lambda})$ are mutually adjoint in H. Let μ be a normal (classical) eigenvalue of $A(\lambda)$ with a corresponding canonical system of Jordan chains

$$y_j^0, \dots, y_j^{p_j}, \qquad j = 1, \dots, N. \tag{3.2}$$

Then μ is a normal eigenvalue of the pencil $\tilde{A}(\lambda)$ and

$$S^2 y_j^0, \dots, S^2 y_j^{p_j}, \qquad j = 1, \dots, N, \tag{3.3}$$

is a canonical system of Jordan chains of the pencil $\tilde{A}(\lambda)$ corresponding to μ. In this case, obviously, $Sy_j^0, \dots, Sy_j^{p_j}$, are eigen and associated elements of the pencil $\hat{A}(\lambda)$. Therefore, $\mu \in \sigma_p(\hat{A})$ and we can define a canonical system of Jordan chains

$$z_j^0, \dots, z_j^{q_j}, \qquad j = 1, \dots, K, \tag{3.4}$$

of the pencil $\hat{A}(\lambda)$ corresponding to the eigenvalue μ. It follows from the definition of a canonical system that $K \geqslant N$ and $q_j \geqslant p_j$ for $j = 1, \dots, N$. If in addition $\mu \in \rho_m(\hat{A})$ (as claimed by assumption) then the adjoint system

$$x_j^0, \dots, x_j^{q_j}, \qquad j = 1, \dots, K,$$

with respect to (3.4) (in the sense of the Laurent expansion for $\hat{A}^{-1}(\lambda)$ at the point μ) is well defined. Now, we observe that $S^{-1}x_j^0, \dots, S^{-1}x_j^{q_j}$ are the Jordan chains of the pencil $\tilde{A}^*(\lambda)$ corresponding to the eigenvalue $\bar{\mu}$. Hence any canonical system of $\tilde{A}^*(\lambda)$ corresponding to $\bar{\mu}$ consists of $l \geqslant q_1 + \dots + q_K$ elements. On the other hand, the system which is adjoint to (3.3) is a canonical system of $\tilde{A}^*(\lambda)$ corresponding to $\bar{\mu}$ and consists of $p_1 + \dots + p_N$ elements. Therefore, $N = K$ and $p_j = q_j$ for $j = 1, \dots, N$.

The same arguments can be applied to show that if $\mu \in \rho(A) \cap \rho_m(\hat{A})$ then $\mu \in \rho(\hat{A})$. □

Proposition 3.3. *Let*

$$A_\pm(\lambda) = \lambda^2 F + (D \pm iG)\lambda + T \qquad (A_+(\lambda) = A(\lambda)). \tag{3.5}$$

Let Ω be a domain in $\mathbb{C}$ which is symmetric with respect to the real axis and such that $\Omega \subset \rho_m(A_+) \cap \rho_m(A_-)$. Then $\Omega \subset \rho_m(\hat{A}_+) \cap \rho_m(\hat{A}_-)$ and all the eigenvalues of the pencils $A_\pm(\lambda)$ and $\hat{A}_\pm(\lambda)$ in Ω as well as the structures of the corresponding canonical systems coincide.

Proof. Denote

$$\tilde{A}_\pm(\lambda) = A_\pm(\lambda) S^{-2}.$$

Obviously, $\zeta \in \rho_m(A_+) \cap \rho_m(A_-)$ if and only if $\zeta \in \rho_m(\tilde{A}_+) \cap \rho_m(\tilde{A}_-)$ (see also [Ma], Lemma 20.1). Hence the functions

$$\tilde{A}_\pm^{-1}(\lambda) = S^2 A_\pm^{-1}(\lambda), \qquad [\tilde{A}_\pm^{-1}(\bar{\lambda})]^* = A_\mp^{-1}(\lambda) S^2$$

are finitely meromorphic in the domain Ω. From this we find that the functions $SA_\pm^{-1}(\lambda)S$ are finitely meromorphic in Ω, i.e. $\Omega \in \rho_m(\hat{A}_+) \cap \rho_m(\hat{A}_-)$. □

Corollary 3.4. *Suppose that the operator G is T-compact and the set $\rho_m(A)$ is connected, i.e. $\rho_m(A)$ is a domain in $\mathbb{C}$. Then $\rho_m(\hat{A}) \supset \rho_m(A)$ and the eigenvalues of the pencils $A(\lambda)$ and $\hat{A}(\lambda)$ in $\rho_m(A)$ as well as the structures of the corresponding canonical systems coincide.*

Proof. Let us consider the pencils $A_\pm(\lambda)$ defined by (3.5). Since $0 \in \rho(A_\pm)$ (as T is boundedly invertible), G is a T-compact operator and $\rho_m(A_+)$ is a domain in $\mathbb{C}$, we have

$$\rho_m(A_+) = \rho_m(A_-) = \rho_m(A_0), \quad \text{where } A_0(\lambda) = \lambda^2 F + \lambda D + T.$$

(see [Ka], Ch.4, Theorems 5.26 and 5.31). It is easy to check that the set $\rho_m(A_0)$ is symmetric with respect to the real axis. Now apply proposition 3.3. $\square$

We shall associate a linear pencil with the quadratic pencil $A(\lambda)$. Let us consider the following linear pencil

$$(3.6) \qquad -\begin{pmatrix} D+iG & T \\ -J & 0 \end{pmatrix} - \lambda \begin{pmatrix} F & 0 \\ 0 & J \end{pmatrix} =: \mathbf{T} - \lambda \mathbf{W} =: \mathbf{A}(\lambda).$$

We can consider $\mathbf{T}$ as an operator acting in the space $H \times H$ with the domain $\mathcal{D}(\mathbf{T}) = H_2 \times H_2$ or as an operator acting in $H_{-1} \times H_{-1}$ with the domain $\mathcal{D}(\mathbf{T}) = H_1 \times H_1$. However, in both these spaces $\mathbf{T}$ is not a dissipative operator. The situation is changed if we define $\mathbf{T}$ as an operator acting in the space $\mathbf{H} = H \times H_1$ with the domain

$$(3.7) \qquad \mathcal{D}(\mathbf{T}) = \left\{ \mathbf{x} = \begin{pmatrix} x_1 \\ x_2 \end{pmatrix}, \; x_1, x_2 \in H_1, \; (D+iG)x_1 + Tx_2 \in H \right\}.$$

We observe that $H_2 \times H_2 \subset \mathcal{D}(\mathbf{T})$, therefore, T is densely defined in $\mathbf{H} = H \times H_1$.

Proposition 3.5. *The operator $-i\mathbf{T}$ with domain (3.7) is dissipative in the space $H \times H_1$.*

Proof. Let $\mathbf{x} = \begin{pmatrix} x_1 \\ x_2 \end{pmatrix} \in \mathcal{D}(\mathbf{T})$. Then $y_j := Sx_j \in H$, for $j = 1, 2$, and

$$\begin{aligned} i(\mathbf{Tx}, \mathbf{x})_{\mathbf{H}} &= -i((D+iG)x_1 + Tx_2, x_1) + i(SJx_1, Sx_2) \\ &= -i(\hat{D}y_1, y_1) + (\hat{G}y_1, y_1) - i(Jy_2, y_1) + i(y_1, Jy_2). \end{aligned}$$

Therefore, $Im\,(i\mathbf{Tx}, \mathbf{x})_{\mathbf{H}} = -(Dx_1, x_1) \leqslant 0$ for all $\mathbf{x} \in \mathcal{D}(\mathbf{T})$. $\square$

Proposition 3.6. *Let $\rho(\mathbf{A})$ be the resolvent set of the linear pencil $\mathbf{A}(\lambda)$ defined by (3.6). Then $\rho(\mathbf{A}) \supset \rho(\hat{A})$. Moreover, $\sigma_p(\mathbf{A}) = \sigma_p(\hat{A})$ and the Jordan structures of the root subspaces corresponding to each eigenvalue μ of $\mathbf{A}(\lambda)$ and $\hat{A}(\lambda)$ coincide.*

Proof. Let us solve the equation

$$\mathbf{A}(\lambda)\begin{pmatrix} x_1 \\ x_2 \end{pmatrix} = \begin{pmatrix} f_1 \\ f_2 \end{pmatrix}, \qquad \text{where } \begin{pmatrix} f_1 \\ f_2 \end{pmatrix} \in \mathbf{H}, \quad \begin{pmatrix} x_1 \\ x_2 \end{pmatrix} \in \mathcal{D}(\mathbf{T}). \tag{3.8}$$

After simple calculations we obtain

$$\begin{pmatrix} x_1 \\ x_2 \end{pmatrix} = -\begin{pmatrix} \lambda A^{-1}(\lambda) & \lambda - J - A^{-1}(\lambda)T \\ A^{-1}(\lambda) & \lambda - A^{-1}(\lambda)T \end{pmatrix} \begin{pmatrix} f_1 \\ f_2 \end{pmatrix}.$$

If $\lambda \in \rho(\hat{A})$ then $A^{-1}(\lambda) : H_{-1} \to H_1$ is a bijection, and so is $A^{-1}(\lambda)T : \ H_1 \to H_1$. Therefore $\begin{pmatrix} x_1 \\ x_2 \end{pmatrix} \in H_1 \times H_1$ if $\begin{pmatrix} f_1 \\ f_2 \end{pmatrix} \in \mathbf{H} = H \times H_1$. Moreover, it follows from (3.8) that

$$(D + iG)x_1 + Tx_2 = -f_1 \ - \lambda F x_1 \in H, \text{ i.e. } \begin{pmatrix} x_1 \\ x_2 \end{pmatrix} \in \mathcal{D}(\mathbf{T}).$$

Hence, the inclusion $\rho(\mathbf{A}) \supset \rho(\hat{A})$ is proved.

Suppose that system (3.2) forms a canonical system of Jordan chains corresponding to an eigenvalue μ of $A(\lambda)$ acting in the space H_{-1}. Then it is easily seen that the elements

$$\begin{pmatrix} \mu y_j^0 \\ y_j^0 \end{pmatrix}, \begin{pmatrix} \mu y_j^0 + y_j^0 \\ y_j^0 \end{pmatrix}, \dots, \begin{pmatrix} \mu y_j^{p_j} + y_j^{p_j - 1} \\ y_j^{p_j} \end{pmatrix} \tag{3.9}$$

belong to $\mathcal{D}(\mathbf{T})$ and form Jordan chains of $\mathbf{A}(\lambda)$ corresponding to $\mu \in \sigma_p(\mathbf{A})$. The converse assertion can also be easily verified, namely: all Jordan chains of $\mathbf{A}(\lambda)$ have representation (3.9), and if (3.9) is a canonical system of $\mathbf{A}(\lambda)$ then (3.2) is a canonical system of $A(\lambda)$ in H_{-1}. □

Let μ be a normal pure imaginary eigavalue of the pencil $A(\lambda)$ in the space H_{-1}. Then according to Proposition 3.6 $\zeta = i\mu$ is a normal real eigenvalue of the linear dissipative pencil $i\mathbf{A}(-i\zeta)$. Using the results of section 2 we can choose a regular canonical system (3.9) of the pencil $i\mathbf{A}(-i\zeta)$ corresponding to the eigenvalue $\zeta = i\mu$. A canonical system (3.2) is said to be regular if the corresponding system (3.9) is regular. Further *we define the sign characteristics ε_j of regular system (3.2) to be equal to those of the corresponding system (3.9).* As in section 2 we define the numbers $\varepsilon_j^+ = \max(0, \varepsilon_j)$ for all $\mu_j \in \sigma_d(\mathbf{A})$ and assume $\varepsilon_j^+ = 0$ for all $\mu_j \in \sigma_p(\mathbf{A}) \setminus \sigma_d(\mathbf{A})$. Recall also that $\nu(F) = rank\, F_-$ and $\nu(T) = rank\, T_-$ are equal to the numbers of negative eigenvalues counting with multiplicities of the operators F and T, respectively.

Theorem 3.7. *Let the numbers $\nu(F)$ and $\nu(T)$ be finite. Then the generalized spectrum of $A(\lambda)$ in the open right-half plane $\mathbb{C}_r$ consists of only normal eigenvalues, and hence coincides with the spectrum of the linear pencil $\mathbf{A}(\lambda)$. If $\kappa(\hat{A})$ is the total algebraic multiplicity of all the eigenvalues lying in $\mathbb{C}_r$ then*

$$\kappa(\hat{A}) + \sum_{\mu_k \in i\mathbb{R}\cap\sigma_p} (\varepsilon_k^+ + [(p_k - 1)/2]) \leqslant \nu(T) + \nu(F). \tag{3.10}$$

Here the numbers ε_k are defined as above and $p_k + 1$ are the lengths of Jordan chains of regular canonical systems corresponding to the pure imaginary eigenvalues μ_k. If $A(\lambda)$ (and hence $\mathbf{A}(\lambda)$) has only normal eigenvalues on the imaginary axis then equality holds in (3.10).

Proof. It was already noticed that the spectrum of $A(\lambda)$ in the space H_{-1} coincides with the spectrum of the pencil

$$\hat{A}(\lambda) = \lambda^2 \hat{F} + \lambda(\hat{D} + i\hat{G}) + J$$

in the space H, where $\hat{F}, \hat{D}, \hat{G}$ are bounded operators in H, while $J = I - 2J_-$ is a finite rank perturbation of the identity operator. Let

$$Q(\lambda) = \lambda|\hat{F}| + (\hat{D} + i\hat{G}) + \lambda^{-1} I, \qquad \lambda = \eta + i\tau.$$

Then

$$Re\,(Q(\lambda)x,\, x) = \eta(|\hat{F}|x,\, x) + (\hat{D}x,\, x) + \frac{\eta}{\eta^2 + \tau^2}(x,\, x) > \frac{\eta}{\eta^2 + \tau^2}(x,\, x)$$

for all $\eta = Re\,\lambda > 0$. Therefore, $Q(\lambda)$ is boundedly invertible in the open right-half plane (see, for example, [Ma], Theorem 26.2). It follows from our assumptions that $\lambda^{-1}\hat{A}(\lambda)$ is a finite rank pertubation of $Q(\lambda)$. Hence, by virtue of the theorem on holomorphic operator function (see [GS], for example) the spectrum of $\hat{A}(\lambda)$ in the open right-half plane $\mathbb{C}_r$ consists only of normal eigenvalues. According to Proposition 3.6 the linear pencil $\mathbf{A}(\lambda) = \mathbf{T} - \lambda\mathbf{W}$ has the same spectrum in $\mathbb{C}_r$. Now, notice that $\nu(\mathbf{W}) = \nu(T) + \nu(F)$ and apply formula (2.19) with respect to the pencil $i\mathbf{T} - \zeta\mathbf{W}$, $\zeta = i\lambda$. □

Corollary 3.8. *Let the numbers $\nu(F)$ and $\nu(T)$ be finite and let the operators F and G be T-compact. Then the generalized spectrum of $A(\lambda)$ consists of normal eigenvalues with the possible exeption of a closed subset lying on the negative semiaxis and the following formula is valid*

$$\kappa(\hat{A}) + \sum_{\mu_k \in i\mathbb{R}\cap\sigma_d} \varepsilon_k^+ + [(p_k - 1)/2]) = \nu(T) + \nu(F). \tag{3.11}$$

If in addition the classical spectrum of the linear pencil $\lambda D + T$ in the closed right-half plane consists of only normal eigenvalues then $\kappa(\hat{A})$ can be replaced by $\kappa(A)$.

Proof. Let us consider the linear pencil $\lambda\hat{D} + I$ in the space H. Obviously, its nondiscrete spectrum is a closed subset, say Δ, belonging to $\mathbb{R}_-$ ($\Delta = \varnothing$ if $\hat{D}$ is a compact operator). It follows from the interpolation theorem (see Proposition 3.1) that $\hat{F}$ and $\hat{G}$ are compact operators in H if F and G are T-compact. Hence the pencil $\hat{A}(\lambda)$ is a compact perturbation of the pencil $\lambda\hat{D} + I$. Now apply the theorem on holomorphic operator functions and Theorem 3.7.

As the operators F and G are T-compact, the complement to the nondiscrete spectrum of the linear pencil $\lambda D + T$ in H coincides with the set $\rho_m(A)$ (this follows as before from the theorem on holomorphic operator functions). Hence we may apply Proposition 3.2 and replace $\kappa(\hat{A})$ by $\kappa(A)$. □

Suppose that $A(\lambda)$ has only semisimple eigenvalues on the imaginary axis. Then the sign characteristics ε_k are defined as follows ($\zeta_k = i\mu_k$)

$$\begin{aligned}\varepsilon_k = &\left(W\begin{pmatrix}\zeta_k y_k\\ y_k\end{pmatrix}, \begin{pmatrix}\zeta_k y_k\\ y_k\end{pmatrix}\right)_{\mathbf{H}} = \\ &= -\mu_k^2(Fy_k, y_k) + (Ty_k, y_k) = (A(\mu_k)y_k, y_k)\\ &- \mu_k(A'(i\mu_k)y_k, y_k) = -\mu_k(A'(i\mu_k)y_k, y_k).\end{aligned}$$

An eigenvalue $i\mu \in i\mathbb{R}$ is said to be of the first (the second) type if $\varepsilon = -\mu(A'(i\mu)y, y) > 0 (< 0)$ for all $y \in Ker A(\mu)$. It is well-known fact that all eigenvalues of definite type are semisimple.

Corollary 3.9. *If assumptions of Theorem 3.7 are fulfilled and all the pure imaginary eigenvalues of $A(\lambda)$ are of definite type then*

$$\kappa(\hat{A}) = \nu(T) + \nu(F) - \varepsilon^+(\hat{A}). \tag{3.12}$$

where $\varepsilon^+(A)$ is the number of the first type eigenvalues counting with multiplicities belonging to the imaginary axis. In particular, if $D > 0$ then

$$\kappa(\hat{A}) = \nu(T) + \nu(F). \tag{3.13}$$

Proof. We have only to notice that $\hat{A}(\lambda)$ has no pure imaginary eigenvalues if the condition $D > 0$ is fulfilled. □

Let us consider the operator differential equation

$$A\left(\frac{du}{dt}\right) = F\frac{d^2u}{dt^2} + (D + iG)\frac{du}{dt} + Tu = 0, \qquad u = u(t). \tag{3.14}$$

If (3.2) are Jordan chains of $A(\lambda)$ corresponding to an eigenvalue μ_k then the functions

$$u_k^h(t) = e^{\mu_k t}\left(y_k^h + \frac{t}{1!}y_k^{h-1} + \ldots + \frac{t^h}{h!}y_k^0\right), \quad h = 0, \ldots, p_k,$$

are called elementary solution of (3.14). Under the assumptions of Corollary 3.9 the number $\kappa(\hat{A})$ coincides with the number of linearly independent elementary solutions of equation (3.14) which are not bounded when $t \to \infty$. Hence the number $\kappa(\hat{A})$ characterizes the index of instability of equation (3.14). Strictly speaking, the index of instability $\kappa'(\hat{A})$ has to be defined as the number of linearly independent generalized solutions of (3.14) which are not bounded when $t \to \infty$. Generally, $\kappa'(\hat{A}) \geqslant \kappa(\hat{A})$. We know abstract examples when $\kappa'(\hat{A}) > \kappa(\hat{A})$ even if there are no pure imaginary eigenvalues (see [M1], for example). We can show that $\kappa'(\hat{A}) = \kappa(\hat{A})$ if in addition to the assumptions of Theorem 3.7 the whole spectrum of $\hat{A}(\lambda)$ is discrete. However, the rigorous proof of this fact (and more general ones) requires additional preparations and is left for a future occasion. With these reservations we may consider (3.12) as the instability index formula. The relation (3.11) may be considered as the generalized instability index formula.

In the end of this section we would like to make some historical remarks concerning formula (3.11). Apparently, the first investigation of the pencil (3.1) with matrix coefficients was carried out by Kelvin and Tait [KT]. They considered the case $F = I$ and made the following interesting observations.

1. *If matrix T is positive then the problem is stable for all gyroscopic matrices G and all $D \geqslant 0$.*

2. *The condition $T > 0$ is not necessary for stability. Even if $T < 0$ the motion can be stabilized by gyroscopic forces (an example was given). However, if rank $T_- = \kappa$ is odd then the problem can not be stabilized by the action of gyroscopic forces.*

3. *If rank $T_- = \kappa > 0$ and $D > 0$ (complete dissipative forces) then the problem can not be stable for any gyroscopic forces G.*

All these observations were rigorously proved by Chetaev [Ch] by introducing the Lyapunov function. The next step was made by Zajac [Z]. He considered the matrix pencil

(3.1) with $F = I$, $D > 0$ and proved the formula $\kappa(A) = \nu(T)$. Wimmer [W] and later Lancaster and Tismenetsky [LT] studied matrix pencil (3.1) and admitted an indefinite leading coefficient F. In particular, they proved the relations

$$\begin{aligned} &\kappa(A) \leqslant \nu(F) + \nu(T) \qquad \text{if } D \geqslant 0, \\ &\kappa(A) = \nu(F) + \nu(T) \qquad \text{if } D > 0. \end{aligned} \tag{3.15}$$

The second relation in (3.15) for the case $D = 0$ has been more closely investigated recently by Barkwell, Lancaster, and Markus [BLM]. The pencils of the form (3.1) with unbounded operator coefficients were studied by Miloslavsky, Pivovarchik et al. in the papers [M1], [M2],[ZKM], [P3], [P4]. The main aim of these papers was to obtain relations (3.15) under the assumptions $F = I$ and various hypotheses on the operator coefficients D, G, T, which were essentially stronger than our assumptions *i) -iii)*. As far as we know, formulas (3.11), (3.12) presented in this paper are new even for matrix pencils.

4. Applications

In this section we shall apply the obtained abstract results to concrete problems considered in Section 1.

Theorem 4.1. *Formula (3.11) or its simplifications (3.12) or (3.13) are valid for operator pencil (1.7) associated with the problem of small oscillations of ideal incompressible fluid in a pipe of finite length if the condition $KerT = \{0\}$ is fulfilled ($T := A + C$). For a pipe of infinite length the assertion of Theorem 3.7 is valid if $g(x)$ is such a function that $KerT = \{0\}$ and $\nu(T) < \infty$.*

Proof. The conditions *i)–ii)* and *iv)* of Section 1 imply conditions *i)–iii)* of Section 3 if it is assumed in addition that $KerT = \{0\}$. Moreover, for a pipe of finite length the conditions *iii)* and *v)* of Section 1 hold. Hence, for a pipe of finite length the assumptions of Corollary 3.8 are fulfilled. For a pipe of infinite length the operators G and I are not T-compact and we must use Theorem 3.7. In the last case we can not guarantee the absence of pure imaginary eigenvalues belonging to the nondiscrete spectrum. □

If $KerT \neq \{0\}$ then $\lambda = 0$ is an eigenvalue of pencil (3.1). In this case the analogue of formula (3.11) can also be obtained. For this purpose one has to modify the results of Section 2 for the case $KerW \neq \{0\}$. Technically this is not a trivial work. However, the estimates for the number $\kappa(\hat{A})$ can be obtained easily if $KerT \neq \{0\}$.

Theorem 4.2. *Suppose that a pencil $A(\lambda)$ is defined by (3.1) and its operator coefficients satisfy the assumptions i)-iii) of Section 3 with the possible exception that the operators F and T are not necessarily boundedly invertible. Suppose that there exists a point μ, $Re\,\mu > 0$ such that $\hat{A}(\mu)$ is boundedly invertible. Then*

$$\kappa(\hat{A}) \leqslant \nu(F) + \nu(T). \tag{4.1}$$

Proof. Let us consider the pencil

$$A_\tau(\lambda) = \lambda^2(F + \tau I) + (D + iG)\lambda + T + \tau I, \qquad \tau > 0.$$

Obviously, $\nu(T + \tau I) = \nu(T)$, $\nu(F + \tau I) = \nu(F)$, if $\tau \in (0, \tau_0$ and τ_0 is sufficiently small. By virtue of Theorem 3.7 we have

$$\kappa(\hat{A}_\tau) \leqslant \nu(F) + \nu(T) \qquad \text{for all} \quad 0 < \tau < \tau_0. \tag{4.2}$$

Repeating the arguments from the proof of theorem 3.7 and .taking into account that $\mu \in \rho(\hat{A})$ for some μ with $Re\,\mu > 0$ we obtain that the spectrum of $\hat{A}(\lambda) := \hat{A}_0(\lambda)$ in the open right half plane consists only of normal eigenvalues. These eigenvalues continuously depend on τ (see [Ka], Ch. 7). Then (4.2) implies (4.1). □

The results of Sections 2 and 3 can also be applied to selfadjoint pencils. Lancaster and Shkalikov [LS] considered an operator pencil $L(\lambda)$ defined by (1.5) with $C = 0, D_\alpha = \alpha A$ and obtained the following estimate

$$\eta/2 \leqslant \min_{k\in\mathbb{R}} \pi(L(k)), \qquad \pi(L) := \nu(-L), \tag{4.3}$$

where η is the number of nonreal eigenvalues of the pencil $L(\lambda)$ counting with algebraic multiplicities. Using an analytic approach Shkalikov and Griniv proved a sharper estimate for the case $C = 0$ and reproved (4.3) for $C \neq 0$ (if C is an A-compact operator). Here we refine the corresponding results from [LS] and [SG].

Theorem 4.3. *Let*

$$L(\lambda) = \lambda^2 F + \lambda D + T,$$

where $T = T^*$ and F, D are symmetric and T-bounded operators. Let $S^2 = |T| + I$ and the scale of Hilbert spaces H_θ be generated by the operator $S \gg 0$. Suppose that there exist real points a and b belonging to $\rho(\hat{L})$ such that

$$\pi(L(a)) < \infty, \qquad \nu(L(b)) < \infty.$$

Then the nonreal spectrum of $L(\lambda)$ *in the space* H_{-1} *consists of finitely many, say* η, *nonreal eigenvalues, and the following estimate is valid*

$$\eta/2 \leqslant \pi(L(a)) + \nu(L(b)) - \delta^{+}(L), \tag{4.4}$$

where $\delta^{+}(L)$ *is the number of real eigenvalues* μ_k *of* $L(\lambda)$ *counting with multiplicities such that*

$$(b-a)\left(\frac{\mu_k - a}{b - \mu_k}\right)(L'(\mu_k)y,\, y) > 0 \qquad \text{for all} \quad y \in Ker L(\mu_k).$$

Proof. We use the same idea as in [LS] where estimate (4.4) was obtained in a slightly different situation not taking into account the number $\delta^{+}(L)$. It was shown in Section 3 that the spectrum of $L(\lambda)$ in the space H_{-1} coincides with the spectrum of $\hat{L}(\lambda) = S^{-1}L(\lambda)S^{-1}$ in the space H. The pencil $\hat{L}(\lambda)$ has the bounded operator coefficients $\hat{F}, \hat{D}, \hat{T}$. After the substitution $\lambda = (b\xi + a)(\xi + 1)^{-1}$ we obtain the quadratic pencil

$$\tilde{L}(\xi) := (\xi+1)^2 \hat{L}(\lambda(\xi)) = \xi^2 \tilde{F} + \xi \tilde{D} + \tilde{T}, \qquad \tilde{F} = \hat{L}(b), \quad \tilde{T} = \hat{L}(a).$$

Let us consider the linearization of $\tilde{L}(\xi)$

$$\mathbf{L}(\xi) = -\begin{pmatrix} \tilde{D} & \tilde{T} \\ \tilde{T} & 0 \end{pmatrix} - \xi \begin{pmatrix} \tilde{F} & 0 \\ 0 & -\tilde{T} \end{pmatrix}.$$

Suppose that ξ_k is a simple (or semisimple) real eigenvalue of $\tilde{L}(\xi)$ with a corresponding eigenvector y_k. Then the sign characteristic ε_k (see section 2) is defined as follows

$$\begin{aligned}\epsilon_k &= \left(\begin{pmatrix} \tilde{F} & 0 \\ 0 & -\tilde{T} \end{pmatrix} \begin{pmatrix} \xi_k y_k \\ y_k \end{pmatrix}, \begin{pmatrix} \xi_k y_k \\ y_k \end{pmatrix} \right)_{H \times H} = \xi(\tilde{L}'(\xi_k) y_k,\, y_k) \\ &= \xi_k \lambda'(\xi_k)(\hat{L}'(\lambda_k) y_k,\, y_k)) = (b-a)(\mu_k - a)(b-\mu_k)^{-1}(\hat{L}'(\mu_k) y_k,\, y_k).\end{aligned}$$

Now apply Corollary 2.11. □

We note that the estimate (4.4) is also new for matrix pencils.

Acknowledgement. The author expresses his gratitude to P.Lancaster and R.O.Griniv who looked through the manuscript, marked a number of misprints and made some useful remarks.

REFERENCES

[A] T.Ja. Azizov, *Dissipative operators in Hilbert space with indefinite metric*, Izv. Acad. Nauk SSSR Ser. Mat. **37** (1973), no. 3 (Russian); English trans. in Math USSR Izv. **7** (1973).

[AI] T. Ja. Azizov and I.S. Iohvidov, *Linear operators in spaces with indefinite metric*, John Wiley, Chichester, 1989.

[BLM] L.Barkwell, P.Lancaster, and A.S.Markus, *Gyroscopically stabilized systems: a class of quadratic eigenvalue problems with real spectrum*, Canadian J.Math. **44** (1992), 42-53.

[Ch] N.G. Chetaev, *The stability of motion*, Pergamon Press, 1961.

[G] A.M. Gomilko, *Invariant subspaces of J-dissipative operators*, J.Funct. Anal. and Appl. **19** (1985), no. 3, 213-214.

[GLR] I. Gohberg, P. Lancaster and L. Rodman, *Matrices and indefinite scalar product,* Operator theory: Advances and Applications, Vol. 8, Birkhäuser Verlag, Basel–Boston–Stuttgart, 1983.

[GS] I. Gohberg and E.Sigal, *An Operator Generalization of the Logarithmic Residue Theorem and the Theorem of Rouché*, Mat. Sbornik **84** (1971); English transl. in Math. USSR Sbornik **13** (1971), 603–625.

[Gr] R.O.Griniv, *On operator pencils arising in the problem of semiinfinite beam oscillations with internal damping*, Moscow Univ. Math. Bulletin (to appear).

[Ka] T. Kato, *Perturbation theory for linear operators (2-nd edition)*, Springer-Verlag, New York, 1976.

[Ke] M.V. Keldysh, *On the completeness of eigenfunctions of certain classes of nonselfadjoint linear operators.*, Russian Math. Surveys **26** (1971), no. 4, 295–305.

[KO] A.G. Kostyuchenko amd M.B. Orazov, *On certain properties of the roots of a selfadjoint quadratic pensil*, J. Funct. Anal Appl. **9** (1975), 28–40.

[KS] A.G. Kostyuchenko and A.A. Shkalikov, *Selfadjoint quadratic operator pencils and elliptic problems*, J.Funct. Anal. and Appl. **17** (1983), 109–128.

[KL] M.G.Krein and H.Langer, *On Definite Subspaces and Generalized Resolvents of Hermitian Operators in Spaces* Π_κ, Funkz. Anal. i Prilozh. vol 5 (1971), no. 2, 59–71; vol 5 (1971), no. 3, 54–69 (Russian); English transl in Funct. Anal. and Appl. **5** (1971).

[KT] W. Tompson (Lord Kelvin) and P. Tait, *Treatise on Natural Philosophy, Part 1*, Cambrige Univ. Press, 1869.

[LM] J.L. Lions and E.Magenes, *Problems aux Limites Nonhomogenes et Applications. Vol.1*, Dunod, Paris, 1968; English transl. in Springer Verlag, 1972.

[LS] P. Lancaster and A.A. Shkalikov, *Damped vibrations of beams and related spectral problems*, Can. Appl. Math. Quart. **2** (1994), no. 1, 45–90.

[LT] P. Lancaster and M. Tismenetsky, *Inertia characteristics of selfadjoint matrix polynomials*, Lin. Algebra and Appl. **52/53** (1983), 479–496.

[Ma] A.S. Markus, *Introduction to the Spectral Theory of Polynomial Operator Pencils*, Amer. Math. Soc., Providence, 1988.

[M1] A.I. Miloslavskii, *Foundation of the spectral approach in nonconservative problems of the theory of elastic stability*, J. Funct. Anal. Appl. **17** (1983), no. 3, 233–235.

[M2] ———, *On stability of some classes of evolutionary equations*, Siberian Math. J. **26** (1985), no. 5, 723–735.

[P] L.S. Pontrjagin, *Hermitian operators in spaces with indefinite metric*, Izv. Acad. Nauk SSSR Ser. Mat. **8** (1944), 243–280. (Russian)

[PI] M.P. Paidoussis and N.T. Issid, *Dynamic stability of pipes conveying fluid*, J. Sound Vibration **33** (1974), 267–294.

[P1] V.N. Pivovarchik, *A boundary value problem connected with the oscillation of elastic beams with internal and viscous damping*, Moscow Univ. Math. Bulletin **42** (1987), 68–71.

[P2] ———, *On oscillations of a semiinfinite beam with internal and external damping*, Prikladnaya Mathem. and Mech. **52** (1988), no. 5, 829-836 (Russian); English transl. in J. Appl. Math. and Mech. (1989).

[P3] ———, *On the spectrum of quadratic operator pencils in the right half plane*, Matem. Zametki **45** (1989), no. 6, 101–103 (Russian); English transl. in Math. Notes **45** (1989).

[P4] ———, *On the total algebraic multiplicity of spectrum in the right half plane for one class of quadratic operator pencils*, Algebra and Analysis **3** (1991), no. 2, 223–230.

[RT] A.C.M. Ran and D. Temme, *Dissipative matrices and invariant maximal semidefinite subspaces*, Linear Algebra Appl. (to appear).

[S1] A.A. Shkalikov, *Selection principles and properties of some parts of eigen and associated elements of operator pencils*, Moscow Univ. Math. Bulletin **43** (1988), no. 4, 16–25.

[S2] ———, *Operator pencils and operator equations in Hilbert space*, (Unpublished manuscript, University of Calgary), 1992.

[S3] ———, *Elliptic equations in Hilbert space and associated spectral problems*, J. Soviet Math. **51** (1990), no. 4, 2399–2467.

[SG] A.A. Shkalikov and R.O. Griniv, *On operator pencils arising in the problem of beam oscillation with internal damping*, Matem. Zametkii **56** (1994), no. 2, 114–131 (Russian); English transl. in Math. Notes **56** (1994).

[W] H.K. Wimmer, *Inertia theorems for matricies, controllability and linear vibrations*, Linear Algebra Appl. (1974), no. 8, 337–343.

[Z] E.E. Zajac, *The Kelvin–Tait–Chetaev theorem and extentions*, J. Aeronaut.. Sci. vol 11 (1964), no. 2, 46–49.

[ZKM] V.N. Zefirov, V.V. Kolesov and A.I. Miloslavskii, *On eigenfrequences of a strightline pipe*, Izv. Acad. Nauk SSSR, Ser. Mech. Tverdogo Tela (1985), no. 1, 179–188 (Russian); English transl. in Math. USSR Izv. Ser. Mech (1985).

Moscow Lomonosov State University
Department of Mechanics and Mathematics
Moscow,199899, Russia

AMS Subject Classifications. 47A70, 70J20.

Operator Theory:
Advances and Applications, Vol. 87

TOEPLITZ-LIKE OPERATORS AND THEIR FINITE SECTIONS

Bernd Silbermann

This article is aimed at the study of norm-stability of finite sections for Toeplitz-like operators with piecewise continuous operator-valued coefficients. Further asymptotic problems are discussed.

1 INTRODUCTION

Let $\mathcal{H}$ be a separable Hilbert space and let $l^2(\mathcal{H})$ stand for the Hilbert space of all sequences $f = (f_n)_{n\in\mathbb{Z}}$ with values $f_n \in \mathcal{H}$ for which

$$||f||^2 := \sum_{n\in\mathbb{Z}} ||f_n||^2_{\mathcal{H}} < \infty.$$

A function a defined on the complex unit circle $\mathbb{T}$ and taking on values in $\mathcal{L}(\mathcal{H})$, the C^*-algebra of all bounded linear operators on $\mathcal{H}$, is said to belong to $L^\infty(\mathcal{L}(\mathcal{H}))$ if it is weakly measurable and

$$||a||_\infty := \operatorname{ess\,sup}_{t\in\mathbb{T}} ||a(t)||_{\mathcal{L}(\mathcal{H})} < \infty.$$

Each function $a \in L^\infty(\mathcal{L}(\mathcal{H}))$ induces a multiplication operator $M(a)$ on $l^2(\mathcal{H})$: if we denote by $(a_n)_{n\in\mathbb{Z}}$ the sequence of Fourier coefficients of a,

$$a_n := \frac{1}{2\pi} \int_{-\pi}^{\pi} a(e^{i\theta}) e^{-in\theta} d\theta,$$

then $M(a)$ is a bounded operator on $l^2(\mathcal{H})$ given by the infinite matrix $(a_{j-k})_{j,k\in\mathbb{Z}}$. Further, define operators P, $P_n (n \in \mathbb{Z}_+)$ acting on $l^2(\mathcal{H})$ by

$$\begin{aligned} P(f_n)_{n\in\mathbb{Z}} &= (\cdots, 0, 0, f_0, f_1, \cdots), \\ P_n(f_n)_{n\in\mathbb{Z}} &= (\cdots, 0, f_{-n}, f_{-n+1}, \cdots, f_{n-2}, f_{n-1}, 0, \cdots). \end{aligned}$$

Let $\mathcal{F}$ denote the collection of all sequences $(A_n)_{n\in\mathbb{Z}_+}$ of linear bounded operators on $l^2(\mathcal{H})$ satisfying

$$(1.1) \qquad ||(A_n)|| := \sup ||A_n|| < \infty.$$

Provided with elementwise operations and the norm (1.1), $\mathcal{F}$ becomes a Banach algebra, which is actually a C^*-algebra if we indroduce the involution by $(A_n)^* := (A_n^*)$. Let $\mathcal{N}$ denote the subset of the algebra $\mathcal{F}$ consisting of all sequences (A_n) with $||A_n|| \to 0$. Evidently $\mathcal{N}$ is a closed two-sided $*$-ideal of the Algebra $\mathcal{F}$. We are mainly interested in the norm-stability of sequences $(A_n) \in \mathcal{F}$. Recall that a sequence $(A_n) \in \mathcal{F}$ is called norm-stable if the operators A_n are invertible for all $n \geq n_0$ and if $\sup\limits_{n\geq n_0} ||A_n^{-1}|| < \infty$. It is not hard to see, that a sequence $(A_n) \in \mathcal{F}$ is norm-stable if and only if the coset $(A_n) + \mathcal{N}$ is invertible in the quotient algebra $\mathcal{F}/\mathcal{N}$. Notice that norm-stability is one of the corner stones in numerical analysis: If the sequence (A_n) is norm-stable and converges strongly to some invertible operator A, $A = s - \lim A_n$, then $A^{-1} = s - \lim A_n^{-1}$. This also can be expressed saying that A is asymptotically invertible by means of the sequence (A_n). Moreover there is known a variety of problems in analysis closely related to norm-stability of some operator sequences. However the algebra $\mathcal{F}$ is essentially to large in order to decide whether a given sequence $(A_n) \in \mathcal{F}$ is norm-stable. On the other hand the algebra $\mathcal{F}$ contains many interesting subalgebras for which there is some hope to answer the above quotet question. We will restrict ourselves to the following important case: Let $\mathcal{A}$ be the smallest closed subalgebra of $\mathcal{F}$ containing the constant sequences (P) and $(M(a))$ for all piecewise continuous functions $PC(\mathbb{C} + \mathcal{K}(\mathcal{H}))$ on $\mathbb{T}$ with values in $\{\lambda I + \mathcal{K}(\mathcal{H})\} \subset \mathcal{L}(\mathcal{H})$ ($\lambda \in \mathbb{C}$, I being the identity operator and $\mathcal{K}(\mathcal{H})$ is the collection of all compact operators) and the sequences $(P_{[rn]})$ for all positive real numbers r (as usual, $[x]$ denotes the greatest integer $\leq x$). Since $P = P^*$ and $P_n = P_n^*$ and $M(a)^* = M(a^*) \quad (a^*(t) := a(t)^*)$, $\mathcal{A}$ is even a C^*-subalgebra of $\mathcal{F}$.

Since $\mathcal{A} + \mathcal{N}$ is a C^*-subalgebra of $\mathcal{F}$ and since $(\mathcal{A} + \mathcal{N})/\mathcal{N}$ can be viewed as a subalgebra of $\mathcal{F}/\mathcal{N}$, a sequence $(A_n) \in \mathcal{A}$ is norm-stable if and only if the coset $(A_n) + \mathcal{N}$ is invertible in $\mathcal{A} + \mathcal{N}/\mathcal{N}$. (Note that the problem of inverse closedness does not appear for C^*-algebras.) But, abbreviating the ideal $\mathcal{A} \cap \mathcal{N}$ with G, the C^*-algebras $\mathcal{A} + \mathcal{N}/\mathcal{N}$ and $\mathcal{A}/G$ are isometrically isomorphic and, thus, norm-stability of a sequence $(A_n) \in \mathcal{A}$ is equivalent to the invertibility of $(A_n) + G$ in $\mathcal{A}/G$. This is at least the reason why we will give a complete description of the algebra $\mathcal{A}/G$ in terms of an algebra of operator-valued functions. This description then yields nessesary and sufficient conditions for a sequence

$(A_n) \in \mathcal{A}$ being norm-stable. Further important consequences will be discussed.

The algebra $\mathcal{A}$ contains for instance the following sequences:

1. $(P_nPM(a)PP_n + \hat{Q}_n)$ with $\hat{Q}_n = I - PP_n$.

The norm-stability of this sequence is completely equivalent to the norm-stability of sequence

$$(R_nT(a)R_n),$$

where $T(a)$ is the Toeplitz operator acting on

$$l^2_+(\mathcal{H}) := \{(f_n)_{n\in\mathbb{Z}_+} : f_n \in \mathcal{H} \text{ and } \sum_{n\in\mathbb{Z}_+} \|f_n\|^2 < \infty\}$$

by the infinite matrix $(a_{i-j})^{\infty}_{i,j\in\mathbb{Z}_+}$ and R_n is given by $R_n(f_k) = (f_0, \cdots, f_{n-1}, 0, 0, \cdots)$. The matrices $T_n(a) := (a_{i-j})^{n-1}_{i,j=0}$ are called finite sections and their norm-stability means the following: The sequence $(T_n(a))$ is norm-stable if $T_n(a) : \text{im } R_n \to \text{im } R_n$ is invertible for all $n \geq n_0$ and $\sup_{n\geq n_0} \|T_n(a)^{-1}R_n\| < \infty$.

2. $(P_n(M(a)P + M(b)Q)P_n + Q_n)$ with $Q_n = I - P_n$.

The operator $M(a)P + M(b)Q$ is a singular integral operator (in its spectral description). Again the norm-stability of $(P_n(M(a)P + M(b)Q)P_n + Q_n)$ is completely aquivalent to the norm-stability of the finite sections, that means of the sequence $(P_n(M(a)P + M(b)Q)P_n)$ in the sense of Example 1.

The norm-stability of finite sections of Toeplitz and singular integral operators has been the subject of many investigations. The first deep result was obtained by Baxter [Ba] who proved the norm-stability of the finite sections of invertible Toeplitz operators generated by elements of the Wiener algeba ($\mathcal{H} = \mathbb{C}$). This paper was the starting point of numerous investigations by many authors. The development of the following years culminated with Gohberg and Feldman's book [G/F], in which a first systematic and comprehensive theory of projection methods for convolution equations was given. Some further comments and historical remarks to investigations carried out by Krupnik and Verbitski [K/V], Ambartsumyan [A], Widom [W], Kosak [K] and Verbitski [V] in the 70-ties can be found in [Bö/S1] or in [Pr/S]. Beginning with the papers [S1] and [S2] Banach algebra techniques have been widely used in the theory of finite sections and other approximation processes. For finite sections of single Toeplitz operators the most general results were likely obtained in [S3]. The next step was then the study of algebras the elements of which are operator sequences. One has to mention first of all the papers [Bö/S2], [S3], [Ro/S1] and the monograph [Bö/S1]. The investigations of Rathsfeld [Ra] are in some sense intermediate: He considered single sequences, but some baby versions of studying algebras are implicitely touched.

The investigations of this paper are strongly based on the recent paper of Roch/ Silbermann [R/S] where the problem of the norm-stability was completely studied for sequences of the above defined algebra $\mathcal{A}$, however in the special case $\mathcal{H} = \mathbb{C}$ (or $\mathbb{C}^N$). That paper has actually a forerunner, namely the article of Roch [R] in which a sketch of the proof (for norm-stability) was already given.

2 THE MAIN RESULTS

Let $L^2_{\mathbb{R}}(\mathcal{H})$ be the Hilbert space of all weakly measurable square integrable functions on $\mathbb{R}$ with values in the separable Hilbert space $\mathcal{H}$. Then the singular integral operator $S_{\mathbb{R}}$, $\rho \mapsto \frac{1}{\pi i}\int\limits_{\mathbb{T}} \frac{\rho(t)}{t-\tau}dt$ is bounded on $L^2_{\mathbb{R}}(\mathcal{H})$ and

$$P_{\mathbb{R}} = \frac{1}{2}(\mathrm{I} + S_{\mathbb{R}}), \quad Q_{\mathbb{R}} = \frac{1}{2}(\mathrm{I} - S_{\mathbb{R}})$$

are orthogonal projections (see [St,§7]).

We shall construct two families of unital algebra homomorphisms, W^s with s running through the one-point compactification $\dot{\mathbb{R}}$ of te real numbers $\mathbb{R}$ and W_t with t running through the unit circle $\mathbb{T}$, acting from $\mathcal{A}$ into certain operator algebras and having the property that the sequence $(A_n) \in \mathcal{A}$ is norm-stable if and only if the operators $W^s((A_n)_{n\in\mathbb{Z}^+})$ and $W_t((A_n)_{n\in\mathbb{Z}^+})$ are invertible for all $s \in \dot{\mathbb{R}}$ and $t \in \mathbb{T}$.

Let π refer to the canonical homomorphism from $\mathcal{L}(l^2(\mathcal{H}))$, the algebra of all bounded linear operators acting on $l^2(\mathcal{H})$ onto the Calkin algebra $\mathcal{L}(l^2(\mathcal{H}))/\mathcal{K}(l^2(\mathcal{H}))$, where $\mathcal{K}(l^2(\mathcal{H}))$ stands for the ideal of all compact operators acting on $l^2(\mathcal{H})$.

PROPOSITION 2.1

(a) For each sequence $(A_n) \in \mathcal{A}$ and each number $s \in \mathbb{R}$ there is a unital algebra homomorphism $W^s : \mathcal{A} \to \mathcal{L}(l^2(\mathcal{H}))$ such that

$$W^s((M(a))) = M(a),$$

$$W^s((P)) = \begin{cases} I, & \text{if } s > 0 \\ P, & \text{if } s = 0 \\ 0, & \text{if } s < 0, \end{cases}$$

and

$$W^s((P_{[rn]})) = \begin{cases} I\,, & \text{if } |s| < r \\ P\,, & \text{if } s = -r \\ I - P\,, & \text{if } s = r \\ 0\,, & \text{if } |s| > r. \end{cases}$$

(b) There is a further unital algebra homomorphism

$$W^\infty : \mathcal{A} \to \mathcal{L}(l^2(\mathcal{H}))/\mathcal{K}(l^2(\mathcal{H}))$$

such that

$$\begin{aligned} W^\infty((M(a))) &= \pi(M(a)), \\ W^\infty((P)) &= \pi(P), \\ W^\infty((P_{[rn]})) &= 0. \end{aligned}$$

(c) The mappings

$$W^s : \mathcal{A} \to \begin{cases} \mathcal{L}(l^2(\mathcal{H}))\,, & \text{if} \quad s \in \mathbb{R} \\ \mathcal{L}(l^2(\mathcal{H}))/\mathcal{K}(l^2(\mathcal{H}))\,, & \text{if} \quad s = \infty \end{cases}$$

are C^*-algebra homomorphisms.

Write χ_M for the characteristic function of the measurable set $M \subset \mathbb{R}$.

PROPOSITION 2.2 *For each $t \in \mathbb{T}$ there is a C^*-algebra homomorphism $W_t : \mathcal{A} \to \mathcal{L}(L^2_{\mathbb{R}}(\mathcal{H}))$ such that*

$$\begin{aligned} W_t((M(a))) &= a(t-0)P_{\mathbb{R}} + a(t+0)Q_{\mathbb{R}}\,, \\ W_t((P)) &= \chi_{\mathbb{R}+} I \end{aligned}$$

and

$$W_t((P_{[rn]})) = \chi_{[-r,r]} I\,.$$

Propositions 2.1 and 2.2 can be accomplished by

REMARK 2.1 If $(A_n) \in G$, then each of the homomorphisms W^s and W_t takes (A_n) into the zero element.

Now we introduce a further algebra smb$\mathcal{A}$, called symbol algebra assigned to $\mathcal{A}$. Namely, let M be the "disjoint union" of the sets $\mathbb{T}$ and $\dot{\mathbb{R}}$ and consider the collection of all functions f on M,

$$m \mapsto f(m) = \begin{cases} W^s((A_n)) & \text{if} \quad s = m, \\ W_t((A_n)) & \text{if} \quad t = m. \end{cases} \tag{2.1}$$

Since W^s and W_t are C^*-homomorphisms, smb$\mathcal{A}$ actually forms an algebra, which is even a unital C^*-algebra under the norm

$$||f|| := \sup_{m \in M} ||f(m)||.$$

Therefore, to each sequence $(A_n) \in \mathcal{A}$ there can be assigned an element smb$(A_n) \in$ smb$\mathcal{A}$ where smb(A_n) is given by (2.1).

THEOREM 2.1 *A sequence $(A_n) \in \mathcal{A}$ (moreover, any subsequence of (A_n)) is norm-stable if and only if smb(A_n) is invertible in smb$\mathcal{A}$.*

REMARK 2.2 This theorem gives raise to call smb(A_n) the *stability symbol* of (A_n). Recall that the notion of symbol is widely used in the theory of singular integral operators and pseudodifferential operators in order to formulate Fredholm properties of these operators.

REMARK 2.3 Remark 2.1 tells us that the homomorphisms W^s and W_t actually depends on the coset $(A_n) + G$ only, so they are defined on $\mathcal{A}/G$ and we denote them again by W^s and W_t, respectively.

THEOREM 2.2 *The algebras $\mathcal{A}/G$ and smb$\mathcal{A}$ are isometrically isomorphic, and this isomorphism is given by*

$$(A_n) + G \mapsto smb(A_n).$$

A special case of this theorem previously occured in the author's paper [S4] (see also [S3]).

THEOREM 2.3 *For each sequence $(A_n) \in \mathcal{A}$, the limit $\lim_{n\to\infty} ||A_n||$ exists and is equal to $||\,smb(A_n)||$. In particular, if the sequence $(A_n) \in \mathcal{A}$ is norm-stable then $\lim_{n\to\infty} ||A_n^{-1}|| = ||(smb(A_n))^{-1}||$, and the condition numbers $||A_n||\,||A_n^{-1}||$ converge to $||smb(A_n)||\,||(smb(A_n))^{-1}||$.*

A. Böttcher recognized (see [Bö]) this result again in a special case as an immediate consequence of the author's paper [S4].

Our next and last result concerns the asymptotic behaviour of the so-called ε-pseudospectra of the operators A_n, $(A_n) \in \mathcal{A}$. The ε-pseudospectrum of an element a which belongs to some unital C^*-algebra is defined to be the set

$$\Lambda_\varepsilon(a) = \{\lambda \in \mathbb{C} : ||(a-\lambda)^{-1}|| \geq \frac{1}{\varepsilon}\},$$

where te convention is used that $||C^{-1}|| = \infty$ in case C is not invertible.

The limit set $\lim\limits_{n\to\infty} \Lambda_\varepsilon(A_n)$ of the sequence $(\Lambda_\varepsilon(A_n))$ is by definition the collection of all complex numbers which are the limit of some sequence (λ_k) of numbers $\lambda_k \in \Lambda_\varepsilon(A_{n_k})$ with $n_1 < n_2 < \cdots$.

THEOREM 2.4 *If* $(A_n) \in \mathcal{A}$, *then the limit set of* $(\Lambda_\varepsilon(A_n))$ *exists and is equal to* $\Lambda_\varepsilon(smb(A_n))$.

In the case of the finite sections method for Toeplitz operators this result goes back to L. Reichel and L. N. Trefethen [Re/T] (generating functions from the Wiener algebra) and A. Böttcher [Bö] (piecewise continuous generating functions). The result for this special case reads as follows:

$$\lim_{n\to\infty} \Lambda_\varepsilon(T_n(a)) = \Lambda_\varepsilon(T(a)).$$

Notice that for the spectra of $T_n(a)$ a related result is in general not true. A discussion of such problems can be found, for instance, in [Bö].

3 PROOFS

The proof of Theorem 2.1 is our first concern. There are two possibilities to proceed. First one can repeat the arguments (with related modifications) of [R/S] in order to get a proof of Theorem 2.1. We will, however, offer a second possibility which is of its own interest. The idea is to extend the assertion of Theorem 2.1 for $\mathcal{H} = \mathbb{C}^N$ (which is proved in [R/S]) by some approximation argument to the general case. For this aim we have to use Theorem 2.2 for $\mathcal{H} = \mathbb{C}_N$ which is not proved (even not formulated) in [R/S]. A particular case however was already treated in [S4].

Let us denote the algebra $\mathcal{A}$ and the ideal G by $\mathcal{A}_N$ and G_N in case $\mathcal{H} = \mathbb{C}^N$.

PROPOSITION 3.1 *The algebra $\mathcal{A}_N/G_N$ is isometrically isomorphic to smb$\mathcal{A}_N$. This isomorphism is given by $(A_n) + G_N \mapsto smb(A_n)$*

PROOF As it was mentioned in the introduction, the norm-stability of a sequence $(A_n) \in \mathcal{A}_N$ is equivalent to the invertibility of the coset $(A_n) + G_N$ in $\mathcal{A}_N/G_N$. Therefore, using Theorem 2.1 for $\mathcal{H} = \mathbb{C}^N$ we get the following claim: $(A_n) + G_N$ is invertible if and only if smb$(\mathcal{A}_N)$ is invertible. Now we use that $\mathcal{A}_N/G_N$ and smb$\mathcal{A}_N$ are C^*-algebras. Because $(A_n) + G_N \mapsto$ smb(A_n) is a C^*-algebra homomorphism which preserves spectra it preserves also norms. ∎

Now we give a proof of Theorem 2.2 which yields immediately a proof of Theorem 2.1.

A dense subset of $\mathcal{A}$ is the set $\mathcal{A}_0$ of all sequences (A_n) which are finite sums of finite products of the form

$$(A_n) = \sum_i \prod_j (A_n^{(i,j)}),$$

where $(A_n^{(i,j)})$ are generators of $\mathcal{A}$ (see the introduction). Furthermore, it can be assumed that for $(A_n^{(i,j)}) = (M(a))$ the function a has only a finite number of jumps.

Let $\{e_1, e_2, \cdots\}$ be some orthogonal basis of $\mathcal{H}$ and denote by S_N the orthogonal projection of $\mathcal{H}$ onto the linear hull of $\{e_1, e_2, \cdots, e_N\}$. Introduce the following C^*-subalgebras $\mathcal{B}_N$ of $\mathcal{A}$ defined as follows: Given $M(a)$, $a = b\mathrm{I} + h$ with $b \in PC(\mathbb{C})$ and $h \in PC(\mathcal{K}(\mathcal{H}))$, set $M_N(a) = M(b) + S_N M(h) S_N$ (more precisely, $S_N M(h) S_N$ denotes the operator of multiplication by the function with values of the form $S_N h(t) S_N$). The algebra $\mathcal{B}_N$ is by definition the smallest C^*-subalgebra of $\mathcal{A}$ containing the sequences (P), $(P_{[rn]})$ and $M_N(a)$, $a \in PC(\mathbb{C}\mathrm{I} + \mathcal{K}(\mathcal{H}))$. It is easely seen that $\mathcal{B}_N$ decomposes into the direct sum of two C^*-subalgebras $\mathcal{B}_N^{(0)}$ and $\mathcal{B}_N^{(1)}$, which consists of such sequences the members of which are operators acting on $l^2(S_N\mathcal{H})$ and $l^2((\mathrm{I} - S_N)\mathcal{H})$, respectively. Thereby, the algebra $\mathcal{B}_N^{(0)}$ is isometrically isomorphic to $\mathcal{A}_N$, and $\mathcal{B}_N^{(1)}$ to $\mathcal{A}_1$. This shows that the assertions of the Propositions 2.1 and 2.2 are in force for the algebras $\mathcal{B}_N$; moreover, the algebra $\mathcal{B}_N/G \cap \mathcal{B}_N$ is isometrically isomorphic to the algebra smb$\,\mathcal{B}_N$. Now use

$$(\mathcal{B}_n + G)/G \cong \mathcal{B}/\mathcal{B}_N \cap G$$

in order to get

$$(\mathcal{B}_n + G)/G \cong .\mathrm{smb}\,\mathcal{B}_N$$

(all isomorphism being isometric). Take now any element $(A_n) \in \mathcal{A}_0$. Replace this element by $(A_n^{(N)})$, where $(A_n^{(N)})$ is given in the following way: each operator $M(a)$ of multiplication occuring in (A_n) is replaced by $M_N(a) = M(b) + S_N M(h) S_N$, where $a = b\mathrm{I} + h$, $b \in PC(\mathbb{C})$ and $h \in PC(\mathcal{K}(\mathcal{H}))$. Since h has at most finitely many jumps, the essential range of h is a compact subset of $\mathcal{K}(\mathcal{H})$. Lemma 4.1 from [Bö/S3] implies that $M_N(a)$ converges in the Norm of $\mathcal{L}(l^2(\mathcal{H}))$ to $M(a)$.

Consequently, the Propositions 2.1 and 2.2 are valid for elements from $\mathcal{A}_0$. Since $\mathcal{A}_0$ is dense in $\mathcal{A}$, this is also true for any sequence from $\mathcal{A}$. Further, by the former discussion,

$$||(A_n^{(N)}) + G|| = ||\mathrm{smb}(A_n^{(N)})||,$$

and passing to limit $N \to \infty$ yields

$$||(A_n) + G|| = ||\mathrm{smb}(A_n)||.$$

Hence, $\mathcal{A}/G \cong \mathrm{smb}\,\mathcal{A}$.

Therefore, Propositions 2.1 and 2.2, Theorems 2.1 and 2.2 are completely proved. ■

PROOF of Theorem 2.3.

First we show that, given a sequence $(A_n) \in \mathcal{A}$,

$$||\mathrm{smb}(A_n)|| \le \lim_{n\to\infty} \inf ||A_n||.$$

To this end consider a subsequence (A_{n_k}) such that

$$||A_{n_k}|| \to \lim_{n\to\infty} \inf ||A_n||$$

as $k \to \infty$. Using

$$W^s(A_{n_k}) = W^s(A_n), \quad W_t(A_{n_k}) = W_t(A_n)$$

(for $\mathcal{H} = \mathbb{C}^N$ this is proved in [R/S] and, by the former used approximation argument, it must be also valid in the general case), we get

$$||\mathrm{smb}(A_n)|| \le \lim_{n\to\infty} \inf ||A_n||.$$

Next we prove $\lim_{n\to\infty} \sup ||A_n|| \le ||\mathrm{smb}(A_n)||$, whence the assertion follows. Since $||\mathrm{smb}(A_n)|| = \inf_{(C_n)\in G} ||(A_n) + (C_n)||$, there is for any $\varepsilon > 0$ a sequence $(C_n) \in G$ such that

$$||(A_n) + (C_n)|| \le ||\mathrm{smb}(A_n)|| + \varepsilon.$$

Using $||A_n|| - ||C_n|| \le ||A_n + C_n|| \le ||(A_n) + (C_n)||$, we get $||A_n|| - ||C_n|| \le ||\mathrm{smb}(A_n)|| + \varepsilon$. Passing to lim sup it results that

$$\lim_{n\to\infty} \sup ||A_n|| \le ||\mathrm{smb}(A_n)|| + \varepsilon$$

that is,

$$\lim_{n\to\infty} \sup ||A_n|| \le ||\mathrm{smb}(A_n)||. \quad \blacksquare$$

PROOF of Theorem 2.4

First we need a proposition, which is taken from [Bö].

PROPOSITION 3.2 *Let $\mathcal{B}$ be a C^*-algebra with unit element e. Suppose $a - \lambda e$ (a in $\mathcal{B}$) is invertible for all λ in some open subset $U \subset \mathbb{C}$ and $||(a - \lambda e)^{-1}|| \le M$ for all $\lambda \in U$. Then $||(a - \lambda e)^{-1}|| < M$ for all $\lambda \in U$.*

The further considerations are adaptations of ideas from [Bö]. We start with verifying that

$$\Lambda_\varepsilon(W_t(A_n)) \subset \lim \Lambda_\varepsilon(A_n) \tag{3.1}$$

for all $t \in \mathbb{T}$ and

$$\Lambda_\varepsilon(W^s(A_n)) \subset \lim \Lambda_\varepsilon(A_n) \tag{3.2}$$

for all $s \in \dot{\mathbb{R}}$.

Let first $\lambda \in sp(W_t(A_n))$. Then, by Theorem 2.1, the sequence $(A_n - \lambda \mathrm{I})$ cannot be norm-stable. Then there is a subsequence $(A_{n_k} - \lambda \mathrm{I})$ such that

$$||(A_{n_k} - \lambda \mathrm{I})^{-1}|| \to \infty \quad \text{as } k \to \infty$$

(recall that if $A_{n_k} - \lambda \mathrm{I}$ is not invertible then we set $||(A_{n_k} - \lambda \mathrm{I})^{-1}|| = \infty$) which implies $\lambda \in \lim_{n\to\infty} \Lambda_\varepsilon(A_n)$.

Now suppose $\lambda \in \Lambda_\varepsilon(W_t(A_n)) \backslash \mathrm{sp}(W_t(A_n))$. Then $W_t(A_n) - \lambda \mathrm{I}$ is invertible and $||(W_t(A_n) - \lambda \mathrm{I})^{-1}|| \ge \frac{1}{\varepsilon}$. Let U be an open neighborhood of λ. Proposition 3.2 shows that there is an r in U such that

$$||(W_t(A_n) - r\mathrm{I})^{-1}|| > \frac{1}{\varepsilon}.$$

Therefore

$$||(W_t(A_n) - r\mathrm{I})^{-1}|| \leq \frac{1}{(\varepsilon - 1/k)}$$

for all k large enough. Consequently, there is a sequence (λ_k), $\lambda_k \in \Lambda_{\varepsilon-(1/k)}(W_t(A_n))$, such that $\lim \lambda_k = \lambda$.

Now Theorem 2.3 gives

$$\lim_{n\to\infty} ||(A_n - \lambda_k \mathrm{I})^{-1}|| \geq \frac{1}{(\varepsilon - 1/k)}.$$

Thus, $||(A_n - \lambda_k\mathrm{I})^{-1}|| \geq \frac{1}{\varepsilon}$ for all sufficiently large n which yields that $\lambda_k \in \Lambda_\varepsilon(A_n)$ and $\lambda = \lim\limits_{n\to\infty} \lambda_k \in \lim\limits_{n\to\infty} \Lambda_\varepsilon(A_n)$. The inclusion (3.2) can be proved analogously. Using that there is an $s \in \dot{\mathbb{R}}$ or a $t \in \mathbb{T}$ such that

$$||\mathrm{smb}(A_n)|| = ||W^s(A_n)|| \qquad (\text{ or } ||\mathrm{smb}(A_n)|| = ||W_t(A_n)||)$$

(this follows from the elementary theory of C^*-algebras, namely from the fact that the spectral radius of a selfadjoint element a is equal to $||a||$) we obtain

$$\Lambda_\varepsilon(\mathrm{smb}(A_k)) \subset \lim \Lambda_\varepsilon(A_n)\ .$$

In order to prove the reverse inclusion,

$$\lim \Lambda_\varepsilon(A_n) \subset \Lambda_\varepsilon(\mathrm{smb}(A_n)),$$

assume $\lambda \notin \Lambda_\varepsilon(\mathrm{smb}(A_n))$. Then there is a $\delta > 0$ such that

$$||(\mathrm{smb}(A_n) - \lambda e)^{-1}|| = \frac{1}{\varepsilon} - 2\delta < \frac{1}{\varepsilon}.$$

Moreover, all operators $W_t(A_n) - \lambda\mathrm{I}$ and $W^s(A_n) - \lambda\mathrm{I}$ are invertible and the norms of their inverses are bounded by $(\frac{1}{\varepsilon}) - 2\delta$. Applying Theorem 2.2, we obtain

$$||(A_n - \lambda\mathrm{I})^{-1}|| < (\frac{1}{\varepsilon}) - \delta$$

for all $n \geq n_0$. If $n \geq n_0$ and $|\lambda - r| \leq \varepsilon\delta(\frac{1}{\varepsilon} - \delta)^{-1}$ then

$$||(A_n - r\mathrm{I})^{-1}|| \leq \frac{||(A_n - \lambda\mathrm{I})^{-1}||}{1 - |r - \lambda|\,||(A_n - \lambda\mathrm{I})^{-1}||} < \frac{1}{\varepsilon}$$

and thus $r \notin \Lambda_\varepsilon(A_n)$. Therefore, $\lambda \notin \lim\limits_{n\to\infty} \Lambda_\varepsilon(A_n)$ and the assertion is proved. ■

REMARK 3.1 It is easy to see that

$$\bigcup_{t\in\mathbb{T}} \Lambda_\varepsilon(W_t(A_n)) \bigcup_{s\in\dot{\mathbb{R}}} \Lambda_\varepsilon(W^s(A_n)) = \Lambda_\varepsilon(\mathrm{smb}(A_n)).$$

REFERENCES

[A] Ambartsumyan, G. V.: On the reduction method for a class of Toeplitz matrices. Mat. Issled. 8:2, 161–168 (1973)(Russian).

[Ba] Baxter, G.: A norm inequality for a finite-section Wiener-Hopf equation. Illionis J. Math. 7, 97–103 (1963).

[Bö] Böttcher, A.: Pseudospectra and singular values of large convolution operators (to appear)

[Bö/S1] Böttcher, A., and Silbermann, B.: The finite section method for Toeplitz operators on the quarter-plane with piecewise continuous symbols. Math. Nachr. 110, 279–291 (1983).

[Bö/S2] Böttcher, A. and Silbermann, B.: Analysis of Toeplitz operators, Akademie-Verlag, Berlin 1990, and Springer-Verlag, Berlin 1990.

[Bö/S3] Böttcher, A. and Silbermann, B.: Operator-valued Szegö-Widom Limit Theorems. Operator Theory: Advances and Applications, Vol. 71, Birkhäuser Verlag, Basel 1994

[G/F] Gohberg, I., and Feldmann, I.A.: Convolution equations and projection methods for their solution. Amer. Math. Soc. Transl. of Math. Monographs 41, Providence, R. I., 1974.

[Ko] Kozak, A.V.: A local principle in the theory of projection methods. Dokl. Akad. Nauk SSSR 212:6, 1287–1289 (1973) (Russian); also in: Soviet Math. Dokl. 14, 1580–1583 (1974).

[Kr/V] Krupnik, N. Ya., and Verbitzki, I. E.: On the applicability of the reduction method to discrete Wiener-Hopf equations with piecewise continuous symbol. In: Spectr. svoista Oper. (mat. issled. 45) Shtiintsa, Kishinev 1977, 17-28 (Russian).

[Pr/S] Prößdorf, S., and Silbermann, B.: Numerical analysis for Integral and related Operator Equations. Akademie-Verlag, Berlin 1991, and Birkhäuser-Verlag, Basel-Boston-Stuttgard 1991.

[Ra] Rathsfeld, A.: Über das Redukionsverfahren für singuläre Integraloperatoren mit stückweise stetigen Koeffizienten. Math. Nachr. 127, 125–143 (1986).

[Re/T] Reichel, L., and Trefethen, L. N.: Eigenvalues and pseudo-eigenvalues of Toeplitz matrices. Linear Alg. Appl. 162, 153–185 (1992).

[R] Roch, S., and Silbermann, B.: Finite sections of operators belonging to the closed algebra of singular integral operators. Seminar Analysis: Operator equations and numerical analysis 1986/1987, 139–148, Berlin 1987.

[R/S] Roch., and Silbermann, B.: Limiting sets of eigenvalues and singular values of Toeplitz matrices. Asymptotic Analysis 8, 293–309 (1994).

[S1] Silbermann, B.: Lokale Theorie des Reduktionsverfahrens für Toeplitzoperatoren. Math. Nachr. 104, 137–146 (1981).

[S2] Silbermann, B.: Lokale Theorie des Reduktionsverfahrens für singuläre Integralgleichungen. ZfAA 1, 45–56 (1982).

[S3] Silbermann, B.: Local objects in the theory of Toeplitz operators. IEOT 9, 706–738 (1986).

[S4] Silbermann, B.: On the limiting set of singular values of Toeplitz matrices. Lin. Alg. Appl. 182, 35–43 (1993).

[St] Stein, N.: Singular integrals and differentiability properties of functions. Princeton Univ. Press, New Jersey 1970.

[V] Verbitski, I. E.: Projection method for the solution of singular integral equation with piecewise continuous coefficients. In: Oper. v. Banach. Prostr. (Math. Issl-. Vyp.47), 12–24, Shtiintsa, Kishinev 1978 (Russian).

[W] Widom, H.: Asymptotic behaviour of block Toeplitz matrices and determinants, II. Adv. Math. 21, 1–29 (1976).

TU Chemnitz–Zwickau
Fakultät für Mathematik
09107 Chemnitz
Germany

AMS subject classification: 47 B35

Operator Theory:
Advances and Applications, Vol. 87

SPECTRAL REPRESENTATIONS AND SPECTRAL FUNCTIONS OF SYMMETRIC OPERATORS

Abraham V.Strauss

In the memory of my parents
Rebecca and Wilhelm Strauss,
my brother Henry, my sister Sonia,
martyrs of Przemysl Ghetto.

The spectral representation of a symmetric nonmaximal operator transforms it into a multiplication operator by a complex variable acting in a linear space of pairs of vector-valued functions holomorphic on the upper and lower half- planes. Each generalized spectral function of the same symmetric operator leads to its realization in form of a multiplication operator by a real variable in some Hilbert space. We study the connection between such realizations of the given operator and its spectral representation.

0. INTRODUCTION

The present paper is related to well-known articles of M.A.Naimark [N1], [N2], M.G.Krein [Kr1], [Kr2], and M.S.Livšic [L1], [L2]. Some our earlier results [S1]- [S9] are applied and supplemented.

For a symmetric nonmaximal operator A acting in a Hilbert space $\mathcal{H}$ a linear mapping of $\mathcal{H}$ onto a linear space of pairs is defined [S9]. The components of each pair are vector-valued functions holomorphic on the upper and lower half-planes, respectively. In this way the operator A is represented as a multiplication operator by a complex variable. Such mapping is called *the spectral representation* of A. This construction may be considered as a development of the approach proposed by M.G.Krein [Kr1] for a densely defined symmetric operator with finite defect numbers equal to n. Such operator was realized there as a multiplication operator by a complex variable in a space of C^n -valued functions meromorphic on the upper and lower half-planes.

The linear representation space which we construct is generated in some sense by an operator-valued function holomorphic on the upper half-plane [S9],[S2], [S5]. If the operator A is densely defined and its defect numbers are finite and equal, this operator-valued function is near to the characteristic function of A introduced by M.S.Livšic [L1], [L2].

According to M.A.Naimark [N2] for a symmetric nonmaximal operator A there exists an infinite set of generalized spectral functions. Each of them leads to a realization of A as a multiplication operator by a real variable in some Hilbert space of vector-valued functions. Our aim is to clarify the connection between such realizations of A and its spectral representation.

1. SPECTRAL REPRESENTATIONS OF LINEAR OPERATORS

Recall some notions and assertions concerning the representation theory of linear operators [S7]-[S9]. For symmetric densely defined operators with finite equal defect numbers the foundations of this theory are due to M.G.Krein [Krl], [Kr2].

Let A be a linear closed operator in Hilbert space $\mathcal{H}$ with $DomA \neq \{0\}$. Let $\Lambda(A)$ be the set of all $\lambda \in C$ for which the operator $(A - \lambda I)^{-1}$ exists and is bounded, but $Ran(A - \lambda I) \neq \mathcal{H}$. $\Lambda(A)$ is an open set. Suppose that $\Lambda(A) \neq \emptyset$.

Let $\mathcal{N}$ be a subspace of $\mathcal{H}$ such that the set

$$\Omega(\mathcal{N}) = \{\lambda \in \Lambda(A) : \mathcal{H} = Ran(A - \lambda I) + \mathcal{N},\ Ran(A - \lambda I) \cap \mathcal{N} = \{0\}\}$$

is not empty. Then $\mathcal{N}$ is called *the module subspace* for A.

For each $\lambda \in \Omega(\mathcal{N})$ we define the operator $Q(\lambda) : \mathcal{H} \to \mathcal{N}$ as a projection of $\mathcal{H}$ onto $\mathcal{N}$ parallel to $Ran(A - \lambda I)$ corresponding to the direct sum decomposition $\mathcal{H} = Ran(A - \lambda I)\dot{+}\mathcal{N}$.

$\Omega(\mathcal{N})$ is an open set, and the operator-valued function $\lambda \mapsto Q(\lambda)$ is holomorphic on each connected component of $\Omega(\mathcal{N})$.

The operator A is called *regular* if $\Lambda(A) = C$. A regular operator A is called *entire* if there exists a module subspace $\mathcal{N}$ for A such that $\Omega(\mathcal{N}) = C$. In this case $\mathcal{N}$ is called *an entire module subspace.*

The operator A is entire if and only if $0 \in \Lambda(A)$, and for A^{-1} there exists a quasinilpotent extension T such that $RanT = DomA$. In this case $KerT$ is an entire module subspace for A.

Every reqular operator in a finite-dimensional space is entire.

The first-order differentiation operator in $\mathcal{L}_2(0,1)$ corresponding to the boundary conditions $f(0) = f(1) = 0$ is regular, but not entire.

Let $\mathcal{N}$ be a module subspace for A, and let $\Omega^0(\mathcal{N})$ be a domain in C which coincides with some connected component of $\Omega(\mathcal{N})$ or is a nonempty part of such component. Now for each $f \in \mathcal{H}$ we have the $\mathcal{N}$ - valued function $\lambda \mapsto Q(\lambda)f$ holomorphic on $\Omega^0(\mathcal{N})$. So we obtain a mapping Φ of $\mathcal{H}$ onto a linear space of such functions. In this way the operator A is represented as a multiplication operator by the complex variable λ. Indeed, if $f \in DomA$, then for every $\lambda \in \Omega^0(\mathcal{N})$ $\quad Q(\lambda)(A - \lambda I)f = 0$, whence $Q(\lambda)Af = \lambda Q(\lambda)f$.

Such mapping Φ is called *the spectral representation* of A corresponding to $\mathcal{N}$ and $\Omega^0(\mathcal{N})$.

The spectral representation Φ is called *exact* if $Ker\Phi = \{0\}$.

Suppose that we have an ordered finite set of module subspaces $\mathcal{N}^1, \ldots, \mathcal{N}^k$ for A together with nonempty domains $\Omega^{01} \subset \Omega(\mathcal{N}^1), \ldots, \Omega^{0k} \subset \Omega(\mathcal{N}^k)$ and corresponding spectral representations $\Phi_1, \ldots, \Phi_k$. Then we can define a direct sum $\Phi_1 + \ldots + \Phi_k$ as a mapping which transforms each $f \in \mathcal{H}$ into an ordered set of holomorphic vector-functions $\Phi_1 f, \dots, \Phi_k f$ with domains $\Omega^{01}, \dots, \Omega^{0k}$, respectively. This mapping also will be called *a spectral representation* of A.

Now let A be a closed symmetric nonmaximal operator in $\mathcal{H}$. For each nonreal γ let $\mathcal{N}_\gamma$ be the corresponding defect subspace of A, that is,

$$\mathcal{N}_\gamma = \mathcal{H} \ominus Ran(A - \gamma I).$$

Denote by C_+ and C_- the upper and lower half-planes, respectively. It is known that $dim\mathcal{N}_\gamma$ is constant on C_+ and C_-. The cardinal numbers $dim\mathcal{N}_i$ and $dim\mathcal{N}_{-i}$ are called *the defect numbers* of A. Since A is not maximal, $\mathcal{N}_i \neq \{0\}$ and $\mathcal{N}_{-i} \neq \{0\}$.

Choose $\mathcal{N}_i$ and $\mathcal{N}_{-i}$ as module subspaces for A. Then, according to [S2] and [S5], $\Omega(\mathcal{N}_i) \supset C_+$, $\Omega(\mathcal{N}_{-i}) \supset C_-$, and

$$\mathcal{R} \cap \Lambda(A) \subset \Omega(\mathcal{N}_i) \cap \Omega(\mathcal{N}_{-i}). \tag{1.1}$$

If $\mathcal{R} \subset \Lambda(A)$, then $\Lambda(A) = C$ and the operator A is regular. In this case by (1.1)

$$\mathcal{R} \subset \Omega(\mathcal{N}_i) \cap \Omega(\mathcal{N}_{-i}).$$

Let Ω^0 and Ω^0_* be connected components of $\Omega(\mathcal{N}_i)$ and $\Omega(\mathcal{N}_{-i})$, respectively, such that $\Omega^0 \supset C_+$ and $\Omega^0_* \supset C_-$. Let Φ and Φ_* be the spectral representations of A corresponding to $\mathcal{N}_i$, Ω^0 and $\mathcal{N}_{-i}$, Ω^0_*, respectively. $\Phi\mathcal{H}$ is a linear space of $\mathcal{N}_i$-valued functions holomorphic on Ω^0, and $\Phi_*\mathcal{H}$ is a linear space of $\mathcal{N}_{-i}$-valued functions holomorphic on Ω^0_*. The spectral representation $\Phi + \Phi_*$ of A is a linear mapping of $\mathcal{H}$ onto a subspace of the direct sum $\Phi\mathcal{H} \dot{+} \Phi_*\mathcal{H}$. For $\zeta \in \Omega^0$ and $\eta \in \Omega^0_*$ we denote by $Q(\zeta)$ and $Q_*(\eta)$ the projections on $\mathcal{N}_i$ and $\mathcal{N}_{-i}$ parallel to $(A - \zeta I)$ and $(A - \eta I)$, respectively. Then for each $f \in \mathcal{H}$

$$(\Phi f)(\zeta) = Q(\zeta) f, \tag{1.2}$$

$$(\Phi_* f)(\eta) = Q_*(\eta) f. \tag{1.3}$$

Put $\quad K(\zeta) = Q(\zeta) | \mathcal{N}_{-i} \quad (\zeta \in \Omega^0), \qquad K_*(\eta) = Q_*(\eta) | \mathcal{N}_i \quad (\eta \in \Omega^0_*).$

According to [S2], [S5] for each $\zeta \in C_+$, $\eta \in C_-$

$$\| K(\zeta) \| \leq 1, \quad \| K_*(\eta) \| \leq 1, \quad \text{and} \quad K^*(\zeta) = K_*(\bar{\zeta}).$$

The operator-valued function

$$\zeta \mapsto C(\zeta) = \frac{\zeta - i}{\zeta + i} K(\zeta)$$

on the domain $\Omega(\mathcal{N}_i) \setminus \{-i\}$ coincides with the characteristic function of A in sense of the definition proposed in [S2], [S5]. The characteristic function for a densely defined symmetric operator with finite equal defect numbers was first introduced, studied, and applied by M.S.Livsic [L1], [L2]. Our approach to this notion is some different.

Let us consider the elements of the direct sum $\Phi\mathcal{H} \dot{+} \Phi_*\mathcal{H}$ as columns with two components. According to [S9] the linear set $(\Phi + \Phi_*) \sum_{Im\gamma \neq 0} \mathcal{N}_\gamma$ is described in terms of the operator-valued function $\zeta \mapsto K(\zeta) \quad (\zeta \in C_+)$ as follows:

$$(\Phi + \Phi_*) \sum_{Im\gamma \neq 0} \mathcal{N}_\gamma =$$

$$= \left\{ \begin{pmatrix} u \\ K_*(\eta)u \end{pmatrix} : u \in \mathcal{N}_i \right\} + \left\{ \begin{pmatrix} K(\zeta)v \\ v \end{pmatrix} : v \in \mathcal{N}_{-i} \right\} +$$

$$+ \sum_{\alpha \in C_+} \left\{ \begin{pmatrix} \frac{1}{\zeta-\alpha}(K(\zeta) - K(\alpha))v \\ \frac{1}{\eta-\alpha}(1 - K_*(\eta)K(\alpha))v \end{pmatrix} : v \in \mathcal{N}_{-i} \right\} +$$

$$+ \sum_{\beta \in C_-} \left\{ \begin{pmatrix} \frac{1}{\zeta-\beta}(1 - K(\zeta)K_*(\beta))u \\ \frac{1}{\eta-\beta}(K_*(\eta) - K_*(\beta))u \end{pmatrix} : u \in \mathcal{N}_i \right\}.$$

Here $\zeta \in C_+$, $\eta \in C_-$ are considered as complex variables. For $\zeta = \alpha$ and $\eta = \beta$ the corresponding divided differences must be replaced by derivatives at α and β, respectively.

Note that if the operator A is completely nonselfadjoint, then the linear sum $\sum_{Im\gamma \neq 0} \mathcal{N}_\gamma$ is dense in $\mathcal{H}$.

2. SELFADJOINT EXTENSION IN A LARGER HILBERT SPACE

As before, let A be a closed symmetric operator in Hilbert space $\mathcal{H}$. Let B be a selfadjoit extension of A going out into a larger Hilbert space $\mathcal{K} \supset \mathcal{H}$. Such extensions have been studied by M. A. Naimark [N1], [N2] for a densely defined A in $\mathcal{H}$.

Put $\mathcal{H}' = \mathcal{K} \ominus \mathcal{H}$. Designate by A' the closed symmetric operator induced in $\mathcal{H}'$ by B so that

$$Dom A' = \{f' \in \mathcal{H}' \cap Dom B \, : \, Bf' \in \mathcal{H}'\}$$

and $A' \subset B$. For every nonreal γ we set $\mathcal{N}'_\gamma = \mathcal{H}' \ominus Ran(A' - \gamma I)$.

We introduce the unitary operator

$$U = (B + iI)(B - iI)^{-1}.$$

Note that $U(\mathcal{N}_i \oplus \mathcal{N}'_i) = \mathcal{N}_{-i} \oplus \mathcal{N}'_{-i}$.

LEMMA 2.1 (cf.[N2], Theorem 3).

$$U\mathcal{N}'_i \cap \mathcal{N}'_{-i} = \{0\}.$$

Proof. Suppose that for some $g \in \mathcal{N}'_i$ $Ug \in \mathcal{H}'$. Then for $f = Ug - g$ we have: $f \in Dom B \cap \mathcal{H}'$ and $Bf = i(Ug + g) \in \mathcal{H}'$. Hence $f \in Dom A'$ and $(A' - iI)f = 2ig$. It follows that $g \in Ran(A' - iI) \cap \mathcal{N}'_i$ and therefore $g = 0$. □

Designate by P and P' the orthogonal projections of $\mathcal{K}$ onto $\mathcal{H}$ and $\mathcal{H}'$, respectively.

LEMMA 2.2. $\mathcal{N}'_i = \overline{P'U^*\mathcal{N}_{-i}}$.

Proof. Suppose that $g \in \mathcal{N}'_i$ and $g \perp P'U^*\mathcal{N}_{-i}$. Then $g \perp U^*\mathcal{N}_{-i}$ and $Ug \perp UU^*\mathcal{N}_{-i} = \mathcal{N}_{-i}$. Therefore $Ug \in \mathcal{N}'_{-i}$ and by *Lemma 2.1* $g = 0$. □

LEMMA 2.3. $$\mathcal{N}'_i \subset U^*\mathcal{N}_{-i} \vee \mathcal{N}_i. \tag{2.1}$$

Proof. For every $h \in \mathcal{N}_{-i}$ $\quad P'U^*h = U^*h - PU^*h.$
Therefore $P'U^*\mathcal{N}_{-i} \subset U^*\mathcal{N}_i + \mathcal{N}_i$.

Hence (2.1) follows by *Lemma 2.2.* □

Following [N2], we call the selfadjoint extension B of A minimal if B has no nonzero reducing subspace in $\mathcal{H}'$. The selfadjoint extension B of A is minimal if and only if the operator A' induced in $\mathcal{H}'$ by B is completely nonselfadjoint.

Let $\quad \mathcal{K}_0 = \bigvee_{k\in Z} U^k\mathcal{H}.$

$\mathcal{K}_0$ reduces B, and the operator B_0 induced in $\mathcal{K}_0$ by B is a minimal selfadjoint extension of A. This proposition in case of a densely defined operator A is due to M.A.Naimark ([N2], Theorem 7), but the proof remains valid also in the general case. Thus the selfadjoint extension B of A is minimal if and only if $\mathcal{K}_0 = \mathcal{K}$.

Let
$$\mathcal{K}_1 = \bigvee_{k\in Z} U^k\mathcal{N}_i. \tag{2.2}$$
$$\mathcal{K}_2 = \bigvee_{k\in Z} U^k\mathcal{N}_{-i}. \tag{2.3}$$

THEOREM 2.4. *If the operator A is completely nonselfadjoint, then*
$$\mathcal{K}_0 = \mathcal{K}_1 \vee \mathcal{K}_2. \tag{2.4}$$

Proof. We may suppose that $\mathcal{K}_0 = \mathcal{K}$. So B is a minimal selfadjoint extension of A. Then the operator A' induced in $\mathcal{H}'$ by B is completely nonselfadjoint. Therefore, if A is completely nonselfadjoint, then the operator $A \oplus A'$ in $\mathcal{K}$ is also completely nonselfadjoint. It follows that the defect subspace $\mathcal{N}_i \oplus \mathcal{N}'_i$ of $A \oplus A'$ is a generating subspace for B, that is,
$$\bigvee_{k\in Z} U^k(\mathcal{N}_i \oplus \mathcal{N}'_i) = \mathcal{K}_0. \tag{2.5}$$

Taking into account (2.2), (2.3), and the equality $U^* = U^{-1}$, we have
$$\mathcal{K}_1 \vee \mathcal{K}_2 = (\bigvee_{k\in Z} U^k\mathcal{N}_i) \vee (\bigvee_{k\in Z} U^{k-1}\mathcal{N}_{-i}) = (\bigvee_{k\in Z} U^k\mathcal{N}_i) \vee (U^k(U^*\mathcal{N}_{-i} + \mathcal{N}_i)).$$

Hence by *Lemma 2.3* we obtain
$$\mathcal{K}_1 \vee \mathcal{K}_2 \supset (\bigvee_{k\in Z} U^k\mathcal{N}_i) \vee (\bigvee_{k\in Z} U^k\mathcal{N}'_i) \supset \bigvee_{k\in Z} U^k(\mathcal{N}_i \oplus \mathcal{N}'_i).$$

So, in view of (2.5) $\quad \mathcal{K}_1 \vee \mathcal{K}_2 \supset \mathcal{K}_0.$
On the other hand obviously $\quad \mathcal{K}_0 \supset \mathcal{K}_1 \vee \mathcal{K}_2.$
Thus (2.4) follows. □

The next theorem is a consequence of *Theorem 2.4*.

THEOREM 2.5. *If the operator A is completely nonselfadjoint, then*
$$\mathcal{H} \subset \mathcal{K}_1 \vee \mathcal{K}_2.$$

3. THE GENERALIZED SPECTRAL FUNCTION OF A SYMMETRIC OPERATOR AND THE CORRESPONDING SPECTRAL TRANSFORMATION

Let A be a completely nonselfadjoint closed symmetric operator in $\mathcal{H}$, and let B be his minimal selfadjoint extension in $\mathcal{K} \supset \mathcal{H}$. Let $\lambda \to E^+(\lambda)$ $(\lambda \in \mathcal{R})$ be the spectral function of B. Taking into account (2.2) and (2.3), we have

$$\mathcal{K}_1 = \bigvee_{\lambda \in \mathcal{R}} E^+(\lambda)\mathcal{N}_i, \tag{3.1}$$

$$\mathcal{K}_2 = \bigvee_{\lambda \in \mathcal{R}} E^+(\lambda)\mathcal{N}_{-i}. \tag{3.2}$$

The operator-valued function

$$\lambda \mapsto PE^+(\lambda)\,|\,\mathcal{H} = E(\lambda) \quad (\lambda \in \mathcal{R})$$

is called *the generalized spectral function* of A defined B. As before, here P denotes the orthogonal projection of K onto H.

The generalized spectral functions of a densely defined closed symmetric operator were studied by M. A. Naimark in his fundamental article [N2], the case of an arbitrary closed symmetric operator was considered by the author [S1], [S3].

Denote by Q_1^+ and Q_2^+ the orthogonal projections of $\mathcal{K}$ onto $\mathcal{N}_i$ and $\mathcal{N}_{-i}$, respectively. Let $Q_1 = Q_1^+\,|\,\mathcal{H}$, $Q_2 = Q_2^+\,|\,\mathcal{H}$. So, Q_1 and Q_2 are orthogonal projections of $\mathcal{H}$ onto $\mathcal{N}_i$ and $\mathcal{N}_{-i}$, respectively.

Define the following operator-valued functions of $\lambda \in \mathcal{R}$ by formulas:

$$E_{11}(\lambda) = Q_1^+E^+(\lambda)\,|\,\mathcal{N}_i = Q_1E(\lambda)\,|\,\mathcal{N}_i\,, \tag{3.3}$$

$$E_{22}(\lambda) = Q_2^+E^+(\lambda)\,|\,\mathcal{N}_{-i} = Q_2E(\lambda)\,|\,\mathcal{N}_{-i}\,, \tag{3.4}$$

$$E_{12}(\lambda) = Q_1^+E^+(\lambda)\,|\,\mathcal{N}_{-i} = Q_1E(\lambda)\,|\,\mathcal{N}_{-i}\,, \tag{3.5}$$

$$E_{21}(\lambda) = Q_2^+E^+(\lambda)\,|\,\mathcal{N}_i = Q_2E(\lambda)\,|\,\mathcal{N}_i\,. \tag{3.6}$$

Note that $E_{12}^*(\lambda) = E_{21}(\lambda)$ for each $\lambda \in \mathcal{R}$.

We also introduce the Hilbert spaces

$$\mathcal{F}_1 = \mathcal{L}_2(\mathcal{N}_i\,,\, dE_{11}(\lambda))\,, \tag{3.7}$$

$$\mathcal{F}_2 = \mathcal{L}_2(\mathcal{N}_{-i}\,,\, dE_{22}(\lambda)) \tag{3.8}$$

(cf.[B]). $\mathcal{F}_1$ is a completion of the pre-Hilbert space obtained as a factor space from the span of the set of vector-valued functions

$$\lambda \mapsto \left(\frac{\lambda+i}{\lambda-i}\right)^k g = g^{[k]}(\lambda) \quad (\lambda \in \mathcal{R})\,,$$

where $g \in \mathcal{N}_i$ and $k \in Z$. This span is endowed with a semi-definite inner product defined by the formula:

$$\left\langle g_1^{[k]}, g_2^{[l]} \right\rangle = \int_{-\infty}^{+\infty} \left(\frac{\lambda+i}{\lambda-i}\right)^{k-l} d(E_{11}(\lambda)g_1\,, g_2)\,,$$

where $g_1, g_2 \in \mathcal{N}_i$, $k, l \in Z$. In a usual way the semi-definite inner product leads to a inner product in $\mathcal{F}_1$. The description of $\mathcal{F}_1$ may be nontrivial if $\mathcal{N}_i$ is infinite-dimensional. Similarly we define $\mathcal{F}_2$.

Now we introduse linear continuons operators

$$\Psi_1 : \mathcal{K}_1 \to \mathcal{F}_1,$$
$$\Psi_2 : \mathcal{K}_2 \to \mathcal{F}_2.$$

such that for $g \in \mathcal{N}_i$, $h \in \mathcal{N}_{-i}$, $k \in Z$

$$\Psi_1 U^k g = g^{[k]} \quad \text{and} \quad \Psi_2 U^k h = h^{[k]}.$$

These operators are isometric, $Ran\Psi_1 = \mathcal{F}_1$ and $Ran\Psi_2 = \mathcal{F}_2$.

Consider the direct sum $\mathcal{F}_1 \dot{+} \mathcal{F}_2$ and agree to write its elements as columns with two components. We define on $\mathcal{F}_1 \dot{+} \mathcal{F}_2$ a semi-definite inner product by the formula

$$\left\langle \begin{pmatrix} u_1 \\ u_2 \end{pmatrix}, \begin{pmatrix} v_1 \\ v_2 \end{pmatrix} \right\rangle = \int_{-\infty}^{+\infty} \left(d \begin{pmatrix} E_{11}(\lambda) & E_{12}(\lambda) \\ E_{21}(\lambda) & E_{22}(\lambda) \end{pmatrix} \begin{pmatrix} u_1(\lambda) \\ u_2(\lambda) \end{pmatrix}, \begin{pmatrix} v_1(\lambda) \\ v_2(\lambda) \end{pmatrix} \right)_{(\mathcal{N}_i \oplus \mathcal{N}_{-i})} \tag{3.9}$$

At first we apply this formula to elements whose components are vector-valued functions $g^{[k]}$, $h^{[l]}$ as above, where $g \in \mathcal{N}_i$, $h \in \mathcal{N}_{-i}$, k, $l \in Z$, then in a usual way we extend the formula onto the corresponding span, its factor space, and onto the completion of this factor space concidered as a pre-Hilbert space. So we obtain a Hilbert space $\mathcal{F}$. According to (3.9), $\mathcal{F}_1$ and $\mathcal{F}_2$ may be regarded as subspaces of $\mathcal{F}$. Moreover, $\mathcal{F} = \mathcal{F}_1 \vee \mathcal{F}_2$.

Finally we introduce a linear continuons operator

$$\Psi : \mathcal{K} \to \mathcal{F}$$

such that $\Psi | \mathcal{K}_1 = \Psi_1$ and $\Psi | \mathcal{K}_2 = \Psi_2$.

By these conditions together with the requirements of linearity and continuity the operator Ψ is uniquely defined. Moreover, Ψ is isometric, $\Psi\mathcal{K} = \mathcal{F}$, and

$$\Psi B = M \Psi,$$

where M denotes the multiplication operator by the real variable in $\mathcal{F}$.

We are interested in the restriction $\Psi | \mathcal{H}$ and in the part of M corresponding to A.

We shall say that $\Psi | \mathcal{H}$ is the spectral transformation defined by B and the pair $\mathcal{N}_i$, $\mathcal{N}_{-i}$, or by the operator-valued function

$$\lambda \mapsto \begin{pmatrix} E_{11}(\lambda) & E_{12}(\lambda) \\ E_{21}(\lambda) & E_{22}(\lambda) \end{pmatrix}.$$

Our aim is to clarify the connection between the spectral transformation $\Psi | \mathcal{H}$ and the spectral representation $\Phi + \Phi_*$ considered in Section 1.

4. Generalized resolvents of a symmetric operator

It is known that a closed symmetric operator A in $\mathcal{H}$ with the defect subspaces $\mathcal{N}_i$, $\mathcal{N}_{-i}$ is densely defined if and only if $DomA$, $\mathcal{N}_i$ and $\mathcal{N}_{-i}$ are linearly independent ([N1],

Theorem 8). If $g \in \mathcal{N}_i$, $h \in \mathcal{N}_{-i}$ and $g - h \in DomA$, then $\| g \| = \| h \|$ ([N1, Theorem 9). By setting $h = Xg$ for all such pairs g, h the isometric operator X is defined with $DomX \subset \mathcal{N}_i$, $RanX \subset \mathcal{N}_{-i}$. $DomX = \{0\}$ if and only if $\overline{DomA} = \mathcal{H}$. M.A.Krasnoselskii [K] introduced this operator X in a some another way, and applying X described the set of all symmetric extensions of a nondensely defined closed symmetric operator A.

We call a linear bounded operator $F : \mathcal{N}_i \to \mathcal{N}_{-i}$ *admissible* with respect to the given symmetric operator A if $Fg = Xg$ only for $g = 0$. For an operator F admissible with respect to A we denote by A_F the extension of A such that

$$DomA_F = DomA + (F - I)\mathcal{N}_i,$$

$$A_F(f + Fg - g) = Af + i(Fg + g), \quad \text{where} \quad f \in DomA,\ g \in \mathcal{N}_i.$$

This operator A_F is densely defined, closed, $i \in \rho(A_F)$, and

$$F = (A_F + iI)(A_F - iI)^{-1}|\mathcal{N}_i.$$

According to [S2], the operator A_F is accumulative, in the sense that $Im(A_F h, h) \leq 0$ for every $h \in DomA_F$, if and only if $\| F \| \leq 1$. In this case the operator $F^* : \mathcal{N}_{-i} \to \mathcal{N}_i$ satisfies the condition $F^* h = X^{-1}h$ only for $h = 0$, the operator A_{F^*} defined by formulas

$$DomA_{F^*} = DomA + (F^* - I)\mathcal{N}_{-i},$$

$$A_{F^*}(f + F^*g - g) = Af - i(F^*g + g), \quad \text{for} \quad f \in DomA,\ g \in \mathcal{N}_{-i}$$

is dissipative, in the sense that $Im(A_{F^*}h, h) \geq 0$ for every $h \in DomA_{F^*}$, and

$$(A_F)^* = A_{F^*}. \tag{4.1}$$

If $\| F \| \leq 1$, then $\rho(A_F) \supset C_+$ and $\rho(A_{F^*}) \supset C_-$. As it was shown in [S2], the operator X^{-1} is connected with the operator-valued function $\zeta \mapsto K(\zeta)$ $(\zeta \in C)$ considered in Section 1 by following formulas:

$$DomX^{-1} = \{h \in \mathcal{N}_{-i} : \varliminf_{\substack{\zeta \to \infty \\ \epsilon < arg\,\zeta < \pi - \epsilon}} (|\zeta|\,(\|h\| - \|K(\zeta)h\|)) < +\infty\},$$

$$X^{-1}h = \lim_{\substack{\zeta \to \infty \\ \epsilon < arg\,\zeta < \pi - \epsilon}} K(\zeta)h \text{ (strong convergence)} \quad \text{for} \quad h \in DomX^{-1}.$$

Let the operator B in $\mathcal{K} \supset \mathcal{H}$ be a selfadjoint extension of A. The generalized resolvent of A corresponding to B is given by the formula

$$R(z) = P(B - zI)^{-1}|\mathcal{H}$$

for each nonreal z. Here, as in Section 3, P denotes the orthogonal projection of $\mathcal{K}$ onto $\mathcal{H}$. For this generalized resolvent of A and for the generalized spectral function $\lambda \mapsto E(\lambda)$ $(\lambda \in R)$ of A considered in Section 3 and defined by the same selfadjoint extension B we have the formula

$$R(z) = \int_{-\infty}^{+\infty} \frac{1}{\lambda - z}\, dE(\lambda) \quad (Imz \neq 0). \tag{4.2}$$

Note that $$R^*(z) = R(\bar{z}) \tag{4.3}$$
for each nonreal z.

Let us consider the set of all generalized resolvents of A. Each of them corresponds to some selfadjoint extension B of A. According to [S1], [S3], this set is described as follows. The formula

$$R(\zeta) = (A_{F(\zeta)} - \zeta I)^{-1} \qquad (\zeta \in C_+) \tag{4.4}$$

defines a one-to-one correspondence between the set of all generalized resolvents of A and the set of all operator-valued function $\zeta \to F(\zeta)$ $(\zeta \in C_+)$ such that: 1) for each $\zeta \in C_+$ $F(\zeta) : \mathcal{N}_i \to \mathcal{N}_{-i}$ is a linear operator with norm $\|F(\zeta)\| \leq 1$; 2) the function $\zeta \mapsto F(\zeta)$ is holomorphic on C_+; 3) if for some $g \in \mathcal{N}_i$

$$\lim_{\substack{\zeta\to\infty \\ \epsilon<\arg\zeta<\pi-\epsilon}} F(\zeta)g = Xg \quad \text{(strong convergence)} \qquad \text{and}$$

$$\varliminf_{\substack{\zeta\to\infty \\ \epsilon<\arg\zeta<\pi-\epsilon}} (|\zeta|\,(\|g\| - \|F(\zeta)g\|)) < +\infty, \qquad \text{then } g = 0$$

Conditions 1)-3) and the maximum modulus principle imply that $F(\zeta)$ for each $\zeta \in C_+$ is an admissible operator with respect to A. If the operator A is densely defined, the condition 3) becomes trivial. By (4.1), (4.3), (4.4)

$$R(\eta) = (A_{F^*(\bar{\eta})} - \eta I)^{-1} \quad \text{for each } \eta \in C_-. \tag{4.5}$$

Let $z \mapsto R(z)$ $(Im(z) \neq 0)$ be an arbitrary generalized resolvent of A, and let $\zeta \mapsto F(\zeta)$ $(\zeta \in C)$ be the corresponding by (4.4) operator-valued function. Suppose that the operator-valued function $\zeta \mapsto Q(\zeta)$, $\zeta \mapsto K(\zeta)$ $(\zeta \in C_+)$, $\eta \mapsto Q_*(\eta)$ $(\eta \in C_-)$. are the same as in Section 1. As in Section 3 denote by Q_1 and Q_2 the orthogonal projections of $\mathcal{H}$ onto $\mathcal{N}_i$ and $\mathcal{N}_{-i}$, respectively. Note that $Q_1 = Q(i)$ and $Q_2 = Q_*(-i)$.

THEOREM 4.1.

$$\begin{aligned} Q_1R(\zeta) &= (K(\zeta)F(\zeta) - I)((i-\zeta)K(\zeta)F(\zeta) + (i+\zeta)I)^{-1}Q(\zeta) + \\ &+ \frac{1}{\zeta - i}(Q(\zeta) - Q(i)) \end{aligned} \tag{4.6}$$

for each $\zeta \in C_+$ (at $\zeta = i$ the divided difference on the right must be replaced by the derivative $Q'(i)$),

$$\begin{aligned} Q_1R(\eta) &= \frac{1}{i-\eta}(Q_1 - 2iF^*(\bar{\eta}) \times \\ &\times ((i+\eta)K^*(\bar{\eta})F^*(\bar{\eta}) + (i-\eta)I)^{-1}Q_*(\eta)) \quad \textit{for each } \eta \in C_-, \end{aligned} \tag{4.7}$$

$$\begin{aligned} Q_2R(\zeta) &= \frac{1}{i+\zeta}(-Q_2 + 2iF(\zeta) \times \\ &\times ((i-\zeta)K(\zeta)F(\zeta) + (i+\zeta)I)^{-1}Q(\zeta)) \quad \textit{for each } \zeta \in C_+, \end{aligned} \tag{4.8}$$

$$\begin{aligned} Q_2R(\eta) &= (K^*(\bar{\eta})F^*(\bar{\eta}) - I)(-(i+\eta)K^*(\bar{\eta})F^*(\bar{\eta}) - (i-\eta)I)^{-1}Q_*(\eta) + \\ &+ \frac{1}{\eta + i}(Q_*(\eta) - Q_*(-i)) \quad \textit{for each } \eta \in C_- \end{aligned} \tag{4.9}$$

(at $\eta = -i$ the divided difference on the right must be replaced by the derivative $Q'_*(-i)$).

Proof. For $\zeta \in C_+$ and $h \in \mathcal{H}$ put $g = R(\zeta)h$. Then by (4.4)

$$g = f + F(\zeta)u - u, \tag{4.10}$$

where $f \in Dom A$, $u \in \mathcal{N}_i$, and

$$h = (A - \zeta I)f + (i - \zeta)F(\zeta)u + (i + \zeta)u. \tag{4.11}$$

Applying the operator $Q(\zeta)$ to both sides, we obtain

$$Q(\zeta)h = (i - \zeta)K(\zeta)F(\zeta)u + (i + \zeta)u. \tag{4.12}$$

Hence

$$u = ((i - \zeta)K(\zeta)F(\zeta) + (i + \zeta)I)^{-1}Q(\zeta)h; \tag{4.13}$$

here the inverse operator on the right side exists, it is bounded, and is defined on the whole $\mathcal{N}_i$ because

$$\|K(\zeta)F(\zeta)\| \leq 1 \quad \text{and} \quad |i - \zeta| < |i + \zeta|.$$

By (4.10) and (4.11)

$$h + (\zeta - i)g = (A - iI)f + 2iu.$$

Applying the orthogonal projection Q_1 to both sides, we obtain

$$Q_1h + (\zeta - i)Q_1g = 2iu. \tag{4.14}$$

According to (4.12), (4.14),

$$Q(\zeta)h - Q_1h - (\zeta - i)Q_1g = (i - \zeta)(K(\zeta)F(\zeta) - I)u.$$

Hence for $\zeta \neq i$

$$Q_1g = (K(\zeta)F(\zeta) - I)u + \frac{1}{\zeta - i}(Q(\zeta) - Q(i))h.$$

Thus, on account of (4.13), we obtain

$$Q_1R(\zeta)h = (K(\zeta)F(\zeta) - I)((i - \zeta)K(\zeta)F(\zeta) + (i + \zeta)I)^{-1}Q(\zeta)h +$$
$$+ \frac{1}{\zeta - i}(Q(\zeta) - Q(i))h.$$

This proves (4.6).

Now for $\eta \in C_-$, $h \in \mathcal{H}$ put $g = R(\eta)h.$. Then, according to (4.5),

$$g = f + F^*(\bar{\eta})v - v, \tag{4.15}$$

where $f \in Dom A$, $v \in \mathcal{N}_{-i}$, and

$$h = (A - \eta I)f - (i + \eta)F^*(\bar{\eta})v - (i - \eta)v. \tag{4.16}$$

Applying the operator $Q_*(\eta)$ to both sides, we obtain

$$Q_*(\eta)h = -(i+\eta)K^*(\bar{\eta})F^*(\bar{\eta})v - (i-\eta)v,$$

whence

$$v = -((i+\eta)K^*(\bar{\eta})F^*(\bar{\eta}) + (i-\eta)I)^{-1}Q_*(\eta)h. \tag{4.17}$$

By (4.15), (4.16)

$$h + (\eta - i)g = (A - iI)f - 2iF^*(\bar{\eta})v$$

and, consequently,

$$Q_1h + (\eta - i)Q_1g = -2iF^*(\bar{\eta})v.$$

Therefore, on account of (4.17),

$$Q_1g = \frac{1}{i-\eta}(Q_1 - 2iF^*(\bar{\eta})((i+\eta)K^*(\bar{\eta})F^*(\bar{\eta}) + (i-\eta)I)^{-1})h$$

and (4.7) follows.

Formulas (4.8) and (4.9) are analogous to (4.7) and (4.6), respectively, we must only transpose i, $-i$, and ζ, η. □

For each nonreal z we put

$$R_{11}(z) = Q_1R(z)|\mathcal{N}_i,$$

$$R_{22}(z) = Q_2R(z)|\mathcal{N}_{-i},$$

$$R_{12}(z) = Q_1R(z)|\mathcal{N}_{-i},$$

$$R_{21}(z) = Q_2R(z)|\mathcal{N}_i.$$

By (3.3)-(3.6) and (4.2)

$$R_{jk}(z) = \int_{-\infty}^{+\infty} \frac{1}{\lambda - z} dE_{jk}(\lambda) \quad (j,k = 1,2).$$

Note that $(R_{jk}(z))^* = R_{kj}(\bar{z})$. On account of *Theorem 4.1* the following proposition is valid.

PROPOSITION 4.2. *For each* $\zeta \in C_+$

$$R_{11}(\zeta) = (K(\zeta)F(\zeta) - I)((i-\zeta)K(\zeta)F(\zeta) + (i+\zeta)I)^{-1}, \tag{4.18}$$

$$R_{22}(\zeta) = \frac{1}{i+\zeta}(-I + 2iF(\zeta)((i-\zeta)K(\zeta)F(\zeta) + (i+\zeta)I)^{-1}K(\zeta)),$$

$$R_{12}(\zeta) = (K(\zeta)F(\zeta) - I)((i-\zeta)K(\zeta)F(\zeta) + (i+\zeta)I)^{-1}K(\zeta) + \\ + \frac{1}{\zeta - i}(K(\zeta) - K(i))$$

(for $\zeta = i$ the divided difference must be replaced by $K'(i)$),

$$R_{21}(\zeta) = \frac{1}{i+\zeta}(-K^*(i) + 2iF(\zeta)((i-\zeta)K(\zeta)F(\zeta) + (i+\zeta)I)^{-1}).$$

Note that formula (4.18) was obtained in our earlier article [S6].

Let us return to constructions considered in *Section 2.* Now they will concern the selfadjoint extension B of A in $\mathcal{K} \supset \mathcal{H}$ which defines the generalized spectral function and the generalized resolvent of A corresponding to the operator-valued function $\zeta \mapsto F(\zeta) \quad \zeta \in C_+$. As before, $\lambda \mapsto E^+(\lambda) \quad (\lambda \in (\mathcal{R}))$ denotes the spectral function of B. Also we shall use other notations introduced in Sections 2 and 3.

For $g_1 \in \sum_{k \in Z} U^k \mathcal{N}_i$ and $g_2 \in \sum_{k \in Z} U^k \mathcal{N}_{-i}$ we have

$$g_n = \int_{-\infty}^{+\infty} (dE(\lambda))(\Psi_n g_n)(\lambda) \quad (n = 1, 2). \tag{4.19}$$

We shall say that $g_n \in \mathcal{K}_n \quad (n = 1, 2)$ is an ordinary element of $\mathcal{K}_n$ if the equality (4.19) holds with the righthand side maintaining its usual meaning, that is, $\Psi_1 g_1$ or $\Psi_2 g_2$ is a $\mathcal{N}_i$-valued or, respectively, $\mathcal{N}_{-i}$-valued function for which the integral exists as a limit of the corresponding sum.

Denote by P_1^+ and P_2^+ the orthogonal projections of $\mathcal{K}$ onto $\mathcal{K}_1$ and $\mathcal{K}_2$, respectively.

PROPOSITION 4.3. *If for* $f \in \mathcal{H} \quad P_n^+ f \quad (n = 1, 2)$ *is an ordinary element of* $\mathcal{K}_n$*, then for each nonreal* z

$$Q_n R(z) f = \int_{-\infty}^{+\infty} \frac{1}{\lambda - z} (dE_{nn}(\lambda))(\Psi_n P_n^+ f)(\lambda). \tag{4.20}$$

Proof. If for $g_n = P_n^+ f$ (4.19) remains valid in an elementary sense, then for each nonreal z

$$(B - zI)^{-1} P_n^+ f = \int_{-\infty}^{+\infty} \frac{1}{\lambda - z} (dE^+(\lambda))(\Psi_n P_n^+ f)(\lambda).$$

Applying the operator Q_n^+ to both sides of this equality, and recalling (3.3), (3.4). we obtain

$$Q_n^+ (B - zI)^{-1} P_n^+ f = \int_{-\infty}^{+\infty} \frac{1}{\lambda - z} (dE_{nn}(\lambda))(\Psi_n P_n^+ f)(\lambda). \tag{4.21}$$

The operators $(B - zI)^{-1}$ and P_n^+ commute because $\mathcal{K}_n$ reduces B. Therefore

$$\begin{aligned} Q_n^+ (B - zI)^{-1} P_n^+ f &= Q_n^+ P_n^+ (B - zI)^{-1} f \\ &= Q_n^+ (B - zI)^{-1} f \\ &= Q_n P (B - zI)^{-1} f = Q_n R(z) f. \end{aligned}$$

This and (4.21) imply (4.20). □

Theorem 4.1, Proposition 4.2 and 4.3, together with the Stieltjes' inversion formula and formulas (1.2) and (1.3), clarify the connection between the spectral transformation $\Psi | \mathcal{H}$ and the spectral representation $\Phi + \Phi_*$.

We have only to explain how for $f \in \mathcal{H}$ Ψf can be found if the both orthogonal projections of Ψf onto the subspaces $\mathcal{F}_1$ and $\mathcal{F}_2$ defined by (3.7) and (3.8) are known. The Aronszajn's projection formula [A] solves this problem. We refer to [S10] for more details concerning this question.

Elsewhere we shall consider the situation with the spectral transformation $\Psi|\mathcal{H}$ when the operator-valued function $\zeta \mapsto F(\zeta)$ is identically equal to zero. Note that this case is closely connected with the theory of dilations of dissipative operators and corresponding function models [SzNF], [dBR], [P], [NV].

References

[A] N.Aronszajn. *Theory of reproducing kernels.* Trans. Amer.Math.Soc., **68** (1950), 337-404.

[B] Berezanskii Ju. M. *Expansions in Eigenfunctions of Selfadjoint Operators.* Transl. Math. monographs, Amer. Math. Soc. Providence, 1968, **17**, 809.

[dBR] L.de Branges,J.Rovnyak. *Canonical models in quantum scattering theory.* In: Perturbation theory and its applications in quantum mechanics, N.Y.,Wiley, 1966, 295-391.

[K] M.A.Krasnoselskii, *On selfadjoint extensions of Hermitian operators.* Ukr. mat. z., **1** (1949), 21-38 (Russian).

[Kr1] M.G.Krein. *Fundamental aspects of the representation theory of Hermitian operators with deficiency index (m,m).* Ukr. mat. zh., **1** (1949), 3-66 (Russian) (English translation: Amer. Math. Soc. Transl. (2), **97** (1970), 75-143).

[Kr2] M.G.Krein. *Analytic problems and results in the theory of linear operators in Hilbert space.* Proc. Int. Congress Math. (Moscow 1966), Mir, Moscow, 1968, 189-216 (Russian) (English translation: Amer. Math. Soc. Transl. (2), **90** (1970), 181-210).

[L1] M.S.Livšic. *On a class of linear operators in Hilbert space.* Mat. Sbornik N.S., **19** (1946), 239-262 (Russian) (English translation: Amer. Math. Soc. Transl. (2), **13** (1960), 61-83).

[L2] M.S.Livšic. *On the spectral decomposition of linear non-selfadjoint operators.* Mat. Sbornik N.S., **34** (1954), 145-199 (Russian) (English translation: Amer. Math. Soc. Transl. (2), **5** (1957), 67-114).

[N1] M.A.Naimark. *On selfadjoint extensions of second kind of a symmetric operator.* Izv.Akad Nauk SSSR, Ser. Mat., **4** (1940), 53-104 (Russian).

[N2] M.A.Naimark. *Spectral functions of a symmetric operator.* Izv.Akad Nauk SSSR, Ser. Mat., **4** (1940), 277-318 (Russian).

[NV] N.K.Nikolskii, V.I.Vasyunin. *A unified approach to function models and the transcription problem.* The Gohberg Anniversary Collection, Birkhäuser, Basel, 1989, 405-434.

[P] B.S.Pavlov. *On separability conditions for spectral components of a dissipative operator.* Izv.Akad Nauk SSSR, Ser.Mat., **39** (1975), 123-148 (Russian).

[S1] A.V.Strauss. *Generalized resolvents of symmetric operators.* Izv. Akad. Nauk SSSR, Ser. Mat., **18** (1954), 51-86 (Russian).

[S2] A.V.Strauss. *On the extensions and the characteristic function of a symmetric operator.* Izv. Akad. Nauk SSSR, Ser. Mat., **32** (1968),186-207 (Russian) (English translation: Math. USSR-Izvestija, **2** (1968) 181-204).

[S3] A.V.Strauss. *On the extensions and generalized resolvents of a symmetric operator which is not densely defined.* Izv. Akad. Nauk SSSR, Ser. Mat., **34** (1970), 175-202 (Russian) (English translation: Math.USSR-Izvestija, **4** (1970) 179-208).

[S4] A.V.Strauss. *On spectral decompositions of a regular symmetric operator.* Dokl. Akad. Nauk SSSR, **204** (1972), 52-55 (Russian) (English translation: Soviet Math. Dokl., **13** (1972) 614-618).

[S5] A.V.Strauss. *On resolvents of extensions of a symmetric operator.* Funct. Analysis, Ulyanovsk, **8** (1977), 162-173 (Russian).

[S6] A.V.Strauss. *On spectral theory of regular symmetric operators.* Funct.Analysis, Ulyanovsk, **10** (1978), 145-153 (Russian).

[S7] A.V.Strauss. *Pseudo resolvents and representation of linear operators.* Funct. Analysis, Ulyanovsk, **29** (1989), 141-152 (Russian).

[S8] A.V.Strauss. *On the theory of regular and entire operators.* Funct. Analysis, Ulyanovsk, **33** (1992), 65-83 (Russian).

[S9] A.V.Strauss. *Spectral representations of linear operators.* Funct. Analysis, Ulyanovsk, **34** (1993), 80-93 (Russian).

[S10] A.V.Strauss. *On Aronszajn's projection formula.* Funct. Analysis, Ulyanovsk, **35** (1994), 123-136 (Russian).

[SzNF] B.Sz.-Nagy, C.Foias. *Harmonic Analysis of Operators on Hilbert Space.* North-Holland, Amsterdam, 1970.

Department of Physics and Mathematics
Ulyanovsk State Pedagogical University
Ulyanovsk, 432700 Russia

AMS Classification Numbers: 47A10, 47A20,
47A45, 47A67, 47B25.

Operator Theory:
Advances and Applications, Vol. 87

HANKEL TYPE OPERATORS, BOURGAIN ALGEBRAS, AND ISOMETRIES

T. TONEV and K. YALE

Isometries and the complete continuity of Hankel type operators together with certain known Bourgain algebras are used to give a simple proof of the Poincaré theorem concerning the biholomorphic inequivalence of the unit ball and the unit polydisc in $\mathbf{C}^n$.

1. Hankel type operators and Bourgain algebras. Let B be a commutative Banach algebra with norm $\|\cdot\|$ and let $A \subset B$ be a linear subspace (not necessarily closed). We let $c_o^w(A)$ denote the space of weakly null sequences in A. The notion of the Bourgain algebra of A with respect to B was introduced by J. Cima and R. Timoney [2] in their study of Dunford-Pettis property of uniform algebras and it is based on a construction of J. Bourgain involving operators of Hankel type.

Definition 1. (i) *The Hankel type operator S_f on A generated by the element $f \in B$ is the mapping $S_f : A \to B/A$ defined as $S_f(g) = \pi_A(fg)$, where $\pi_A : B \to B/A$ is the natural projection.*

(ii) *The Bourgain algebra $(A, B)_b$ of A with respect to B is the set of all elements $f \in B$ for which S_f is completely continuous.*

Thus $f \in (A, B)_b$ in case $S_f(c_o^w(A)) \subset c_o^{\|\cdot\|}(B/A)$. In other words $(A, B)_b$ consists of all $f \in B$ such that for every $\{\varphi_n\} \in c_o^w(A)$ there exists a sequence $\{g_n\}$ in A for which

$$\lim_{n\to\infty} \|\varphi_n f - g_n\| = 0.$$

It is known that $(A, B)_b$ is a commutative Banach algebra; moreover, $A \subset (A, B)_b$ if A is an algebra (e.g. [2]). We mention the following simple, but useful fact about Bourgain algebras: if $A \subset B \subset C$ are commutative Banach algebras then $(A, B)_b = (A, C)_b \cap B$. We call B the *enveloping algebra* of the Bourgain algebra $(A, B)_b$. Note that the structure of $(A, B)_b$ is quite sensitive to B.

Research partially supported by grants from the University of Montana and from the National Science Foundation.

The Bourgain algebra construction can be applied to a more general situation. Let $c_o^w(M)$ denote the space of weakly null sequences in an arbitrary linear subspace M of B. We can regard c_o^w as a function on the set of linear subspaces of B. Denote the Bourgain algebra more precisely as $(A,B;c_o^w)_b$. We can replace c_o^w by a different assignment $\mathcal{S}$ of sequences to subspaces and define the space $(A,B;\mathcal{S})_b$ in a way analogous to $(A,B;c_o^w)_b$ simply by requiring $\{\varphi_n\} \in \mathcal{S}_A$ instead of $\{\varphi_n\} \in c_o^w(A)$, i.e. $(A,B;\mathcal{S})_b = \{f \in B : S_f(\mathcal{S}_A) \subset c_o^{\|\cdot\|}(B/A)\}$. If $\mathcal{S}$ is smaller than c_o^w (i.e. $\mathcal{S}_M \subset c_o^w(M)$ for all M) then $(A,B;\mathcal{S})_b \supset (A,B;c_o^w)_b = (A,B)_b$ is larger. If $\mathcal{S}$ is larger than c_o^w then $(A,B;\mathcal{S})_b$ is smaller than $(A,B)_b$. For example, if $\mathcal{S} = \mathcal{B}^w$ where $\mathcal{B}_M^w$ is the space of weakly bounded sequences in M and if A is an algebra then $(A,B;\mathcal{B}^w)_b$ is simply the norm closure of A.

The argument in [2] can be applied to this general setting to show that $(A,B;\mathcal{S})_b$ is a closed subalgebra of B whenever the assignment $\mathcal{S}$ of sets of sequences $\mathcal{S}_M$ to linear subspaces M of B satisfies the following four properties:

(i) $\mathcal{S}_M \subset \mathcal{S}_N$ for all $M \subset N$,

(ii) $\mathcal{S}_M \subset \mathcal{B}_M^w$ for all M,

(iii) if $\{\varphi_n\} \in \mathcal{S}_A$ and if $f \in B$ then $\{f\varphi_n\} \in \mathcal{S}_B$, and

(iv) if $\{\varphi_n\} \in \mathcal{S}_B$ and if $\psi_n \in A$ with $\|\varphi_n - \psi_n\| \to 0$ as $n \to \infty$, then $\{\psi_n\} \in \mathcal{S}_A$.

Properties (i) and (ii) are required for all subspaces M and N while (iii) and (iv) are needed only for the fixed A and B under consideration. Observe that by (ii) the elements of the sequences in $\mathcal{S}_M$ are in M.

The argument is actually valid in the more general setting of commutative Frechét algebras. The uniform boundedness principle and property (ii) are needed to show that $(A,B;\mathcal{S})_b$ is closed. Very little is known concerning Bourgain algebras in this setting.

The Bourgain algebra $(A,B)_b$ contains important information about A. If A is an algebra of continuous functions on a set Ω then $(A,B)_b$ contains information about Ω as well and we apply this fact to the problem of biholomorphic equivalency in the next section.

The generalized Bourgain algebras $(A,B;\mathcal{S})_b$ become less closely associated to A and B as $\mathcal{S}$ becomes smaller. Is every closed algebra lying between A and B of the form $(A,B;\mathcal{S})_b$ for some $\mathcal{S}$? When $A = \{0\}$ and B has an unit then $A \cup \{1\} \subset (A,B;\mathcal{S})_b$ implies $(A,B;\mathcal{S})_b = B$ because $\mathcal{S}_A$ consists only of the zero sequence. A similar argument applies whenever A is an *ideal*. The question is mainly of interest when A is *not* an ideal in B. For example, the closed algebras between $A = H^\infty(\mathbf{T})$ and $B = L^\infty(\mathbf{T})$ on the circle $\mathbf{T}$ are characterized in terms of interpolating Blaschke products and it would be interesting to have another type of characterization. The closed subalgebras between $H^\infty(\mathbf{D})$ and $L^\infty(\mathbf{D})$ on the unit disc $\mathbf{D}$ have not been determined.

2. Complete continuity of Hankel type operators and isometries. Our next proposition says that the complete continuity is an invariant property of Hankel type operators under *algebraic isometries.* This result is not surprizing but is sufficient for our purpose.

Proposition 2. *Let $A \subset C$ and $B \subset D$ be two pairs of commutative Banach algebras. If $T : C \to D$ is an isometric algebra isomorphism with $TA = B$, then the Hankel type operator $S_{Tf} : B \to D/B$ is completely continuous if and only if $S_f : A \to C/A$ is completely continuous.*

Proof: First note that T carries the set $c_o^w(A)$ *onto* the set $c_o^w(B)$ because T^* is an isometry. Secondly, for $\psi_n, g_n \in A$ and $f \in C$ we have that

$$\|\psi_n f - g_n\| = \|T(\psi_n f - g_n)\| = \|(T\psi_n)(Tf) - Tg_n\|.$$

Thus, for example, if $\{\psi_n\} \in c_o^w(A)$ and S_f is completely continuous, then $\{T\psi_n\} \in c_o^w(B)$ and $Tg_n \in B$. Consequently, S_{Tf} is completely continuous since *all* weakly null sequences in B are of type $\{T\psi_n\}$, where $\{\psi_n\} \in c_o^w(A)$. The argument is readily reversible. ∎

Note that Proposition 2 holds for the case when T is a Banach algebra isomorphism, i.e. a continuous isomorphism of the algebraic structure that has a continuous inverse. Moreover, if T is an algebra homomorphism, while $T|_A$ is a topological linear isomorphism, one can show that S_{Tf} is completely continuous together with S_f.

Corollary 3. *Under the hypothesis of Proposition 2*

$$T((A, C)_b) = (TA, TC)_b = (B, D)_b.$$

Corollary 3 is quite natural since Bourgain algebras are defined solely in terms of algebraic and metric conditions. Let ∂B denote the Shilov boundary of B. The restriction map $T : B \to B|_{\partial B}$ is an isometric algebra isomorphism and from Proposition 2 and Corollary 3 we obtain:

Corollary 4. *Let X be a compact Hausdorff space and $A \subset B \subset C(X)$ are uniform algebras. Then*

(i) *The Hankel type operator $S_{f|_{\partial B}} : A|_{\partial B} \to B|_{\partial B}/A|_{\partial B}$ is completely continuous if and only if $S_f : A \to B/A$ is completely continuous.*

(ii) $(A|_{\partial B}, B|_{\partial B})_b = ((A, B)_b)|_{\partial B}$.

Observe that Proposition 2 and Corollary 3 do not hold for an isometry T between A and B that is not be extendable as an isometry between the enveloping algebras C and D. For example the Bourgain algebras of H^∞ relative to L^∞ for the unit disc $\mathbf{D}$ and the unit circle $\mathbf{T} = \partial\mathbf{D}$ are given by

$$(H^\infty(\mathbf{T}), L^\infty(\mathbf{T}))_b = H^\infty(\mathbf{T}) + C(\mathbf{T}), \tag{1}$$

$$(H^\infty(\mathbf{D}), L^\infty(\mathbf{D}))_b = H^\infty(\mathbf{D}) + \mathrm{UC}(\mathbf{D}) + V(\mathbf{D}), \tag{2}$$

where $\mathrm{UC}(\mathbf{D})$ is the space of uniformly continuous functions on $\mathbf{D}$ and $V(\mathbf{D})$ is the ideal of functions in $L^\infty(\mathbf{D})$ that vanish near the boundary (namely $f \in V(\mathbf{D})$ if for every $\varepsilon > 0$

there is a compact set $K \subset \mathbf{D}$ for which $\operatorname{ess\,sup}_{z\in K}|f(z)| < \varepsilon$) [1, 5]. Here the boundary value mapping $f \to f^*$ from $H^\infty(\mathbf{D})$ to $H^\infty(\mathbf{T})$ is an isometry that does not extend to the corresponding L^∞ enveloping algebras nor does it even extend to the corresponding Bourgain algebras of H^∞. However, the boundary value mapping extends isometrically from $H^\infty(\mathbf{D})$ to the algebra $\mathcal{U}(\mathbf{D}) = [H^\infty(\mathbf{D}), \overline{H^\infty}(\mathbf{D})]$ generated by $H^\infty(\mathbf{D}) \cup \overline{H^\infty}(\mathbf{D})$. Indeed, it extends to the generators of $\mathcal{U}(\mathbf{D})$ and a closure argument provides a further extension to $\mathcal{U}(\mathbf{D})$. Appropriate analogs of these facts hold for several complex variables and are used in the next section.

The algebra $\mathcal{U}(\mathbf{D}) = [H^\infty(\mathbf{D}), \overline{H^\infty}(\mathbf{D})]$ is a natural enveloping algebra for $H^\infty(\mathbf{D})$ since it can be identified with $C(\mathrm{sp}H^\infty(\mathbf{D}))$. There is a second reason for the algebra $\mathcal{U}(\mathbf{D})$ to be a natural enveloping algebra for $H^\infty(\mathbf{D})$. Namely isometries on $H^\infty(\mathbf{D})$ induced by automorphisms of $\mathbf{D}$ extend naturally to isometries of $\mathcal{U}(\mathbf{D})$.

For $\widehat{\ }$, the Gelfand map into $C(\mathcal{M})$, where $\mathcal{M} = \partial H^\infty$ is the maximal ideal space of L^∞ we have

Corollary 5.

(i) $(\widehat{H^\infty(\mathbf{T})}|_{\partial H^\infty(T)}, \widehat{L^\infty(\mathbf{T})}|_{\partial H^\infty(T)})_b = \widehat{H^\infty(\mathbf{T})}|_{\partial H^\infty(T)} + \widehat{C(\mathbf{T})}|_{\partial H^\infty(T)}.$

(ii) $(\widehat{H^\infty(\mathbf{D})}|_{\partial H^\infty(D)}, \widehat{L^\infty(\mathbf{D})}|_{\partial H^\infty(D)})_b = \widehat{H^\infty(\mathbf{D})}|_{\partial H^\infty(D)} + \widehat{\mathrm{UC}(\mathbf{D})}|_{\partial H^\infty(D)}.$

This follows immediately from (1), (2) and Corollary 4. Note that Corollary 5(ii) implies in particular that $\widehat{H^\infty(\mathbf{D})}|_{\partial H^\infty(D)} + \widehat{\mathrm{UC}(\mathbf{D})}|_{\partial H^\infty(D)}$ is a closed subalgebra of $\widehat{L^\infty(\mathbf{D})}|_{\partial H^\infty(D)}$ since Bourgain algebras are automatically closed.

3. Biholomorphic equivalence and Bourgain algebras. Here we apply completely continuous Hankel type operators to the problem of biholomorphic equivalence of domains in $\mathbf{C}^n$. Our arguments are functional analytic in character and involve Bourgain algebras of the space H^∞ relative to $\mathcal{U} = [H^\infty, \overline{H^\infty}] \subset L^\infty$.

Proposition 6. *If U_1 and U_2 are biholomorphically equivalent domains in $\mathbf{C}^n$ then the corresponding Bourgain algebras $(H^\infty(U_1), \mathcal{U}(U_1))_b$ and $(H^\infty(U_2), \mathcal{U}(U_2))_b$ are isometrically isomorphic.*

Proof: Let U_1 and U_2 be biholomorphically equivalent and $\tau : U_2 \to U_1$ is a biholomorphic mapping. Define the map $T : \mathrm{BC}(U_1) \to \mathrm{BC}(U_2)$ by

$$(Tf)(\mathbf{z}) = f(\tau(\mathbf{z}))$$

for all $f \in \mathrm{BC}(U_1)$ and $\mathbf{z} \in U_2$. The mapping T is an isometric algebra isomorphism with respect to the sup norms on U_1 and U_2. Moreover, $T(H^\infty(U_1)) = H^\infty(U_2)$, $T(\mathcal{U}(U_1)) = \mathcal{U}(U_2)$. Corollary 3 now implies that $T((H^\infty(U_1), \mathcal{U}(U_1))_b) = (H^\infty(U_2), \mathcal{U}(U_2))_b$. ∎

We can apply Proposition 6 to $\mathbf{B}^n$ and $\mathbf{D}^n$, the unit ball and the unit polydisc in $\mathbf{C}^n$, respectively, to obtain a functional-analytic proof of the famous Poincaré theorem which states that $\mathbf{B}^n$ and $\mathbf{D}^n$ are biholomorphically inequivalent for $n \geq 2$. From this viewpoint the sourse of the inequivalence resides in the peak point arguments used to determine the relevant Bourgain algebras.

Recall that every function $f \in H^\infty(\mathbf{D}^n)$ has *radial limits*

$$\lim_{r\to 1^-} f(r\mathbf{z}) = f^*(\mathbf{z})$$

at almost every point $\mathbf{z} \in \mathbf{T}^n$; The *radial boundary value function* $f^*(\mathbf{z})$ of every function $f \in H^\infty(\mathbf{D}^n)$ belongs to $H^\infty(\mathbf{T}^n)$ (e.g. [4, Theorem 2.3.2]). We need the following

Lemma 7. *Let $g \in \mathcal{U}(\mathbf{D}^n)$ generate a completely continuous Hankel type operator $S_g : H^\infty(\mathbf{D}^n) \to \mathcal{U}(\mathbf{D}^n)/H^\infty(\mathbf{D}^n)$. If g^* is the boundary value function of g on $\mathbf{T}^n$ then $S_{g^*} : H^\infty(\mathbf{T}^n) \to \mathcal{U}(\mathbf{T}^n)/H^\infty(\mathbf{T}^n)$ is completely continuous.*

This follows directly from the fact that every weakly null sequence $\{f_n\}$ in $H^\infty(\mathbf{T}^n)$ is of the form $f_n = \varphi_n^*$ where $\{\varphi_n\}$ is weakly null in $H^\infty(\mathbf{D}^n)$.

Suppose that $\mathbf{D}^n$ and $\mathbf{B}^n$ are biholomorphically equivalent and let $\tau : \mathbf{D}^n \to \mathbf{B}^n$ be a biholomorphic mapping. Define $T : \mathrm{BC}(\mathbf{B}^n) \to \mathrm{BC}(\mathbf{D}^n)$ by

$$(Tf)(\mathbf{z}) = f(\tau(\mathbf{z}))$$

for all $f \in \mathrm{BC}(\mathbf{B}^n)$ and $\mathbf{z} \in \mathbf{D}^n$ as before. Proposition 6 implies that

$$T((H^\infty(\mathbf{B}^n), \mathcal{U}(\mathbf{B}^n))_b) = (H^\infty(\mathbf{D}^n), \mathcal{U}(\mathbf{D}^n))_b,$$

which will yield a contradiction.

Let f be a fixed nonconstant function in the algebra $A(\mathbf{B}^n)$. Observe that $(T\overline{f})^*$ exists because $\overline{f} \in \overline{H^\infty}(\mathbf{B}^n)$ and so $T\overline{f} \in \overline{H^\infty}(\mathbf{D}^n)$. If we can show that the Hankel type operator

$$S_{T\overline{f}} : (H^\infty(\mathbf{D}^n)) \to \mathcal{U}(\mathbf{D}^n)/H^\infty(\mathbf{D}^n) \text{ is completely continuous,} \tag{3}$$

then from Lemma 7 it follows that for $n \geq 2$ the radial boundary value function $(T\overline{f})^*$ of the nonconstant *antiholomorphic* function $T\overline{f}$ belongs to $H^\infty(\mathbf{T}^n) = (H^\infty(\mathbf{T}^n), L^\infty(\mathbf{T}^n))_b = (H^\infty(\mathbf{T}^n), \mathcal{U}(\mathbf{T}^n))_b$ (see [3, 5]), which is impossible.

It remains to prove (3). Note that $\overline{f} \in C(\overline{\mathbf{B}^n})|_{\mathbf{B}^n}$ and also $\overline{f} \in \mathcal{U}(\mathbf{B}^n)$. An argument of Izuchi [3] shows that

$$C(\overline{\mathbf{B}^n})|_{B^n} \subset (H^\infty(\mathbf{B}^n), L^\infty(\mathbf{B}^n))_b.$$

Consequently, by the remark immediately following Definition 1 we have

$$\overline{f} \in (H^\infty(\mathbf{B}^n), L^\infty(\mathbf{B}^n))_b \cap \mathcal{U}(\mathbf{B}^n) = (H^\infty(\mathbf{B}^n), \mathcal{U}(\mathbf{B}^n))_b,$$

and hence

$$T\overline{f} \in T((H^\infty(\mathbf{B}^n), \mathcal{U}(\mathbf{B}^n))_b) = (H^\infty(\mathbf{D}^n), \mathcal{U}(\mathbf{D}^n))_b,$$

by Proposition 6, i.e. $S_{T\overline{f}}$ is completely continuous, as desired.

Note that the boundary value technique avoids the need of a direct reference to the more complicated Bourgain algebras of $H^\infty(\mathbf{D}^n)$ and $H^\infty(\mathbf{B}^n)$.

The authors thank the referees for their helpful remarks and comments.

4. References.

1. J. Cima, K. Stroethoff and K. Yale, *Bourgain algebras on the unit disk,* Pacific J. Math. **160** (1993), 27-41.
2. J. Cima and R. Timoney, *The Dunford-Pettis property for certain planar uniform algebras,* Michigan Math. J. **34** (1987), 99-104.
3. K. Izuchi, *Bourgain algebras of the disk, polydisk and ball algebras,* Duke Math. J. **66** (1992), 503-519.
4. W. Rudin, *Function Theory in the Polydiscs*, Benjamin, New York-Amsterdam, 1969.
5. K. Yale, *Bourgain algebras,* pp. 413-422 in "Proceedings Conference on Function Spaces" (edited by K. Jarosz), Lecture Notes in Pure and Appl. Math., Vol. 136, Marcel Dekker, 1992.

Department of Mathematical Sciences
The University of Montana
Missoula, Montana, USA

AMS classification numbers: 46 J 10, 46 J 15, 32 E 25, 32 H 02

Operator Theory:
Advances and Applications, Vol. 87

EFFECTIVE COMPUTATION OF OPERATORS DEFINED BY LINE INTEGRALS

EZIO VENTURINO

We extend the recent results obtained for the calculation of line integrals over smooth curves, for which an explicit parametrization is not known. The original parametrization of the curve is replaced by another one, interpolating to the curve at a set of given knots. We then investigate the use of Gaussian quadrature formulae specifically devised for this problem.

1 INTRODUCTION

Some recent researches of the author have dealt with the numerical calculation of line integrals, when an explicit parametrization is not explicitly known. In such cases it is necessary to interpolate the points at which the curve is known. This step introduces an extra error in the procedure that needs to be taken into account. It turns out however that under favorable circumstances, the convergence rate is higher than can be a priori expected. Our analyses point out in fact that for an interpolant of odd order q, cancellations lead to a rate of convergence of order $q+1$, if a sufficiently high order quadrature formula is used. These results, originally based on Newton-Cotes quadratures [1], have been extended to the case of Cauchy principal value integrals [2], and to the inverse problem [7] where also the possible use of classical Gaussian quadrature is addressed.

The underlying assumption of all these studies is that, apart possibly from the Cauchy principal value singularity, the integrand does not possess other forms of singularities. This assumption is quite restrictive, since it rules out an important class of singular integral equations, those for which the unknown function possesses endpoint integrable singularities.

Our aim in this paper is to address this problem and to find correctives to the standard Gaussian rules. Indeed the latter do not lead to any reasonable convergence behavior, if applied naively in the above described procedure when endpoint singularities are present. We intend to suitably adapt a scheme devised in [6] for Newton Cotes quadrature formulae, to obtain modified Gaussian quadratures, and use these in the scheme for the calculation of the singular integrals. These investigations are considered a preliminary step toward the construction of an automated package for the calculation of Cauchy transforms over lines in the complex domain.

The paper is organized as follows: in the next section we describe the problem and outline the general discretization procedure, in section 3 we study the modified Gaussian quadratures, providing also their convergence analysis. The latter are then applied for solving integrals over the interval $[-1,1]$ in section 4, and then extended to consider the direct problem for smooth integrands over lines in section 5. Finally the problem consisting in the effective calculation of the Cauchy transform over lines is treated in the last section.

2 NUMERICAL PROCEDURE

We begin with some preliminaries. To illustrate the problem, let $r(t)$, $-1 \leq t \leq 1$, be a smooth parametrization of the curve L in $\mathbf{R}^2$, with $r'(t) \neq 0$. We want to evaluate the line integral

$$I \equiv \int_L f(r)ds = \int_{-1}^{1} f(r(t))|r'(t)|dt.$$

Here f is a given smooth function, defined on the interior of L, with possible endpoint integrable singularities. In other words, we assume that near $+1$,

$$f(r(t)) \cong |r(1) - r(t)|^{\alpha} g(r(t)),$$

for some $\alpha > -1$, and g smooth. Similarly at the other endpoint -1,

$$f(r(t)) \cong |r(-1) - r(t)|^{\beta} k(r(t)),$$

for some $\beta > -1$ and k smooth. Incorporating these behaviors into the integrand we can rewrite it in the form

$$f(r(t)) \cong w(t) H(r(t)), \ w(t) = |r(1) - r(t)|^{\alpha} |r(-1) - r(t)|^{\beta}$$

with H smooth on $[-1,1]$. This clearly entails that it is also a bounded function, so that the case of the Cauchy principal value integrals is here excluded. It will be examined in the last section of the paper. Our task therefore reduces to the evaluation of the integral

$$I \equiv \int_{-1}^{1} w(t) H(r(t)) |r'(t)| dt. \tag{1}$$

Usually an explicitly differentiable parametrization $r(t)$ is supposed to be given. The above integral can then be calculated by evaluating the analytical derivative of $r(t)$, and by discretizing it by means of a suitable Gaussian quadrature, thus yielding

$$I \cong \sum_{i=1}^{M} w_i H(r(x_i)) |r'(x_i)|.$$

We want instead to follow a different approach, using an interpolant for $r(t)$, i.e. approximating the curve L. In terms of our integral, this is not a major problem, since it can be shown that in many instances f can be extended to a differentiable function that is defined in a domain Ω containing the curve L in its interior, so that f can be evaluated at nearby points of L. It follows that the same property holds for the newly introduced function H.

Consider now a uniform partition of the interval $\Delta \equiv [-1,1]$, given by the breakpoints t_j, $j = 0(1)n$, with $t_j = -1 + jh$, $h = 2/n$. On each subinterval $\Delta_j \equiv [t_{j-1}, t_j]$, $j = 1(1)n$, the parametrization $r(t)$ is replaced by the interpolation polynomial $r_p(t)$ of order $p \geq 2$ in each component of $r(t)$. This newly introduced interpolating function replaces $r(t)$ in the above integral. Finally the suitable Gaussian quadrature is applied to the rewritten integral.

A possible procedure we have attempted consisted in the following. On the first subinterval, $[-1, -1+h]$, we could use a Gauss-Jacobi formula with weight given by

$$w_1(t) = (1+t)^{\beta},$$

since this incorporates the singular behavior at the endpoint $t = -1$. For the intermediate subintervals $[t_{j-1}, t_j]$, $j = 2(1)n-1$, the integrand now being smooth, the appropriate quadrature is given by a Gauss-Legendre formula, i.e.

$$w_j(t) = 1, \ j = 2(1)n-1,$$

since no singularity can be present at any node t_j, $j = 2(1)n-1$. Finally on the last interval $[1-h, 1]$, the symmetric argument of the first subinterval holds, for which the appropriate weight now is

$$w_n(t) = (1-t)^{\alpha}.$$

This approach however suffers of the problem that its rate of convergence is essentially determined by the worst endpoint singular behavior of the integrand. Indeed, as the number of subintervals n increases, the rate of convergence will only be $O(h^{\min\{\alpha,\beta\}})$. The problem essentially lies in the fact that as n increases, the distance of the "central" subintervals Δ_j, for $j = 2(1)n-1$, from the endpoints decreases, and this fact is not taken into consideration by the method. To correct it, one could impose that as n increases also for the two subintervals next to the extreme subintervals Gaussian quadrature of Jacobi type be used, and then also for the next two inner subintervals and so on. But it seems very difficult a priori to determine for the innermost subintervals how appropriately to switch from the strategy of Gauss-Legendre quadrature to Gauss-Jacobi.

Rather than pursuing the modifications to the previous approach, we prefer to propose and investigate an alternative method, which could be described as a "smooth" transition from the former type of quadrature to the latter. The nodes and weights relative to each subinterval are based on modified Gaussian quadratures, incorporating the influence of nearby singularities even on subintervals where the integrand is supposedly smooth, as the interior subintervals.

We can summarize this discussion for the evaluation of line integrals with the formula,

$$\int_{t_{j-1}}^{t_j} w_j(t)g(t)dt \cong \sum_{k=1}^{q} \nu_k g(\eta_k),$$

where η_k denote the quadrature nodes and ν_k the respective weights. The original integral then is replaced by

$$\sum_{j=1}^{n} \int_{t_{j-1}}^{t_j} w_j(t)H(r(t))|r'(t)|dt \cong \sum_{j=1}^{n}\sum_{k=1}^{m} w_{kj}H(r_p(s_{kj}))|r'_p(s_{kj})| \equiv I_N,$$

with $N = nm$. The w_{kj}'s represent the quadrature weights appropriate for the quadrature in use in each subinterval $[t_{j-1}, t_j]$, and the s_{kj}'s the relative quadrature nodes. These shall be described in detail in the next section.

3 MODIFIED GAUSSIAN QUADRATURES

In this section we consider integrals of the form

$$I_\gamma = \int_p^q t^\gamma \, g(t) \, dt, \;\; 0 \le p < q \le 1, \;\; -1 < \gamma < 1,$$

where $g(t)$ is a smooth function. Observe that here the "nearby" singularity is located at the origin, so that to apply the present machinery to the problem outlined in section 2, it will be necessary to transform the intervals $[-1, 0]$ and respectively $[0, 1]$ into $[0, 1]$ itself, but in the latter case with reversed orientation. These tasks can be easily accomplished by means of the transformations

$$x = t - 1, \;\; y = 1 - t. \tag{2}$$

We want to devise Gauss-type quadratures for the calculation of I_γ. Our approach follows the standard development of the Gaussian quadratures, except that for the determination of weights and nodes we will explicitly solve the resulting nonlinear system. In order to devise formulae that are exact for a polynomial of degree m as high as possible, consider the moments

$$I_i = \int_p^q t^{\gamma+i} \, dt, \;\; i = 0(1)2m - 1.$$

In what follows we are going to concentrate on the cases $m = 1(1)4$, for two main reasons. The first one is that the algebraic approach proposed here cannot be extended beyond $m = 4$, since we need the explicit solution of algebraic equations of degree m. Secondly and most importantly, we are satisfied with the orders of convergence that are attained by these formulae. Indeed as it will be apparent later, on line integrals the quadrature needs to be coupled with an interpolation procedure on the line where the integration is to be performed. To be consistent, we need an interpolation error of the same order of the quadrature error. However, when m exceeds 4 this would lead to the use of interpolation of a moderately high a degree, which we prefer to avoid.

In our approach we consider t^γ as the weight of the quadrature to be constructed. Upon using m quadrature nodes, we will have $2m$ unknowns, the nodes x_k and the corresponding weights w_k. We thus obtain the system

$$I_i = \sum_{k=1}^{m} w_k x_k^i, \ i = 0(1)2m - 1. \tag{3}$$

Let us now examine in detail the various cases.

- For the case $m = 1$ the system is linear, so that the node and weight are easily found:

$$x_1 = \frac{\gamma+1}{\gamma+2}\frac{t^{\gamma+2}|_p^q}{t^{\gamma+1}|_p^q}, \quad w_1 = \frac{t^{\gamma+1}|_p^q}{\gamma+1}. \tag{4}$$

- For $m = 2$, the system can be reduced by eliminating w_1 from the last three equations, using the first equation

$$w_1 = I_0 - w_2; \tag{5}$$

this yields

$$w_2 = \frac{I_1 - x_1 I_0}{x_2 - x_1}. \tag{6}$$

From the remaining last two equations, eliminating $(I_1 - x_1 I_0)x_2$ we then have

$$x_2 = \frac{I_2 - x_1 I_1}{I_1 - x_1 I_0}, \tag{7}$$

from which we obtain a quadratic equation for x_1

$$(I_1^2 - I_0 I_2)x_1^2 + x_1(I_0 I_3 - I_2 I_1) + I_2^2 - I_3 I_1 = 0. \tag{8}$$

Upon its solution, working backwards on equations (7), (6), (5), the values of x_2, w_2, w_1 are easily determined.

The remaining cases are more complicated. The weights are obtained after elimination of the nodes. The nodes are instead found by solving a suitable algebraic equation.

- For $m = 3$, the weights are given by the formulae

$$w_3 = \frac{I_2 - (x_1 + x_2)I_1 + x_1 x_2 I_0}{(x_3 - x_1)(x_3 - x_2)}, \quad w_2 = \frac{I_1 - x_1 I_0 - w_3(x_3 - x_1)}{x_2 - x_1}, \tag{9}$$

$$w_1 = I_0 - w_2 - w_3. \tag{10}$$

The nodes are instead the roots of the cubic equation

$$x^3 - 2ux^2 + (u^2 + \chi)x - \chi u + v = 0 \tag{11}$$

where

$$u = -(D_1 D_3 + D_5 D_2)/D_0, \; v = (D_1 D_5 - D_3 D_4)/D_0$$

and

$$D_1 = I_2^2 - I_1 I_3, \; D_2 = I_0 I_2 - I_1^2, \; D_3 = I_3 I_2 - I_1 I_4,$$
$$D_4 = I_3^2 - I_4 I_2, \; D_5 = I_3 I_4 - I_2 I_5, \; D_0 = -(D_1^2 + D_4 D_2)$$
$$\chi = \left(I_4^2 - I_5 I_3 + D_3 v\right)/D_4.$$

- For the case $m = 4$ the weights are obtained from

$$w_4 = \frac{I_3 - (x_1 + x_2 + x_3)I_2 + (x_1 x_2 + x_1 x_3 + x_2 x_3)I_1 - x_1 x_2 x_3 I_0}{(x_4 - x_1)(x_4 - x_2)(x_4 - x_3)}, \tag{12}$$

$$w_3 = \frac{I_2 - (x_1 + x_2)I_1 + x_1 x_2 I_0 - w_4(x_4 - x_1)(x_4 - x_2)}{(x_3 - x_1)(x_3 - x_2)}, \tag{13}$$

$$w_2 = \frac{I_1 - x_1 I_0 - w_3(x_3 - x_1) - w_4(x_4 - x_1)}{x_2 - x_1}, \; w_1 = I_0 - w_2 - w_3 - w_4. \tag{14}$$

The nodes are the roots of the quartic equation

$$x^4 - yx^3 + zx^2 - tx - s = 0, \tag{15}$$

where the coefficients are obtained by solving the linear system

$$I_3 y - I_2 z + I_1 t - I_0 s = I_4$$
$$I_4 y - I_3 z + I_2 t - I_1 s = -I_5$$
$$I_5 y - I_4 z + I_3 t - I_2 s = I_6$$
$$I_6 y - I_5 z + I_4 t - I_3 s = -I_7.$$

The error analysis for the above formulae follows once again from the standard theory. In summary we have, denoting by $q_m(t)$ the $m-th$ polynomial of the family of orthogonal polynomials over $[p, q]$ with respect to the weight t^γ, see [5] p. 334,

Theorem 1 *For the above described quadrature formulae the following error estimates hold*

$$E_m\{g\} \equiv \int_p^q t^\gamma g(t)\, dt - \sum_{j=1}^m w_j g(x_j) = \frac{g^{(2m)}(\xi)}{(2m)!} \int_p^q q_m^2(t) t^\gamma dt. \tag{16}$$

The problem that is left then consists in estimating the integral term, in terms of the interval length $q - p = h$. Now observing that theorem 4, p. 203 of [5] holds for any generic family of orthogonal polynomials, we see that the zeros x_k, $k = 1(1)m$, of $q_m(t)$ must lie in $[p, q]$. Hence the following estimate is established observing that in (16) $q_m(t)$ is a monic polynomial

$$q_m^2(t) = \prod_{k=1}^m (t - x_k)^2 = O\left(h^{2m}\right), \ t \in [p, q].$$

On using this result in (16),

Theorem 2 *For the proposed modified Gaussian quadrature, the error estimate for some* $\zeta \in [p, q]$, *is given by*

$$|E_m\{g\}| \leq \frac{|g^{(2m)}(\zeta)|}{(2m)!} \frac{h^{2m+\gamma+1}}{\gamma + 1} = O(h^{2m+1+\gamma}). \tag{17}$$

Remark. The above result tells us that in the best possible case the quadrature is essentially of order almost $O(h^{2m+1})$, for $0 > \gamma > -1$. In the worst situation however, γ can be arbitrarily close to its lower bound, thus possibly giving only a convergence rate of order just slightly higher than $2m$.

4 PIECEWISE MODIFIED GAUSSIAN QUADRATURE ON AN INTERVAL

In this section we apply the formulae derived in the previous section to the calculation of integrals possessing endpoint singularities over the finite interval, here normalized to be $\Delta \equiv [-1, 1]$. By means of breakpoints we partition Δ into subintervals and apply the modified quadrature rules to each subinterval using a fixed number m of quadrature nodes. The convergence of the method is obtained as the number n of subintervals increases, and therefore its rate will be given in terms of this very same parameter. On using the results of the previous section on each subinterval $\Delta_k = [t_{k-1}, t_k]$, $k = 1(1)n$, by summing the estimates (17) over all the subintervals of the partition, observing that $nh = 2$, it is easy to prove the following result.

Theorem 3 *For the proposed composite rule based on modified piecewise Gaussian quadrature of order m, the quadrature for the integral is*

$$Q \equiv \int_{-1}^{1} w(t)g(t)dt \doteq \sum_{i=1}^{n}\sum_{k=1}^{m} w_k g(x_{ki}) \equiv Q_N^m \tag{18}$$

and has the convergence rate

$$E_N \equiv Q - Q_N^m = O(h^{2m+\gamma}), \tag{19}$$

where $\gamma = \min(\alpha, \beta)$, $w(t) = (1-t)^\alpha(1+t)^\beta$, and $N = nm$.

Remark. The bound (19) shows that the convergence rate is exponential in m, the number of Gaussian quadrature nodes, but since in the method m is fixed, (19) says that the convergence rate is a power of the number n of partitions.

To illustrate the empyrical behavior of these estimates, we provide a sample of results of our computations in tables 1-3. Specifically, we have calculated the integral (18), with $g(t) = \exp(t)$. The calculations show that the empyrical order of convergence, $\log_2(\frac{Q_{n-1}^m - Q_{n-2}^m}{Q_n^m - Q_{n-1}^m})$, is quite close to its theoretical value derived above.

In order to apply the formulae developed in section 3 to the case of singular integrals, we need to suitably reformulate the problem. We rewrite the task as follows

$$\begin{aligned} G(x) &= \int_{-1}^{1} \frac{w(t)g(t)}{t-x} dt \\ &= \int_{-1}^{1} w(t)\frac{g(t)-g(x)}{t-x} dt + g(x)\int_{-1}^{1} \frac{w(t)}{t-x} dt, x \in [-1,1]. \end{aligned}$$

The evaluation of the last term can be performed analytically, but also its numerical calculation has been discussed, see e.g. [4]. As this approach will be used once again in the last section to deal with Cauchy principal value integrals over a line, we do not provide here any further numerical evidence, and just refer to the set of results included in section 6.

5 THE CALCULATION OF LINE INTEGRALS

We now apply the formulae developed in sections 3 and 4 to the case of line integrals, by coupling them with suitable polynomial interpolants. The curve L is approximated by means of a suitable piecewise interpolant $r_p(t)$ of L, whose construction over each subinterval generated by the breakpoints $t_j = -1 + jh$, $j = 0, ..., n$, $h = 2/n$ is explained below. The line integral (1) is evaluated by means of the formula

$$I_N^m \equiv \sum_{j=1}^{n} I_j^m \equiv \sum_{j=1}^{n}\sum_{k=1}^{m} w_{kj}\Psi\left(r\left(s_{kj}\right)\right) H_j(r_p(s_{kj}))|r_p'(s_{kj})|.$$

Define indeed on the generic subinterval $[t_{j-1}, t_j]$ the functions

$$\Psi(r(t)) = \left(\frac{|r(t)| - |r(0)|}{t}\right)^{\gamma}$$

and

$$w_j(t) = \begin{matrix} |r(-1)-r(t)|^{\beta} \ for \ j=1(1)\left\lfloor \frac{n}{2} \right\rfloor \\ |r(1)-r(t)|^{\alpha} \ for \ j=\left\lfloor \frac{n+1}{2} \right\rfloor(1)n \end{matrix}, \quad H_j(t) = \begin{cases} |r(1)-r(t)|^{\alpha} H(r(t)) \ for \ j=1(1)\left\lfloor \frac{n}{2} \right\rfloor \\ |r(-1)-r(t)|^{\beta} H(r(t)) \ for \ j=\left\lfloor \frac{n+1}{2} \right\rfloor(1)n \end{cases}.$$

The integral can be rewritten to take into account the appropriate weight on each subinterval, namely $|r(-1) - r(t)|^{\beta}$ over $[-1, 0]$ and $|r(1) - r(t)|^{\beta}$ over $[0, 1]$

$$\sum_{j=1}^{n} \int_{t_{j-1}}^{t_j} w_j(t)\, \Psi(r(s_{kj}))\, H_j(r(t))\, |r'(t)|\, dt$$

and where H_j represent smooth functions. Let $E_N = I - I_N^m$ denote the error term. For the error analysis, we need to modify a little the approach of [1]. Using the mappings (2), we can just consider integrals of smooth functions $K = H_j$ of the type

$$\int_{t_{j-1}}^{t_j} t^{\gamma} \Psi(r(t))\, K(r(t))\, |r'(t)|\, dt, \ j = 1(1)n.$$

We break the error term into the following parts

$$\begin{aligned} E_j^m &= \int_{t_{j-1}}^{t_j} t^{\gamma} \Psi(r(t))\ K(r(t))\ |r'(t)|\, dt - I_j^m \\ &= E_{j,0} + E_{j,1} + E_{j,2} + E_{j,3}^m, \ j = 1(1)n \end{aligned}$$

with for $j = 1(1)n$

$$E_{j,0} = \int_{t_{j-1}}^{t_j} t^{\gamma} \left[\Psi(r(t)) - \Psi(r_p(t))\right] K(r(t))\, |r'(t)|\, dt,$$

$$E_{j,1} = \int_{t_{j-1}}^{t_j} t^{\gamma} \Psi(r_p(t)) \left[K(r(t)) - K(r_p(t))\right] |r'(t)|\, dt,$$

$$E_{j,2} = \int_{t_{j-1}}^{t_j} t^{\gamma} \Psi(r_p(t))\, K(r_p(t)) \left[|r'(t)| - \left|r_p'(t)\right|\right] dt,$$

$$E_{j,3}^m = \int_{t_{j-1}}^{t_j} t^{\gamma} \Psi(r_p(t))\, K(r_p(t)) \left|r_p'(t)\right| dt - I_j^m,$$

where w_{kj} and s_{kj} represent quadrature weights and nodes in section 4. We will estimate the interpolation errors $E_{j,0}$, $E_{j,1}$, $E_{j,2}$, while for $E_{j,3}$ the estimates of section 4 hold, in view of the fact that Ψ and K are smooth. Notice indeed that $\Psi(r(0)) = \gamma |r'(0)| \neq 0$, $\frac{d}{dt}\Psi(r(0)) = 0$. We will later need also an estimate of its gradient

$$\nabla\Psi(r(t)) = \frac{\gamma}{t^{\gamma}} \frac{r(t) - r(0)}{|r(t) - r(0)|^{2-\gamma}} \simeq t^{-1}, \ as\ t \longrightarrow 0. \tag{20}$$

For the error estimates we use Taylor series expansions; denoting by Φ either of the functions K or Ψ, for $j = 1\,(1)\,n$

$$\begin{aligned}\Phi(r(t)) - \Phi(r_p(t)) &= \nabla\Phi(r(t)) \cdot (r(t) - r_p(t)) + o(h^p) \\ &= (\nabla\Phi(r(t_j)) + O(t - t_j)) \cdot (r(t) - r_p(t)) + o(h^p).\end{aligned}$$

Also we need

$$F(r(t)) = F(r(t_j)) + O(t - t_j), \; j = 1\,(1)\,n$$

where here F can represent either F, Ψ, or r'. Finally, let us denote the error as it is standard in interpolation theory,

$$r(t) - r_p(t) = \omega_p(t)\, r[v_1, ..., v_p, t] \equiv \omega_p(t)\, r_1[t]$$

where $\omega_p(t) = \prod_{i=1}^{p}(t - v_i)$, the v_i's being the equispaced interpolation knots in the subinterval $[t_{j-1}, t_j]$, $j = 1\,(1)\,n$

$$v_j = \frac{j-1}{p-1}h, \; j = 1(1)p.$$

Taylor expansion allows us to write also

$$r(t) - r_p(t) = \omega_p(t)\, r_1[t] = \omega_p(t)\,(r_1[t_{j-1}] + (t - t_{j-1})\, r_2[\zeta]), \; \zeta \in [t_{j-1}, t_j], \; j = 1\,(1)\,n.$$

Using these formulae then for $j = 1\,(1)\,n$

$$E_{j,0} = \nabla\Psi(r(t_{j-1}))\, r_1[t_{j-1}]\, K(r(t_{j-1}))\, |r'(t_{j-1})| \int_{t_{j-1}}^{t_j} t^\gamma \omega_p(t)\, dt + o\left(h^{p+\gamma+1}\right), \quad (21)$$

$$E_{j,1} = \Psi(r(t_{j-1}))\, r_1[t_{j-1}]\, \nabla K(r(t_{j-1}))\, |r'(t_{j-1})| \int_{t_{j-1}}^{t_j} t^\gamma \omega_p(t)\, dt + o\left(h^{p+\gamma+1}\right). \quad (22)$$

For the last term instead recall from [1] that

$$\begin{aligned}\left|r'(t) - r_p'(t)\right| &= \frac{|r'(t)|^2 - \left|r_p'(t)\right|^2}{|r'(t)| + \left|r_p'(t)\right|} \\ &= \left(\frac{1}{2\,|r'(t)|} + O\left(h^{p-1}\right)\right) \{2\omega'(t)\, r'(t) \cdot r_1[t] + 2\omega(t)\, r'(t) \cdot r_2[t] \\ &\quad + O\left(h^{2p-2}\right)\},\end{aligned}$$

so that again using Taylor expansions about t_{j-1}, $j = 1\,(1)\,n$ for r', r_1, r_2 and for suitable constants C_k, $k = 1, 2, 3$

$$E_{j,2} = C_1 \int_{t_{j-1}}^{t_j} t^\gamma \omega_p'(t)\, dt + C_2 \int_{t_{j-1}}^{t_j} t^\gamma \omega_p(t)\, dt + C_3 \int_{t_{j-1}}^{t_j} t^\gamma (t - t_{j-1})\, \omega_p'(t)\, dt + o\left(h^{p+\gamma+1}\right). \quad (23)$$

Evidently, the dominant term here is the first one, so that

$$E_{j,2} = O\left(h^{p+\gamma}\right), \ j = 1\,(1)\,n. \tag{24}$$

However, forgetting for now for n odd the last interval $\Delta_n = [1-h, 1]$, if we consider contributions over two neighboring intervals $[t_{2j-1}, t_{2j}]$ and $[t_{2j}, t_{2j+1}]$, $j = 1(1)\hat{n}$ with $\hat{n} = \left\lfloor \frac{n-1}{2} \right\rfloor$, omitting higher order terms we have

$$\begin{aligned}
E^1_{2j,2} &= \int_{t_{2j-1}}^{t_{2j}} t^\gamma \omega_p'(t)\, dt + \int_{t_{2j}}^{t_{2j}\ +1} t^\gamma \omega_p'(t)\, dt \\
&= (t_{2j})^\gamma \left[\int_{t_{2j-1}}^{t_{2j}} \left(1 - \frac{t_{2j}-t}{t_{2j}}\right)^\gamma \omega_p'(t)\, dt + \int_{t_{2j}}^{t_{2j}\ +1} \left(1 + \frac{t-t_{2j}}{t_{2j}}\right)^\gamma \omega_p'(t)\, dt\right] \\
&= (t_{2j})^\gamma \left[\int_{t_{2j-1}}^{t_{2j}} \omega_p'(t)\, dt + \int_{t_{2j}}^{t_{2j}\ +1} \omega_p'(t)\, dt\right] + \gamma\,(t_{2j})^\gamma \left[\int_{t_{2j-1}}^{t_{2j}} + \int_{t_{2j}}^{t_{2j}\ +1}\right] \frac{t-t_{2j}}{t_{2j}} \omega_p'(t)\, dt.
\end{aligned}$$

Since $\omega_p(t_k) = 0$, for $k = 0, ..., n$, the only surviving terms are of the form

$$h^\gamma \left[\int_{t_{2j-1}}^{t_{2j}} + \int_{t_{2j}}^{t_{2j}\ +1}\right] t\omega_p'(t)\, dt.$$

Thus combining with (23),

$$E^1_{2j,2} = \begin{cases} O\left(h^{p+\gamma+1}\right) & \text{for } p \text{ even} \\ O(h^{p+\gamma+2}) & \text{for } p \text{ odd} \end{cases}, \ j = 1\,(1)\,\hat{n}. \tag{25}$$

In view of (21), (22) and (25), for $j \geq 2$ and assuming also the worst case scenario, namely p even,

$$E_{j,0} = O\left(h^{p+\gamma+1}\right), \ E_{j,1} = O\left(h^{p+\gamma+1}\right), \ E_{j,2} = O\left(h^{p+\gamma+1}\right).$$

Summing over all these subintervals, and taking into account (24) for $j = n$ and assuming the worst endpoint singularity, i.e. $\gamma < 0$,

$$\sum_{j=2}^{n/2} [E_{j,0} + E_{j,1} + E_{j,2}] + O\left(h^{p+\gamma}\right) = O\left(h^{p+\gamma}\right). \tag{26}$$

In the first subinterval instead we have to remember (20), so that

$$E_{1,0} = O\left(h^{p+\gamma}\right), \ E_{1,1} = O\left(h^{p+\gamma+1}\right), \ E_{1,2} = O\left(h^{p+\gamma+1}\right). \tag{27}$$

But this is compatible with (26), so that from (26) and (27)

$$\sum_{j=1}^{n} [E_{j,0} + E_{j,1} + E_{j,2}] = O\left(h^{p+\gamma}\right). \tag{28}$$

Remark. In view of (27), it is now clear why a more refined analysis on (21), (22), (25) is useless in this situation, contrary to what is observed in ([1]).

Coupling (28) with the quadrature error estimate, we easily establish the following convergence result.

Theorem 4 *Let $p \geq 2$ and $m \geq 1$ be integers as used in the above definitions of the interpolation and quadrature rules. Assume the curve L has a parametrization $r \in C^{p+2}[a,b]$ with $r'(t) \neq 0$. For $f = wH$, $H \in C(L)$, assume the composition map $H \circ r \in C^{2m}[a,b]$. Furthermore, without loss of generality, assume H is the restriction to L of a twice continuously differentiable function of several variables, also called H, which is defined in an open neighborhood U of L. Then for the error we have*

$$E_N(f) = O(h^{Min\{p+\gamma, 2m+\gamma\}}).$$

Remark. By taking $m \geq p/2$, we see that it is the order of approximation of r_p to r that governs the convergence rate, rather the quadrature error.

In tables 4-7 we give numerical evidence of the performance of these rules. In all these examples, the integral (1) is evaluated, where $H(r(t)) = \exp(-(x+y))$, with $r(t) = (x(t), y(t)) = (\cos(t), t^3)$, $t \in [-1,1]$.

6 EXTENSION FOR THE CALCULATION OF OPERATORS DEFINED BY CAUCHY PRINCIPAL VALUE INTEGRALS

For the Cauchy principal value integrals, we can proceed in a similar way as done on the real line. Following the approach of [2], we consider the problem formulated as

$$C(s) = \int_{-1}^{1} g(r(t))(\ell(t) - \ell(s))^{-1}|r'(t)|dt, \; s \in (-1,1). \tag{29}$$

We assume than that the endpoint behavior of the integrand is expressed by the weight function of the Jacobi polynomials, so that we can redefine the integrand as the product $g(r(t)) = w(r(t))f(r(t))$. The quadrature we use is given by

$$\begin{aligned} C_N &= \sum_{j=1}^{n}\sum_{k=1}^{q} w_{kj}[g(r_p(s_{kj})) - g(r_p(s))][\ell_p(s_{kj}) - \ell_p(s)]^{-1}|r_p'(s_{kj})| \\ &\quad + g(r_p(s)) \ln|1 - \ell(1)/\ell_p(s)|, \end{aligned} \tag{30}$$

with

$$\ell_p(t) = \int_0^t |r_p'(u)|du,$$

where $N = nq$ and the nodes and weights are the ones obtained from the modified Gaussian quadrature. The error can then be studied by defining the functions

$$f(r(t)) = [g(r(t)) - g(r(s))][\ell(t) - \ell(s)]^{-1}, \; f(r_p(t)) = [g(r_p(t)) - g(r_p(s))][\ell_p(t) - \ell_p(s)]^{-1}.$$

and following the steps of [2], just appropriately using the results of sections 3, 4 and 5. We then obtain

Theorem 5 *If the hypotheses of theorem 1 on the parametrization of the curve are satisfied, and if $g \in C^{2q+1}[0,1]$, then for (30) a composite method using a q-point Gaussian quadrature rule and a p order interpolatory formula for the parametrization of the line has convergence rate p, if $q \geq p/2$.*

In the tables 8-11 we summarize the numerical evidence for the above described scheme of integration. The computations are done for the evaluation of the integral (29), with the parametrization $r(t) = (\log(2+t^3), \sin(t)), t \in [-1,1]$, and for the integrand $g(t) = (1-t)^\alpha (1+t)^\beta \exp[-(x(t)+y(t))]$. The singularity is located in all the examples at $s = -.83$, but from other examples not here reported, the results are independent of its location within the interval $[-1,1]$, as well as of the endpoint singularities.

REFERENCES

[1] K.E.Atkinson, E.Venturino, (1993), Numerical evaluation of line integrals, *SIAM Journal on Numerical Analysis*, vol. 30, p. 882-888.

[2] C.Barone, E.Venturino, (1993), On the numerical evaluation of Cauchy transforms, *Numerical Algorithms*, vol. 5, p. 429-436.

[3] D. Chien (1991), *Piecewise Polynomial Collocation for Integral Equations on Surfaces in Three Dimensions*, PhD thesis, University of Iowa, Iowa City.

[4] W. Gautschi, J. Wimp, (1987), Computing the Hilbert transform of a Jacobi weight function, BIT, 27, p. 203-215.

[5] E. Isaacson, H.B. Keller, (1966), *Analysis of numerical methods*, Wiley.

[6] F.Potra, E.Venturino, (1993), Low order methods for Cauchy principal value integrals with endpoint singularities, *IMA Journal of Numerical Analysis*, vol. 14, p. 295-310.

[7] E. Venturino, (1995), On the solution of discretely given Fredholm integral equations over lines, *Approximation Theory, Wavelets and Applications,* (S.P. Singh, Ed.), Kluwer, p. 513-522.

Mathematics Department,
University of Iowa,
Iowa City, IA 52242, USA

AMS MSC: 65D30, 65D05

APPENDIX

n	Q_n	$Q_n - Q_{n-1}$	$\log_2(\frac{Q_{n-1}-Q_{n-2}}{Q_n-Q_{n-1}})$
2	1386.8834638159790		
4	1386.9335475260600	.50084E-01	
8	1386.9417715113740	.82240E-02	2.61
16	1386.9422706736970	.49916E-03	4.04
32	1386.9423028626700	.32189E-04	3.95
64	1386.9423049465910	.20839E-05	3.95
128	1386.9423050817310	.13514E-06	3.95
256	1386.9423050905020	.87714E-08	3.95
512	1386.9423050910710	.56889E-09	3.95
1024	1386.9423050911080	.37062E-10	3.94

Table 1: Calculation of the integral (18), with endpoint singularities given by $\alpha = -.999, \beta = -.99$, number of quadrature nodes used $q = 2$

n	Q_n	$Q_n - Q_{n-1}$	$\log_2(\frac{Q_{n-1}-Q_{n-2}}{Q_n-Q_{n-1}})$
2	1386.9419226498920		
4	1386.9420546375550	.13199E-03	
8	1386.9423011605880	.24652E-03	-.90
16	1386.9423050251670	.38646E-05	6.00
32	1386.9423050900440	.64877E-07	5.90
64	1386.9423050910940	.10500E-08	5.95
128	1386.9423050911100	.16598E-10	5.98
256	1386.9423050911110	.22737E-12	6.19

Table 2: Calculation of the integral (18), with endpoint singularities given by $\alpha = -.999, \beta = -.99$, number of quadrature nodes used $q = 3$

n	Q_n	$Q_n - Q_{n-1}$	$\log_2(\frac{Q_{n-1}-Q_{n-2}}{Q_n-Q_{n-1}})$
2	1386.9423035459670		
4	1386.9422977719610	-.57740E-05	
8	1386.9423050577030	.72857E-05	-.34
16	1386.9423050909540	.33251E-07	7.78
32	1386.9423050911100	.15598E-09	7.74
64	1386.9423050911110	.45475E-12	8.42

Table 3: Calculation of the integral (18), with endpoint singularities given by $\alpha = -.999, \beta = -.99$, number of quadrature nodes used $q = 4$

n	I_n	$I_n - I_{n-1}$	$\log_2(\frac{I_{n-1}-I_{n-2}}{I_n-I_{n-1}})$
2	16.4014878927241		
4	43.9700629909325	.27569E+02	
8	50.7670657454148	.67970E+01	2.02
16	52.9405075020835	.21734E+01	1.64
32	53.5928031466468	.65230E+00	1.74
64	53.7757366122800	.18293E+00	1.83
128	53.8248700782486	.49133E-01	1.90
256	53.8377583724809	.12888E-01	1.93
512	53.8410962010888	.33378E-02	1.95
1024	53.8419543248481	.85812E-03	1.96

Table 4: Calculation of the integral (1), with endpoint singularities given by $\alpha = -.99, \beta = -.89$, number of quadrature nodes used $q = 1$, interpolation order $p = 3$

n	I_n	$I_n - I_{n-1}$	$\log_2(\frac{I_{n-1}-I_{n-2}}{I_n-I_{n-1}})$
2	13.7954789725467		
4	53.4866886977736	.39691E+02	
8	53.7822822844566	.29559E+00	7.07
16	53.8355668943302	.53285E-01	2.47
32	53.8416748608500	.61080E-02	3.12
64	53.8422069169540	.53206E-03	3.52
128	53.8422462741797	.39357E-04	3.76
256	53.8422488720122	.25978E-05	3.92
512	53.8422490230435	.15103E-06	4.10
1024	53.8422490297576	.67141E-08	4.49

Table 5: Calculation of the integral (1), with endpoint singularities given by $\alpha = -.99, \beta = -.89$, number of quadrature nodes used $q = 2$, interpolation order $p = 4$

n	I_n	$I_n - I_{n-1}$	$\log_2(\frac{I_{n-1}-I_{n-2}}{I_n-I_{n-1}})$
2	53.0838147322906		
4	53.8133700930639	.72956E+00	
8	53.8406167308681	.27247E-01	4.74
16	53.8421933275570	.15766E-02	4.11
32	53.8422476789435	.54351E-04	4.86
64	53.8422490024784	.13235E-05	5.36
128	53.8422490292070	.26729E-07	5.63
256	53.8422490297017	.49474E-09	5.76
512	53.8422490296965	-.52793E-11	6.55

Table 6: Calculation of the integral (1), with endpoint singularities given by $\alpha = -.99, \beta = -.89$, number of quadrature nodes used $q = 3$, interpolation order $p = 6$

n	I_n	$I_n - I_{n-1}$	$\log_2(\frac{I_{n-1}-I_{n-2}}{I_n-I_{n-1}})$
2	17.9398456275961		
4	53.8406370002045	.35901E+02	
8	53.8422151513609	.15782E-02	14.47
16	53.8422486972661	.33546E-04	5.56
32	53.8422490275462	.33028E-06	6.67
64	53.8422490296935	.21473E-08	7.27

Table 7: Calculation of the integral (1), with endpoint singularities given by $\alpha = -.99, \beta = -.89$, number of quadrature nodes used $q = 4$, interpolation order $p = 7$

n	C_n	$C_n - C_{n-1}$	$\log_2(\frac{C_{n-1}-C_{n-2}}{C_n-C_{n-1}})$
4	-24.8161873408085		
8	-21.4558813207764	.33603E+01	
16	-20.3920638629499	.10638E+01	1.66
32	-20.0701424661158	.32192E+00	1.72
64	-20.0258171229097	.44325E-01	2.86
128	-20.0155631700582	.10254E-01	2.11
256	-20.0133626526278	.22005E-02	2.22
512	-20.0128417444683	.52091E-03	2.08
1024	-20.0127141917267	.12755E-03	2.03

Table 8: Calculation of the singular integral (29), using (30), where the endpoint singularities are given by $\alpha = -.93, \beta = -.07$, location of singularity at $t = -.83$ number of quadrature nodes used $q = 1$, interpolation order $p = 5$

n	C_n	$C_n - C_{n-1}$	$\log_2(\frac{C_{n-1}-C_{n-2}}{C_n-C_{n-1}})$
4	-20.9321769875053		
8	-20.5298650672099	.40231E+00	
16	-19.8534867182626	.67638E+00	-.75
32	-20.0224105732950	-.16892E+00	2.00
64	-20.0106751247412	.11735E-01	3.85
128	-20.0127697725793	-.20946E-02	2.49
256	-20.0126688031879	.10097E-03	4.37
512	-20.0126690542661	-.25108E-06	8.65

Table 9: Calculation of the singular integral (29), using (30), where the endpoint singularities are given by $\alpha = -.93, \beta = -.07$, location of singularity at $t = -.83$ number of quadrature nodes used $q = 2$, interpolation order $p = 4$

n	C_n	$C_n - C_{n-1}$	$\log_2(\frac{C_{n-1}-C_{n-2}}{C_n-C_{n-1}})$
4	-20.2797422964244		
8	-19.9941118664595	.28563E+00	
16	-20.0142236811867	-.20112E-01	3.83
32	-20.0126354086122	.15883E-02	3.66
64	-20.0126725104673	-.37102E-04	5.42
128	-20.0126725295965	-.19129E-07	10.92
256	-20.0126725591574	-.29561E-07	-.63
512	-20.0126725626454	-.34880E-08	3.08
1024	-20.0126725626755	-.30106E-10	6.86

Table 10: Calculation of the singular integral (29), using (30), where the endpoint singularities are given by $\alpha = -.93, \beta = -.07$, location of singularity at $t = -.83$ number of quadrature nodes used $q = 3$, interpolation order $p = 6$

n	C_n	$C_n - C_{n-1}$	$\log_2(\frac{C_{n-1}-C_{n-2}}{C_n-C_{n-1}})$
4	-20.1025755913295		
8	-20.0078167547070	.94759E-01	
16	-20.0122881899902	-.44714E-02	4.41
32	-20.0126710957889	-.38291E-03	3.55
64	-20.0126725119544	-.14162E-05	8.08
128	-20.0126725628427	-.50888E-07	4.80

Table 11: Calculation of the singular integral (29), using (30), where the endpoint singularities are given by $\alpha = -.93, \beta = -.07$, location of singularity at $t = -.83$ number of quadrature nodes used $q = 4$, interpolation order $p = 7$

Titles previously published in the series

OPERATOR THEORY: ADVANCES AND APPLICATIONS

BIRKHÄUSER VERLAG

Edited by **I. Gohberg,**

School of Mathematical Sciences, Tel-Aviv University, Ramat Aviv, Israel

74. **R. Hagen, S. Roch, B. Silbermann**: Spectral Theory of Approximation Methods for Convolution Equations, 1994, (3-7643-5112-8)

75. **C.B. Huijsmans, M.A. Kaashoek, B. de Pagter** (Eds): Operator Theory in Function Spaces and Banach Lattices. The A.C. Zaanen Anniversary Volume, 1994, (ISBN 3-7643-5146-2)

77. **J. Lindenstrauss, V.D. Milman** (Eds): Geometric Aspects of Functional Analysis Israel Seminar GAFA 1992-94, 1995, (ISBN 3-7643-5207-8)

78. **M. Demuth, B.-W. Schulze** (Eds): Partial Differential Operators and Mathematical Physics: International Conference in Holzhau (Germany), July 3-9, 1994, 1995, (ISBN 3-7643-5208-6)

79. **I. Gohberg, M.A. Kaashoek, F. van Schagen**: Partially Specified Matrices and Operators: Classification, Completion, Applications, 1995, (ISBN 3-7643-5259-0)

80. **I. Gohberg, H. Langer** (Eds): Operator Theory and Boundary Eigenvalue Problems International Workshop in Vienna, July 27-30, 1993, 1995, (ISBN 3-7643-5259-0)

81. **H. Upmeier**: Toeplitz Operators and Index Theory in Several Complex Variables, 1995, (ISBN 3-7643-5280-5)

82. **T. Constantinescu**: Schur Parameters, Factorization and Dilation Problems, 1996, (ISBN 3-7643-5285-X)

83. **A.B. Antonevich**: Linear Functional Equations. Operator Approach, 1995, (ISBN 3-7643-2931-9)

84. **L.A. Sakhnovich**: Integral Equations with Difference Kernels on Finite Intervals, 1996, (ISBN 3-7643-5267-1)

85. **Y.M. Berezansky, G.F. Us, Z.G. Sheftel**: Functional Analysis, **Vol. I**, 1996, (ISBN 3-7643-5344-9)

86. **Y.M. Berezansky, G.F. Us, Z.G. Sheftel**: Functional Analysis, **Vol. II**, 1996, (ISBN 4-7643-5345-7)